Selected Titles in This Series

Volume

7 **Robert Steinberg**
Robert Steinberg Collected Papers
1997

6 **Solomon Feferman, Editor**
The Collected Works of Julia Robinson
1996

5 **Freeman Dyson**
Selected Papers of Freeman Dyson with Commentary
1996

4 **Krystyna Kuperberg, Editor**
Collected Works of Witold Hurewicz
1995

3.2 **Richard E. Block, Nathan Jacobson, J. Marshall Osborn, David J. Saltman, and Daniel Zelinsky, Editors**
A. Adrian Albert Collected Mathematical Papers: Nonassociative Algebras and Miscellany, Part 2
1993

3.1 **Richard E. Block, Nathan Jacobson, J. Marshall Osborn, David J. Saltman, and Daniel Zelinsky, Editors**
A. Adrian Albert Collected Mathematical Papers: Associative Algebras and Riemann Matrices, Part 1
1993

2 **Robert C. Gunning, Editor**
Collected Papers of Salomon Bochner, Parts 1–4
1992

1 **Sukhjit Singh, Steve Armentrout, and Robert J. Daverman, Editors**
The Collected Papers of R. H. Bing, Parts 1 and 2
1988

ROBERT STEINBERG

ROBERT STEINBERG

COLLECTED PAPERS

ROBERT STEINBERG
COLLECTED PAPERS

Robert Steinberg

with Foreword by
J-P.Serre

American Mathematical Society
Providence, Rhode Island

1991 *Mathematics Subject Classification.* Primary 20-xx; Secondary 22-xx.

Library of Congress Cataloging-in-Publication Data

Steinberg, Robert, 1922–
[Works. 1997]
Robert Steinberg, collected papers / Robert Steinberg: with foreword by J-P.Serre
p. cm.
ISBN 0-8218-0576-2
1. Group theory. I. Title.
QA3.S725 1997
512′.2—dc21 96-51039
CIP

A complete list of acknowledgments can be found at
the back of this publication.

♾ The paper used in this book is acid-free and falls within the guidelines
established to ensure permanence and durability.

10 9 8 7 6 5 4 3 2 1 02 01 00 99 98 97

Contents

Foreword

The present volume contains the Collected Papers of Robert Steinberg.

It includes all his published papers on group theory, with commentaries by the author himself.

Among these, one will find those on the "special representations" (now called Steinberg representations), on tensor products of representations, on finite reflection groups, on regular elements of algebraic groups, on Galois cohomology, on universal central extensions, etc.

I have always been an avid reader of Steinberg's papers. It gives me great pleasure to thank him, and the AMS, for making them easily available to a large audience.

Jean-Pierre Serre

Preface

Collected in this volume are most of my published papers. All of them are reproduced directly from the journals in which they first appeared with only cosmetic changes, except for (23) and (35) (of the list starting on page xvii) which have been retyped with only minor changes and (40) which is still in press. Not included are eight published papers and two sets of lecture notes, one of which has appeared in mimeographed form and the other in book form; all of these are listed on pages xix–xx. I have also listed (as (41)) a published solution to a problem in the *American Mathematical Monthly*, one of many in my early days, since it represents the first appearance of my work in print, and it gives, in my opinion, the "book proof" of a significant, although elementary, result. Also I still remember fondly how this "publication" came into being. I was taking a class in projective geometry at the University of Toronto at the time. It was being taught by Professor H. S. M. Coxeter who seemed to have a never-ending supply of interesting problems. One of these was to show that if finitely many points in the real projective plane have the property that the line through two of them always contains a third then there exists a line containing all of them. I solved it and showed my solution to Professor Coxeter. He was obviously delighted with it and asked my permission to type it up and send it in to the *Monthly*, where, it turned out, the problem had come from, and I of course agreed. I had not heard of the *Monthly* until then, but soon after I began sending in solutions on my own.

Also included in this volume is a section, "Comments on the Papers," which comes directly after the papers themselves. It contains minor corrections and clarifications of the papers and indications of how their main ideas and results have evolved and been used since they first appeared. My remarks here are far from complete and there are undoubtedly many relevant works that I should have mentioned and did not. To the authors of all of these I offer my sincere apologies. At the suggestion of several of my colleagues, I have included, in my comments on (4), an extended discussion of what has come to be known as the Steinberg representation.

When Donald Babbitt, the publisher of the AMS (American Mathematical Society) and my former colleague at U.C.L.A., first suggested, in a telephone call, the possibility of doing this book, my first thought was that he had made a mistake and reached the wrong party. But no, he assured me that he had dialed the right number. Needless to say, I am delighted that he had, and I thank him and the AMS for this great honor. I also thank Christine Thivierge and Thomas Costa of the publishing office of the AMS for their work in the production of this book. J-P. Serre, the official editor of this book, has also been very helpful, especially in matters mathematical, and I thank him warmly.

ROBERT STEINBERG

Many people have contributed, directly and indirectly, to my mathematical work and my continuing mathematical education. I thank all of them and two of them especially: Richard Brauer, who was my teacher at the University of Toronto, and Claude Chevalley, whose mathematics long ago struck a resonant chord which is with me still.

I also thank my colleagues, past and present, at my home institution U.C.L.A., where most of my work has been done, for providing a stimulating, yet relaxed and unhurried, atmosphere, an atmosphere ideally suited to my temperament.

Most of all I thank my wife Maria who over the years has supported all of my endeavors, mathematical and other, in every possible way.

Robert Steinberg
November, 1996
Pacific Palisades, California

Curriculum Vitae

1922	Born on May 25 in Soroki, Romania, to David (a fruit dealer) and Cecelia, née Yalonetsky
1924	Family moved to Edmonton, Alberta
1926	Family moved to Toronto, Ontario
1935	Graduated from King Edward Public School
1940	Graduated from Harbord Collegiate (a high school)
1944	B.A. in Mathematics from University of Toronto
1948	Ph.D. in Mathematics from University of Toronto, advisor Richard Brauer; Thesis: Representations of the Linear Fractional Groups
1952	Married on December 19 to Maria Weber, no children
1955–56	Sabbatical at I.A.S. (Institute for Advanced Study)
1961–62	On leave at I.A.S.
1963–64	Sabbatical in Paris, loose affiliation with Institute Henri Poincaré
1966	Invited speaker at I.C.M. (International Congress of Mathematicians), Moscow
1967–68	Visited Yale University, gave course Lectures on Chevalley Groups
1968–69	Visited M.I.T. (fall, winter) and I.A.S. (spring)
1972–73	Sabbatical spent partly at T.I.F.R., Bombay (course Conjugacy Classes in Algebraic Groups), partly at University of Paris VII, and partly at University of Warwick
1979–80	Sabbatical at Queen Mary College, University of London
1985	Steele Prize of AMS for papers (18), (20) and (23) of this volume
1985	Elected to U.S. National Academy of Sciences
1986–87	Sabbatical at Mathematical Sciences Research Institute, Berkeley
1992	Retired as Emeritus at U.C.L.A.

Bibliography

(a) Papers Appearing in Volume

1. A geometric approach to the representations of the full linear group over a Galois field, *Trans. Amer. Math. Soc.* **71** (1951), 274–282.

2. The representations of $GL(3,q), GL(4,q), PGL(3,q)$, and $PGL(4,q)$, *Canad. J. Math.* **3** (1951), 225–235.

3. Prime power representations of finite linear groups, *Canad. J. Math.* **8** (1956), 580–591.

4. Prime power representations of finite linear groups II, *Canad. J. Math.* **9** (1957), 347–351.

5. On the number of sides of a Petrie polygon, *Canad. J. Math.* **10** (1958), 220–221.

6. (with A. Horn) Eigenvalues of the unitary part of a matrix, *Pacific J. Math.* **9** (1959), 541–550.

7. Finite reflection groups, *Trans. Amer. Math. Soc.* **91** (1959), 493–504.

8. Variations on a theme of Chevalley, *Pacific J. Math.* **9** (1959), 875–891.

9. The simplicity of certain groups, *Pacific J. Math.* **10** (1960), 1039–1041.

10. Automorphisms of finite linear groups, *Canad. J. Math.* **12** (1960), 606–615.

11. Invariants of finite reflection groups, *Canad. J. Math.* **12** (1960), 616–618.

12. Automorphisms of classical Lie algebras, *Pacific J. Math.* **11** (1961), 1119–1129.

13. A general Clebsch-Gordan theorem, *Bull. Amer. Math. Soc.* **67** (1961), 406–407.

14. Generators for simple groups, *Canad. J. Math.* **14** (1962), 277–283.

15. A closure property of sets of vectors, *Trans. Amer. Math. Soc.* **105** (1962), 118–125.

16. Complete sets of representations of algebras, *Proc. Amer. Math. Soc.* **13** (1962), 746–747.

17. Générateurs, relations et revêtements de groupes algébriques, *Colloquium on Algebraic Groups*, Brussels (1962), 113–127.

18. Representations of algebraic groups, *Nagoya Math. J.* **22** (1963), 33–56.

19. Differential equations invariant under finite reflection groups, *Trans. Amer. Math. Soc.* **112** (1964), 392–400.

20. Regular elements of semisimple algebraic groups, *Publ. Sci. I.H.E.S.* **25** (1965), 49–80.

21. On the Galois cohomology of linear algebraic groups, *Proc. International Conf. on the Theory of Groups*, Canberra (1965), Gordon and Breach (1967), 315–319.

22. Classes of elements of semisimple algebraic groups, *Proc. International Congress of Mathematicians*, Moscow (1966), 277–284.

23. Endomorphisms of linear algebraic groups, *Mem. Amer. Math. Soc.* **80** (1968), 1–108.

24. Algebraic groups and finite groups, *Illinois J. Math.* **13** (1969), 81–86.

25. (with T. A. Springer) Conjugacy classes, Seminar in algebraic groups and related finite groups, *Lecture Notes in Math.*, Springer-Verlag **131** (1970), 167–266; Addendum to second corrected printing (1986), 322–323.

26. Abstract homomorphisms of simple algebraic groups (after A. Borel and J. Tits), Séminaire Bourbaki, *Lecture Notes in Math.*, Springer-Verlag **383** (1972), 307–326.

27. Torsion in reductive groups, *Advances in Math.* **15** (1975), 63–92.

28. On a theorem of Pittie, *Topology* **14** (1975), 173–177.

29. On the desingularization of the unipotent variety, *Invent. Math.* **36** (1976), 209–224.

30. On theorems of Lie-Kolchin, Borel and Lang, *Contributions to Algebra*, Academic Press (1977), 349–354.

31. Conjugacy in semisimple algebraic groups, *J. Algebra* **55** (1978), 348–350.

32. Kleinian singularities and unipotent elements, *Proc. Sympos. Pure Math. of Amer. Math. Soc.* **37** (1980), 265–270.

33. Generators, relations and coverings of algebraic groups II, *J. Algebra* **71** (1981), 527–543.

34. Finite subgroups of SU_2, Dynkin diagrams and affine Coxeter elements, *Pacific J. Math.* **118** (1985), 587–598.

35. Some consequences of the elementary relations of SL_n, *Contemp. Math.* **45** (1985), 335–350.

36. Tensor product theorems, *Proc. Sympos. Pure Math. of Amer. Math. Soc.* **47** (1987), 331–338.

37. On Dickson's theorem on invariants, *J. Faculty of Sciences*, Univ. of Tokyo **34** (1987), 699–707.

38. An occurrence of the Robinson-Schensted correspondence, *J. Algebra* **113** (1988), 523–528.

39. (with R. Richardson and G. Röhrle) Parabolic subgroups with Abelian unipotent radical, *Invent. Math.* **110** (1992), 649–671.

40. Nagata's example, in Algebraic groups and Lie groups, *Australian Math. Soc. Lecture Series 9*, Cambridge University Press, 1997, 375–384.

(b) Papers not Appearing in Volume

41. Solution to problem 4065, *Amer. Math. Monthly* **51** (1944), 169–170.

42. (with R. M. Redheffer) The Laplacian and mean values, *Quart. Appl. Math.* **9** (1951), 315–317.

43. (with R. M. Redheffer) Analytic proof of the Lindemann theorem, *Pacific J. Math.* **2** (1952), 231–242.

44. (with R. M. Redheffer) Simultaneous trigonometric approximations, *J. Math. Phys.* **31** (1953), 260–266.

45. An example in functional equations, *Math. Magazine* **29** (1956),129–130.

46. Note on a theorem of Hadwiger, *Pacific J. Math.* **6** (1956), 775–777.

47. Note on linear differential equations, *Amer. Math. Monthly* **64** (1957), 35–36.

48. Generalizations of the theorem of Chasles, *Amer. Math. Monthly* **64** (1957), 352–353.

(c) Paper in Preparation

49. The isomorphism and isogeny theorems for the reductive algebraic groups.

(d) Books not Appearing in Volume

50. Lectures on Chevalley groups, previously available from Yale Univ. Math. Dept. (1968), 1–277, now being prepared as a book; Russian translation by J. N. Bernstein and N. N. Yakovlev, Mir, Moscow, 1975.

51. Conjugacy classes in algebraic groups, based on lectures given at the Tata Institute in Bombay, *Lecture Notes in Math.*, Springer-Verlag **366** (1974), 1–159.

A GEOMETRIC APPROACH TO THE REPRESENTATIONS OF THE FULL LINEAR GROUP OVER A GALOIS FIELD

BY

R. STEINBERG

1. **Introduction.** In this paper, methods are given for obtaining a large number of representations of $G \equiv GL(n,q)$, the group of all nondegenerate linear transformations of $S(n, q)$, the n-dimensional vector space over the Galois field $GF(q)$ of $q = p^r$ elements; this group will be considered equivalently as the group of all nondegenerate n by n matrices over the Galois field.

In §2, through a favorable comparison of G and H, the symmetric group on n symbols, we obtain a basic set of $p(n)$ irreducible(1) characters of G closely related to those of H. Among the characters obtained is one of degree $q^{n(n-1)/2}$ which is of particular interest from the group theoretical and the modular representations [1](2) points of view since $q^{n(n-1)/2}$ is the highest power of a prime p dividing the order g of G. In §3, the characters of this representation are computed explicitly.

In §4, by making use of linear characters of suitably chosen subgroups of G, a large number of irreducible characters of G is obtained.

The methods used involve the elementary properties of finite group representations(3) and characters, especially of permutation representations, and the Frobenius formula for induced characters which enables one to find a character of a group if he knows one for a subgroup [2].

2. We shall first determine $p(n)$ irreducible representations of G by considering simple geometric properties of $S(n, q)$ or, briefly, of $S(n)$. Corresponding to a fixed partition of n, $n = \nu_1 + \nu_2 + \cdots + \nu_n \equiv (\nu)$, say, with $0 \leqq \nu_1 \leqq \nu_2 \leqq \cdots \leqq \nu_n$, let $e(\nu)$ be a sequence of subspaces $S(\nu_1)$, $S(\nu_1+\nu_2)$, $\cdots$, $S(\nu_1+\nu_2+\cdots+\nu_{n-1})$, $S(n)$ which are such that each subspace is contained in the following one. The $e(\nu)$'s are permuted by the elements of G, and hence we get a permutation representation $C(\nu)$ of G of degree

$$(2.1) \qquad c(\nu) = \{n\}/\{\nu_1\}\{\nu_2\} \cdots \{\nu_n\},$$

the number of $e(\nu)$'s, where the notation used is

$$(2.2) \qquad [r] = q^{r-1} + q^{r-2} + \cdots + q + 1, \qquad \{r\} = \prod_{i=1}^{r} [i].$$

Presented to the Society, February 26, 1949; received by the editors December 18, 1948 and November 23, 1949 and, in revised form, February 3, 1951.

(1) $p(n)$ denotes the number of ways of dividing n into non-negative integers.

(2) Numbers in brackets refer to the references at the end of the paper.

(3) An account of these properties and methods can be found in [3].

We shall find it convenient to define $[r]$ for all integers by

$$[-r] = -q^{-r}[r]. \tag{2.3}$$

Then, for any two integers r and s, we have

$$[r] - [s] = q^s[r - s]. \tag{2.4}$$

Corresponding to the $p(n)$ partitions of n, we thus get $p(n)$ representations of G.

Now, let $\lambda_i = \nu_i + (i-1)$. Then, the λ's satisfy the relations

$$\begin{gathered} 0 \leqq \lambda_1 < \lambda_2 < \cdots < \lambda_n, \\ \sum_{i=1}^{n} \lambda_i = n(n+1)/2. \end{gathered} \tag{2.5}$$

Also, let us define sgn $(\kappa_1, \kappa_2, \cdots, \kappa_n)$ to be 0 if two of the κ's are equal and otherwise 1 or -1 according as $\kappa_1, \kappa_2, \cdots, \kappa_n$ form an even or an odd permutation of the κ's written in ascending order of magnitude.

Then our main result of this section is the following theorem.

THEOREM 2.1. $\Gamma(\nu) = \sum_\kappa \operatorname{sgn}(\kappa_1, \kappa_2, \cdots, \kappa_n) C(\lambda_1 - \kappa_1, \lambda_2 - \kappa_2, \cdots, \lambda_n - \kappa_n) = \sum_\kappa \operatorname{sgn}(\kappa) C(\lambda - \kappa)$ *is an irreducible representation of* G. *(The summation is made as* $\kappa_1, \kappa_2, \cdots, \kappa_n$ *run through the* $n!$ *permutations of the numbers* $0, 1, 2, \cdots, n-1$; *and* $C(\lambda - \kappa)$ *is defined to be* 0 *if any* $\lambda_i - \kappa_i$ *is negative.) Moreover, the* $p(n)$ *representations* $\Gamma(\nu)$ *so obtained are all distinct.*

We shall first prove two lemmas of a geometric nature on which the proof of Theorem 2.1 depends.

LEMMA 2.1. *If* $e^{(1)}$ *and* $e^{(2)}$ *are two*(4) $e(1^n)$*'s, there exist n vectors* $V_1, V_2, \cdots, V_n$ *which, when taken in the order* $V_1 V_2 \cdots V_n$, *span* $e^{(1)}$, *and, taken in some other order* $V_{p_1} V_{p_2} \cdots V_{p_n}$, *span* $e^{(2)}$.

We say that the ordered set of vectors $V_1 V_2 \cdots V_n$ *spans* $e^{(1)} = S^{(1)}(1), S^{(1)}(2), \cdots, S^{(1)}(n)$ *if* $S^{(1)}(i) = \{V_1, V_2, \cdots, V_i\}$ *for* $i = 1, 2, \cdots, n$.

Proof of Lemma 2.1. The proof will be by induction, and, since the lemma is trivially true for $n=1$ or 2, we shall assume $n \geqq 3$, and that the lemma is true for $n-1$.

Now $e^{(1)}$ and $e^{(2)}$ intersect $S^{(1)}(n-1)$ in two $e(1^{n-1})$'s which we shall denote by $e^{(1)\prime}$ and $e^{(2)\prime}$. By the induction assumption, we can choose $n-1$ vectors $V_1, V_2, \cdots, V_{n-1}$ such that $V_1 V_2 \cdots V_{n-1}$ and $V_{p_1} V_{p_2} \cdots V_{p_{n-1}}$ span $e^{(1)\prime}$ and $e^{(2)\prime}$ respectively. If $S^{(1)}(n-1)$ contains $S^{(2)}(i)$ but not $S^{(2)}(i+1)$, choose V_n to be any vector in $S^{(2)}(i+1)$ but not in $S^{(2)}(i)$. Then $V_1 V_2 \cdots V_n$ and $V_{p_1} V_{p_2} \cdots V_{p_i} V_n V_{p_{i+1}} \cdots V_{p_{n-1}}$ span $e^{(1)}$ and $e^{(2)}$ respectively, and the lemma is proved.

(4) (1^n) denotes the partition of n into n ones.

We next note an analogy between G and H, the symmetric group on n symbols, and make use of this analogy. Corresponding to the general partition (ν) of n, we define an $s(\nu)$ to be an entity consisting of the symbols 1 to n in any order, the first ν_1 symbols being bracketed together, the next ν_2 being bracketed together, and so forth. Two $s(\nu)$'s are considered to be the same if one can be obtained from the other by permuting the symbols in their separate brackets; for example, a typical $s(2^2)$ is $12, 34 \equiv 21, 34 \equiv 12, 43 \equiv 21, 43$. The $s(\nu)$'s are permuted by the elements of H, and hence furnish a permutation representation $D(\nu)$ of H of degree

$$d(\nu) = n!/\nu_1!\nu_2! \cdots \nu_n!, \tag{2.6}$$

the number of $s(\nu)$'s (cf. (2.1)).

It should be noted here that $D(\nu)$ is the representation of H induced by the unit representation of the subgroup $H(\nu) = H\nu_1 \times H\nu_2 \times \cdots \times H\nu_n$, where H_i is a symmetric group on i symbols, and that $C(\nu)$ is the representation of G induced by the unit representation of the subgroup $G(\nu)$ consisting of matrices with square blocks of degrees $\nu_1, \nu_2, \cdots, \nu_n$ (in this order) down the main diagonal and zeros above them.

LEMMA 2.2. *The number of classes of $(e(\nu), e(\mu))$'s is equal to the number of classes of $(s(\nu), s(\mu))$'s.*

Here, we mean that $(e^{(1)}(\nu), e^{(1)}(\mu))$ and $(e^{(2)}(\nu), e^{(2)}(\mu))$, for example, belong to the same class if there is an element of G taking $e^{(1)}(\nu)$ into $e^{(2)}(\nu)$ and $e^{(1)}(\mu)$ into $e^{(2)}(\mu)$.

Proof of Lemma 2.2. For any $(e(\nu), e(\mu))$, we supply the missing subspaces, if necessary, and then by Lemma 2.1 we choose n vectors $V_1, V_2, \cdots, V_n$ such that $V_1 V_2 \cdots V_n$ and $V_{p_1} V_{p_2} \cdots V_{p_n}$ span $e(\nu)$ and $e(\mu)$ respectively. We then associate the class containing $(e(\nu), e(\mu))$ with the class containing the $(s(\nu), s(\mu))$ given by

$$(1\ 2 \cdots \nu_1, \nu_1 + 1 \cdots \nu_1 + \nu_2, \cdots, n;$$
$$p_1\ p_2 \cdots p_{\mu_1}, p_{\mu_1+1} \cdots p_{\mu_1+\mu_2}, \cdots p_n).$$

Since there is a nondegenerate linear transformation taking any n vectors which span $S(n)$ into any other n vectors which span $S(n)$, we see that two $(e(\nu), e(\mu))$'s belong to the same class if and only if the corresponding $(s(\nu), s(\mu))$'s do, and this proves the lemma.

Geometrically, we have the following result.

COROLLARY. *There is a nondegenerate linear transformation taking any $(e(\nu), e(\mu))$ into any other $(e(\nu), e(\mu))$ with the same degrees of intersection of all corresponding pairs of subspaces.*

Now, let $\psi(\nu)$, $\phi(\nu)$, and $\chi(\nu)$ be the characters of $C(\nu)$, $D(\nu)$, and $\Gamma(\nu)$

respectively. Then, the permutation representations of the $(e(\nu), e(\mu))$'s and the $(s(\nu), s(\mu))$'s are given by the Kronecker products $C(\nu)\times C(\mu)$ and $D(\nu)\times D(\mu)$, and thus their characters by $\psi(\nu)\psi(\mu)$ and $\phi(\nu)\phi(\mu)$ respectively. But, for any permutation representation of character χ of a group of order g, $g^{-1}\sum\chi(x)$ is the number of times χ contains the unit character, that is, the number of classes of transitivity. (The summation is made over all elements x of G.) This remark together with Lemma 2.2 proves the following lemma.

LEMMA(5) 2.3. $g^{-1}\sum_{x\in G}\psi(\nu, x)\psi(\mu, x) = h^{-1}\sum_{y\in H}\phi(\nu, y)\phi(\mu, y)$.

Proof of Theorem 2.1. Let $\lambda_i=\nu_i+i-1$ and $\sigma_i=\mu_i+i-1$, and let $\kappa_1, \kappa_2, \cdots, \kappa_n$ and $\rho_1, \rho_2, \cdots, \rho_n$ be permutations of the numbers $0, 1, 2, \cdots, n-1$.

Then (6),

$$\begin{aligned} g^{-1}\sum_{x\in G}\chi(\nu, x)\chi(\mu, x) &= g^{-1}\sum_{x\in G}\sum_{\kappa}\operatorname{sgn}(\kappa)\psi(\lambda-\kappa, x)\sum_{\rho}\operatorname{sgn}(\rho)\psi^*(\sigma-\rho, x) \\ &= h^{-1}\sum_{y\in H}\sum_{\kappa}\operatorname{sgn}(\kappa)\phi(\lambda-\kappa, y)\sum_{\rho}\operatorname{sgn}(\rho)\phi^*(\sigma-\rho, y), \end{aligned}$$

by the definitions and Lemma 2.3 respectively.

But, $\sum_\kappa \operatorname{sgn}(\kappa)\phi(\lambda-\kappa)$ is the character of $\Delta(\nu)=\sum_\kappa \operatorname{sgn}(\kappa)D(\lambda-\kappa)$, and it was proved by Frobenius [3] that the $p(n)$ representations $\Delta(\nu)$ of H are irreducible and distinct. Thus

$$g^{-1}\sum_{x\in G}\chi(\nu, x)\chi^*(\mu, x) = \delta_{(\nu)(\mu)}. \tag{2.7}$$

Thus, if in addition $\chi(\nu, E)>0$, where E is the identity element of G, $\chi(\nu)$, being an integral linear combination of characters, will be an irreducible character.

Now,

$$\begin{aligned} \chi(\nu, E) &= \sum_{\kappa}\operatorname{sgn}(\kappa)\{n\}/\{\lambda_1-\kappa_1\}\{\lambda_2-\kappa_2\}\cdots\{\lambda_n-\kappa_n\} \\ &= \{n\}\det|\{\lambda_i-(j-1)\}^{-1}| \qquad (\text{where } \{r\}^{-1}\equiv 0 \text{ if } r<0) \\ &= \frac{\{n\}\det|[\lambda_i][\lambda_i-1]\cdots[\lambda_i-(j-2)]|}{\{\lambda_1\}\{\lambda_2\}\cdots\{\lambda_n\}}. \end{aligned} \tag{2.8}$$

To evaluate this, let us consider the determinant

$$\left|X_i\left(\frac{X_i-[1]}{q}\right)\left(\frac{X_i-[2]}{q^2}\right)\cdots\left(\frac{X_i-[j-2]}{q^{j-2}}\right)\right|.$$

It is of degree $n(n-1)/2$ in the n indeterminates $X_1, X_2, \cdots, X_n$ and van-

(5) $\psi(\nu, x)$ denotes the character in $\psi(\nu)$ of the element x of G.

(6) * denotes conjugate complex.

ishes if $X_i = X_j$, $i \neq j$. Hence, it is a constant(7) times $\Delta(X_1, X_2, \cdots, X_n) = \prod_{i<j}(X_j - X_i)$, and a perusal of the coefficient of $X_2 X_3^2 \cdots X_n^{n-1}$ shows this constant to be $q^{-n(n-1)(n-2)/3}$. Since this determinant reduces to the one in (2.8) if $X_i = [\lambda_i]$, we finally get

$$\begin{aligned}\gamma(\nu) &= \chi(\nu, E) \\ &= q^{-n(n-1)(n-2)/3}\{n\}\Delta([\lambda_1], [\lambda_2], \cdots, [\lambda_n])/\{\lambda_1\}\{\lambda_2\}\cdots\{\lambda_n\}.\end{aligned} \tag{2.9}$$

This is positive since $\lambda_i < \lambda_j$ for $i < j$. This completes the proof of Theorem 2.1, and the equation (2.9) furnishes us with the degree $\gamma(\nu)$ of $\Gamma(\nu)$.

As an immediate consequence of Theorem 2.1 and Lemma 2.3, we get the following corollaries.

COROLLARY 1. *$C(\nu)$ and $D(\nu)$ split into irreducible representations in exactly the same manner; that is, if $C(\nu) = \sum_{(\mu)} k(\nu, \mu)\Gamma(\mu)$, then $D(\nu) = \sum_{(\mu)} k(\nu, \mu)\Delta(\mu)$.*

COROLLARY 2. *Let $(\nu_1), (\nu_2), \cdots, (\nu_n)$ be further partitions of $\nu_1, \nu_2, \cdots, \nu_n$, let C be the representation of G induced by the representation $\Gamma(\nu_1) \times \Gamma(\nu_2) \times \cdots \times \Gamma(\nu_n)$ of $G(\nu)$, and let D be the corresponding representation of H. Then C and D split into irreducible representations in exactly the same manner.*

We can give an alternate representation to that given by (2.5) for the partitions of n, and this will lead to a simpler formula for the degree $\gamma(\nu)$ of $\Gamma(\nu)$. Among the numbers $\lambda_1, \lambda_2, \cdots, \lambda_n$ there are a certain number, say r, which are not less than n. Denote these by $n+b_1, n+b_2, \cdots, n+b_r$. Denote the remaining λ's by $n-1-a_{r+1}, n-1-a_{r+2}, \cdots, n-1-a_n$. Let $n-1-a_1, n-1-a_2, \cdots, n-1-a_r$ be the rest of the numbers $0, 1, \cdots, n-1$ so that the a's are a permutation of the numbers $0, 1, \cdots, n-1$. Then, we can represent the partition by the two sets of integers $a_1, a_2, \cdots, a_r$ and $b_1, b_2, \cdots, b_r$. These a's and b's can be ordered so that

$$0 \leqq a_1 < a_2 < \cdots < a_r, \qquad 0 \leqq b_1 < b_2 < \cdots < b_r. \tag{2.10}$$

Because of (2.5), we have also

$$\sum_{i=1}^{r} a_i + \sum_{i=1}^{r} b_i = n - r. \tag{2.11}$$

Conversely, it can be shown that two sets of r integers satisfying the relations (2.10) and (2.11) correspond to a unique partition of n. The partition obtained from a given one by interchanging the a's and b's we shall call the conjugate partition(8).

(7) Hereafter we shall call this the Vandermonde determinant.

(8) One can represent a partition diagrammatically by the lattice points (i, j) which satisfy $1 \leqq j \leqq \nu_{n-i}$ and alternately define the conjugate partition to be the one represented by the points (j, i). These two definitions are equivalent.

Then, by a computation similar to that used by Frobenius [3], and making use of (2.4), we can show that

$$\gamma(\nu) = q^{r(r-1)/2+\sum_{i=1}^{r}(a_i+1)a_i/2} \cdot \frac{\{n\}\Delta([a_1], [a_2], \cdots, [a_r])\Delta([b_1], [b_2], \cdots, [b_r])}{\{a_1\}\{a_2\}\cdots\{a_r\}\{b_1\}\{b_2\}\cdots\{b_r\}\prod_{\alpha,\beta=1}^{r}[a_\alpha + b_\beta + 1]}.$$

From this a direct computation shows that the degrees $\gamma(\nu)$, $\gamma(\nu')$ of representations corresponding to conjugate partitions (ν), (ν') are polynomials in q related by the following formula:

$$\gamma(\nu', q) = q^{n(n-1)/2}\gamma(\nu, q^{-1}). \tag{2.13}$$

As an important special case, if $(\nu)=(0^{n-1}n)$ and $(\nu')=(1^n)$, then $\gamma(\nu)=1$ (since $\Gamma(\nu)$ in this case is the unit representation) and $\gamma(\nu')=q^{n(n-1)/2}$. As previously stated, this last representation is of group-theoretical importance and in the next section we shall compute its characters explicitly.

In closing this section, a remark on the analogy between G and H seems to be in order. Instead of considering G as a group of linear transformations of a vector space, we could consider G as a collineation group of a finite $(n-1)$-dimensional geometry. If $q=1$, the vector space fails to exist but the finite geometry does exist and, in fact, reduces to the n vertices of a simplex with a collineation group isomorphic to H. Hence, if we put $q=1$, $[r]=r$, and $\{r\}=r!$ in (2.1), (2.9), (2.12), and (2.13), we get corresponding results for H.

3. In this section, the characters of the representation of degree $q^{n(n-1)/2}$ of G mentioned in the previous section will be found. These characters are given by Theorem 3.1 at the end of this section.

The determination will depend on a number of lemmas which we proceed to state and prove.

By the generating function of an element x of G, we shall mean the symmetric polynomial

$$f(q, t) = \sum f(\nu)t^{(\nu)},$$

where the sum is taken over all partitions (ν) of n, $f(\nu)$ is the number of $e(\nu)$'s left fixed by x, and $t^{(\nu)} = t_1^{\nu_1}t_2^{\nu_2}\cdots t_n^{\nu_n}$.

LEMMA 3.1. *Let A be an irreducible matrix of degree m and En the unit matrix of degree n (over $GF(q)$). Then, if the generating function of En is $f(q, t)$, that of $En\times A$ is $f(q^m, t^m)$.*

Proof. Let the field $GF(q)$ be extended to $GF(q^m)$. In this field A can be transformed to a diagonal matrix R whose main-diagonal elements are conjugate relative to $GF(q)$. Then, after a further simple transformation, we get $En \times A$ similar to $R \times En$. The latter matrix leaves fixed all of the vectors of m disjoint conjugate (relative to $GF(q)$) $S(n)$'s and all subspaces spanned by such vectors. If we take m conjugate vectors, one from each $S(n)$, we get a fixed real $S(m)$. Similarly, the only fixed real $S(rm)$'s are spanned by conjugate $S(r)$'s. Since each $S(n)$ is a $GF(q^m)$ vector space, the lemma is proved.

LEMMA 3.2. *Let f_1, f_2, f be the generating functions for $A_1, A_2, A_1 \dotplus A_2$, where A_1 and A_2 are matrices which, when written in canonical form, have no common constituents. Then $f = f_1 f_2$.*

This lemma follows from the definitions and from the fact that any fixed $S(i)$ of $A_1 \dotplus A_2$ is obtained as the join of a fixed $S(j)$ in the space of A_1 and a fixed $S(i-j)$ in the space of A_2, this representation being unique.

LEMMA 3.3. *Let f_1, f_2 be homogeneous polynomials of degree m, n respectively, symmetric in the $m+n$ variables t_i, and let $\Delta_1, \Delta_2, \Delta$ be the Vandermonde determinants in $t_1, t_2, \cdots, t_m; t_{m+1}, t_{m+2}, \cdots, t_{m+n}; t_1, t_2, \cdots, t_{m+n}$ respectively. Let c_1, c_2, c be the coefficients of $t_1 t_2^2 \cdots t_m^m$, $t_{m+1} t_{m+2}^2 \cdots t_{m+n}^n$, $t_1 t_2^2 \cdots t_{m+n}^{m+n}$ in $f_1\Delta_1, f_2\Delta_2, f\Delta$ respectively. Then $c_1 c_2 = c$.*

Proof. If we consider $f_2\Delta$, it is evident that any term with a repeated index will vanish, and, since we are interested only in those terms in which each t_i occurs to a power of at most $m+n$, we need consider only terms of the sort $t_2 t_3^2 \cdots t_m^{m-1} t_{m+1}^{m+1} \cdots t_{m+n}^{m+n}$. The coefficient of this term is c_2. If we multiply by f_1, we readily see that the terms involving $t_{m+1}^{m+1} \cdots t_{m+n}^{m+n}$ are $c_2 f_1 \Delta_1 t_{m+1}^{m+1} \cdots t_{m+n}^{m+n}$. Hence the lemma is established.

LEMMA 3.4. *Let $f(t)$ be a homogeneous polynomial of degree n, symmetric in the n variables $t_1, \cdots, t_n$, and let Δ_1, Δ be the Vandermonde determinants in $t_1, t_2, \cdots, t_n$ and $t_1, t_2, \cdots, t_{mn}$ respectively. Then the coefficient of $t_1 t_2^2 \cdots t_n^n$ in $f(t)\Delta_1$ is equal to $(-1)^{(m-1)n}$ times the coefficient of $t_1 t_2^2 \cdots t_{mn}^{mn}$ in $f(t^m)\Delta$.*

This lemma is evident if one considers how the $t_1 t_2^2 \cdots t_{mn}^{mn}$ term is obtained when $f(t^m)$ is multiplied by Δ.

LEMMA 3.5. *For the matrix En, the coefficient of $t_1 t_2^2 \cdots t_n^n$ in $f\Delta$ is $q^{n(n-1)/2}$.*

This follows from §2 and the following statement.

This coefficient, in the case of En, gives the degree of the representation under consideration, and, in general, the character of the element in this representation.

LEMMA 3.6. *For a matrix which cannot be reduced to diagonal form in any extension of $GF(q)$, the coefficient of $t_1 t_2 \cdots t_n$ in $f\Delta$ is 0.*

This follows from a theorem due to R. Brauer and C. Nesbitt [1] to the effect that the character of any p-singular element[9] is 0 in an irreducible representation of order divisible by p^r, where p^r is the highest power of a prime p which divides the order of the group.

From these lemmas and the statement in Lemma 3.5, we get the following theorem.

THEOREM 3.1. *For the element $En \times A$, where A is irreducible and of degree m, the character is $(-1)^{(m-1)n}q^{mn(n-1)/2}$. For any element x of G which is made up of blocks of the type $En \times A$, the character is obtained by multiplying the $(-1)^{(m-1)n}q^{mn(n-1)/2}$'s for the various blocks. For any p-singular element, the character is* 0.

4. We can make use of the $p(n)$ basic irreducible representations $\Gamma(\nu)$ to determine a large set of irreducible representations of G.

We first note that there are $q-1$ linear representations of G corresponding to the powers of the determinants of the elements x of G. Thus if ρ is a primitive $(q-1)$th root of unity in $GF(q)$ and ϵ is one in the field of complex numbers, and if $\det x=\rho^\alpha$, these representations are given by $x \to \epsilon^{\alpha u}$, $u=1, 2, \cdots, q-1$.

If we multiply $\Gamma(\nu)$ by each of these representations, we get $q-1$ irreducible representations which we shall denote by $\Gamma_u(\nu)$.

Let us now return to our general partition $n=\nu_1+\nu_2+ \cdots +\nu_n$ and partition each ν_i further: $\nu_i=\sum \nu_{ij}$. We shall denote these partitions by (ν_i). Let the partition of ν_i have k_{i1} ones, k_{i2} twos, and so forth, and let $k=\sum k_{ij}$. Corresponding to these partitions, we can take the Kronecker product of representations of the type $\Gamma_u(\nu)$ of the constituents of the subgroup $G(\nu)$ of G and get a representation of G which we can write as $\times \prod_{i=1}^n \Gamma_{u_i}(\nu_i)$. This representation of $G(\nu)$ induces one in G which we shall call Γ.

THEOREM 4.1. *If no two u's are equal, Γ, is irreducible. Moreover, the number of representations so obtained (by varying the u's and keeping the ν's fixed) is $(q-1)(q-2) \cdots (q-k)/\prod k_{ij}!$. The degree of each is*

$$\gamma = \frac{\gamma(\nu_1)\gamma(\nu_2) \cdots \gamma(\nu_n)\{n\}}{\{\nu_1\}\{\nu_2\} \cdots \{\nu_n\}}. \tag{4.1}$$

Due to the complexities of notation, the proof will not be given here. It depends on a computation similar to that used by Frobenius [3] in his determination of the characters of the symmetric group.

A combinatorial argument shows that the number of different degrees of characters obtained by this method has as generating function $h(t) = \prod_{i=1}^\infty (1-t^i)^{-p(i)}$. I conjecture that the total number of different degrees has as generating function $\prod_{i=1}^\infty h(t^i)$. For $n=1, 2, 3, 4$, this method gives

(9) A p-singular element is one of order divisible by p.

characters of 1, 3, 6, 14 distinct degrees, and the total[10] number of distinct degrees is 1, 4, 8, 22 respectively.

References

1. R. Brauer and C. Nesbitt, *On the modular characters of groups*. Ann. of Math. vol. 42 (1941) pp. 556–590.

2. G. Frobenius, *Über Relationen zwischen den Charakteren einer Gruppe und denen ihrer Untergruppen*, Preuss. Akad. Wiss. Sitzungsber. (1898) pp. 501–515.

3. ———, *Über die Charaktere der symmetrische Gruppe*, Preuss. Akad. Wiss. Sitzungsber. (1900) pp. 516–534.

4. F. D. Murnaghan, *The theory of group representations*, Baltimore, 1938.

5. R. Steinberg, *Representations of the linear fractional groups*, Thesis, 1948, University of Toronto Library.

University of California,
Los Angeles, Calif.

(10) All of the characters of G for $n=1, 2, 3, 4$ have been determined [5].

THE REPRESENTATIONS OF GL(3, q), GL(4, q), PGL(3, q), AND PGL(4, q)

ROBERT STEINBERG

1. Introduction. This paper is a result of an investigation into general methods of determining the irreducible characters of $GL(n, q)$, the group of all non-singular linear substitutions with marks in $GF(q)$, and of the related groups, $SL(n, q)$, $PGL(n, q)$, $PSL(n, q)$, the corresponding group of determinant unity, projective group, projective group of determinant unity, respectively. This investigation is not complete, but the general problem was answered partially in [9]. In [3], [7], [6], [1], Frobenius, Schur, Jordan, and Brinkmann gave the characters of $PSL(2, p)$; $SL(2, q)$, $GL(2, q)$; $SL(2, q)$, $GL(2, q)$; $PSL(3, q)$, respectively. In this paper in §2 and §3, the characters of $GL(2, q)$ and $GL(3, q)$ are determined, and, from them, those of $PGL(2, q)$ and $PGL(3, q)$ deduced. In §4, an outline of the determination of the characters of $GL(4, q)$ is given together with the degrees and frequencies of the characters of $GL(4, q)$ and $PGL(4, q)$ and a table of the rational characters of $GL(4, q)$.

The simple properties of the underlying geometry, $PG(n-1, q)$, of which $PGL(n, q)$ is the collineation group, are used throughout the work. The most powerful and frequent tool used in the determination of the characters is the Frobenius method[1] of induced representations [5] which enables one to construct a representation of a group if a representation of a subgroup is known. The explicit formula for the character in this case is $\chi(G) = \frac{m}{g_G} \Sigma\psi(G')$, where m is the index of the subgroup, g_G is the number of elements of the group similar to G, ψ is the character of the subgroup, and the summation is made over all elements G' which are similar to G and lie in the subgroup. Of fundamental use in the application of this method are the $q-1$ linear characters of $GL(n, q)$ which correspond to the powers of the determinants of the matrices which define the elements of $GL(n, q)$. Also very useful are pseudo-characters—linear combinations of irreducible characters with negative coefficients permissible—and the fact that a pseudo-character, $\chi(G)$, is an irreducible character if and only if $\Sigma|\chi(G)|^2 = g$ and $\chi(E) > 0$, where E is the unit element of the group.

The descent from the characters of $GL(n, q)$ to those of $PGL(n, q)$ is immediate because of the following two theorems due to Frobenius [4], [5]:

If $\mathfrak{H}$ is a normal subgroup of a group $\mathfrak{G}$, then every character of $\mathfrak{G}/\mathfrak{H}$ is also a character of $\mathfrak{G}$.

Received March 13, 1950.

This paper is part of a Ph.D. thesis written at the University of Toronto under the direction of Professor Richard Brauer.

[1] See [8] for a complete account of the properties of group characters used here.

In order that a character of $\mathfrak{G}$ may belong to the group $\mathfrak{G}/\mathfrak{H}$, it is necessary and sufficient that it have the same value for all elements of $\mathfrak{H}$. Then, it has also equal values for every two elements of $\mathfrak{G}$ which are equivalent mod $\mathfrak{H}$.

In our case, $\mathfrak{G}$ is the group GL(n, q), $\mathfrak{H}$ is the cyclic group of the $q - 1$ scalar matrices, and $\mathfrak{G}/\mathfrak{H}$ is the group PGL(n, q). For this reason, and also because the group GL(n, q) is easier to handle, its characters are first determined and then those of PGL(n, q) obtained from them.

In what follows, $\chi_q^{(r)}$ for example, will denote a character of degree q, the superscript being used to distinguish between two characters of the same degree. GL$(1, 2; q)$ denotes the subgroup $\begin{pmatrix} A_1 & 0 \\ * & A_2 \end{pmatrix}$ of GL$(3, q)$; $\rho, \sigma, \tau, \omega$ are primitive elements of GF(q), GF(q^2), GF(q^3), GF(q^4) respectively, such that $\rho = \sigma^{q+1} = \tau^{q^2+q+1} = \omega^{q^3+q^2+q+1}$ and $\sigma = \omega^{q^2+1}$.

2. The characters of GL$(2, q)$ **and** PGL$(2, q)$. The group GL$(2, q)$ is of order $q(q-1)^2(q+1)$ and each of its elements is similar to a matrix of one of the following four types [2]:

$$A_1: \begin{pmatrix} \rho^a & \\ & \rho^a \end{pmatrix}, \quad A_2: \begin{pmatrix} \rho^a & \\ 1 & \rho^a \end{pmatrix}, \quad A_3: \begin{pmatrix} \rho^a & \\ & \rho^b \end{pmatrix}_{a \neq b}, \quad B_1: \begin{pmatrix} \sigma^a & \\ & \sigma^{aq} \end{pmatrix}_{a \neq \text{mult. } (q+1)}.$$

The number of classes of each type and the number of elements in each class is given by Table I. The total number of classes is $(q-1)(q+1) = k$.

TABLE I

Element	Number of classes	Number of elements in each class
A_1	$q-1$	1
A_2	$q-1$	$(q-1)(q+1)$
A_3	$\frac{1}{2}(q-1)(q-2)$	$q(q+1)$
B_1	$\frac{1}{2}q(q-1)$	$q(q-1)$

Now, if we consider each matrix as a linear transformation of PG$(1, q)$, we get a representation of degree $q + 1$ representing the permutation of the points of PG$(1, q)$. The character of any element of GL$(2, q)$ is just the number of points left fixed by it. This permutation group is doubly transitive and hence splits into the unit representation and an irreducible representation [9] of degree q. Multiplication of each of these characters by each of the $q - 1$ linear characters given by the powers of the determinants gives us $q - 1$ irreducible characters of degree 1 and $q - 1$ of degree q. (See Table II.)

We next consider the subgroup GL$(1, 1; q) = \begin{pmatrix} A_1 & 0 \\ * & B_1 \end{pmatrix}$ of index $q + 1$. Clearly, any character of A_1 or GL$(1, q)$ multiplied by any character of B_1 or GL$(1, q)$ is a character of GL$(1, 1; q)$. If we use the linear characters of GL$(1, 1; q)$ obtainable in this way as a basis for Frobenius's method of induced characters, we get $\frac{1}{2}(q-1)(q-2)$ irreducible characters of degree $q+1$ of GL(2, q). (See Table II.)

Finally, the linear characters of the cyclic subgroup $\begin{pmatrix}\sigma & \\ & \sigma^q\end{pmatrix}^a$ of index $q(q-1)$ induce in GL(2, q) the following representations $\Psi_{q(q-1)}{}^{(n)}$ of degree q^2-q, all of which are reducible:

$$A_1\colon (q^2-q)\,\epsilon^{na(q+1)},\quad A_2\colon 0,\quad A_3\colon 0,\quad B_1\colon \epsilon^{na}+\epsilon^{naq},$$

where $\epsilon^{q^2-1}=1$ and $n=1,2,\ldots,q-1$. But, if we form $\chi_q{}^{(o)}\chi_{q+1}{}^{(o,n)}-\chi_{q+1}{}^{(o,n)}-\psi_{q(q-1)}{}^{(n)}$, we get an irreducible character provided $n \neq$ mult. $(q+1)$. We thus have $\frac{1}{2}q(q-1)$ irreducible characters of degree $q-1$ and this completes the list since we now have in all $(q-1)(q+1)=k$ characters. They are shown in Table II.

TABLE II

Characters of GL(2, q)

Element	$\chi_1{}^{(n)}$	$\chi_q{}^{(n)}$	$\chi_{q+1}{}^{(m,n)}$	$\chi_{q-1}{}^{(n)}$
	$n=1,2,\ldots,q-1$; $\epsilon^{q-1}=1$	$n=1,2,\ldots,q-1$; $\epsilon^{q-1}=1$	$m,n=1,2,\ldots,q-1$; $m\neq n$; $(m,n)\equiv(n,m)$; $\epsilon^{q-1}=1$	$n=1,2,\ldots,q^2-2$; $n\neq$ mult. $(q+1)$; $\epsilon^{q^2-1}=1$
A_1	ϵ^{2na}	$q\epsilon^{2na}$	$(q+1)\epsilon^{(m+n)a}$	$(q-1)\epsilon^{na(q+1)}$
A_2	ϵ^{2na}	0	$\epsilon^{(m+n)a}$	$-\epsilon^{na(q+1)}$
A_3	$\epsilon^{n(a+b)}$	$\epsilon^{n(a+b)}$	$\epsilon^{ma+nb}+\epsilon^{na+mb}$	0
B_1	ϵ^{na}	$-\epsilon^{na}$	0	$-(\epsilon^{na}+\epsilon^{naq})$

The theorems of Frobenius [4], [5] mentioned in the introduction immediately give us the characters of PGL(2, q). For q even they are as in Table III. For q odd, there are in addition the two characters

$$A_1\colon 1,\quad A_2\colon 1,\quad A_3\colon (-1)^{a+b},\quad B_1\colon (-1)^a,$$

and

$$A_1\colon q,\quad A_2\colon 0,\quad A_3\colon (-1)^{a+b},\quad B_1\colon (-1)^{a+1}.$$

TABLE III

Characters of PGL(2, q)

Element	χ_1	χ_q	$\chi_{q+1}{}^{(n)}$	$\chi_{q-1}{}^{(n)}$
			$n=1,2,\ldots,[\frac{1}{2}(q-1)]$; $\epsilon^{q-1}=1$	$n=1,2,\ldots,[\frac{1}{2}(q+1)]$; $\epsilon^{q+1}=1$
A_1	1	q	$q+1$	$q-1$
A_2	1	0	1	-1
A_3	1	1	$\epsilon^{n(b-a)}+\epsilon^{-n(b-a)}$	0
B_1	1	-1	0	$-(\epsilon^{na}+\epsilon^{naq})$

3. The characters of GL(3, q) **and** PGL(3, q). The group GL(3, q) is of order $q^3(q-1)^3(q+1)(q^2+q+1)$ and each of its elements similar to one of the following types [2]:

$$A_2: \begin{pmatrix} \rho^a & & \\ & \rho^a & \\ & & \rho^a \end{pmatrix}, \quad A_2: \begin{pmatrix} \rho^a & & \\ 1 & \rho^a & \\ & & \rho^a \end{pmatrix}, \quad A_3: \begin{pmatrix} \rho^a & & \\ 1 & \rho^a & \\ & 1 & \rho^a \end{pmatrix}, \quad A_4: \begin{pmatrix} \rho^a & & \\ & \rho^a & \\ & & \rho^b \end{pmatrix},$$

$$A_5: \begin{pmatrix} \rho^a & & \\ 1 & \rho^a & \\ & & \rho^b \end{pmatrix}, \quad A_6: \begin{pmatrix} \rho^a & & \\ & \rho^b & \\ & & \rho^c \end{pmatrix}, \quad B_1: \begin{pmatrix} \rho^a & & \\ & \sigma^b & \\ & & \sigma^{bq} \end{pmatrix}, \quad C_1: \begin{pmatrix} \tau^a & & \\ & \tau^{aq} & \\ & & \tau^{aq^2} \end{pmatrix},$$

where $a \neq$ mult. (q^2+q+1) in C. The number of elements in each class and the number of classes of each type are given in Table IV. The total number of classes is $q(q-1)(q+1)=k$.

TABLE IV

Element	Number of Classes	Elements in each Class
A_1	$q-1$	1
A_2	$q-1$	$(q-1)(q+1)(q^2+q+1)$
A_3	$q-1$	$q(q-1)^2(q+1)(q^2+q+1)$
A_4	$(q-1)(q-2)$	$q^2(q^2+q+1)$
A_5	$(q-1)(q-2)$	$q^2(q-1)(q+1)(q^2+q+1)$
A_6	$\frac{1}{6}(q-1)(q-2)(q-3)$	$q^3(q+1)(q^2+q+1)$
B_1	$\frac{1}{2}q(q-1)$	$q^3(q-1)(q^2+q+1)$
C_1	$\frac{1}{3}q(q-1)(q+1)$	$q^3(q-1)^2(q+1)$

Here, as before, the permutation of the points of the underlying geometry gives us a double-transitive permutation group, in this case of degree q^2+q+1. We thus get the unit representation and an irreducible representation of degree q^2+q. The geometric entities each of which consists of a point and a line through it are also permuted by the elements of GL(3, q), and this furnishes us with a representation of degree $(q+1)(q^2+q+1)$. The orthogonality properties of group characters tell us that the character of this representation contains the unit character χ_1 once and χ_{q^2+q} twice and an irreducible character [9] of degree q^3. Multiplying each of the characters of degrees 1, q^2+q, q^3 by each of the $q-1$ linear characters given by the powers of the determinants, we obtain $q-1$ irreducible characters of each of these degrees, as in Table V.

TABLE V

Element	$\chi_1^{(n)}$	$\chi_{q^2+q}^{(n)}$	$\chi_{q^3}^{(n)}$
A_1	ϵ^{3na}	$(q^2+q)\epsilon^{3na}$	$q^3\epsilon^{3na}$
A_2	ϵ^{3na}	$q\epsilon^{3na}$	0
A_3	ϵ^{3na}	0	0
A_4	$\epsilon^{n(2a+b)}$	$(q+1)\epsilon^{n(2a+b)}$	$q\epsilon^{n(2a+b)}$
A_5	$\epsilon^{n(2a+b)}$	$\epsilon^{n(2a+b)}$	0
A_6	$\epsilon^{n(a+b+c)}$	$2\epsilon^{n(a+b+c)}$	$\epsilon^{n(a+b+c)}$
B_1	$\epsilon^{n(a+b)}$	0	$-\epsilon^{n(a+b)}$
C_1	ϵ^{na}	$-\epsilon^{na}$	ϵ^{na}

(where $n = 1, 2, \ldots q-1$ and $\epsilon^{q-1} = 1$).

We next consider the subgroup of index $q^2 + q + 1$:

$$\mathrm{GL}(1, 2; q) = \begin{pmatrix} A_1 & 0 & 0 \\ * & A_2 & \\ * & & \end{pmatrix}.$$

It is clear that any character of A_1 (or GL(1, q)) multiplied by any character of A_2(or GL(2, q)) is a character of GL(1, 2; q). By multiplying linear characters of GL(1, q) by the characters of degree 1, q, $q+1$, $q-1$ of GL(2, q) determined in §2, we get characters of these degrees of GL(1, 2; q). These characters induce in GL(3, q) a set of characters from which we can extract $(q-1)(q-2)$ irreducible characters of degree q^2+q+1, $(q-1)(q-2)$ of degree $q(q^2+q+1)$, $\frac{1}{6}(q-1)(q-2)(q-3)$ of degree $(q+1)(q^2+q+1)$, $\frac{1}{2}q(q-1)^2$ of degree $(q-1)(q^2+q+1)$. See Table VI and Table VII.

TABLE VI

Element	$\chi_{q^2+q+1}^{(m,n)}$	$\chi_{q(q^2+q+1)}^{(m,n)}$
A_1	$(q^2+q+1)\,\epsilon^{(m+2n)a}$	$q(q^2+q+1)\,\epsilon^{(m+2n)a}$
A_2	$(q+1)\epsilon^{(m+2n)a}$	$q\epsilon^{(m+2n)a}$
A_3	$\epsilon^{(m+2n)a}$	0
A_4	$(q+1)\epsilon^{(m+n)a+nb}+\epsilon^{2na+mb}$	$(q+1)\epsilon^{(m+n)a+nb}+q\epsilon^{2na+mb}$
A_5	$\epsilon^{(m+n)a+nb}+\epsilon^{2na+mb}$	$\epsilon^{(m+n)a+nb}$
A_6	$\Sigma_{(a,b,c)}\epsilon^{ma+n(b+c)}$	$\Sigma_{(a,b,c)}\epsilon^{ma+n(b+c)}$
B_1	ϵ^{ma+nb}	$-\epsilon^{ma+nb}$
C_1	0	0

(where $m, n = 1, 2, \ldots q-1$; $m \neq n$ and $\epsilon^{q-1} = 1$).

TABLE VII

Element	$\chi_{(q+1)(q^2+q+1)}{}^{(l,m,n)}$ $l, m, n, = 1, 2, \ldots, q-1; l \neq m \neq n \neq l;$ $\epsilon^{q-1} = 1$	$\chi_{(q-1)(q^2+q+1)}{}^{(m,n)}$ $m = 1, 2, \ldots, q-1; n = 1, 2, \ldots, q^2-2;$ $n \neq$ mult. $(q+1)$ $\epsilon^{q^2-1} = 1$
A_1	$(q+1)(q^2+q+1)\epsilon^{(l+m+n)a}$	$(q-1)(q^2+q+1)\epsilon^{(m+n)a(q+1)}$
A_2	$(2q+1)\epsilon^{(l+m+n)a}$	$-\epsilon^{(m+n)a(q+1)}$
A_3	$\epsilon^{(l+m+n)a}$	$-\epsilon^{(m+n)a(q+1)}$
A_4	$(q+1)\Sigma_{(l,m,n)}\epsilon^{(l+m)a+nb}$	$(q-1)\epsilon^{(na+mb)(q+1)}$
A_5	$\Sigma_{(l,m,n)}\epsilon^{(l+m)a+nb}$	$-\epsilon^{(na+mb)(q+1)}$
A_6	$\Sigma_{(l,m,n)}\epsilon^{la+mb+nc}$	0
B_1	0	$-\epsilon^{ma(q+1)}(\epsilon^{nb}+\epsilon^{nbq})$
C_1	0	0

By $\Sigma_{(l,m,n)}\epsilon^{(l+m)a+nb}$, we mean the symmetric function in l, m, and n which has $\epsilon^{(l+m)a+nb}$ as its typical term.

Finally, we turn to the cyclic subgroup of order $(q-1)(q^2+q+1)$:

$$\begin{pmatrix} \tau & & \\ & \tau^q & \\ & & \tau^{q^2} \end{pmatrix}^a .$$

The linear characters of this subgroup induce the following in the group GL$(3,q)$:

A_1: $q^3(q-1)^2(q+1)\ \epsilon^{na(q^2+q+1)}$, A_2: 0, A_3: 0, A_4: 0,
A_5: 0, A_6: 0, B_1: 0, C_1: $\epsilon^{na}+\epsilon^{naq}+\epsilon^{naq^2}$.

If from this character we subtract $[\chi_{q^3}{}^{(0)} - \chi_{q^2+q}{}^{(0)} + \chi_1{}^{(0)}]\ \chi_{(q-1)(q^2+q+1)}{}^{(0,n)}$, we get:

A_1: $(q-1)^2(q+1)\epsilon^{na(q^2+q+1)}$, A_2: $-(q-1)\epsilon^{na(q^2+q+1)}$, A_3: $\epsilon^{na(q^2+q+1)}$
A_4: 0, A_5: 0, A_6: 0, B_1: 0, C_1: $\epsilon^{na}+\epsilon^{naq}+\epsilon^{naq^2}$.

This is an irreducible character if $n \neq$ mult. (q^2+q+1). Since $(n) \equiv (nq) \equiv (nq^2)$, we thus get $\frac{1}{3}q(q-1)(q+1)$ irreducible characters of degree $(q-1)^2(q+1)$.

This completes the list of characters since we have now obtained $q(q-1)(q+1) = k$ irreducible characters.

In obtaining the characters of PGL$(3, q)$, again two cases must be distinguished: $q = 3t+1$ or $q \neq 3t+1$. The revision of classes and characters in each case is straightforward and we shall content ourselves with a list of the number of characters of each degree. (See Table VIII.)

TABLE VIII
Characters of PGL$(3, q)$

Degree	1	q^2+q	q^3	q^2+q+1	$q(q^2+q+1)$	$(q+1)\times$ (q^2+q+1)	$(q-1)\times$ (q^2+q+1)	$(q-1)^2\times$ $(q+1)$
Frequency								
$q=3t+1$	3	3	3	$q-4$	$q-4$	$\frac{1}{6}(q^2-5q+10)$	$\frac{1}{2}q(q-1)$	$\frac{1}{3}(q-1)(q+2)$
$q\neq 3t+1$	1	1	1	$q-2$	$q-2$	$\frac{1}{6}(q-2)(q-3)$	$\frac{1}{2}q(q-1)$	$\frac{1}{3}q(q+1)$

4. The characters of GL(4, q) **and** PGL(4, q). The group GL(4, q) is of order $q^6(q-1)^4(q+1)^2(q^2+1)(q^2+q+1)$ and each of its elements is similar to one of the following twenty-two types [2]:

$$A_1: \begin{pmatrix} \rho^a &&& \\ & \rho^a && \\ && \rho^a & \\ &&& \rho^a \end{pmatrix}, \quad A_2: \begin{pmatrix} \rho^a &&& \\ 1 & \rho^a && \\ && \rho^a & \\ &&& \rho^a \end{pmatrix}, \quad A_3: \begin{pmatrix} \rho^a &&& \\ 1 & \rho^a && \\ && \rho^a & \\ && 1 & \rho^a \end{pmatrix},$$

$$A_4: \begin{pmatrix} \rho^a &&& \\ 1 & \rho^a && \\ & 1 & \rho^a & \\ &&& \rho^a \end{pmatrix}, \quad A_5: \begin{pmatrix} \rho^a &&& \\ 1 & \rho^a && \\ & 1 & \rho^a & \\ && 1 & \rho^a \end{pmatrix}, \quad A_6: \begin{pmatrix} \rho^a &&& \\ & \rho^a && \\ && \rho^a & \\ &&& \rho^b \end{pmatrix},$$

$$A_7: \begin{pmatrix} \rho^a &&& \\ 1 & \rho^a && \\ && \rho^a & \\ &&& \rho^b \end{pmatrix}, \quad A_8: \begin{pmatrix} \rho^a &&& \\ 1 & \rho^a && \\ & 1 & \rho^a & \\ &&& \rho^b \end{pmatrix}, \quad A_9: \begin{pmatrix} \rho^a &&& \\ & \rho^a && \\ && \rho^b & \\ &&& \rho^b \end{pmatrix},$$

$$A_{10}: \begin{pmatrix} \rho^a &&& \\ 1 & \rho^a && \\ && \rho^b & \\ &&& \rho^b \end{pmatrix}, \quad A_{11}: \begin{pmatrix} \rho^a &&& \\ 1 & \rho^a && \\ && \rho^b & \\ && 1 & \rho^b \end{pmatrix}, \quad A_{12}: \begin{pmatrix} \rho^a &&& \\ & \rho^a && \\ && \rho^b & \\ &&& \rho^c \end{pmatrix},$$

$$A_{13}: \begin{pmatrix} \rho^a &&& \\ 1 & \rho^a && \\ && \rho^b & \\ &&& \rho^c \end{pmatrix}, \quad A_{14}: \begin{pmatrix} \rho^a &&& \\ & \rho^b && \\ && \rho^c & \\ &&& \rho^d \end{pmatrix}, \quad B_1: \begin{pmatrix} \rho^a &&& \\ & \rho^a && \\ && \sigma^b & \\ &&& \sigma^{bq} \end{pmatrix},$$

$$B_2: \begin{pmatrix} \rho^a &&& \\ 1 & \rho^a && \\ && \sigma^b & \\ &&& \sigma^{bq} \end{pmatrix}, \quad B_3: \begin{pmatrix} \rho^a &&& \\ & \rho^b && \\ && \sigma^c & \\ &&& \sigma^{cq} \end{pmatrix}, \quad C_1: \begin{pmatrix} \sigma^a &&& \\ & \sigma^{aq} && \\ && \sigma^a & \\ &&& \sigma^{aq} \end{pmatrix},$$

$$C_2: \begin{pmatrix} \sigma^a &&& \\ & \sigma^{aq} && \\ 1 && \sigma^a & \\ & 1 && \sigma^{aq} \end{pmatrix}, C_3: \begin{pmatrix} \sigma^a &&& \\ & \sigma^{aq} && \\ && \sigma^b & \\ &&& \sigma^{bq} \end{pmatrix}, D_1: \begin{pmatrix} \rho^a &&& \\ & \tau^b && \\ && \tau^{bq} & \\ &&& \tau^{bq^2} \end{pmatrix}, E_1: \begin{pmatrix} \omega^a &&& \\ & \omega^{aq} && \\ && \omega^{aq^2} & \\ &&& \omega^{aq^3} \end{pmatrix}.$$

Now, we shall make use of the underlying geometry to obtain five irreducible characters. To do this, we consider the following five geometric entities: the PG(3, q); a point; a line; a point and a line through it; a point, a line through it, and a plane through the line. It will be noted that these five entities correspond to the five partitions of 4: (4), (13), (2^2), $(1^2 2)$, (1^4), respectively. In fact, GL(4, q), GL(1, 3; q), GL(2, 2; q), GL(1, 1, 2; q) and GL(1, 1, 1, 1; q) are the subgroups of GL(4, q) which leave fixed one of each of these entities, respectively. Each of these sets of entities will be permuted by the elements of

TABLE IX

Characters of GL(4 q) and PGL(4, q)

Element	Unit	Point	$q(q^2+q+1)$	Line	$q^2(q^2+1)$	Point–line	$q^3(q^2+q+1)$	Point–Line–Plane	q^6
A_1	1	$(q+1)(q^2+1)$	$q(q^2+q+1)$	$(q^2+1)(q^2+q+1)$	$q^2(q^2+1)$	$(q+1)(q^2+1)(q^2+q+1)$	$q^3(q^2+q+1)$	$(q+1)^2(q^2+1)(q^2+q+1)$	q^6
A_2	1	q^2+q+1	q^2+q	$2q^2+q+1$	q^2	q^3+3q^2+2q+1	q^3	$3q^3+5q^2+3q+1$	0
A_3	1	$q+1$	q	q^2+q+1	q^2	q^2+2q+1	0	$2q^2+3q+1$	0
A_4	1	$q+1$	q	$q+1$	0	$2q+1$	0	$3q+1$	0
A_5	1	1	0	1	0	1	0	1	0
A_6	1	q^2+q+2	q^2+q+1	$2q^2+2q+2$	q^2+q	q^3+4q^2+4q+3	q^3+q^2+q	$4q^3+8q^2+8q+4$	q^3
A_7	1	$q+2$	$q+1$	$2q+2$	q	$4q+3$	q	$8q+4$	0
A_8	1	2	1	2	0	3	0	4	0
A_9	1	$2q+2$	$2q+1$	q^2+2q+3	q^2+1	$2q^2+6q+4$	q^2+2q	$6q^2+12q+6$	q^2
A_{10}	1	$q+2$	$q+1$	$q+3$	1	$3q+4$	q	$6q+6$	0
A_{11}	1	2	1	3	1	4	0	6	0
A_{12}	1	$q+3$	$q+2$	$2q+4$	$q+1$	$5q+7$	$2q+1$	$12q+12$	q
A_{13}	1	3	2	4	1	7	1	12	0
A_{14}	1	4	3	6	2	12	3	24	1
B_1	1	$q+1$	q	2	$-q+1$	$q+1$	-1	0	$-q$
B_2	1	1	0	2	1	1	-1	0	0
B_3	1	2	1	2	0	2	-1	0	-1
C_1	1	0	-1	q^2+1	q^2+1	0	$-q^2$	0	q^2
C_2	1	0	-1	1	1	0	0	0	0
C_3	1	0	-1	2	2	0	-1	0	1
D_1	1	1	0	0	-1	0	0	0	1
D_2	1	0	-1	0	0	0	1	0	-1

GL(4, q) and in this way permutation representations of degree 1, $(q+1)(q^2+1)$, $(q^2+1)(q^2+q+1)$, $(q+1)(q^2+1)(q^2+q+1)$ and $(q+1)^2(q^2+1)(q^2+q+1)$ will be obtained. All except the first of these five characters are reducible, but they can be combined to give five irreducible characters as follows [9]:

$$1 = 1;\ (q+1)(q^2+1) - 1 = q(q^2+q+1);$$
$$(q^2+1)(q^2+q+1) - (q+1)(q^2+1) = q^2(q^2+1);$$
$$(q+1)(q^2+1)(q^2+q+1) - (q^2+1)(q^2+q+1) - (q+1)(q^2+1) + 1 = q^3(q^2+q+1);$$
$$(q+1)^2(q^2+1)(q^2+q+1) - 3(q+1)(q^2+1)(q^2+q+1) + (q^2+1)(q^2+q+1) + 2(q+1)(q^2+1) - 1 = q^6.$$

Multiplication of each of these characters by the $q-1$ linear characters given by the powers of the determinants gives $q-1$ irreducible characters of each of these degrees. Table IX lists the basic characters and shows the "fixed entity" situation.

We next consider characters induced by those of subgroup GL(1, 3; q) of index $(q+1)(q^2+1)$. In a manner analogous to those obtained of GL(3, q) from GL(1, 2; q), we get irreducible characters of the degrees and frequencies[2] shown in Table X:

TABLE X

Degree	Frequency
$(q+1)(q^2+1)$	$(q-1)(q-2)$
$q(q+1)^2(q^2+1)$	$(q-1)(q-2)$
$q^3(q+1)(q^2+1)$	$(q-1)(q-2)$
$(q+1)(q^2+1)(q^2+q+1)$	$\frac{1}{2}(q-1)(q-2)(q-3)$
$q(q+1)(q^2+1)(q^2+q+1)$	$\frac{1}{2}(q-1)(q-2)(q-3)$
$(q+1)^2(q^2+1)(q^2+q+1)$	$\frac{1}{24}(q-1)(q-2)(q-3)(q-4)$
$(q-1)(q+1)(q^2+1)(q^2+q+1)$	$\frac{1}{4}q(q-1)^2(q-2)$
$(q-1)^2(q+1)^2(q^2+1)$	$\frac{1}{3}q(q-1)^2(q+1)$

In the same way, the subgroup GL(2, 2; q) yields the irreducible characters shown in Table XI:

TABLE XI

Degree	Frequency
$(q^2+1)(q^2+q+1)$	$\frac{1}{2}(q-1)(q-2)$
$q^2(q^2+1)(q^2+q+1)$	$\frac{1}{2}(q-1)(q-2)$
$q(q^2+1)(q^2+q+1)$	$(q-1)(q-2)$
$(q-1)(q^2+1)(q^2+q+1)$	$\frac{1}{2}q(q-1)^2$
$q(q-1)(q^2+1)(q^2+q+1)$	$\frac{1}{2}q(q-1)^2$
$(q-1)^2(q^2+1)(q^2+q+1)$	$\frac{1}{8}q(q-1)(q+1)(q-2)$.

[2]The actual characters of GL(4, q) with a more detailed account of the methods are available in [10].

As a bi-product of the set of characters of degree $(q-1)^2(q^2+1)(q^2+q+1)$ we obtain $\frac{1}{2}q(q-1)$ characters of this degree each of which is the sum of two irreducible characters which are not among those that we have already obtained. Let us denote them by $\chi^{(n)}$, $n = 1, 2, \ldots, \frac{1}{2}q(q-1)$.

Finally, the linear characters of the cyclic subgroup of order q^4-1,

$$\begin{pmatrix} \omega & & & \\ & \omega^q & & \\ & & \omega^{q^2} & \\ & & & \omega^{q^3} \end{pmatrix}^a,$$

induce in GL$(4, q)$ a set of characters of degree $q^6(q-1)^3(q+1)(q^2+q+1)$. Each of these is reducible, but by a suitable use of the characters already obtained, i.e., by multiplication, addition and subtraction, a set of $\frac{1}{4}q^2(q-1)(q+1)$ irreducible characters of degree $(q-1)^3(q+1)(q^2+q+1)$ can be extracted from them. Again there is a bi-product: $\frac{1}{2}q(q-1)$ pseudocharacters of degree $(q-1)^3(q+1)(q^2+q+1)$ each of which is the difference of two irreducible characters. Denote them by $\psi^{(n)}$. Then, if the proper correlation is made be-

TABLE XII
Characters of PGL$(4, q)$

Degrees	Frequencies		
	$q = 4t$ or $4t+2$	$q = 4t+1$	$q = 4t+3$
1	1	4	2
(10) (111)	1	4	2
$(10)^2(101)$	1	4	2
$(10)^3(111)$	1	4	2
$(10)^6$	1	4	2
(11)(101)	$1-2$	$1-5$	$1-3$
(10) $(11)^2(101)$	$1-2$	$1-5$	$1-3$
$(10)^3(11)$ (101)	$1-2$	$1-5$	$1-3$
(11) (101) (111)	$\frac{1}{2}(1-2)(1-3)$	$\frac{1}{2}(1-6-13)$	$\frac{1}{2}(1-3)^2$
(10)(11)(101)(111)	$\frac{1}{2}(1-2)(1-3)$	$\frac{1}{2}(1-6-13)$	$\frac{1}{2}(1-3)^2$
$(11)^2(101)(111)$	$\frac{1}{24}(1-2)(1-3)(1-4)$	$\frac{1}{24}(1-5)(1-49)$	$\frac{1}{24}(1-3)(1-6-11)$
$(1-1)(11)(101)(111)$	$\frac{1}{4}(10)(1-1)(1-2)$	$\frac{1}{4}(1-1)^3$	$\frac{1}{4}(1-1)^3$
$(1-1)^2(11)^2(101)$	$\frac{1}{3}(10)(1-1)(11)$	$\frac{1}{3}(10)(1-1)(11)$	$\frac{1}{3}(10)(1-1)(11)$
(101)(111)	$\frac{1}{2}(1-2)$	$1-3$	$1-2$
$(10)^2(101)(111)$	$\frac{1}{2}(1-2)$	$1-3$	$1-2$
(10) (101)(111)	$1-2$	$2-6$	$2-4$
$(1-1)(101)(111)$	$\frac{1}{2}(10)(1-1)$	$\frac{1}{2}(1-1)^2$	$\frac{1}{2}(1-1)^2$
$(10)(1-1)(101)(111)$	$\frac{1}{2}(10)(1-1)$	$\frac{1}{2}(1-1)^2$	$\frac{1}{2}(1-1)^2$
$(1-1)^2(101)(111)$	$\frac{1}{8}(10)(11)(1-2)$	$\frac{1}{8}(1-1)(10-3)$	$\frac{1}{8}(11)(1-2-1)$
$(1-1)^2(111)$	$\frac{1}{2}(10)$	$1-1$	10
$(10)^2(1-1)^2(111)$	$\frac{1}{2}(10)$	$1-1$	10
$(1-1)^3(11)(111)$	$\frac{1}{4}(10)^2(11)$	$\frac{1}{4}(1-1)(11)^2$	$\frac{1}{4}(1-1)(11)^2$

tween the $\chi^{(n)}$'s and the $\psi^{(n)}$'s, it turns out that $\frac{1}{2}(\chi^{(n)}+\psi^{(n)})$ and $\frac{1}{2}(\chi^{(n)}-\psi^{(n)})$ are irreducible characters. In this way we obtain $\frac{1}{2}\,q(q-1)$ irreducible characters of each of the degrees $q^2(q-1)^2(q^2+q+1)$ and $(q-1)^2(q^2+q+1)$. This completes the character list since we have now obtained $q^4-q=k$ of them.

In cutting down the characters of GL(4, q) to get those of PGL(4, q), three cases are distinct: q even, $q=4t+1$, $q=4t+3$. Table XII gives the degrees and frequencies in each of these cases. For convenience in notation, we shall mean by $\frac{1}{3}$ (10-11), for example, $\frac{1}{3}\,(q^3-q+1)$, etc.

References

[1] H. W. Brinkmann, Bull. Amer. Math. Soc., vol. 27 (1921), 152.

[2] L. E. Dickson, *Linear Groups in Galois Fields* (Leipzig, 1901).

[3] G. Frobenius, *Über Gruppencharaktere*, Berliner Sitz. (1896), 985.

[4] ———, *Über die Darstellung der endlichen Gruppen durch lineare Substitutionen*, Berliner Sitz. (1897), 994.

[5] ———, *Über Relationen zwischen den Charakteren einer Gruppe und denen ihrer Untergruppen*, Berliner Sitz. (1898), 501.

[6] H. Jordan, *Group-Characters of Various Types of Linear Groups*, Amer. J. of Math., vol. 29 (1907), 387.

[7] I. Schur, *Untersuchungen über die Darstellung der endlichen Gruppen durch gebrochene lineare Substitutionen*, J. für Math., vol. 132 (1907), 85.

[8] A. Speiser, *Die Theorie der Gruppen von endlicher Ordnung* (Berlin, 1937).

[9] R. Steinberg, *A Geometric Approach to the Representations of the Full Linear Group over a Galois Field*, submitted to Trans. Amer. Math. Soc.

[10] ———, *Representations on the Linear Fractional Groups*, Thesis, University of Toronto Library.

University of California at Los Angeles

PRIME POWER REPRESENTATIONS OF FINITE LINEAR GROUPS

ROBERT STEINBERG

1. Introduction. There are five well-known, two-parameter families of simple finite groups: the unimodular projective group, the symplectic group,[1] the unitary group,[2] and the first and second orthogonal groups, each group acting on a vector space of a finite number of elements **(2**; **3)**. If k is the dimension of this space, we denote these groups by $\mathfrak{L}_k$, $\mathfrak{S}_k$, $\mathfrak{U}_k$, $\mathfrak{O}_k$ and $\mathfrak{O}_k'$, respectively. By analogy, groups $\mathfrak{O}_2$, $\mathfrak{O}_4$ and $\mathfrak{O}_2'$ (which are not simple) can be defined. Our main conern then is the proof of the following result:

THEOREM. *Let $\mathfrak{G}$ be one of the groups $\mathfrak{L}_k$, $\mathfrak{S}_k$, $\mathfrak{U}_k$, $\mathfrak{O}_k$ or $\mathfrak{O}_k'$ with $k \geqslant 2$. Let p be the characteristic of the base field, let d be the order of a p-Sylow subgroup $\mathfrak{P}$ of $\mathfrak{G}$, and let m be the index of the normalizer of $\mathfrak{P}$ in $\mathfrak{G}$. Let Σ be any vector space of dimension d over a field of characteristic* 0 *or prime to m. Then $\mathfrak{G}$ has an irreducible representation of degree d with Σ as the representation space.*

The special case $\mathfrak{G} = \mathfrak{L}_2$ was proved by Jordan **(7)** and Schur **(12)**, independently; the case $\mathfrak{G} = \mathfrak{L}_3$ by Brinckmann **(1)**; and the case $\mathfrak{G} = \mathfrak{L}_n$ first by the present author **(13)** and then later by Green **(5)**. In **(4)**, Frame proved the theorem when $\mathfrak{G} = \mathfrak{U}_3$. All of these authors dealt only with the character of the representation, not with the representation itself. The methods of the present paper are constructive and yield the representation space and the representing matrices explicitly. It is hoped that the geometric ideas introduced in this construction may be of independent interest.

In §§2, 3, 4, and 5, the group $\mathfrak{L}_n$ is dealt with. In §6, the other groups are considered. In §7, a few observations are added.

As a general reference to the definitions and properties of the spaces and groups to be considered, we cite **(2)** and **(3)**.

2. Preliminary definitions and notations. Throughout §§2, 3, 4 and 5, V denotes a vector space of dimension n over a field of q elements and of characteristic p. The symbol S^r denotes an r-dimensional subspace of V; if the superscript is omitted, the dimension is to be taken as 1; subscripts are used to distinguish subspaces of the same dimension. The symbol $\{S^i, S^j, \ldots\}$ denotes the subspace spanned by $S^i, S^j, \ldots$.

Definition 1. An *r-simplex* is an ordered set of r linearly independent 1-spaces: $[S_1, S_2, \ldots, S_r]$. Each S_j is called a vertex of the simplex. An n-simplex is more briefly called a simplex.

Received January 16, 1956.

[1]Sometimes called the abelian group.

[2]Sometimes called the hyperorthogonal group.

Definition 2. A *composition sequence* (abbreviated to c.s.) is a sequence of n subspaces $[S^1, S^2, \ldots, S^n]$ such that $S^j \subset S^{j+1} (j = 1, 2, \ldots, n-1)$.

Definition 3. Let $\Delta = [S_1, S_2, \ldots, S_n]$ be a simplex and $\nabla = [S^1, S^2, \ldots, S^n]$ a c.s. Suppose that there exists a permutation σ of the numbers $1, 2, \ldots, n$ such that

$$S^j = \{S_{\sigma(1)}, S_{\sigma(2)}, \ldots, S_{\sigma(j)}\} \qquad (j = 1, 2, \ldots, n). \tag{1}$$

Then ∇ is called a *face* of Δ: a positively or negatively oriented face according as σ is even or odd. Each of the $n!$ faces of Δ determines an opposite face $\nabla_1 = [S_1^1, S_1^2, \ldots, S_1^n]$ defined by

$$S_1^j = \{S_{\sigma(n)}, S_{\sigma(n-1)}, \ldots, S_{\sigma(n-j+1)}\} \qquad (j = 1, 2, \ldots, n). \tag{2}$$

Our first result is a useful characterization of opposite faces:

LEMMA 1. *If $\nabla = [S^1, S^2, \ldots, S^n]$ and $\nabla_1 = [S_1^1, S_1^2, \ldots, S_1^n]$ are two faces of a simplex $\Delta = [S_1, S_2, \ldots, S_n]$ then a necessary and sufficient condition for ∇ and ∇_1 to be opposite is that*

$$S^j \cap S_1^{n-j} = 0 \qquad (j = 1, 2, \ldots, n-1). \tag{3}$$

If ∇ and ∇_1 are two c.s. for which (3) *holds, there is a simplex Δ, uniquely determined to within an ordering of its vertices, which has ∇ and ∇_1 as* (*opposite*) *faces.*

Proof. The assumption that ∇ and ∇_1 are opposite faces of Δ implies the existence of a permutation σ such that (1) and (2) hold. But then, since the S_j are linearly independent, (3) holds. If ∇ and ∇_1 are faces which are not opposite, there exists a permutation τ, different from σ, such that

$$S_1^j = \{S_{\tau(n)}, S_{\tau(n-1)}, \ldots, S_{\tau(n-j+1)}\} \qquad (j = 1, 2, \ldots, n).$$

If j is the first index such that $\sigma(j) \neq \tau(j)$, then $S_{\sigma(j)} \subset S^j \cap S_1^{n-j}$, contradicting (3). Suppose finally that ∇ and ∇_1 are c.s. for which (3) holds. Then it follows that

$$S_j = S^j \cap S_1^{n-j+1}$$

is 1-dimensional for each j. Thus these S_j are the only possible choices for vertices of a simplex Δ relative to which ∇ and ∇_1 are opposite faces. To complete the proof, we note that, for each j, $S_j \subset S^j$ but $S_j \not\subset S^{j-1}$. Thus the S_j are linearly independent so that $\Delta = [S_1, S_2, \ldots, S_n]$ is a simplex and the equations (1) and (2) hold with σ the identity.

3. The spaces Σ and Σ^*. We proceed now to define representation spaces and to develop some of their properties. If Δ is a simplex and ∇ a c.s., we introduce an inner product (Δ, ∇), defined to be 1, -1 or 0 according as ∇ is a positive face, a negative face or not a face of Δ. If F is an arbitrary but fixed field, we can extend this inner product, by linearity, to linear combinations of simplexes and to linear combinations of c.s. over F. In this way, relative to this inner product, dual spaces Σ and Σ^* are determined. Thus an element of Σ (Σ^*) is a linear combination of simplexes (c.s.), and it is defined to be 0 if and only if it is orthogonal to all elements of Σ^* (Σ).

If $\epsilon(\sigma)$ is defined to be 1 or -1 according as σ is even or odd, an immediate consequence of the definitions is the following:

LEMMA 2. *If* $[S_1, S_2, \ldots, S_n]$ *is a simplex and* σ *a permutation, then*

$$[S_{\sigma(1)}, S_{\sigma(2)}, \ldots, S_{\sigma(n)}] = \epsilon(\sigma)[S_1, S_2, \ldots, S_n].$$

If Δ is an $(n-1)$-simplex and S a linearly independent 1-dimensional subspace, then $[S, \Delta]$ is used in our next result to denote the n-simplex whose vertices are obtained by taking first S and then the vertices of Δ in a positive order.

LEMMA 3. *Let* $\{\Delta\}$ *be a set of* $(n-1)$*-simplexes, all contained in one* $(n-1)$*-dimensional subspace* S^{n-1} *of* V. *Let* S *be a* 1*-space not in* S^{n-1}. *Then* $\sum \Delta = 0$ *implies* $\sum [S, \Delta] = 0$.

Proof. To each face $\nabla = [S^1, S^2, \ldots, S^{n-1}]$ of Δ we make correspond n faces $\nabla_1, \nabla_2, \ldots, \nabla_n$ of $[S, \Delta]$ defined by

$\nabla_k = [S^1, S^2, \ldots, S^{k-1}, \{S, S^{k-1}\}, \{S, S^k\}, \ldots, \{S, S^{n-1}\}]$; $(k = 1, 2, \ldots, n)$. Then one sees that, for each k, $([S, \Delta], \nabla_k) = (-1)^{k-1}(\Delta, \nabla)$. The required result now follows by summation on Δ with ∇ and ∇_k held fast.

LEMMA 4. *Let* $S_1, S_2, \ldots, S_{n+1}$ *be* 1*-spaces, and, for* $k = 1, 2, \ldots, n+1$, *let* $\Delta_k = [S_1, S_2, \ldots, S_{k-1}, \tilde{S}_k, S_{k+1}, \ldots, S_{n+1}]$, *where* $\tilde{S}_k$ *denotes that this vertex is to be omitted. Then*

$$\sum{}' (-1)^k \Delta_k = 0, \tag{4}$$

the summation being over those Δ_k, *which are simplexes.*

Proof. Suppose first that no n of the S_j are linearly dependent. Let $\nabla = [S^1, S^2, \ldots, S^n]$ be any face of Δ_{n+1}. Thus there is a permutation σ such that (1) holds. It is easy to see that

$$(\Delta_{n+1}, \nabla) = \epsilon(\sigma), (\Delta_{\sigma(n)}, \nabla) = (-1)^{n-\sigma(n)}\epsilon(\sigma), (\Delta_j, \nabla) = 0 \text{ if } j \neq n+1, \sigma(n).$$

Thus the left side of (4) is orthogonal to each face of Δ_{n+1}. Similarly, it is orthogonal to each face of $\Delta_1, \Delta_2, \ldots, \Delta_n$ (since an interchange of two S_j changes the sign of this sum); hence it is 0.

In proving the general case, we may assume that n of the S_j, say $S_1, S_2, \ldots, S_n$, are linearly independent and that S_{n+1} is linearly dependent on $S_{n-k+1}, S_{n-k+2}, \ldots, S_n$ but on no smaller number of S_j $(j = 1, 2, \ldots, n)$. Then the k-dimensional case of the first part of the proof shows that the analogue of (4) holds for the $k+1$ k-simplexes formed from the vertices $S_{n-k+1}, \ldots, S_{n+1}$. By Lemma 3 (applied $n-k$ times), we may prefix each of these simplexes with the vertices $S_1, S_2, \ldots, S_{n-k}$ and get our result.

We now introduce convenient bases for Σ and Σ^*.

THEOREM 1. *Let* $\nabla_0 = [S_0^1, S_0^2, \ldots, S_0^n]$ *be a fixed c.s. Let* $\mathfrak{B}$ *be the set of simplexes* Δ *such that* $(\Delta, \nabla_0) = (-1)^{[\frac{1}{2}n]}$. *For each* Δ_j *in* $\mathfrak{B}$, *let* ∇_j *be the face opposite to* ∇_0. *Let* $\mathfrak{B}^*$ *be the set of such faces. Then the sets* $\mathfrak{B}$ *and* $\mathfrak{B}^*$ *are dual bases of* Σ *and* Σ^*.

Proof. We first prove that $\mathfrak{B}$ spans Σ. Let $\mathfrak{B}_r$ $(r = 0, 1, 2, \ldots, n-1)$ be the set of simplexes $\Delta = [S_1, S_2, \ldots, S_n]$ such that

$$S_0^j = \{S_1, S_2, \ldots, S_j\} \qquad (j = 1, 2, \ldots, r).$$

Thus $\mathfrak{B}_0$ consists of all simplexes, and $\mathfrak{B}_{n-1} = \mathfrak{B}$. We now show that any member of $\mathfrak{B}_r$ is the signed sum of at most $n - r$ members of $\mathfrak{B}_{r+1}$. Let

$$\Delta = [S_1, S_2, \ldots, S_r, \; T_{r+1}, \ldots, T_n]$$

be in $\mathfrak{B}_r$. Let S_{r+1} be any 1-space in $S_0^{r+1} \cap \{T_{r+1}, \ldots, T_n\}$. Then, by Lemma 4, applied to the $(n-r)$-space $\{T_{r+1}, \ldots, T_n\}$, the $(n-r)$-simplex $[T_{r+1}, \ldots, T_n]$ is a signed sum of at most $n - r$ $(n-r)$-simplexes, each of which has S_{r+1} as a vertex. By Lemma 3, Δ is a signed sum of at most $n - r$ members of $\mathfrak{B}_{r+1}$.

To complete the proof, we invoke Lemma 1, which implies that, if Δ_j, ∇_k are in $\mathfrak{B}, \mathfrak{B}^*$, then $(\Delta_j, \nabla_k) = \delta_{jk}$. Thus $\mathfrak{B}$ is linearly independent, hence is a basis of Σ, and $\mathfrak{B}^*$ is the dual basis of Σ^*.

COROLLARY. *A simplex* Δ *is the signed sum of those members* Δ_j *of* $\mathfrak{B}$ *which have a face* ∇_j *in common with* Δ *and which have* ∇_0 *as the opposite face, the signature being positive or negative according as the common face* ∇_j *does or does not have the same orientations on* Δ *and* Δ_j. *The sum consists of at most* $n!$ *terms. If* Δ *is not a member of* $\mathfrak{B}$, *the sum of these signatures is* 0.

Proof. The first two statements follow from the equation $\Delta = \sum (\Delta, \nabla_j) \Delta_j$ which is valid since $\mathfrak{B}$ and $\mathfrak{B}^*$ are dual bases. The equations $(\Delta_j, \nabla_0) = (-1)^{[\frac{1}{2}n]}$ then imply the third statement.

4. The Sylow subgroup $\mathfrak{P}$. We now turn to the group $\mathfrak{L}_n$ of unimodular projective transformations. Since we are concerned only with the permutations of simplexes effected by members of $\mathfrak{L}_n$, and since a scalar transformation leaves all simplexes fixed, we may work with $\mathfrak{L}_n$ via representative elements of the unimodular group. Similar considerations apply to the other groups dealt with in §6.

The order and existence of a useful p-Sylow subgroup of $\mathfrak{L}_n$ is given by the next two lemmas:

LEMMA 5. *The order of a p-Sylow subgroup of* $\mathfrak{G} = \mathfrak{L}_n$ *is* $d = q^{\frac{1}{2}n(n-1)}$.

LEMMA 6. *Let* ∇_0 *be a given c.s. Let* $\mathfrak{N}$ *be the set of elements of* $\mathfrak{G}$ *which leave* ∇_0 *fixed, and let* $\mathfrak{P}$ *be the subset of* $\mathfrak{N}$ *composed of elements whose orders are powers of p. Then* $\mathfrak{P}$ *is a p-Sylow subgroup of* $\mathfrak{G}$ *and* $\mathfrak{N}$ *is its normalizer in* $\mathfrak{G}$.

Proof. For Lemma 5, the order of $\mathfrak{G}$ is available **(2; 3)**. To prove Lemma 6, we let $\nabla_0 = [S_0^1, S_0^2, \ldots, S_0^n]$ and then choose an ordered basis $X = (x_1, x_2, \ldots, x_n)$ of V such that

$$S_0^j = \{x_1, x_2, \ldots, x_j\} \qquad (j = 1, 2, \ldots, n). \tag{5}$$

Then, relative to the basis X, $\mathfrak{N}$ consists of all subdiagonal matrices, and $\mathfrak{P}$ of those which in addition have only 1's on the main diagonal. All conclusions now easily follow.

We proceed to set up a 1-1 correspondence between the elements of the p-Sylow subgroup $\mathfrak{P}$ defined in Lemma 6 and the members of the basis $\mathfrak{B}$ of Σ defined in Theorem 1. Again let X be a basis of V satisfying (5). Relative to X, each element P of $\mathfrak{P}$ is represented by a matrix whose rows $s_1, s_2, \ldots, s_n$ may be interpreted as vectors in V. It is easy to see that the simplex

$$\Delta = (-1)^{[\frac{1}{2}n]}[\{s_1\}, \{s_2\}, \ldots, \{s_n\}]$$

is a member of $\mathfrak{B}$ and that the correspondence θ defined by $\theta P = \Delta$ is 1-1 from $\mathfrak{P}$ onto $\mathfrak{B}$. From the fact that each row in the product of two matrices is the image of the corresponding row of the first matrix under the transformation corresponding to the second matrix, it follows that $(\theta P_1)P_2 = \theta(P_1P_2)$, where P_1 and P_2 are any two elements of $\mathfrak{P}$. Thus the right multiplication by P on the set $\mathfrak{P}$ is mapped by θ onto the application of P to the set $\mathfrak{B}$; and this mapping is an isomorphism since θ is 1-1. We may sum up the results of this paragraph in the following theorems:

THEOREM 2. *The dimension of Σ (or Σ^*) is equal to the order of a p-Sylow subgroup of $\mathfrak{G}$.*

THEOREM 3. *If θ is the mapping from $\mathfrak{P}$ onto $\mathfrak{B}$ defined in the preceding paragraph, then θ induces an isomorphism between the right regular representation of $\mathfrak{P}$ and the group $\mathfrak{P}$ considered as acting on the set $\mathfrak{B}$. The group $\mathfrak{P}$ is simply transitive on the members of $\mathfrak{B}$.*

5. The representation $\mathfrak{R}$.

Two final geometric results are necessary for the proof of the main theorem.

LEMMA 7. *Let m be the index of the normalizer of a p-Sylow subgroup of $\mathfrak{G} = \mathfrak{L}_n$. Then*

(i) *m is the number of c.s.;*

(ii) $m = \left(\prod_{j=1}^{n} (q^j - 1)\right) \Big/ (q-1)^n.$

Proof. The first statement follows from Lemma 6 and the fact that the group G is transitive on the c.s. For the second statement, see **(13)**.

LEMMA 8. *Let $\Delta_0 = [S_1, S_2, \ldots, S_n]$ be a simplex. Let ∇_σ be the face of Δ_0 corresponding to the permutation σ. For each σ, let $\{\Delta_{\sigma j}\}$ $(j = 1, 2, \ldots)$ be the set* *of simplexes which have ∇_σ as a positive face. Let m be the integer defined by Lemma* 7. *Then*

$$\sum_{\sigma, j} \epsilon(\sigma)\, \Delta_{\sigma j} = m\, \Delta_0. \tag{6}$$

Proof. An arbitrary simplex Δ makes one appearance on the left side of (6) for each face that Δ has in common with Δ_0, the signature being positive or negative according as this face does or does not have the same orientation on Δ and Δ_0. This face determines a unique opposite face ∇_0 of Δ. If we keep ∇_0 fixed and sum over those terms of (6) which give rise to ∇_0 in this way, then, by the corollary to Theorem 1, we get Δ_0. We then sum over ∇_0 to get the stated result.

In the case that $\mathfrak{G} = \mathfrak{L}_n$, we now state our main result:

THEOREM 4. *Let $\mathfrak{G} = \mathfrak{L}_n$ and let $\mathfrak{R}$ be the representation induced*[3] *in the space Σ by $\mathfrak{G}$. Further suppose that $\mathfrak{P}$ is a p-Sylow subgroup in $\mathfrak{G}$, that $\mathfrak{B}$ is the basis of Σ defined by Theorem* 1, *and that d and m are the order of $\mathfrak{P}$ and the index of the normalizer of $\mathfrak{P}$, respectively.* (These numbers are given by Lemma 5 and Lemma 7.) *Then*

(i) *in the sense of Theorem* 3, *$\mathfrak{R}$ restricted to $\mathfrak{P}$ is equivalent to the right regular representation of $\mathfrak{P}$; the degree of $\mathfrak{R}$ is thus d;*

(ii) *relative to $\mathfrak{B}$, $\mathfrak{R}$ is represented by a set of matrices each of which has only entries of* 0, 1 *or* -1; *in each row, at most $n!$ non-zero entries occur, and their sum is* 0, *if the row has more than one such entry;*

(iii) *if the base field F of the space Σ has characteristic prime to m, then $\mathfrak{R}$ is irreducible—in particular, this is so if the characteristic is* 0 *or p.*

Proof. Statements (i) and (ii) follow from Theorem 3 and the corollary to Theorem 1. To prove (iii), we show that the enveloping algebra of $\mathfrak{R}$ consists of all linear transformations from Σ to Σ. First, choose a basis X of V such that (5) holds, and set $S_j = \{x_j\}$ $(j = 1, 2, \ldots, n)$. We now note that, corresponding to each permutation σ of the numbers $1, 2, \ldots, n$, there exists an element Q_σ of $\mathfrak{L}_n$ such that $S_j Q_\sigma = S_{\sigma(j)}$. If σ is even, Q_σ may be defined by $x_j Q_\sigma = x_{\sigma(j)}$; if σ is odd, Q_σ may be defined by $x_1 Q_\sigma = -x_{\sigma(1)}$, $x_j Q_\sigma = x_{\sigma(j)}$, $j \neq 1$. If we now let Δ_0 be the simplex $[S_1, S_2, \ldots, S_n]$, it follows, by Theorem 3 and Lemma 8, that, for each Δ_i in $\mathfrak{B}$,

$$\Delta_i\, Q = m\, \Delta_0, \tag{7}$$

where $Q = \sum \epsilon(\sigma) P_j Q_\sigma$, the summation being over all permutations σ and all elements P_j of $\mathfrak{P}$. Now, let ∇' be the face of Δ_0 opposite to ∇_0 and let $\mathfrak{B}' = \{\Delta_i'\}$ (with $\Delta_0' = (-1)^{[\frac{1}{2}n]}\Delta_0$) be the corresponding basis of Σ, as given by Theorem 1. By Lemma 1, the only member of $\mathfrak{B}'$ which has ∇_0 as a face is Δ_0'. By the corollary to Theorem 1 and by (7),

[3]Since the elements of $\mathfrak{G}$ leave (Δ, ∇) invariant, they induce well-defined linear transformations in Σ and Σ^*. A similar remark applies in the case of Theorem 4′.

(8) $$\Delta_0' Q = m\,\Delta_0',\ \Delta_i' Q = 0, \qquad i \neq 0.$$

By Theorem 3, applied to the basis $\mathfrak{B}'$, there exists, for each i, an element P_i' of $\mathfrak{G}$ such that

(9) $$\Delta_0' P_i' = \Delta_i'.$$

Now, for each pair i, j, we set

$$T_{ij} = \frac{1}{m} P_i'^{-1}\, Q P_j'.$$

By (8) and (9), it follows that $\Delta_i' T_{ij} = \Delta_j'$; $\Delta_k' T_{ij} = 0$, $k \neq i$. Since the T_{ij} form a basis for the linear transformations from Σ to Σ, the proof of irreducibility is complete.

6. The symplectic, unitary and orthogonal groups. In this section, we consider the modifications necessary in the preceding development if the group $\mathfrak{L}_n$ is replaced by the other classical linear groups.

In the case of the unitary group, V denotes a vector space over a field of q^2 elements and of characteristic p; in the other cases, the field is to have q elements.

The symplectic group, $\mathfrak{S}_{2n}$, has an invariant, skew, bilinear form of pairs of vectors, $x = (\alpha_i)$, $y = (\beta_i)$, which may be taken as

(*) $$(x, y) = \sum_{j=1}^{n} (\alpha_j \beta_{n+j} - \alpha_{n+j} \beta_j).$$

For the unitary group, $\mathfrak{U}_{2n}$, this is to be replaced by

(*) $$(x, y) = \sum_{j=1}^{n} (\alpha_j \bar{\beta}_{n+j} + \alpha_{n+j} \bar{\beta}_j),$$

with $\bar{\beta} = \beta^q$; for $\mathfrak{U}_{2n+1}$, a term $\alpha_{2n+1}\bar{\beta}_{2n+1}$ is added.

For the first orthogonal group, $\mathfrak{O}_{2n}$, we choose the quadratic form

(*) $$Q(x) = \sum_{j=1}^{n} \alpha_j \alpha_{n+j};$$

a term α^2_{2n+1} is to be added in the case of $\mathfrak{O}_{2n+1}$; an irreducible quadratic form in α_{2n+1} and α_{2n+2} is to be added for $\mathfrak{O}'_{2n+2}$, the second orthogonal group. In these three cases, we introduce the inner product

$$(x, y) = Q(x + y) - Q(x) - Q(y).$$

Thus, in all cases, the concept of orthogonality of pairs of vectors exists.

Unless the contrary is stated, it is assumed in what follows that $\mathfrak{G}$ is any one of the groups S_{2n}, $\mathfrak{U}_{2n}$, $\mathfrak{U}_{2n+1}$, $\mathfrak{O}_{2n}$, $\mathfrak{O}_{2n+1}$ or $\mathfrak{O}'_{2n+2}$ with $n \geqslant 1$, the group $\mathfrak{G} = \mathfrak{O}_{2n+1}$ being excluded if q is even, since then $\mathfrak{G}$ is isomorphic to $\mathfrak{S}_{2n}$.

The symbol $c(S^r)$ denotes the subspace orthogonal to S^r.

Definition 4. If V underlies an orthogonal group, and if q is even, a subspace is *isotropic* if each of its vectors annuls the quadratic form Q. In all other cases, a subspace is isotropic if every two of its vectors are orthogonal.

Definition 5. A special $2r$-simplex is an ordered set of $2r$ isotropic 1-spaces $[S_1, S_2, \ldots, S_{2r}]$ for which there exist vectors s_j in S_j such that

$$(s_j, s_k) = 0, \quad (s_{r+j}, s_{r+k}) = 0, \quad (s_j, s_{r+k}) = \delta_{jk} \qquad (j, k = 1, 2, \ldots, r).$$

It is clear that the vertices of such a simplex are linearly independent. The vertices S_j and S_{r+j} are termed *opposite*. We shorten "special $2n$-simplex" to "simplex".

The existence of isotropic n-spaces and of simplexes follows at once from the equations (*). In each case, the first n basis vectors span an isotropic n-space and the $2n$ 1-spaces generated by the first $2n$ basis vectors are the vertices of a simplex.

Definition 6. A *special composition sequence* (s.c.s.) is a sequence of n isotropic subspaces $[S^1, S^2, \ldots, S^n]$ such that $S^j \subset S^{j+1}$ $(j = 1, 2, \ldots, n-1)$.

Definition 7. An *admissible permutation* (a.p.) of the numbers $1, 2, \ldots, 2n$ is a permutation σ such that $\sigma(n+j) \equiv n + \sigma(j) \pmod{2n}$ $(j = 1, 2, \ldots, n)$.

It is to be noted that an a.p. is determined by its effect on $1, 2, \ldots, n$. The a.p. form a group of order $2^n\, n!$ isomorphic to the hyper-octahedral group **(15)**. Each a.p. σ induces a permutation $\bar{\sigma}$ of the n pairs $(j, n+j)$. We set $\epsilon'(\sigma) = \epsilon(\sigma)\,\epsilon(\bar{\sigma})$.

Definition 8. Let $\Delta = [S_1, S_2, \ldots, S_{2n}]$ be a simplex and $\nabla = [S^1, S^2, \ldots, S^n]$ an s.c.s. Suppose that there exists an a.p. σ such that

$$S^j = \{S_{\sigma(1)}, S_{\sigma(2)}, \ldots, S_{\sigma(j)}\} \qquad (j = 1, 2, \ldots, n).$$

Then ∇ is termed a face of Δ: a positively or negatively oriented face according as $\epsilon'(\sigma)$ is 1 or -1. The face ∇_1 of Δ which is opposite to ∇ is defined by

$$\nabla_1 = [S_1{}^1, S_1{}^2, \ldots, S_1{}^n],\ S_1^j = \{S_{\sigma(n+1)}, S_{\sigma(n+2)}, \ldots, S_{\sigma(n+j)}\} \quad (j = 1, 2, \ldots, n).$$

The spaces Σ and Σ^* are defined as in §3.

Lemmas 1, 2, 3 and 4 have analogues which are:

LEMMA 1′. *If $\nabla = [S^1, S^2, \ldots, S^n]$ and $\nabla_1 = [S_1{}^1, S_1{}^2, \ldots, S_1{}^n]$ are two faces of a simplex $\Delta = [S_1, S_2, \ldots, S_{2n}]$, then a necessary and sufficient condition for ∇ and ∇_1 to be opposite is that*

$$(3') \qquad S^j \cap c(S_1^j) = 0, \qquad j = 1, 2, \ldots, n.$$

If ∇ and ∇_1 are two s.c.s. for which (3′) holds, there exists a simplex Δ, uniquely determined to within an ordering of its vertices, which has ∇ and ∇_1 as (opposite) faces.

LEMMA 2′. *If $[S_1, S_2, \ldots, S_{2n}]$ is a simplex and σ is an a.p., then*

$$[S_{\sigma(1)}, S_{\sigma(2)}, \ldots, S_{\sigma(2n)}] = \epsilon'(\sigma)[S_1, S_2, \ldots, S_{2n}].$$

LEMMA 3′. *Let* $[S_1, S_2]$ *be a special* 2-*simplex contained in a* 2-*space* S^2. *Let* $\{\Delta\}$ *be a collection of special* $(2n-2)$-*simplexes contained in* $c(S^2)$. *For each* Δ, *let* Δ' *be the special* $2n$-*simplex which has* S_1 *and* S_2 *as its first and* $(n+1)$*st vertices and the vertices of* Δ, *taken in positive order, as its remaining vertices. Then* $\sum \Delta = 0$ *implies* $\sum \Delta' = 0$.

LEMMA 4′. *Let* Δ *be a simplex and* S *an isotropic* 1-*space. Then* Δ *can be expressed as a sum of simplexes each of which has* S *as a vertex.*

Proof. The proofs of Lemmas 1′, 2′ and 3′ are virtually the same as those of Lemmas 1, 2 and 3, and so may be omitted. As a first step in the proof of Lemma 4′, we check two special cases. If $n = 1$, $\Delta = [S_1, S_2]$, $S \neq S_1$, $S \neq S_2$, then it is easy to verify that $[S_1, S]$ and $[S, S_2]$ are simplexes (see Definition 5), and that

$$[S_1, S_2] = [S_1, S] + [S, S_2].$$

Next suppose that $n = 2$, $\Delta = [S_1, S_2, S_3, S_4]$, and that S is orthogonal to exactly one vertex of Δ, say to S_1. Then, if

$$T = c(S) \cap \{S_2, S_3\}, \quad U = c(S) \cap \{S_3, S_4\},$$

the required conclusion may be drawn from the equation

$$[S_1, S_2, S_3, S_4] = [S, T, S_3, U] + [S_1, S_2, T, S] + [S_1, S, U, S_4].$$

The rest of the proof consists in showing that any other case can be reduced to one of these two cases. We may suppose that $n \geqslant 2$ and that S is not orthogonal to a pair of opposite vertices, say S_1 and S_{n+1}, since, then, the restriction to $c(\{S_1, S_{n+1}\})$ and an application of Lemma 3′ effectively replaces n by $n-1$. Thus we may suppose that the two vertices S_{n+1} and S_{n+2} are not orthogonal to S. Now set

$$T = c(S) \cap \{S_{n+1}, S_{n+2}\}, \quad U = c(T) \cap \{S_1, S_2\}.$$

Then the following is a relation among special 4-simplexes, all in one 4-space:

$$[S_1, S_2, S_{n+1}, S_{n+2}] = [S_1, U, T, S_{n+2}] + [U, S_2, S_{n+1}, T]. \tag{10}$$

By Lemma 3′, if the vertices $S_3, \ldots, S_n, S_{n+3}, \ldots, S_{2n}$ are adjoined to these 4-simplexes, an expression is obtained for Δ as a sum of two simplexes each of which has at least one vertex orthogonal to S. If $n \geqslant 3$, this construction can be repeated, with the indices 1 and 2 replaced by 2 and 3, to yield a second vertex orthogonal to S. Finally, if $n \geqslant 2$ and S is orthogonal to two vertices of Δ (which may be taken as non-opposite), say to S_1 and S_2, and not orthogonal to S_{n+1} and S_{n+2}, then the same construction yields, on the right side of (10), two simplexes, each of which has a pair of opposite vertices orthogonal to S; and this case has already been considered.

In the statement of Theorem 1, the number $(-1)^{[\frac{1}{2}n]}$ is to be replaced by $(-1)^n$, in the present case; in the corollary to Theorem 1, $n!$ by $2^n\, n!$. No changes are required in the proof.

The analogue of Lemma 5 is:

LEMMA 5′. *The order of a p-Sylow subgroup of* $\mathfrak{G}$ *is*

$$d = q^{n^2},\quad q^{n(2n-1)},\quad q^{n(2n+1)},\quad q^{n(n-1)},\quad q^{n^2} \quad \text{or} \quad q^{n(n+1)}$$

according as

$$\mathfrak{G} = \mathfrak{S}_{2n},\quad \mathfrak{U}_{2n},\quad \mathfrak{U}_{2n+1},\quad \mathfrak{O}_{2n},\quad \mathfrak{O}_{2n+1} \quad \text{or} \quad \mathfrak{O}'_{2n+2}.$$

Proof. See (**2**; **3**) for the order of $\mathfrak{G}$.

The statement of Lemma 6 goes over intact and the proof is similar; so both may be omitted. The same remark applies to Theorems 2 and 3.

We now note an exception that occurs (only) in the case that $\mathfrak{G} = \mathfrak{O}_{2n}$. Then the isotropic n-spaces form two families such that two members of the same family (of opposite families) intersect in a space of dimension $n - r$ with r even (odd), and such that the elements of $\mathfrak{O}_{2n}$ permute these n-spaces within their separate families (**3**, p. 48). The first property implies that at most one-half of the 2^n isotropic n-spaces spanned by sets of n vertices of a given simplex can fail to intersect a given isotropic n-space; thus, in the corollary to Theorem 1, the number $n!$ may be replaced by $2^{n-1}\, n!$, in this case. The second property implies that the group $\mathfrak{O}_{2n}$ is not transitive on all of the s.c.s., only on one-half of them. Thus the analogue of Lemma 7 takes the following form:

LEMMA 7′. *Let m be the index of the normalizer of a p-Sylow subgroup of* $\mathfrak{G}$. *Then*

(i) *if* $\mathfrak{G} = \mathfrak{O}_{2n}$, *m is one-half the number of s.c.s.; if* $\mathfrak{G} = \mathfrak{S}_{2n}, \mathfrak{U}_{2n}, \mathfrak{U}_{2n+1}, \mathfrak{O}_{2n+1}$ *or* $\mathfrak{O}'_{2n+2}$, *m is the number of s.c.s.;*

(ii) *if* $\mathfrak{G} = \mathfrak{S}_{2n}, \mathfrak{U}_{2n}, \mathfrak{U}_{2n+1}, \mathfrak{O}_{2n}, \mathfrak{O}_{2n+1}$ *or* $\mathfrak{O}'_{2n+2}$, *then*

$$m = \Big(\prod_{j=1}^{n} (q^{2j} - 1)\Big) \Big/ (q-1)^n,\ \Big(\prod_{j=1}^{2n} (q^j - (-1)^j)\Big) \Big/ (q^2-1)^n,$$
$$\Big(\prod_{j=1}^{2n+1} (q^j - (-1)^j)\Big) \Big/ (q^2-1)^n,$$
$$(q^n - 1)\Big(\prod_{j=1}^{n-1} (q^{2j} - 1)\Big) \Big/ (q-1)^n,$$
$$\Big(\prod_{j=1}^{n} (q^{2j} - 1)\Big) \Big/ (q-1)^n \quad \text{or} \quad (q^{n+1} + 1)\Big(\prod_{j=2}^{n} (q^{2j} - 1)\Big) \Big/ (q-1)^{n-1}.$$

Proof. Part (ii) is easily established by counting the number of s.c.s. using induction on n. If S is an isotropic 1-space, one may invoke the induction hypothesis on the quotient space $c(S)/S$ with the induced definition of isotropy We omit the details.

In the modified statement of Lemma 8, only admissible permutations are to be considered; if $\mathfrak{G} = \mathfrak{O}_{2n}$, a further restriction is to be made to even permutations. No essential change occurs in the proof. The analogue of Theorem 4 may be stated as follows:

THEOREM 4′. *Let* $\mathfrak{G}$ *be one of the groups* $\mathfrak{S}_{2n}$, $\mathfrak{U}_{2n}$, $\mathfrak{U}_{2n+1}$, $\mathfrak{O}_{2n}$, $\mathfrak{O}_{2n+1}$ *or* $\mathfrak{O}'_{2n+2}$. *Let* $\mathfrak{R}$ *be the representation induced*[3] *in the space* Σ *by* $\mathfrak{G}$. *Further suppose that* $\mathfrak{P}$ *is a p-Sylow subgroup in* $\mathfrak{G}$, *that* $\mathfrak{B}$ *is the basis of* Σ *defined by Theorem* 1′, *and that d and m are the order of* $\mathfrak{P}$ *and the index of the normalizer of* $\mathfrak{P}$, *respectively.* (These numbers are given by Lemma 5′ and Lemma 7′). *Then all conclusions of Theorem* 4 *are valid if, in* (ii), *the number* $n!$ *is replaced by* $2^{n-1}\,n!$ *if* $\mathfrak{G} = \mathfrak{O}_{2n}$, *and by* $2^n\,n!$ *in all other cases.*

The proof of Theorem 4 carries over without essential change.

Theorems 4 and 4′ imply the theorem stated in the introduction.

7. Concluding remarks. Our first remarks take the form of two conjectures which, if true, provide converses to the theorem of the introduction:

CONJECTURE 1. *The group* $\mathfrak{G}$ *does not have an irreducible representation of degree d over a field whose characteristic divides m.*

We are able to prove the following weaker result:

THEOREM 5. *Using the notations of Theorems* 4 *and* 4′, *if the characteristic of F divides m, then the representation* $\mathfrak{R}$ *is reducible.*

Proof. It is convenient to introduce "boundary" operators b and b^* on Σ and Σ^*: for each simplex Δ, let $b\Delta$ be the signed sum of the faces of Δ, the sign of a face being that of its orientation on Δ; for each c.s. ∇, let $b^*\nabla$ be the sum of those simplexes which have ∇ as a positive face; then extend b and b^* to all of Σ and Σ^* by linearity. Lemmas 8 and 8′ may now be rewritten as: $b^*b\Delta_0 = m\Delta_0$. Thus, if Δ is a simplex and ∇ a c.s., it follows that $(b^*b\,\Delta, \nabla) = m(\Delta, \nabla)$, and this is easily seen to be equivalent to $(b^*\nabla, b\Delta) = m(\Delta, \nabla)$. The assumption $m = 0$ then implies that $b^*\,\Sigma^*$ and $b\Sigma$ are orthogonal. It is easy to see that neither of them is 0. Hence $b^*\,\Sigma^*$ is a proper non-zero subspace of Σ. This subspace is invariant under $\mathfrak{G}$: if G is an element of $\mathfrak{G}$ and ∇ is a c.s., then $(b^*\nabla)G = b^*(\nabla G)$. Thus $\mathfrak{R}$ is reducible.

CONJECTURE 2. *The notation being that of Theorems* 4 *and* 4′, *any proper subgroup* $\mathfrak{H}$ *of* $\mathfrak{G}$ *does not have an irreducible representation of degree d. In particular, the restriction of* $\mathfrak{R}$ *to* $\mathfrak{H}$ *is reducible.*

If $\mathfrak{G} = \mathfrak{L}_2, \mathfrak{L}_3, \mathfrak{U}_3$ or $\mathfrak{S}_4$, this statement follows from results of Moore **(11)**, Wiman **(14)**, Hartley **(6)** and Mitchell **(9; 10)**, who have shown that, in these cases, every proper subgroup $\mathfrak{H}$ of $\mathfrak{G}$ has order less than d^2.

In **(13)**, an alternative method is used to derive the character of $\mathfrak{R}$ in the case that $\mathfrak{G} = \mathfrak{L}_n$. There, use is made of a correspondence between $\mathfrak{L}_n$ and the symmetric group of degree n. If $\mathfrak{G}$ is one of the other groups considered in this paper, a similar correspondence exists between $\mathfrak{G}$ and the hyper-octahedral group of the appropriate degree, and yields the character of $\mathfrak{R}$. However, this method leans heavily on a previous determination of the characters of the symmetric and hyper-octahedral groups and does not deal with the representation itself.

Our final observation is that the special case $n = 3$ of the corollary to Theorem 1 also follows from a theorem on graphs (**8**, p. 126).

References

1. H. W. Brinckmann, *The group characteristics of the ternary linear fractional group and of various other groups*, Bull. Amer. Math. Soc., *27* (1921), 152.

2. L. E. Dickson, *Linear groups in Galois fields* (Leipzig, 1901).

3. J. Dieudonné, *La géométrie des groupes classiques*, Ergeb. Math. (Berlin, 1955).

4. J. S. Frame, *Some irreducible representations of hyperorthogonal groups*, Duke Math. J., *1* (1935), 442–448.

5. J. A. Green, *The characters of the finite linear groups*, Trans. Amer. Math. Soc., *80* (1955), 402–447.

6. R. W. Hartley, *Determination of the ternary collineation groups whose coefficients lie in the* $GF(2^n)$, Ann. of Math., ser. 2, *29* (1925–26), 140–158.

7. H. Jordan, *Group-characters of various types of linear groups*, Amer. J. of Math., *29* (1907), 387–405.

8. D. König, *Theorie der endlichen und unendlichen Graphen* (Chelsea, New York, 1950).

9. H. H. Mitchell, *Determination of the ordinary and modular ternary linear groups*, Trans. Amer. Math. Soc., *12* (1911), 207–242.

10. ———, *The subgroups of the quaternary abelian linear group*, Trans. Amer. Math. Soc., *15* (1914), 379–396.

11. E. H. Moore, *The subgroups of the generalized finite modular group*, Dec. Publ. Univ. of Chicago, *9* (1904), 141–190.

12. I. Schur, *Untersuchungen über die Darstellung der endlichen Gruppen durch gebrochene lineare Substitutionen*, J. für Math., *132* (1907), 85–137.

13. R. Steinberg, *A geometric approach to the representations of the full linear group over a Galois field*, Trans. Amer. Math. Soc., *71* (1951), 274–282.

14. A. Wiman, *Bestimmung aller Untergruppen einer doppelt unendlichen Reihe von einfachen Gruppen*, Handl. Svenska Vet.-Akad., *25* (1899), 1–47.

15. A. Young, *On quantitative substitutional analysis* (fifth paper), Proc. London Math. Soc., ser. 2, *31* (1930), 273–288.

Institute for Advanced Study

PRIME POWER REPRESENTATIONS OF FINITE LINEAR GROUPS II

ROBERT STEINBERG

1. Introduction. The aim of this paper is two-fold: first, to extend the results of **(4)** to the exceptional finite Lie groups recently discovered by Chevalley **(1)**, and, secondly, to give a construction which works simultaneously for the groups A_n, B_n, C_n, D_n, E_n, F_4 and G_2 (in the usual Lie group notation), and which depends only on intrinsic structural properties of these groups. It seems likely that the statements of this paper, especially (1) to (14) upon which the later work is based, also hold for the other known simple linear groups, namely the unitary and second orthogonal groups **(4)**.

Throughout this paper, the phrase "finite Lie group" and the symbol L refer to any of the groups in the first list above. We lean heavily on the properties of these groups developed by Chevalley **(1)**, and use his notations, slightly modified. The symbols u, h, etc., always denote elements of the groups U, H, etc., respectively, these groups being defined in §2.

2. Basic properties of finite Lie groups. In this section we set forth the properties of the groups $G = L$ to be used in the sequel. Statements (1) to (10) are extracted from **(1)**, and the others are proved in §4.

(1) *There exist two subgroups U and H of G such that $U \cap H = 1$, $U H$ is a group, and U is normal in $U H$* (**1**, *p.* 40, Lemma 9).

(2) *There exists a group W (the Weyl group) and for each $w \in W$ an element of G which is also denoted by w such that the union of the sets $H w$ is a group, H is a normal subgroup, and the quotient group is the isomorphic image of W under the map $w \to H w$* (**1**, p. 37 Lemma 3).

(3) *Corresponding to each $w \in W$, U has two subgroups U_w' and U_w* (this is U_w'' in **(1)**) *such that*

(4) $U = U_w' U_w$,

(5) $w U_w' w^{-1} \subseteq U$ *and*

(6) $U_{w_0} = U$ *for some* $w_0 \in W$ (**1**, pp. 41–43).

(7) *G is the union of the sets $U H w U_w$, and*

$$u\, h\, w_1\, u_{w_1} = \bar{u}\, \bar{h}\, w_2\, u_{w_2}$$

implies that $u = \bar{u}$, $h = \bar{h}$, $w_1 = w_2$, and

$$u_{w_1} = u_{w_2}$$

(**1**, p. 42, Theorem 2).

Received November 26, 1956.

(8) *W contains a set of elements* $\{w_\alpha\}$ *(the fundamental reflections) such that*

(9) $w_\alpha^2 = 1$ *for each* α *and*

(10) $\{w_\alpha\}$ *generates* W (**3**, p. 16–05, Theorem 1).

(11) *For each* α, *if* $w = w_\alpha$, *we set* $U_w = U_\alpha$ *and* $U_w' = U_\alpha'$; *then the union of* $U_\alpha H$ *and* $U_\alpha H w_\alpha U_\alpha$ *is a group.*

(12) *For each* w *and* α, *at least one of*

$$U_\alpha \subseteq U_w', \; U_\alpha \subseteq U_{ww_\alpha}'$$

holds.

(13) *There is a homomorphism* ϵ *of* W *onto the group* $\{1, -1\}$ *of two elements such that* $\epsilon(w_\alpha) = -1$ *for each* α.

(14) *There is an element* u *such that* $u \notin U_w'$ *for all* $w \neq 1$.

These properties are not independent. For example, the condition $U \cap H = 1$ of (1) follows from (7), and (9) follows from (4), (5), (7) and (11). The complete list is given here for ready reference in what follows.

3. The ideal I and the representation R. In this section, we assume that G is a group for which (1) to (14) hold. Let F be any field and A the group algebra of G over F. For any subset S of G, let the symbol S also denote the sum of the members of S considered as elements of A, and $|S|$ the cardinality of S. Let e be the element of A defined by

$$e = U H \Sigma \epsilon(w)\, w, \tag{15}$$

the summation being over the elements of W. Then we can state the following fundamental result.

LEMMA 1. *Let e be defined by* (15). *Then* (i) $e\, w_\alpha = -e$; (ii) *if* $u_\alpha \neq 1$, *then*

$$w_\alpha u_\alpha w_\alpha^{-1} = \tilde{u}_\alpha h\, w_\alpha \bar{u}_\alpha \tag{16}$$

for some $\tilde{u}_\alpha$, $\bar{u}_\alpha$ *and* h, *and*

$$e\,(u_\alpha w_\alpha^{-1} - \bar{u}_\alpha + 1) = 0. \tag{17}$$

Proof. Equation (i) follows from (2) and (13). By (7), $w_\alpha u_\alpha \notin U_\alpha H w_\alpha$. Thus $w_\alpha u_\alpha w_\alpha^{-1} \notin U_\alpha H$, and (11) implies (16). By (9) and (12), each right coset of W relative to the group $\{1, w_\alpha\}$ consists of two elements v and vw_α such that $U_\alpha \subseteq U_v'$. Hence, by (1), (2), (5), (9), (13) and (16), we get

$$U H v\, u_\alpha w_\alpha^{-1} = U H v\, w_\alpha, \tag{18}$$

$$U H v = U H v\, \bar{u}_\alpha \text{ and} \tag{19}$$

$$U H v\, w_\alpha u_\alpha w_\alpha^{-1} = U H v\, w_\alpha \bar{u}_\alpha. \tag{20}$$

Now to get (17), multiply (18), (19), and (20) by $\epsilon(v)$, $\epsilon(v)$ and $\epsilon(vw_\alpha)$, add, and then sum on the right cosets of W relative to the group $\{1, w_\alpha\}$.

THEOREM 1. *Let e be given by* (15), *and let I be the right ideal of A generated by e. Then the set* $B = \{e\, u,\, u \in U\}$ *is a vector space basis for I. The dimension of I is* $|U|$.

Proof. For a fixed u, by (4), (5), (6), and (7), the coset $U\,H\,w_0\,u$ is present in $e\,u$ with a coefficient $\epsilon(w_0)$ and is not present in $e\,u_1$ if $u_1 \neq u$. Thus B is linearly independent. Let I_0 be the linear span of B. Then $e \in I_0 \subseteq e\,A = I$. Thus $I_0 = I$ if it can be shown that I_0 is an ideal. For any u and h, $I_0\,u \subseteq I_0$ and $I_0\,h \subseteq I_0$, by (1) and (2). Also for any u and α, by (4) we can write $u = u_\alpha\,u_\alpha'$ and then $e\,u\,w_\alpha^{-1} = (e\,u_\alpha\,w_\alpha^{-1})\,(w_\alpha\,u_\alpha'\,w_\alpha^{-1}) \in I_0$, using Lemma 1 and (5); hence $I_0\,w_\alpha^{-1} \subseteq I_0$. By (2), (7), and (10), A is generated by elements of the form u, h and w_α^{-1}. Thus $I_0\,A \subseteq I_0$ and Theorem 1 is proved.

COROLLARY. *If $s \in I$ and $s = \Sigma\,\gamma_u\,e\,u$, $\gamma_u \in F$, then γ_u is $\epsilon(w_0)$ times the coefficient of $U\,H\,w_0\,u$ in the right coset decomposition of s relative to $U\,H$, and $\Sigma\,\gamma_u$ is the coefficient of $U\,H$. If $x \in G$ and $e\,x = \Sigma\,\gamma_u\,e\,u$, then $\gamma_u = 1, -1$ or 0, and at most $|W|$ of the γ_u are non-zero.*

Proof. The first statement follows from (6), (7) and Theorem 1. Since e is the signed sum of $|W|$ right cosets relative to $U\,H$, the same is true of $e\,x$, and the last statement now follows from the first.

LEMMA 2. *Let m be the index of $U\,H$ in G: $m = \Sigma|U_w|$. Then*

$$e\,U\,\Sigma\,\epsilon\,(w)\,w = m\,e. \tag{21}$$

Proof. For each $x \in G$, write $e\,x = \Sigma\,\gamma(x;\,u)\,e\,u$ so that

$$e = \Sigma\,\gamma(x;\;u)\;\;e\,u\,x^{-1}. \tag{22}$$

Here $\gamma(x;\,u)$ is $\epsilon(w\,w_0)$ if

$$U\,H\,w_0\,u = U\,H\,w\,x \tag{23}$$

for some w and is 0 otherwise, by the corollary to Theorem 1. Now (23) is equivalent to $u\,x^{-1} \in w_0^{-1}\,U\,H\,w$. Thus summation of (22) on $x \in G$ gives

$$|G|\;e = \Sigma_u\,\Sigma_w\epsilon(w\,w_0)\;e\,w_0^{-1}\;U\,H\,w = \Sigma_u e\;U\,H\;\Sigma_w\epsilon(w)\;w = |U\,H| e\;U\;\Sigma\epsilon(w)w$$

by (1), (2) and (13). If the base field F has characteristic 0, division by $|\;U\,H\;|$ yields (21). Since only integral coefficients occur in (21), it remains valid for any field.

We now state the principal result of this section.

THEOREM 2. *Let G be a group for which* (1) *to* (13) *hold. Let I and B be as in Theorem* 1, *let m be the index of $U\,H$ in G, and let R be the representation of G on I by right multiplication. Then*

(i) *R restricted to U is equivalent to the right regular representation of U; the degree of R is $|\;U\;|$;*

(ii) *relative to B, R is represented by a set of matrices each of which has only entries of* 1, $-$ 1 *or* 0 *with at most $|\;W\;|$ non-zero entries in each row;*

(iii) *if the base field F has characteristic* 0 *or prime to m, then R is irreducible.*

Proof. Theorem 1 and its corollary imply (i) and (ii). We prove (iii) by showing that any element $s \neq 0$ of I generates all of I under right multiplication by the elements of A, the linear span of G. If $s \neq 0$, there is a coset $U\,H\,x$ represented in s with a coefficient $\gamma \neq 0$. Then $U\,H$ has the coefficient γ in $s\,x^{-1}$. By the corollary to Theorem 1, $s\,x^{-1} = \Sigma\,\gamma_u\,e\,u$ with $\Sigma\,\gamma_u = \gamma$. Hence, by (1) and Lemma 2,

$$s\,x^{-1}\,(\gamma\,m)^{-1}\,U\,\Sigma\,\epsilon(w)\,w = (\gamma\,m)^{-1}\,\Sigma_u\,\gamma_u\,e\,u\,U\,\Sigma_w\,\epsilon(w)\,w = e.$$

Thus $I = e\,A \subseteq s\,A \subseteq I$, and (iii) is proved.

We use the condition (14) for the first time in the proof of the following converse to Theorem 2.

THEOREM 3. *In the notation of Theorem* 2, *if* (1) *to* (14) *hold for G and if the characteristic of F divides m, then R is reducible.*

Proof. By (6) and (7), $e\,U \neq 0$ so that $e\,U$ generates a non-zero invariant subspace I_1 of I. We complete the proof by showing that $I_1 \neq I$ or, equivalently, that $e \notin I_1$. Lemma 2, with the elements of G replaced by their inverses, implies, because of (1), (2), and (13), that $\Sigma\,\epsilon(w)\,w$ is a left annihilator of $e\,U$ and hence of I_1. But it is not a left annihilator of e. Indeed we show that an element u for which (14) holds appears in $\Sigma\,\epsilon(w)\,w\,e$ with a coefficient 1. Now $u \in w_1\,U\,H\,w_2$ implies that $w_1^{-1}\,u \in U\,H\,w_2$. By (2) (7), and (14), $w_1 = 1$ and then $w_2 = 1$. Again by (2) and (7), u can be written uniquely as $1\,\bar{u}\,\bar{h}\,1$, so that all assertions are proved.

4. Representations of finite Lie groups. In order to apply the results of §3 to the groups L, we first prove:

LEMMA 3. *Each group L has properties* (1) *to* (14).

Proof. As has already been stated in §2, (1) to (10) are extracted from **(1)**. In the proof of (11) to (14), we use the standard Lie algebra terminology **(2)**, and note that the index set $\{\alpha\}$ of (8) refers to a simple system of positive roots and the w_α are the reflections in these roots. It is proved in **(3**, p. 19–01, Lemma 1**)** that w_α maps α upon $-\alpha$ and permutes the remaining positive roots. Then (11) follows by **(1**, pp. 36, 41**)**. Also

$$(w\,w_\alpha)(\alpha) = w(-\alpha) = -w(\alpha).$$

Thus either $(w\,w_\alpha)\,(\alpha)$ or $w(\alpha)$ is positive, and (12) is true. For each w, let $n(w)$ be the number of roots r such that $r > 0$ and $w(r) < 0$. Then $\epsilon(w) = (-1)^{n(w)}$ fulfils the requirements of (13). Finally, if $w \neq 1$, then $w(\alpha) < 0$ for at least one α **(3**, p. 16–08, Theorem 2**)**, and so, by the same reasoning as above, $U_w' \subseteq U_\alpha'$ for some α, and similarly, $U_\beta \subseteq U_\gamma'$ if $\beta \neq \gamma$. Thus, if we choose $u_\alpha \neq 1$ for each α and set $u = \Pi u_\alpha$, then $u \notin U_\alpha'$ for each α, hence $u \notin U_w'$ for each $w \neq 1$, and (14) is proved.

If the group L is defined over a field of q elements and of characteristic p, it is further proved in **(1)** that U can be chosen as a p-Sylow subgroup of L,

UH as the normalizer of U, and then $|U_w| = q^{n(w)}$, with $n(w)$ as above. Thus we have:

THEOREM 4. *Let L be a finite Lie group over a field of q elements and of characteristic p. Let U be a p-Sylow subgroup of L, and let m be the index of the normalizer of U in L, this number being given by $m = \Sigma q^{n(w)}$. Let I be any vector space of dimension $|U|$ over a field of characteristic 0 or prime to m. Then L has an irreducible representation R of degree $|U|$ which has I as the representation space and for which the results of Theorem 2 are valid.*

REFERENCES

1. C. Chevalley, *Sur certains groupes simples*, Tohoku Math. J., *7* (1955), 14–66.
2. E. B. Dynkin, *The structure of semi-simple algebras*, Amer. Math. Soc. Translation No. 17.
3. Séminaire "Sophus Lie" (Paris, 1954–5).
4. R. Steinberg, *Prime power representations of finite linear groups*, Can. J. Math., *8* (1956), 580–591.

University of California, Los Angeles

ON THE NUMBER OF SIDES OF A PETRIE POLYGON

ROBERT STEINBERG

Let $\{p, q, r\}$ be the regular 4-dimensional polytope for which each face is a $\{p, q\}$ and each vertex figure is a $\{q, r\}$, where $\{p, q\}$, for example, is the regular polyhedron with p-gonal faces, q at each vertex. A Petrie polygon of $\{p, q\}$ is a skew polygon made up of edges of $\{p, q\}$ such that every two consecutive sides belong to the same face, but no three consecutive sides do. Then a Petrie polygon of $\{p, q, r\}$ is defined by the property that every three consecutive sides belong to a Petrie polygon of a bounding $\{p, q\}$, but no four do. Let $h_{p,q,r}$ be the number of sides of such a polygon, and $g_{p,q,r}$ the order of the group of symmetries of $\{p, q, r\}$. Our purpose here is to prove the following formula:

$$\frac{h_{p,q,r}}{g_{p,q,r}} = \frac{1}{64}\left(12 - p - 2q - r + \frac{4}{p} + \frac{4}{r}\right). \tag{1}$$

We use the following result of Coxeter (**1**, p. 232; **2**):

$$\frac{h_{p,q,r}}{g_{p,q,r}} = \frac{1}{16}\left(\frac{6}{h_{p,q} + 2} + \frac{6}{h_{q,r} + 2} + \frac{1}{p} + \frac{1}{r} - 2\right), \tag{2}$$

where $h_{p,q}$, for example, denotes the number of sides of a Petrie polygon of $\{p, q\}$. Both proofs referred to depend on the fact that the number of hyperplanes of symmetry of $\{p, q, r\}$ is $2h_{p,q,r}$. This is proved in a more general form in (**3**). Clearly (1) is a consequence of (2) and the following result:

If h is the number of sides of a Petrie polygon of the polyhedron $\{p, q\}$, then

$$h + 2 = \frac{24}{10 - p - q}. \tag{3}$$

Proof of (3). The planes of symmetry of $\{p, q\}$ divide a concentric sphere into congruent spherical triangles each of which is a fundamental region for the group $\mathfrak{G}$ of symmetries of $\{p, q\}$ (**1**, p. 81). The number of triangles is thus g, the order of $\mathfrak{G}$. The vertices of one of these triangles can be labelled P, Q, R so that the corresponding angles are π/p, π/q, $\pi/2$. There are $g/2p$ images of P under $\mathfrak{G}$, since the subgroup leaving P fixed has order $2p$. At each of these points there are $p(p-1)/2$ intersections of pairs of circles of symmetry. Counting intersections at the images of Q and R in a similar fashion, one gets for the total number of intersections of pairs of circles of symmetry the number

Received October 21, 1957.

$g(p + q - 1)/4$. However, the number of such circles is $3h/2$ (**1**, p. 68), and every two intersect in two points. Hence

$$\frac{g(p + q - 1)}{4} = \frac{3h}{2}\left(\frac{3h}{2} - 1\right). \tag{4}$$

Dividing (4) by the relation $g = h(h + 2)$ of Coxeter (**1**, p. 91), and solving for h, one obtains (3).

REFERENCES

1. H. S. M. Coxeter, *Regular polytopes* (London, 1948).
2. ———, *The product of the generators of a finite group generated by reflections*, Duke Math. J. *18* (1951), 765–782.
3. R. Steinberg, *Finite reflection groups*, submitted to Trans. Amer. Math. Soc.

University of California

EIGENVALUES OF THE UNITARY PART OF A MATRIX

ALFRED HORN AND ROBERT STEINBERG

1. Introduction. It is well known that every matrix A (square and with complex entries) has a polar decomposition $A = P_1U_1 = U_2P_2$, where U_i are unitary and P_i are unique positive semi-definite Hermitian matrices. If A is non-singular then $U_1 = U_2 = U$, where U is also unique. In this case we call U the unitary part of A. The eigenvalues of P_1 are the same as those of P_2.

In [2] the following problem was solved. Given the eigenvalues of P_1, what is the exact range of variation of the eigenvalues of A? The answer shows that a knowledge of the eigenvalues of P_1 puts restrictions only on the moduli of the eigenvalues of A. In this paper we are going to consider the corresponding question for the unitary part U of A. In turns out that a knowledge of the eigenvalues of U restricts only the arguments of the eigenvalues of A.

Before stating the result, we need some definitions. An ordered pair of n-tuples (λ_i), (α_i) of complex numbers is said to be *realizable* if there exists a non-singular matrix A of order n with eigenvalues λ_i such that the unitary part of A has eigenvalues α_i. If (γ_j) is an n-tuple of complex numbers of modulus 1, and if two of the γ_j are of the form e^{ib}, e^{ic} with $0 < b - c < \pi$ and $0 \leqq d \leqq (b-c)/2$, then the operation of replacing e^{ib}, e^{ic} by $e^{i(b-d)}$, $e^{i(c+d)}$ is called a *pinch* of (γ_j). In other words, a pinch of (γ_j) consists in choosing two of the γ_j which do not lie on the same line through 0 and turning them toward each other through equal angles.

If (a_i), (b_i) are n-tuples of real numbers, and if (a'_i), (b'_i) are their rearrangements in non-decreasing order, then we write $(a_i) \prec (b_i)$ when $\sum_r^n a'_i \leqq \sum_r^n b'_i$, $r = 2, \cdots, n$ and $\sum_1^n a'_i = \sum_1^n b'_i$. It is easily seen that the conditions are equivalent to the conditions $\sum_1^r a'_i \geqq \sum_1^r b'_i$, $r = 1, \cdots, n-1$, and $\sum_1^n a'_i = \sum_1^n b'_i$.

Our main theorem is the following.

THEOREM 1. *Let (λ_i), (α_i) be n-tuples of complex numbers such that $\lambda_i \neq 0$ and $|\alpha_i| = 1$. Then the following statements are equivalent:*

(1) *the pair (λ_i), (α_i) is realizable;*

(2) *(α_i) can be reduced to $(\lambda_i/|\lambda_i|)$ by a finite sequence of pinches;*

(3) *$\prod_1^n \alpha_i = \prod_1^n (\lambda_i/|\lambda_i|)$, and exactly one of the following holds:*

(a) *there is a line through 0 containing all the α_i and $(\lambda_i/|\lambda_i|)$ is a rearrangement of (α_i);*

(b) *there is no line through 0 containing all α_i but there is*

Received September 26, 1958.

a closed half plane H with 0 *on its boundary containing all* α_i, *and, if we choose a branch of the argument function which is continuous in* $H - \{0\}$, *then* $(\arg \lambda_i) \prec (\arg \alpha_i)$;

(c) *there is no closed half plane with* 0 *on its boundary which contains all* α_i.

The proof of Theorem 1 will be given at the end of the paper.

2. Definitions and preliminary results. Two matrices A and B are said to be *congruent* if there exists a non-singular matrix X such that $B = X^*AX$. A *triangular* matrix is a matrix such that all entries below the main diagonal are 0. If P is a positive definite matrix, then $P^{1/2}$ denotes the unique positive definite matrix whose square is P. We will use the symbol diag $(a_1, \cdots, a_n)$ to denote the diagonal matrix with diagonal elements $a_1, \cdots, a_n$.

LEMMA 1. *If* $\lambda_i \neq 0$ *and* $|\alpha_i| = 1$, *then the pair* (λ_i), (α_i) *is realizable if and only if there exists a matrix A with eigenvalues* λ_i *which is congruent to* $D = \text{diag}(\alpha_1, \cdots, \alpha_n)$.

Proof. We use the fact that for any two matrices B and C, BC and CB have the same eigenvalues. If (λ_i), (α_i) is realizable, there exists a unitary matrix U with eigenvalues α_i and a positive definite matrix P such that PU has eigenvalues λ_i. Let V be a unitary matrix such that $U = V^*DV$. Then PU has the same eigenvalues as $P^{1/2}V^*DVP^{1/2}$, which is congruent to D. Conversely, if X^*DX has eigenvalues λ_i, then so does $A = XX^*D$, and D is the unitary part of A since XX^* is positive definite.

LEMMA 2. *If* (λ_i), (α_i) *is realizable and* $\rho_i > 0$ *for each* i, *then* $(\rho_i\lambda_i)$, (α_i) *is realizable.*

Proof. Suppose $D = \text{diag}(\alpha_1, \cdots, \alpha_n)$ is congruent to a matrix A with eigenvalues λ_i. Then A is congruent to a triangular matrix B with diagonal elements λ_i. If $X = \text{diag}(\rho_1^{1/2}, \cdots, \rho_n^{1/2})$, then X^*BX obviously has eigenvalues $\rho_i\lambda_i$ and is congruent to D.

LEMMA 3. *If* (λ_i), (α_i) *is realizable and z is any complex number of modulus* 1, *then* $(z\lambda_i)$, $(z\alpha_i)$ *is realizable.*

LEMMA 4. *If* (μ_1, μ_2) *results from* (λ_1, λ_2) *by a pinch and T is a triangular matrix with diagonals elements* λ_1, λ_2, *then T is congruent to a matrix with eigenvalues* μ_1, μ_2.

Proof. By multiplication by a suitable constant, we may suppose that $\lambda_1 = e^{i\theta}$, $\lambda_2 = e^{-i\theta}$, and $\mu_1 = e^{i\phi}$, $\mu_2 = e^{-i\phi}$, where $0 \leqq \phi \leqq \theta < \pi/2$. It suffices to find a positive matrix P such that PT has eigenvalues $e^{\pm i\phi}$. Suppose

$$T = \begin{pmatrix} e^{i\theta} & a \\ 0 & e^{-i\theta} \end{pmatrix}.$$

Let

$$P = \begin{pmatrix} x & \bar{y} \\ y & x \end{pmatrix},$$

where $x \geqq 1$, $|y|^2 = x^2 - 1$ and $ya = |a|(x^2 - 1)^{1/2}$. Since P has determinant 1, we need only choose x so that the trace of PT is $2 \cos \phi$. The trace of PT is $f(x) = xe^{i\theta} + xe^{-i\theta} + ya = 2x \cos \theta + |a|(x^2 - 1)^{1/2}$. When $x = 1$, this is $2 \cos \theta$, and for $x \geqq 1$, $f(x)$ increases to infinity.

LEMMA 5. *If (α_i) can be reduced to $(\lambda_i / |\lambda_i|)$ by a finite number of pinches, then (λ_i), (α_i) is realizable.*

Proof. By Lemma 2 we may assume $|\lambda_i| = 1$. We need only prove the following: if (λ_i), (α_i) is realizable, if $|\lambda_i| = 1$ and if (μ_i) is a pinch of (λ_i), then (μ_i), (α_i) is realizable. We may suppose that the pinch consists in replacing λ_1, λ_2 by μ_1, μ_2. By hypothesis there exists a triangular matrix A with eigenvalues λ_i which is congruent to diag $(\alpha_1, \cdots, \alpha_n)$. By Lemma 4 there exists a two rowed non-singular matrix Z such that

$$B = Z^* \begin{pmatrix} \lambda_1 & a_{12} \\ 0 & \lambda_2 \end{pmatrix} Z$$

has eigenvalues μ_1, μ_2. Here a_{12} is the (1, 2) entry of A. If we set

$$Y = \begin{pmatrix} Z & 0 \\ 0 & I \end{pmatrix},$$

where I is the identity matrix of order $n - 2$, then

$$Y^*AY = \begin{pmatrix} B & C \\ 0 & D \end{pmatrix},$$

where D is triangular with diagonal elements $\lambda_3, \cdots, \lambda_n$. But this last matrix obviously has eigenvalues $(\mu_1, \mu_2, \lambda_3, \cdots, \lambda_n) = (\mu_1, \cdots, \mu_n)$.

LEMMA 6. *If $(a_1, \cdots, a_k) \prec (b_1, \cdots, b_k)$ and $(c_1, \cdots, c_p) \prec (d_1, \cdots, d_p)$ then $(a_1, \cdots, a_k, c_1, \cdots, c_p) \prec (b_1, \cdots, b_k, d_1, \cdots, d_p)$.*

Proof. A proof is given in [1; 63].

LEMMA 7. *If A is a matrix such that $(Ax, x) \neq 0$ and $0 < \arg (Ax, x) < \pi$ for all $x \neq 0$, then A is congruent to a unitary matrix.*

Proof. Let $H = (A + A^*)/2$, $K = (A - A^*)/2i$. Then $A = H + iK$, and H, K are Hermitian. Since $(Ax, x) = (Hx, x) + i(Kx, x)$, the hypothesis implies that $(Kx, x) > 0$ for all $x \neq 0$, so that K is positive definite. Therefore by [3; 261] H and K are simultaneously congruent to real diagonal matrices. Hence $A = H + iK$ is congruent to a diagonal unitary matrix.

LEMMA 8. *If A is congruent to a unitary matrix U with eigenvalues α_i, and if $0 < \arg \alpha_1 < \cdots < \arg \alpha_n < \pi$, then $(Ax, x) \neq 0$ for all $x \neq 0$ and*

$$\arg \alpha_j = \inf_{\substack{\dim S \\ = j}} \sup_{\substack{x \in S \\ x \neq 0}} \arg (Ax, x) = \sup_{\substack{\dim S \\ = n-j+1}} \inf_{\substack{x \in S \\ x \neq 0}} \arg (Ax, x)$$

where S ranges over subspaces of n-dimensional complex Euclidean space.

Proof. Let (u_i) be an ortho-normal sequence of eigenvectors of U corresponding to (α_i). If $A = X^*UX$, then $(Ax, x) = \sum_1^n \alpha_i |(Xx, u_i)|^2$. If S is the space spanned by $X^{-1}u_1, \cdots, X^{-1}u_j$, then

$$\sup_{\substack{x \in S \\ x \neq 0}} \arg (Ax, x) = \arg \alpha_j .$$

Now let S be any subspace of dimension j. Let M be the space spanned by $X^{-1}u_j, \cdots, X^{-1}u_n$. Then there exists a non-zero vector x in $M \cap S$. But

$$\arg (Ax, x) \geqq \inf_{y \neq 0} \arg \sum_j^n \alpha_i |(y, u_i)|^2 = \arg \alpha_j .$$

Therefore

$$\sup_{\substack{x \in S \\ x \neq 0}} \arg (Ax, x) \geqq \arg \alpha_j .$$

The proof of the second statement is analogous.

Lemma 8 is of course the analogue of the minimax principle for Hermitian matrices. The generalization due to Wielandt [4] also has an analogue for unitary matrices, which we mention without proof since it will not be used.

If A and U satisfy the hypotheses of Lemma 8 and $1 \leqq i_1 < \cdots < i_k \leqq n$, then

$$\arg \alpha_{i_1} + \cdots + \arg \alpha_{i_k} = \inf_{\substack{M_1 \subset \cdots \subset M_k \\ \dim M_p = i_p}} \sup_{x_p \in M_p} (\arg \beta_1 + \cdots + \arg \beta_k)$$

where $(x_1, \cdots, x_k)$ *ranges over linearly independent sequences of vectors, and the* β_j *are the eigenvalues of the matrix of order* k *whose* (i, j) *entry is* (Ux_i, x_j). *The number* $\arg \beta_1 + \cdots + \arg \beta_k$ *depends only on the subspace generated by* $x_1, \cdots, x_k$.

LEMMA 9. *If* $(\lambda_i), (\alpha_i)$ *is realizable and* $0 \leqq \arg \alpha_1 \leqq \cdots \leqq \arg \alpha_n \leqq \pi$, *then* $(\arg \lambda_i) \prec (\arg \alpha_i)$.

Proof. By Lemma 1, λ_i are the eigenvalues of X^*DX, where X is non-singular and $D = \text{diag}(\alpha_1, \cdots, \alpha_n)$. Since the eigenvalues of X^*DX vary continuously with the α_i, we need only prove the theorem for the case where $0 < \arg \alpha_1$, $\arg \alpha_n < \pi$. We proceed by induction on n. The statement being obvious when $n = 1$, suppose $n > 1$ and the theorem holds for matrices of order $n - 1$. Let A be a triangular matrix with eigenvalues λ_i which is congruent to D. Suppose the λ_i are arranged so that $\arg \lambda_1 \leqq \cdots \leqq \arg \lambda_n$. Let B be the principal minor of A formed from the first $n - 1$ rows and columns of A. If $x = (x_1, \cdots, x_{n-1})$ is a vector with $n - 1$ components and $y = (x_1, \cdots, x_{n-1}, 0)$ then $(Bx, x) = (Ay, y)$. Therefore for any such $x \neq 0$, $(Ax, x) \neq 0$ and

$$0 < \arg \alpha_1 \leqq \arg (Ay, y) = \arg (Bx, x) \leqq \arg \alpha_n < \pi ,$$

by Lemma 8, since A is congruent to D.

By Lemma 7, B is congruent to a unitary matrix V. Let the eigenvalues of V be β_i, where $\arg \beta_1 \leqq \cdots \leqq \arg \beta_{n-1}$. Since the quadratic form (Bx, x) associated with B is a restriction of the quadratic form associated with A, it follows from Lemma 8 that $\arg \alpha_{j+1} \geqq \arg \beta_j \geqq \arg \alpha_j$, $j = 1, \cdots, n-1$. Also by the induction hypothesis $(\arg \lambda_1, \cdots, \arg \lambda_{n-1}) \prec (\arg \beta_1, \cdots, \arg \beta_{n-1})$. Therefore
$\arg \lambda_1 + \cdots + \arg \lambda_r \geqq \arg \beta_1 + \cdots + \arg \beta_r \geqq \arg \alpha_1 + \cdots + \arg \alpha_r$,
$r = 1, \cdots, n-1$
and

$$\begin{aligned} \arg \alpha_2 + \cdots + \arg \alpha_n &\geqq \arg \lambda_1 + \cdots + \arg \lambda_{n-1} \\ &\geqq \arg \alpha_1 + \cdots + \arg \alpha_{n-1} . \end{aligned}$$

Hence

$$-\pi < \arg \lambda_n - \arg \alpha_n \leqq \sum_1^n (\arg \lambda_i - \arg \alpha_i) \leqq \arg \lambda_n - \arg \alpha_1 < \pi .$$

But

$$\prod_1^n \lambda_i = |\det X|^2 \cdot \prod_1^n \alpha_i .$$

Therefore

$$\sum_1^n \arg \lambda_i = \sum_1^n \arg \alpha_i \ .$$

The proof is complete.

LEMMA 10. *If* (β_i), (α_i) *are n-tuples of complex numbers of modulus* 1 *which lie on a line through* 0, *and if* (β_i), (α_i) *is realizable, then* (β_i) *must be a rearrangement of* (α_i).

Proof. By Lemma 3 we may suppose that the α_i and β_i are all real. Let A be a matrix with eigenvalues β_i which is congruent to $\operatorname{diag}(\alpha_1, \cdots, \alpha_n)$. Then A is Hermitian and therefore A is also congruent to $\operatorname{diag}(\beta_1, \cdots, \beta_n)$. But by Lemma 1 it follows that (α_i), (β_i) is realizable. Therefore by Lemma 9 we have $(\arg \beta_i) \prec (\arg \alpha_i) \prec (\arg \beta_i)$, from which the present theorem follows immediately.

LEMMA 11. *Suppose* (β_i), (α_i) *are n-tuples of complex numbers of modulus* 1 *such that* $\prod_1^n \beta_i = \prod_1^n \alpha_i$. *Then there exist determinations of* $\arg \alpha_i$, $\arg \beta_i$ *such that*

$$\max \arg \alpha_i - \min \arg \alpha_i \leqq 2\pi$$

and

$$(\arg \beta_i) \prec (\arg \alpha_i) \ .$$

Proof. The statement is obvious for $n = 1$. Suppose $n > 1$ and it holds for n-1-tuples. If any of the β_i is equal to any of the α_i, say $\beta_1 = \alpha_1$, then by the induction hypothesis, we can find determinations of the remaining $\arg \alpha_i$, $\arg \beta_i$ as stated. If we now choose a value of $\arg \alpha_1$ which lies between μ and $\mu + 2\pi$, where $\mu = \min_{i>1} \arg \alpha_i$, and set $\arg \beta_1 = \arg \alpha_1$, then the conditions of our theorem will be satisfied, by Lemma 6. So henceforth we may assume that $\beta_i \neq \alpha_j$ for all i, j.

As another special case, suppose the α_i are all equal, say to 1. If we assign arguments to the β_i such that $0 < \arg \beta_i < 2\pi$, then $\sum_1^n \arg \beta_i = 2\pi k$, where k is some positive integer $< n$. We need only assign arguments to the α_i such that exactly k of them have argument 2π and the remaining ones have argument 0.

Now assume the previous two cases do not occur. The α_i divide the unit circle into arcs. At least one of them must contain more than one of the β_i, for if not the α_i would be all distinct and each of the n arcs determined by them would contain exactly one of the β_i. We could then assign arguments to arrangements of the α_i, β_i so that

$$\arg \alpha_1 < \arg \beta_1 < \arg \alpha_2 < \cdots < \arg \alpha_n < \arg \beta_n < \arg \alpha_1 + 2\pi \ .$$

But then $0 < \sum_1^n \arg \beta_i - \sum_1^n \arg \alpha_i < 2\pi$, contradicting the hypothesis $\prod_1^n \alpha_i = \prod_1^n \beta_i$.

Let C be an arc containing more than one of the β_i. By changing subscripts, we may assume that the endpoints of C when described counterclockwise are α_1 and α_2. Let β_1 be one of the β_i in C which is nearest to α_1 and β_2 be one of the β_i with subscript $\neq 1$ which is nearest to α_2. Note that β_1 may equal β_2, but $\alpha_1 \neq \alpha_2$. As will be seen from the following argument, we may assume the subarc $\alpha_1\beta_1$ of $C \leqq$ the subarc $\beta_2\alpha_2$ of C, (all arcs are described counterclockwise). Let $\beta_1' = \alpha_1$ and let β_2' be the point in $\beta_2\alpha_2$ such that $\beta_2\beta_2' = \alpha_1\beta_1 = \delta$. By the first case of the proof, we may assign arguments to $\beta_1', \beta_2', \beta_3, \cdots, \beta_n$ and $\alpha_1, \cdots, \alpha_n$ so that

(1) $\max \arg \alpha_i - \min \arg \alpha_i \leqq 2\pi$

and

(2) $(\arg \beta_1', \arg \beta_2', \arg \beta_3, \cdots, \arg \beta_n) \prec (\arg \alpha_1, \cdots, \arg \alpha_n)$.

If $\arg \alpha_1$ happens to be the largest of $\arg \alpha_i$, and therefore $\arg \alpha_2$ is the smallest of $\arg \alpha_i$, then none of $\beta_1', \beta_2', \beta_3, \cdots, \beta_n$ can lie in the interior of C. Therefore $\beta_2' = \alpha_2$, and if we decrease $\arg \alpha_1$ and $\arg \beta_1$ by 2π, then (1) and (2) will still hold. Thus we may assume $\arg \alpha_1 < \arg \alpha_2$, and therefore $\arg \beta_1' < \arg \beta_2'$. Now assign to β_1 the argument $\beta_1' + \delta$ and to β_2 the argument $\arg \beta_2' - \delta$. Since

$$(\arg \beta_1' + \delta, \arg \beta_2' - \delta) \prec (\arg \beta_1', \arg \beta_2') ,$$

we have by Lemma 6,

$$\begin{aligned}(\arg \beta_1, \cdots, \arg \beta_n) &\prec (\arg \beta_1', \arg \beta_2', \arg \beta_3, \cdots, \arg \beta_n) \\ &\prec (\arg \alpha_1, \cdots, \arg \alpha_n) .\end{aligned}$$

This completes the proof.

LEMMA 12. *If* (β_i), (α_i) *are* n*-tuples of complex numbers of modulus* 1 *which can be assigned arguments such that*

$$\arg \alpha_1 \leqq \cdots \leqq \arg \alpha_n \leqq \arg \alpha_1 + 2\pi ,$$
$$\arg \beta_1 \leqq \cdots \leqq \arg \beta_n ,$$
$$(\arg \beta_i) \prec (\arg \alpha_i) ,$$

and

$$\arg \alpha_{i+1} - \arg \alpha_i < \pi, \ i = 1, \cdots, n-1 ,$$

then a finite number of pinches will reduce (α_i) to (β_i).

Proof. We proceed by induction on n. When $n = 2$, we have $\arg \alpha_1 \leqq \arg \beta_1 \leqq \arg \beta_2 \leqq \arg \alpha_2$, $\arg \alpha_1 + \arg \alpha_2 = \arg \beta_1 + \arg \beta_2$ and $\arg \alpha_2 - \arg \alpha_1 < \pi$. Therefore $\arg \beta_1 - \arg \alpha_1 = \arg \alpha_2 - \arg \beta_2$ and so

(β_1, β_2) is a pinch of (α_1, α_2).

Suppose $n > 2$ and the theorem holds for all m-tuples, $m < n$. Let

$$\delta = \min_{1 \leq p \leq n-1} \sum_1^p (\arg \beta_i - \arg \alpha_i) .$$

There exists k such that $\sum_1^k \arg \beta_i - \sum_1^k \arg \alpha_i = \delta$. It is easy to verify that

$$(\arg \beta_1, \cdots, \arg \beta_k) \prec (\arg \alpha_1 + \delta, \arg \alpha_2, \cdots, \arg \alpha_k)$$

and

$$(\arg \beta_{k+1}, \cdots, \arg \beta_n) \prec (\arg \alpha_{k+1}, \cdots, \arg \alpha_{n-1}, \arg \alpha_n - \delta) .$$

Also

$$\arg \alpha_1 + \delta \leqq \arg \beta_1 \leqq \arg \beta_n \leqq \arg \alpha_n - \delta .$$

By the induction hypothesis, we can reduce $(\alpha_1 e^{i\delta}, \alpha_2, \cdots, \alpha_k)$ to $(\beta_1, \cdots, \beta_k)$ and $(\alpha_{k+1}, \cdots, \alpha_{n-1}, \alpha_n e^{-i\delta})$ to $(\beta_{k+1}, \cdots, \beta_n)$ by a finite number of pinches. We need only show that $(\alpha_1, \cdots, \alpha_n)$ can be reduced to $(\alpha_1 e^{i\delta}, \alpha_2, \cdots, \alpha_{n-1}, \alpha_n e^{-i\delta})$ by a finite number of pinches. This will follow from the next lemma if we consider only the distinct α_i.

If the α_i all coincide, then so do the β_i and the statement of our theorem is trivial.

LEMMA 13. *If (α_i) is an m-tuple of numbers of modulus 1 with assigned arguments such that*

$$\arg \alpha_1 < \cdots < \arg \alpha_m \leqq \arg \alpha_1 + 2\pi$$

and

$$\arg \alpha_{i+1} - \arg \alpha_i < \pi, \quad i = 1, \cdots, m-1 ,$$

and if δ is a positive number such that $\arg \alpha_1 + \delta \leqq \arg \alpha_m - \delta$, then (α_i) can be reduced to $(\alpha_1 e^{i\delta}, \alpha_2, \cdots, \alpha_{m-1}, \alpha_m e^{-i\delta})$ by a finite number of pinches.

Proof. This is obvious for $m = 2$. Assume $m > 2$ and the lemma holds for $m - 1$ – tuples. If

$$\begin{aligned} \eta = \min(&\arg \alpha_2 - \arg \alpha_1, \pi - (\arg \alpha_3 - \arg \alpha_2), \cdots, \\ &\pi - (\arg \alpha_m - \arg \alpha_{m-1})) , \end{aligned}$$

and $0 < \varepsilon < \eta$, then each sequence in the following list is a pinch of the preceeding sequence:

$$\alpha_1, \cdots, \alpha_m$$

$$\alpha_1 e^{i\varepsilon},\ \alpha_2 e^{-i\varepsilon},\ \alpha_3,\ \cdots,\ \alpha_m$$
$$\alpha_1 e^{i\varepsilon},\ \alpha_2,\ \alpha_3 e^{-i\varepsilon},\ \cdots,\ \alpha_m$$
$$\cdots$$
$$\alpha_1 e^{i\varepsilon},\ \alpha_2,\ \cdots,\ \alpha_{m-2},\ \alpha_{m-1} e^{-i\varepsilon},\ \alpha_m$$
$$\alpha_1 e^{i\varepsilon},\ \alpha_2,\ \cdots,\ \alpha_{m-1},\ \alpha_m e^{-i\varepsilon}\ .$$

Note that $\arg\alpha_1 + \varepsilon$ need not be $\leqq \arg\alpha_2 - \varepsilon$, and $\arg\alpha_2$ need not be $\leqq \arg\alpha_3 - \varepsilon$, etc.

We may repeat this cycle of m pinches $k-1$ more times to pass from

$$\alpha_1 e^{i\varepsilon},\ \alpha_2,\ \cdots,\ \alpha_{m-1},\ \alpha_m e^{-i\varepsilon} \text{ to } \alpha_1 e^{ki\varepsilon},\ \alpha_2,\ \cdots,\ \alpha_{m-1},\ \alpha_m e^{-ki\varepsilon}$$

as long as $\arg\alpha_1 + k\varepsilon \leqq \arg\alpha_2$, since

$$\arg\alpha_2 + p\varepsilon - \arg\alpha_1 > \arg\alpha_2 - \arg\alpha_1$$

and

$$\pi - (\arg\alpha_n - p\varepsilon - \arg\alpha_{m-1}) > \pi - (\arg\alpha_n - \arg\alpha_{m-1})$$

for $p < k$. Therefore if $\delta \leqq \arg\alpha_2 - \arg\alpha_1$, we need only choose $\varepsilon = \delta/k$, where k is an integer so large that $\delta/k < \eta$. If $\delta > \arg\alpha_2 - \arg\alpha_1$, choose $\varepsilon = (\arg\alpha_2 - \arg\alpha_1)/k$, where k is so large that $\varepsilon < \eta$. Then $(\alpha_1, \cdots, \alpha_m)$ is reduced to $(\alpha_2, \alpha_2, \cdots, \alpha_{m-1}, \alpha_m e^{-ik\varepsilon})$ by the above sequence of pinches. By the induction hypothesis, $(\alpha_2, \alpha_3, \cdots, \alpha_{m-1}, \alpha_m e^{-ik\varepsilon})$ can by a finite number of pinches be reduced to $(\alpha_1 e^{i\delta}, \alpha_3, \cdots, \alpha_{m-1}, \alpha_m e^{-i\delta})$. (The fact that $\alpha_m e^{-ik\varepsilon}$ might be equal to one of the α_j is clearly unimportant.) Therefore $(\alpha_1, \cdots, \alpha_m)$ can be reduced to $(\alpha_1 e^{i\delta}, \alpha_2, \cdots, \alpha_{m-1}, \alpha_m e^{-i\delta})$, and the proof is complete.

3. Proof of Theorem 1.

(2) → (1): This is the statement of Lemma 5.

(1) → (3): If (λ_i), (α_i) is realizable, then by Lemma 1 there exists a matrix A and a non-singular matrix X such that $A = X^* \operatorname{diag}(\alpha_1, \cdots, \alpha_n) X$ and A has eigenvalues λ_i. Therefore $\prod\lambda_i = \prod\alpha_i \cdot |\det X|^2$ and hence $\prod\lambda_i/|\lambda_i| = \prod\alpha_i$. If the α_i lie on a line through 0, then $(\lambda_i/|\lambda_i|)$ is a rearrangement of (α_i) by Lemmas 2 and 10. If the α_i lie in a closed half plane through 0, then by Lemma 3 we may assume they lie in the upper half plane. By Lemma 9 it follows that $(\arg\lambda_i) \prec (\arg\alpha_i)$.

(3) → (2): In case (a), the statement is obvious. In case (c), Lemma 11 and the fact that the α_i do not lie in any closed half plane with 0 on its boundary show that the hypotheses of Lemma 12 are satisfied by arrangements of $(\lambda_i/|\lambda_i|)$, (α_i). In case (b), the hypotheses of

Lemma 12 also are satisfied by arrangements of $(\lambda_i/|\lambda_i|)$, (α_i). Thus an application of Lemma 12 completes the proof.

REFERENCES

1. G. H. Hardy, J. E. Littlewood, and G. Polya, *Inequalities*, Cambridge, 1952.

2. A. Horn, *On the eigenvalues of a matrix with prescribed singular values*, Proc. Amer. Math. Soc., **5** (1954), 4-7.

3. R. R. Stoll, *Linear algebra and matrix theory*, New York, 1952.

4. H. Wielandt, *An extremum property of sums of eigenvalues*, Proc. Amer. Math. Soc., **6** (1944), 106-110.

UNIVERSITY OF CALIFORNIA, LOS ANGELES

FINITE REFLECTION GROUPS

BY
ROBERT STEINBERG

1. **Introduction.** The finite groups generated by reflections (g.g.r.) of real Euclidean space of n dimensions (E^n) have been classified by Coxeter [4]. He has noticed a number of properties common to these groups, but has been able to prove them only by verification in the individual cases. Our prime purpose here is to give general proofs of some of these results (1.1 to 1.4 below).

If $\mathfrak{G}$ is a finite g.g.r. on E^n, the reflecting hyperplanes (r.h.) all pass through one point, which may be taken as the origin 0, and partition E^n into a number of chambers each of which is a fundamental region of $\mathfrak{G}$; further $\mathfrak{G}$ is generated by the reflections in the walls of any one of these chambers. The group $\mathfrak{G}$ is *irreducible* in the usual algebraic sense if and only if there are n linearly independent r.h. and there is no partition of the r.h. into two nonempty sets which are orthogonal to each other [7, p. 403]. In this case each chamber is a simplicial cone with vertex at 0 [3, p. 254; 4, p. 590].

This leads us to the first result of Coxeter [4, p. 610]:

1.1. THEOREM. *If $\mathfrak{G}$ is a finite irreducible g.g.r. on E^n and if h is the order of the product of the reflections in the walls of one of the fundamental chambers, then the number of reflecting hyperplanes is $nh/2$.*

Associated with each simple Lie algebra (or Lie group) of rank n over the complex field there is a finite irreducible g.g.r. $\mathfrak{G}$ on E^n and a set of vectors (roots) normal to the corresponding r.h. [1; 13]. There then exists a *fundamental set of roots* and a so-called *dominant root* relative to this set (definitions in §§6 and 8).

Then Coxeter's second observation [6, p. 234] is this:

1.2. THEOREM. *If $\alpha_1, \alpha_2, \cdots, \alpha_n$ is a fundamental set of roots for a simple Lie algebra of rank n, and if $\sum y^j\alpha_j$ is the dominant root, then the number of reflecting hyperplanes of the corresponding group $\mathfrak{G}$ (or one-half the number of roots) is $n(1+\sum y^j)/2$.*

From 1.2 (see [6, p. 212]) one immediately gets:

1.3. THEOREM. *The dimension of the Lie algebra (or Lie group) is $n(2+\sum y^j)$.*

As Coxeter [6, p. 212] has remarked, this is an interesting analogue to the formula of Weyl for the order of $\mathfrak{G}$, namely, $g=f\cdot n!\prod y^j$, with $f-1$ denoting the number of y's equal to 1.

Received by the editors, October 19, 1957.

The proof of 1.2 (and hence also of 1.3) is rather short and is independent of 1.1. If 1.1 and 1.2 are combined, the result is the formula [6, p. 234]:

1.4. THEOREM. $\sum y^i = h-1$.

Following the proof of 1.1, 1.2 and some related results concerning root systems and Petrie polygons, in the last section there is set forth another curious property of root systems which the author has recently discovered but has been unable to prove by general methods.

The abbreviations g.g.r. and r.h. introduced above are used throughout the paper.

2. **Preliminary lemmas.** Recall that a *tree* is a finite connected graph with no circuits and that an *end node* is one linked to at most one other node.

2.1. LEMMA. *A tree has an end node.*

2.2. LEMMA. *If each circuit of a graph has even length, then the nodes can be placed in two classes so that each link joins two nodes belonging to distinct classes.*

2.3. LEMMA. *Let $p_1, p_2, \cdots, p_n$ be the nodes of a tree T. Then any circular arrangement of the numbers $1, 2, \cdots, n$ can be obtained from any given one by a sequence of moves each of which consists of the interchange of a pair ij which satisfy the condition that i and j are adjacent on the circle and p_i and p_j are not directly linked on T.*

Proof. See [10, pp. 49, 151] for proofs of the first two results. Since 2.3 (cf. [4, p. 602]) is clearly true for $n=1$ or 2, we assume $n \geqq 3$ and proceed by induction. By 2.1 there is a node p_n joined to exactly one other node of T. If p_i and p_j are nodes other than p_n, at least one of them, say p_i, is not linked to p_n. Thus if the numbers inj occur in this order on the circle, and if p_i and p_j are not linked, the following moves can be made: $inj \to nij \to nji$; here i and j have been interchanged in the circular arrangement restricted to $1, 2, \cdots, n-1$. Thus by the inductive hypothesis applied to $T-p_n$ one can transform the numbers $1, 2, \cdots, n-1$ into any given arrangement, and one can then move the number n around the circle to its required position.

As is customary, we associate with each spherical $(n-1)$-simplex (or each simplicial cone in E^n) a graph by choosing a node for each wall of the simplex (or cone) and linking two nodes if and only if the corresponding walls are not orthogonal. One can also associate a graph with a symmetric $n \times n$ matrix (a_{ij}) by introducing n nodes $p_1, \cdots, p_n$ and linking p_i and p_j if and only if $a_{ij} \neq 0$.

2.4. LEMMA. *Let F be a spherical $(n-1)$-simplex on a sphere of radius 1. Assume that the interior dihedral angles of F are nonobtuse and that the graph* *corresponding to F is connected. Then the spherical distance between any two points of F is less than $\pi/2$.*

This result can easily be proved by the methods used by Cartan [3, p. 253] to establish similar results.

2.5. Lemma. *If (a_{ij}) is a symmetric matrix containing only non-negative real entries and if the corresponding graph is connected, then (a_{ij}) has a characteristic value which is positive, strictly larger than all other characteristic values, and such that the corresponding characteristic vector can be chosen to have only positive entries.*

This is a special case of a theorem of Frobenius [9, p. 471].

3. **Spherical simplexes.** Throughout this section the following conventions and assumptions are made. Let F denote a spherical simplex on the unit sphere in E^n such that the corresponding graph is connected and has only even circuits. By 2.2 the walls of F can be so labeled that

3.1. *$W_1, W_2, \cdots, W_s$ are mutually orthogonal as also are $W_{s+1}, \cdots, W_n$.*

Set $E^{(2)} = W_1 \cdot W_2 \cdot \cdots \cdot W_s$ and $E^{(1)} = W_{s+1} \cdot \cdots \cdot W_n$, these being "opposite edges" of F. In the space E^n in which F is embedded let $\epsilon_1, \epsilon_2, \cdots, \epsilon_n$ be inwardly directed unit normals to $W_1, W_2, \cdots, W_n$, and let $\epsilon^1, \epsilon^2, \cdots, \epsilon^n$ be the dual basis, so that $(\epsilon_i, \epsilon^j) = \delta_i^j$, where $(\cdot, \cdot)$ denotes the inner product of E^n. Let $G = (g_{ij}) = ((\epsilon_i, \epsilon_j))$, so that $G^{-1} = (g^{ij}) = ((\epsilon^i, \epsilon^j))$. Corresponding to each characteristic value x of $1-G$ we introduce a characteristic vector $(l_1, l_2, \cdots, l_n)$ and the notations

$$\bar{\sigma} = \sum_1^s l_i \epsilon^i, \quad \bar{\tau} = \sum_{s+1}^n l_i \epsilon^i, \bar{\rho} = \bar{\sigma} + \bar{\tau}, \sigma = \bar{\sigma}/(\bar{\sigma}, \bar{\sigma})^{1/2}, \tau = \bar{\tau}/(\bar{\tau}, \bar{\tau})^{1/2}, \rho = \bar{\rho}/(\bar{\rho}, \bar{\rho})^{1/2}.$$

3.2. Lemma. *If the assumptions of the preceding paragraph are made, then* (1) *corresponding to the characteristic value $x=0$ of $1-G$ the point σ is orthogonal to the edge $E^{(2)}$, and τ is orthogonal to $E^{(1)}$*; (2) *corresponding to each $x \neq 0$ the spherical line joining σ and τ cuts the edges $E^{(1)}$ and $E^{(2)}$ orthogonally, has ρ as one of its mid-points, and is of length* $\cos^{-1} x$; (3) *the largest characteristic value x is positive and has multiplicity* 1, *the corresponding l's may be taken as positive, so that one of the open segments $\sigma\tau$ is completely interior to F, and the length $\sigma\tau = \cos^{-1} x$ in this case is the absolutely minimum distance between $E^{(1)}$ and $E^{(2)}$.*

Proof. Because of 3.1 the matrix G can be written in partitioned form

$$G = \begin{pmatrix} 1 & A \\ A' & 1 \end{pmatrix}$$

with the 1's denoting identity matrices and A' the transpose of A. Let

$$G^{-1} = \begin{pmatrix} B & C \\ C' & D \end{pmatrix}$$

be the corresponding form of G^{-1}. Then $G^{-1}\cdot G=1$ implies

3.3 $$BA + C = 0, \qquad C' + DA' = 0.$$

Let $l^{(1)}$ and $l^{(2)}$ be the column vectors with co-ordinates $l_1, l_2, \cdots, l_s$ and $l_{s+1}, \cdots, l_n$ respectively. Then since $(l_1; l_2, \cdots, l_n)$ and x are corresponding characteristic vector and value of $1-G$, we have

3.4 $$xl^{(1)} + Al^{(2)} = 0, \qquad A'l^{(1)} + xl^{(2)} = 0.$$

Because of 3.3 these equations yield:

3.5 $$xBl^{(1)} - Cl^{(2)} = 0, \qquad C'l^{(1)} - xDl^{(2)} = 0.$$

If $x=0$, the equation $C'l^{(1)}=0$ in co-ordinate form reads

$$0 = \sum_{j=1}^{s} g^{ij}l_j = \left(\epsilon^i, \sum_{j=1}^{s} l_j\epsilon^j\right) = (\epsilon^i, \bar{\sigma}), \qquad i = s+1, \cdots, n,$$

so that $\bar{\sigma}$ (or σ) is orthogonal to $E^{(2)}$; similarly τ is orthogonal to $E^{(1)}$. Next suppose $x \neq 0$. Now 3.5 yields $l^{(1)\prime}Bl^{(1)} = l^{(2)\prime}Dl^{(2)}$, or $\sum\sum_1^s g^{ij}l_il_j = \sum\sum_{s+1}^n g^{ij}l_il_j$, or $(\bar{\sigma}, \bar{\sigma}) = (\bar{\tau}, \bar{\tau})$. Thus ρ is one of the mid-points of $\sigma\tau$. We can normalize the l's so that

$$(\bar{\sigma}, \bar{\sigma}) = (\bar{\tau}, \bar{\tau}) = 1, \qquad \bar{\sigma} = \sigma, \qquad \bar{\tau} = \tau.$$

Then the equations 3.5 say that the vectors $x\sigma-\tau$ and $\sigma-x\tau$ are respectively orthogonal to $E^{(1)}$ and $E^{(2)}$. Thus the line $\sigma\tau$ cuts $E^{(1)}$ and $E^{(2)}$ orthogonally. Another application of 3.5 yields $\cos\sigma\tau = l^{(1)\prime}Cl^{(2)} = xl^{(1)\prime}Bl^{(1)} = x$. Finally suppose that x is the largest characteristic value of $1-G$. Since F has no obtuse dihedral angles, the entries of $1-G$ are non-negative; since the graph of F is connected, so also is the graph of $1-G$. Hence 2.5 implies that x is positive and of multiplicity 1, and that the l's can be chosen all positive. It is easily seen that the problem of finding the minimum distance from a point $\sigma = \sum_1^s l_i\epsilon^i$ of $E^{(1)}$ to a point $\tau = \sum_{s+1}^n l_i\epsilon^i$ of $E^{(2)}$, because of the restriction $(\sigma, \sigma) = (\tau, \tau) = 1$, gives Lagrange equations which lead to 3.5. Thus the minimum distance $\cos^{-1} x$ corresponds to the largest value of x, and 3.2 is proved.

From 3.2 we easily deduce:

3.6. LEMMA. *In addition to the assumptions made in* 3.1, *let* $R_1, R_2, \cdots, R_n$ *denote the reflections in the walls* $W_1, W_2, \cdots, W_n$, *and set* $S = R_1R_2 \cdots R_s$, $T = R_{s+1} \cdots R_n$ *and* $R = ST$ *(operations performed from right to left). Then* R *is a product of translations along the lines* $\sigma\tau$ *of* 3.2 *through distances of* $2\sigma\tau = 2\cos^{-1} x$ *corresponding to nonzero values of* x *together with the central inversion in the space spanned by the* σ*'s and* τ*'s corresponding to* $x=0$.

Proof. By 3.1, S leaves each point of $E^{(2)}$ fixed and maps each point orthogonal to $E^{(2)}$ on its antipode. Hence if x is a nonzero characteristic value of $1-G$ and if $L=\sigma\tau$ is the corresponding spherical line given by 3.2, the restriction of S to L is equal to the reflection in τ; similarly T effects the reflection in σ and $R=ST$ the translation of length $2\sigma\tau$ along L. The case $x=0$ is treated similarly.

In regard to the results above, it is to be noted that Coxeter [5; 8, p. 766], by a somewhat different method and under the more restrictive assumption that the graph of F is a tree, proved that the characteristic values of $1-G$ give the lengths of the basic translations of which R is a product according to 3.6. The present development, which also gives a geometric interpretation to the characteristic vectors of $1-G$, was inspired by a remark of Coxeter [6, p. 233].

4. **Finite irreducible g.g.r.** Assume that $\mathfrak{G}$ is a finite irreducible g.g.r. leaving 0 fixed, that F is a fundamental chamber, and that F also denotes the corresponding spherical simplex on the unit sphere with center at 0. Then the graph of F is a tree [6, p. 195]; and, since reflections in orthogonal hyperplanes are commutative, 2.3 implies the following result due to Coxeter [4, p. 602]:

4.1. *The products of the reflections in the n walls of a fundamental chamber of a finite irreducible g.g.r. taken in the various orders are all conjugate.*

Since also the dihedral angles of F are submultiples of π, the various notations, assumptions (in particular 3.1), and conclusions of §3 are applicable and will be used henceforth to refer to this specific situation.

It is now easy to prove the first main result stated in §1:

4.2. THEOREM. *If the product of the reflections in the walls of F has order h, then the number of reflecting hyperplanes is $nh/2$.*

Proof. Let $L=\sigma\tau$ be the line corresponding to the largest characteristic value of $1-G$ as given by 3.2, and let $R=ST$ be the product of the reflections in the walls of F as in 3.6. Let $\mathfrak{H}$ be the restriction to L of the group generated by S and T; that is, $\mathfrak{H}$ is the group generated by reflections in the points σ and τ of L. Then R also has order h in $\mathfrak{H}$: if R^k leaves L fixed pointwise, it leaves fixed points interior to F by 3.2, and so is the identity in $\mathfrak{G}$. Thus $\mathfrak{H}$ has order $2h$. Since $\mathfrak{H}$ contains the reflections in σ and in τ, and since the open segment $\sigma\tau$ is interior to F, it follows that $\sigma\tau$ is a fundamental region for $\mathfrak{H}$. Hence $\sigma\tau=\pi/h$ and there are h transforms of each of the points σ and τ alternating around L. Clearly L meets r.h. at these points only; it meets $n-s$ r.h. at each transform of σ and s r.h. at each transform of τ; however, each hyperplane is met twice at antipodes. Hence the number of r.h. is $nh/2$, and 4.2 is proved.

In the following corollaries, the same assumptions as in 4.2 are made.

4.3. Corollary. *The minimum distance between* $E^{(1)}$ *and* $E^{(2)}$ *is* π/h; *all other extreme distances between* $E^{(1)}$ *and* $E^{(2)}$ *are integral multiples of* π/h. *One of the characteristics values of* R *is* $\exp(2\pi i/h)$; *all other characteristics values are integral powers of it.*

Because of what has already been said, this is clear (cf. [6, p. 233]).

4.4. Corollary. *If* h *is odd, then* $s=n-s$ *and* $W_1, \cdots, W_s$ *are mapped onto* $W_{s+1}, \cdots, W_n$ *by an element of* $\mathfrak{G}$.

Proof. Indeed in this case $R^{(h-1)/2}$ maps τ onto $-\sigma$ and the hyperplanes meeting at τ onto those meeting at $-\sigma$ (or at σ).

4.5. Corollary. *If* $\mathfrak{G}$ *contains the central inversion* I, *then* h *is even and* $I=R^{h/2}$.

Proof. If it exists, I maps F onto $-F$. If h is odd, the transformation $U=R^{(h-1)/2}S$ maps σ and τ onto $-\tau$ and $-\sigma$ respectively and hence F onto $-F$ since the segment $\sigma\tau$ has points interior to F; clearly $U\neq I$. If h is even, one sees similarly that $R^{h/2}$ maps F on $-F$ and hence is the central inversion if it exists.

This result also has been verified by Coxeter [4, p. 606]. By 4.1 it remains valid if $R_1, R_2, \cdots, R_n$ are multiplied in an arbitrary order to give R.

4.6. Corollary. *Assume again the order* 3.1 *for the walls of* F. *Define* $W_k=W_j$ *and* $R_k=R_j$ *if* $k\equiv j \pmod{n}$. *Then the r.h. are*

$$R_1R_2\cdots R_{k-1}W_k \tag{4.7}$$

and the reflections of $\mathfrak{G}$ *are*

$$R_1R_2\cdots R_{k-1}R_kR_{k-1}\cdots R_1 \tag{4.8}$$

for $k=1, 2, \cdots, nh/2$.

Proof. If $1\leqq k\leqq s$, then $R_1R_2\cdots R_{k-1}W_k=W_k$, so that the first s hyperplanes of 4.7 are those met by the line L of 4.2 at the point τ. Similarly one sees that the r.h. of 4.7 are listed in the order in which they are met in a trip along L for a point interior to F to its antipode in $-F$. Clearly 4.8 is the reflection in 4.7.

5. **Petrie polygons.** Here we assume $n\geqq 3$ and consider the spherical honeycomb C in which the unit sphere is cut by the r.h. This consists of a number of spherical $(n-1)$-simplexes bounded by simplexes of lower orders. In this case the following definition of a *Petrie polygon* (p.p.) exists (cf. [6, p. 223]): any n consecutive vertices of a p.p. belong to one simplex of C, but no $n+1$ consecutive vertices do.

It easily follows that any n consecutive vertices of a p.p. are distinct, that any ordered set of n vertices of a simplex of C determines a unique p.p., that $n-1$ ordered vertices of a simplex of C determine two p.p., and that, if $\epsilon^1, \epsilon^2, \cdots, \epsilon^{n+1}$ are $n+1$ consecutive vertices of a p.p., then ϵ^{n+1} is the reflection of ϵ^1 in the hyperplane determined by $\epsilon^2, \epsilon^3, \cdots, \epsilon^n$.

5.1. THEOREM. *Let $W_1, W_2, \cdots, W_n$ be the walls of F taken in an arbitrary order, and let $\epsilon^1, \epsilon^2, \cdots, \epsilon^n$ be the respectively opposite vertices. Set $R_k = R_j$ and $\epsilon^k = \epsilon^j$ if $k \equiv j \pmod n$. Let P be the Petrie polygon of which $\epsilon^1, \epsilon^2, \cdots, \epsilon^n$ are the first n vertices. Then* (1) *the vertices of P in order are given by $\beta^k = R_1R_2 \cdots R_{k-1}\epsilon^k$, $k=1, 2, \cdots, nh$;* (2) *the vertices of P are distinct;* (3) *the nh $(n-2)$-simplexes determined by sets of $n-1$ consecutive vertices of P lie 2 in each of the $nh/2$ r.h.*

Proof. Since $R_j\epsilon^k = \epsilon^k$ if $j \not\equiv k \pmod n$, we see that $\beta^k = \epsilon^k$ for $1 \leqq k \leqq n$. Also $(\beta^j, \beta^{j+1}, \cdots, \beta^{j+n-1}) = R_1R_2 \cdots R_{j-1}(\epsilon^j, \cdots, \epsilon^{j+n-1})$, so that these n consecutive β's are the vertices of the simplex $R_1R_2 \cdots R_{j-1}F$. But β^{j+n} can not be a vertex of this simplex since $R_1R_2 \cdots R_{j-1}F \neq R_1R_2 \cdots R_jF$. Hence (1) is proved.

Next (2) and (3) are proved under the assumption that the W's are ordered as in 3.1. If $k \not\equiv j \pmod n$, the points β^k and β^j are distinct since they are transforms of distinct points of F. Thus if $\beta^k = \beta^j$, we can normalize so that $1 \leqq j \leqq n$, $k = j + rn$, $0 \leqq r < h$; then $R^r\beta^j = \beta^k = \beta^j$. If β^j lies on the line L of 4.2, then $r=0$ and $k=j$ since R effects a translation of order h along L. If β^j does not lie on L, then either $\beta^j\sigma$ or $\beta^j\tau$ is orthogonal to L; say the first is and set $\sigma' = R^r\sigma$. Then $\beta^j\sigma'$ is also orthogonal to L. But $\beta^j\sigma' = \beta^j\sigma < \pi/2$ by 2.4. Hence $\sigma = \sigma' = R^r\sigma$ from which we conclude $r=0$ and $k=j$ as before. Since $\epsilon^{k+1}, \epsilon^{k+2}, \cdots, \epsilon^{k+n-1}$ lie on the wall W_k of F, it follows on application of $R_1R_2 \cdots R_{k-1}$ that $\beta^{k+1}, \cdots, \beta^{k+n-1}$ lie on the hyperplane $R_1R_2 \cdots R_{k-1}W_k$. Then a slight modification of 4.6 yields (3).

The first step in removing the restriction on the order of the W's is to modify the order adopted in the previous paragraph by a simple cyclic permutation thus: set $W_1' = W_2$, $W_2' = W_3, \cdots, W_n' = W_1$ and define corresponding R', ϵ' and β'. Then $\beta^{k\prime} = R_1'R_2' \cdots R_{k-1}'\epsilon^{k\prime} = R_2R_3 \cdots R_k\epsilon^{k+1} = R_1\beta^k$ for each k, so that (2) and (3) are still valid for the new order. Next suppose that we have an order for which (2) and (3) hold and that two consecutive W's, say W_1 and W_2, are orthogonal so that R_1 and R_2 commute. Adopting the new order $W_1' = W_2$, $W_2' = W_1$, $W_3' = W_3, \cdots, W_n' = W_n$, one sees that $\beta^{j\prime} = \beta^{j+1}$ and $\beta^{j+1\prime} = \beta^j$ if $j \equiv 1 \pmod n$ while $\beta^{j\prime} = \beta^j$ if $j \not\equiv 1, 2 \pmod n$. The hyperplanes determined by sets of $n-1$ consecutive β's are permuted in a similar manner. Thus (2) and (3) hold for the new order of the W's, and, because of 2.3, for an arbitrary order.

This result and the following corollary are suggested by Coxeter [6, p. 231], but with a general proof only in the case $n=3$.

5.2. COROLLARY. *Let g be the order of $\mathfrak{G}$. Then the number of $(n-2)$-simplexes in each hyperplane of the honeycomb C is g/h.*

Proof. There are $n!/2$ p.p. which contain the vertices of a given simplex, say F, as n consecutive vertices. Since each p.p. determines nh simplexes by its sets of n consecutive vertices, the number of p.p. is $g\cdot(n-1)!/2h$. By 5.1 each p.p. leads to nh $(n-2)$-simplexes which lie in pairs in the $nh/2$ hyperplanes. Since each $(n-2)$-simplex is counted $(n-1)!$ times in this enumeration, the number of $(n-2)$-simplexes in each hyperplane is

$$(g\cdot(n-1)!/2h)2/(n-1)! \text{ or } g/h.$$

Coxeter has shown that the symmetry group $\mathfrak{G}$ of each regular polytope Π in E^n is an irreducible group generated by reflections in the hyperplanes of symmetry of Π and that every p.p. of Π (for definition, see [4, p. 605]) can be obtained by taking every nth vertex of a p.p. of the spherical honeycomb corresponding to $\mathfrak{G}$ [4, p. 605]. Hence by 4.2 we have:

5.3. Corollary to 4.2. *If a Petrie polygon of a regular polytope* Π *in* E^n *has* h *vertices, then the number of hyperplanes of symmetry of* Π *is* $nh/2$.

6. **Roots.** We leave the spherical honeycomb and introduce two nonzero normal vectors for each r.h. in such a way that each element of $\mathfrak{G}$ permutes the resulting nh vectors. These vectors will be called *roots*. Thus if ρ is a root, so is $-\rho$. The n roots lying along the inwardly directed normals to a fixed fundamental chamber F are called *fundamental roots*. Since no r.h. has points interior to F, every root ρ is a linear combination of fundamental roots in which all coefficients are non-negative or all are nonpositive; in the first (second) case ρ is called *positive* (*negative*) and we write $\rho>0$ $(\rho<0)$. From the equation for the reflection R in the hyperplane orthogonal to ρ,

$$R\eta = \eta - 2\rho(\eta, \rho)/(\rho, \rho), \tag{6.1}$$

we get the following result, important for our purposes [11, p. 19–01]:

6.2. Lemma. *The reflection in the hyperplane orthogonal to a fundamental root* α *maps* α *upon* $-\alpha$ *and permutes the remaining positive roots.*

For roots a somewhat sharper analogue of 4.6 exists.

6.3. Theorem. *Let the fundamental roots be ordered similarly to* 3.1 *so that* $\alpha_1, \alpha_2, \cdots, \alpha_s$ *are mutually orthogonal as also are* $\alpha_{s+1}, \cdots, \alpha_n$. *Let* R_j *be the reflection in the wall* W_j *orthogonal to* α_j. *Let* $\alpha_k=\alpha_j$ *and* $R_k=R_j$ *if* $k\equiv j \pmod n$, *and then let* $\rho_k=R_1R_2\cdots R_{k-1}\alpha_k$, $k=1, 2, \cdots$. *Then* (1) *the positive roots are the* ρ'*s given by* $k=1, 2, \cdots, nh/2$; (2) *the negative roots are given by* $k=nh/2+1, \cdots, nh$.

Proof. Suppose that $1\leqq k\leqq nh/2$. Then if $1\leqq j<k$, it follows by 4.6 that $\rho_k\neq\pm\rho_j$, so that $R_{j-1}R_{j-2}\cdots R_1\rho_k\neq\pm\alpha_j$. Thus by 6.2 the roots $R_{j-1}R_{j-2}\cdots R_1\rho_k$ and $R_jR_{j-1}\cdots R_1\rho_k$ have the same sign. Taking $j=1, 2, \cdots, k-1$, we see that ρ_k has the same sign as $R_{k-1}\cdots R_1\rho_k=\alpha_k$ and hence is positive. Sup-

pose next that $nh/2 < k \leqq nh$. Since $R_1R_2 \cdots R_{nh} = 1$, it now follows that $\rho_k = -R_{nh}R_{nh-1} \cdots R_{k+1}\alpha_k$; applying (1) to the case in which the α's and R's are relabeled so as to appear in reverse order, we get (2).

There exists the result that *all* roots are given by ρ_k, $k=1, 2, \cdots, nh$, for an arbitrary initial ordering of the fundamental roots. However 4.2 is valid for all initial orderings if and only if $\mathfrak{G}$ contains the central inversion. These statements can be proved by modifications of the methods already introduced.

By a simple computation using the definition of the ρ's and 6.1, one can also prove the recursion formula

$$6.4 \qquad \rho_k = \alpha_k + \sum_{j=1}^{k-1} a_{kj}\rho_j, \qquad a_{kj} = -2(\alpha_k, \alpha_j)/(\alpha_j, \alpha_j).$$

6.5. Corollary. *Suppose that the fundamental roots are partitioned under the action of $\mathfrak{G}$ into transitive sets of $n_1, n_2, \cdots, n_r$ elements. Then the set of all roots is partitioned into transitive sets of $n_1h, n_2h, \cdots, n_rh$ elements.*

Proof. This is clear from 6.2.

6.6. Corollary. *If h is even, the set of all roots can be partitioned into h fundamental sets, each consisting of the roots which lie along the inwardly directed normals to the walls of a fundamental chamber.*

Proof. $R_1R_2 \cdots R_s(\alpha_1, \alpha_2, \cdots, \alpha_n) = (-\rho_1, -\rho_2, \cdots, -\rho_s, \rho_{s+1}, \cdots, \rho_n)$ is a fundamental set. By 6.3 one can apply $1, R, \cdots, R^{h/2-1}$ in turn to this set to get $h/2$ fundamental sets of which the union contains exactly one of ρ and $-\rho$ for each root ρ. Hence one gets the required result by adjoining to the sets already obtained their negatives.

Consulting the known list of values of h [4, p. 618], one sees that h is odd only if $\mathfrak{G} = \mathfrak{A}_{h-1}$, the symmetry group of the regular $(h-1)$-simplex. Representing $\mathfrak{G}$ by the permutations of the h vertices of the simplex, one sees that the question left open by 6.6 can be phrased as follows:

6.7. *If h is odd, is it possible to order the numbers* $1, 2, \cdots, h$ *in h ways so that the resulting h $(h-1)$ ordered consecutive pairs are distinct?*

If $h=3$ or 5, the answer is *no*, but the general question seems to be open.

7. **Traces.** Here we prove two lemmas to be used in the proof of 1.2.

Recall that $\alpha_1, \alpha_2, \cdots, \alpha_n$ denote the fundamental roots relative to a fixed fundamental chamber F. If $\mu = \sum_1^n x^j\alpha_j$ is an arbitrary vector, we set $\sum x^j = \text{tr}\,\mu$, the *trace* of μ.

7.1. Lemma. *If μ is an arbitrary vector, then* $\text{tr}\,\mu = \sum(\mu, \rho)/(\rho, \rho)$, *the sum being on the positive roots ρ.*

Proof. Set $\eta = \sum \rho/(\rho, \rho)$. The reflection in the hyperplane orthogonal to a fundamental root α maps η onto $\eta - 2\alpha(\eta, \alpha)/(\alpha, \alpha)$ by 6.1. By 6.2 this last vector is equal to $\eta - 2\alpha/(\alpha, \alpha)$. Hence $(\eta, \alpha) = 1 = \operatorname{tr} \alpha$, for each fundamental α. It follows by linearity that for an arbitrary μ we have $\operatorname{tr} \mu = (\eta, \mu)$, that is 7.1. The argument used here occurs in [11, p. 19–01] in a similar context.

7.2. LEMMA. *If μ is an arbitrary vector, and if k is the number of r.h. in $\mathfrak{G}$, then*

$$\sum_{\rho>0} (\mu, \rho)^2/(\mu, \mu)(\rho, \rho) = k/n.$$

Proof. $Q(\mu) = \sum(\mu, \rho)^2/(\rho, \rho)$ and (μ, μ) are quadratic forms in μ which are invariant under $\mathfrak{G}$. Since $\mathfrak{G}$ is algebraically irreducible over the reals, $Q(\mu) = c(\mu, \mu)$ for some real c. Letting μ run through an orthonormal basis of E^n, one gets $c = k/n$, as required.

8. **Crystallographic restriction, dominant root.** Assume that $\overline{\mathfrak{G}}$ is an infinite irreducible discrete g.g.r. on E^n, that 0 is a "special point" which lies on one member of each family of parallel r.h. of $\overline{\mathfrak{G}}$, and that $\mathfrak{G}$ is the subgroup of $\overline{\mathfrak{G}}$ leaving 0 fixed (see [4]).

8.1. LEMMA. *Under the assumptions of the preceding paragraph, the roots corresponding to $\mathfrak{G}$ can be so chosen that $2(\rho, \sigma)/(\sigma, \sigma)$ is an integer for each pair of roots ρ, σ.*

Proof. Because $\overline{\mathfrak{G}}$ is irreducible, it easily follows that each r.h. through 0 has other r.h. parallel to it. Now choose each root ρ to be equal in length to the distance between consecutive r.h. orthogonal to ρ. Let ρ and σ be any two roots. The point 2ρ is a transform of 0 by $\overline{\mathfrak{G}}$ and so lies on a r.h. orthogonal to σ. Since the distance of this r.h. from 0 is $2(\rho, \sigma)/(\sigma, \sigma)^{1/2}$, we conclude that the latter number is an integral multiple of the length $(\sigma, \sigma)^{1/2}$ of σ, as desired [7, p. 404].

Henceforth we impose these *crystallographic restrictions* on $\mathfrak{G}$ and the root system. It is in this case that there exists a corresponding Lie group which is closely related through its algebraic and topological properties with $\overline{\mathfrak{G}}$, $\mathfrak{G}$ and the root system [1; 2; 3; 13].

From 6.1, 6.3 and 8.1, it follows that each root has integral coefficients in terms of the fundamental roots. One can also easily conclude from 8.1 the following fact:

8.2. LEMMA. *If ρ and σ are two roots such that $\rho \neq \pm\sigma$ and $(\rho, \rho) \leqq (\sigma, \sigma)$, then $2(\sigma, \rho)/(\sigma, \sigma) = 0$, 1 or -1.*

Cartan [3, p. 256] has proved the existence of a *dominant root* $\mu = \sum y^j\alpha_j$ with the property that, if $\rho = \sum x^j\alpha_j$ is any root, then $y^j \geqq x^j$ for all j. It follows that μ is in F, or, equivalently, that $(\mu, \alpha_j) \geqq 0$ for each fundamental root α_j, since otherwise one could increase a coefficient of μ by reflection in some wall of F.

8.3. LEMMA. *If μ is the dominant root and ρ is any root, then $(\rho, \rho) \leqq (\mu, \mu)$.*

Proof. Since there is an element of $\mathfrak{G}$ mapping ρ onto a root in F of the same length, we may assume that ρ is in F. Since $\mu-\rho$ has only non-negative coefficients, and ρ and μ are in F, it follows that $(\mu-\rho, \mu)\geqq 0$ and $(\rho, \mu-\rho)\geqq 0$. Thus $(\mu, \mu)\geqq(\rho, \mu)\geqq(\rho, \rho)$, with equality only if $\mu=\rho$.

The stage is now set for the proof of the second main theorem.

8.4. THEOREM. *If μ is the dominant root, then the number of r.h. is $k=n(1+\operatorname{tr}\mu)/2$.*

Proof. If $\rho>0$, then $(\mu, \rho)\geqq 0$, and then 8.2 and 8.3 imply that $2(\mu, \rho)/(\mu, \mu)=0$ or 1 if $\rho\neq\mu$. Hence using 7.1 and 7.2 and noting that the range of summation below is the set of positive roots, we have

$$\begin{aligned}\operatorname{tr}\mu &= \sum (\mu, \rho)/(\rho, \rho) = \sum (2(\mu, \rho)/(\mu, \mu))((\mu, \mu)/2(\rho, \rho))\\ &= \sum (2(\mu, \rho)/(\mu, \mu))^2((\mu, \mu)/2(\rho, \rho)) - 1\\ &= 2\sum (\mu, \rho)^2/(\mu, \mu)(\rho, \rho) - 1 = 2k/n - 1.\end{aligned}$$

A similar computation also explains an analogous formula encountered by Coxeter in his work on extreme quadratic forms [7, p. 413].

9. **Trace distribution.** Following is another easily verified property of root systems which the author has not been able to explain in general terms. By 4.3 the characteristic values of R are of the form $\exp(2\pi i m_j/h)$ with $1=m_1\leqq m_2\leqq\cdots\leqq m_n=h-1$. The numbers m_j have been called the *exponents* of $\mathfrak{G}$, and enter into many questions concerning $\mathfrak{G}$; for example, the numbers m_j+1 are the degrees of a basic set of invariants for $\mathfrak{G}$, and, if $\mathfrak{G}$ is crystallographically restricted, then the Poincaré polynomial of the corresponding Lie group is $\prod(1+t^{2m_j+1})$ (see [12] for other properties). Under this restriction one can also prove by verification the following curious fact.

9.1. *Let the number of positive roots of trace* $1, 2, \cdots, h-1$ *be* $p_1, p_2, \cdots, p_{h-1}$, *respectively. Then* $p_1\geqq p_2\geqq\cdots\geqq p_{h-1}$ *and the partition conjugate to that determined by the* p'*s consists of the* m'*s, the exponents of* $\mathfrak{G}$.

For example, if $\mathfrak{G}=\mathfrak{D}_4$ (in the usual notation), then the roots of trace 1, 2, 3, 4, 5 occur with multiplicities of 4, 3, 3, 1, 1, respectively. The partition of 12 conjugate to the last list of numbers consists of the numbers 1, 3, 3, 5, the exponents of $\mathfrak{D}_4$.

REFERENCES

1. E. Cartan, Thèse, Paris, Nony, 1894.

2. ———, *La géométrie des groupes simples*, Annali Mat. Pura Appl. vol. 4 (1927) pp. 209–256.

3. ———, *Complément au mémoire "sur la géométrie des groupes simples,"* Annali Mat. Pura Appl. vol. 5 (1928) pp. 253–260.

4. H. S. M. Coxeter, *Discrete groups generated by reflections*, Ann. of Math. vol. 35 (1934) pp. 588–621.

5. ———, *Lösung der Aufgabe* 245, Jber. Deutsch. Math. Verein. vol. 49 (1939) pp. 4–6.

6. ———, *Regular polytopes*, New York, 1949.

7. ———, *Extreme forms*, Canad. J. Math. vol. 3 (1951) pp. 391–441.

8. ———, *The product of the generators of a finite group generated by reflections*, Duke Math. J. vol. 18 (1951) pp. 765–782.

9. G. Frobenius, *Über Matrizen aus positiven Elementen*, Preuss. Akad. Wiss. Sitzungsber. (1908) pp. 471–476.

10. D. König, *Theorie der endlichen und unendlichen Graphen*, New York, 1950.

11. Séminaire "Sophus Lie," Paris, 1954–1955.

12. G. C. Shepard, *Some problems on finite reflection groups*, Enseignement Math. vol. 2 (1956) pp. 42–48.

13. H. Weyl, *Über die Darstellungen halbeinfacher Gruppen durch lineare Transformationen*, Math. Z. vol. 23 (1925) pp. 271–309, vol. 24 (1926) pp. 328–395

UNIVERSITY OF CALIFORNIA,
LOS ANGELES, CALIF.

VARIATIONS ON A THEME OF CHEVALLEY

ROBERT STEINBERG

1. Introduction. In this paper we use the methods of C. Chevalley to construct some simple groups and to gain for them the structural theorems of [3]. Among the groups obtained there are two new families of finite simple groups[1], not to be found in the list of E. Artin [1]. Whether the infinite groups constructed are new has not been settled yet.

Section 5 contains statements of the main results of [3]. In §§ 2, 3, 4 and 7, we define analogues of certain real forms of the Lie groups of type A_l, D_l and E_6 (in the usual notation), and extend to them the structural properties of the groups of Chevalley. Sections 6 and 9 treat some identifications, and § 8 deals with the question of simplicity. In §§ 10 and 11, using the extra symmetry inherent in a Lie algebra of type D_4, we consider two modifications of the first construction which are, perhaps, of more interest since they produce groups which have no analogue in the classical complex-real case: in fact, a basic ingredient of each of these variants is a field automorphism of order 3. In Sections 12 and 13, it is proved that new finite simple groups are obtained[1], and their orders are given. Section 14 deals with an application to the theory of group representations, and § 15 with some concluding observations.

The notation is cumulative. We denote by $|S|$ the cardinality of the set S, by K^* the multiplicative group of the field K, and by C the complex field. An introduction to the standard Lie algebra terminology together with statements of the principal results in the classical theory can be found in [3, p. 15–19]. (Proofs are available in [8] or [10]).

2. Roots and reflections. We first introduce some notations. Relative to a Cartan decomposition of a simple complex Lie algebra of rank l, let E be the real space generated by the roots, made into an Euclidean space in the usual way, and normalized as in [3, p. 17–18]. Relative to an ordering $\prec$ of the additive group generated by the roots, let Π be the set of positive roots, and $a(1), a(2), \cdots, a(l)$ the fundamental roots. For each root $r = \Sigma z_i a(i)$, set $\Sigma z_i = ht\ r$, the *height* of r. The ordering $\prec$ can always be chosen so that $ht\ r < ht\ s$ implies $r \prec s$ (see [3, p. 20, l. 35–40]); suppose this is done. Assume now the existence of an automorphism σ of E of order 2 such that $\sigma\Pi = \Pi$. This restricts the type of algebra to A_l, D_l ($l \geqq 4$) or E_6 (see [3, p. 18]), and hence

Received October 2, 1958, in revised form January 8, 1959.

[1] Since the preparation of this paper, the author has learned that these groups have also been discovered by D. Hertzig [**6**], who has shown that they complete the list of finite simple algebraic groups.

implies that all roots have the same length. We also denote σr by $\bar{r}$. Clearly σ permutes the fundamental roots. Thus $ht\,\bar{r} = ht\,r$ for each root r. Finally, let W be the Weyl group, W^1 the subgroup of elements commuting with σ, and for each $w \in W$ denote by $n(w)$ the number of roots r for which $r \succ 0$ and $wr \prec 0$.

Consider now subsets S of Π of the following three types:

(1) S consists of one root r, which is self-conjugate ($\bar{r} = r$), and which can not be written as a sum of a conjugate pair of roots;

(2) S consists of a conjugate pair r, $\bar{r}$ such that $r + \bar{r}$ is not a root;

(3) S consists of three roots of the form r, $\bar{r}$, $r + \bar{r}$.

Note that in case (2) one has $r \perp \bar{r}$ because $ht\,r = ht\,\bar{r}$ implies that $r - \bar{r}$ is not a root. Shortly we prove the important fact:

2.1 LEMMA. *If Π^1 denotes the collection of sets of types* (1), (2) *and* (3) *above, then Π^1 is a partition of Π.*

In any case, the *fundamental sets* of Π^1 - those which contain fundamental roots - are disjoint because the fundamental roots are linearly independent. If w_r denotes the reflection in the hyperplane orthogonal to r, we set $w_s = w_r$, $w_r w_{\bar{r}}$ or $w_{r+\bar{r}} (= w_r w_{\bar{r}} w_r)$ according as S is of type (1), (2) or (3) above. Note that $w_s \in W^1$.

2.2 LEMMA. *For each fundamental $S \in \Pi^1$, w_s maps S onto $-S$ and permutes the positive roots not in S. Hence $n(w_s) = |S|$.*

Proof. Since $n(w_a) = 1$ for each fundamental root a [8, p. 19-01, Lemma 1], and since w_s can be written as a product of $|S|$ such reflections, it follows that $n(w_s) \leqq |S|$. By direct verification one sees that $w_s S = -S$. Hence the lemma is proved.

2.3 LEMMA. *The group W^1 is generated by the w_s corresponding to fundamental $S \in \Pi^1$.*

Proof. Using induction on $n(w)$, we show that each $w \in W^1$ is a product of elements of the given form. If $n(w) = 0$, $w = 1$, the statement is clearly true. If $n(w) > 0$, $w \neq 1$, there is a fundamental root a such that $a \succ 0$ and $wa \prec 0$. Since $\bar{a} \succ 0$ and $w\bar{a} = \overline{wa} \prec 0$, it follows that $r \succ 0$, $wr \prec 0$ for each root r in the set $S \in \Pi^1$ which contains a. Hence $n(ww_s^{-1}) = n(w) - n(w_s)$ by 2.2, and the induction hypothesis can be applied to ww_s^{-1} to complete the proof.

2.4 LEMMA. *W is a normal subgroup of the group generated by W and σ.*

Proof. One has $\sigma w_r \sigma^{-1} = w_{\bar{r}}$ for each root r. Since σ permutes the roots, and the root reflections generate W, one gets $\sigma W \sigma^{-1} = W$, and hence 2.4.

2.5 LEMMA. *The element* w_0 *of* W *defined by* $w_0 \Pi = -\Pi$ *is in* W^1.

Proof. By 2.4, $\sigma w_0 \sigma^{-1} \in W$. Since $\sigma w_0 \sigma^{-1} \Pi = -\Pi$, one concludes that $\sigma w_0 \sigma^{-1} = w_0$ and that $w_0 \in W^1$.

2.6 LEMMA. *Each* $S \in \Pi^1$ *is congruent under* W^1 *to a fundamental set.*

Proof. Write the element w_0 of 2.5 in the form $w_0 = w_k \cdots w_2 w_1$ guaranteed by 2.3. Since $S \succ 0$ and $w_0 S \prec 0$, there is an index i such that $w_{i-1} \cdots w_1 S \succ 0$ and $w_i \cdots w_1 S \prec 0$. If $T \in \Pi^1$ corresponds to w_i, it follows from 2.2 that $w_{i-1} \cdots w_1 S \subseteq T$, and clearly equality must hold.

By using 2.6 and examinining the fundamental root systems for groups of type A_l, D_l and E_6 (see [3, p. 18] or [8, p. 13-08]), one sees that a set in Π^1 of type (3) can occur only in the case A_l (l even). This turns out to be the most troublesome case in the sequel. Note however that sets of types (1) and (3) do not occur simultaneously.

Proof of 2.1. This follows from 2.6 and the fact that the fundamental sets of Π^1 are non-overlapping.

We now associate with W^1 a reflection group. Let E^+ and E^- respectively denote the positive and negative subspaces of E under σ, and for each $w \in W^1$ let $\tilde{w}$ and $\widetilde{W}^1$ denote the restrictions of w and W^1 to E^+. Also denote by $\tilde{S}$ the vector r, $r + \bar{r}$ or $r + \bar{r}$ in the respective cases (1), (2) or (3) of 2.1.

2.7 LEMMA. *The restriction of* W^1 *to* $\widetilde{W}^1$ *is faithful.* $\widetilde{W}^1$ *is a reflection group of type* $C_{[(l+1)/2]}$, B_{l-1} *or* F_4 *in the respective cases that* W *is of type* A_l, D_l *or* E_6, *and, to within a change of scale,* $\{\tilde{S} \mid S \in \Pi^1\}$ *is a corresponding system of positive root vectors.*

Proof. First if $w \in W^1$, $\tilde{w} = 1$, then w maps each positive root onto another one. Hence $w = 1$, and the restriction is faithful. Those $\tilde{S}$ which correspond to the fundamental $S \in \Pi^1$ form a new fundamental root system (to within a change of scale) of the listed type, as one sees by considering the separate cases (see [3, p. 18]). Becuse of 2.3 and 2.6, the proof is complete if it can be shown that, for each fundamental $S \in \Pi^1$, $\tilde{w}_s$ is the reflection in the hyperplane orthogonal to $\tilde{S}$. If

$|S| = 2$ and $S = \{a, \bar{a}\}$, then w_s has -1 as a characteristic value of multiplicity 2. Since $w_s(a + \bar{a}) = -(a + \bar{a})$, $w_s(a - \bar{a}) = -(a - \bar{a})$, $a + \bar{a} \in E^+$, and $a - \bar{a} \in E^-$, it follows that $\tilde{w}_s$ has -1 as a characteristic value of multiplicity 1, and then that $\tilde{w}_s$ is the required reflection. If $|S| = 1$ or $|S| = 3$, the result follows from the definitions.

2.8 COROLLARY. *Any two sets of the same type in the partition 2.1 are congruent under* W^1.

Proof. Since sets of types (1) and (3) do not occur simultaneously, and since $\tilde{W}^1$ is transitive on its root vectors of a given length, 2.8 follows from 2.7.

A new ordering $<$ of the positive roots is now introduced. First if $R, S \in \Pi^1$, then $R < S$ means that $\min r \in R \prec \min s \in S$. Then if $r, s \in \Pi$, define $r < s$ to mean that either r and s belong to distinct sets R and S of Π^1 and $R < S$, or r and s belong to the same set of Π^1 and $r \prec s$,

2.9 LEMMA. *The roots in each set S of* Π^1 *occur consecutively in the ordering of the roots of* Π *relative to* $<$. *If* r, s *and* $r + s$ *are positive roots, then* $r + s > \min(r, s)$.

Proof. The first statement follows from the definition. Since $\prec$ respects heights, the second assertion is true if $r + s$ has minimum height in the set S of Π^1 containing it. Thus one may assume that there is a root t such that $r + s = t + \bar{t}$, $r \neq t$, $r \neq \bar{t}$, and that W is of type A_l (l even). Then each positive root is a sum of a string of distinct fundamental roots, and the strings corresponding to r and s are necessarily of different lengths. Thus $ht\, t = ht\, \bar{t} > \min(ht\, r, ht\, s)$. Since $\prec$ respects heights, this implies that $r + s > \min(t, \bar{t}) > \min(r, s)$.

3. Construction of an involution. Suppose that $\mathfrak{g}$ is a simple complex Lie algebra with a generating system $(X_r, X_{-r}, H_r, r \in \Pi)$ chosen to satisfy the conditions of Theorem 1 of [3]. Assume also that $\mathfrak{g}$ is restricted to type A_l, D_l ($l \geqq 4$) or E_6 so that the results of § 2 can be applied. Set $r(H_s) = r(s)$. Then, all roots being of the same length, it follows that:

$$3.1 \qquad X_r X_s = N_{rs} X_{r+s}\,;\ N_{rs} = 0,\ \pm 1;\ r,\ s \in \Pi\,.$$

For the same reason $r(s) = s(r)$ and $r(r) = 2$. By the uniqueness theorem for a simple Lie algebra with a given root structure (see [8, p. 11-04] or [10, p. 94]), there exists an automorphism σ_C of $\mathfrak{g}$ such that $\sigma_C H_r = H_{\bar{r}}$ and $\sigma_C X_r = c_r X_{\bar{r}}$, $c_r \in C^*$, $r \in \Pi$ or $-\Pi$, with $c_a = 1$ for each fundamental root a. Then each $c_{-a} = 1$, and by induction on the height one gets each $c_r = \pm 1$. Next let K be a field on which an automorphism σ of order 2 acts, let K_0 be the fixed field, and write $\sigma k = \bar{k}$, $k \in K$. Then following the procedure of [3, p. 32], one can transfer the base field of $\mathfrak{g}$ from C to K, and thus gain a Lie algebra $\mathfrak{g}_K$ over K and a semi-automorphism σ of $\mathfrak{g}_K$ such that $\sigma(kH_r) = \bar{k}H_{\bar{r}}$ and $\sigma(kX_r) = \pm \bar{k}X_{\bar{r}}$, $k \in K$, $r \in \Pi$ or $-\Pi$. Note [3, p. 32] that the field is not transferred for roots (or weights) and that the expression $r(s)$ retains its original meaning.

3.2 Lemma. *The order of σ is 2. By appropriate sign changes of the X_r one can arrange things so that in the equations $\sigma X_r = k_r X_{\bar{r}}$, $r \in \Pi$, one has:*

(a) $k_{\bar{r}} = k_r$;

(b) *if $\bar{r} \neq r$, then $k_r = 1$;*

(c) *if $\bar{r} = r$, then k_r is 1 or -1 according as r belongs to an $S \in \Pi^1$ of 1 or 3 elements.*

Proof. One has $\sigma^2 X_a = X_a$, $\sigma^2 X_{-a} = X_{-a}$ for each fundamental root a. Thus $\sigma^2 = 1$, and this implies (a). If r, $\bar{r}$ is a conjugate pair in Π, if $r < \bar{r}$, and if $k_r = -1$, replace X_r by $-X_r$. Then (b) holds. If $|S| = 3$ in (c), there is a root s such that $r = s + \bar{s}$, and one gets (c) by applying σ to the equation $X_s X_{\bar{s}} = kX_r$. If $|S| = 1$, assume $ht\, r > 1$. Then there is either a self-conjugate fundamental root a such that $r - a$ is a root, or a conjugate pair of orthogonal fundamental roots b, $\bar{b}$ such that $r - b$, $r - \bar{b}$ and $r - b - \bar{b}$ are all roots. One then applies σ to the equation $X_{r-a}X_a = k_1 X_r$ or $(X_{r-b-\bar{b}}X_b)X_{\bar{b}} = k_2 X_r$, respectively, and completes the proof of (c) by induction on the height.

We assume henceforth that the normalization indicated by 3.2 has been made and that the corresponding treatment has been given to the negative roots, so that one has once again the equations of structure of Theorem 1 of [3] (in particular, $X_r X_{-r} = H_r$).

4. Some nilpotent groups. As in [3], we set $x_r(t) = \exp(t\, ad\, X_r)$, $t \in K$, $r \in \Pi$, denote by $\mathfrak{X}_r$ the one-parameter group $\{x_r(t) \mid t \in K\}$, and by $\mathfrak{U}$ the group generated by all $\mathfrak{X}_r$, $r \in \Pi$.

4.1 Lemma. *For r, $s \in \Pi$ and t_1, $t_2 \in K$, one has the commutator relation $(x_r(t_1), x_s(t_2)) = x_{r+s}(N_{rs}t_1t_2)$.*

Proof. This follows from [3, p. 33, *l.* 22] and the fact that all roots have the same length.

A straightforward computation yields:

$$4.2 \qquad \sigma \exp(t\, ad\, X_r)\, \sigma^{-1} = \exp(\bar{t}\, ad\, \sigma X_r)\ .$$

4.3 LEMMA. *Let Σ be a subset of Π satisfying the condition*

4.4 $$r,\ s \in \Sigma,\ r+s \in \Pi \text{ imply } r+s \in \Sigma.$$

Then each $x \in \mathfrak{U}_\Sigma$, the group generated by all $\mathfrak{X}_r$, $r \in \Sigma$, can be written uniquely in the form $x = \prod x_r(t_r)$, the product being over the roots of Σ arranged in increasing order relative to $<$ (see § 2).

Proof. Using the formulas 4.1 repeatedly, one sees that the set of elements of the given form is closed under multiplication; thus each $x \in \mathfrak{U}_\Sigma$ has an expression of the given form. Uniqueness is proved by induction on $|\Sigma|$. If $|\Sigma| = 1$ and $\Sigma = \{r\}$, this follows from $x_r(t)X_{-r} = X_{-r} + tH_r - t^2X_r$ (see [3, p. 36, *l*. 15]). If $|\Sigma| > 1$, let r be the least element of Σ (relative to $<$), and set $\Sigma' = \Sigma - r$. Let $x \in \mathfrak{U}_\Sigma$ be written as $x = x_r(t_1)x_1$ and $x = x_r(t_2)x_2$ with $t_i \in K$ and $x_i \in \mathfrak{U}_{\Sigma'}$. Then $x_r(t_2 - t_1) = x_1x_2^{-1}$. Since $x_r(t_2 - t_1)X_{-r} = X_{-r} + (t_2 - t_1)H_r - (t_2 - t_1)^2X_r$, since $x_r(t_2 - t_1) \in \mathfrak{U}_{\Sigma'}$, and since r can not be written as a sum of roots larger than r by 2.9, it follows that the coefficient of H_r, namely $t_2 - t_1$, must be 0. Thus $x_1 = x_2$, and the induction hypothesis can be applied to Σ' to complete the proof.

The result 4.3 can be applied in the cases $\Sigma = \Pi$ and $\Sigma = S \in \Pi^1$. Because of 2.9, one gets:

4.5 COROLLARY. *Each $x \in \mathfrak{U}$ can be written uniquely in the form $x = \prod x_s$, $x_s \in \mathfrak{U}_s$, the product being over the sets S of Π^1 arranged in increasing order.*

Denote now by $\mathfrak{U}^1$, $\mathfrak{U}_s^1$, etc. the subgroups of elements of $\mathfrak{U}$, $\mathfrak{U}_s$, etc. commuting with σ.

4.6 LEMMA. *If $x \in \mathfrak{U}$ is written in the form 4.5, then $x \in \mathfrak{U}^1$ if and only if each $x_s \in \mathfrak{U}^1$. A necessary and sufficient condition for $x_s \in \mathfrak{U}_s$ to be in $\mathfrak{U}^1$ is that, in the cases (1), (2) or (3) of 2.1, x_s has the respective form (1) $x_r(t), \bar{t} = t$, (2) $x_r(t)x_{\bar{r}}(v)$, $v = \bar{t}$, or (3) $x_r(t)x_{\bar{r}}(v)x_{r+\bar{r}}(w)$, $v = \bar{t}$, $w + \bar{w} = N_{\bar{r}r}t\bar{t}$.*

Proof. If $x \in \mathfrak{U}^1$ commutes with σ, one has $x = \sigma x\sigma^{-1} = \prod(\sigma x_s\sigma^{-1})$. Since $\sigma x_s\sigma^{-1} \in \mathfrak{U}_s$ by 4.2, one gets $\sigma x_s\sigma^{-1} = x_s$ by the uniqueness in 4.5. Thus each $x_s \in \mathfrak{U}^1$. The converse is clear. In the cases listed in the second statement, one has

(1) $\sigma x_r(t)\sigma^{-1} = x_{\bar{r}}(\bar{t})$,

(2) $\sigma x_r(t)x_{\bar{r}}(v)\sigma^{-1} = x_r(\bar{v})x_{\bar{r}}(\bar{t})$, and

(3) $\sigma x_r(t)x_{\bar{r}}(v)x_{r+\bar{r}}(w)\sigma^{-1} = x_r(\bar{v})x_{\bar{r}}(\bar{t})x_{r+\bar{r}}(-\bar{w} + N_{\bar{r}r}\bar{t}\bar{v})$ by 3.2, 4.1 and 4.2. The required results now follow from 4.3.

4.7 LEMMA. *Let Π be the union of the disjoint sets Σ and Σ',* *each invariant under* σ, *and each satisfying* 4.4. *Then* $\mathfrak{U}^1 = \mathfrak{U}^1_\Sigma \mathfrak{U}^1_{\Sigma'}$ *and* $\mathfrak{U}^1_\Sigma \cap \mathfrak{U}^1_{\Sigma'} = 1$.

Proof. By [3, p. 41, Lemma 11], one can write $x \in \mathfrak{U}^1$ uniquely in the form $x = yy'$, $y \in \mathfrak{U}_\Sigma$, $y' \in \mathfrak{U}_{\Sigma'}$. The proof that y and y' are in $\mathfrak{U}^1$ is the same as that for the first part of 4.6.

If $\mathfrak{B}$ denotes the group generated by all $\mathfrak{X}_r$, $r < 0$, then one can define $\mathfrak{B}^1$, $\mathfrak{B}_s$, etc., and gain for these groups corresponding results.

5. Main results of Chevalley. For each simple complex Lie algebra $\mathfrak{g}$ (not necessarily one for which σ exists), consider the groups $\mathfrak{U}$ and $\mathfrak{B}$ and also the group G (denoted in [3] by G') which they generate. For each $w \in W$, if Σ consists of the roots r for which $r > 0$ and $wr < 0$, we set $\mathfrak{U}_\Sigma = \mathfrak{U}_w$ (denoted in [3] by $\mathfrak{U}''_w$). Let P_r and P, respectively, denote the additive groups generated by the roots and by the weights. Corresponding to each character χ of P_r into K^*, there is an automorphism $h = h(\chi)$ of $\mathfrak{g}_K$ defined by $hX_r = \chi(r)X_r$, $r \in \Pi$ or $-\Pi$. Let $\mathfrak{H}$ (denoted in [3] by $\mathfrak{H}'$) be the group generated by those automorphisms which correspond to characters which can be extended to P. For $h(\chi) \in \mathfrak{H}$, one has

5.1 $$hx_r(t)h^{-1} = x_r(\chi(r)t) .$$

The main results of [3] are as follows:

5.2 G contains $\mathfrak{H}$.

5.3 Corresponding to each $w \in W$ there is $\omega(w) \in G$ such that $\omega(w)X_r = c_r X_{wr}$, $\omega(w)H_r = H_{wr}$, $c_r \in K^*$, $r \in \Pi$ or $-\Pi$. The union of the sets $\mathfrak{H}\omega(w)$ is a group $\mathfrak{W}$ and the map $w \to \mathfrak{H}\omega(w)$ is an isomorphism of W on $\mathfrak{W}/\mathfrak{H}$.

Parenthetically, we remark that here one has:

5.4 $$\omega(w)\mathfrak{X}_r\omega(w)^{-1} = \mathfrak{X}_{wr} .$$

5.5 G is the union of the sets $\mathfrak{U}\mathfrak{H}\omega(w)\mathfrak{U}_w$, $w \in W$. These sets are disjoint and each element of G has a unique expression of the indicated form.

5.6 G is simple if one excludes the case (1) $|K| = 2$ and $\mathfrak{g}$ of type A_1, B_2 or G_2, and (2) $|K| = 3$ and $\mathfrak{g}$ of type A_1.

Before proving corresponding results for the group G^1 generated by $\mathfrak{U}^1$ and $\mathfrak{B}^1$, we identify G^1 in the case that $\mathfrak{g}$ is of type A_l.

6. Some unitary groups. Consider the form

6.1 $$f(\alpha, \beta) = \sum_1^{l+1} (-1)^i \alpha_i \bar{\beta}_{l+2-i}$$

on a space of $l+1$ dimensions over K. Let $U_{l+1}(f)$ denote the corresponding unimodular unitary group and $C_{l+1}(f)$ its center. Then one has:

6.2 *If* $\mathfrak{g}$ *is of type* A_l, $G^1 \cong U_{l+1}(f)/C_{l+1}(f)$.

Proof. *If* $\mathfrak{g}$ *is of type* A_l, one can identify $\mathfrak{g}_K$ with $\mathfrak{sl}_{l+1}(K)$, the algebra of $(l+1)$th order matrices of trace 0, in such a way that, for each fundamental root $a(i)$, $X_{a(i)} \in \mathfrak{g}_K$ corresponds to $E_{i,i+1}$, the matrix with 1 in the $(i, i+1)$ position and 0 elsewhere [7, p. 393]. If $m = ((-1)^i \delta_{i,l+2-j})$ is the matrix corresponding to f, one can then verify that σ is the product of the transformations $Y \to mYm^{-1}$ (matrix multiplication) and $Y \to -\overline{Y}^t$ (t = transpose). According to a recent identification of R. Ree [7], $\mathfrak{U}$ and $\mathfrak{V}$, respectively, consist of the superdiagonal matrices (0 below and 1 on the diagonal) and the subdiagonal matrices, acting on $\mathfrak{sl}_{l+1}$ via inner automorphisms, so that the group G of Chevalley is in this case the projective unimodular group. Now it follows from material in [4, p. 66-69] that $U_{l+1}(f)$ is generated by its superdiagonal and subdiagonal elements and that $C_{l+1}(f)$ consists of scalar matrices. Thus to prove 6.2 it is enough to prove:

6.3 Let x be a superdiagonal matrix. Then $x \in \mathfrak{U}^1$ if and only if $x \in U_{l+1}(f)$.

A simple calculation using the concrete form of σ given above shows that $x\sigma = \sigma x$ if and only if $\bar{x}^t m^{-1} x m$ commutes with each $Y \in \mathfrak{sl}_{l+1}$. This is equivalent to $xm\bar{x}^t = km$, $k \in K$. If x is superdiagonal, k must be 1, because the $(1, l+1)$ entries of the matrices $xm\bar{x}^t$ and m are both -1. Thus 6.3 and 6.2 are proved.

It is to be observed that the form f has index $[(l+1)/2]$.

7. **Structure of G^1.** Recall that G^1 is the group generated by $\mathfrak{U}^1$ and $\mathfrak{V}^1$. For each $w \in W^1$, set $\mathfrak{U}^1_w = \mathfrak{U}^1 \cap \mathfrak{U}_w$. For each $S \in \Pi^1$, let G^1_s be the group generated by $\mathfrak{U}^1_s$ and $\mathfrak{V}^1_s$. Denote by X^1 the group of those characters of P_r into K^* which can be extended to characters χ of P which are selfconjugate in the sense that $\chi(\bar{a}) = \overline{\chi(a)}$ for all $a \in P$, and by $\mathfrak{H}^1$ the corresponding subgroup of $\mathfrak{H}$. For $S \in \Pi^1$, set $\mathfrak{H}^1_s = \mathfrak{H}^1 \cap G_s$. Finally, for each root r and each $k \in K^*$, denote by $\chi_{r,k}$ the character on P_r defined by $\chi_{r,k}(s) = k^{s(r)}$.

It is assumed until further notice that $\mathfrak{g}$ is not of type A_l (l even). We aim to prove:

7.1 Lemma. *For each* $w \in W^1$, $\mathfrak{H}\omega(w) \cap G^1$ *is not empty.*

Once this is established, it can (and will) be assumed that $\omega(w) \in G^1$ for each $w \in W^1$. Then:

7.2 Theorem. *G^1 is the union of the sets $\mathfrak{U}^1\mathfrak{H}^1\omega(w)\mathfrak{U}^1_w$, $w \in W^1$. The sets are disjoint and each element of G^1 has a unique expression of the indicated form.*

The steps of the proof are quite analogous to those in the proof of 5.5 in view of the following:

7.3 LEMMA. *Assume* $S \in \Pi^1$. *Then* (1) *if* $S = \{r\}$, *there is a homomorphism* φ_1 *of* $SL_2(K_0)$, *the unimodular group, onto* G_s^1 *such that*

$$\varphi_1\begin{pmatrix}1 & t\\ 0 & 1\end{pmatrix} = x_r(t),\ \varphi_1\begin{pmatrix}1 & 0\\ t & 1\end{pmatrix} = x_{-r}(t),\ \varphi_1\begin{pmatrix}k & 0\\ 0 & k^{-1}\end{pmatrix} = h(\chi_{r,k}) ,$$

and

$$\varphi_1\begin{pmatrix}0 & 1\\ -1 & 0\end{pmatrix} \equiv \omega(w_r) \pmod{\mathfrak{H}} ;$$

(2) *if* $S = \{r, \bar{r}\}$, *there is a homomorphism* φ_2 *of* $SL_2(K)$ *onto* G_s^1 *such that*

$$\varphi_2\begin{pmatrix}1 & t\\ 0 & 1\end{pmatrix} = x_r(t)x_{\bar{r}}(\bar{t}),\ \varphi_2\begin{pmatrix}1 & 0\\ t & 1\end{pmatrix} = x_{-r}(t)x_{-\bar{r}}(t),\ \varphi_2\begin{pmatrix}k & 0\\ 0 & k^{-1}\end{pmatrix} = h(\chi_{r,k}\chi_{\bar{r},\bar{k}}) ,$$

and

$$\varphi_2\begin{pmatrix}0 & 1\\ -1 & 0\end{pmatrix} \equiv \omega(w_r w_{\bar{r}}) \pmod{\mathfrak{H}} .$$

Proof. The existence of φ_1 is established in [3, p. 29, p. 36]. Since $\mathfrak{X}_r$ and $\mathfrak{X}_{-r}$ commute elementwise with $\mathfrak{X}_{\bar{r}}$ and $\mathfrak{X}_{-\bar{r}}$, it is clear that φ_2 also exists.

Proof of 7.1. By 7.3, $\mathfrak{H}\omega(w_s) \cap G^1$ is non-empty for each $S \in \Pi^1$. Thus 7.1 follows from 2.3.

Now we choose $\omega(w) \in G^1$ for each $w \in W^1$, and denote by $\mathfrak{W}^1$ the union of the sets $\mathfrak{H}^1\omega(w)$. Then the analogue of 5.3 holds.

7.4 LEMMA. G^1 *contains* $\mathfrak{H}^1$.

Proof. G^1 contains all $h(\chi) \in \mathfrak{H}^1$ such that χ is of the form $\chi_{a,k}$, $\bar{a} = a$, $\bar{k} = k$, or $\chi_{a,j}\chi_{\bar{a},\bar{j}}$ by 7.3. These characters generate X^1 (see [3, p. 48, Lemma 2]). Thus $G^1 \supset \mathfrak{H}^1$.

7.5 LEMMA. *For each* $S \in \Pi^1$, G_s^1 *is the the union of the sets* $\mathfrak{U}_s^1\mathfrak{H}_s^1$ *and* $\mathfrak{U}_s^1\mathfrak{H}_s^1\omega(w_s)\mathfrak{U}_s^1$.

Proof. Because of 7.3, this follows from the corresponding properties of the groups $SL_2(K_0)$ and $SL_2(K)$ (see [3, p. 34, Lemma 2]).

7.6 LEMMA. G^1 *is generated by the groups* $\mathfrak{U}_s^1$ *and* $\mathfrak{V}_s^1$ *which correspond to fundamental sets* $S \in \Pi^1$.

Proof. This follows from 7.1, 5.4 and 4.5.

7.7 LEMMA. $G^1 = \mathfrak{U}^1\mathfrak{W}^1\mathfrak{U}^1$.

Proof. This follows from 7.6, 7.5, 7.1 and 4.7 as in [3, p. 40, Lemma 10].

Proof of 7.2. That G^1 is the union of the given sets follows from 7.7 and 4.7 as in [3, p. 42, Theorem 2]. The disjointness and uniqueness follow from 5.5.

7.8 COROLLARY. $\mathfrak{H}^1 = \mathfrak{H} \cap G^1$.

Proof. Because of 7.2, this is clear.

7.9 COROLLARY. $\mathfrak{U}^1\mathfrak{H}^1$ *is the normalizer of* $\mathfrak{U}^1$ *in* G^1.

Proof. The normalizer contains $\mathfrak{U}^1\mathfrak{H}^1$ by 5.1, and equality follows from 7.2.

One also concludes from the preceding results:

7.10 COROLLARY. *The sets of* 7.2 *are the double cosets of* G^1 *relative to* $\mathfrak{U}^1\mathfrak{H}^1$.

7.11 COROLLARY. *If* K *is a finite field of characteristic* p, *then* $\mathfrak{U}^1$ *and* $\mathfrak{V}^1$ *are* p*-Sylow subgroups of* G^1.

In regard to 7.11, one sees from 4.5 and 4.6 that, if $|K| = q^2$ and $|\Pi| = N$, then $|\mathfrak{U}^1| = q^N$.

We now remove the restriction on $\mathfrak{g}$ and remark that the results of this section remain valid even if $\mathfrak{g}$ is of type A_l (l even). The key point here is that, if $S \in \Pi^1$ and $|S| = 3$, then there exists a homomorphism of $U_3(f)$ (see 6.1) onto G^1_s with properties like those of φ_1 and φ_2 in 7.3. We omit the proof which can be made to depend on the representation of G^1 by unitary matrices given in § 6.

8. Proof of simplicity.

Our aim here is to prove:

8.1 THEOREM. *If* K_0 *has at least* 5 *elements, then* G^1 *is simple.*

The simplicity of the group SL_2 over its center is assumed to be known. It is further assumed that $\mathfrak{g}$ is not of type A_l (l even) and that $l \geqq 3$. The proof to be given can be adapted with minor modifications to the missing groups, which are in any case adequately covered by 6.2 and [4, p. 70, Theorem 5].

8.2 LEMMA. *Assume R, $T \in \Pi^1$, $R \neq T$, and that r, t are elements of R, T, respectively. Then there is $\chi \in X^1$ such that $\chi(r) = 1$, $\chi(t) \neq 1$.*

Proof. Let $\tilde{R}$ (or more simply R) denote r or $r + \bar{r}$ in the cases $\bar{r} = r$ or $\bar{r} \neq r$, respectively, and then set $\chi_{R,k} = \chi_{r,k}$ or $\chi_{R,k} = \chi_{r,k}\chi_{\bar{r},\bar{k}}$ accordingly. Treat t and T similarly. If $R(T) = 0$, set $\chi = \chi_{T,k}$, $k \in K_0^*$, $k^2 \neq 1$. If $R(T) = \pm 1$, or if $R(T) = \pm 2$ and $|R| = |T| = 2$, set $\chi = \chi_{T,k^2}\chi_{R,k}^{-r(T)}$, $k \in K_0^*$, $k^3 \neq 1$. In the other cases of $R(T) = \pm 2$, set $\chi = \chi_{t,k}\chi_{\bar{t},k}\chi_{R,k}^{-r(t)}$, $k \in K_0^*$, $k^2 \neq 1$. Finally if $R(T) = \pm 4$, set $\chi = \chi_{t,k}\chi_{\bar{t},\bar{k}}$, $k = \bar{k}_1/k_1$, $k_1 \in K$, $\bar{k}_1 \neq \pm k_1$. One can check that these cases are exhaustive and that $\chi(r) = 1$ and $\chi(t) \neq 1$ in each case.

8.3 LEMMA. *If $w \in W^1$ and $w \neq 1$, there is $h \in \mathfrak{H}^1$ such that $\omega(w)h \neq h\omega(w)$.*

Proof. We first show that there exist $\chi \in X^1$ and $r \in \Pi$ such that $\chi(wr) \neq \chi(r)$. If there is an $R \in \Pi^1$ such that $wR \neq \pm R$, then χ and r exist by 8.2. If $wR = \pm R$ for all $R \in \Pi^1$, then, since $w \neq 1$, one has $wR = -R$ for all $R \in \Pi^1$. Since $l \geqq 3$, one can readily choose $r, t \in \Pi$ so that $r \perp \bar{r}$, $t = \bar{t}$ and $r(t) < 0$. If $k \in K_0^*$, $k^2 \neq 1$, then $\chi = \chi_{t,k}$ and r have the required property. If $h = h(\chi)$, a simple calculation now shows that X_r has different images under $\omega(w)h$ and $h\omega(w)$.

Assume now that H is a normal subgroup of G^1 and that $|H| > 1$.

8.4 LEMMA. $|H \cap \mathfrak{U}^1\mathfrak{H}^1| > 1$.

Proof. By 7.2 there is $x \in H$ such that $x \neq 1$ and $x = uh_1\omega(w)$ with $u \in \mathfrak{U}^1$, $h_1 \in \mathfrak{H}^1$ and $w \in W^1$. If $w \neq 1$, then by 8.3 there is $h \in \mathfrak{H}^1$ such that $\omega(w)h \neq h\omega(w)$. Then $y = hxh^{-1}x^{-1} \in H \cap \mathfrak{U}^1\mathfrak{H}^1$, and we assert that $y \neq 1$. Indeed, if $y = 1$, then

$$x = hxh^{-1} = huh^{-1}(hh_1\omega(w)h^{-1}\omega(w)^{-1})\omega(w) ,$$

and by 7.2 one gets $h\omega(w)h^{-1}\omega(w)^{-1} = 1$, a contradiction. Thus the assertion and the lemma are proved.

8.5 LEMMA. $|H \cap \mathfrak{U}^1| > 1$.

Proof. By 8.4, there is $x \in H \cap \mathfrak{U}^1\mathfrak{H}^1$ such that $x \neq 1$. Write $x = uh$, $u \in \mathfrak{U}^1$, $h \in \mathfrak{H}^1$, and suppose $h \neq 1$. Then there is a fundamental root r such that $hX_r = cX_r$, $c \in K$, $c \neq 1$. If $r \in S \in \Pi^1$, let y be the commutator of x with $x_r(1)$ or $x_r(1)x_{\bar{r}}(1)$ according as $|S| = 1$ or 2. Then $y \in H \cap \mathfrak{U}^1$, and it remains to show that $y \neq 1$. If $y = 1$, then, for the case $|S| = 1$, one has $x_r(1) = uhx_r(1)h^{-1}u^{-1} = ux_r(c)u^{-1}$. Now it follows easily from 4.1 that the subgroup $\mathfrak{U}_2$ of $\mathfrak{U}$ generated by those $\mathfrak{X}_r$ for which $ht\ r > 1$ contains the commutator subgroup of $\mathfrak{U}$. Thus $x_r(1-c) = x_r(1)x_r(c)^{-1} \in \mathfrak{U}_2$, whence $1 - c = 0$ by 4.3. This contradiction establishes $y \neq 1$. The case $|S| = 2$ can be treated similarly.

8.6 LEMMA. *For some* $R \in \Pi^1$, $|H \cap U^1_R| > 1$.

Proof. Among all $x \in H \cap \mathfrak{U}^1$ with $x \neq 1$, choose one which maximizes the minimum $S \in \Pi^1$ for which $x_s \neq 1$ in the representation 4.5. If this minimum is R, we show $x = x_R$. Assuming the contrary, one can write $x = x_R x_T x_1$ with $x_R \neq 1$, $x_T \neq 1$, and x_1 denoting the remaining terms in 4.5. By 8.2, 5.1 and 4.6, there is $h \in \mathfrak{H}^1$ such that $hx_Rh^{-1} = x_R$ and $hx_Th^{-1} \neq x_T$. Thus $hxh^{-1} \neq x$ by 4.5. But then $y = x^{-1}hxh^{-1} \neq 1$, $y \in H \cap \mathfrak{U}^1$, and y provides a contradiction to the choice of x.

Using 8.6, one can deduce as in [3, p. 62, Lemma 15]:

8.7 LEMMA. *If* $|H \cap \mathfrak{U}^1_R| > 1$ *for* $R \in \Pi^1$, *then* $H \supset \mathfrak{U}^1_R$.

Proof of 8.1. As in 8.3 choose (fundamental) roots r, t such that $r \perp \bar{r}$, $t = \bar{t}$ and $r(t) < 0$. Since $r \perp \bar{r}$, this implies that $r + t$, $\bar{r} + t$ and $r + \bar{r} + t$ are all roots. Set $R = \{r, \bar{r}\}$, $T = \{t\}$, $U = \{r + t, \bar{r} + t\}$, $V = \{r + \bar{r} + t\}$, $x_R(1) = x_r(1)x_{\bar{r}}(1)$, $x_T(1) = x_t(1)$, etc.. Then by 4.1 (used several times), one gets:

$$(x_R(1),\ x_T(1)) = x_U(N_{rt})x_V(N_{rt}N_{r,\bar{r}+t}) . \tag{8.8}$$

By 8.7, 5.4 and 2.8, either $x_R(1)$ or $x_T(1)$ is in H; hence so is their commutator. For the same reason one of the elements on the right of 8.8 is in H; hence so is the other. Thus, by 8.7, 5.4, 3.1 and 2.8, H contains all $\mathfrak{U}^1_S$, hence also $\mathfrak{U}^1$ by 4.5. Similarly H contains $\mathfrak{V}^1$, whence $H = G^1$. Thus G^1 is simple.

9. Some identifications. If $\mathfrak{g}$ is of type A_l, then G^1 has been identified in § 6 as a projective unitary group in $l + 1$ dimensions. Similarly, if $\mathfrak{g}$ is of type D_l ($l \geqq 4$), then using the representation of G given by Ree [7], one can show that G^1 is isomorphic to a projective orthogonal group corresponding to a form in $2l$ variables which has index $l - 1$ relative to K_0 and index l relative to K. The details in the complex-real case can be found in [2, p. 422]. If $\mathfrak{g}$ is of type E_6, then, again in the complex-real case, one can identify G^1 with a real form of E_6, the one characterized by Cartan [2, p. 493] by the fact that its Killing form, when written as a sum of real squares, contains a surplus of 2 positive terms. If $\mathfrak{g}$ is of type E_6 and K is finite, we show in § 12 that new groups are obtained[1], not isomorphic to any appearing in the list of finite simple groups given by Artin [1].

10. Second variation for D_4. A root system for D_4 has a fundamental basis consisting of roots a, b, c, d of the same length such that b, c, d are mutually orthogonal and each makes an angle of $2\pi/3$ with a. Let τ be the automorphism of order 3 of the underlying Euclidean space defined by $a, b, c, d \to a, c, d, b$, and let W^2 be the subgroup of elements of W commuting with τ. One can then obtain the analogues of the results of § 2 without essential change in the proofs. For example: W^2 is generated by the elements w_a and $w_b w_c w_d$, and is of type G_2. The roots are partitioned into sets of the types (1) $S = \{r\}$, $\tau r = r$, and (2) $S = \{r, \tau r, \tau^2 r\}$. Any 2 sets of the same type are congruent under W^2. One then introduces a field K on which an automorphism τ of order 3 acts, and defines a semi-automorphism τ of $\mathfrak{g}_K$ by $\tau(kX_r) = (\tau k)X_{\tau r}$. Then $\mathfrak{U}^2$ and $\mathfrak{V}^2$ are the subgroups of $\mathfrak{U}$ and $\mathfrak{V}$, respectively, made up of elements commuting with τ and G^2 is the group they generate. The whole previous developement goes through. It turns out that in the proof of simplicity it is enough to assume that the fixed field K_0 has at least 4 elements. In § 12, it is shown that once again new finite groups[1] are obtained.

11. Third variation for D_4. Assume now that K is a field admitting automorphisms σ and τ which are of orders 2 and 3 respectively, and which generate a group isomorphic to S_3, the symmetric group on 3 objects. Define corresponding semi-automorphisms σ and τ of the Lie algebra $\mathfrak{g}_K$ of type D_4 as in §§ 3 and 10. Then set $\mathfrak{U}^3 = \mathfrak{U}^1 \cap \mathfrak{U}^2$, $\mathfrak{V}^3 = \mathfrak{V}^1 \cap \mathfrak{V}^2$, and let G^3 be the group generated by $\mathfrak{U}^3$ and $\mathfrak{V}^3$. Again everything goes through. It need only be remarked that the present construction is possible only if K is infinite, and that all groups of type G^3 are simple.

12. Some new groups. The list L of known finite simple groups consists of the cyclic, alternating and Mathieu groups, and the "Lie groups", namely the groups G of Chevalley over A_l $(l \geqq 1)$, B_l $(l \geqq 2)$, C_l $(l \geqq 3)$, D_l $(l \geqq 4)$, E_6, E_7, E_8, F_4 and G_2, the groups G^1 over A_l $(l \geqq 2)$, D_l $(l \geqq 4)$ and E_6, and the groups G^2 over D_4, all constructed on a finite field. By the type of one of these latter groups we mean a combination consisting of the general mode of construction (G or G^1 or G^2), the underlying complex Lie algebra $\mathfrak{g}$, and the field K. We adopt the notation: $E_6^1(r)$ is the group of type G^1 over E_6 on a field of r elements. Our aim is to prove:

12.1 THEOREM. *If G is one of the groups $E_6^1(q^2)$ or $D_4^2(q^3)$, then $\hat{G}$*

is not isomorphic to a cyclic, alternating or Mathieu group, and two representations of $\hat{G}$ as Lie groups necessarily have the same type.

In other words the groups $E_6^1(q^2)$ and $D_4^2(q_1^3)$ are new[1] and distinct among themselves. We need some preliminary results. Let $\hat{G}$ be a Lie group over a field K of q, q^2 or q^3 elements in the cases G, G^1 or G^2, respectively, and set $\hat{W} = W$, W^1 or W^2 accordingly. The *Poincaré sequence* of $\hat{G}$ shall mean the list of numbers $q^{n(w)}(w \in \hat{W})$ arranged in non-decreasing order. Thus the first term is 1 and the last term is q^N, the integer N being the number of positive roots of $\mathfrak{g}$ (see 2.5, 4.5 and 4.6).

12.2 Lemma. *The Poincaré sequence of $A_l^1(q^2)$, $D_l^1(q^2)$, $E_6^1(q^2)$ or $D_4^2(q^3)$ is obtained by writing the respective polynomial* $\prod_1^{l+1} \frac{t^i - (-1)^i}{t - (-1)^i}$, $(t^l + 1) \prod_2^{l-1} \frac{t^{2i} - 1}{t - 1}$, $\frac{t^2 - 1}{t - 1} \cdot \frac{t^5 + 1}{t + 1} \cdot \frac{t^6 - 1}{t - 1} \cdot \frac{t^8 - 1}{t - 1} \cdot \frac{t^9 + 1}{t + 1} \cdot \frac{t^{12} - 1}{t - 1}$ *or* $(t + 1)(t^3 + 1)(t^8 + t^4 + 1)$ *as a sum of non-decreasing powers of t and then replacing t by q in the individual terms.*

To avoid interruption of the present development we give the proof in the next section. We also need the polynomials for the groups of Chevalley. As one sees from considerations in [3, p. 44, p. 64], these polynomials take the form $\prod[(t^{a(i)} - 1)/(t - 1)]$, the $a(i)$ being given in [3, p. 64]. Since $q^{n(w)} = |\mathfrak{U}_w^1|$ by 4.6 and 4.7, one can use 12.2 in conjunction with 7.2 and the definition of $\mathfrak{H}^1$ to compute $|G^1|$. In the same way, one can find $|G^2|$. Thus:

12.3 Lemma. *If u is the g.c.d. of 3 and $q + 1$, the orders of $E_6^1(q^2)$ and $D_4^2(q^3)$ are $u^{-1}q^{36}(q^2 - 1)(q^5 + 1)(q^6 - 1)(q^8 - 1)(q^9 + 1)(q^{12} - 1)$ and $q^{12}(q^2 - 1)(q^6 - 1)(q^8 + q^4 + 1)$, respectively.*

The orders of the other Lie groups can be found in [1]. It is interesting to note that, if in the expressions in 12.2 and 12.3 which relate to the group $E_6^1(q^2)$ one replaces all plus signs by minus signs, then one obtains the corresponding properties of $E_6(q)$. A similar phenomenon occurs for each of the groups $A_l^1(q^2)$ and $D_l^1(q^2)$.

12.4 Lemma. *The Poincaré sequence of a finite Lie group $\hat{G}$ is determined by the abstract group and the characteristic p of the base field K. The type of a finite Lie group is determined by its Poincaré sequence except that $B_l(q)$ and $C_l(q)$ have the same sequence, as do $A_1(q^3)$ and $A_2^1(q)$ also.*

Proof. If $\hat{G}$ is of type G, then, to within an inner automorphism, G and p determine $\mathfrak{U}$ as a p-Sylow subgroup, then $\mathfrak{U}\mathfrak{H}$ as the normalizer of $\mathfrak{U}$, and finally the numbers $|\mathfrak{U} \cap x\mathfrak{U}x^{-1}|$ as x runs through a system of representatives of the double coset decomposition of G relative to $\mathfrak{U}\mathfrak{H}$. These latter numbers are just the terms of the Poincaré sequence by the analogue of 7.10, since $|\mathfrak{U} \cap \omega(w)\mathfrak{U}\omega(w)^{-1}| = q^{n(w_0 w)}$ by 4.3. A similar proof of the first statement holds for groups of type G^1 or G^2. One proves the second statement by inspection of the Poincaré sequences for the various Lie groups.

By checking their orders, one sees that $A_1(q^3)$ and $A_2^1(q)$ can not be isomorphic. Thus the two statements of 12.4 can be combined to yield:

12.5. *The type of a finite Lie group is determined by the abstract group and the characteristic of the base field except that $B_l(q)$ and $C_l(q)$ may be isomorphic.*

This result has been obtained previously (for the previously known finite simple Lie groups) by Artin [**1**] and Dieudonné [**5**, p. 71–75] by different, more detailed methods. Artin actually draws the conclusion under the weak assumption that only $|\hat{G}|$ and p are known.

One also concludes from 12.4 the well-known fact that $A_2(4)$ and $A_3(2)$, both of order 20160, are not isomorphic.

An inspection of the results of 12.3 yields:

12.6 LEMMA. *Let $\hat{G}$ be either $E_6^1(q^2)$ or $D_4^2(q^3)$ over a field of characteristic p, and let Q be the largest power of p which divides $|\hat{G}|$. Let Q' be any prime power which divides $|\hat{G}|$. Then $Q^3 > |\hat{G}|$ and $Q \geqq Q'$.*

Proof of 12.1. Clearly $\hat{G}$ is not cyclic. Since $|\hat{G}| > 10^8$ and $Q^3 > |\hat{G}|$, it follows that $\hat{G}$ is not an alternating group (see [**1**]). $D_4^2(8)$ does not have the order of a Mathieu group and all other values of $|\hat{G}|$ are too large. $\hat{G}$ is not isomorphic to either of the groups $A_1(p_1)$ with $p_1 = 2^r - 1 =$ prime, or $A_1(2^s)$ with $2^s + 1 =$ prime, since in each case one has a prime p_2 such that p_2 divides $|\hat{G}|$ and $p_2^3 > |\hat{G}|$, and this is readily seen to be impossible by 12.3. But except for these two types, every simple finite Lie group verifies 12.6 (see [**1**] where the other groups are considered). Thus any representation of $\hat{G}$ as a Lie group must be over a field of characteristic p. An application of 12.4 completes the proof.

13. Proof of 12.2. By 2.2, 2.3 and 2.6, $n(w) = \sum |S|$, summed over those $S \in \Pi^1$ for which $wS < 0$. By 2.7, one can compute $n(w)$ within the framework of $\tilde{W}^1$ and its root system, but each root is to be counted with the right multiplicity (1, 2 or 3). Assume first that the group under consideration is $E_6^1(q^2)$. Then $\tilde{W}^1$ is of type F_4 and, in terms of coordinates relative to an orthonormal basis, its roots can be taken as $\pm x_i$, $(\pm x_1 \pm x_2 \pm x_3 \pm x_4)/2$, each of multiplicity 1, and $\pm x_i \pm x_j$ $(i \neq j)$, each of multiplicity 2 (see [8, p. 13-08]). The inequalities $x_1 - x_2 - x_3 - x_4 > 0$, $x_2 - x_3 > 0$, $x_3 - x_4 > 0$, and $x_4 > 0$ determine a fundamental region F of $\tilde{W}^1$ by [10, p. 160]. The last 3 inequalities determine a region L whose intersection with the unit sphere is lune-shaped with $(1, 0, 0, 0)$ as one of its vertices. The subgroup V of $\tilde{W}^1$ leaving $(1, 0, 0, 0)$ fixed is of type C_3 and has L as a fundamental region. Let $P(t)$ be the polynomial sought, let $P_1(t)$ be the corresponding polynomial for the group V, and let $P_2(t)$ be $\sum t^{n(w)}$, the sum being over those $w \in W^1$ for which $\tilde{w}F \subset L$. A simple geometric argument shows that $P = P_1 P_2$. We next find P_2. The point $a = (16, 8, 4, 2)$ is in F. It has 24 transforms in L corresponding to the 24 elements $\tilde{w} \in \tilde{W}^1$ for which $\tilde{w}F \subset L$. These are a, $b = (15, 5, 3, 9)$, $c = (13, 11, 7, 1)$ and the points in L obtained from these by coordinate permutations. One can now find $n(w)$ for each of the 24 elements above. For example, if w maps a on b, then the roots positive at a and negative at b are $(x_1 - x_2 - x_3 - x_4)/2$, of multiplicity 1, and $x_2 - x_4$ and $x_3 - x_4$, each of multiplicity 2. Hence $n(w) = 5$. Thus P_2 is determined, and the original problem of rank 4 is reduced to one of rank 3. A similar reduction to rank 2 is possible, whence P can be determined. If one starts with $A_l^1(q^2)$ or $D_l^1(q^2)$ instead, the same inductive procedure can be carried through, and for $D_4^2(q^3)$ the polynomial P can be found rather quickly by enumerating $n(w)$ for the 12 elements of W^2. The results are those listed in 12.2.

14. Prime power representations. In [9], 14 assumptions on a finite group are made, and then some properties concerning the representations of the group are deduced. It is then verified that the groups of Chevalley satisfy the basic assumptions. The verification for G^1 or G^2 is virtually the same as for G because of the structure theorems of the present paper. Thus one gains the results of [9] (in particular Theorem 4) simultaneously for all known finite simple Lie groups.

15. Concluding remarks. We first note that it is possible to cover somewhat more ground than was indicated in the main development given here by allowing certain degeneracies to occur. For example, if σ on E is of order 2, if σ on K is of order 1, and if $\mathfrak{g}$ is of type A_{2l} or A_{2l-1}, then the construction of §§ 3, 4 and 5 yields a group of type B_l or C_l, respectively. Thus B_l, C_l and also A_m may be regarded as degenerate cases of A_m^1. Similarly D_l^1 degenerates to B_{l-1} and D_l; E_6^1 to F_4 and E_6; and D_4^3 to G_2, B_3, D_4, D_4^1 and D_4^2. It is easily verified that no other groups can be obtained by the present method of combining automorphisms of E and of K in various ways[1].

In regard to the construction given for G^1, it is to be noted that $\mathfrak{g}_K^1$, the set of fixed points of σ, is the Lie algebra (over K_0) of G^1 in many cases. We could have defined G^1 on $\mathfrak{g}_K^1$ in view of the easily proved facts that an automorphism x of $\mathfrak{g}_K$ commutes with σ if and only if $x\mathfrak{g}_K^1 = \mathfrak{g}_K^1$, and that, in this case, the restriction of x to $\mathfrak{g}_K^1$ is 1 only if $x = 1$; but this would have led to a much more complicated development. It is also to be noted that one can not define G^1 as the subgroup G^σ of G made up of elements which commute with σ. The difference, roughly speaking, lies in $\mathfrak{H}$: a self-conjugate character on P_r may be extendable to a character on P but not to a self-conjugate one, as is proved by the following example. Let $\mathfrak{g}$ be of type A_1, and let w and $a = 2w$ be fundamental weight and root, respectively. Then χ defined by $\chi(a) = k^2$, $k^2 \in K_0^*$, $\bar{k} \neq k$, has the given property. One sees rather easily, however, that G^σ/G^1 is always isomorphic to a subgroup of P/P_r.

The proof of simplicity given in §8 is considerably shorter than the one given in [3], but this is at the expense of the assumption that K has enough elements: left open is the question of simplicity for the groups $E_6^1(q^2)$ with $q \leqq 4$, and $D_4^2(q^3)$ with $q \leqq 3$. The answer quite likely requires rather detailed methods such as those of [3].

More important, perhaps, and probably more difficult is the identification of the infinite groups constructed. An infinite analogue of 12.4 would go a long way in this direction. Finally, it seems likely that there is some sort of description of D_4^2 and D_4^3 by Cayley numbers.

References

1. E. Artin, *Orders of classical simple groups*, Comm. Pure Appl. Math., **8** (1955), 455.
2. É. Cartan, *Oeuvres complètes*, Paris, 1952.
3. C. Chevalley, *Sur certains groupes simples*, Tôhoku Math. J., (2) **7** (1955), 14.
4. J. Dieudonné, *Sur les groupes classiques*, Paris, 1948.
5. ———, *On the automorphisms of the classical groups*, Mem. Amer. Math. Soc. 2, 1955.
6. D. Hertzig, *On simple algebraic groups*, Short communications, Int. Congress of Math., Edinburgh, 1958.
7. R. Ree, *On some simple groups defined by C. Chevalley*, Trans. Amer. Math. Soc. **84** (1957), 392.
8. Séminaire "*Sophus Lie*", Paris, 1954-1955.
9. R. Steinberg, *Prime power representations of finite linear groups* II, Canad. J. Math., **9** (1957), 347.
10. H. Weyl, *The structure and representation of continuous groups*, I.A.S. notes, Princeton, 1934-35.

UNIVERSITY OF CALIFORNIA, LOS ANGELES

THE SIMPLICITY OF CERTAIN GROUPS

ROBERT STEINBERG

The purpose of this note is to give a proof of the simplicity of certain "Lie groups" considered in [2]. The main feature of the present development is the proof of Lemma 2 below: it is superior to the corresponding proof given in [2], because no assumption on the number of elements of the base field is required, and is very much shorter than the one given by Chevalley [1] for the direct analogues, over arbitrary fields, of the simple (complex) Lie groups. Thus it turns out that the groups $E_6^1(q^2)$ with $q \leqq 4$, and $D_4^2(q^3)$ with $q \leqq 3$, to which the proof in (2) is not applicable, are simple.

Assuming the notations of [1] and [2] to be in effect, we shall prove:

1. THEOREM. *If $\hat{G}$ is one of the groups of type G^1, G^2 or G^3, defined in* [2], *and the rank l of the corresponding Lie algebra is at least* 3, *then $\hat{G}$ is simple.*

It will be noticed that the case A_2^1 is excluded by the assumption on l. This is of necessity, since the simplicity of A_2^1 is not universal, but depends on the base field. The same is true of groups of type A_1.

2. MAIN LEMMA. *Let $\hat{G}$ be a group of type G, that is, one of the direct analogues of the ordinary simple Lie groups, or a group of type G^1, G^2* or *G^3, but assume $\hat{G}$ is not of type A_1 or A_2^1. Let $\hat{\mathfrak{U}}$ be the nilpotent subgroup of $\hat{G}$ corresponding to the positive roots of the underlying Lie algebra. Let H be a normal subgroup of $\hat{G}$ such that $|H| > 1$. Then $|H \cap \hat{\mathfrak{U}}| > 1$.*

Proof. Assume first that G is of type G^1. By 7.2 of [2], there is $x = uh\omega(w) \in H$ with $u \in \mathfrak{U}^1$, $h \in \mathfrak{H}^1$.

If $w = 1$, then [2, Lemma 8.5] yields the required conclusion.

If $w \neq 1$, consider first the case in which $w = w_S$ with S a fundamental element of Π^1. Then there is a fundamental $A \in \Pi^1$ such that $B = wA > 0$ and $wA \neq A$ (because A_1 and A_2^1 are excluded). Choose $y \in \mathfrak{U}_A^1$ so that $y \neq 1$ and $y \notin \mathfrak{U}_2$, the subgroup of $\mathfrak{U}$ generated by those $\mathfrak{X}_r$ for which $ht\ r \geqq 2$. Then we assert that the commutator $z = (x, y)$ is in $H \cap \mathfrak{U}^1$ and that $z \neq 1$. In fact, $z = uh\omega(w)y\omega(w)^{-1}h^{-1}u^{-1}y^{-1} = utu^{-1}y^{-1}$ with $t \in \mathfrak{U}_B^1$; hence $z \in H \cap \mathfrak{U}^1$, and, since $\mathfrak{U}/\mathfrak{U}_2$ is Abelian, we have $z \equiv ty^{-1} \not\equiv 1 \bmod \mathfrak{U}_2$, by 4.3 of [2], whence $z \neq 1$.

Finally, consider the general case in which $w \neq 1$. Choose $R \in \Pi^1$

Received July 31, 1959.

so that $-wR = S$ is fundamental in Π^1, and then $y \in \mathfrak{U}_R^1$ so that $y \neq 1$. Again form $z = (x, y)$. In the present case, $\omega(w)y\omega(w)^{-1} \in \mathfrak{U}_S^1 \mathfrak{H}^1 \omega(w_S)\mathfrak{U}_S^1$ by 7.3 of [2], so that z is conjugate to an element x_1 of the form $u_1 h_1 \omega(w_S)$ with $u_1 \in \mathfrak{U}^1$, $h_1 \in \mathfrak{H}^1$. Clearly $x_1 \neq 1$ and $x_1 \in H$. Thus the situation is that at the beginning of the preceding paragraph, and Lemma 2 is proved for groups of type G^1.

Now to get a proof for groups of type other than G^1, we need only delete all superscripts or replace them all by 2 or all by 3, depending on the group under consideration.

From this point on, we assume that $\hat{G}$ is of type G^1, but not of type A_l^1 (l even), and the ensuing discussion refers explicitely to this case. For groups of type A_l^1 (l even), G^2 or G^3, the changes to be made are quite clear: a prototype for these changes is the replacement of (*) below by an appropriate analogue. For groups of type G, the rest of the proof of Theorem 1 is given in [**1**].

3. LEMMA. *If G^1 is not of type A_l^1 (l even) and H is a normal subgroup of G^1 such that $|H| > 1$, then, for some $R \in \Pi^1$, $|H \cap \mathfrak{U}_R^1| > 1$.*

It is convenient to precede the proof of this lemma by some preparatory results.

4. LEMMA. *If $s, a, s + a$ and t are roots such that $\bar{a} \neq a$ and $s + a = t + \bar{a}$, then $t = \bar{s}$.*

Proof. We have $s(a) < 0$ and $s(\bar{a}) = (s + a)(\bar{a}) > 0$. Hence $\bar{s} \neq s$, and a simple calculation shows that $t - \bar{s} = s + a - \bar{s} - \bar{a}$ has length 0, since all roots have the same length and the only possible angles are the multiples of $\pi/3$ and $\pi/2$. Hence $t = \bar{s}$.

Let us recall that, for each positive integer m, $\mathfrak{U}_m$ denotes the subgroup of $\mathfrak{U}$ generated by those $\mathfrak{X}_r$ for which $ht\ r \geqq m$.

5. LEMMA. *Let s be a positive root, a a fundamental root, and S and A the elements of Π^1 which contain them. Assume $s(a) < 0$, $x \in \mathfrak{U}_S^1$, $y \in \mathfrak{U}_A^1$, and set $ht\ s = n$. Then*

(a) *(x, y) is congruent,* mod $\mathfrak{U}_{n+2}$, *to an element of $\mathfrak{U}^1$ whose representation 4.3 of* [2] *has all components other than those from $\mathfrak{X}_{s+a}$ and $\mathfrak{X}_{\bar{s}+\bar{a}}$ equal to* 1, *and*

(b) *if x is given and $x \neq 1$, then y can be chosen so that the $\mathfrak{X}_{s+a}$ component is not* 1.

Proof. Assume first $|S| = |A| = 2$. Then $(s, a) < 0$, whence $(s, \bar{a}) \geqq 0$, because the contrary assumption yields the false conclusion that $s + \bar{s} + a + \bar{a}$ has length 0. Thus $\mathfrak{X}_s$ and $\mathfrak{X}_a$ commute elementwise with $\mathfrak{X}_{\bar{s}}$ and $\mathfrak{X}_{\bar{a}}$, and 4.1 of [2] yields

$$(*) \qquad (x_s(k)x_{\bar{s}}(\bar{k}),\ x_a(l)x_{\bar{a}}(\bar{l})) = x_{s+a}(N_{sa}kl)x_{\bar{s}+\bar{a}}(N_{sa}\overline{kl}) .$$

Thus (a) is true. If $k \neq 0$, we can choose l so that $kl + \overline{kl} \neq 0$, and then coalesce the terms on the right of (*) if $\bar{s} + \bar{a} = s + a$. Thus (b) is also true. If $|S| = 1$ or $|A| = 1$, we replace (*) in the above argument by an appropriate analogue (see 4.1 and 8.8 of [2]).

Let us recall that a root d is dominant if $d(a) \geqq 0$ for each fundamental root a. Since these inequalities define a fundamental region for W, and all roots are congruent under W in the present case, it follows that there is a unique dominant root d. If s is any other root, then $(s, a) < 0$ for some fundamental root a, and then $s + a$ is also a root. Thus the dominant root d may also be described as the unique root of maximum height; and one has $\bar{d} = d$ and $d > s$ for each root $s \neq d$.

We now turn to the proof of Lemma 3. Among all $x \in H \cap \mathfrak{U}^1$ for which $x \neq 1$, choose one which maximizes the minimum $S \in \Pi^1$ for which $x_S \neq 1$ in the representation 4.5 of [2]. If this minimum is R, we show $x = x_R$. Assuming the contrary, one can write $x = x_R x_T \cdots$ with $x_T \neq 1$. Set $ht\ R = n$. If $r \in R$, then r is not dominant, since $R < T$. Thus $r(a) < 0$ for some fundamental root a, and $r + a$ is a root. If $a \in A \in \Pi^1$, we conclude from Lemma 5 that there is $y \in \mathfrak{U}^1_A$ such that (x_R, y) is congruent, mod $\mathfrak{U}_{n+2}$, to an element of $\mathfrak{U}^1$ with the $\mathfrak{X}_{r+a}$ component not 1. Since $z = (x, y) \in H \cap \mathfrak{U}_{n+1}$, and $>$ respects heights, we need only show $z \neq 1$ to reach a contradiction. We have $(x, y) = (x_R, y)(x_T, y) \cdots \bmod \mathfrak{U}_{n+2}$. Here the elements on the right are in $\mathfrak{U}_{n+1}$. By choice of y, the $\mathfrak{X}_{r+a}$ component of (x_R, y) is not 1, and by Lemmas 4 and 5, the $\mathfrak{X}_{r+a}$ component of each of $(x_T, y) \cdots$ is 1. Thus we conclude from 4.3 of [2] and the fact that $\mathfrak{U}_{n+1}/\mathfrak{U}_{n+2}$ is Abelian that $(x, y) \not\equiv 1 \bmod \mathfrak{U}_{n+2}$. Therefore $(x, y) \neq 1$, and Lemma 3 is proved.

The proof of Theorem 1 can now be completed, just as in [2].

References

1. C. Chevalley, *Sur certains groupes simples*, Tôhoku Math. J. (2) **7** (1955), p. 14.
2. R. Steinberg, *Variations on a theme of Chevalley*, Pacific J. Math. **9** (1959), p. 875.

UNIVERSITY OF CALIFORNIA AT LOS ANGELES

AUTOMORPHISMS OF FINITE LINEAR GROUPS

ROBERT STEINBERG

1. Introduction. By the methods used heretofore for the determination of the automorphisms of certain families of linear groups, for example, the (projective) unimodular, orthogonal, symplectic, and unitary groups **(7, 8)**, it has been necessary to consider the various families separately and to give many case-by-case discussions, especially when the underlying vector space has few elements, even though the final results are very much the same for all of the groups. The purpose of this article is to give a completely uniform treatment of this problem for all the known finite simple linear groups (listed in §2 below). Besides the "classical groups" mentioned above, these include the "exceptional groups," considered over the complex field by Cartan and over an arbitrary field by Dickson, Chevalley, Hertzig, and the author **(3, 4, 5, 6, 10, 15)**. The automorphisms of the latter groups are given here for the first time. The unifying principles come from the theory of Lie algebras: each group is a group of automorphisms of a corresponding Lie algebra and this leads to structural properties shared by all of the groups. These centre around the so-called Bruhat decomposition (see **(2)** and 4.8 below), which, in case the underlying field is complex, reduces to the decomposition of the group into double cosets relative to a maximal solvable connected subgroup (see also **(13)** where much use is made of this decomposition). Stated roughly, the final result is that the outer automorphisms of these groups are generated by field automorphisms, graph automorphisms, which come from symmetries of the Schlaefli (or Coxeter) graph of the root structure of the corresponding Lie algebra, and diagonal automorphisms, a prototype of which is an automorphism of the unimodular group produced by conjugation by a (diagonal) matrix of determinant other than 1. Exact statements of these results (3.2 to 3.6 below) follow a description of the groups and automorphisms to be considered.

An introduction to the standard Lie algebra terminology together with statements of the principal results in the classification of the simple Lie algebras over the complex field can be found in **(4**, pp. 15–19**)**. (Proofs are available in **(3**: thesis**)**, **(9)**, **(14)**, or **(16)**.)

2. The groups. Let us start with a Cartan decomposition of a simple Lie algebra over the complex field and denote by Π and Σ respectively the sets of positive and fundamental roots relative to a fixed ordering of the additive group generated by the roots. Then, as in **(4)**, one can replace the complex field by an arbitrary base field K after choosing a generating set $\{X_r, X_{-r},$

Received July 7, 1959.

$H_r, r \in \Pi\}$ to fulfil the conditions of Theorem 1 of **(4)**, and then define: $x_r(k) = \exp(\text{ad } kX_r)$; $\mathfrak{X}_r = \{x_r(k), k \in K\}$; $\mathfrak{U}(\mathfrak{V})$ is the group generated by those $\mathfrak{X}_r$ for which r is positive (negative); and finally G (denoted G' in **(4)**) is the group generated by $\mathfrak{U}$ and $\mathfrak{V}$. The various groups G obtained in this way are A_l $(l \geqslant 1)$, B_l $(l \geqslant 2)$, C_l $(l \geqslant 3)$, and D_l $(l \geqslant 4)$, which are identified in **(11)** as suitable (projective) unimodular, orthogonal, symplectic, and orthogonal groups acting on spaces of $l+1$, $2l+1$, $2l$ and $2l$ dimensions respectively, as well as the exceptional groups E_6, E_7, E_8, F_4 and G_2. The groups G of this paragraph are called *normal types.*

If the additive group generated by the roots admits an automorphism $r \to \bar{r}$ of order 2 such that $\bar{\Sigma} = \Sigma$ and if the field K admits an automorphism $k \to \bar{k}$ of order 2, one can define an automorphism σ of the normal type of group such that $x_a(k)^\sigma = x_{\bar{a}}(\bar{k})$ for all $a \in \Sigma$ or $-\Sigma$, $k \in K$, then restrict each of $\mathfrak{U}$ and $\mathfrak{V}$ to the subgroup of elements invariant under σ, and finally restrict G to the group generated by these restrictions **(15)**. In this way one gets subgroups of A_l $(l \geqslant 2)$, D_l, and E_6 which we denote A_l^1 (unitary group in $l+1$ dimensions), D_l^1 (a second orthogonal group in $2l$ dimensions), and E_6^1 respectively. Similarly, automorphisms of order 3 yield a second subgroup D_4^2 of D_4. The groups of this paragraph are called *twisted types* and are also denoted generically by G.

3. The automorphisms. Since each of the groups G above is centreless (to be proved in 4.4; actually with 5 exceptions the groups are all simple **(4, 15)**), we can identify G with its group of inner automorphisms.

For each normal type let $\hat{\mathfrak{H}}$ (this is essentially $\mathfrak{H}$ in **(4)**) denote the group of homomorphisms $r \to h(r)$ of the additive group generated by the roots into K^*, the multiplicative group of K, with multiplication in $\hat{\mathfrak{H}}$ defined by $(h_1h_2)(r) = h_1(r)h_2(r)$, and let $\mathfrak{H}$ (this is $\mathfrak{H}'$ in **(4)**) denote the subgroup consisting of those homomorphisms which can be extended to the group of weights. Each $h \in \hat{\mathfrak{H}}$ leads to an automorphism of the Lie algebra and then to one of G (also denoted h) such that:

3.1 $$x_r(k)^h = x_r(h(r)k) \qquad (r \in \Pi \text{ or } -\Pi, k \in K).$$

If G is a twisted type, then $\hat{\mathfrak{H}}$ is to be restricted to those elements which are self-conjugate in the sense that $h(\bar{r}) = \overline{h(r)}$ and $\mathfrak{H}$ to those which have self-conjugate extensions to the group of weights. The elements of $\hat{\mathfrak{H}}$ considered as acting on G are called *diagonal automorphisms.*

Each group G as a linear group admits *field automorphisms* induced by automorphisms of K (which must be restricted to commute with $k \to \bar{k}$ in the twisted cases).

Finally, symmetries of the corresponding graph lead to automorphisms of G. If $r \to \bar{r}$ is an automorphism of the group generated by the roots such that $\bar{\Sigma} = \Sigma$, there exists an automorphism σ of G such that $x_a(k)^\sigma = x_{\bar{a}}(k)$, $a \in \Sigma$ or $-\Sigma$, $k \in K$ (see **(14**, pp. 11–104**)** or **(16**, p. 94**)**). This yields extra automorphisms of A_l ($l \geqslant 2$), D_l and E_6 (5 extra for D_4 and 1 extra for each other group). Also if K is perfect and of characteristic 3 and if G is of type G_2 with fundamental roots a and b such that $2a + 3b$ is also a root, there is an automorphism σ of G such that $x_a(k)^\sigma = x_b(k)$, $x_b(k)^\sigma = x_a(k^3)$, $k \in K$, with similar equations for $-a$ and $-b$. If K is perfect and of characteristic 2 and if G is of type B_2 or F_4, a similar automorphism exists (**13**, Exposés 21 to 24). The automorphisms of this paragraph as well as the identity are called *graph automorphisms.* Note that distinct graph automorphisms effect distinct permutations of the groups $\mathfrak{X}_a$, $a \in \Sigma$.

Our aim is to prove first:

3.2. *If G is one of the groups defined in §2 and if G is finite, each automorphism σ of G can be written $\sigma = gfdi$, with i, d, f, and g being inner, diagonal, field and graph automorphisms respectively. In this representation f and g are uniquely determined by σ.*

Then denoting by $\hat{G}$ (this is G in (**4**)) the group of automorphisms of G generated by $\hat{\mathfrak{H}}$ and G, by $\hat{A}$ the group generated by $\hat{G}$ and the group of field automorphisms F, and by A the group of all automorphisms of G, and assuming that K has q^3, q^2, or q elements in the respective cases that G is of type D_4^2, one of the other twisted types, or a normal type, we show:

3.3. $G \subseteq \hat{G} \subseteq \hat{A} \subseteq A$ *is a normal sequence for* A.

3.4. *$\hat{G}/G$ is isomorphic to $\hat{\mathfrak{H}}/\mathfrak{H}$, hence is Abelian. Thus $\hat{G} = G$ for the groups E_8, F_4, G_2 and D_4^2; $\hat{G}/G$ has order $(l+1, q-1)$, $(2, q-1)$, $(2, q-1)$, $(4, q^l - 1)$, $(3, q-1)$, $(2, q-1)$, $(l+1, q+1)$, $(4, q^l+1)$, or $(3, q+1)$ for the respective group A_l, B_l, C_l, D_l, E_6, E_7, A_l^1, D_l^1 or E_6^1; $\hat{G}/G$ is cyclic with the sole exception: G of type D_l (l even) and q odd.*

3.5. *$\hat{A}/\hat{G}$ is isomorphic to F, hence is cyclic if K is finite.*

3.6. *The graph automorphisms form a system of coset representatives of A over $\hat{A}$. Thus $A = \hat{A}$ with the exceptions: $A/\hat{A}$ has order 2 if G is A_l ($l \geqslant 2$), D_l ($l \geqslant 5$) or E_6, or if G is B_2 or F_4 and K has characteristic 2, or if G is G_2 and K has characteristic 3; $A/\hat{A}$ is isomorphic to the symmetric group on 3 objects if G is D_4.*

An immediate consequence of 3.3 to 3.6 is that each of the above groups which is simple verifies the Schreier conjecture (**12**, p. 303): *if A is the automorphism group of a finite simple non-Abelian group G, then A/G is solvable.*

Before starting the proofs of the above statements, we shall examine the groups under consideration a bit more closely.

4. Structure of the groups. In this section G need not be finite. However, until the last paragraph it is assumed that G is a normal type. Using the notation of §§2 and 3, one has:

4.1. *Each $x \in \mathfrak{U}$ can be written uniquely $x = \Pi x_r$, $x_r \in \mathfrak{X}_r$, the product being over the positive roots in increasing order.*

The proof of 4.1 as well as 4.2, 4.3, 4.5, 4.6, 4.7, 4.8, and 4.9 below can be found in **(4)**.

4.2. *$\mathfrak{H}$ is a subgroup of G, $\mathfrak{U}\mathfrak{H}$ is the normalizer of $\mathfrak{U}$ and $\mathfrak{U}\mathfrak{H} \cap \mathfrak{B} = 1$.*

4.3. *If K is finite and has characteristic p, then $\mathfrak{U}$ is a p-Sylow subgroup of G.*

4.4. *The centre of G is* 1.

Proof. If x is in the centre of G, then $x \in \mathfrak{U}\mathfrak{H}$ by 4.2. Similarly $x \in \mathfrak{B}\mathfrak{H}$, whence $x \in \mathfrak{H} = \mathfrak{U}\mathfrak{H} \cap \mathfrak{B}\mathfrak{H}$ by 4.2. But then 3.1 with $x = h$ yields $h(r) = 1$ for each $r \in \Pi$, whence $x = h = 1$.

Let W denote the Weyl group and w_r the reflection in W corresponding to the root r. One has:

4.5. *For each $w \in W$ there is $\omega(w) \in G$ such that $\omega(w)x_r(k)\omega(w)^{-1} = x_{wr}(\eta k)$, $r \in \Pi$ or $-\Pi$, with $\eta = \pm 1$ depending on w and r but not on k.*

4.6. *The union of the sets $\mathfrak{H}\omega(w)$ is a group $\mathfrak{W}$ and the map $w \to \mathfrak{H}\omega(w)$ is an isomorphism of W on $\mathfrak{W}/\mathfrak{H}$.*

Next for each $w \in W$ define $\mathfrak{U}_w = \mathfrak{U} \cap \omega(w)^{-1}\mathfrak{B}\omega(w)$, $\mathfrak{U}_w' = \mathfrak{U} \cap \omega(w)^{-1}$ $\mathfrak{U}\omega(w)$ so that $\mathfrak{U}_w$ $(\mathfrak{U}_w')$ is the group generated by those $\mathfrak{X}_r$ for which $r > 0$ and $wr < 0$ $(wr > 0)$. Thus if $a \in \Sigma$ and $w = w_a$ one has $\mathfrak{U}_w = \mathfrak{X}_a$.

4.7. $\mathfrak{U} = \mathfrak{U}_w\mathfrak{U}_w' = \mathfrak{U}_w'\mathfrak{U}_w$.

4.8. *The sets $\mathfrak{U}\mathfrak{H}\omega(w)\mathfrak{U}_w$, $w \in W$, are the distinct double cosets of G relative to $\mathfrak{U}\mathfrak{H}$, and each element of G has a unique expression of the indicated form.*

Analogous results hold with $\mathfrak{U}$ replaced by $\mathfrak{B}$.

4.9. *For each $r \in \Pi$ there is a homomorphism ϕ of $SL_2(K)$, the unimodular group, onto G_r, the group generated by $\mathfrak{X}_r$ and $\mathfrak{X}_{-r}$ such that*

$$\phi\begin{pmatrix} 1 & k \\ 0 & 1 \end{pmatrix} = x_r(k),$$

$$\phi\begin{pmatrix} 1 & 0 \\ k & 1 \end{pmatrix} = x_{-r}(k),$$

$$\phi\begin{pmatrix} 0 & 1 \\ -1 & 0 \end{pmatrix} \equiv \omega(w_r) \bmod \mathfrak{H},$$

and $\phi^{-1}(\mathfrak{H} \cap G_r)$ consists of the diagonal matrices. The kernel of ϕ is contained in the centre of $SL_2(K)$.

We may (and do) normalize so that

$$\phi\begin{pmatrix} 0 & 1 \\ -1 & 0 \end{pmatrix} = \omega(w_a)$$

for each $a \in \Sigma$ and then define each $\omega(w)$ to be a product of elements $\omega(w_a)$, $a \in \Sigma$. Then 4.9 implies:

4.10. *For each* $a \in \Sigma$ *the equation* $x_a(1)x_a(k)x_a(1) = x_{-a}(k)x_a(1)x_{-a}(k)$ *holds only if* $k = -1$ *and then both sides are equal to* $\omega(w_a)$. *Given* $k, l \in K^*$, $a \in \Sigma$, *then* $x_a(k)x_{-a}(l)x_a(m)x_{-a}(t)$ *is in* $\mathfrak{H}\omega(w_a)$ *only if* $m = -l^{-1}$ *and* $t = l + kl^2$.

4.11. *If* $w = w_a$, $a \in \Sigma$, *then* (1) $T =$ *union* $\mathfrak{B}\mathfrak{H}$, $\mathfrak{B}\mathfrak{H}\omega(w)\mathfrak{B}_w$ *is a group and* (2) $T \cap \mathfrak{U} = \mathfrak{U}_w$.

Proof. By 4.8, T is closed under right (or left) multiplication by $\mathfrak{B}\mathfrak{H}$. A consequence of 4.9 (see **(4**, p. 34, Lemma 2**)**) is $\omega(w)\mathfrak{B}_w\omega(w)^{-1} \subseteq$ union $\mathfrak{B}_w\mathfrak{H}$, $\mathfrak{B}_w\mathfrak{H}\omega(w)\mathfrak{B}_w$. Thus $T\omega(w)^{-1} \subseteq T$, hence $TT^{-1} \subseteq T$, and T is a group. Next write $\mathfrak{B} = \mathfrak{B}_w\mathfrak{B}_w'$. Then $\omega(w)\mathfrak{B}_w\omega(w)^{-1} = \mathfrak{U}_w$ and $\omega(w)\mathfrak{B}_w'\omega(w)^{-1} = \mathfrak{B}_w'$ by 4.5 so that $T = \omega(w)T\omega(w)^{-1} =$ union $\mathfrak{U}_w\mathfrak{B}_w'\mathfrak{H}$, $\mathfrak{U}_w\mathfrak{B}_w'\mathfrak{H}\omega(w)\mathfrak{U}_w$. Now $\mathfrak{U}_w\mathfrak{B}_w'\mathfrak{H} \cap \mathfrak{U} = \mathfrak{U}_w$ by 4.2; and if $w_0 \in W$ is defined by $w_0\Pi = -\Pi$, then $\mathfrak{U}\mathfrak{H}\omega(w_0w) \cap \omega(w_0)\mathfrak{U} = 0$ by 4.8, then left multiplication by $\omega(w_0)^{-1}$ yields $\mathfrak{B}\mathfrak{H}\omega(w) \cap \mathfrak{U} = 0$ and then $\mathfrak{U}_w\mathfrak{B}_w'\mathfrak{H}\omega(w)\mathfrak{U}_w \cap \mathfrak{U} = 0$. Thus $T \cap \mathfrak{U} = \mathfrak{U}_w$.

4.12. *Among the double cosets* $\mathfrak{B}\mathfrak{H}\omega(w)\mathfrak{B}_w$ *for which* $w \neq 1$ *and* (1) *of* 4.11 *holds, those for which* w *has the form* $w = w_a$, $a \in \Sigma$, *are characterized by the fact that* $T \cap \mathfrak{U}$ *is minimal.*

Proof. If w does not have the form $w = w_a$, $a \in \Sigma$, then $T \supseteq \omega(w)\mathfrak{B}_w\omega(w)^{-1} = \mathfrak{U}_{w^{-1}} \supset \mathfrak{X}_b$ for some $b \in \Sigma$, the last inclusion being proper; thus $T \cap \mathfrak{U}$ is not minimal by 4.11. If $a, b \in \Sigma$, $a \neq b$, then $\mathfrak{X}_a \not\supseteq \mathfrak{X}_b$; hence $T \cap \mathfrak{U}$ with $w = w_a$, $a \in \Sigma$, is minimal by 4.11.

Because of the results of **(15)**, the twisted types of groups have corresponding properties whose proofs are entirely analogous to those given above and in **(4)**.

5. Proof of 3.2 for the normal types. Throughout this section and the next assume that G is a normal type. The method of proof is as follows: we start with an arbitrary automorphism σ of G and multiply in turn by an inner, a diagonal, a graph and a field automorphism, referring at each stage to a normalization of σ; the final normalization yields $\sigma = 1$, whence 3.2 soon follows. Only in the first step is the finiteness of K used. Hence the rest of the argument is phrased so as to be applicable even if K is infinite.

5.1. *If K is finite, the automorphism σ can be normalized by an inner automorphism of G so that* $\mathfrak{U}^\sigma = \mathfrak{U}$ *and* $\mathfrak{B}^\sigma = \mathfrak{B}$. *If this is done, then* $\mathfrak{H}^\sigma = \mathfrak{H}$ *and there is a permutation* ρ *of the fundamental roots such that* $\mathfrak{X}_a^\sigma = \mathfrak{X}_{\rho a}$ *and* $\mathfrak{X}_{-a}^\sigma = \mathfrak{X}_{-\rho a}$ *for each fundamental root* a.

Proof. By 4.3, $\mathfrak{U}$, $\mathfrak{B}$, $\mathfrak{U}^\sigma$, and $\mathfrak{B}^\sigma$ are all p-Sylow subgroups of G, hence are conjugate. Thus one can normalize σ by an inner automorphism to fulfil $\mathfrak{U}^\sigma = \mathfrak{U}$. Now $\mathfrak{B}^\sigma = x^{-1}\mathfrak{U}x$ for some x in G; Thus $\mathfrak{B}^\sigma = u^{-1}\omega(w)^{-1}\mathfrak{U}\omega(w)u$

with $u \in \mathfrak{U}$, $w \in W$, by 4.8 and 4.2. Since $\mathfrak{V} \cap \mathfrak{U} = 1$ by 4.2, one has $\mathfrak{V}^\sigma \cap \mathfrak{U} = 1$, whence $w = w_0$ (defined by $w_0\Pi = -\Pi$) by 4.7, and then $\mathfrak{V}^\sigma = u^{-1}\mathfrak{V}u$ by 4.5. A second normalization, the inner automorphism effected by u, now yields $\mathfrak{V}^\sigma = \mathfrak{V}$. Then $\mathfrak{U}\mathfrak{H}$ and $\mathfrak{V}\mathfrak{H}$ are invariant under σ by 4.2, and so is $\mathfrak{H}$, since $\mathfrak{H} = \mathfrak{U}\mathfrak{H} \cap \mathfrak{V}\mathfrak{H}$ by 4.2. The double cosets of G relative to $\mathfrak{V}\mathfrak{H}$ ($\mathfrak{U}\mathfrak{H}$) are thus permuted by σ, and by 4.12 there exists a permutation ρ (a permutation τ) of the fundamental roots (of their negatives) such that $\mathfrak{X}_a{}^\sigma = \mathfrak{X}_{\rho a}$ and $\mathfrak{X}_{-a}{}^\sigma = \mathfrak{X}_{\tau(-a)}$ for each $a \in \Sigma$. If b and c are in Σ and $b \neq c$, then $b+(-c)$ is not a root, hence $X_bX_{-c} = 0$ (in the Lie algebra) and $\mathfrak{X}_b$ commutes with $\mathfrak{X}_{-c}$; if $b = c$, then $\mathfrak{X}_b$ does not commute with $\mathfrak{X}_{-c}$ by 4.9. Setting $b = \rho a$, $c = -\tau(-a)$, one concludes $\rho a = -\tau(-a)$. Thus 5.1 is proved.

5.2. *The normalization of σ attained in* 5.1 *can be refined by application of a diagonal automorphism of G so that in addition $x_a(1)^\sigma = x_{\rho a}(1)$ for each fundamental root a. It is then true that $x_{-a}(1)^\sigma = x_{-\rho a}(1)$, $\omega(w_a)^\sigma = \omega(w_{\rho a})$, and the orders of w_aw_b and $w_{\rho a}w_{\rho b}$ are equal for any fundamental roots a and b.*

Proof. Let $x_a(1)^\sigma = x_{\rho a}(k_a)$, $a \in \Sigma$. Then there exists a homomorphism h of the additive group generated by the roots into K^* such that $h(a) = k_a^{-1}$, $a \in \Sigma$. Application of the corresponding diagonal automorphism now yields the refinement $x_a(1)^\sigma = x_{\rho a}(1)$, $a \in \Sigma$, by 3.1. Applying σ to the first equation of 4.10, one then gets $x_{-a}(-1)^\sigma = x_{-\rho a}(-1)$ (so that $x_{-a}(1)^\sigma = x_{-\rho a}(1)$) and $\omega(w_a)^\sigma = \omega(w_{\rho a})$ for each $a \in \Sigma$. Lastly, the orders of w_aw_b and $w_{\rho a}w_{\rho b}$ are respectively equal to the orders of $\omega(w_a)\omega(w_b)$ and $\omega(w_{\rho a})\omega(w_{\rho b})$ mod $\mathfrak{H}$ by 4.6, hence are equal to each other because σ is an automorphism.

The last conclusion can be interpreted geometrically. If the order of w_aw_b is n, then the angle between a and b is $\pi - \pi/n$. Thus ρ effects an angle preserving permutation of the fundamental roots, hence is the identity unless the corresponding graph has extra "angular" symmetries (see (**3**, p. 18)).

5.3. *The normalization of σ in* 5.2 *can be refined by application of a graph automorphism of G so that ρ is the identity.*

Proof. Suppose first that G is of type F_4 or B_2 and that ρ is not the identity. Then there are $a, b \in \Sigma$ such that $a+b$ and $a+2b$ are roots, $\rho a = b$ and $\rho b = a$, by the remarks above. Let α and β be the maps of K defined by $x_a(k)^\sigma = x_b(k^\alpha)$ and $x_b(l)^\sigma = x_a(l^\beta)$. One has $x_{a+b}(l) = \omega(w_a)x_b(l)\omega(w_a)^{-1}$ and $x_{a+2b}(k) = \omega(w_b)x_a(k)\omega(w_b)^{-1}$ by 4.5 (in which the normalization $\eta = 1$ is achieved by replacing X_{a+b} or X_{a+2b} by its negative if necessary). Applying σ to these equations, one gets $x_{a+b}(l)^\sigma = x_{a+2b}(l^\beta)$ and $x_{a+2b}(k)^\sigma = x_{a+b}(k^\alpha)$. Consider the commutator equation

5.4. $$(x_a(k), x_b(l)) = x_{a+b}(\delta kl)x_{a+2b}(\epsilon kl^2),$$

with each of δ and ϵ equal to ± 1 and independent of k and l by (**4**, p. 27, ll. 22–26). Apply σ to 5.4:

5.5. $$(x_b(k^\alpha), x_a(l^\beta)) = x_{a+2b}((\delta kl)^\beta)x_{a+b}((\epsilon kl^2)^\alpha).$$

Now let us replace k and l by l^{β} and k^{α} respectively in 5.4 and take inverses:

5.6. $$(x_b(k^{\alpha}), x_a(l^{\beta})) = x_{a+2b}(-\epsilon l^{\beta}(k^{\alpha})^2)x_{a+b}(-\delta l^{\beta}k^{\alpha}).$$

Comparing the $\mathfrak{X}_{a+2b}$ components of 5.5 and 5.6 with $l = 1$, one gets $\delta k^{\beta} = -\epsilon(k^{\alpha})^2$ by 4.1 and 5.2. Setting first $k = 1$ and then $k = -1$, one gets $\delta = -\epsilon$ and $-\delta = -\epsilon$, whence $\delta + \delta = 0$, so that K is of characteristic 2. Then since β is onto, $k^{\beta} = (k^{\alpha})^2$ implies that K is perfect. Hence (see the paragraph preceding 3.2) there exists a graph automorphism which normalizes σ so that $\rho a = a$ and $\rho b = b$. As is easily seen, ρ is now the identity. If G is of type G_2, one proceeds similarly and finds that a normalization is not required unless K is perfect and of characteristic 3. If G is of type A_l, D_l, or E_6 and K is arbitrary, then again there is a graph automorphism to normalize ρ to the identity. In all other cases, due to the lack of "angular" symmetry of the graph, ρ is already the identity. Thus 5.3 is proved.

5.7. *The normalization of σ in* 5.3 *can be refined to $\sigma = 1$ by application of a field automorphism of G.*

That is, if σ satisfies $\mathfrak{U}^{\sigma} = \mathfrak{U}$, $\mathfrak{V}^{\sigma} = \mathfrak{V}$, and $x_a(1)^{\sigma} = x_a(1)$ for each $a \in \Sigma$, then σ is a field automorphism.

Proof. Choose $a \in \Sigma$ and define α by $x_a(k)^{\sigma} = x_a(k^{\alpha})$. We first show that α is an automorphism of K by the method of Schreier and van der Waerden (**12**, p. 318). By 5.2 we know that α maps K onto K and that $1^{\alpha} = 1$. The equation $x_a(k+l) = x_a(k)x_a(l)$ implies that $(k+l)^{\alpha} = k^{\alpha} + l^{\alpha}$. The equation 4.5 with $r = a$ and $w = w_a$ yields $x_{-a}(k)^{\sigma} = x_{-a}(k^{\alpha})$, and then the second part of 4.10 implies $(l + kl^2)^{\alpha} = l^{\alpha} + k^{\alpha}(l^{\alpha})^2$, whence $(kl^2)^{\alpha} = k^{\alpha}(l^{\alpha})^2$ and $(l^2)^{\alpha} = (l^{\alpha})^2$. If K is of characteristic 2, then $((kl)^{\alpha})^2 = ((kl)^2)^{\alpha} = (k^2l^2)^{\alpha} = (k^2)^{\alpha}(l^{\alpha})^2 = (k^{\alpha})^2(l^{\alpha})^2 = (k^{\alpha}l^{\alpha})^2$, whence $(kl)^{\alpha} = k^{\alpha}l^{\alpha}$; if K is of characteristic other than 2, then polarization of the equation $(l^2)^{\alpha} = (l^{\alpha})^2$ yields $(kl)^{\alpha} = k^{\alpha}l^{\alpha}$. Thus in either case α is an automorphism of K. Now choose a second root $b \in \Sigma$ (if one exists) such that $a + b$ is a root, and let β be the corresponding automorphism of K. By labelling appropriately a and b one may assume that $\omega(w_a)\mathfrak{X}_b\omega(w_a)^{-1} = \mathfrak{X}_{a+b}$. Then applying σ to the equation $(x_a(k), x_b(1)) = x_{a+b}(\delta k) \ldots$ as in 5.3 (but now with ρ the identity), one gets $k^{\beta} = k^{\alpha}$, whence $\alpha = \beta$. Since any 2 roots of Σ are the end terms of a sequence of roots of Σ such that the sum of each consecutive pair is a root (in other words, the graph is connected), it follows that there is a single automorphism γ of K such that $x_c(k)^{\sigma} = x_c(k^{\gamma})$ for each c in Σ. Normalization of σ by the field automorphism of G corresponding to γ^{-1} now yields $x_c(k)^{\sigma} = x_c(k)$, $c \in \Sigma$. One has also $\omega(w_c)^{\sigma} = \omega(w_c)$, $c \in \Sigma$, by 5.3. However, since W is generated by the elements w_c, $c \in \Sigma$, and each root has the form wc with $w \in W$, $c \in \Sigma$, it follows from 4.5 that G is generated by the elements $x_c(k)$ and $\omega(w_c)$ with $k \in K$, $c \in \Sigma$. Hence $\sigma = 1$, and 5.7 is proved.

Let us now prove 3.2. Let σ be an automorphism of G and let i, d, g, and f' be the respective inner, diagonal, graph, and field automorphisms used in

5.1, 5.2, 5.3, and 5.7 to achieve the normalization of σ^{-1}. One has $f'gdi\sigma^{-1} = 1$, thus $\sigma = f'gdi$. Since $g^{-1}f'g = f$ is in F by 5.7, one gets $\sigma = gfdi$, and the first statement of 3.2 is proved. Now suppose $\sigma = g_1f_1d_1i_1$ is a second representation of σ in the indicated form. Then $d^{-1}f^{-1}g^{-1}g_1f_1d_1 = ii_1^{-1}$. The left side of this equation maps $\mathfrak{U}$ onto $\mathfrak{U}$ and $\mathfrak{B}$ onto $\mathfrak{B}$. Hence $ii_1^{-1} \in \mathfrak{U}\mathfrak{H} \cap \mathfrak{B}\mathfrak{H} = \mathfrak{H}$ by 4.2. Then $f^{-1}g^{-1}g_1f_1 = dii_1^{-1}d_1^{-1} \in \mathfrak{H}$. This element leaves fixed each $x_a(1)$, $a \in \Sigma$, hence $dii_1^{-1}d_1^{-1} = 1$ by 3.1; that is, $g^{-1}g_1 = ff_1^{-1}$. This implies that g and g_1 effect the same permutation of the groups $\mathfrak{X}_a$, $a \in \Sigma$. Hence $g = g_1$, then $f = f_1$, and 3.2 is proved completely.

6. Proof of 3.3 to 3.6 for the normal types. The group G of inner automorphisms is clearly a normal subgroup of each of $\hat{G}$, $\hat{A}$, and A. This implies $\hat{G} = \hat{\mathfrak{H}}G$. One has also $\hat{\mathfrak{H}} \cap G = \mathfrak{H}$ since $\mathfrak{H} \subseteq \hat{\mathfrak{H}} \cap G$ by 4.2, whereas $h \in \hat{\mathfrak{H}} \cap G$ implies $\mathfrak{U}^h = \mathfrak{U}$, $\mathfrak{B}^h = \mathfrak{B}$ and then $h \in \mathfrak{U}\mathfrak{H} \cap \mathfrak{B}\mathfrak{H} = \mathfrak{H}$ by 4.2. Thus $\hat{G}/G \cong \hat{\mathfrak{H}}/(\hat{\mathfrak{H}} \cap G) = \hat{\mathfrak{H}}/\mathfrak{H}$. The specific results of 3.4 can now be verified from the fact that the Cartan integers $2(a, b)/(a, a)$ $(a, b \in \Sigma)$, taken mod $(q - 1)$, build a relation matrix for $\hat{\mathfrak{H}}/\mathfrak{H}$ (**4**, p. 48, ll. 13–18). It is easily verified (from the definitions) that $f\hat{\mathfrak{H}}f^{-1} = \hat{\mathfrak{H}}$ for each f in F. Hence $\hat{G}$ is normal in $\hat{A}$ and $\hat{A} = F\hat{\mathfrak{H}}G$, whence the uniqueness feature of 3.2 implies 3.5. Finally, if g is a graph automorphism, one verifies (by considering the effect on each $x_a(k)$) that $gFg^{-1} = F$ and $g\hat{\mathfrak{H}}g^{-1} = \hat{\mathfrak{H}}$, whence $\hat{A}$ is normal in A, and 3.3 is completely proved. The uniqueness feature of 3.2 then implies the first statement of 3.6; the last statement follows from the definition of graph automorphism given in the paragraph before 3.2.

7. The twisted types. The proofs of 3.2 to 3.6 for the twisted types are virtually the same as those given above for the normal types and to a large extent involve little more than a change of notation in view of the structural properties developed for the twisted types in (**15**). A comparison of 4.1 and 5.4 with their analogues 4.5 and 8.8 in (**15**) should make completely clear what modifications are to be made, if G is not A_l^1 (l even), and even in the latter case if $l \geqslant 4$. This leaves the group A_2^1 to be considered. Although the proofs in this case are also of the same genre as those given above for the normal types, the details are sufficiently more complicated to warrant a separate exposition, especially since the case in which K has few elements has not been completely treated elsewhere.

Let us recall that A_2^1 is a subgroup of A_2 and may be identified with a 3-dimensional projective unimodular unitary group (**15**). The positive roots of A_2 can be written as a, $\bar{a}$, and b (with $b = a + \bar{a}$), and then the elements of $\mathfrak{U}$ take the form $x_a(k)x_{\bar{a}}(\bar{k})x_b(l)$, subject to $k\bar{k} = l + \bar{l}$ (**15**, Lemma 4.6). For given k this last equation is always solvable for l: choose m so that $m + \bar{m} \neq 0$, and then set $l = k\bar{k}m(m + \bar{m})^{-1}$. For convenience, we denote $x_a(k)x_{\bar{a}}(\bar{k})x_b(l)$ by $(k|l)$, so that the rule of multiplication is $(k|l)(m|n) = (k + m|l + n + \bar{k}m)$ (see (**4**, p. 27, ll. 22–26) or use the unitary identification). From this it follows that the elements $(0|t)$, subject to $t + \bar{t} = 0$, build the centre $\mathfrak{C}$ of $\mathfrak{U}$. Let us now turn to the proof of 3.2.

If σ is an automorphism of G, it can be normalized by an inner automorphism, just as before, so that $\mathfrak{U}^\sigma = \mathfrak{U}$ and $\mathfrak{V}^\sigma = \mathfrak{V}$; assume this is done. Then $\mathfrak{C}^\sigma = \mathfrak{C}$, and since $(k|l_1)^{-1}(k|l_2) \in \mathfrak{C}$, the map α defined by $(k|l)^\sigma = (k^\alpha|m)$ is single-valued; clearly α is also onto.

The normalization of σ can now be refined by application of a diagonal automorphism of G so that $1^\alpha = 1$, and the next thing to be proved is that α becomes an automorphism of K. If K has 4 elements, then α leaves fixed 0 and 1 and permutes the other 2 elements of K, hence is an automorphism. Thus in the rest of the proof we may assume that K has more than 4 elements. Because σ is an automorphism we have $(k + l)^\alpha = k^\alpha + l^\alpha$; $k, l \in K$. Next if $h \in \mathfrak{H}$ and $h(a) = k$, we have $h(l|*)h^{-1} = (kl|*)$, whence $(h^\sigma)(l^\sigma|*)(h^\sigma)^{-1} = ((kl)^\alpha|*)$. Setting $l = 1$, we get $h^\sigma(a) = k^\alpha$, and then from the equation, $(kl)^\alpha = k^\alpha l^\alpha$. Here l is arbitrary, but k is restricted to the set S of numbers of the form $m^2\bar{m}^{-1}$ (see the definition of $\mathfrak{H}$ in §3). The field K_0 of numbers left fixed by $k \to \bar{k}$ is contained in S, and this inclusion is proper: if K is of characteristic 3 and $m \notin K_0$, then $m^2\bar{m}^{-1} \notin K_0$; if K is of characteristic other than 3 and $k \notin K_0$, $r \in K_0$, $r \neq 0, 1$, then not all three of $m_1 = k$, $m_2 = 1 + k$, $m_3 = r + k$ can have cubes in K_0 because, in the contrary case, differencing yields $k + k^2$, $rk + k^2 \in K_0$ and then $k \in K_0$, a contradiction, and so $m_i{}^2\bar{m}_i{}^{-1} \notin K_0$ for some $i = 1, 2$, or 3. There is thus $k \in S$ such that $k \notin K_0$, and each element of K can be written as $rk + s$ $(r, s \in K_0)$, that is, as the sum of 2 elements of S. Since α is additive on K and multiplicative on S, this implies that α is multiplicative on all of K and hence is an automorphism.

Thus the normalization of σ can be refined by a field automorphism of G so that α is the identity, and what remains to be shown is that now $\sigma = 1$. Choose $k \notin K_0$, and set $j = k - \bar{k}$. Then σ applied to $((1|*), (k|*)) = (0|k - \bar{k}) = (0|j)$ yields $(0|j)^\sigma = (0|j)$, that is, σ leaves fixed $x_b(j)$. A slight extension of the first statement in 4.10 shows that $x_{-b}(-j^{-1})$ and $x_b(j)x_{-b}(-j^{-1})x_b(j)$ are also left fixed by σ, and that the latter element is in $\omega(w)\mathfrak{H}$ and so may be denoted $\omega(w)$ after a normalization. A final calculation shows that for given $(k|l)$, $l \neq 0$, one has $(k|l)\omega(w)(m|n) \in \mathfrak{V}\mathfrak{H}$ if and only if $m = \bar{j}k\bar{l}^{-1}$ and $n = j\bar{j}\bar{l}^{-1}$. If this condition is met, if $(k|l)^\sigma = (k|l_1)$ and if $(m|n)^\sigma = (m|n_1)$, then application of σ yields $m = \bar{j}k\bar{l}_1{}^{-1}$, whence $l_1 = l$ and $(k|l)^\sigma = (k|l)$. Thus $\sigma = 1$ on $\mathfrak{U}$. Since $\omega(w)^\sigma = \omega(w)$ and $\omega(w)\mathfrak{U}\omega(w)^{-1} = \mathfrak{V}$, we get $\sigma = 1$. The first statement of 3.2 is hereby proved, and the other statement as well as 3.3 to 3.6 follow from it, just as before.

8. Final observations. As we have already stated, the finiteness of K is used above only in the proof that an automorphism of G necessarily maps $\mathfrak{U}$ onto one of its conjugates. If K is algebraically closed, this fact is proved in **(13)** by rather advanced methods of topology and algebraic geometry. It is hoped that an elementary proof, along the lines of the present article, can be found to handle all fields K simultaneously. In regard to **(13)**, we also mention that the proof of existence of graph automorphisms for B_2, F_4, and G_2 is quite long and that a shortened self-contained treatment is desirable.

An interesting special case of such an automorphism occurs when G is $B_2(2)$, the group of type B_2 over a field K of 2 elements. This group is isomorphic to S_6, the only symmetric group which admits outer automorphisms: one of these shows up as a graph automorphism which owes its existence to the fact that K has characteristic 2. It is also interesting to compare the groups $A_2(4)$ and $A_3(2)$. For the first the order of A/G is 12 and for the second it is 2. Thus one has another proof of the well-known fact that these groups, both of order 20160, are not isomorphic.

Finally let us remark that a companion problem to that treated here, namely the determination of the isomorphisms among the various finite groups of §2 is handled in **(1)** by uniform number-theoretic methods. In this connection we also refer the reader to 12.5 of **(15)** which can be used to eliminate some of the computations of **(1)**.

References

1. E. Artin, *Orders of classical simple groups*, Comm. Pure Appl. Math., *8* (1955), 455.

2. F. Bruhat, *Représentations induites des groupes de Lie semi-simples connexes*, C. R. Acad. Sci. Paris, *238* (1954), 437.

3. E. Cartan, *Oeuvres complètes* (Paris, 1952).

4. C. Chevalley, *Sur certains groupes simples*, Tôhoku Math. J., *7* (1955), 14.

5. L. E. Dickson, *Linear groups* (Leipzig, 1901).

6. ——— *A new system of simple groups*, Math. Ann., *60* (1905), 137.

7. J. Dieudonné, *On the automorphisms of the classical groups*, Mem. Amer. Math. Soc., *2* (1951).

8. ——— *La géométrie des groupes classiques*, Ergeb. der Math. u. i. Grenz. (1955).

9. E. B. Dynkin, *The structure of semi-simple algebras*, Amer. Math. Soc. Translation No. 17.

10. D. Hertzig, *On simple algebraic groups*, Short communications, Int. Congress of Math. (Edinburgh, 1958).

11. R. Ree, *On some simple groups defined by C. Chevalley*, Trans. Amer. Math. Soc., *84* (1957), 392.

12. O. Schreier and B. L. van der Waerden, *Die Automorphismen der projektiven Gruppen*, Abh. Math. Sem. Univ. Hamburg, *6* (1928), 303.

13. Séminaire C. Chevalley, *Classification des Groupes de Lie Algébriques* (Paris, 1956–8).

14. Séminaire "Sophus Lie" (Paris, 1954–1955).

15. R. Steinberg, *Variations on a theme of Chevalley*, Pac. J. Math., *9* (1959), 875.

16. H. Weyl, *The structure and representations of continuous groups*, I. A. S. notes (Princeton, 1934–1935).

University of California, Los Angeles

INVARIANTS OF FINITE REFLECTION GROUPS

ROBERT STEINBERG

Let us define a reflection to be a unitary transformation, other than the identity, which leaves fixed, pointwise, a (reflecting) hyperplane, that is, a subspace of deficiency 1, and a reflection group to be a group generated by reflections. Chevalley **(1)** (and also Coxeter **(2)** together with Shephard and Todd **(4)**) has shown that a reflection group G, acting on a space of n dimensions, possesses a set of n algebraically independent (polynomial) invariants which form a polynomial basis for the set of all invariants of G. Our aim here is to prove:

THEOREM. *Let G be a finite reflection group, acting on a space V of finite dimension. Let J be the Jacobian (matrix) of a basic set of invariants of G, computed relative to any basis of V. Let p be any point of V. Then the following numbers are equal:*

(a) *the maximum number of linearly independent reflecting hyperplanes containing p;*

(b) *the maximum rank of $1 - x$ for all x in G for which $xp = p$;*

(c) *the nullity of J at p.*

The equality of the numbers defined in (b) and (c) is the essence of a conjecture of Shephard **(3)**.

Throughout the paper, G is a reflection group, of finite order g, acting on a space V of n dimensions. The symbols $L_1, \ldots, L_v$ denote the hyperplanes in which reflections of G take place, as well as non-zero linear forms which vanish on the corresponding hyperplanes, and for each i, a_i is a corresponding non-zero normal vector, r_i is the order of the (cyclic) subgroup of G which leaves L_i fixed pointwise, and R_i is a generator of this subgroup. Finally, $I_1, \ldots, I_n$ are basic invariants of G; $d_1, \ldots, d_n$ are their degrees; and J generically denotes their Jacobian, relative to whatever basis is at hand.

LEMMA. *For some non-zero scalar c,*

$$\det J = c \prod_{i=1}^{v} L_i^{r_i - 1}.$$

A proof of this well-known result will be included because it and the corollary below play a key role in the proof of the theorem. Choose an orthonormal basis of V so that the first co-ordinate x_1 is a multiple of L_1. If I is any invariant of G, the equation $R_1 I = I$ implies that I is a polynomial in $x_1{}^{r_1}$, whence

$$x_1^{r_1 - 1} \text{ divides } \partial I / \partial x_1.$$

Received July 15, 1959.

Thus the first row of J, and hence also det J, is divisible by

$$x_1^{r_1-1}, \text{ and hence also by } L_1^{r_1-1}.$$

Similarly, det J is divisible by each $L_i^{r_i-1}$. Using the formula

$$\sum_{j=1}^{n} (d_j - 1) = \sum_{i=1}^{v} (r_i - 1) ,$$

proved in **(4**, p. 290, l. 12**)**, a comparison of degrees shows that the factor c in the statement of the lemma is a scalar, non-zero because the I_j are algebraically independent.

From the first part of the proof we have:

COROLLARY. *The determinant of the Jacobian of any n invariants of G is divisible by* $\prod L_i^{r_i-1}$.

Proof of the theorem. If k, l, and m denote the respective numbers defined by (a), (b), and (c), we prove in turn that $m \leqslant k$, $k \leqslant l$, and $l \leqslant m$.

First label the L's so that $L_1, \ldots, L_u$ are those which contain p, and then choose an orthonormal basis $p_1, \ldots, p_n$ of V so that $p_1, \ldots, p_k$ span the same subspace as $a_1, \ldots, a_u$, the normals to the L's. Let G' be the (reflection) group generated by $R_1, \ldots, R_u$. The co-ordinates $x_{k+1} = I_{k+1}', \ldots, x_n = I_n'$ are invariants of G'. If $I_1', \ldots, I_k'$ are any invariants of G, they are also invariants of G', and the corollary above shows that

$$\prod_1^u L_i^{r_i-1}$$

divides

$$\partial(I_1', \ldots, I_n')/\partial(x_1, \ldots, x_n),$$

that is, divides

$$\partial(I_1', \ldots, I_k')/\partial(x_1, \ldots, x_k).$$

Consider now the expansion of det J across the first k rows:

$$\det J = \sum \pm J'(i_1, \ldots, i_k) J''(i_{k+1}, \ldots, i_n),$$

with $J'(i_1, \ldots, i_k)$ denoting the minor corresponding to the rows $1, \ldots, k$ and columns $i_1, \ldots, i_k$ of J, $J''(i_{k+1}, \ldots, i_n)$ denoting the minor corresponding to the rows $k+1, \ldots, n$ and columns $i_{k+1}, \ldots, i_n$, and the sum being over all permutations $i_1, \ldots, i_n$ of $1, \ldots, n$ for which $i_1 < \ldots < i_k$ and $i_{k+1} < \ldots < i_n$. By what has just been shown, each J' is divisible by

$$\prod_1^u L_i^{r_i-1},$$

so that, by the lemma, there are polynomials $M(i_1, \ldots, i_k)$ such that

$$\prod_{u+1}^{v} L_i^{r_i-1} = \sum M(i_1, \ldots, i_k) J''(i_{k+1}, \ldots, i_n).$$

Since the left side of this equation is not 0 at p, we conclude that some J'' is not 0 at p, whence J has rank $n - k$ at least and nullity k at most at p. Thus $m \leqslant k$.

Next, assume that the labelling is such that $L_1, \ldots, L_k$ contain p and are linearly independent. Set $x = R_1 R_2 \ldots R_k$. Suppose $xq = q$, with $q \in V$. Then $R_1^{-1} q = R_2 \ldots R_k q$ implies that

$$q + c_1 a_1 = q + c_2 a_2 + \ldots + c_k a_k$$

for suitable scalars c_j, whence, because of the linear independence of the a_j, we conclude that $c_1 = 0$ and $R_1 q = q$. Similarly $R_2 q = q, \ldots, R_k q = q$, hence q lies in each of $L_1, \ldots, L_k$, and the solution space of the equation $xq = q$ has dimension $n - k$. Thus $1 - x$ has rank k, and the inequality $k \leqslant l$ has been established.

Finally choose $x \in G$ so that $1 - x$ has rank l and $xp = p$, and then an orthonormal basis $p_1, \ldots, p_n$ of V so that $xp_j = c_j p_j$ with $c_j \neq 1$ for $1 \leqslant j \leqslant l$ and $c_j = 1$ for $l + 1 \leqslant j \leqslant n$. If I is an invariant of G, the equation $xI = I$ implies that each term of I has a total exponent in the co-ordinates $x_1, \ldots, x_l$ which is either 0 or at least 2. Thus for each j such that $1 \leqslant j \leqslant l$, $\partial I/\partial x_j$ is 0 at any point at which $x_1, \ldots, x_l$ are all 0, in particular, at p. This implies that the first l rows of J vanish at p, whence $l \leqslant m$.

Thus the theorem is completely proved.

References

1. C. Chevalley, *Invariants of finite groups generated by reflections*, Amer. J. Math., *77* (1955), 778.
2. H. S. M. Coxeter, *The product of the generators of a finite group generated by reflections*, Duke Math. J., *18* (1951), 765.
3. G. C. Shephard, *Some problems of finite reflection groups*, Enseignement Math., *II* (1956), 42.
4. G. C. Shephard and J. A. Todd, *Finite unitary reflection groups*, Can. J. Math., *6* (1954), 274.

University of California, Los Angeles

AUTOMORPHISMS OF CLASSICAL LIE ALGEBRAS

ROBERT STEINBERG

1. Introduction. Starting with a simple Lie algebra over the complex field C, Chevalley [2] has given a procedure for replacing C by an arbitrary field K. Under mild restrictions on the characteristic of K, the algebra so obtained is simple over its center, and it is our purpose here to determine the automorphisms of each such quotient algebra $\mathfrak{g}$. In terms of the group G defined in [2] and also in § 3 below and the group A of all automorphisms of $\mathfrak{g}$, the principal result is that, with some exceptions, which occur only at characteristic 2 or 3, A/G is isomorphic to the group of symmetries of the corresponding Schläfli diagram. As might be expected, the main step in the development is the proof of a suitable conjugacy theorem for Cartan subalgebras (4.1 and 7.1 below). The final result then quickly follows.

Definitions of the algebras and automorphisms to be considered are given in § 2 and § 3. Sections 4, 5 and 6 contain the main development and § 7 treats some special cases. The last section contains some remarks on the extension of the preceding results to other algebras. In 4.6, 4.7, 4.8, 7.2 and 7.3 the results are interpreted for the various types of algebras occurring in the Killing-Cartan classification, thereby yielding results of other authors [**4, 5, 6, 7, 8, 9, 12, 14, 15, 16, 18**] who have worked on various types of algebras from among those usually denoted A, B, C, D, G and F. For other treatments in which all types are considered simultaneously, the reader is referred to [**4**; **16**, Exp. 16] where the problem is solved over the complex field, however by topological methods which can not be used for other fields, and to [**14**] where general fields occur but only partial results are obtained. General references to the classical theory of Lie algebras over the complex field are [**1**, thesis; **3**; **16**; **19**].

2. The algebras. Let us start with a simple Lie algebra $\mathfrak{g}_C$ over the complex field C, a Cartan subalgebra $\mathfrak{h}_C$, the (ordered) system Σ of (nonzero) roots relative to $\mathfrak{h}_C$, the set Φ of fundamental positive roots, and for each pair of roots r and s, define c_{rs} to be the Cartan integer $2(r, s)/(s, s)$, and p_{rs} to be 0 if $r + s$ is not a root and otherwise to be the least positive integer p for which $r - ps$ is not a root. Then Chevalley [2, Th. 1] has shown that there exists a set of root elements $\{X_r\}$ and a set $\{H_r\}$ of elements of $\mathfrak{h}_C$ such that the equations of structure of $\mathfrak{g}_C$ are:

Received July, 15, 1960.

2.1 *$H_{-r} = -H_r$, and if r, s and t are roots such that $r+s+t=0$ and r is at most as long as s or t, then*

$$H_r + (s, s)/(r, r)H_s + (t, t)/(r, r)H_t = 0 .$$

2.2 $H_r H_s = 0$.

2.3 $H_r X_s = c_{sr} X_s$.

2.4 $X_r X_{-r} = H_r$.

2.5 $X_r X_s = \pm p_{rs} X_{r+s}$ *if* $r + s \neq 0$.

The equations 2.1 imply that each H_r is an integral linear combination of the elements $H_a (a \in \Phi)$, which form a basis for $\mathfrak{h}_C$. Just as in [2] the base field C can now be replaced by an arbitrary field K (because the structural constants are all integers), yielding an algebra $\bar{\mathfrak{g}}$ over K, an Abelian subalgebra $\bar{\mathfrak{h}}$, a set of numbers $\{\bar{p}_{rs}, \bar{c}_{rs}\}$ in K, and a set of roots relative to $\bar{\mathfrak{h}}$ defined by $\bar{r}(H_s) = \bar{c}_{rs}$. We use the notation $\{X_r, H_r\}$ for the generating set of $\bar{\mathfrak{g}}$, the subscript r referring to a root of the original system Σ.

2.6 *Assume that $\bar{\mathfrak{g}}$ is one of the algebras just constructed, but if Σ has roots of unequal length or if Σ is of type A_1 assume that K is not of characteristic 2, and if Σ is of type G_2 assume further that K is not of characteristic 3. Then*

(1) *if $p_{rs} \neq 0$, then $\bar{p}_{rs} \neq 0$, whereas if $c_{rs} \neq 0$, then $\bar{c}_{rs} \neq 0$ unless $r = \pm s$ and K is of characteristic 2;*

(2) *no H_r is in the center of $\bar{\mathfrak{g}}$;*

(3) *the center $\bar{\mathfrak{c}}$ of $\bar{\mathfrak{g}}$ consists of those H in $\bar{\mathfrak{h}}$ such that $\bar{r}(H) = 0$ for all r in Σ;*

(4) *if $\mathfrak{h} = \bar{\mathfrak{h}}/\bar{\mathfrak{c}}$ and $\mathfrak{g} = \bar{\mathfrak{g}}/\bar{\mathfrak{c}}$, then $\mathfrak{h}$ is a Cartan subalgebra of $\mathfrak{g}$;*

(5) *$\mathfrak{g}$ is simple.*

Proof. (1) From known properties of root systems if $r \neq \pm s$ then p_{rs} and c_{rs} take on values other than 0, ± 1 only if Σ has roots of different lengths: the values ± 2 and ± 3 if Σ is of type G_2 and ± 2 if Σ is one of the other types. The possibility of these numbers becoming 0 in K has been ruled out by the assumptions. On the other hand $\bar{c}_{rr} = \bar{2}$ which is 0 if and only if K is of characteristic 2.

(2) If K is not of characteristic 2, then $H_r X_r = 2X_r \neq 0$, and if K is of characteristic 2, then there is a root s not orthogonal to r, whence $H_r X_s = c_{sr} X_s \neq 0$ by (1). Thus H_r is not in the center.

(3) Assume that $X = H + \Sigma c_s X_s$ is in the center. Then multiplication by X_{-r} yields $c_r = 0$ because of 2.4, whence

$$0 = XX_r = HX_r = \bar{r}(H)X_r$$

so that $\bar{r}(H) = 0$. The converse is easily checked.

(4) $\mathfrak{h}$ is Abelian, and if $X = H + \Sigma c_s X_s$ is in the normalizer of $\mathfrak{h}$, then $c_r = 0$ just as before and X is in $\mathfrak{h}$. Hence $\mathfrak{h}$ is a Cartan subalgebra of $\mathfrak{h}$.

(5) Let $\mathfrak{m}$ be an ideal in $\mathfrak{g}$ and Y a nonzero element of $\mathfrak{m}$. Then by repeated multiplication by elements of the form X_a $(a \in \Phi)$ (now considered to be in $\mathfrak{g}$) we arrive at a nonzero X which is in $\mathfrak{m}$ and commutes with all X_a. By (1) this implies that X is a scalar multiple of X_d, d being the unique root such that $d + a$ is not a root for each a in Φ. Thus X_d is in $\mathfrak{m}$ and by repeated multiplication by elements of the form X_{-a} $(a \in \Phi)$ we get all X_r in $\mathfrak{m}$, whence $\mathfrak{m} = \mathfrak{g}$. Hence $\mathfrak{g}$ is simple.

In regard to the cases excluded by the assumptions of 2.6, let us observe first that if K is of characteristic 2 and Σ is of type A_1 then $\bar{\mathfrak{g}}$ is nilpotent while if Σ is of type G_2 then $\bar{\mathfrak{g}}$ is isomorphic to the algebra $\mathfrak{g}$ of type D_3, as is seen by an examination of the multiplication tables. In the other cases $\mathfrak{g}$ is not simple because those X_r and H_r for which r is a short root span an ideal as is seen from 2.1 to 2.5 and the following properties of Σ: if r is a long root and s is a short one, then c_{rs} is 0 or $\pm(r, r)/(s, s)$; if $r + s$ is also a root then it is a short one (because $(r + s, r + s) = (1 + c_{sr})(r, r) + (s, s)$ which is not a multiple of (r, r)); if r and s are short roots and $r + s$ is a long root, then

$$p_{rs} = (r + s, r + s)/(r, r)$$

(check for Σ of type B_2 or G_2).

In the sequel, each algebra $\mathfrak{g}$ of 2.6 is called a classical Lie algebra, and the algebra $\mathfrak{h}$ and the set of elements $\{X_r, H_r \mid r \in \Sigma\}$, now considered to be in $\mathfrak{g}$, which occur in the explicit mode of construction described are called standard Cartan subalgebra and standard set of generators, respectively. (Actually the subset $\{X_a \mid \pm a \in \Phi\}$ is enough to generate $\mathfrak{g}$.) In addition the notations $\bar{p}_{rs}$, $\bar{c}_{rs}$ and $\bar{r}$ are used in reference to $\mathfrak{g}$ rather than $\bar{\mathfrak{g}}$. Observe that $\bar{r}$ is defined in a natural way on $\mathfrak{h}$ because of (3) of 2.6.

A consequence of (4) of 2.6 which should be borne in mind is that $\bar{r} \neq 0$ if $r \neq 0$, although it may happen that $\bar{r} = \bar{s}$ with $r \neq s$.

3. The groups. Following Chevalley [2], let us now describe certain automorphisms of classical Lie algebras. Let $\mathfrak{g}$ be such an algebra and $\{X_r H_r \mid r \in \Sigma\}$ a standard set of generators. For each r in Σ and each k in K, let $x_r(k)$ be the automorphism of $\mathfrak{g}$ which has the same effect as $\exp \operatorname{ad} kX_r$ on each generator, with the sole exception: if K is of characteristic 2, then $x_r(k)X_{-r} = X_{-r} + kH_r + k^2X_r$ (see [2, p. 24]), and then let G' be the group generated by all such automorphisms as r runs through Σ and k through K. Then for each $w \in W$, the Weyl group of Σ, there is $\omega(w)$ in G' such that $\omega(w)X_r = \pm X_{wr}$ and $\omega(w)H_r = H_{wr}$ for each r in Σ [2, p. 35]. If χ is a homomorphism of the additive group generated by the roots of Σ into the multiplicative group K^* of K, then there is an automorphism h of $\mathfrak{g}$ such that $hX_r = \chi(r)X_r$ for each r in Σ. The group of such automorphisms is denoted $\mathfrak{H}$, and the subgroup corresponding to those homomorphisms which can be extended to the group of weights relative to Σ is denoted $\mathfrak{H}'$. Let G be the group generated by G' and $\mathfrak{H}$. One has [2]:

3.1. *G' is normal in G, $\mathfrak{H}' = \mathfrak{H} \cap G'$, $G = G'\mathfrak{H}$, and G/G' is isomorphic to $\mathfrak{H}/\mathfrak{H}'$.*

Finally, if $r \to r'$ is a permutation of Φ, the set of fundamental roots, such that $c_{a'b'} = c_{ab}$ for all a and b in Φ, then there is a *graph* automorphism g of $\mathfrak{g}$ defined by: $gX_a = X_{a'}$ if $\pm a$ is in Φ (see one of [3, p. 116; **16**, p. 11–04; **19**, p. 94] for the proof of existence and [1, p. 361] for an interesting discussion). Although the automorphisms of this paragraph are defined in [2] to act on $\bar{\mathfrak{g}}$, we can (and shall) think of them as acting on $\mathfrak{g}$. Nothing is lost in the passage from $\bar{\mathfrak{g}}$ to $\mathfrak{g}$: if x is an automorphism of $\bar{\mathfrak{g}}$ which induces the identity on $\mathfrak{g}$ then, in the notation prior to 2.6, $xX_r \equiv X_r \bmod \bar{\mathfrak{c}}$ for each r, whence $xH_r = H_r$ by 2.4 and then $xX_r = X_r$ by 2.3, implying that x is the identity.

The following observation will be used later:

3.2. *Let S be a standard set of generators of $\mathfrak{g}$ and x an automorphism of $\mathfrak{g}$. Let G' be the group defined above relative to S, and let G'' be the corresponding group defined relative to the standard set xS. Then $G'' = xG'x^{-1}$.*

Proof. Let B be a subset of S which is also a vector space basis for $\mathfrak{g}$. Then the matrices representing G' relative to B are the same as those representing G'' relative to xB, whence $G'' = xG'x^{-1}$.

4. Principal results. Throughout the next three sections, $\mathfrak{g}$ denotes a classical Lie algebra with a fixed standard set of generators

$$S = \{X_r, H_r \mid r \in \Sigma\}$$

and corresponding Cartan subalgebra $\mathfrak{h}$, K is the underlying field, the symbols G', G, $\mathfrak{H}'$, and "graph" refer to the automorphisms of $\mathfrak{g}$ defined relative to S as in § 3, and A denotes the group of all automorphisms of $\mathfrak{g}$. It is assumed that Σ is not of type A_2 if K is of characteristic 3 and not of type D_n if K is of characteristic 2. These exceptional cases are considered in § 7.

4.1. *Conjugacy theorem. If* $\mathfrak{h}_1$ *and* $\mathfrak{h}_2$ *are standard Cartan subalgebras of* $\mathfrak{g}$, *there is* x *in* G' *such that* $x\mathfrak{h}_1 = \mathfrak{h}_2$.

4.2. *Each* x *in* A *can be written uniquely* $x = ihg$, *with* i *in* G', h *in* $\mathfrak{H}$ *and* g *a graph automorphism.*

4.3. G' *and* G *are normal subgroups of* A.

4.4. G/G' *is isomorphic to* $\mathfrak{H}/\mathfrak{H}'$, *hence is Abelian.*

4.5. *The graph automorphisms form a system of coset representatives for* A *over* G. *Hence* A/G *is isomorphic to the group of symmetries of the Schläfli graph.*

These results may be amplified thus:

4.6. $G = G'$ *if* Σ *is of type* E_8, F_4 *or* G_2 *or if* Σ *is of arbitrary type and* K *is algebraically closed.* G/G' *is isomorphic to* K^*/K^{*f}, *with* $f = n+1, 2, 2, 4, 3, 2$ *in the respective cases that* Σ *is of type* A_n, B_n, C_n, D_n *(n odd),* E_6, E_7, *and is isomorphic to the direct product of* 2 *copies of* K^*/K^{*2} *if* Σ *is of type* D_n *(n even).*

4.7. $A = G$ *with the exceptions:* A/G *is of order* 2 *if* Σ *if of type* $A_n (n \geqq 2)$, $D_n (n \geqq 5)$ *or* E_6, *and is isomorphic to the symmetric group on* 3 *objects if* Σ *is type* D_4.

4.8. $A = G'$, *hence is simple, if* Σ *is of type* E_8, F_4 *or* G_2 *and* K *is arbitrary or if* Σ *is of type* B_n, C_n *or* E_7 *and every element of* K *is a square.* $A \neq G'$ *otherwise.*

5. The theorem of conjugation. We first show that the group G' depends only on $\mathfrak{h}$, not on all of S.

5.1. *If* r *and* s *are in* Σ *and* $r \neq s$, *then* $\bar{r} = \bar{s}$ *if and only if both* $r = -s$ *and* K *is of characteristic* 2.

Proof. Let r and s be roots such that $r \neq s$ and $\bar{r} = \bar{s}$. Assume first that K is of characteristic other than 2. The equations

$$\overline{(-r)}(H_r) = -2 = -\bar{r}(H_r)$$

show that $r \neq -s$. Then since $c_{rs}c_{sr} = 0, 1, 2$ or 3 and $\bar{c}_{rs}\bar{c}_{sr} = \bar{c}_{rr}\bar{c}_{ss} = \bar{4}$, the only possibility is that $c_{rs}c_{sr} = 1$ and K is of characteristic 3. From $\bar{c}_{rs} = \bar{c}_{ss} = \bar{2} = -\bar{1}$, we see that r and s have the same length and form an angle of $2\pi/3$. Since Σ is not of type G_2, this implies that

r and s can be incorporated into a fundamental set [2, p. 19], and since Σ is not of type A_2, this set contains a third root t which can be taken orthogonal to one of r, s and not to the other. But then $\bar{c}_{rt} \neq \bar{c}_{st}$ by (1) of 2.6, contradicting $\bar{r} = \bar{s}$. Now assume that K is of characteristic 2. Then all roots have the same length. Thus if r is orthogonal to s, then r and s can be incorporated into a fundamental set, and since Σ is not of type D_n, one reaches a contradiction just as before. On the other hand if r is not orthogonal to s, then the equation $\bar{c}_{rs} = \bar{c}_{ss} = 0$ implies that $c_{rs} = \pm 2$, whence $r = \pm s$ because r and s have the same length. Since $\overline{(-r)} = \bar{r}$ if K is of characteristic 2, 5.1 is proved.

5.2. *Let $S = \{X_r, H_r \mid r \in \Sigma\}$ be the standard set of generators of $\mathfrak{g}$ introduced in* § 4 *and let $S' = \{X_q, H_q \mid q \in \Sigma'\}$ be a second standard set such that S and S' determine the same Cartan subalgebra $\mathfrak{h}$. Then there exists a bijective mapping $r \to r'$ of Σ onto Σ' such that*

(1) *if r, s and $r + s$ are in Σ, then $r' + s'$ is in Σ', $(r + s)' = r' + s'$, and $(-r)' = -r'$, and*

(2) *for each r in Σ, $H_{r'} = H_r$ and $X_{r'} = c_r X_r$ with c_r in K and $c_r c_{-r} = 1$.*

Proof. The nonzero root spaces of $\mathfrak{g}$ relative to $\mathfrak{h}$ are determined by 5.1 as $\{KX_r\}$ if K is not of characteristic 2 and $\{KX_r + KX_{-r}\}$ if K is of characteristic 2. In the latter case, if $X = kX_r + lX_{-r}$, then ad X is nilpotent only if either k or l is 0, as one sees by choosing a root s of Σ such that $r + s$ is also a root and then computing $(\operatorname{ad} X)^2 X_s = klX_s$. Thus in all cases $\mathfrak{h}$ determines $\{KX_r\}$ (and $\{KX_q\}$) and there exist a bijective mapping $r \to r'$ and scalars c_r such that $X_{r'} = c_r X_r$. Since $X_s X_r$ is a nonzero element of $\mathfrak{h}$ if and only if $s = -r$, one has $(-r)' = -r'$, and if r, s and $r + s$ are in Σ, then $X_r X_s$ is a nonzero element of KX_{r+s} by (1) of 2.6, which implies that $r' + s'$ is in Σ' and $(r + s)' = r' + s'$. Next $H_{r'} = c_r c_{-r} H_r$ by 2.4. Now one can find a root s such that $\bar{s}(H_{r'}) = \bar{s}(H_r) \neq 0$: if K is of characteristic 2, choose for s any root not orthogonal to r, and if K is not of characteristic 2, choose $s = r$. Thus $c_r c_{-r} = 1$, $H_{r'} = H_r$, and 5.2 is proved.

5.3. *Under the assumptions of* 5.2 *if G'' is the group defined relative to S' in the same way that G' is defined relative to S then $G'' = G'$.*

Proof. If either $r \neq -s$ or K is not of characteristic 2, then $x_{r'}(k) X_s = (\exp \operatorname{ad} kX_{r'})(c_s^{-1} X_{s'}) = (\exp \operatorname{ad} kc_r X_r) X_s = x_r(kc_r) X_s$, while if K is of characteristic 2, then

$$\begin{aligned} x_{r'}(k) X_{-r} &= x_{r'}(k)(c_{-r}^{-1} X_{-r'}) = c_{-r}^{-1}(X_{-r'} + kH_{r'} + k^2 X_{r'}) \\ &= X_{-r} + (kc_r) H_r + (kc_r)^2 X_r = x_r(kc_r) X_{-r} , \end{aligned}$$

by 5.2. Hence $x_{r'}(k) = x_r(kc_r)$ and $G'' = G'$.

Let us now turn to the proof of 4.1. Clearly it is enough to prove that $\mathfrak{h}_1$ is conjugate to $\mathfrak{h}$ under G'. For then by symmetry $\mathfrak{h}_2$ is also conjugate to $\mathfrak{h}$ and then to $\mathfrak{h}_1$. Let S_1 be a standard set of generators corresponding to $\mathfrak{h}_1$. Assume first that K is algebraically closed (so that $G' = G$ by 3.1) and let G_1 be the group defined relative to S_1 in the same way that G is defined relative to S. By a familiar argument of Harish-Chandra (see [16, Exp. 15] or [13]), there exist y in G and y_1 in G_1 such that $y\mathfrak{h} = y_1\mathfrak{h}_1$. Set $x = y_1^{-1}y$. Then $x\mathfrak{h} = \mathfrak{h}_1$, and by 3.2 and 5.3, $G_1 = xGx^{-1}$, whence $G_1 = y_1G_1y_1^{-1} = yGy^{-1} = G$ and x is in G. Now assume that K is not algebraically closed. Let $\hat{K}$ be its algebraic closure and let $\hat{\mathfrak{g}}$, etc., be the objects corresponding to $\mathfrak{g}$, etc., when K is replaced by $\hat{K}$. As has just been shown, there is y in $\hat{G}$ such that $y\hat{\mathfrak{h}} = \hat{\mathfrak{h}}_1$. By 5.2 the elements of yS are multiples of those of S_1. One can normalize y by multiplication by an element of $\hat{\mathfrak{H}}$ so that yX_a is in S_1 for each fundamental root a, and then by 5.2, yX_{-a} and yH_a are also in S_1. Since $\mathfrak{h}$ is generated over K by the elements H_a, and $\mathfrak{g}$ is generated by the elements X_a, X_{-a}, it follows that $y\mathfrak{h} = \mathfrak{h}_1$ and $y\mathfrak{g} = \mathfrak{g}$. Since y is in $\hat{G}$ and y induces an automorphism of $\mathfrak{g}$, a result of Ono [10] implies that y is is G. By 3.1 one can write $y = xh$ with x in G' and h in $\mathfrak{H}$. Thus $x\mathfrak{h} = xh\mathfrak{h} = y\mathfrak{h} = \mathfrak{h}_1$, and 4.1 is completely proved.

By combining 3.2, 4.1 and 5.3 we get:

5.4. *The group G' is independent of the standard set of generators used to define it.*

Finally, let use observe that the word standard may be omitted from 4.1 if K is algebraically closed and not of characteristic 2, 3 or 5 because then every Cartan subalgebra is standard (see [1, thesis; 12; 2]).

6. Proofs of 4.2 to 4.8. If x is an automorphism of $\mathfrak{g}$, then $x\mathfrak{h}$ is a standard Cartan subalgebra of $\mathfrak{g}$. Hence by 4.1 there is j in G' such that $j^{-1}x\mathfrak{h} = \mathfrak{h}$. Then 5.2 implies that there is a permutation $r \to r'$ on Σ such that (1′) if r, s and $r + s$ are in Σ, then $(r + s)' = r' + s'$ and $(-r)' = -r'$, and (2′) $j^{-1}xX_r = c_rX_{r'}$ and $c_rc_{-r} = 1$ for each r in Σ. By (1′), Φ' is a fundamental set of roots since Φ is. Hence [16, p. 16–05] there is w in W, the Weyl group, such that $w\Phi = \Phi'$. Then replacing j by $i = j\omega(w)$ we see that the refinement $\Phi' = \Phi$ is achieved. We can now choose h in $\mathfrak{H}$ so that $hX_{a'} = c_aX_{a'}$ for each a in Φ, whence $h^{-1}i^{-1}xX_a = X_{a'}$ and then $h^{-1}i^{-1}xX_{-a} = X_{-a'}$, because $c_ac_{-a} = 1$. Thus by 2.3 and 2.4 and the fact that $h^{-1}i^{-1}x$ is an automorphism $c_{a'b'} = c_{ab}$ for a and b in Φ. That is, $h^{-1}i^{-1}x$ is a graph automorphism, and 4.2 is proved.

From the definitions, it is easily checked that $hG'h^{-1} = G'$, $gG'g^{-1} = G'$ and $g\mathfrak{H}g^{-1} = \mathfrak{H}$ if h is in $\mathfrak{H}$ and g is a graph automorphism. Thus 4.2 implies 4.3.

As a restatement of part of 3.1, 4.4 is true.

Next assume that the graph automorphism g is in G. Let $\mathfrak{u}$ and $\mathfrak{U}$, respectively, be the subalgebra of $\mathfrak{g}$ and subgroup of G' generated by those X_r and $x_r(k)$ for which r is positive. Then by [2, Th. 2] there are u, u'' in $\mathfrak{U}$, h in $\mathfrak{H}$ and w in W such that $g = uh\omega(w)u''$, whence $\omega(w)\mathfrak{u} = h^{-1}u^{-1}gu''^{-1}\mathfrak{u} \subseteqq \mathfrak{u}$. This implies that w maps positive roots onto positive roots, whence $w = 1$. Then the equation

$$gX_r = uhu''X_r = c_rX_r + \sum_{s>r} c_sX_s \,,$$

$c_r \neq 0$, in conjunction with the definition of graph automorphism, implies that $g = 1$, that 4.5 is true.

Let P and P_r be the additive groups generated by the weights and by the roots relative to Σ. By a basic theorem for free modules, there exist bases $\{b_i\}$ and $\{b'_i\}$ of P and P_r and a set of positive integers $\{f_i\}$ such that $b'_i = f_ib_i$ for each i. Then from the definitions $\mathfrak{H}/\mathfrak{H}'$ is isomorphic to the direct product of the groups K^*/K^{*f_i}. Now since Φ is a basis for P_r and $\{a' \mid a \in \Phi, (2a', b)/(b, b) = \delta_{ab}, b \in \Phi\}$ is a basis for P, the numbers f_i can be found by reducing the matrix $(c_{ab})(a, b \in \Phi)$ to diagonal form. In this way 4.6 is proved.

Finally, an examination of the various root systems yields 4.7, and then 4.6 and 4.7 imply 4.8.

7. The other algebras. Continuing with the previous notation, but dropping the assumption in the second sentence of § 4, we define G'' to be the group generated by the automorphisms of type $x_r(k)$ constructed relative to all standard sets of generators for which $\mathfrak{h}$ is the corresponding Cartan subalgebra. By 5.3, $G'' = G'$ for the algebras treated there, but this is not the case for the algebras yet to be considered.

7.1. *If $\mathfrak{h}_1$ and $\mathfrak{h}_2$ are standard Cartan subalgebras of $\mathfrak{g}$, there is x in G'' such that $x\mathfrak{h}_1 = \mathfrak{h}_2$.*

7.2. *In the respective cases that Σ is of type A_2, D_4 or D_n $(n \neq 4)$ and K is of characteristic 3, 2 or 2, the group G'' is isomorphic to the group G' of type G_2, F_4 or C_n.*

7.3. *In the first two cases above $A = G''$ and in the third A/G'' is isomorphic to K^*/K^{*2}.*

The proofs of these results require suitable analogues of 5.1 and 5.2:

7.4. *In the respective cases of* 7.2, *the nonzero root spaces of* $\mathfrak{g}$ *relative to* $\mathfrak{h}$ *have dimensions* 3, 8 *or* 4.

7.5. *If* S *and* S'' *are standard sets of generators both of which have* $\mathfrak{h}$ *as the corresponding Cartan subalgebra, then there is* x *in* G'' *such that* S *and* $S' = xS''$ *satisfy the properties* (1) *and* (2) *of* 5.2.

The ideas in the proofs of these results are the same for all three types of algebras. However, the details are somewhat different. Hence we shall restrict ourselves to a discussion of the algebra of type A_2 over a field of characteristic 3.

Now the roots of a system of type G_2 may be so labelled that the set of short ones is $\Sigma = \{\pm a, \pm b, \pm(a+b)\}$ and the set of long ones is $\Lambda = \{\pm(a-b), \pm(a+2b), \pm(2a+b)\}$ (see any of [**1**, p. 93; 3, p. 141; **16**, p. 14–06]). As has already been mentioned, the construction of 2.6 does not yield a simple algebra if a root system of type G_2 is combined with a field K of characteristic 3: the set $S = \{X_r, H_r \mid r \in \Sigma\}$ spans an ideal which is easily seen to be a classical Lie algebra of type A_2 with S as a standard set of generators. Let $\mathfrak{g}$ denote the ideal and $\mathfrak{m}$ the full algebra. First we observe that an automorphism of $\mathfrak{m}$ which is the identity on $\mathfrak{g}$ is the identity on $\mathfrak{m}$ because the adjoint action of $\mathfrak{m}$ on $\mathfrak{g}$ is faithful by 2.3 and 2.5. Thus the automorphisms of $\mathfrak{m}$ may be considered to act on $\mathfrak{g}$ (the unique minimal ideal) without any ambiguity. Now $\bar{c}_{ab} = \bar{c}_{ba} = -1 = 2 = \bar{c}_{aa} = \bar{c}_{bb}$. Hence $\bar{a} = \bar{b}$, and then $-\bar{a} - \bar{b} = -\bar{a} - \bar{a} = \bar{a}$. Thus $\bar{a}$ corresponds to a root space R^+ spanned by those X_r for which r is in $\Sigma^+ = \{a, b, -a-b\}$; a similar statement for $-\bar{a}$ establishes 7.4. Now each r in Λ can be written uniquely $r = t - s$ with t and s in Σ^+. Hence if u denotes the third element of Σ^+, $x_r(k)$ maps X_s, X_t, X_u onto $X_s \pm kX_t, X_t, X_u$, respectively. Here s, t, u run through the permutations of Σ^+ as r runs through Λ. Hence the group generated by $\{x_r(k) \mid r \in \Lambda, k \in K\}$ induces in R^+ the three-dimensional unimodular group. Now if $S'' = \{Y_q, J_q, q \in \Sigma'\}$ is a second standard set of generators of $\mathfrak{g}$ corresponding to the same Cartan subalgebra $\mathfrak{h}$ as S, then the root spaces, as determined by $\mathfrak{h}$, are three dimensional and Σ' is of type A_2, whence its roots can be labelled so that R^+ is spanned by $Y_{a'}$, $Y_{b'}$ and $Y_{-a'-b'}$. Thus by what has just been said there is x in G''', the group of type G' for $\mathfrak{m}$ such that, if we set $xY_r = X_r$ and $xJ_r = H_r$ for each r in Σ', then $X_{a'}, X_{b'}, X_{-a'-b'}$ are scalar multiples of X_a, X_b, X_{-a-b}, respectively. But then also $X_{a'+b'} = \pm X_{a'}X_{b'}$ is a scalar multiple of $X_aX_b = \pm X_{a+b}$, with similar statements for X_{-a} and X_{-b}, whence the properties $H_{r'} = H_r$ and $c_rc_{-r} = 1$ are proved as before. Now consider the identity [**2**, p. 63, 1.7]

$$x_{r+3s}(k)x_{-s}(1)x_{r+3s}(k)^{-1} = x_{-s}(1)x_{r+2s}(\pm k)x_{r+s}(\pm k)x_r(\pm k)x_{2r+3s}(\pm k^2)$$

which is valid if s and $r+s$ are in Σ, r is in Λ and k is in K. By 3.2 the left side is in G'' as are the first three terms on the right. Thus the product of the last two is also, and replacing k by $-k$, we conclude that $x_r(k)$ is in G''. Thus $G''' \subseteqq G''$, completing the proof of 7.5. We see by 3.2 that G'' is generated by elements of the form $xx_r(k)x^{-1}$, with r in Σ and x in G'''. Hence $G'' \subseteqq G'''$, whence $G'' = G'''$ and 7.2 is proved. The deduction of 7.1 and 7.3 now proceeds as before and details are left to the reader.

8. Classification theorem. By 4.1, 5.2, 7.1 and 7.5, if two classical Lie algebras are isomorphic, then they can be identified so that specified standard sets of generators satisfy conditions (1) and (2) of 5.2, whence the root systems are of the same type. Hence (see [**13**]).

8.1. *Two classical Lie algebras are isomorphic if and only if they have the same type.*

9. Extensions. If $\hat{\mathfrak{g}}$ is obtained from an algebra $\mathfrak{g}$ by extension of the base field, then any automorphism of $\mathfrak{g}$ has a unique extension to $\hat{\mathfrak{g}}$, whence the automorphisms of $\mathfrak{g}$ may be described as the restrictions to $\mathfrak{g}$ of those automorphisms of $\hat{\mathfrak{g}}$ which fix $\mathfrak{g}$. Thus if $\hat{\mathfrak{g}}$ turns out to be a direct sum of classical Lie algebras, the results above enable us to determine the automorphisms of $\mathfrak{g}$. For example, using well-known identifications [**11**], we infer from 4.2 to 4.8 for $\mathfrak{g}$ of type B_n or D_n that each automorphism of the Lie algebra of those linear transformations of a vector space of dimension not 8 over an algebraically closed field of characteristic not 2 which are skew relative to a non singular symmetric bilinear form is induced by an orthogonal transformation of the underlying space, and we then easily deduce if the field is not necessarily algebraically closed that every automorphism is induced by a similitude.

A procedure often used to construct a Lie algebra $\mathfrak{g}$ is to start with $\hat{\mathfrak{g}}$, a direct sum of classical Lie algebras, to then prescribe a group F of semiautomorphisms of $\hat{\mathfrak{g}}$, and finally to define $\mathfrak{g}$ as the set of fixed points of F. Let us assume that F is so chosen that $\hat{\mathfrak{g}}$ can be regarded as a field extension of $\mathfrak{g}$. Then the device stated above is applicable in the following easily proved form: the automorphisms of $\mathfrak{g}$ are the restrictions to $\mathfrak{g}$ of those automorphisms of $\hat{\mathfrak{g}}$ which commute with the elements of F. Examples here are the analogues over general fields of the real forms of Cartan [**1**, p. 399], and the algebras which can be constructed from those classical ones which admit graph automorphisms by naturally defined semiautomorphisms. For these latter algebras one can thus obtain explicit statements such as 4.2 to 4.5 with the rôle of G' taken by the simple groups considered in [**17**].

References

1. E. Cartan, *Oeuvres complètes*, Paris, 1952.
2. C. Chevalley, *Sur certains groupes simples*, Tôhoku Math. J., (2) **7** (1955), 14.
3. E. B. Dynkin, *The structure of semi-simple algebras*, A. M. S. Translation no. 17.
4. F. Gantmacher, *Canonical representation of automorphisms of a complex semi-simple Lie group*, Math. Sb., **5** (1939), 101.
5. N. Jacobson, *Simple Lie algebras over a field of characteristic O*, Duke Math. J., **4** (1938), 534.
6. ———, *Cayley numbers and normal simple Lie algebras of type G*, Duke Math. J., **5** (1939), 775.
7. ———, *Classes of restricted Lie algebras*, Amer. J. Math., **73** (1941), 481.
8. ———, *Exceptional Lie algebras*, Multilithed, New Haven, 1959.
9. W. Landherr, *Über einfache Liesche Ringe*, Hrmburger Abh., **11** (1935), 41.
10. T. Ono, *Sur les groupes de Chevalley*, J. Math. Soc. Japan, **10** (1958), 307.
11. R. Ree, *On some simple groups defined by C. Chevalley*, Trans. A.M.S., **84** (1957), 347.
12. G. B. Seligman, *On Lie algebras of prime characteristic*, Men. A.M.S., **19** (1956).
13. ———, *Some remarks on classical Lie algebras*, J. Math. and Mech. **6** (1957), 549.
14. ———, *On automorphisms of Lie algebras of classical type I*, Trans. A.M.S., **92** (1959), 430.
15. ———, *idid. II*, Trans. A.M.S., **93** (1960), 452.
16. Séminaire "*Sophus Lie*", Paris, 1954-1955.
17. R. Steinberg, *Variations on a theme of Chevalley*, Pacific J. Math., **9** (1959), 875.
18. M. L. Tomber, *Lie algebras of type F*, Proc. A.M.S., **4** (1953), 759.
19. H. Weyl, *The structure and representation of continuous groups*, I.A.S. notes, Princeton, 1934-1935.
20. J. Dieudonné, *Les algèbres de Lie simples associées aux groupes simples algébriques sur un corps de caractéristique* $p > 0$, Rend. Circ. Mat. Palmero, (2) **6** (1957), 198.

UNIVERSITY OF CALIFORNIA, LOS ANGELES

Reprinted from the
BULLETIN OF THE AMERICAN MATHEMATICAL SOCIETY
July 1961, Vol. 67, No. 4
Pp. 406-407

A GENERAL CLEBSCH-GORDAN THEOREM

BY ROBERT STEINBERG

Communicated by N. Jacobson, March 17, 1961

Relative to a Cartan decomposition of a simple Lie algebra over the complex field and an ordering of the roots, let W be the Weyl group, ϕ half the sum of the positive roots, and $P(\beta)$ the number of partitions of β as a sum of positive roots. In a fairly complicated way, Kostant [2] has proved that the multiplicity of μ as a weight in the irreducible representation with highest weight λ is

$$m_\lambda(\mu) = \sum_{s \in W} \det s \; P(s(\phi + \lambda) - (\phi + \mu)). \tag{1}$$

Cartier [1] and the present author have noticed, independently, that Weyl's character formula and (1) are simple formal consequences of each other. (Incidentally, Cartier seems to be wrong in saying that Kostant's work thus provides another algebraic proof of Weyl's formula, since the latter is Kostant's starting point for the proof of (1).) In this note we deduce from (1) the following explicit formula for the multiplicity of an irreducible representation in the tensor product of two others. If the algebra is of type A_1, the result is the classical Clebsch-Gordan Theorem.

THEOREM. *Let π_λ be the irreducible representation with highest weight λ. Then the multiplicity of π_δ in $\pi_\beta \otimes \pi_\gamma$ is*

$$m(\beta, \gamma; \delta) = \sum_{r,s \in W} \det rs\, P(r(\phi + \beta) + s(\phi + \gamma) - (2\phi + \delta)).$$

To prove this, we use Weyl's formula to write

$$\sum_{\mu} m_\beta(\mu) \exp \mu \sum_{s \in W} \det s \exp[s(\phi + \gamma)]$$
$$= \sum_{\eta} m(\beta, \gamma; \eta) \sum_{s \in W} \det s \exp[s(\phi + \eta)].$$

Here μ runs over the set of weights and η over the set of highest weights. A comparison of the coefficients of $\exp(\phi+\delta)$ yields

$$m(\beta, \gamma; \delta) = \sum_{s \in W} \det s\; m_\beta(\phi + \delta - s(\phi + \gamma)),$$

and then by (1), the theorem.

References

1. P. Cartier, *On H. Weyl's character formula*, Bull. Amer. Math. Soc. vol. 67 (1961) p. 228–230.

2. B. Kostant, *A formula for the multiplicity of a weight*, Trans. Amer. Math. Soc. vol. 93 (1959) pp. 53–73.

University of California, Los Angeles

GENERATORS FOR SIMPLE GROUPS

ROBERT STEINBERG

1. Introduction. The list of known finite simple groups other than the cyclic, alternating, and Mathieu groups consists of the classical groups which are (projective) unimodular, orthogonal, symplectic, and unitary groups, the exceptional groups which are the direct analogues of the exceptional Lie groups, and certain twisted types which are constructed with the aid of Lie theory (see §§ 3 and 4 below). In this article, it is proved that each of these groups is generated by two of its elements. It is possible that one of the generators can be chosen of order 2, as is the case for the projective unimodular group **(1)**, or even that one of the generators can be chosen as an arbitrary element other than the identity, as is the case for the alternating groups. Either of these results, if true, would quite likely require methods much more detailed than those used here.

As a model on which the construction for all groups is based, the situation is now described for the group G of $(n+1)$th order unimodular matrices taken modulo the scalar multiples of the identity. Let k be a generator of the multiplicative group K^* of the finite field K; h the diagonal matrix with entries $k, k^{-1}, 1, 1, \ldots$; x the matrix with 1 in all diagonal positions and the $(1, 2)$ position and 0 in all other positions; and w the matrix with 1 in the $(i, i+1)$ position for $1 \leqslant i \leqslant n$, $(-1)^n$ in the $(n+1, 1)$ position, and 0 elsewhere. Then if K has more than three elements, G is generated by the elements represented by h and xw, while if K has two or three elements, x and w will do.

The two-element generation of all of the above groups is covered by 3.11, 3.13, 3.14 and 4.1 below.

With the exception of the complex field, all fields considered in this paper are assumed to be finite.

2. Roots and reflections. Let $\sum = \{a_1, a_2, \ldots, a_n\}$ be a simple (also called fundamental) system of roots corresponding to a simple Lie algebra over the complex field. Throughout the paper we assume that the elements of $\sum$ for the various possible root systems are so labelled that $(a, a) = 2$ and $(a, b) = 0$ for each pair of roots in $\sum$ with the following exceptions:

A_n: $(a_i, a_{i+1}) = -1$ for $1 \leqslant i \leqslant n-1$
B_n: $(a_1, a_1) = 1$, $(a_i, a_{i+1}) = -1$ for $1 \leqslant i \leqslant n-1$
C_n: $(a_i, a_i) = 1$ and $(a_i, a_{i+1}) = -1/2$ for $1 \leqslant i \leqslant n-2$,
$(a_{n-1}, a_{n-1}) = -(a_{n-1}, a_n) = 1$

Received January 10, 1961.

D_n: $(a_1, a_3) = (a_i, a_{i+1}) = -1$ for $2 \leqslant i \leqslant n-1$
E_n: $(a_i, a_{i+1}) = (a_{n-3}, a_n) = -1$ for $1 \leqslant i \leqslant n-2$
F_4: $(a_1, a_1) = (a_2, a_2) = 1$, $(a_1, a_2) = -1/2$, $(a_2, a_3) = (a_3, a_4) = -1$
G_2: $(a_1, a_1) = 2/3$, $(a_1, a_2) = -1$.

Whenever it is convenient, the notation $a, b, \ldots$ is also used for $a_1, a_2, \ldots$. The reflections w_r reversing the various roots r generate a finite group W (the Weyl group) which is at the same time generated by the reflections w_i corresponding to the simple roots a_i. As is well known, any two roots of the same length are congruent under W.

2.1 *Let* $w = w_1 w_2 \ldots w_n$ *(operations from right to left).* (a) *W is generated by w and w_1 with the exceptions: type B_n ($n \geqslant 3$) or D_n (n even) when w, w_1 and w_2 will do, C_n ($n \geqslant 3$) when w, w_{n-1} and w_n will do, F_4 when w, w_2 and w_3 will do.* (b) *W contains the central reflection -1 (defined by $(-1)r = -r$ for each r in $\sum$) if W is not of type A_n ($n \geqslant 2$), D_n (n odd) or E_6.* (c) *If -1 is in W, it is a power of w.*

Proof. Let V be the subgroup of W generated by the given elements. If W is of type A_n, then V contains each $w_i = w^{i-1}w_1w^{1-i}$, hence all of W. If W is of type B_2 or G_2, then V contains w_1 and $w_2 = w_1 w$, hence all of W. If W is of type B_n ($n \geqslant 3$), V contains w_1 and each $w_i = w^{i-2}w_2w^{2-i}$ for $2 \leqslant i \leqslant n$, hence all of W. If W is of type C_n ($n \geqslant 3$), the situation is similar. If W is of type D_n, V contains w_2 even if n is odd since then $w_2 = w^{n-1}w_1w^{1-n}$; thus V contains $w_3 = w_2ww_1w^{-1}w_2$, $w_i = w^{i-3}w_3w^{3-i}$ for $3 \leqslant i \leqslant n$, hence all of W. If W is of type E_n, V contains $w_i = w^{i-1}w_1w^{1-i}$ for $1 \leqslant i \leqslant n-3$, then $w_{n-2} = w_{n-3}w^{-2}w_1w^2w_{n-3}$, $w_{n-1} = ww_{n-2}w^{-1}$ and $w_n = w_{n-1} \ldots w_2w_1w$, hence all of W. Finally, if W is of type F_4, V contains w_2, $w_1 = w^{-1}w_2w$, w_3 and $w_4 = ww_3w^{-1}$, hence all of W. Thus (a) is true. Now if w_0 is the element of W such that $w_0 \sum = -\sum$, then $-w_0$ is an orthogonal transformation which permutes the roots of $\sum$. If $\sum$ is not of type A_n, D_n or E_6, the only possibility is that $-w_0$ is the identity, whence $-1 = w_0$ is in W. If W is of type D_n (n even), one can verify that $w^{n-1}a_i = -a_i$ for each i, whence $-1 = w^{n-1}$ is in W. Thus (b) is proved. For the proof of (c), see 4.1 and 4.5 of **(6)**.

3. The normal types. Following Chevalley, let us consider a Cartan decomposition of a simple Lie algebra over the complex field, choose a generating set $\{X_r, H_r | r = \text{root}\}$ to fulfil the conditions of Theorem 1 of **(2)** (so that the structural constants are all integers), transfer the base field to a finite field K, and then define $x_r(k) = \exp(\text{ad } kX_r)$ for each root r and each k in K, $\mathfrak{X}_r = \{x_r(k) | k \text{ in } K\}$, and G as the group generated by all $\mathfrak{X}_r$. Excluding the cases in which the corresponding simple system of roots $\sum$ is of type A_1, B_2 or G_2 and K has two elements and the case in which $\sum$ is of type A_1 and K has three elements, we obtain a simple group G and call it a *normal type.* Henceforth we also exclude explicit mention of the group G of type C_n constructed over a field of characteristic 2 since it is isomorphic to the corresponding group of type B_n.

The following properties are shared by the normal types.

3.1. *G is generated by those $\mathfrak{X}_r$ for which $\pm r$ is in $\sum$.*

3.2. *If r and s are roots such that $r + s$ is not a root, then $\mathfrak{X}_r$ and $\mathfrak{X}_s$ commute elementwise.*

3.3. *Let r and s be roots such that $r + s$ is a root and $(r + s, r + s) = e(r, r) = e(s, s)$ with $e \geqslant 1$. Then there holds the commutator relation $(x_r(k), x_s(l)) = x_{r+s}(\epsilon ekl)$ with $\epsilon = \pm 1$ depending only on r and s.*

3.4. *For each root r and each k in K^*, the multiplicative group of K, there exists $h = h_{r,k}$ in G such that $hx_s(l)h^{-1} = x_s(k^{2(s,r)/(r,r)}l)$ for each root s and each l in K. The elements $h_{r,k}$ generate an Abelian subgroup $\mathfrak{H}$.*

For each h in $\mathfrak{H}$, we also use h to denote the character on the roots defined by $hx_s(1)h^{-1} = x_s(h(s))$. Thus $hx_s(l)h^{-1} = x_s(h(s)l)$ for every root s and every l in K.

3.5. *For each w in W, there is $\omega(w)$ in G such that $\omega(w)x_r(k)\omega(w)^{-1} = x_{wr}(\epsilon k)$ for each k in K and each root r with $\epsilon = \pm 1$ independent of k.*

3.6. *$\mathfrak{H}\omega(W)$ is a group $\mathfrak{W}$ which contains $\mathfrak{H}$ as a normal subgroup and $\omega(W)$ as a system of coset representatives relative to $\mathfrak{H}$. Further, the map $w \rightarrow \mathfrak{H}\omega(w)$ is an isomorphism of W on $\mathfrak{W}/\mathfrak{H}$.*

3.7. *For each positive root r, we can (and do) choose $\omega(w_r) = x_r(1)x_{-r}(-1)x_r(1)$.*

For the proof of these results, see **(2)**.

3.8. *If G is a normal type, then G is generated by any system of coset representations for $\mathfrak{W}$ over $\mathfrak{H}$ together with $\mathfrak{X}_a$ except when G is of type B_n over a field of characteristic 2, or of type F_4 over a field of characteristic 2, or of type G_2 over a field of characteristic 3, in which case "$\mathfrak{X}_a$" is to be replaced by "$\mathfrak{X}_a$ and $\mathfrak{X}_b$", or "$\mathfrak{X}_b$ and $\mathfrak{X}_c$", or "$\mathfrak{X}_a$ and $\mathfrak{X}_b$", respectively.*

Proof. Since W is transitive on roots of the same length, the result is clear from 3.4, 3.5 and 3.6 if all roots have the same length. For the same reason if G is of type B_n, a system of representatives for $\mathfrak{W}$ over $\mathfrak{H}$ and $\mathfrak{X}_a$ and $\mathfrak{X}_b$ generate G. But if the characteristic is not 2 in the latter case, then $\mathfrak{X}_b$ may be omitted since the other elements generate $\mathfrak{X}_{-a}$, $\mathfrak{X}_{a+b}$ and then $(\mathfrak{X}_{-a}, \mathfrak{X}_{a+b}) = \mathfrak{X}_b$ by 3.3 with $e = 2$. The argument is similar in the other exceptional cases.

3.9. *Let r be a root, l in K^*, and h in $\mathfrak{H}$ such that $h(r)$ is either a generator or the square of a generator of K^*. Then h and $x_r(l)$ generate $\mathfrak{X}_r$.*

Proof. By repeated conjugation by h, we get from $x_r(l)$ all elements of the form $x_r(lk^2)$, and then by multiplication, $x_r(l \sum k_i^2)$. The numbers inside the last brackets form an additive subgroup which contains more than half the elements of K, hence must be K.

3.10. *Let r be a root, w in W, and h in $\mathfrak{H}$ such that $h(r)$ and $h(w^{-1}r)$ are generators or squares of generators of K^* and different from 1. Then h and $x_r(1)\ \omega(w)$ generate $\mathfrak{X}_r$ and $\omega(w)$.*

Proof. Set $h(r) = k$, $h((w^{-1}r) = l$, $x_r(1)\ \omega(w) = x$. Then $y = xhx^{-1} = x_r(1-l)h_1$ with h_1 in $\mathfrak{H}$ by 3.4 and 3.5. Since $(y, h) = x_r(1-l)\ (1-k))$ $= x_r(m)$ with $m \neq 0$, the desired result follows from 3.9.

We can now prove our first principal result.

3.11. *Let G be a normal type, but assume that G is not of type D_n (n even), or of type B_n or F_4 if the underlying field K is of characteristic 2, or of type G_2 if K is of characteristic 3. Let k be a generator of K^*, $a = a_1$, $h = h_{a,k}$ except that for type B_n $h = h_{r,k}$ with $r = 2a_1 + a_2 + \ldots + a_n$, and $w = w_1w_2 \ldots w_n$. Then G is generated by n and $x_a(1)\ \omega(w)$ if K has more than three elements and by $x_a(1)$ and $\omega(w)$ if K has not.*

Proof. Let F be the group generated by the given elements. By 3.10, F contains $\mathfrak{X}_a$ and $\omega(w)$. By 2.1, 3.5, 3.6 and 3.8, it suffices to prove that F also contains an element congruent to $\omega(w_1)$ mod $\mathfrak{H}$, unless G is of type B_n, C_n, or F_4 in which respective cases elements must be produced which are congruent to $\omega(w_1)$ and $\omega(w_2)$, to $\omega(w_{n-1})$ and $\omega(w_n)$, or to $\omega(w_2)$ and $\omega(w_3)$. If G is of type A_n, F contains $\mathfrak{X}_a$, $\mathfrak{X}_b = \omega(w)\mathfrak{X}_a\omega(w)^{-1}, \ldots$, and then by commutation, $\mathfrak{X}_r$ with $r = a + b + \ldots$ and $\mathfrak{X}_{-a} = \omega(w)\mathfrak{X}_r\omega(w)^{-1}$, hence also $\omega(w_a) = \omega(w_1)$ by 3.7. If G is of type B_n, F contains $\mathfrak{X}_a$, $\mathfrak{X}_{a+b} = \omega(w)\mathfrak{X}_a\omega(w)^{-1}$, $\mathfrak{X}_{2a+b} = (\mathfrak{X}_a, \mathfrak{X}_{a+b})$ by 3.3, and $\mathfrak{X}_{-a}$ and $\mathfrak{X}_{-2a-b}$ by 2.1 and 3.5, thus also $\omega(w_a)$ and $\omega(w_{2a+b})$ by 3.7, and $\omega(w_a)\omega(w_{2a+b})\omega(w_a)^{-1}$ which is congruent to $\omega(w_b)$ mod $\mathfrak{H}$. If G is of type C_n, set $s = a_1 + a_2 + \ldots + a_{n-1}$, $t = a_{n-1}$, $u = a_n$. Then F contains $\mathfrak{X}_r$ for $r = a_1, a_2, \ldots, a_{n-1}$ and then for $r = s$, the first by conjugation of $\mathfrak{X}_a$ by $\omega(w)$ and the second by commutation. Thus F also contains $\omega(w)^{-1}\mathfrak{X}_s\omega(w) = \mathfrak{X}_{-t-u}$, $\mathfrak{X}_{-u} = (\mathfrak{X}_t, \mathfrak{X}_{-t-u})$, $\mathfrak{X}_{-t}$ and $\mathfrak{X}_u$ by 2.1 and 3.5, and then $\omega(w_t) = \omega(w_{n-1})$ and $\omega(w_u) = \omega(w_n)$ by 3.7. If G is of type D_n (n odd), F contains $\mathfrak{X}_a$, $\mathfrak{X}_{b+c} = \omega(w)\mathfrak{X}_a\omega(w)^{-1}$, $\mathfrak{X}_{-b} = \omega(w)^{n-1}\mathfrak{X}_a\omega(w)^{1-n}$, $\mathfrak{X}_{-a-c} = \omega(w)\mathfrak{X}_{-b}\omega(w)^{-1}$, $\mathfrak{X}_c = (\mathfrak{X}_{b+c}, \mathfrak{X}_{-b})$, $\mathfrak{X}_{-a} = (\mathfrak{X}_c, \mathfrak{X}_{-a-c})$, hence also $\omega(w_a)$ by 3.7. If G is of type E_6, F contains $\mathfrak{X}_a$ and $\mathfrak{X}_{-a} = (\omega(w)^4\mathfrak{X}_a\omega(w)^{-4}, \omega(w)^8\mathfrak{X}_a\omega(w)^{-8})$, hence also $\omega(w_a)$ by 3.7. If G is of type E_7 or E_8, F contains $\mathfrak{X}_{-a}$ by 2.1 and 3.5, hence also $\omega(w_a)$ by 3.7. If G is of type F_4, F contains $\mathfrak{X}_b = \omega(w)\mathfrak{X}_a\omega(w)^{-1}$, $\mathfrak{X}_{a+b+c} = \omega(w)\mathfrak{X}_b\omega(w)^{-1}$, $\mathfrak{X}_{-a}$ and $\mathfrak{X}_{-b}$ by 2.1 and 3.5, $\mathfrak{X}_{-a-b} = (\mathfrak{X}_{-a}, \mathfrak{X}_{-b})$, $\mathfrak{X}_c = (\mathfrak{X}_{a+b+c}, \mathfrak{X}_{-a-b})$ by 3.3 with $e = 2$, $\mathfrak{X}_{-c}$ by 2.1 and 3.5, and then $\omega(w_b)$ and $\omega(w_c)$ by 3.7. Finally, if G is of type G_2, F contains $\mathfrak{X}_a$ and $\mathfrak{X}_{-a}$ by 2.1 and 3.5, and then $\omega(w_a)$ by 3.7.

In order to treat the normal types excluded by 3.11, we require the following statement.

3.12. *Assume that r and s are roots such that $\mathfrak{X}_r$ and $\mathfrak{X}_s$ commute elementwise,* *and that w in W and h in $\mathfrak{H}$ are such that $h(r) = 1$ and, setting $h(s) = k$, $h(w^{-1}r) = l$, $h(w^{-1}s) = m$, $h(wr) = n$, that each of k, l, m, n is either a generator or the square of a generator of K^* and different from 1. Then h and $x = x_r(1)x_s(1)\omega(w)$ generate $\mathfrak{X}_r$, $\mathfrak{X}_s$ and $\omega(w)$.*

Proof. If F is the subgroup generated by h and x, then F contains $y = xhx^{-1} = x_r(1-l)x_s(1-m)h_1$ with h_1 in $\mathfrak{H}$, then also $(y, h) = x_s((1-m)(1-k))$ and all of $\mathfrak{X}_s$ by 3.9. Thus F contains $t = x_r(1)\omega(w)$, $h_2 = t^{-1}ht = \omega(w)^{-1}h\omega(w)$ with $h_2(r) = h(wr) = n$ by 3.4 and 3.5, $u = tht^{-1} = x_r(1-l)h_1$, and $x_r((1-l)(1-n)) = (u, h_2)$, thus all of $\mathfrak{X}_r$ by 3.9 (with h replaced by h_2).

We can now give two-element generations for the remaining normal types.

3.13. *Let G be of normal type D_n (n even), k a generator of K^*, and set $h = h_{b,k}$ and $w = w_1w_2 \ldots w_n$. Then h and $x_{-a}(1)x_c(1)\omega(w)$ generate G if K has more than 2 elements, while $x_a(1)x_c(1)$ and $\omega(w)$ do if K has not.*

Proof. Let F be the group generated by the given elements. If K has two elements, then F contains $x_{a+c}(1) = (x_a(1)x_c(1))^2$, $x_b(1) = \omega(w)^{-1}x_{a+c}(1)\omega(w)$, $x_{b+c}(1) = ((x_a(1)x_c(1))^{-1}, x_b(1))$, $x_a(1) = \omega(w)^{-1}x_{b+c}(1)\omega(w)$, hence $x_{-b}(1)$ and $x_{-a}(1)$ by 2.1 and 3.5, $\omega(w_a)$ and $\omega(w_b)$ by 3.7, and all of G by 2.1, 3.5 and 3.8. If K has more than two elements, F contains $\mathfrak{X}_{-a}$, $\mathfrak{X}_c$ and $\omega(w)$ by 3.12 with $r = -a$ and $s = c$, hence also $\mathfrak{X}_{-b-c} = \omega(w)\mathfrak{X}_{-a}\omega(w)^{-1}$, $\mathfrak{X}_{-b} = (\mathfrak{X}_{-b-c}, \mathfrak{X}_c)$, and then all of G just as before.

3.14. *Let G be of normal type B_n, F_4, G_2, and in these respective cases let K be of characteristic 2, 2, 3, and define $r = b + c + \ldots$, $s = -a$; $r = c$, $s = -b$; $r = b$, $s = -a$. Let k be a generator of K^*, $t = r - 2s$, $h = h_{t,k}$ and $w = w_1w_2 \ldots w_n$. Then G is generated by h and $x_r(1)x_s(1)\omega(w)$ if K has more than two elements and by $x_r(1)x_s(1)$ and $\omega(w)$ if it has not.*

Proof. Let F be the group generated by the given elements. If K has more than two elements, F contains $\mathfrak{X}_r$, $\mathfrak{X}_s$ and $\omega(w)$ by 3.12. Thus if G is of type F_4 or G_2, F contains $\mathfrak{X}_{-r}$, $\mathfrak{X}_{-s}$, $\omega(w_r)$ and $\omega(w_s)$ by 2.1, 3.5 and 3.7, thus all of G by 3.5 and 3.8; whereas if G is of type B_n, F contains $\mathfrak{X}_{-a}$, then $\omega(w_a)$ by 2.1, 3.5 and 3.7, then $\mathfrak{X}_{-b} = \omega(w_a)\omega(w)\mathfrak{X}_r\omega(w)^{-1}\omega(w_a)^{-1}$, thus all of G as before. If K has two elements, and G is of type B_n, then $n \geqslant 3$, and F contains $x = x_r(1)x_s(1)$, $x_b(1) = (x, (\omega(w)^n x\omega(w)^{-n}, (x, \omega(w)^{n+1}x\omega(w)^{-n-1})))$, thus $x_c(1) = \omega(w)x_b(1)\omega(w)^{-1}, \ldots$, by commutation $x_r(1)$, then $x_s(1) = x_{-a}(1)$ and again all of G by 2.1, 3.5 and 3.8; whereas if G is of type F_4, F contains $x = x_{-b}(1)x_c(1)$, $y = x_{c+d}(1) = (x, (\omega(w)^{-2}x\omega(w)^2, \omega(w)x\omega(w)^{-1}))$, $x_c(1) = (\omega(w)^2y\omega(w)^{-2}, \omega(w)^{-3}y\omega(w)^3)$, $x_{-b}(1) = xx_c(1)$ and all of G once again.

4. The twisted types. Each of the groups yet to be considered occurs as a subgroup of a normal type and will be treated as such. Let the simple root system $\sum$ possess a permutation $r \to \bar{r}$ such that $(\bar{r}, \bar{s}) = (r, s)$ for each pair r, s in $\sum$, and let the field K possess an automorphism $k \to \bar{k}$ of the same period. Then the normal type G constructed from $\sum$ and K has an automorphism α such that $x_a(k)^\alpha = x_{\bar{a}}(\bar{k})$ whenever $\pm a$ is in $\sum$ and k is in K. We then define: $\mathfrak{U}$ (respectively $\mathfrak{V}$) is the subgroup of G generated by those $\mathfrak{X}_r$ for which r is positive (respectively negative), $\mathfrak{U}^1$ (respectively $\mathfrak{V}^1$) is the subgroup of $\mathfrak{U}$ (respectively $\mathfrak{V}$) consisting of the elements invariant under α, and G^1 is the group generated by $\mathfrak{U}^1$ and $\mathfrak{V}^1$. If the period of α is 2, the groups G^1 obtained in this way are $A_n{}^1$ $(n \geqslant 2)$, $D_n{}^1$ $(n \geqslant 4)$ and $E_6{}^1$ (in the notation of **(7)** and **(8)**; see also **(3)**, **(11)**, **(12)**, while if it is 3, one obtains $D_4{}^2$, a second subgroup of D_4; these groups are all simple except for the type $A_2{}^1$ over a field of four elements. Next, the normal type C_2 over a field of $2^{2f+1} = 2e^2$ elements has an automorphism α such that $x_a(k)^\alpha = x_b(k^{2e})$ and $x_b(k)^\alpha = x_a(k^e)$ with similar equations for $-a$ and $-b$ (**5**, Exposés 21 to 24), and one constructs as before a subgroup G^1 (see also (**10**)). A similar construction is possible if the normal type is F_4 over a field of 2^{2f+1} elements or G_2 over a field of 3^{2f+1} elements (see **4**). If $f \geqslant 1$, we get simple groups $C_2{}^1$, $F_4{}^1$ and $G_2{}^1$ in this way and call them, as well as the other simple groups constructed in this paragraph, *twisted types*.

For each twisted type, a *simple set* (of roots) is one which contains a simple root, is closed under addition and the permutation $a \to \bar{a}$ used in the construction, and is minimal relative to these properties. We label the various simple sets S_i thus:

$A_{2n}{}^1$: $S_1 = \{a_n, a_{n+1}, a_n + a_{n+1}\}$, $S_i = \{a_{n+1-i}, a_{n+i}\}$, $2 \leqslant i \leqslant n$
$A_{2n-1}{}^1$: $S_i = \{a_i, a_{2n-i}\}$, $S_n = \{a_n\}$, $1 \leqslant i \leqslant n-1$
$D_n{}^1$: $S_1 = \{a_1, a_2\}$, $S_i = \{a_{i+1}\}$, $2 \leqslant i \leqslant n-1$
$E_6{}^1$: $S_1 = \{a_1, a_5\}$, $S_2 = \{a_2, a_4\}$, $S_3 = \{a_3\}$, $S_4 = \{a_6\}$
$D_4{}^2$: $S_1 = \{a_1, a_2, a_4\}$, $S_2 = \{a_3\}$
$C_2{}^1$: $S_1 = \{a, b, a+b, 2a+b\}$
$F_4{}^1$: $S_1 = \{b, c, b+c, 2b+c\}$, $S_2 = \{a, d\}$
$G_2{}^1$: $S_1 = \{a, b, a+b, 2a+b, 3a+b, 3a+2b\}$.

For each simple set S_i, let $w_i{}^1$ be the unique element of W which maps S_i on $-S_i$ and is in the group generated by those w_r for which r is in S_i (cf. **7**, 2.2), and then set $w = w_1{}^1 w_2{}^1 \ldots$. Further, define h thus: if k is a generator of K^* and r is a simple root in S_1, then $h = h_{r,k} h_{r,k}{}^\alpha$ unless the type is $D_4{}^2$ in which case $h = h_{r,k} h_{r,k}{}^\alpha h_{r,k}{}^{\alpha\alpha}$. Finally, define x thus: for type $A_{2n-1}{}^1$, $D_n{}^1$ or $E_6{}^1$, $x = x_a(1) x_a(1)^\alpha$ with $a = a_1$; for type $D_4{}^2$, $x = x_a(1) x_a(1)^\alpha x_a(1)^{\alpha\alpha}$ with $a = a_1$; for type $A_{2n}{}^1$, $x = x_r(1) x_s(1) x_{r+s}(k)$ with $r = a_n$, $s = a_{n+1}$ and $k + \bar{k} = 1$ (this is $(1|k)$ in (**9**)); for type $C_2{}^1$, $x = x_a(1) x_b(1) x_{2a+b}(1)$ (this is $S(1, 0)$ in (**10**)); for type $F_4{}^1$, $x = x_b(1) x_c(1) x_{2b+c}(1)$; for type $G_2{}^1$, $x = x_a(1) x_b(1) x_{a+b}(1) x_{2a+b}(1)$ (this is $\alpha(1)$ in (**4**)). We can now state our results on the generation of the twisted types.

4.1. *Let G^1 be a twisted type and let w, h and x be defined as in the preceding paragraphs. Then G^1 is generated by h and $x\omega(w)$.*

The properties 2.1 and 3.1 to 3.7 for the normal types have analogues for the twisted types (see **7** and **4**). For this reason, a proof of 4.1 can be patterned after that of 3.11. The details are omitted.

Added in proof. Since the preparation of this paper, I have learned that the symplectic groups (groups of type C_n in the above notation) have been considered by several other authors. In **(13)** and **(14)** a two element generation is given in case the underlying field has a prime number of elements, and in **(15)** the general case is dealt with.

REFERENCES

1. A. A. Albert and J. Thompson, Illinois J. Math., *3* (1959), 421.
2. C. Chevalley, *Sur certains groupes simples*, Tôhoku Math. J., *7* (1955), 14.
3. D. Hertzig, *On simple algebraic groups*, Short communications, Int. Cong. Math. (Edinburgh, 1958).
4. R. Ree, *A family of simple groups associated with the simple Lie algebra of type* G_2, Bull. Amer. Math. Soc., *66* (1960), 508.
5. Séminaire C. Chevalley, *Classification des groupes de Lie algébriques* (Paris, 1956–8).
6. R. Steinberg, *Finite reflection groups*, Trans. Amer. Math. Soc., *91* (1959), 493.
7. ——— *Variations on a theme of Chevalley*, Pacific J. Math., *9* (1959), 875.
8. ——— *The simplicity of certain groups*, Pacific J. Math., *10* (1960), 1039.
9. ——— *Automorphisms of finite linear groups*, Can. J. Math., *12* (1960), 606.
10. M. Suzuki, *A new type of simple groups of finite order*, Proc. Nat. Acad. Sci., *46* (1960), 868.
11. J. Tits, *Les "formes réelles" des groupes de type* E_6, Séminaire Bourbaki, Exposé 162 (Paris, 1958).
12. ——— *Sur la trialité et certains groupes qui s'en déduisent*, Publ. Math. Inst. Hautes Etudes Sci., *2* (1959), 14.
13. T. G. Room and R. J. Smith, *A generation of the symplectic group*, Quart. J. Math., *9* (1958), 177.
14. T. G. Room, *The generation by two operators of the symplectic group over* $GF(2)$, J. Austr. Math. Soc., *1* (1959), 38.
15. P. F. G. Stanek, *Two element generation of the symplectic group*, Bull. Amer. Math. Soc., *67* (1961), 225.

University of California

A CLOSURE PROPERTY OF SETS OF VECTORS

BY
ROBERT STEINBERG

1. **Introduction and statement of results.** If the system S of roots of a simple Lie algebra over the complex field is imbedded in a real Euclidean space in the usual way, the following important property, which we call property P, is true:

(P) *If* $x,\ y \in S,\ x \neq -y$ *and* $(x,y) < 0$, *then* $x + y \in S$; $\quad 0 \notin S$.

Our aim here is to study this property. The exclusion of the vector 0 is not essential in what follows. It turns out that many of the other properties of root systems of Lie algebras depend only on property P and are thus shared by a much wider class of vector systems. In the statements to follow it is assumed that all vectors considered come from a real Euclidean space V of finite dimension n.

1.1 *Let S be finite and have property* P, *and let A be a subset with property* P. *For each x in S, assume that at most one (resp. exactly one, at least one) of x and $-x$ is in A. Then there is an ordering of the space V such that A is contained in (resp. A is, A contains) the set of positive elements of S.*

Harish-Chandra [6, Lemma 4] proves the second part of this result for Lie algebra root systems, however, using nontrivial properties of Lie algebras. Borel and Hirzebruch [2, pp. 471–473] give a geometric proof of the second and third parts, but then revert to Lie algebra techniques to prove the first part, all for Lie algebra root systems. All of these authors make a somewhat stronger assumption on A than property P, namely, if $x,\ y \in A$ and $x + y \in S$, then $x + y \in A$. Our proof of 1.1 depends on a preliminary result which may have some independent interest.

1.2. *Let B be finite and have property* P, *and assume that $x \in B$ implies $-x \notin B$. Then there is an ordering of the space V such that all elements of B are positive.*

The real numbers of the form $k - l\sqrt{2}$ (k, l positive integers) show that the assumption of finiteness in 1.1 (or 1.2) can not be dropped. It can, however, be weakened thus.

1.1′. *In* 1.1 *replace the assumption of finiteness by the assumption that* 0 *is not a point of accumulation of S.*

In 1.2 we can go further, in terms of a weakening of property P.

(P_n) *If* $x,y \in S,\ x \neq -y$ *and* $(x,y) \leqq -1/n\ |x|\ |y|$, *then* $x + y \in S$; $\quad 0 \notin S$.

Received by the editors September 15, 1961.

1.2′. *If B does not have* 0 *as a point of accumulation, if B has property* P_n, *and if $x \in B$ implies $-x \notin B$, the conclusion of* 1.2 *holds.*

In this statement the number $-1/n$ which enters via P_n can not be replaced by a smaller one, as we see by taking B to be the set of vertices of a regular simplex. We also remark that the change P to P_n renders false the third part of 1.1′ as well as all of the results to follow.

We call a set of vectors S *symmetric* if $x \in S$ implies $-x \in S$, and *indecomposable* if it is not contained in the union of two lower-dimensional orthogonal subspaces of V. Observe that if S is indecomposable then it generates V.

1.3. *Let S be finite, symmetric and have property* P, *and relative to a fixed ordering of V, let B consist of those positive elements of S which can not be written as sums of other positive elements of S. Then* (a) *B is linearly independent*; (b) *every positive element of S is a sum of elements of B. If further S is indecomposable (so that B is a basis of V), then* (c) *B cannot be split into subsets B_1, B_2 such that $x_1 \in B_1$, $x_2 \in B_2$ implies $x_1 + x_2 \notin S$*; (d) *the sum of all elements of B is in S*; (e) *there exists a (dominant) positive element d of S, $d = \sum k(b)b$ ($b \in B$, $k(b)$ = nonnegative integer), such that if $r = \sum l(b)b$ is any element of S then $k(b) \geqq l(b)$ for all b.*

The proofs of these results can be patterned after those of Cartan [3] and Dynkin [4, p. 106], and are not given here. The existence of the dominant element d is easily seen to be equivalent to the fact that the inequalities $(d,x) < 1$ and $(b,x) > 0$ ($b \in B$) define a simplex which is not pierced by any of the hyperplanes $(r,x) =$ integer ($r \in S$). More generally, we show:

1.4. *Let S be finite, symmetric, indecomposable and have property* P. *Then the regions into which V is partitioned by the hyperplanes $(r,x)=k$ ($r \in S$, $k = 0, \pm 1, \pm 2, \ldots$) are all simplexes.*

An example which yields a nice pattern of triangles of various sizes and shapes (in contrast to the Lie algebra case when all regions are congruent under a discrete group generated by reflections) is obtained by letting S consist of the vectors $(\pm 3, 0)$, $(0, \pm 3)$, (i, j), with i,j integers, $|i| \leqq 2$, $|j| \leqq 2$, $(i,j) \neq (0,0)$ (coordinates relative to an orthonormal basis of V). Another example, $S = \{\pm(1,0), \pm(0,1), \pm(1,1), \pm(1/2,-1/2)\}$, shows that property P is not necessary for the conclusion of 1.4 and raises the interesting question: *what condition is*?

One can easily classify the one-dimensional finite symmetric indecomposable sets with property P: $S = \{\pm r, \pm 2r, \ldots, \pm kr;\ r \neq 0,\ k$ an arbitrary positive integer$\}$. In higher dimensions, however, arbitrarily large multiples can not occur if the set is to remain finite.

1.5. *Let S be symmetric, indecomposable, and have property* P *and dimension $n \geqq 2$. If r and $5r$ are in S, then S is infinite, in fact all nonzero integral multiples* *of r are in S. If it is assumed instead that r and $4r$ are in S, then S need not be infinite.*

But for infinite sets of the type under consideration there is a rather drastic consequence.

1.6. *Let S be infinite, symmetric, indecomposable, and have property* P *and dimension $n \geqq 2$. Then $S \cup \{0\}$ is a group (under the addition of V).*

Combining 1.5 and 1.6, we get:

1.7. *If S is symmetric, indecomposable, has property* P *and dimension $n \geqq 2$, and if S contains r and $5r$, then $S \cup \{0\}$ is a group.*

A second consequence of 1.6 is a characterization of vector lattices.

1.8. *If $0 \notin S$, then $S \cup \{0\}$ is an n-dimensional lattice in V if and only if S is infinite, symmetric, indecomposable, has property* P, *and does not have* 0 *as a point of accumulation.*

A third consequence is that there is no infinite analogue of 1.4. In order to refine 1.7 (see 4.1), we require a final basic result.

1.9. *Let S be symmetric, indecomposable and have property* P. *Let s be in S, and let S_1 be the nonzero part of the projection of S on s^0, the orthogonal complement of s. Then S_1 is symmetric, indecomposable and has property* P.

2. **Proof of 1.1, 1.2, 1.1′ and 1.2′.** The deductions of 1.1 from 1.2 and of 1.1′ from 1.2′ are identical and as follows. The choice $B = A$ in 1.2 yields the first part of 1.1 and half of the second part. To get the other half, observe that if $x \in S$, and $x \notin A$, then in turn $-x \in A$, $-x$ is positive, and x is negative. To prove the third part, we choose B as the complement of A in S. Then B has property P: the assumptions $x \in B$, $y \in B$, $x \neq -y$, $(x,y) < 0$, $x + y \notin B$ imply $x + y \in A$, $-x \in A$, $-y \in A$, either $(-x, x + y) < 0$ or $(-y, x + y) < 0$, so that either $(-x) + (x + y) = y$ is in A or $(-y) + (x + y) = x$ is in A, a contradiction. Further $x \in B$ implies $-x \notin B$. Thus by 1.2 there is an ordering of V in which all elements of B are positive. If this ordering is reversed, the vectors of B become negative, and A contains all positive elements of S as required.

To establish 1.2, we first show that there is a nonzero vector z such that $(z,x) \geqq 0$ for all x in B. If this is false, then because B has property P, there is a sequence $x_1, x_2, \ldots$ of elements of B such that $x_1 + x_2$, $x_1 + x_2 + x_3, \ldots$ are also in B. Since B is finite, two of these sums must be equal, so that a nontrivial relation $\sum y_j = 0$ $(y_j \in B)$ exists. Among all such relations, pick a shortest one. Its length is at least 3. Now $(y_1, y_2 + y_3 + \ldots) = (y_1, -y_1) < 0$, so that $(y_1, y_j) < 0$ for some j, and $y_1 + y_j \in B$. But then the relation can be shortened, a contradiction. This proves our assertion, from which we quickly deduce 1.2 by induction on the dimension of V. The result being clear if this dimension is 0, assume it is positive. Then by the inductive assumption there is an ordering of z^0, the orthogonal complement of z, such that all elements of $z^0 \cap B$ are positive. We extend this ordering (lexicographically) to V by defining x to be positive if in the representation $x = kz + u$ (k scalar, $u \in z^0$) either k is positive or k is 0 and u is positive. Clearly all elements of B are now positive.

Our proof of 1.2′ requires a preliminary result, proved in two steps.

2.1 *If* 0 *is in the convex closure of a set* B *of unit vectors, then* $(x,y) \leqq -1/n$ *for some* x, y *in* B.

By [1, p. 9], there is a relation $\sum_{j=1}^m c_j x_j = 0$ ($c_j > 0$, $x_j \in B$, $m \leqq n+1$). If c_1 is the largest of the c_j, and 2.1 is false, we have $0 = c_1(x_1,x_1) + \sum_2^m c_j(x_1,x_j) > c_1 - \sum_2^m c_j/n \geqq c_1(1-(m-1)/n) \geqq 0$, a contradiction.

2.2 *If* B *is a finite set of unit vectors and* b *is a number such that* $b \geqq -1/n$ *and* $(x,y) > b$ *for all* x,y *in* B, *there is a unit vector* z *such that* $(z,x) > ((1+(n-1)b)/n)^{\frac{1}{2}}$ *for all* x *in* B.

We use only the case $b = -1/n$, $((1+(n-1)b)/n)^{\frac{1}{2}} = 1/n$, but 2.2 in general is no more difficult to prove. Let z be a unit vector that maximizes $\min_{x \in B}(z,x)$. If this maximum is a, than $a \geqq 0$ since 0 is not in the convex closure of B by 2.1 (see [1, p. 9]). Further if B_1 is the subset of B on which the maximum occurs, then 0 is in the convex closure of the projection of B_1 on the orthogonal complement of z, since otherwise there would be a unit vector y orthogonal to z such that $(y,x) > 0$ ($x \in B_1$), and then the choice $z_1 = (z+ey)/(1+e^2)^{\frac{1}{2}}$ (e positive, sufficiently small) would contradict the definition of z. Thus there is a relation $\sum_1^m c_j x_j = kz$ ($c_j > 0$, $\sum c_j = 1$, $x_j \in B_1$, $m \leqq n$). Taking the inner product of this equation with z and then with $\sum x_j$, we get $k = a$ and then

$$1 + \textstyle\sum_j c_j \sum_{i \neq j}(x_i, x_j) = ma^2.$$

Since $(x_i, x_j) > b$, this yields $a^2 > (1+(m-1)b)/m \geqq (1+(n-1)b)/n$, as required.

Now to start the proof of 1.2′, we assume there is no nonzero vector z such that $(z,x) \geqq 0$ for all x in B. This implies that 0 is in the convex closure of B, so that a nontrivial relation $\sum c_j x_j = 0$ ($c_j > 0$, $x_j \in B$) exists. If l is a positive lower bound to the lengths of the elements of B, then by [5, Theorem 201] there exist positive integers m_j and m such that if $c_j = (m_j + r_j)/m$ then the r_j are so small that $|\sum r_j x_j| \leqq 2l/n$. Then $\sum m_j x_j$ is a nontrivial sum of elements of B, and its length is at most $2l/n$. Let $\sum y_j$ be such a sum with a minimum number of terms. Since B has property P_n, $(y_i,y_j) > -1/n\,|y_i|\,|y_j|$, all i,j, and thus by 2.2 there is a unit vector z such that $(z,y_j) > |y_j|/n$, all j. But then $2l/n \geqq |\sum y_j| \geqq (z, \sum y_j) > \sum |y_j|/n \geqq 2l/n$, a contradiction. Thus there is a nonzero vector z such that $(z,x) \geqq 0$ for all x in B, and we can complete the proof by induction on n, just as for 1.2.

3. **Proof of 1.4.** The conclusion of 1.4 can be stated (and is proved) in the following form: *if an integer* $p(r)$ *is associated with each* r *in* S *so that* $p(-r) = -p(r) - 1$ *and the inequalities* $(x,r) > p(r)$ ($r \in S$) *are consistent, (that is, have a solution for* x*), then these inequalities are a consequence of a subset of* $n+1$ *of them.*

Let $x = x_0$ be a solution of the inequalities and let A be a minimal subset of elements of S such that the inequalities $(x,a) > p(a)$ $(a \in A)$ imply all inequalities. The proof that A has $n + 1$ elements is given in several steps.

(1) *If* $b \in A$, $r \in S$, $r - b \in S$ *and* r *is independent of* b, *then* $p(r) = p(r - b) + p(b)$. First note that $(x,-s) > p(-s)$ is equivalent to $(x,s) < p(s) + 1$, for every s in S. Thus $p(r - b) < (x_0, r - b) < p(r - b) + 1$, $p(b) < (x_0,b) < p(b) + 1$, by addition $p(r - b) + p(b) < (x_0,r) < p(r - b) + p(b) + 2$, whence $p(r) = p(r-b) + p(b)$ or $p(r) = p(r - b) + p(b) + 1 = p(b) - p(b - r)$. The last equation, however, yields a contradiction as follows. Because of the minimal nature of A, we can choose x_1 so that $(x_1,a)>p(a)$ $(a \in A,\ a \neq b)$ and $(x_1,b) = p(b)$ (we can first achieve the inequality $(x_1,b) \leqq p(b)$, and then by moving x_1 towards x_0, the equality), and because r is independent of b, so that also $(x_1,r) \neq p(r)$. Now at a point x_2 of the open segment x_0x_1 we have $(x_2,a) > p(a)$ for every a in A, so that, by the definition of A, $(x_2,r) > p(r)$ and $(x_2,b - r) > p(b - r)$. If x_2 approaches x_1, this yields $(x_1,r) \geqq p(r)$ and $(x_1,b - r) \geqq p(b - r)$, whence $(x_1,r) \geqq p(r) = p(b) - p(b - r) \geqq (x_1,b) - (x_1,b - r) = (x_1,r)$, so that $(x_1,r) = p(r)$, a contradiction.

(2) *If* b *and* c *are in* A *and independent, then* $c - b \notin S$. For otherwise $p(b) = p(b-c)+p(c)$ and $p(c)=p(c-b)+p(b)$ by (1), then $0 = -1$ by addition.

(3) *If* x *is a nonzero vector, there is* a *in* A *such that* $(x,a) < 0$. For, if $(x,a) \geqq 0$ for every a in A, then $(x_0 + kx,a) > n(a)$ for every positive number k, whence $(x_0 + kx,r) > n(r)$ and $(x,r) \geqq 0$ for every r in S. Since $-r \in S$, this yields $(x,r) = 0$ and then $x = 0$, a contradiction.

(4) *A is not the union of two nonempty subsets B and C which are orthogonal to each other.* Assume the contrary. Since S is indecomposable, there is r in S not orthogonal to B or C. By (3) applied to the projection of r on the subspace generated by B there is b_1 in B such that $(r,b_1) > 0$, whence $r - b_1$ is in S. If $r - b_1$ is not orthogonal to B, then there is b_2 in B such that $r - b_1 - b_2$ is in S, and so on. This process can not continue indefinitely since there would then be a repetition $r - b_1 - \dots - b_k = r - b_1 - \dots - b_m$ $(k < m)$, whence $p(r - b_1 - \dots - b_i) = p(r - b_1 - \dots - b_{i+1}) + p(b_{i+1})$ $(k \leqq i < m)$ by (1), and $p(b_{k+1}) + \dots + p(b_m) = 0$ by addition, yielding the contradiction $0 = (x_0,b_{k+1} + \dots + b_m) = (x_0,b_{k+1}) + \dots + (x_0,b_m) > p(b_{k+1}) + \dots + p(b_m) = 0$. Thus for some t, $r - b_1 - \dots - b_t$ is orthogonal to B. Starting with $s = -r + b_1 + \dots + b_{t-1}$, we can repeat the above procedure to get $c_1,c_2 \dots, c_u$ in C such that $s - c_1$, $s - c_1 - c_2, \dots, s - c_1 - \dots - c_u$ are all in S and the last vector is orthogonal to C. But then $s - c_1 - \dots - c_u + b_t$ is orthogonal to B and C, hence it is 0 by (3), so that $c_u - b_t = s - c_1 - \dots - c_{u-1}$, an element of S in contradiction to (2).

(5) *Completion of proof.* Let m be the number of elements of A. Since the region $(x,a) > p(a)$ $(a \in A)$ is bounded, $m \geqq n + 1$. Thus there is a linear relation among the elements of A, and in fact one of shortest nonzero length q, say

$k_1a_1 + \ldots + k_qa_q = 0$. Clearly $q \leqq n + 1$. If r (resp. s) denotes the sum of the terms with positive (resp. negative) coefficients, then $(r,s) \geqq 0$ by (2), whence $(r,r) \leqq (r,r + s) = 0$, so that $r = 0$ and all terms have the same sign which may be taken as positive. Now assume that b is in A and is not one of the a_j. Then b is not a multiple of any a_j since this would imply that $q = 2$, that b is a positive multiple of some a_j, and then that one of the inequalities $(x,b) > p(b)$, $(x,a_j) > p(a_j)$ is a consequence of the other, contradicting the minimality of A. By (4) we can choose b to be nonorthogonal to some a_j; then $(b,a_j) < 0$ by (2) and the fact that S has property P, so that $(b,a_i) > 0$ for some i since all coefficients in the above relation are positive, and $b - a_i$ is in S, contradicting (2). Thus b does not exist, $m = q \leqq n + 1$, $m = n + 1$, and 1.4 is proved.

4. **Proofs of 1.5 to 1.9 and related results.** Assume that 1.5 is false, that k is a positive integer such that kr is in S but $(k + 1)r$ is not. Now $(5r,r) > 0$, whence $4r$ is in S, and similarly so are $3r$ and $2r$. Thus $k \geqq 5$. By the assumptions on S, there is s in S such that s is independent of r and not orthogonal to r. Since r may be replaced by $-r$, there is no restriction in assuming that also $(s,r) < 0$, so that $s + r$ is in S. Then if $(s + r,r) < 0$, $s + 2r$ is in S, and so on. After a finite number of steps we arrive at $t = s + jr$ in S such that $-(r,r) \leqq (t,r) < 0$. Then $t + kr$ and $t+2r$ are in S, and since $(t + 2r, -r) < 0$, so are $t - r$ and $t - 2r$. Now $(t + kr,t - r) \leqq 0$ since $(k + 1)r$ is not in S; hence $(t + kr,\ t - 2r) < 0$ and $2t + (k - 2)r$ is in S. This last vector, say s_1, is twice as far from the line of r as s (or t) is, and it is not orthogonal to r since $(2t + (k-2)r,r) = 2(t + r,r) + (k - 4)(r,r) > 0$. Since the step from s to s_1 can be repeated indefinitely, our original vector s can be chosen so that s (and hence t) is so far from the line of r that $(t,t) > (2k - 1)(r,r)$. But now $(t + kr,t - r) = (t,t) + (k - 1)(t, r) - k(r,r) > (2k - 1)(r,r) - (k - 1)(r,r) - k(r,r) = 0$, so that $(k + 1)r$ is in S, a contradiction. The second part of 1.5 is proved by the (unique) examples $S = \{(\pm i,0), (\pm j,\pm\sqrt{6}), (0,\pm 2\sqrt{6}); 1 \leqq |i| \leqq 4, |j| \leqq 3\}$.

Next we prove 1.6. For convenience, we use *multiple* to mean nonzero integral multiple.

(1) *There is an element of S which has all its multiples in S.* If S is bounded, it has 0 as a point of accumulation: otherwise, there is a point of accumulation $x \neq 0$, then for elements s and t of S which are distinct and close enough to x, $(s,t) > 0$, $t - s \in S$, and 0 is a point of accumulation anyway. Since V can be covered by a finite number of cones with vertex 0 and verticle angle $\pi/3$, there thus exists such a cone containing a sequence of distinct elements of S tending to 0 or to ∞, and hence also containing elements r and s such that $|s| > 8|r|$. It follows that $s - r, s - 2r, \ldots, s - 5r$ are all in S, the last inclusion, for example, coming from $(s - 4r,r) = (s,r) - 4(r,r) \geqq 1/2\,|s|\,|r| - 4|r|^2 > 0$. Since

$$(s,s - 5r) = (s,s) - 5(s,r) \geqq |s|^2 - 5|s|\,|r| > 0,$$

the vector $5r$ belongs to S, and so do all multiples of r by 1.5.

(2) *If s and t are elements of S such that all multiples of s are in S and t is not orthogonal to s, then* (a) *all vectors $ks + t$ (k integer) are in $S \cup \{0\}$*; (b) *all multiples of t are in S.* Replacing s by $-s$ if necessary, we may assume that $(s,t) < 0$, whence $ls + t \in S \cup \{0\}$ for any positive integer l. Now if l is large enough, $(s, ls + t) > 0$, so that $(l - m)s + t \in S \cup \{0\}$ for any positive integer m, which is (a). Also for l large enough $(ls + t, -ls + t) < 0$, whence $2t \in S$. Similarly $4t$, $8t \in S$. But then $5t \in S$, and (b) follows from 1.5.

(3) *If r and t are in S and orthogonal, there is s in S not orthogonal to r or t.* Since S is indecomposable, it contains a sequence $x_1, x_2, \ldots, x_j$ such that $x_1 = r$, $x_j = t$, and consecutive terms are not orthogonal. If a shortest such sequence is chosen, then nonconsecutive terms are orthogonal, and $j = 3$, since otherwise the sequence could be shortened by replacing the pair x_2, x_3 by $x_2 + x_3$ if $(x_2,x_3) < 0$ or by $x_2 - x_3$ if $(x_2,x_3) > 0$. The choice $s = x_2$ yields (3).

(4) *If $r,t \in S$, $r \neq t$, then $r - t \in S$.* First note that all multiples of all elements of S are again in S by (1), (2b) and (3). Thus if r is not orthogonal to t, $r - t \in S$ by (2a); while if r is orthogonal to t, there is $s \in S$ not orthogonal to r or t by (3), then $ks + t$, $ks + r \in S$ for all integers k by (2a), and, since $(ks + t, ks + r) > 0$ for k large enough, $r - t \in S$ in this case also.

By (4), $S \cup \{0\}$ is closed under subtraction, hence is a group, and 1.6 is proved.

From 1.6 we get 1.7. The nonzero real (or rational) numbers of absolute value less than 1 show that neither of these results is true if $n = 1$.

Since the *only if* part of 1.8 is easily verified, we turn to the *if* part. If $n \geqq 2$, $S \cup \{0\}$ is a group by 1.6, and since it is discrete (because 0 is not a point of accumulation) and contains n independent elements, it must be an n-dimensional lattice. If $n = 1$, it is easily proved that S has a smallest element r and consists of the multiples of r.

To start the proof of 1.9, assume u, $v \in S_1$, $u \neq -v$ and $(u,v) < 0$. Then there are scalars k, l such that $u + ks$, $v + ls \in S$. Now if $k > 0$, then $(u + ks,s) > 0$ and $u + ks$ can be replaced by $u + (k - 1)s$ which is also in S. Repeating this procedure as often as necessary, we may assume $k \leqq 0$, and similarly $l \geqq 0$. Then $(u + ks,v + ls) = (u,v) + kl(s,s) < 0$, $u + v + (k + l)s \in S$, and $u + v \in S_1$. Thus S_1 has property P. Now let Q and R be two lower-dimensional orthogonal subspaces of s^0. Since S is indecomposable, then by the same reasoning as in step (3) of the proof of 1.6, there is x in S orthogonal to neither Q nor R. Since x projects onto an element of S_1 not in Q or R and the pair Q, R is arbitrary, S_1 is indecomposable.

We can now refine 1.7.

4.1. *Let S be symmetric, indecomposable, have property* P *and contain r. In the respective cases that the dimension n of S is* 2, 3, *at least* 4, *assume that S contains $5r$, $4r$, $3r$. Then $S \cup \{0\}$ is a group. If a smaller positive multiple* *of r is assumed to be in S, then $S \cup \{0\}$ need not be a group, in fact need not be infinite.*

If $n = 2$, we use 1.7. Assume $n = 3$. By the methods used at the beginning of the proof 1.5, there is s in S such that s is independent of r, and $s + r$, $s + 2r, \ldots,$ $s + 5r$ are all in S. The projection of S on s^0 contains r_1, the projection of r, and also $5r_1$. By 1.9 and 1.5 we conclude that S is infinite, and then by 1.6 that $S \cup \{0\}$ is a group. The corresponding proof for $n \geqq 4$ is similar. To prove the final part of 4.1, we add to the example used in the proof of 1.5 two further examples: if $n = 3$, $S = \{(i,0,0),\ (j,\pm\sqrt{3},\pm 1),\ (j,0,\pm 2),\ (0,\pm 2\sqrt{3},0),$ $(0,\pm\sqrt{3},\pm 3);\ 1 \leqq |i| \leqq 3,\ |j| \leqq 2\}$, and if $n \geqq 4$ (in fact, if n is arbitrary), $S = \{(x_1,x_2,\ldots,x_n);\ x_j$ an integer, $\sum |x_j| = 1$ or $2\}$.

References

1. T. Bonnesen and W. Fenchel, *Theorie der konvexen Koerper*, Springer, Berlin, 1927; or New York, Chelsea, 1948.

2. A. Borel and F. Hirzebruch, *Characteristic classes and homogeneous spaces*. I, Amer. J. Math. **80** (1958), 458–538.

3. E. Cartan, *Complément au mémoire "sur la géométrie des groupes simples"*, Ann. Mat. **5** (1928), 253–260.

4. E. B. Dynkin, *The structure of semi-simple Lie algebras*, Amer. Math. Soc. Transl. (2) **17** (1950).

5. G. H. Hardy and E. M. Wright, *Theory of numbers*, 4th ed., Oxford, 1960.

6. Harish-Chandra, *Representations of semisimple Lie groups*. IV, Amer. J. Math. **77** (1955), 743–777.

University of California,
Los Angeles, California

COMPLETE SETS OF REPRESENTATIONS OF ALGEBRAS

ROBERT STEINBERG

1. Introduction and results. A classical theorem [1, Chapter XV, Theorem IV] states:

(1) *Let G be a finite group and R a faithful representation*[1] *of G over a field K. Then each irreducible representation of G over K is a constituent of some tensor power of R.*

The only proof of this result known to us actually requires the additional assumption that K is of characteristic 0 and involves a calculation with characters which is not very revealing (to us). In an attempt to construct a more conceptual proof we have been led to a considerably more general result.

(2) *Let A be an algebra over a field K. Assume that A has a basis B over K such that $B\cup\{0\}$ is closed under multiplication. Finally, let R be a representation of A which is faithful on $B\cup\{0\}$, and for each $r=1, 2, \cdots$ let $\otimes^r R$ be the representation of A defined by $(\otimes^r R)(b) = \otimes^r R(b)$ $(b\in B)$ together with linearity. Then the representations $\otimes^r R$ $(r=1, 2, \cdots)$ form a complete set of representations of A (in the sense that their direct sum is faithful on A).*

Observe that the assumptions on B imply that each $\otimes^r R$ really is a representation of A and that A is associative, but that there is no restriction on the characteristic of K or the dimension of A or R. The transition from (2) to (1) is immediately effected by applying to the group algebra of G the statement (2) and the following probably well-known result, for which a proof is sketched at the end of this paper.

(3) *If $\{{}^rR \mid r=1, 2, \cdots\}$ is a complete set of representations of a finite-dimensional algebra A, then each irreducible representation of A is a constituent of some ${}^rR$.*

That the finiteness assumptions cannot be dropped in (1) or (3) may be seen from the following example. Let $e(k)$ be the real 2×2 matrix obtained by replacing the 12 entry of the identity matrix by k, G the multiplicative group of all $e(k)$, A the group algebra of G over the reals, B the set G (imbedded in A), and R the defining representation of G extended to A. Then no tensor power of R contains the one-dimensional representation S of A (or G) defined by $S(e(k)) = \exp k$ (k real).

The proof of (2) depends on the following lemma.

Received by the editors September 21, 1961.

[1] Throughout this note all representations are assumed to correspond to left modules and the 0-representation is excluded from the list of irreducible representations.

(4) *If C is a set of nonzero elements of a vector space V, then in the strong direct sum $\sum_{r=1}^{\infty} \otimes^r V$ the vectors $\sum \otimes^r c$ $(c \in C)$ are linearly independent.*

2. **Proofs.** If the conclusion of (4) does not hold, there is a minimal nonempty finite subset D of C such that there are nonzero scalars $k(d)(d \in D)$ for which

$$(*) \qquad \sum_{d \in D} k(d) \otimes^r d = 0 \qquad (r = 1, 2, \cdots).$$

Since D clearly has at least two elements, there is a linear function v^* on V which is not constant on D. Replacing r by $r+s$ in (*), taking the tensor product with $\otimes^s v^*$, and then contracting, we get

$$\sum_{d \in D} (k(d) \otimes^r d) v^*(d)^s = 0 \qquad (r = 1, 2, \cdots; s = 0, 1, 2, \cdots).$$

Thus if $k_1, k_2, \cdots, k_n$ are the distinct values taken by v^* on D, the value k_1 being taken on the subset D_1 of D, then because the van der Monde matrix (k_t^s) $(1 \leqq t \leqq n, 0 \leqq s \leqq n-1)$ is nonsingular, the equations (*) hold with D replaced by D_1, contradicting the minimal nature of D. Thus (4) is established.

Under the assumptions of (2) let $a = \sum k(b) b$ $(b \in B, k(b) \in K)$ be an element of A such that $(\otimes^r R)(a) = 0$ for $r = 1, 2, \cdots$. Then $\sum k(b) \otimes^r R(b) = 0$ for $r = 1, 2, \cdots$, each $k(b)$ is 0 by (4), whence a is also 0. Thus (2) is proved.

For the proof of (3) one may assume that $\{{}^rR\}$ is finite and consists of finite-dimensional representations. Let ${}^rM = {}^rM_0 \supset {}^rM_1 \supset \cdots$ be a composition series for the A-module ${}^rM$ corresponding to ${}^rR$, and let N be an arbitrary irreducible A-module. If A^0 is the radical of A, then A/A^0 is a sum of minimal left ideals. Hence there is a minimal left ideal I/A^0 such that $IN \neq 0$, and then there is a corresponding pair (r, i) such that $I({}^rM_i/{}^rM_{i+1}) \neq 0$, since otherwise I would be nilpotent because $\{{}^rR\}$ is complete and thus would be contained in A^0. If m and n are nonzero elements of ${}^rM_i/{}^rM_{i+1}$ and N respectively, it is then readily verified that the map $im \rightarrow in$ $(i \in I)$ is an A-module isomorphism of ${}^rM_i/{}^rM_{i+1}$ on N. Hence (3).

Reference

1. W. Burnside, *Theory of groups of finite order*, 2nd ed., Cambridge Univ. Press, Cambridge, 1911.

University of California, Los Angeles and
The Institute for Advanced Study

Reprinted from the
Proceedings of the American Mathematical Society
Vol. 13, No. 5, October, 1962
pp 746-747

GÉNÉRATEURS, RELATIONS ET REVÊTEMENTS DE GROUPES ALGÉBRIQUES

PAR

ROBERT STEINBERG (Los Angeles)

I. Introduction et énoncé des résultats

1. *Résumé*

Pour chaque groupe simple G de forme normale (c'est-à-dire un des groupes étudiés par Chevalley dans [1], et désigné là par G′), nous allons étudier les propriétés de deux groupes Γ et Δ qui jouent le rôle de revêtement universel de G, sous des conditions variées. Pour certaines formes non normales de groupes simples, celles étudiées par Hertzig [4], Tits [15, 16], Suzuki [14], Ree [5] et l'auteur [10, 11], il y a des résultats analogues qui seront présentés dans un article ultérieur. Ces résultats sont de première importance dans l'étude des représentations de tous ces groupes [13].

2. *Terminologie et notations*

Si K est un corps, K^* désigne le groupe multiplicatif de K, et $|K|$ la cardinalité de K. Si G est un groupe, G' désigne le sous-groupe des commutateurs de G. Soit V un espace vectoriel. On emploie la notation $GL(V)$, $PL(V)$, ... de [8] pour les groupes classiques associés à V. Soit π un homomorphisme d'un groupe $\widetilde{G}$ sur un groupe G de façon que le noyau de π soit contenu dans le centre de $\widetilde{G}$. On dit dans ce cas que $(\pi, \widetilde{G})$ est une extension centrale de G.

3. *Les groupes G, Γ et Δ*

Soit $\mathfrak{g}_C$ une algèbre de Lie simple sur le corps complexe, Σ l'ensemble des racines relatif à une décomposition de Cartan de $\mathfrak{g}_C$, et $\{X_r, H_r \mid r \in \Sigma\}$ un ensemble de générateurs qui satisfont aux équations de structure de [1, p. 24, Th. 1]. On peut remplacer le corps C des coefficients par un corps arbitraire K [1, p. 32], et obtenir une algèbre $\mathfrak{g}$ pour laquelle on peut définir, de façon naturelle, l'automorphisme $exp\ ad\ tX_r$ $(r \in \Sigma,\ t \in K)$. Le groupe engendré par tous ces automorphismes sera désigné par G (c'est G' dans [1]). A part quatre exceptions, que nous excluons dans ce qui suit, G est simple [1, p. 63, Th. 3]. Si K est algébriquement clos, on obtient un ensemble complet de groupes algébriques simples sur K [8], et si K est le corps complexe, un ensemble complet de groupes de Lie simples avec des paramètres complexes. Parmi les relations de G, on a celles de [1, p. 33] qui se déduisent de celles ci-dessous par la substitution de $exp\ ad\ tX_r$ pour $x_r(t)$.

(A) $$x_r(t)\, x_r(u) = x_r(t+u) \qquad (r \in \Sigma;\ t,\ u \in K)$$

(B) $$(x_r(t), x_s(u)) = {}^{(*)}\Pi x_{ir+js}(c_{ij,rs} t^i u^j) \qquad (r, s \in \Sigma,\ r+s \neq 0).$$

Ici (x, y) désigne le commutateur $xyx^{-1}y^{-1}$, le produit s'étend à tous les entiers i, j pour lesquels $ir + js \in \Sigma$, les termes étant arrangés dans un ordre lexicographique, et les $c_{ij,rs}$ sont certains entiers qui ne dépendent que de la structure de g_C et non de t ou u. Nous définissons $w_r(t) = x_r(t)\, x_{-r}(-t^{-1})\, x_r(t)$ et $h_r(t) = w_r(t)\, w_r(1)^{-1} = w_r(t)\, w_r(-1)$ $(t \in K^*)$, et considérons aussi les relations suivantes imposées aux $x_r(t)$:

(B') $$w_r(t)\, x_r(u)\, w_r(t)^{-1} = x_{-r}(-t^{-2}u) \qquad (r \in \Sigma,\ t \in K^*,\ u \in K).$$

3.1. Théorème. *Soit* Δ *le groupe abstrait engendré par les symboles* $x_r(t)$ $(r \in \Sigma,\ t \in K)$, *sujets aux relations* (A) *et* (B) *si* $rk\ \Sigma > 1$, *et aux relations* (A) *et* (B') *si* $rk\ \Sigma = 1$ (rk veut dire rang). (a) *Soit* π *la projection canonique de* Δ *sur* G. *Alors* (π, Δ) *est une extension centrale de* G. (b) *On a* $\Delta' = \Delta$.

On a aussi une décomposition de Bruhat de Δ (voir 7.5 et 7.6 ci-dessous), mais on ne connaît pas exactement la structure de Δ (excepté si K est une extension algébrique d'un corps fini (voir 3.2 et 3.3)). A cause des propriétés de Δ démontrées ci-dessous (voir n° 4), la nature exacte de sa structure serait d'un grand intérêt (à l'auteur). Nous allons aussi considérer les relations suivantes imposées aux $x_r(t)$:

(C) $$h_r(t)\, h_r(u) = h_r(tu) \qquad (r \in \Sigma;\ t, u \in K^*).$$

(*) le Π majuslule désigne *toujours* un produit.

3.2. THÉORÈME. *Soit Γ le groupe abstrait engendré par les $x_r(t)$ sujets aux relations de Δ (voir 3.1.) et celles de* (C). (a) *Soit π la projection canonique de Γ sur G. Alors (π, Γ) est une extension centrale de G ayant un centre fini. En plus, si K est algébriquement clos Γ est, d'une façon naturelle, isomorphe au groupe simplement connexe, groupe de revêtement de G, ce dernier étant considéré comme groupe algébrique, et si K est arbitraire, Γ a une structure correspondante.* (b) *On a $\Gamma' = \Gamma$.*

Observations. 1. Dickson [3] a employé des relations semblables à (A),(B) et (C) pour définir les groupes Γ du type A_n et C_n, c'est-à-dire $SL(n+1, K)$ et $Sp(2n, K)$. Ses calculs, en particulier pour $Sp(2n, K)$, sont assez compliqués. 2. La définition de Γ donnée dans [13, n° 3], quoique plus restrictive en apparence, est équivalente à celle donnée ci-dessus, comme nous allons le démontrer.

3.3. THÉORÈME. *Si K est une extension algébrique d'un corps fini, les groupes Δ de 3.1 et Γ de 3.2 sont canoniquement isomorphes. C'est-à-dire, le groupe Γ peut être défini par les relations* (A) *et* (B) *seules, si $rk\,\Sigma > 1$, et par les relations* (A) *et* (B′) *seules, si $rk\,\Sigma = 1$.*

Exemple. Si K est une extension algébrique d'un corps fini, le groupe $SL(n, K)$ $(n \geqslant 3)$ est défini abstraitement par les générateurs $x_{ij}(t)$ $(1 \leqslant i, j \leqslant n,\ i \neq j,\ t \in K)$, sujets aux relations $x_{ij}(t)\, x_{ij}(u) = x_{ij}(t+u)$, et sous la condition $i \neq l$, aux relations $(x_{ij}(t), x_{kl}(u)) = x_{il}(tu)$ ou 1, selon que $j = k$ ou non.

Tout au cours de cet exposé nous utiliserons les notations $\Sigma, G, \Gamma, \Delta, \ldots$ introduites ci-dessus. En plus, $c(r, s)$ désigne l'entier de Cartan $2(r, s)/(s, s)$.

4. *Le groupe Δ comme revêtement de G*

4.1. THÉORÈME. *Soit $|K| > 4$, si $rk\,\Sigma > 1$, et soit $|K| \neq 4{,}9$, si $rk\,\Sigma = 1$. Soit π la projection canonique de Δ sur G.* (a) *Alors (π, Δ) est une extension centrale de G possédant la propriété suivante : si (π_1, Δ_1) est une extension centrale d'un groupe G_1, et $\bar{\varrho}$ est un homomorphisme de G dans G_1, il existe un homorphisme unique ϱ de Δ dans Δ_1 tel que $\bar{\varrho}\pi = \pi_1\varrho$.* (b) *La propriété de* (a) *caractérise (π, Δ) : soit $(\tilde{\pi}, \tilde{\Delta})$ une extension centrale de G telle que* (a) *soit satisfait avec $(\tilde{\pi}, \tilde{\Delta})$ au lieu de (π, Δ); alors il existe un isomorphisme unique ϱ de Δ sur $\tilde{\Delta}$ tel que $\tilde{\pi}\varrho = \pi$.*

Au lieu de 4.1 (a) on a un résultat plus général.

4.2. THÉORÈME. *Soit* K *le même que dans* 4.1, *soit* (π_1, Δ_1) *une extension centrale d'un groupe* G_1, *et soit* $\bar{\sigma}$ *un homomorphisme de* Δ *dans* G_1. *Alors il existe un homomorphisme unique* σ *de* Δ *dans* Δ_1 *de façon que* $\bar{\sigma} = \pi_1 \sigma$.

4.3. COROLLAIRE. *Soit* K *le même que dans* 4.1. *Alors chaque extension centrale de* Δ *est triviale : si* (π_1, Δ_1) *est une extension centrale de* Δ, Δ_1 *est le produit direct du noyau de* π_1 *et d'un groupe isomorphe à* Δ *par* π_1.

4.4. COROLLAIRE. *Soit* K *le même que dans* 4.1. *Chaque représentation projective de* G (*ou* Γ *ou* Δ) *est, de façon unique, la projection d'une représentation linéaire de* Δ.

La propriété 4.4 caractérise aussi (π, Δ) parmi toutes les extensions centrales de G. En langage de Schur [6, p. 27, 1.2] le «multiplikator» de G est isomorphe au centre A de Δ, et Δ a un «multiplikator trivial». En language homologique, $H_2(G, \mathbb{Z})$ est isomorphe à A, et $H_2(\Delta, \mathbb{Z})$ est trivial, C étant le corps complexe. Par des méthodes complètement différentes de celles ci-dessous, Shur [6, p. 119, IX] a démontré 4.4 pour K fini et $rk\, \Sigma = 1$, c'est-à-dire, quand $\Delta = \Gamma = SL(2, K)$ et $G = PSL(2, K)$, et a montré que $SL(2, 4)$ (groupe de l'icosaèdre) et $SL(2, 9)$ (une extension centrale du groupe alterné sur six lettres) sont de véritables exceptions [7, p. 170, II]. Si $rk\, \Sigma > 1$ et $|K| = 2$, 3 ou 4, il y a d'autres exceptions, par exemple $\Gamma = SL(3,2) \cong PSL(2,7)$, mais on peut démontrer qu'il n'y en a qu'un nombre fini. Ces exceptions existent dans 4.4 seulement si la caractéristique du corps de représentation est différente de celle de K, car on a :

4.5. THÉORÈME. *Si* K *est arbitraire, chaque représentation projective de* G (*ou* Γ *ou* Δ) *sur un espace de même caractéristique que* K *est, d'une façon unique, la projection d'une représentation linéaire de* Δ.

5. *Le groupe* Γ *comme revêtement de* G

D'abord soit K une extension algébrique d'un corps fini. Alors par 3.3, on a les résultats précédents avec Γ au lieu de Δ.

On dit qu'une extension centrale (π_1, Γ_1) d'un groupe G_1

est de type fini, s'il existe un entier positif n tel que, pour chaque x dans le noyau de π_1, on ait $x^n = 1$.

5.1. Théorème. *Soit* $|K| \neq 2$ *et soit* K *clos pour l'opération d'extraction des racines. Soit* π *la projection canonique de* Γ *sur* G. (a) *Alors* (π, Γ) *est une extension centrale de type fini de* G *ayant la propriété suivante : si* (π_1, Γ_1) *est une extension centrale de type fini d'un groupe* G_1, *et* $\bar{\varrho}$ *est un homomorphisme de* G *dans* G_1, *il existe un homomorphisme unique* ϱ *de* Γ *dans* Γ_1 *de façon que* $\bar{\varrho}\pi = \pi_1\varrho$. (b) *Si* $(\tilde{\pi}, \tilde{\Gamma})$ *est une extension centrale de type fini de* G *tel que* (a) *soit satisfait pour* $(\tilde{\pi}, \tilde{\Gamma})$ *au lieu de* (π, Γ), *il existe un isomorphisme unique* ϱ *de* Γ *sur* $\tilde{\Gamma}$ *de façon que* $\tilde{\pi}\varrho = \pi$.

Le résultat 5.1 (a) sera une conséquence de :

5.2. Théorème. *Soit* K *le même que dans* 5.1, *soit* (π_1, Γ_1) *une extension centrale du type fini d'un groupe* G_1, *et soit* $\bar{\sigma}$ *un homomorphisme de* Γ *dans* G_1. *Alors il existe un homomorphisme unique* σ *de* Γ *dans* Γ_1 *de façon que* $\bar{\sigma} = \pi_1\sigma$.

5.3. Corollaire. *Soit* K *le même que dans* 5.1. *Alors chaque extension centrale de type fini de* Γ *est triviale.*

5.4. Corollaire. *Soit* K *le même que dans* 5.1. *Alors chaque représentation projective de dimension finie de* G (*ou* Γ) *est d'une façon unique la projection d'une représentation linéaire de* Γ.

Ceci nous donne une autre démonstration du résultat 4.4 de [13].

5.5. Corollaire. *Soit* $\tilde{\Gamma}$ *un groupe de Lie semi-simple, simplement connexe avec des paramètres complexes, ou un groupe algébrique, semi-simple, simplement connexe sur un corps algébriquement clos. Alors chaque représentation projective de dimension finie de* $\tilde{\Gamma}$ *est, d'une façon unique, la projection d'une représentation linéaire.*

Notez qu'il n'y ait aucune hypothèse sur la continuité ou rationalité des représentations dans 5.5. La première partie de 5.5 dépend naturellement du fait que Γ est simplement connexe comme groupe topologique quand K est le corps complexe, et en effet nos méthodes peuvent être utilisées à en donner une preuve simple. Mais comme c'est un résultat bien connu, nous l'omettons. D'autre part, si K est le corps réel, Γ n'est en général pas simplement connexe. A cause de 4.2 (cf. aussi [2, p. 50, Prop. 2]), cela prouve que Γ n'est pas toujours isomorphe à Δ.

6. *Algèbres de Lie classiques*

A part quelques exceptions [12], l'algèbre $\mathfrak{g}$ du nº 3 est simple, et elle possède des propriétés tout à fait analogues à celles de Γ et Δ.

6.1. Théorème. *Soit* $\mathfrak{g}$ *l'algèbre simple du nº 3, K de caractéristique* $\neq 2,3$, *et* $rk\,\Sigma > 1$. (a) *Parmi les équations de structure pour* $\mathfrak{g}$, *celles de la forme* $[X_r, X_s] = N_{rs} X_{r+s}$ $(r, s \varepsilon \Sigma, r + s \neq 0)$ *entraînent toutes les autres, et pour cette raison fournissent une définition abstraite de* $\mathfrak{g}$. (b) $\mathfrak{g}' = \mathfrak{g}$. (c) *Si* $(\pi, \mathfrak{h})$ *est une extension centrale d'une algèbre de Lie* $\mathfrak{h}_1$, *et* $\bar{\sigma}$ *est un homomorphisme de* $\mathfrak{g}$ *dans* $\mathfrak{h}_1$, *il existe un homomorphisme unique* σ *de* $\mathfrak{g}$ *dans* $\mathfrak{h}$ *tel que* $\bar{\sigma} = \pi\sigma$. (d) *Chaque extension centrale de* $\mathfrak{g}$ *est triviale.*

Si K est de caractéristique 0, la partie (d) est une conséquence du théorème de réductibilité complète pour les représentations de $\mathfrak{g}$ [9, p. 7-02], mais si K n'est pas de caractéristique 0, on n'a pas ce dernier théorème.

II. Démonstrations

7. *Structure de* Δ

Dans ce numéro, π désigne la projection canonique de Δ sur G.

7.1. *Soit* Σ_1 *un ensemble des racines tel que* (a) *les éléments de* Σ_1 *sont tous positifs relatif à quelque ordination de* Σ, *et* (b) *si* $r, s \varepsilon \Sigma_1$ *et* $r + s \varepsilon \Sigma$, *alors* $r + s \varepsilon \Sigma_1$. *Alors le sous-groupe* U_1 *de* Δ, *engendré par les* $x_r(t)$ $(r \varepsilon \Sigma_1, t \varepsilon K)$, *est isomorphe par* π *au sous-groupe correspondant de* G.

C'est que les relations (A) et (B) permettent de réduire tout élément de U_1 à la forme $\Pi_{r \varepsilon \Sigma_1} x_r(t_r)$, tandis que dans $\pi(U_1)$ la forme correspondante est unique [1, p. 39].

On désigne par W le groupe de Weyl de Σ, et par σ_r la symétrie correspondante à la racine r.

7.2. *Soit* $r, s \varepsilon \Sigma$; $s' = \sigma_r s$; $c = c(s', r)$; $t, u \varepsilon K^*$. *Alors pour*

$\eta = \pm 1$, *indépendant de t et u, on a* $w_r(t)\, x_s(u)\, w_r(-t) = x_{s'}(\eta t^c u)$ *dans* Δ.

Soit d'abord $rk\, \Sigma > 1$. Soit $r \neq \pm s$. Soit Σ_1 l'ensemble des racines de la forme $ir + js$ (i, j entiers; $j > 0$), et U_1 le sous-groupe de Δ engendré par les $x_a(t)$ ($a \,\varepsilon\, \Sigma_1$, $t \,\varepsilon\, K$). De (A) et (B) on déduit $x_r(t)\, U_1 x_r(-t) \subseteq U_1$ et $x_{-r}(-t^{-1})\, U_1 x_{-r}(t^{-1}) \subseteq U_1$, d'où $w_r(t)\, x_s(u)\, w_r(-t) \,\varepsilon\, U_1$. On a aussi $x_{s'}(\eta t^c u) \,\varepsilon\, U_1$. Comme $\pi(w_r(t)\, x_s(u)\, w_r(-t)) = \pi(x_{s'}(\eta t^c u))$ ($\eta = \pm 1$ est indépendant de t et u) est une conséquence des équations de structure de G [1, p. 37], 7.1 entraîne 7.2. Puis soit $r = s$. Parmi les relations (B), il y'en a une pour laquelle le terme $x_r(u)$ apparaît dans le produit Π: $(x_a(t'), x_b(u')) = x_r(u)\, \Pi'$. Soit Σ_2 l'ensemble des racines de la forme $i\sigma_r a + j\sigma_r b$ (i, j entiers positifs), et soit U_2 le sous-groupe correspondant de Δ. Transformant la dernière équation ci-dessus par $w_r(t)$, et employant la première partie de la démonstration, on a $w_r(t)\, x_r(u)\, w_r(-t) \,\varepsilon\, U_2$. Comme on a aussi $x_{-r}(-t^{-2}u) \,\varepsilon\, U_2$, on déduit $w_r(t)\, x_r(u)\, w_r(-t) = x_{-r}(-t^{-2}u)$ à l'aide de 7.1 et de l'équation correspondante dans G. Comme $w_r(t)^{-1} = w_r(-t)$, le cas $r = -s$ de 7.2 découle du cas $r = s$.

Si $rk\, \Sigma = 1$, alors 7.2 est une conséquence des relations (B′).

Grâce aux définitions de $w_s(u)$ et $h_s(u)$, on déduit immédiatement de 7.2 :

7.3. *Employant les mêmes notations qu'au* 7.2 *on a dans* Δ :

(*a*) $\quad w_r(t)\, w_s(u)\, w_r(-t) = w_{s'}(\eta t^c u)$

(*b*) $\quad w_r(t)\, h_s(u)\, w_r(-t) = h_{s'}(\eta t^c u)\, h_{s'}(\eta)^{-1}$.

Soit $d = c(s, r)$; *alors* :

(*c*) $\quad h_r(t)\, x_s(u)\, h_r(t)^{-1} = x_s(t^d u)$

(*d*) $\quad h_r(t)\, w_s(u)\, h_r(t)^{-1} = w_s(t^d u)$

(*e*) $\quad h_r(t)\, h_s(u)\, h_r(t)^{-1} = h_s(t^d u)\, h_s(t^d)^{-1}$.

7.4. Lemme. *Le groupe de Weyl W est engendré par les* σ_r *sujets aux relations* $\sigma_r^2 = 1$ *et* $\sigma_r \sigma_s \sigma_r^{-1} = \sigma_{s'}$ (r, $s \,\varepsilon\, \Sigma$; $s' = \sigma_r s$).

L'on sait [8, p. 14-08] que W est déterminé par les relations $\sigma_r^2 = 1$ et $(\sigma_r \sigma_s)^n = 1$ (r, $s \,\varepsilon\, \Sigma$, n l'ordre de $\sigma_r \sigma_s$ dans W), que l'on peut écrire $(\sigma_r \sigma_s)^{n/2} \sigma_s (\sigma_r \sigma_s)^{-n/2} = \sigma_s$, si n est pair, ou $(\sigma_r \sigma_s)^{(n-1)/2} \sigma_r (\sigma_r \sigma_s)^{-(n-1)/2} = \sigma_s$, si n est impair.

7.5. *Soit* $\mathfrak{W}$ *le sous-groupe de* Δ *engendré par tous les* $w_r(t)$, *et H le sous-groupe engendré par tous les* $h_r(t)$. *Alors H est un sous-*

groupe normal de $\mathfrak{W}$, *et il y a un isomorphisme* φ *de* $\mathfrak{W}/H$ *sur* W, *tel que* $\varphi(Hw_r(t)) = \sigma_r (r \,\varepsilon\, \Sigma,\ t \,\varepsilon\, K^*)$.

H est normal dans $\mathfrak{W}$ à cause de 7.3 (d). A cause de la structure de $\pi(\mathfrak{W})$ [1, p. 37], on a un homomorphisme de $\mathfrak{W}$ sur W et par suite aussi un homomorphisme φ de $\mathfrak{W}/H$ sur W tel que $\varphi(Hw_r(t)) = \sigma_r$. Soit $\psi(\sigma_r) = Hw_r(t)$ $(r \,\varepsilon\, \Sigma)$. Comme $w_r(t) = h_r(t) w_r(1)$ ceci est indépendant de t. On a $\psi(\sigma_r)^2 = Hw_r(1)^2 = Hh_r(-1)^{-1} = H$, et, employant la notation de 7.3 (a), $\psi(\sigma_r)\psi(\sigma_s)\psi(\sigma_r)^{-1}\,\psi(\sigma_{s'})^{-1} = Hw_r(1)\,w_s(1)\,w_r(-1)\,w_{s'}(-1) = H$, par 7.3 (a). Il suit de 7.4 que l'on peut étendre ψ à un homomorphisme de W sur $\mathfrak{W}/H$. On a $\psi\varphi = 1$, d'où φ est un isomorphisme.

Soit U le groupe engendré par les $x_r(t)$ $(r > 0)$ relatif à une ordination de Σ. De plus, pour chaque $\sigma \,\varepsilon\, W$, soit U_σ le groupe engendré par les $x_r(t)$ $(r > 0,\ \sigma r < 0)$, et soit $w(\sigma)$ un élément de W tel que $\varphi(w(\sigma)) = \sigma$.

7.6. Forme normale de Bruhat. *Chaque élément de* Δ *peut être écrit d'une façon unique* $x = uhw(\sigma)\,u_1$ $(u \,\varepsilon\, U,\ h \,\varepsilon\, H,\ \sigma \,\varepsilon\, W,\ u_1 \,\varepsilon\, U_\sigma)$.

Soit S l'ensemble des racines positives, simples. Les relations (A), (B), 7.2 et 7.3 nous permettent de démontrer, exactement comme dans [1, p. 40-42], que l'ensemble des éléments de la forme 7.6 est clos pour la multiplication par $x_a(t)$ et $w_a(1)$ $(a \,\varepsilon\, S)$, qui engendrent Δ par 7.2. Cet ensemble comprend par conséquent tous les éléments de Δ. Si l'on écrit x dans la forme 7.6, alors, grâce à l'unicité de la forme correspondante dans G, x détermine σ, $\pi(u)$ et $\pi(u_1)$ et par suite aussi u et u_1 à cause de 7.1.

On peut simplifier la structure de H.

7.7. *Pour chaque* $a \,\varepsilon\, S$ *(l'ensemble des racines simples), soit* h_a *le sous-groupe de* H *engendré par tous les* $h_a(t)$. *Alors tout* h_a *est normal dans* H, *et* $H = \Pi_{a \varepsilon S}\, h_a$.

A cause de 7.3 (a), h_a est normal dans H. Soit $s \notin S$, $s \,\varepsilon\, \Sigma$, $s > 0$. Il existe $a \,\varepsilon\, S$, tel que $(s, a) > 0$, de façon que $\sigma_a s = s'$ soit de hauteur moins grande que celle de s. Soit $c = c(a, s)$; on a $h_s(t)\,h_a(t^{-c}) = h_s(t)\,w_a(t^{-c})\,w_a(-1) = w_a(1)\,h_s(t)\,w_a(-1)$ (par 7.3 (d)) $= h_{s'}(\eta t)\,h_{s'}(\eta)^{-1}$ (par 7.3 (b)). Par conséquent, par récurrence sur la hauteur de s, on a $h_s(t) \,\varepsilon\, \Pi h_a$. Si $s < 0$, on a par 7.3 (a),

$$h_s(t) = w_s(t)\,w_s(-1) = w_{-s}(1)\,w_{-s}(-t)\,w_{-s}(1)\,w_{-s}(-1) = h_{-s}(t)^{-1} \,\varepsilon\, \Pi h_a.$$

Ceci nous permet de prouver 3.1. (a) Soit $\pi x = 1$, $x = uhw(\sigma)u_1$ de la forme de 7.6. On a $u = 1$, $\sigma = 1$ et $u_1 = 1$ à cause de 7.1 et de la forme normale de G. C'est pourquoi $x \varepsilon H$. Soit $x = \Pi h_s(t_s)$, le produit s'étendant à une suite (s) de racines. Alors $xx_r(u)x^{-1} = x_r(\Pi_s t_s^{c(r,s)} u)$ par 7.3 (c). Comme $\pi x = 1$, on a $\Pi_s t_s^{c(r,s)} = 1$ à cause de 7.1. Alors $xx_r(u)\, x^{-1} = x_r(u)$, x est dans le centre de Δ, et (π, Δ) est une extension centrale de G. (b) Si $|K| > 3$, il existe $k \varepsilon K^*$ tel que $k^2 \neq 1$. Alors $(h_r(k), x_r(u)) = x_r((k^2 - 1)\, u)$, d'où Δ' contient tous les $x_r(t)$. En particulier ceci donne la démonstration du cas $rk\, \Sigma = 1$. Soit $rk\, \Sigma > 1$. Si r peut s'écrire $r = q + s$ avec q, r et s de même longueur et $N_{qs} \neq 0$ (notation de 6.1 (a)), alors une des relations (B) donne $x_r(t) \varepsilon \Delta'$. Restent à considérer les cas pour lesquels r est une longue racine (respectivement courte racine), et Σ est de type C_n (respectivement B_n). Au premier cas, on a $r = a + 2b$ $(a, b \varepsilon \Sigma)$ et $(x_a(t), x_b(u)) = x_{a+b}(\varepsilon tu) x_{a+2b}(\eta tu^2)$ $(\varepsilon, \eta = \pm 1)$, par une des relations (B). Si $rk\, \Sigma > 2$, on a $x_{a+b}(\varepsilon\, tu) \varepsilon \Delta'$ et par conséquent $x_{a+2b}(\eta tu^2) \varepsilon \Delta'$, tandis que si $rk\, \Sigma = 2$, on n'a qu'à considérer $|K| = 3$, et remplaçant $t \to -t$, $u \to -u$, on obtient $x_{a+2b}(2\eta tu^2) \varepsilon \Delta'$. Le second cas peut être traité de façon analogue. Par conséquent $\Delta' = \Delta$.

Pour finir cette section nous démontrons 3.3. Nous prenons $r \varepsilon \Sigma$ fixe et posons $x(t) = x_r(t)$, $y(t) = x_{-r}(t)$, $h(t) = h_r(t)$. Pour chaque t, $u \varepsilon K^*$, $h(tu)\, h(t)^{-1}\, h(u)^{-1}$ est dans le centre de Δ, et t et u engendrent un sous-groupe cyclique de K^*. Par conséquent les $h(t)$ engendrent un groupe abélien. Par 7.3 (e), $h(t) = h(u)\, h(t)\, h(u)^{-1} = h(tu^2)\, h(u^2)^{-1}$, d'où $h(tu^2) = h(t)\, h(u^2)$. Nous allons prouver dans le paragraphe suivant que $h(v - v^2) = h(v)\, h(1 - v)$. Si non pas chaque élément de K est un carré, on peut choisir v tel que v et $1 - v$ ne soient pas carrés : soit $\bar{v}$ le premier parmi 1,2 ... qui ne soit pas un carré, puis posons $v = \bar{v}^{-1}$. Soit $t_1 = v$, $u_1 = 1 - v$. Nous avons déjà démontré (C) si l'un des t ou u est un carré. Si tous les deux ne sont pas des carrés, on a $h(t)\, h(u) = h(t/t_1)\, h(t_1)\, h(u_1)\, h(u/u_1) = h(t/t_1)\, h(t_1 u_1)\, h(u/u_1) = h(tu)$.

Reste à prouver que $h(v - v^2) = h(v)\, h(1 - v)$. Il n'est pas nécessaire de faire l'hypothèse du 3.3 concernant K. Soit $w_r(t) = w(t)$. Alors $w_{-r}(t) = w_r(-t^{-1})$ par 7.3 (a). Donc, grâce à 7.2 et aux définitions, $h(v - v^2)\, h(1 - v)^{-1} = w(v - v^2)\, w(1 - v)^{-1}$
$= x(v)\, y(v^{-1}(v-1)^{-1})\, x(-v(v-1))\, y((v-1)^{-2})\, w(1-v)^{-1}$
$= x(v)\, y(v^{-1}(v-1)^{-1})\, w(1-v)^{-1}\, y(v(v-1)^{-1})\, x(-1)$
$= x(v)\, y(-v^{-1})\, x(v-1)\, y(1)\, x(-1) = w(v)\, w(-1) = h(v)$.

8. *Structure de Γ*

Dans cette section, les symboles $x_r(t)$, $h_r(t)$, etc. désignent des éléments de Γ, et π désigne la projection canonique de Γ sur G. A cause de 7.2, il suffit de supposer que dans la définition de Γ pour chaque longueur de racine possible, il y a une racine r qui satisfait aux relations (C).

8.1. *Si* $s \in \Sigma$, *il existe un homomorphisme* φ_s *de* $SL(2, K)$ *dans* Γ *tel que* $\varphi_s\begin{pmatrix}1 & t\\ 0 & 1\end{pmatrix} = x_s(t)$ et $\varphi_s\begin{pmatrix}1 & 0\\ t & 1\end{pmatrix} = x_{-s}(t)$.

En effet les relations (A), 7.3 (a) et (C) avec $r = \pm s$ donnent une définition de $SL(2\ K)$, car elles permettent la réduction à la forme normale de Bruhat. Donc la définition de Γ donnée dans [13, nº 3] est équivalente à celle du nº 3 ci-dessus.

Par (C) et 7.3 (e), le groupe H engendré par les $h_r(t)$ est abélien, et par 7.7 et (C), chaque élément peut s'écrire $\Pi_{a \in S} h_a(t_a)$. Nous démontrons que les t_a sont déterminés de façon unique. Il suffit pour cela de présenter un groupe qui possède des générateurs satisfaisant aux relations (A), (B) et (C), et tel que les t_a y soient déterminés de façon unique. Il est évident qu'un tel groupe sera canoniquement isomorphe au groupe abstrait Γ. Pour Σ de type A_n, C_n, B_n, D_n respectivement, on a le groupe $SL(n + 1)$, $Sp(2n)$, $Spin\ (2n + 1)$, $Spin\ (2n)$, les deux derniers groupes étant définis relatif à une forme d'index maximal. Pour Σ de type E_8, F_4, G_2, on a le groupe simple G même, tandis que pour Σ de type E_6 ou E_7, on peut prendre un sous-groupe convenable du groupe G de type E_8. Du fait que le noyau de φ_a ($a \in S$) est contenu dans le centre de $SL(2, K)$, et aussi $h_a(-1) \neq 1$, si $-1 \neq 1$, on déduit que les φ_a sont des isomorphismes, et par 7.2 sont donc tous les φ_s.

8.2. (a) *Dans* Γ *on a la décomposition* 7.6 (*mais maintenant* H, h_a, U et U_σ *désignent des sous-groupes de* Γ), *et en plus* : H *est abélien et* $h = \Pi h_a(t_a)$, *les* t_a *étant déterminés par* h *d'une façon unique.* (b) *Chaque* φ_s *dans* 8.1 *est un isomorphisme.*

Nous allons maintenant démontrer 3.2. Le centre de Γ consiste en tous les éléments de la forme $\Pi h_a(t_a)$ tels que $\Pi t_a^{c(r,a)} = 1$ pour tous les $r \in \Sigma$; par suite il est fini; en plus il est le noyau de π, d'où la première partie de 3.2 (a). Pour la seconde, voir la discussion qui précède 8.2. De 3.1 (b) on obtient 3.2 (b).

9. *Démonstration de* 4.2

Soit C le centre de Δ_1. Soit k, $c \varepsilon K^*$ tels que $c = k^2 \neq 1$. On a dans Δ, par 7.3 (c),

9.1. $$x_r(t) = (h_r(k),\ x_r(t/(c-1))).$$

Pour chaque élément $x \varepsilon \Delta$ nous choisissons un élément $\sigma(x)$ dans Δ_1 tel que $\pi_1 \sigma(x) = \bar{\sigma}(x)$, et aussi de façon que

9.2. $$\sigma(x_r(t)) = (\sigma(h_r(k)),\ \sigma(x_r(t/(c-1)))) \quad (r \varepsilon \Sigma,\ t \varepsilon K).$$

Cette normalisation des $\sigma(x_r(t))$ n'est pas circulaire, car un commutateur (x, y) dans Δ_1 ne dépend que des classes mod C de x et y. (Ce fait sera souvent utilisé). Nous allons démontrer les relations (A) et (B) avec les $\sigma(x_r(t))$ au lieu des $x_r(t)$. Nous démontrons ainsi que σ s'étend à un homomorphisme de Δ dans Δ_1 tel que $\bar{\sigma} = \pi_1 \sigma$. L'unicité dans 4.2 est conséquence de 3.1 (b), cette dernière proposition ayant pour résultat que chaque homomorphisme de Δ dans un groupe abélien est trivial (constant). En tout cas σ conserve les relations de Δ à une multiplication par des éléments de C près.

(1) *On a* $\sigma(h)\,\sigma(x_r(t))\,\sigma(h)^{-1} = \sigma(hx_r(t)\,h^{-1})$ $(h \varepsilon H)$. Soit $hx_r(t)\,h^{-1} = x_r(dt)$ $(d \varepsilon K^*)$. Transformant 9.2 par $\sigma(h)$, on a

$$\sigma(h)\,\sigma(x_r(t))\,\sigma(h)^{-1} = (\sigma(hh_r(k)h^{-1}h_r(k)^{-1})\,\sigma(h_r(k)),\ \sigma(x_r(dt/(c-1)))).$$

Par 7.3 (c) on a que $h_1 = hh_r(k)h^{-1}\,h_r(k)^{-1}$ est dans le centre de Δ. C'est pourquoi $\sigma(h_1)$ commute avec tous les $\sigma(x)$ $(x \varepsilon \Delta)$: soit $\sigma(h_1)\,\sigma(x)\,\sigma(h_1)^{-1} = f(x)\,\sigma(x)$ $(f(x) \varepsilon C)$; alors l'application $x \to f(x)$ est un homomorphisme de Γ dans C, qui a pour conséquence $f(x) = 1$ par 3.1 (b). Donc, utilisant 9.2, l'expression ci-dessus peut être mise sous la forme simplifiée suivante, $(\sigma(h_r(k)),\ \sigma(x_r(dt/(c-1)))) = \sigma(x_r(dt))$.

(2) $\sigma(x_r(t))$ *commute avec* $\sigma(x_s(u))$ *si* $r + s \notin \Sigma$ *et* $r + s \neq 0$. Soit $\sigma(x_r(t))\,\sigma(x_s(u))\,\sigma(x_r(t))^{-1} = f(t, u)\,\sigma(x_s(u))$. L'on a

9.3. $$f(t, u) \varepsilon C, \qquad f(t + t', u) = f(t, u)\,f(t', u) \quad et \quad f(t, u + u') = f(t, u)\,f(t, u').$$

(2a) *Soit d'abord* $r \neq s$. Si $(r, s) = 0$, $f(tv^2, u) = f(t, u)$ à cause de 7.3 (c) et (1) avec $h = h_r(v)$. Si $(r, s) > 0$, $f(tv^d, u) = f(t, u)$, avec $d = 4 - c(r, s)\,c(s, r) = 1, 2,$ ou 3, par 7.3 (c) et (1) avec

$h = h_r(v^2)\, h_s(v^{-c(s,r)})$. Donc $f(t(v^d - 1), u) = 1$ par 9.3. Nous pouvons choisir $v \in K^*$ de façon que $v^d - 1 \neq 0$ $(d = 1, 2, 3)$. Comme t et u sont arbitraires, f est identiquement 1, d'où (2).

(2b) *Soit maintenant* $r = s$ *et* $rk\, \Sigma > 1$. S'il existe une racine q telle que $c(r, q) = 1$, par 7.3 (c) et (1) avec $h = h_q(v)$, on a $f(t, u) = f(tv, uv)$. Nous choisissons $v \in K^*$ de façon que $v \neq 1$ et $v^2 - v + 1 \neq 0$, ce qui est possible à cause de $|K| > 4$. Alors par 9.3, $f(t(v - v^2), u) = f(t, u/(v - v^2)) = f(t, u/v) f(t, u/(1-v)) = f(tv, u) f(t - tv, u) = f(t, u)$, d'où $f(t(1 - v + v^2), u) = 1$, et f est identiquement 1. S'il n'existe pas de racine q telle que $c(r, q) = 1$, alors Σ est de type C_n et r est une longue racine. Alors on a $r = a + 2b$ $(a, b \in \Sigma)$, et
$(\sigma(x_a(t)),\ \sigma(x_b(1))) = f \sigma(x_{a+b}(\varepsilon t))\, \sigma(x_{a+2b}(\eta t))$ $(\varepsilon, \eta = \pm 1,$ $f \in C)$. Comme $\sigma(x_{a+2b}(u))$ commute avec $\sigma(x_a(t))$, $\sigma(x_b(1))$ et $\sigma(x_{a+b}(\varepsilon t))$ par (2a), $\sigma(x_{a+2b}(u))$ commute aussi avec $\sigma(x_{a+2b}(\eta t))$.

(2c) *Finalement, soit* $r = s$ *et* $rk\, \Sigma = 1$. Si $|K|$ est un nombre premier, $x_r(t)$ et $x_r(u)$ sont des puissances de $x_r(1)$, ce qui entraîne $f(t, u) = 1$. Supposons que $|K|$ ne soit pas un nombre premier. L'équation (1) avec $h = h_r(v)$ donne $f(t, u) = f(tv^2, uv^2)$. On peut choisir $v \in K^*$ de façon que $v^2 + 1 \neq 0$, $v^4 + v^2 + 1 \neq 0$, et v^2+1 soit un carré : si K est de caractéristique 2, on n'a qu'à satisfaire $v^3 \neq 1$, ce qui est possible parce que $|K| > 4$; si K n'est pas de caractéristique 2, on pose $v = 2w/(w^2 - 1)$; il ne faudra éviter pour le choix de w que 12 éléments de K^*, ce qui est possible parce que $|K| \geqslant 25$ dans ce cas. Alors par 9.3,

$$\begin{aligned} f(tv^2,\ u(v^2 + 1)^2) &= f(tv^2,\ u(v^2 + 1)) f(tv^2,\ u(v^4 + v^2)) \\ &= f(tv^2,\ u(v^2 + 1)) f(t,\ u(v^2 + 1)) = f(t(v^2 + 1),\ u(v^2 + 1)) \\ &= f(t(v^2 + 1)^2, u(v^2 + 1)^2), \text{ d'où } f(t(v^4 + v^2 + 1), u(v^2 + 1)^2) = 1 \end{aligned}$$

Comme t et u sont arbitraires, f est identiquement 1.

(3) *Les relations* (A) *sont préservées par* σ. L'élément $\sigma(x_r(t/(c - 1)))\, \sigma(x_r(u/(c - 1)))\, \sigma(x_r((t + u)/(c - 1)))^{-1}$ est dans C, donc est égal à son transformé par $\sigma(h_1(k))$ (voir 9.1). Par 9.2, (1) et (2), ce transformé est égal à l'élément lui-même multiplié par $\sigma(x_r(t))\, \sigma(x_r(u))\, \sigma(x_r(t + u))^{-1}$, d'où (3).

(4) *Les relations* (B) *sont préservées par* σ. A partir de (B) on obtient $\sigma(x_r(t))\, \sigma(x_s(u))\, \sigma(x_r(t))^{-1}$
$= f(t, u)\, \Pi \sigma(x_{ir+js}(c_{ij,rs} t^i u^j)) \sigma(x_s(u))$, avec $f(t, u) \in C$. On démontre $f(t, u) = 1$ par récurrence sur $n(r, s)$, ce nombre étant le nombre

des racines de la forme $ir + js$ dans (B). Si $n(r, s) = 0$, on a $f(t, u) = 1$ par (2) ci-dessus, tandis que si $n(r, s) > 0$, on obtient 9.3 par l'hypothèse de récurrence, et puis on prouve (4) de la même façon que (2a) ci-dessus.

Finalement, si $rk\,\Sigma = 1$, les relations (B′) sont préservées par σ; la démonstration est l'analogue de celle de (1). Nous avons achevé de démontrer 4.2.

10. *Démonstration des autres résultats du n*° 4

De 3.1 (a) et 4.2 avec $\bar{\sigma} = \bar{\varrho}\pi$, découle 4.1 (a). Employant la notation de 4.1 (b), nous montrons d'abord $\widetilde{\Delta}' = \widetilde{\Delta}$. Soit Δ_1 le groupe $\widetilde{\Delta}/\widetilde{\Delta}'$, et π_1 et $\bar{\varrho}$ les homomorphismes de Δ_1 et G, respectivement, sur le groupe d'un élément, G_1. Comme $(\widetilde{\pi}, \widetilde{\Delta})$ possède la propriété de 4.1 (a), le seul homomorphisme ϱ de $\widetilde{\Delta}$ dans Δ_1 est trivial. Par conséquent $\widetilde{\Delta}' = \widetilde{\Delta}$. Comme (π, Δ) et $(\widetilde{\pi}, \widetilde{\Delta})$ possèdent la propriété de 4.1 (a), il existe un homomorphisme ϱ de Δ dans $\widetilde{\Delta}$ et un homomorphisme $\widetilde{\varrho}$ de $\widetilde{\Delta}$ dans Δ tel que $\widetilde{\pi}\varrho = \pi$ et $\pi\widetilde{\varrho} = \widetilde{\pi}$. Pour $x \in \Delta$, on a $\pi\widetilde{\varrho}\varrho x = \pi x$. Donc $\widetilde{\varrho}\varrho x = f(x)x$, avec $f(x)$ dans le centre de Δ. L'application $x \to f(x)$ est un homomorphisme, et $\Delta' = \Delta$. C'est pourquoi $f(x) = 1$, d'où $\widetilde{\varrho}\varrho = 1$. D'une façon analogue $\varrho\widetilde{\varrho} = 1$. Donc ϱ est un isomorphisme de Δ sur $\widetilde{\Delta}$, et l'on a 4.1 (b).

Nous rapportant à la notation de 4.3, nous avons par 4.2 un homomorphisme σ de Δ dans Δ_1 tel que $\pi_1\sigma$ est l'identité. Ceci entraîne évidemment que σ est un isomorphisme et que Δ_1 est le produit direct de $\sigma\Delta$ et du noyau de π_1.

Soit V un espace vectoriel. Remplaçant Δ_1 dans 4.1 par le groupe $GL(V)$ et π_1 par la projection canonique sur $PL(V)$, nous avons 4.4.

Pour démontrer 4.5, il suffit de considérer des représentations de Δ. Soit $\bar{\sigma}$ une représentation projective de Δ dans $PL(V)$, π la projection canonique de $GL(V)$ sur $PL(V)$, et L le corps de base de V. Par 4.4 on peut supposer que K et L soient de caractéristique $p \neq 0$. Pour chaque $x \in \Delta$, soit $\sigma(x) \in GL(V)$ tel que $\pi\sigma(x) = \bar{\sigma}(x)$. A cause des relations (A), $\sigma(x_r(t))^p$ est un scalaire. Etendant le corps L au corps $\bar{L}$ algébriquement clos, on peut obtenir la normalisation $\sigma(x_r(t))^p = 1$ $(r \in \Sigma,\ t \in K)$. En prenant la p^{ème} puissance de l'équation $\sigma(x_r(t))\,\sigma(x_r(u))\,\sigma(x_r(t))^{-1} = f\sigma(x_r(u))$

125

($f \varepsilon L^*$), on a $f^p = 1, f = 1$, d'où $\sigma(x_r(t))$ commute avec $\sigma(x_r(u))$. On démontre de la même façon que $f = 1$ dans l'équation $\sigma(x_r(t))\,\sigma(x_r(u)) = f\sigma(x_r(t+u))$, et ainsi les relations (A) sont préservées par σ. De manière analogue on voit par récurrence sur le nombre des termes dans Π, que les relations (B) sont préservées par σ, ainsi que les relations (B') pour $rk\ \Sigma = 1$. Donc on peut étendre σ a un homomorphisme de Δ dans $GL(V^{\overline{L}})$. Mais comme $\sigma(\Delta)' = \sigma(\Delta)$, cet homomorphisme est dans $GL(V)$. De $\Delta' = \Delta$ on déduit aussi l'unicité dans 4.5.

11. *Démonstration de* 5.2

Soit C le centre de Γ_1 et n un entier positif tel que $x^n = 1$ pour chaque x dans C. Soit $\sigma(x)$ ($x \varepsilon \Gamma$) tel que $\pi_1\sigma(x) = \bar{\sigma}(x)$, et soit $\sigma(x_r(t))$ normalisé de façon que les équations 9.2 soient satisfaites. A cause de 4.2, les relations (A) et (B) sont préservées par σ. Reste à démontrer qu'il en est de même pour les relations (C). Définissons les $\sigma(h_r(t))$ en fonction des $\sigma(x_r(t))$ de la même façon qu'ont été définis les $h_r(t)$ en fonction des $x_r(t)$, et soit $f(t, u) = \sigma(h_r(tu))\,\sigma(h_r(u))^{-1}\,\sigma(h_r(t))^{-1}$. On a $f(t, u)\ \varepsilon\ C$. Donc $f(t, u) = \sigma(h_r(v))f(t, u)\,\sigma(h_r(v))^{-1} = f(t, uv^2)f(t, v^2)^{-1}$ par 7.3 (e) et (1) de la démonstration de 4.2, d'où $f(t, uv^2) = f(t, u)f(t, v^2)$. Par suite $f(t, v^{2n}) = f(t, v^2)^n = 1$. Grâce aux hypothèses sur K, f est donc identiquement 1, et les relations (C) sont préservées par σ. Donc on a un homomorphisme σ de Γ dans Γ_1, tel que $\pi_1\sigma = \bar{\sigma}$, et il est unique car $\Gamma' = \Gamma$.

12. *Démonstration des autres résultats du n*° 5

La démonstration de 5.1 est la même que celle de 4.1, pourvu qu'on ajoute dans la démonstration de $\tilde{\Gamma}' = \tilde{\Gamma}$ dans 5.1 (b) que la projection de $\tilde{\Gamma}/\tilde{\Gamma}'$ sur le groupe d'un élément est de type fini. Si C est le noyau de $\tilde{\pi}$, on a $\tilde{\Gamma} = \tilde{\Gamma}'\, C$, car $G' = G$, et ceci donne le résultat désiré.

La démonstration de 5.3 est semblable à celle de 4.3.

Pour la démonstration de 5.4, il suffit de prendre pour Γ_1 dans 5.1 le groupe $SL(V)$ sur un espace vectoriel V de dimension finie et pour π_1 la projection canonique sur $PSL(V)$, ce qui est permis à cause de $\Gamma' = \Gamma$. De là l'on obtient 5.5 en prenant pour K

le corps complexe (respectivement un corps algébriquement clos) et en utilisant la classification de [8].

13. *Démonstration de* 6.1

(a) Soit $H_r = [X_r, X_{-r}]$. Grâce à l'identité de Jacobi, $[H_r, X_s] = c(s, r)\, X_s$, d'abord pour $r \neq \pm s$ et puis pour $r = \pm s$, et il n'est pas difficile de vérifier les autres équations [12, p. 1120] de structure de $\mathfrak{g}$. (b) Ceci est une conséquence immédiate des équations (a). (c) On normalise les $\sigma(X_r)$ de façon que $\overline{\sigma}(X) = \pi\sigma(X)$ $(X \,\varepsilon\, \mathfrak{g})$ et $[\sigma(H_r), \sigma(X_r)] = 2\sigma(X_r)$ (cf. 9.2). La démonstration, que σ préserve toutes les relations de (a), est alors analogue à celle dans le n° 9, quoique plus simple. De (c) on déduit (d) comme avant.

BIBLIOGRAPHIE

[1] C. Chevalley : Sur certains groupes simples, *Tôhoku Math. J.*, **7** (1955), 14-66.

[2] C. Chevalley : Theory of Lie Groups, *Princeton* (1946).

[3] L. E. Dickson : The abstract form ..., deux articles, *Quart. J. Math.*, **38** (1907), 141-158.

[4] D. Hertzig : Forms of algebraic groups, *Proc. Amer. Math. Soc.*, **12** (1961), 657-660.

[5] R. Ree : A family of simple groups associated with the simple Lie algebra of type (F_4); 2me article : ... (G_2), *Amer. J.. Math.*, **83** (1961), 401-420; 432-462.

[6] J. Schur : Ueber die Darstellung der endlichen Gruppen durch gebrochene lineare Substitutionen, deux articles, *J. reine u. a. Math.*, **127** (1904), 20-50; **132** (1907), 85-137.

[7] J. Schur : Ueber die Darstellung der symmetrischen und der alternierenden Gruppe durch gebrochene lineare Substitutionen, *J. reine u. a. Math.*, **139** (1911), 155-250.

[8] Séminaire C. Chevalley : Classification des Groupes de Lie Algébriques, deux tomes, Paris (1956-8).

[9] Séminaire «Sophus Lie», Paris (1954-5).

[10] R. Steinberg : Variations on a theme of Chevalley, *Pacific J. Math.*, **9** (1959), 875-891.

[11] R. Steinberg : The simplicity of certain groups, *Pacific J. Math.*, **10** (1960), 1039-1041.

[12] R. Steinberg : Automorphisms of classical Lie algebras, *Pacific J. Math.* **11** (1961), 1119-1129.

[13] R. Steinberg : Representations of algebraic groups, *Nagoya Math. J.*, à paraître.

[14] M. Suzuki : On a class of doubly transitive groups, *Ann. of Math.*, **75** (1962), 105-145.

[15] J. Tits : Les «formes réelles» des groupes de type E_6, *Séminaire Bourbaki*, 162 Paris (1958).

[16] J. Tits : Sur la trialité et certains groupes qui s'en déduisent, *Paris Inst. Hautes Études Sci. Publ. Math.*, **2** (1959), 37-84.

REPRESENTATIONS OF ALGEBRAIC GROUPS

ROBERT STEINBERG*

To Professor Richard Brauer on the occasion of his 60th birthday

§ 1. Introduction

Our purpose here is to study the irreducible representations of semisimple algebraic groups of characteristic $p \neq 0$, in particular the rational representations, and to determine all of the representations of corresponding finite simple groups. (Each algebraic group is assumed to be defined over a universal field which is algebraically closed and of infinite degree of transcendence over the prime field, and all of its representations are assumed to take place on vector spaces over this field.)

To state our first principal result, we observe that relative to a Cartan decomposition of a semisimple algebraic group, there is described in § 5 below (in a somewhat more general context) a standard way of converting an isomorphism on the universal field into one on the group, and that relative to a choice of a set S of simple roots, an irreducible rational projective representation of the group is characterized by a function from S to the nonnegative integers, to be called, together with the corresponding function on the Cartan subgroup of the decomposition, the high weight of the representation [13, Exp. 14 and 15].

1.1 Theorem. *Let G be a semisimple algebraic group of characteristic $p \neq 0$ and rank l, and let $\mathfrak{R}$ denote the set of p^l irreducible rational projective representations of G in each of which the high weight λ satisfies $0 \leq \lambda(a) \leq (p-1)$ $(a \in S)$. Let α_i denote the automorphism $t \to t^{p^i}$ of the universal field as well as the corresponding automorphism (see § 5) of G, and for $R \in \mathfrak{R}$ let R^{α_i} denote the composition of α_i and R. Then every irreducible rational projective representation of G can be written uniquely as $\prod_{i=0}^{\infty} R_i^{\alpha_i}$ (weak tensor product, $R_i \in \mathfrak{R}$).*

Received May 21, 1962.

* This research was supported by the Air Force Office of Scientific Research.

Conversely, every such product yields an irreducible rational projective representation of G.

This follows from 6.1 below. We need only remark here that there is no corresponding phenomenon for groups of characteristic 0, since then the identity is the only rational field automorphism and the tensor product of two rational representations is never irreducible unless one of them is one-dimensional. Related to 1.1 is the following conjecture for which there is much evidence and for which a proof for the group of type A_1 would go a long way.

1.2 CONJECTURE. *If G and $\mathfrak{R}$ are as in* 1.1 *and R is an irreducible, not necessarily rational, projective representation of G, there exist distinct isomorphisms β_i of the universal field into itself and corresponding representations R_i in $\mathfrak{R}$ such that $R = \prod R_i^{\beta_i}$ (see* §5 *for the definition of $R_i^{\beta_i}$).*

That the above product is always irreducible follows from 5.1 below.

Our second main result applies to naturally defined finite simple subgroups of the groups considered above. These include all the "finite simple algebraic groups" (those made up of the rational points of simple algebraic groups suitably defined over finite fields), that is (see Hertzig [8]), the groups considered by Chevalley [3] and those considered by Hertzig [8], Tits [24, 25] and the author [19, 20], and also include the nonalgebraic groups considered by Suzuki [22] and Ree [11], all the known finite simple groups other than the cyclic, alternating and Mathieu groups.

1.3 THEOREM. *If G is a finite simple algebraic group and the rational field has $q = p^n$ elements, then every irreducible projective representation is the restriction of a rational representation of the corresponding infinite algebraic group. If the rank is l, the number of such representations is q^l. Each has a high weight λ for which $0 \leq \lambda(a) \leq q-1$ $(a \in S)$.*

Here we also have the product representation of 1.1 with the upper limit n in place of ∞ (see 7.4 and 9.3). For the nonalgebraic finite groups mentioned above there is a corresponding result (12.2 below), but the relevant representations of the containing infinite algebraic groups are those that satisfy the further condition: $\lambda(a) = 0$ if a is long; hence their number is $q^{l/2}$. A gap in our development is that for finite odd-dimensional unitary groups and finite

Ree groups of type G_2 we have established these results, and also the following (see 8.1, 8.2, 9.6 and 12.5) only for ordinary representations, not for projective representations.

1.4 THEOREM. *Each of the finite groups above, algebraic or not, has an irreducible (ordinary) representation of dimension equal to the order of a p-Sylow subgroup. No other irreducible (projective or ordinary) representation has as high a dimension.*

Among the subsidiary results below, we consider the character of this highest representation (8.4, 9.6, 11.3), and present in §§10 and 11 some results related to those rather special isogenies which give rise to the existence of the groups of Suzuki and Ree.

In addition to [13], to which frequent references will be made, earlier work related to our results is as follows. In [2] Brauer and Nesbitt determine the irreducible representations of finite groups of type $SL(2)$ and prove the appropriate tensor product theorem, while in [13, Exp. 20] Chevalley does the same for rational representations of the corresponding infinite groups. In [10] Mark considers the finite groups of type $SL(3)$, while in [27] Wong considers groups of type $SL(l+1)$ and $Sp(l)$ and proves 1.1, 1.3 and 1.4 for ordinary representations. His methods, however, are quite different from ours, and are not readily extendable to the other types of groups. Our methods are closely related to those of Curtis in [4] where the representations of $\mathfrak{R}$ in 1.1 are constructed by infinitesimal methods and in [5] where they are shown to remain irreducible on restriction to the corresponding finite Chevalley groups (under the assumption $p>7$, which can easily be removed).

§ 2. Classical Lie algebras

Let $\mathfrak{g}_C$ be a simple Lie algebra over the complex field C, $\mathfrak{h}_C$ a Cartan subalgebra, $\sum$ the (ordered) system of roots relative to $\mathfrak{h}_C$, S the set of simple positive roots, and for each pair r, s of roots, set $c_{rs}=2(r, s)/(s, s)$, and define p_{rs} to be 0 if $r+s$ is not a root, otherwise to be the least positive integer p for which $r-ps$ is not a root. Then Chevalley [3, p. 24] has shown that there exists a generating set $\{X_r, H_r \mid r\in\sum\}$ such that the equations of structure of $\mathfrak{g}_C$ are:

2.1. $H_{-r} = -H_r \qquad (r \in \sum)$.

2.2. $H_{r+is} = H_r + H_s$ *if i is a positive integer and $r+is$ and r have the maximum root length.*

2.3. $[H_r, H_s] = 0 \qquad (r, s \in \sum)$.

2.4. $[H_r, X_s] = c_{sr} X_s \qquad (r, s \in \sum)$.

2.5. $[X_r, X_{-r}] = H_r \qquad (r \in \sum)$.

2.6. $[X_r, X_s] = \pm p_{rs} X_{r+s} \qquad (r, s \in \sum, r+s \neq 0)$.

Let $\mathfrak{g}$ and $\mathfrak{h}$ denote the algebras obtained by shifting the coefficients to an arbitrary field K of characteristic p. Then X_r and H_r shall be considered to belong to $\mathfrak{g}$ but the subscript r shall continue to denote an element of $\sum$. For the algebras just constructed, Curtis [4] has developed a theory of irreducible representations quite analogous to the classical theory in characteristic 0. Although he states and proves his results under the assumption that K is algebraically closed and $p > 7$, his proofs can be modified to apply to the present situation. We recall that a representation ρ of $\mathfrak{g}$ is restricted if $\rho(X_r)^p = 0$ and $\rho(H_r)^p = \rho(H_r)$ for each root r.

2.7 CURTIS. *With $\mathfrak{g}$ as above, every irreducible restricted $\mathfrak{g}$-module M contains a nonzero element v_+, uniquely determined to within multiplication by a scalar, such that $X_r v_+ = 0$ if r is positive, and there exist integers $\lambda(a)$, $0 \leq \lambda(a) \leq p-1$, such that $H_a v_+ = \lambda(a) v_+ (a \in S)$. Inequivalent modules yield distinct sequences $\lambda(a)$, and all sequences are realized. Thus there are p^l modules for an algebra of rank l.*

Here and elsewhere in the paper, "$\mathfrak{g}$-module" means vector space over the algebraic closure $\bar{K}$ of K on which $\bar{K}$ and $\mathfrak{g}$ act according to the usual rules, "irreducible" means absolutely irreducible, and $\mathfrak{M}$ denotes the p^l modules given by 2.7. As is easily seen, the modules of $\mathfrak{M}(\mathfrak{g}_{\bar{K}})$ may be viewed as extensions of those of $\mathfrak{M}(\mathfrak{g}_K)$, or equivalently, the latter as restrictions of the former. For each $M \in \mathfrak{M}$, v_+ is called a high vector and the linear function λ on $\mathfrak{h}$ defined by $\lambda(H_a) = \lambda(a)$ the high weight of M. Further for a positive root $r = \sum n(a) a \ (a \in S)$, we set $\operatorname{ht} r = \sum n(a)$, the height of r, then order the positive roots $r_1, r_2, \ldots, r_m$ in a manner consistent with heights (if $\operatorname{ht} r_i < \operatorname{ht} r_j$, then $i < j$), and for the monomial

$$v = X_{-r_m}^{i_m} \cdots X_{-r_2}^{i_2} X_{-r_1}^{i_1} v_+ \qquad (0 \leq i_k \leq p-1) \tag{2.8}$$

set $\operatorname{ht} v = -\sum i_k \operatorname{ht} r_k$, and finally call w homogeneous of height n if it is a linear combination of monomials of height n. We recall that a basis for M can be selected from the monomials.

2.9 LEMMA. *Nonzero vectors of different heights are linearly independent.*

Proof. Given a relation $v_0 + v_1 + \cdots + v_d = 0$ with v_i of height $-i$, we prove by induction on d that each v_i is 0. If $d = 0$, this is clear.
Assume $d > 0$. If r is any positive root, $X_r v_0 + X_r v_1 + \cdots + X_r v_d = 0$, and since $X_r v_i$ is higher than v_i, the induction assumption yields $X_r v_d = 0$. Thus by 2.7, $v_d \in K v_+$, and since the algebra generated by those X_{-r} for which $r > 0$ acts nilpotently on M because M is restricted, $v_d = 0$. Then each v_i is 0 by the induction assumption.

§ 3. Classical algebraic groups

Now set $x_r(t) = \exp \operatorname{ad} tX_r$ $(t \in K,\ r \in \sum)$, and let G (this is G' in [3]) denote the group generated by all of these automorphisms. With 4 exceptions [3, p. 63], which we henceforth exclude, G is simple. In G there are commutator relations [3, p. 36]:

3.1. $$(x_r(t),\ x_s(u)) = \prod x_{ir+js}(C_{ij,rs}\, t^i u^j) \qquad (r,\ s \in \textstyle\sum,\ r+s \neq 0).$$

Here the product is taken over all positive integers i, j for which $ir + js$ is a root, the terms being arranged in some fixed, but arbitrary, order, and the $C_{ij,rs}$ are integers that depend on the order, but not on t, u or the field K. We also have from [3, p. 36]:

3.2. *For each positive root r there is a homomorphism φ_r of $SL(2, K)$ into G such that $\varphi_r\begin{pmatrix}1 & t\\ 0 & 1\end{pmatrix} = x_r(t)$ and $\varphi_r\begin{pmatrix}1 & 0\\ t & 1\end{pmatrix} = x_{-r}(t)$.*

Together with G, we consider a covering group Γ, the abstract group generated by a set of elements $x_r(t)$ $(t \in K,\ r \in \sum)$ subject to the relations 3.1 and those implied by 3.2 with Γ in place of G. That these relations define $SL(l+1, K)$ and $Sp(l, K)$ for $\sum$ of type A_l and C_l respectively was already known to Dickson [7]. The properities of Γ that we require, 3.3 to 3.6 below, are taken from [21].

3.3. *Each φ_r is an isomorphism.*

3.4. *Γ is equal to its commutator subgroup.*

3.5. *If* $h_a(t) = \varphi_a(\mathrm{diag}(t, t^{-1}))$ *and* $h_a = \{h_a(t) | t \in K^*\}$, *the* h_a $(a \in S)$ *generate a subgroup* H *as a direct product.*

For each $r \in \sum$ the symbol r is also used to denote the root on H: $\prod h_a(t_a) \to \prod t_a^{c_{ra}}$ (see 2.4).

3.6. *The center* C *of* Γ *consists of those* $h \in H$ *for which all* $r(h)$ *are* 1; C *is the kernel of the natural projection of* Γ *on* G; Γ/C *is isomorphic to* G.

Thus Γ acts naturally on G-modules, in particular on $\mathfrak{g}$. Let $\Gamma(K)$, $G(K)$, etc. denote the dependence of Γ, G etc. on K.

3.7. *Let* K *be algebraically closed and of infinite degree of transcendence over the prime field. Then* G *and* Γ *may be identified, via isomorphisms, with a simple algebraic group and its simply connected covering group, and both may be defined over the prime field. It is then true that* (a) *the* pth *power automorphism of* Γ *is given by* $x_r(t) \to x_r(t^p)$, (b) *if* k *is a subfield of* K, Γ_k *is naturally isomorphic to* $\Gamma(k)$, *and* (c) H *is a Cartan subgroup of* Γ.

These results, which cover all simple algebraic groups because of the classification in [13], are proved at the end of § 4.

We use ω_a $(a \in S)$ to denote the function (fundamental weight) on H defined by $\prod h_b(t_b) \to t_a$, and set $\omega = \prod \omega_a$.

3.8. $$\omega^2 = \prod_{r>0} r.$$

For a proof of the additive version of this result see [14, p. 19-01].

Finally to close this section we prove a result of fundamental importance in our later discussion of the representations of finite groups. We are indebted to T. A. Springer for the main ideas of the proof.

3.9 Lemma. *Assume that* K *is algebraically closed and of infinite transcendence degree over its prime field* F_p, *that* τ *is a rational automorphism of* Γ *such that* $H^\tau = H$, *that* σ *is the composition of* τ *with the* pth *power automorphism, and that* Γ_σ *is the subgroup of fixed points of* σ. *Then* (a) *the semisimple classes of conjugate elements of* Γ_σ *are in natural one-one correspondence with those orbits of* H *under* W, *the Weyl group, that are invariant under* σ, *and* (b) *if for each* $a \in S$, γ_a *is the sum of the distinct images of* ω_a *under* W, *the orbit space* H/W *is an affine variety with coordinates* γ_a $(a \in S)$.

The proof proceeds in several steps.

(1) *Assume that B is a (connected) algebraic subgroup of Γ and that $B^\sigma = B$. Then for each x in B there is a y in B such that $x = y^{-1}y^\sigma$.* This result is quite close to one of Lang [9], and it does not depend on the simple-connectedness or semisimplicity of Γ. The proof, a straightforward modification of Lang's, is omitted.

(2) *The centralizer of a semisimple element of a simply-connected semisimple algebraic group is connected.* Here are the main steps in a proof due to Springer (unpublished). The semisimple element h is put in a Cartan subgroup H, and then by the Bruhat decomposition [13, p. 13-11], the problem is reduced to showing that an element of the Weyl group that leaves h fixed is a product of reflections (corresponding to roots) that also do. After the problem is shifted from H to an ordinary torus T and then to the covering space of T, the proof is completed by geometric means.

(3) *Two semisimple elements of Γ_σ which are conjugate in Γ are also conjugate in Γ_σ.* Assume $x = zwz^{-1}$ (x, $w \in \Gamma_\sigma$, $z \in \Gamma$). Then $x = z^\sigma wz^{-\sigma}$, whence $z^{-1}z^\sigma$ is in B, the centralizer of w. By 3.7 and (2), B is connected if w is semisimple, and because $w \in \Gamma_\sigma$, $B^\sigma = B$. Thus by (1) we can write $z^{-1}z^\sigma = y^{-1}y^\sigma$ ($y \in B$). Then $zy^{-1} \in \Gamma_\sigma$, and since $x = (zy^{-1})w(zy^{-1})^{-1}$, we have (3).

(4) *An element of Γ is conjugate to an element of Γ_σ if and only if it is conjugate to its image under σ.* For if $z \in \Gamma$, then $z = xz^\sigma x^{-1}$ for some $x \in \Gamma$ if and only if $z = y^{-1}y^\sigma z^\sigma y^{-\sigma}y$ for some $y \in \Gamma$, by (1) with $B = \Gamma$, that is, if and only if $yzy^{-1} \in \Gamma_\sigma$ for some $y \in \Gamma$.

(5) *Two elements of H are conjugate in Γ if and only if they are conjugate under W.* This easily comes from the uniqueness in the Bruhat decomposition.

Since an element of Γ is semisimple if and only if it is conjugate to an element of H [13, p. 6-13], we may combine (3), (4) and (5) to get (a). In [15, p. 57-8] it is proved that H/W is an affine algebraic variety whose coordinate ring is got from that of H by selecting the invariants under W. This means the polynomials in ω_a, ω_a^{-1} ($a \in S$) that are symmetric relative to W. Thus to complete the proof of (b) we need only establish the following result, which in case W is of type A_l reduces to the fundamental theorem for symmetric functions.

(6) *Every polynomial in* ω_a, ω_a^{-1} ($a \in S$) *that is symmetric relative to* W *is a polynomial in the elementary symmetric polynomials* γ_a ($a \in S$). Partially order the monomials $\prod \omega_a^{n_a}$ so that each is higher than those obtained by multiplying it by a product of negative (multiplicative) roots. Thus if β is a nonzero symmetric polynomial and $c\prod \omega_a^{n_a}$ is one of its highest terms, each $n_a \geq 0$ because β is symmetric. Then $\beta - c\prod \gamma_a^{n_a}$ does not contain this term, and the proof of (6) may be completed by induction.

§ 4. Lifting representations from algebras to groups

The notations p, K, $\sum$, S, H, ω_a, etc. introduced in §§ 2 and 3 in connection with the algebra $\mathfrak{g}$ and corresponding groups G and Γ will be used throughout the paper. By a module (or representation) for these groups we mean one over $\overline{K}$, the algebraic closure of K. Following Curtis [4], we first convert each $M \in \mathfrak{M}$ into a projective Γ-module. For $x \in \Gamma$, let M^x be the irreducible $\mathfrak{g}$-module obtained by defining the action of $\mathfrak{g}$ on M by the rule:

4.1 $$(M^x) \quad X.v = X^x v \qquad (X \in \mathfrak{g},\ v \in M).$$

Here X^x is the image of X under x and we use the convention $(X^x)^y = X^{yx}$. The module M^x is equivalent to M [4]. Thus there is a $\mathfrak{g}$-module isomorphism $T(x)$, uniquely determined to within a scalar multiple by Schur's lemma, of M on M^x. This satisfies:

4.2 $$T(x)Xv = X^x T(x)v \qquad (x \in \Gamma,\ X \in \mathfrak{g},\ v \in M).$$

The map $x \to T(x)$ is a projective representation of Γ (or G) on M, again by Schur's lemma. For each positive root r we may (and do) normalize all $T(x_r(t))$ to keep v_+ fixed (see 2.7); since 4.1 and 4.2 imply that $T(x_r(t))v = v + $ *higher terms*, for each monomial v, this amounts to making each $T(x_r(t))$ unipotent. After treating negative roots in a similar way, we want to show that the normalization can be extended to yield an ordinary (not just a projective) representation of Γ. When it is convenient, we write xv for $T(x)v$.

4.3 Lemma. *Let* $M \in \mathfrak{M}$ *have high weight* $\lambda(a)$ ($a \in S$), *fix* $a \in S$ *and set* $\lambda(a) = n$. *Then* (a) $v_+, X_{-a}v_+, \ldots, X_{-a}^n v_+$ *are linearly independent and* $X_{-a}^{n+1}v_+ = 0$; (b) *the normalized action of the* $x_a(t)$ *and* $x_{-a}(t)$ *on* M *can be extended to an ordinary representation of* Z_a, *the group generated by these elements; we then have* (c) $h_a(t)v_+ = t^n v_+$.

Proof. By induction, $X_a X^i_{-a} v_+ = i(n-i+1)X^{i-1}_{-a}v_+$ $(i \geq 1)$, whence the vectors $X^i_{-a}v_+$ $(0 \leq i \leq n)$ are nonzero and then linearly independent by 2.9. Further $X_a X^{n+1}_{-a} v_+ = 0$, and clearly $X_r X^{n+1}_{-a} v_+ = 0$ for $r > 0$, $r \neq a$. Thus $X^{n+1}_{-a} v_+ = 0$ by 2.7 and 2.9. Now set $v_0 = v_+$, $iv_i = X_{-a}v_{i-1}$ $(1 \leq i \leq n)$, so that also $(n-i)v_i = X_a v_{i+1}$. From $x_a(t)v_+ = v_+$, $x_a(t)X_{-a} = X_{-a} + tH_a - t^2 X_a$ and 4.2, we see by induction that $x_a(t)v_i = \sum_{j=0}^{i} \binom{n-j}{n-i} t^{i-j} v_j$. Now interchanging the roles of X_a and X_{-a}, and also of v_0 and v_n, that is, replacing M by M^w with w an element of Z_a corresponding to the Weyl reflection relative to a, we get $x_{-a}(t)v_i = \sum_{j=i}^{n} \binom{j}{i} t^{j-i} v_j$. Introducing a space with coordinates x and y and setting $v'_i = x^{n-i}y^i$, we see that in the space of polynomials of degree n exactly the same equations hold for the transformations $x'_a(t)$: $x, y \to x, y + tx$ and $x'_{-a}(t)$: $x, y \to x + ty, y$. We thus see that the relations on the $x_a(t)$ and $x_{-a}(t)$ $(a \in S)$ implied by 3.2 also hold for the $T(x_a(t))$ and $T(x_{-a}(t))$. Further the relations 3.1 with $r, s > 0$ are also preserved, as we see by applying both sides to v_+ and noting that every term leaves v_+ fixed. Since we may choose an element w in Γ corresponding to an arbitrary element of the Weyl group and apply the above considerations to M^w, we see that all of the relations of 3.1 and 3.2 are preserved. We thus get 4.3(b), 4.3(c) and also:

4.4. *The projective representation of* Γ *on* M *can be lifted in a unique way to an ordinary representation.*

The uniqueness comes from 3.4, which implies that Γ has no nontrivial one-dimensional representation.

From the definitions we see that if v has height n in M then

4.5 $$(x_r(t) - 1)v = tX_r v + \textit{higher (lower) terms, when } r > 0 \ (r < 0).$$

By 2.7 this yields:

4.6. *The vectors* $\bar{K}v_+$ *are the only ones fixed by all* $x_r(t)$ $(r > 0)$.

From this and 4.3, we see that the Γ-module M determines $\bar{K}v_+$, which in turn determines $\lambda(a)$ $(a \in S)$ since $\lambda(a) + 1$ is the dimension of the subspace generated by the elements $x_{-a}(t)$ $(t \in K)$ acting on $\bar{K}v_+$. Thus using also 4.3(c),

4.7. *If M_1 and M_2 in $\mathfrak{M}$ are distinct as $\mathfrak{g}$-modules, they are distinct as Γ-modules. If M_1 has high weight λ as $\mathfrak{g}$-module, it has high weight $\prod \omega_a^{\lambda(a)}$ as Γ-module.*

In order to pass to projective G-modules, we use:

4.8. *Under the natural projection from Γ to G, each irreducible Γ-module leads to an irreducible projective G-module. Distinct Γ-modules yield distinct G-modules.*

We need only observe that the center C of Γ must act via scalars in any irreducible representation of Γ, and then use 3.4.

Now Γ acts faithfully on $\mathfrak{M}$ as a set: since Γ/C is simple, the kernel is contained in C and consists of those h for which all $\omega_a(h)$ $(a \in S)$ are 1. Further, relative to monomial bases, the group $\{x_r(t),\ t \in K\}$ acts via matrices that are polynomials in t with coefficients in F_p, thus acts as an algebraic group defined over F_p with the pth power map given by $x_r(t) \to x_r(t^p)$. The same is thus true of the group Γ. Comparing the structure just put on Γ with the one put on G by T. Ono [J. Math. Soc. Japan 10 (1958)], and using his results and methods and those of [13, Exp. 23], we easily get the assertions of 3.7 and also

4.9. *Γ (hence also G) acts rationally on each M in $\mathfrak{M}$.*

§ 5. Tensor product theorem

Each isomorphism α of K into $\overline{K}$ gives rise to an isomorphism of $\Gamma(K)$ into $\Gamma(\overline{K})$, defined by $x_r(t) \to x_r(t^\alpha)$, and can thus be used to convert each Γ-module $M \in \mathfrak{M}$ into another Γ-module, denoted M^α, by the rule $x \cdot v = x^\alpha v$ $(x \in \Gamma(K),\ v \in M)$.

5.1 THEOREM. (a) *If $M_1, M_2, \ldots, M_k$ are in $\mathfrak{M}$ and $\alpha_1, \alpha_2, \ldots, \alpha_k$ are distinct isomorphisms of K into $\overline{K}$, then $M = M_1^{\alpha_1} M_2^{\alpha_2} \cdots M_k^{\alpha_k}$ (tensor product) is an irreducible Γ-module.* (b) *For a fixed sequence of α's, two Γ-modules M constructed in this way are equivalent if and only if the sequences of M_i's are the same. Or, equivalently, if $N = N_1^{\beta_1} N_2^{\beta_2} \cdots N_l^{\beta_l}$ with the N_j in $\mathfrak{M}$ and the β_j distinct isomorphisms of K into $\overline{K}$, then M is equivalent to N if and only if, after the deletion of all one-dimensional factors, $k = l$ and, for some permutation π of $1, 2, \ldots, k$, M_i is equivalent to $N_{\pi i}$ and $\alpha_i = \beta_{\pi i}$ for $i = 1, 2, \ldots, k$.*

(c) *If the modules in* (a) *and* (b) *are taken to be projective G-modules, the modified statements are also true.*

Proof. Let $X_r^{(i)}$ be the transformation that is X_r on $M_i^{\alpha_i}$ and the identity on the other components of M, let $v_+ = \prod v_+^{(i)}$ be the product of the high vectors in the separate components, and let the height of a product of monomials be defined as the sum of the heights of its terms.

(1) *The set of vectors of M annihilated by all $X_r^{(i)}$ $(r>0,\ i=1, 2, \ldots, k)$ is $\overline{K}v_+$. Nonzero vectors of distinct heights are linearly independent.* For $k=1$, this follows from 2.7 and 2.9. If $k>1$, we can write any v in M as $v=\sum u_j w_j$ $(u_j \in M_1^{\alpha_1} \cdots M_{k-1}^{\alpha_{k-1}},\ w_j \in M_k^{\alpha_k})$ with the u_j and also the w_j linearly independent. Then $X_r^{(k)} v = \sum u_j (X_r w_j)$, which is 0 only if all $X_r w_j$ are 0, because the u_j are linearly independent. Thus $X_r^{(k)} v = 0$ for all $r>0$ only if all w_j are in $\overline{K}v_+^{(k)}$, whence the first part of (1) follows by induction. This implies the second part (see the proof of 2.9) and also:

(2) *If v is a homogeneous vector of M and $r>0$ (resp. $r<0$), then $(x_r(t)-1)v = \sum t^{\alpha_i} X_r^{(i)} v +$ higher (resp. lower) terms.*

(3) *Irreducibility.* Let M' be a Γ-submodule of M and v a nonzero vector of M'. Write $v = v_0 + v_1 + \cdots + v_d$, $v_d \neq 0$, height $v_j = -j$. The α_i are distinct, thus linearly independent. By (2) this implies that for every $r>0$ and every $i = 1, 2, \ldots, k$ there is a vector $X_r^{(i)} v_d +$ *higher terms* in M', whence by (1) and induction on d the vector v_+ is also in M'. Then using negative roots, we see by (downward) induction on the height that for every monomial v of M there is a vector $v +$ *lower terms* in M'. By induction on the height this implies that M' contains all monomials, that $M' = M$, that M is irreducible.

(4) *Uniqueness.* Let λ_i be the high weight of M_i as $\mathfrak{g}$-module. We must show that M as Γ-module intrinsically determines the numbers $\lambda_i(a)$ $(i=1, 2, \ldots, k;\ a \in S)$. First note that M determines $\overline{K}v_+$ by (1). Fix a and set $\lambda_i(a) = a_i$. If all $x_{-a}(t)$ $(t \in K)$ fix v_+, then the a_i are certainly determined by M—they must all be 0 by 4.3(a) and 4.5. Assume henceforth that this is not the case. Then $h_a(t)$ acts on $\overline{K}v_+$ with the characteristic value t^α, $\alpha = \sum a_i \alpha_i$, some $a_i \neq 0$, by 4.3(c). Assuming that M does not determine the a_i uniquely, we are thus led to the existence of a nontrivial identical relation $t^\alpha = t^\beta$ ($\beta = \sum b_i \alpha_i$, some $b_i \neq 0$, some $b_j \neq a_j$, $t \in K$). Among all such relations on the

monomials t^{γ} ($\gamma = \sum c_i \alpha_i$, $0 \leq c_i \leq p-1$), ordered lexicographically, we choose one of minimum degree. The substitution $t \to tu$ shows that this relation has the form $t^{\delta} = t^{\varepsilon}$ ($\delta > \varepsilon$, $\delta = \sum d_i \alpha_i$, $\varepsilon = \sum e_i \alpha_i$). Then using the minimality of the degree and the substitution $t \to t+u$, we get in turn $d_k = 0$ (whence we may assume $e_k \neq 0$ since otherwise the proof is completed by induction), $k > 1$, $\delta = \alpha_1$, $\varepsilon = \alpha_k$, and $\alpha_1 = \alpha_k$, a contradiction. This proves the first statement of (b). The second follows immediately.

(5) *G-modules.* By 4.8 the results we have proved for Γ-modules are equally valid for G-modules, which is (c).

We remark that considering $\mathfrak{g}$ as Lie ring rather than Lie algebra we can interpret $M_i^{\alpha_i}$ above as $\mathfrak{g}$-module and then prove a theorem entirely analogous to 5.1 with $\mathfrak{g}$ in place of Γ.

§ 6. Rational representations

As a first application of 5.1 we have:

6.1 THEOREM. *If K is infinite and perfect and α_i denotes the automorphism $t \to t^{p^i}$ of K, then every irreducible rational Γ-module or irreducible rational projective G-module can be expressed uniquely as a tensor product $M = \prod_{i=0}^{\infty} M_i^{\alpha_i}$ ($M_i \in \mathfrak{M}$, almost all M_i trivial).*

Proof. By 4.8 we need only consider Γ-modules, and by the density theorem of Rosenlicht [12, p. 44] we may assume that K is algebraically closed and of infinite transcendence degree over its prime field. For given, but arbitrary, nonnegative integers $\lambda(a)$ ($a \in S$), we can uniquely write $\lambda(a) = \sum p^i \lambda_i(a)$ ($0 \leq \lambda_i(a) \leq p-1$), choose M_i in $\mathfrak{M}$ as the Γ-module with high weight $\prod \omega_a^{\lambda_i(a)}$ (see 4.7), and thus construct a Γ-module $\prod M_i^{\alpha_i}$ which is rational by 4.9, irreducible by 5.1, and has high weight $\prod \omega_a^{\lambda(a)}$. Using the classification [13, Exp. 14 and 15] of irreducible representations of semisimple algebraic groups in terms of high weights, we see that this construction yields a complete set of irreducible rational Γ-modules, whence 6.1.

By now we have also shown that if K is infinite and perfect every irreducible rational projective representation of Γ or G comes in a unique way from an ordinary representation of Γ.

§ 7. Finite groups, normal forms

If K is finite with q elements, we write Γ_q, G_q for Γ, G. The following fact, not used here, is proved in [21].

7.1. *If the rank is at least* 2, *the relations* 3.1 *alone are enough to define* Γ_q.

What is required here from [21] is:

7.2. *Every irreducible projective G_q-module can be lifted uniquely to a Γ_q-module.*

Now by 3.9 with σ the qth power mapping, the semisimple classes of conjugate elements of Γ_q are characterized by coordinates $\gamma(a)$ subject to the condition $\gamma(a)^q = \gamma(a)$ (see also 3.7(a)). This yields:

7.3 LEMMA. *If the rank is l, the number of semisimple classes of conjugate elements of Γ_q is q^l.*

We can now prove one of the main results of this paper. Observe that the modules considered are not assumed to be rational, throughout this section.

7.4 THEOREM. *Let $q = p^n$ and let α_i denote the field automorphism $t \to t^{p^i}$. Then every irreducible Γ_q-module, also every irreducible projective G_q-module, can be expressed uniquely as a tensor product $M = \prod_{i=0}^{n-1} M_i^{\alpha_i}$ ($M_i \in \mathfrak{M}$). If the rank of Γ_q is l, there are q^l such modules.*

Proof. Again we need only consider Γ_q-modules, this time by 7.2. By 7.3 and a theorem of Brauer and Nesbitt [1, p. 14] the number of inequivalent irreducible Γ_q-modules is q^l. Since the q^l modules $\prod M_i^{\alpha_i}$ are inequivalent and irreducible by 5.1, they thus form a complete set, as required.

7.5 COROLLARY. *If L is an infinite field containing the finite field K, every irreducible representation of $\Gamma(K)$ can be extended to $\Gamma(L)$. Every irreducible representation of $\Gamma(K)$ can be realized over K. The corresponding statements for projective representations of G are also true.*

The first statement is clear. For the second we need only observe that relative to a monomial basis 2.8 for $M_i \in \mathfrak{M}$ each generator $x_r(t)$ is represented by a matrix which is a polynomial in t with coefficients in the prime field.

The modules of 7.4 have high weights $\prod \omega_a^{\lambda(a)}$ with $0 \leq \lambda(a) \leq q-1$. Those in which the center C of Γ (see 3.6) acts trivially, or what is equivalent, fixes

v_+, yield all of the irreducible G_q-modules. Thus

7.6 COROLLARY. *Every irreducible G_q-module is obtained from a Γ_q-module for which the high weight $\Pi\omega_a^{\lambda(a)}$ is* 1 *on the center of Γ.*

In individual cases, when the root system $\sum$ is specfied, more detailed results can be given. Thus if $\sum$ is of type E_8, F_4 or G_2 (and K is arbitrary), $\Gamma = G$, so that 7.6 is superfluous, while, for example, if $\sum$ is of type A_l (so that Γ_q and G_q are respectively isomorphic to $SL(l+1, q)$ and $PSL(l+1, q)$), the irreducible G_q-modules correspond to the sequences $(\lambda_1, \lambda, \ldots, \lambda_l)$ for which $0 \leq \lambda_i \leq q-1$ and $\sum i\lambda_i$ is divisible by the greates common divisor of $l+1$ and $q-1$.

Dually, one can make similar statements concerning the classes of conjugate elements of G in terms of those of Γ.

Finally, we remark that there are results analogous to 7.4 and 7.5 with the finite Lie ring $\mathfrak{g}_q$ in place of Γ_q. The proof of completenteness of the modified 7.4 can be given along the lines of [14; p. 22-11, 22-12].

§ 8. Prime power representations

A general formula in characteristic p, comparable to Weyl's formula in characteristic 0 (cf. [6]), for the characters or dimensions of the above modules does not yet exist (except for groups of type A_1 and A_2 [13, p. 588], [14]). However, for the irreducible Γ_q-module with the greatest of all possible high weights, ω^{q-1} (recall that $\omega = \Pi\omega_a$), that is, the module $M_q = \Pi M_i^{\sigma_i}$ of 7.4 in which each M_i is equivalent to the module M_p of $\mathfrak{M}$ with high weight $\lambda(a) = p-1$ $(a \in S)$ as $\mathfrak{g}$-module, the situation can be described rather completely and is very much as in characteristic 0. The following result is proved in [17].

8.1 LEMMA. *If m is the number of positive roots in $\sum$, there is an irreducible Γ_q-module $\overline{M}_q$ of dimension q^m, that is, the order of a p-Sylow subgroup of Γ_q.*

8.2 THEOREM. *The Γ_q-modules $\overline{M}_q$ of* 8.1 *and M_q of high weight ω^{q-1} are eqnivalent. All other irreducible Γ_q-modules have smaller dimensions than M_q.*

Proof. Because each module of $\mathfrak{M}$ is spanned by the monomials 2.8, the only one that could have a dimension as large as p^m is M_p by 4.3(a), and hence by 7.4 the only possible irreducible Γ-module of dimension as large as

q^m is M_q. From the existence of the module $\overline{M}_q$ with dimension q^m, it follows that M_q is equivalent to $\overline{M}_q$ and that the other irreducible Γ_q-modules have smaller dimensions.

In the course of the argument, we have proved:

8.3 COROLLARY. *The irreducible module M_p with high weight ω^{p-1} has dimension p^m and a basis consisting of all the monomials* 2.8.

A direct proof of this result, within the framework of $\mathfrak{g}$-modules, also exists. Using 8.3 we can compute the (Brauer) character of M_q. To define this we write the order g of Γ_q as $g=p^e g'$, $(p, g')=1$, choose an isomorphism θ of the group of g'th roots of 1 in $\overline{K}$ onto the corresponding group in the complex field, and then for any semisimple element x of Γ_q and any module M for Γ_q, define $\chi(x)$, the character of x on M, to be $\sum\theta(c_i)$, the sum to be taken over the characteristic roots c_i of x on M. Generally χ depends on the choice of θ, but not on M_q where it turns out to be rational.

The following result has been proved previously only for groups of type A_l [16, p. 281] and in somewhat different terms.

8.4 THEOREM. *If Γ_q is of rank l, and x is a semisimple element whose centralizer in the correpsonding algebraic group has dimension $l+2d(x)$, or equivalently, whose action on $\mathfrak{g}$ has fixed point set of dimension $l+2d(x)$, the character of x on M_q is given by $\chi(x) = \pm q^{d(x)}$.*

Proof. In $\Gamma(\overline{K})$, x is conjugate to an h in $H(\overline{K})$. Now since h acts on the monomial 2.8 of M_p by multiplication by $\omega^{p-1}(h)\Pi r_k(h)^{-i_k}$, we see by 8.3 that the character of h on M_p is

$$\psi(h) = \theta(\omega^{p-1}(h))\Pi_{r>0}\textstyle\sum_{i=0}^{p-1}\theta(r(h)^{-i}), \tag{8.5}$$

and then using 3.8, that the character $\chi(h)$ on M_q satisfies $\chi(h)^2 = \Pi_r\sum_{i=0}^{q-1}\theta(r(h)^{(q-1-2i)/2})$, the product over all roots. Since h is conjugate to an element of Γ_q and the roots are permuted by the Weyl group, it follows from (4) and (5) of the proof of 3.9 that the numbers $\theta(r(h))^q$ form a permutation of the numbers $\theta(r(h))$. Thus the roots can be arranged in cycles (of various lengths) $(r_1 r_2 \cdots r_k)$ such that $\theta(r_i(h))^q = \theta(r_{i+1}(h))$, $\theta(r_k(h))^q = \theta(r_1(h))$. If $r(h) \neq 1$, the cycle containing r telescopically contributes 1 to the product for $\chi(h)^2$, since the term for r may be written, subject to a consistent choice of square roots, as $(c^{q/2}-c^{-q/2})/(c^{1/2}-c^{-1/2})$ with $c=\theta(r(h))$. The terms for which $r(h)=1$ contribute q each to the product, $q^{2d(h)}$ together. Thus $\chi(h)^2=q^{2d(h)}$. Since h is conjugate to x, we have 8.4.

8.6 COROLLARY. *If Γ_q is replaced by G_q, 8.2 and 8.4 remain valid.*

We need only remark that M_p, hence M_q, is an ordinary (not just projective) G_q-module: if c is in the center of Γ, $\omega^2(c)=\Pi r(c)=1$ by 3.6 and 3.8, whence $\omega^{p-1}(c)=1$, even if p is 2.

Finally we remark that for semisimple algebraic groups over an algebraically closed field of characteristic 0, results analogous to 8.3 to 8.6, in which p need not be a prime nor q a prime power are true. Here we content ourselves with showing that the formula 8.5 for the character on the irreducible module with high weight ω^{p-1} is essentially unchanged. With $\Delta(j)=\sum(\det w)(w\omega)^j$, the sum over the Weyl group, Weyl's formula for the character [26, p. 389] yields $\Delta(p)/\Delta(1)$, that is, 8.5 with $\theta=1$ because of the basic factorization $\Delta(j)=\omega^j\Pi_{r>0}(1-r^{-j})$ [26, p. 386].

§ 9. Finite groups, nonnormal forms

In this section we treat the simple groups denoted as A_l^1 (a projective unitary group in $l+1$ dimensions), D_l^1 (a second projective orthogonal group in $2l$ dimensions), E_6^1 (a nonnormal "real" form of E_6) and D_4^2 (a "triality" form of D_4) in [19], and their covering groups. Each of the latter groups can be defined in terms of generators and relations derived from the structure of the corresponding simple group, just as Γ is in terms of G in § 3; however, it is more convenient to define them directly as subgroups of the groups Γ. Starting with an automorphism σ, other than the identity, of the root system $\sum$ such that $\sigma S=S$, and an automorphism σ of the same period on the field K, we can construct an automorphism, also called σ, of the corresponding group Γ such that $x_a(t)^\sigma=x_{\sigma a}(t^\sigma)$ for all $a\in\pm S$ and all $t\in K$, and then define Γ^1 to be the group of fixed points of σ. Comparing this definition with the one given in [19] for the corresponding simple groups, and using 3.6, we easily get:

9.1. *Let C^1 be the center of Γ^1. Then $C^1=C\cap\Gamma^1$, and Γ^1/C^1 is naturally isomorphic to the corresponding simple group of* [19].

We write G^1 for Γ^1/C^1, K_0 for the fixed field under σ, and Γ_q^1, etc. when K_0

has q elements. As a subgroup of Γ, Γ^1 acts naturally on each Γ-module.

9.2 THEOREM. *If the M_i are in $\mathfrak{M}$, and the α_i are isomorphisms of K into $\overline{K}$ which are distinct on K_0, then $M = M_1^{\alpha_1} M_2^{\alpha_2} \cdots M_k^{\alpha_k}$ is an irreducible Γ^1-module, and there is uniqueness in this product representation in the sense of the second sentence of* 5.1(b).

Proof. We use the notations of the proof of 5.1 and assume that σ above has period 2. If the period is 3, as it is for one of the groups of type D_4, the argument is similar. If r is a root such that $\sigma r = r$ and there is no root s such that $r = s + \sigma s$, then we may assume $x_r(t) \in \Gamma^1$ for all $t \in K_0$ [19, *p.* 879], and we have 5.1(2) holding. If r is such that $\sigma r \neq r$ and $r + \sigma r$ is not a root, then (see [19]) $x_r(t) x_{\sigma r}(t^\sigma) \in \Gamma^1$ for all t in K, and we have instead $(x_r(t) x_{\sigma r}(t^\sigma) - 1) v = \sum t^{\alpha_i} X_r^{(i)} v + \sum t^{\sigma \alpha_i} X_{\sigma r}^{(i)} v + \cdots$. If r, σr and $r + \sigma r$ are all roots, the situation is similar. Since the α_i act distinctly on K_0 and the α_i and $\sigma \alpha_i$ together act distinctly on K, the proofs of irreducibility and uniqueness in 9.2 are from this point on straightforward modifications of those in 5.1.

For finite groups, we have:

9.3 THEOREM. *If K_0 has $q = p^n$ elements and α_i denotes the field automorphism $t \to t^{p^i}$, then every irreducible Γ_q^1-module can be written uniquely as a tensor product $M = \prod_{i=0}^{n-1} M_i^{\alpha_i}$ $(M_i \in \mathfrak{M})$. If the rank is l, the number of such modules is q^l.*

Thus every irreducible Γ_q^1-module is the restriction of some $\Gamma(\overline{K})$-module, and the largest high weight that occurs is ω^{q-1}. Again, by 9.2, we are reduced to showing that the number of semisimple classes of conjugate elements is q^l. By 3.9, these are characterized by coordinates $\gamma(a)$ $(a \in S)$ subject to the condition $\gamma(a)^q = \gamma(\sigma a)$. Whatever permutation σ effects on S, the number of solutions is q^l (the contribution for each cycle of length d is q^d), as required.

Turning again to [21], we have:

9.4. *If the type A_l (l even) is excluded, every irreducible projective G_q^1-module can be lifted uniquely to a Γ_q^1-module.*

Quite likely this exclusion is unnecessary, but we have not yet shown this. From 9.3 and 9.4 we get:

9.5 COROLLARY. *If the type A_l (l even) is excluded,* 9.3 *also holds for* *projective G_q^1-modules.*

Here also one can get the irreducible G_q^1-modules as those of Γ_q^1 in which the center acts trivially. Since 8.1 is true with Γ_q^1 in place of Γ_q [17, 19], the same is true of 8.2. Also if h in $H(\bar{K})$ is conjugate to an element of Γ_q^1, then by (4) and (5) of the proof of 3.9, it is conjugate under the Weyl group to h^σ, so that the numbers $r(h)^q$ again form a permutation of the numbers $r(h)$ (see the definition of σ at the beginning of this section). Thus the proof 8.4 carries over as is.

9.6 THEOREM. *The statements* 8 1, 8.2 *and* 8.4 *are true with* Γ_q^1 *in place of* Γ_q.

§ 10. Special isogenies, infinitesimal and global

In this section we present a discussion of the rather special isogenies that exist for simple groups of type B_l, C_l and F_4 and characteristic 2, and type G_2 and characteristic 3 (cf. [23, p. 282], [13, Exp. 21-24]). The results will be used in the next sections, where we return to group representations.

In what follows we identify two root systems that are related by a scalar multiplication. Associated with each system Σ, there is a dual system Σ^* and a map of Σ onto Σ^* such that $r^* = 2r/(r, r)$ $(r \in \Sigma)$. When roots of unequal lengths occur, this map preserves angles, sends short roots to long roots and vice versa, maps the simple set S onto another, and puts types B_l and C_l in duality with each other and types C_2, F_4 and G_2 with copies of themselves.

The pair (Σ, p) will be called special if Σ contains roots r and s such that $(s, s)/(r, r) = p$. The possibilities are those listed in the first paragraph of this section. In the corresponding algebra $\mathfrak{g}$ of § 2, those X_r and H_r for which r is short span an ideal, denoted $\mathfrak{g}_1$ in what follows. To see this, observe that if r is short and s is long $c_{sr} = pc_{rs}$, and that if r, s and $r+s$ are roots with r short and $r+s$ long then s is short and $p_{rs} = p$ (check for Σ of type C_2 and G_2). Observe also that in the present case $\Gamma = G$, Γ maps $\mathfrak{g}_1$ onto itself (because each $x_r(t)$ does), and, being simple, Γ acts faithfully on each of $\mathfrak{g}_1$ and $\mathfrak{g}/\mathfrak{g}_1$. To indicate the dependence of $\mathfrak{g}$, etc. on Σ we write $\mathfrak{g}(\Sigma)$, etc.

10.1 (Existence of isogenies). *If (Σ, p) is special, it is possible to normalize* *equations* 2.6 *for* $\mathfrak{g}(\sum)$ *and* $\mathfrak{g}(\sum^*)$ *so that the following hold.* (a) *There exists a homomorphism* $\bar{\theta}$ *of* $\mathfrak{g}(\sum)$ *into* $\mathfrak{g}(\sum^*)$ *such that* $\bar{\theta}X_r = X_{r^*}$ *if* r *is long,* $\bar{\theta}X_r = pX_{r^*} = 0$ *if* r *is short, and similar equations hold for* $\bar{\theta}H_r$. (b) *The kernel of* $\bar{\theta}$ *in* (a) *is* $\mathfrak{g}_1(\sum)$. *Thus* $\bar{\theta}$ *induces an isomorphism, also dented* $\bar{\theta}$, *of* $(\mathfrak{g}/\mathfrak{g}_1)(\sum)$ *onto* $\mathfrak{g}_1(\sum^*)$. (c) *If* $\Gamma(\sum)$ *acts on* $(\mathfrak{g}/\mathfrak{g}_1)(\sum)$ *and* $\Gamma(\sum^*)$ *on* $\mathfrak{g}_1(\sum^*)$, *the map* θ: $x^\theta = \bar{\theta}x\bar{\theta}^{-1}$ $(x \in \Gamma(\sum))$ *is an isomorphism of* $\Gamma(\sum)$ *into* $\Gamma(\sum^*)$ *such that* $x_r(t)^\theta = x_{r^*}(t)$ *if* r *is long,* $x_r(t)^\theta = x_{r^*}(t^p)$ *if* r *is short, and similar equations hold for* $h_r(t)^\theta$.

The proof that we have in mind for $(\sum, p)$ of type $(G_2, 3)$ involves many details and will not be given here. When $p=2$, however, the situation is quite simple since $-1 \equiv 1 \pmod 2$, and no normalization is required.

Proof of 10.1 *for* $p=2$. To show that the equations of (a) define a homomorphism, we must verify that the relations 2.1 to 2.6 are preserved. For this the relations 2.2 and 2.6 will suffice since they together with the relations $[H_r, X_r] = 2X_r = 0$ and $[X_r, X_{-r}] = H_r$, which are clearly preserved, imply all of the others (cf. [21]). We give details only for 2.6. Now if either r or s is short, then 2.6 is preserved (both sides go to 0) because $g_1(\sum)$ is an ideal, while if r and s are long and linearly independent, then either $(r, s) < 0$, whence $p_{rs} = 1 = p_{r^*s^*}$, or $(r, s) \geq 0$, whence $p_{rs} = 0$ and $p_{r^*s^*} = 0$ or 2. Since $p=2$, we have (a), and then (b). For the proof of (c), we fix a long root $s \in \sum$. If r is long, $r \in \sum$, then either $r = -s$ and $x_r(t)^\theta X_{s^*} = \bar{\theta}x_r(t)X_{-r} = \bar{\theta}(X_{-r} + tH_r - t^2X_r) = X_{-r^*} + tH_{r^*} - t^2X_{r^*} = x_{r^*}(t)X_{-r^*} = x_{r^*}(t)X_{s^*}$, or $r \neq -s$ and $x_r(t)^\theta X_{s^*} = \bar{\theta}x_r(t)X_s = \bar{\theta}(X_s + p_{rs}tX_{r+s}) = X_{s^*} + p_{r^*s^*}tX_{r^*+s^*} = x_{r^*}(t)X_{s^*}$; whereas if r is short, $r \in \sum$, then either $r+s \notin \sum$ in which case $r^* + s^* \notin \sum^*$ and $x_r(t)^\theta X_{s^*} = X_{s^*} = x_{r^*}(t^p)X_{s^*}$, or $r+s \in \sum$ in which case $2r+s \in \sum$, $(2r+s)^* = r^* + s^*$ and $x_r(t)^\theta X_{s^*} = \bar{\theta}x_r(t)X_s = \bar{\theta}(X_s + tX_{r+s} + t^2X_{2r+s}) = X_{s^*} + t^2X_{r^*+s^*} = x_{r^*}(t^2)X_{s^*}$. Since the X_{s^*} (s long) generate $\mathfrak{g}_1(\sum^*)$, we have (c).

10.2 Corollary (well known). *Over a perfect field of characteristic* 2 *the groups* Γ *of type* B_l *and* C_l *are isomorphic.*

§ 11. Special algebraic groups

In case $(\sum, p)$ is special our previous results on representations can be refined. In $\mathfrak{M}$ let $\mathfrak{M}'$ $(\mathfrak{M}'')$ be the subset each of whose elements has, as

$\mathfrak{g}$-module, high weight λ vanishing on all long (short) roots of S.

11.1 THEOREM. *Assume that* $(\sum, p)$ *is special and regard the elements of* $\mathfrak{M}$ *either as* $\mathfrak{g}$*-modules or as* Γ*-modules. If* $M' \in \mathfrak{M}'$ *and* $M'' \in \mathfrak{M}''$, *then* $M'M'' \in \mathfrak{M}$, *and conversely every element of* $\mathfrak{M}$ *can be expressed, uniquely, as such a product.*

Proof. Assume $M' \in \mathfrak{M}'$ and $M'' \in \mathfrak{M}''$.

(1) *M' restricted to $\mathfrak{g}_1$ is irreducible.* If M_0 is the space spanned by those monomials 2.8 for which all r_i are short, we show first that $M_0 = M'$. Let r be a long positive root. Then $X_r M_0 \subseteq M_0$ because $\mathfrak{g}_1$ is an ideal and $X_r v_+ = 0$. We use induction on the height of r to show that $X_{-r} M_0 \subseteq M_0$. This is so if r is simple since then $X_{-r} v_+ = 0$ because $\lambda(r) = 0$. If r is not simple, there is a simple root a such that $(r, a) < 0$. If a is long, we may write $X_{-r} = \pm [X_{-a}, X_{-(r-a)}]$ and use induction. If a is short, and w denotes the corresponding Weyl reflection and also a corresponding element of Γ, we may apply the induction hypothesis to the module M'^w (see 4.1) with high vector $X_{-a}^{\lambda(a)} v_+$ (see 4.3(a)) and with M_0^w defined accordingly to get $X_{-wr} \cdot M_0^w \subseteq M_0^w$, that is, $X_{-r} M_0^w \subseteq M_0^w$. By 4.3(a) we have $M_0^w = M_0$. Thus $X_{-r} M_0 \subseteq M_0$, M_0 is a $\mathfrak{g}$-submodule of M', and since M' is irreducible, $M_0 = M'$. Next if $v = v_0 + v_1 + \cdots + v_d$ with v_i of height $-i$ and $v_d \neq 0$, we prove by induction on d that the $\mathfrak{g}_1$-module generated by v contains v_+. If $d > 0$, $X_r v \neq 0$ for some $r > 0$ by 2.7. By the induction hypothesis, $X_{r_1} \cdots X_{r_k} X_r v = c v_+$, $c \neq 0$, for a sequence $r_1, r_2, \ldots, r_k$ of short roots. Thus $\sum_{i=1}^k X_{r_1} \cdots [X_{r_i}, X_r] \cdots X_{r_k} v + X_r X_{r_1} \cdots X_{r_k} v = c v_+$. If a term in the sum is nonzero we are done, while if the last term on the left is nonzero we may finish by imitating the last part of the proof that $M_0 = M'$ to show that if $r > 0$ and $X_r v' = v_+$ then v_+ is in the $\mathfrak{g}_1$-module generated by v'. By the two parts above, an arbitrary nonzero element of M' generates M' as $\mathfrak{g}_1$-module, which is (1).

(2) *M'' restricted to $\mathfrak{g}_1$ is* 0. Let M^* be the $\mathfrak{g}(\sum^*)$-module in $\mathfrak{M}'(\sum^*)$ with high weight λ^* given in terms of the high weight λ of M'' by $\lambda^*(a^*) = \lambda(a)$. We may convert M^* into a $\mathfrak{g}(\sum)$-module by the rule $X.v = (\bar{\theta} X)v$ $(X \in \mathfrak{g}(\sum)$, $v \in M^*)$. As such it is irreducible by (1) and the definition of $\bar{\theta}$, is restricted, and has high weight λ. Thus by 2.7 it is equivalent to M''. From the definition of $\bar{\theta}$, $X_r.v = 0$ if r is short, which is (2).

(3) *Proof of* 11.1 *for* $\mathfrak{g}$-*modules.* Choose a nonzero $u = \sum u_i' u_i''$ with the u_i' linearly independent in M' and the u_i'' in M''. By (1) and (2) we can multiply u by a sequence of X_r (r short, $r>0$) to get a nonzero $v = v_+'v''$, and then v by a sequence of X_r (r long, $r>0$) to get a nonzero multiple of $v_+'v_+''$. Using negative roots instead, first short ones and then long ones, we see that this last vector generates $M'M''$: we first get all $v'v_+''$ (v' monomial in M'), and then using induction on the height of v', all $v'M''$, hence $M'M''$, which is thus irreducible. Since it is also restricted, it is in $\mathfrak{M}$. Since the high weight of $M'M''$ is the sum of those of M' and M'', the uniqueness and completeness in 11.1 follow immediately.

(4) *Proof of* 11.1 *for* Γ-*modules.* As is easily verified, the process 4.2 of lifting the modules of $\mathfrak{M}$ from $\mathfrak{g}$ to Γ is consistent with the two types of tensor products—for algebras and for groups. Thus (4) follows from (3).

11.2 COROLLARY. *If* $(\sum, p)$ *is special, the map* $\bar{\theta}$ (*resp.* θ) *puts the algebra modules* (*resp. group modules*) *of* $\mathfrak{M}''(\sum)$ *in one-one correspondence with those of* $\mathfrak{M}'(\sum^*)$.

In case $(\sum, p)$ is special, 11.1 and 11.2 lead to corresponding refinements of 5.1, 6.1 and 7.4 with $\mathfrak{M}$ replaced by $\mathfrak{M}'$ and $\mathfrak{M}''$. However, there does not seem to be a refinement of the semisimple classes of elements in 7.3. Passing on to 8.2, 8.3 and 8.4, we have:

11.3 COROLLARY. *Assume that* $(\sum, p)$ *is special. Let* ω' $(\omega'') = \prod \omega_a$, *the product over the short* (*long*) *simple roots, let* l' (l'') *be the number of short* (*long*) *simple roots, and let* m' (m'') *be the uumber of short* (*long*) *positive roots. Then* (a) *the* Γ_q-*module* M_q' (M_q'') *with high weight* ω'^{q-1} (ω''^{q-1}) *has dimension* $q^{m'}$ $(q^{m''})$ *and character at a semisimple element* h *given by* $\chi(h) = \pm q^{d(x)}$ *with* $2\,d(x) + l'$ (l'') *the dimension of the set of fixed points of* x *on* $\mathfrak{g}_1$ $(\mathfrak{g}/\mathfrak{g}_1)$. (b) *The* Γ_q-*module* M_p' (M_p'') *has dimension* $p^{m'}$ $(p^{m''})$ *and a basis consisting of all monomials* 2.8 *corresponding to short* (*long*) *roots* r_i.

The proofs are as in §8 and will be omitted. To supplement 11.3 we remark that $M_q = M_q'M_q''$, that M_q' is quite similar to the corresponding module for the group Γ of characteristic 0 (it is of the same dimension, by Weyl's formula), that M_q'' in contrast has lower dimension, that M_q' for $\sum$ of type B_l

and $q=2$ is just the spin module [13, p. 20-04] of dimension 2^l, and that $m'/m''=l'/l''$ in all cases [18, p. 501].

§ 12. Twisted groups

In this, the last, section we extend our results to the finite simple groups associated with the names of Suzuki [22] and Ree [11]. These are defined as follows. A system $\sum$ of type C_2, F_4 or G_2 may be identified with its dual $\sum^*$ in such a way that the map $r\to r^*$ of $\sum$ on $\sum^*$ (see § 10) yields an involutary map, σ, on $\sum$ such that positivity and simplicity of roots is preserved, angles are preserved, but short and long roots are interchanged. If also $(\sum, p)$ is special, K is algebraically closed, and $k\geq 1$, the isomorphism θ of 10.1(c) accordingly yields an automorphism of Γ which combines with the p^kth power map to produce an automorphism, also denoted σ, such that

12.1 $$x_r(t)^\sigma = x_{\sigma r}(t^{ps}),\ x_{\sigma r}(t)^\sigma = x_r(t^s) \qquad (r \text{ short},\ s=p^k).$$

If now $n=2k+1$ and $q=p^n$, the fixed points of σ form the sought simple group, to be denoted Γ_q^1, a subgroup of Γ_q. Observe (see 11.3) that $l'=l''$ and $m'=m''$ here.

12.2 THEOREM. *If α_i denotes the p^ith power map, every irreducible Γ_q^1-module can be written uniquely as $\prod_{i=0}^{n-1} M_i^{\alpha_i}$ $(M_i\in\mathfrak{M}')$. If the rank of $\sum$ is l, there are $q^{l/2}$ such modules. Each can be realized over F_q.*

Proof. If $r>0$ and $t\in F_q$, Γ_q^1 contains an element of the form $x= x_r(t)x_{\sigma r}(t^{ps})\prod x_{r'}(t')$, the product over roots r' which are positive integral linear combinations of r and σr (proof by downward induction on the height of r). Thus if $v\in M\in\mathfrak{M}'$ and v is homogeneous,

12.3 $$(x-1)v = tX_r v + t^{ps}X_{\sigma r}v + \text{higher terms}.$$

If we refine the notion of height so that a positive linear combination of simple roots is taken to be lower than another one of the same height as previously defined if fewer short simple roots are used, then r is lower than σr. In fact, if $r=\sum[c(a)a+d(a)\sigma a]$, the sum over the short simple roots a, then $\sigma r=\sum[pd(a)a+c(a)\sigma a]$: since σ preserves angles, the map $r'\to p^{-1/2}\sigma r'$, $\sigma r'\to p^{1/2}r'$ (r' short) comes from an isometry, which maps r to $\sum[p^{-1/2}c(a)\sigma a + p^{1/2}d(a)a] = p^{-1/2}\sum[pd(a)a+c(a)\sigma a]$. With a corresponding refinement in the notion of homogeneity, we thus get 12.3 with the second term of the right missing, and a similar equation for negative roots. Combining these equations with the key fact, proved as 11.1(1), that each $M \in \mathfrak{M}'$ is irreducible as $\mathfrak{g}_1$-module, we can now prove, almost exactly as in 5.1, that $\Pi M_i^{a_i}$ $(M \in \mathfrak{M}')$ is irreducible. The proof that this product determines its components M_i can also be taken from 5.1. As for completeness of the set of $q^{l/2}$ product modules, the semisimple classes of Γ_q^1 are characterized by coordinates $\gamma(a)$ and $\gamma(\sigma a)$ $(a \in S,\ a$ short) for which $\gamma(a)^{ps} = \gamma(\sigma a)$ and $\gamma(\sigma a)^s = \gamma(a)$ (see 3.9 and 12.1); their number is thus $q^{l/2}$, as required. Finally, as the restriction of a Γ_q-module, each irreducible Γ_q^1-module can be realized over F_q by 7.5.

As a supplement to 12.2 we have:

12.4. *For $\sum$ of type C_2 or F_4 and $p = 2$, every irreducible projective representation of Γ_q^1 can be lifted to an ordinary one.*

This result, proved in [21], quite likely also holds for the remaining case, $\sum$ of type G_2 and $p = 3$.

Also since the axioms of [17] are easily verified for the groups Γ_q^1 there is an analogue of 8.1, and modifying the development of § 8, we have no trouble in proving:

12.5 Theorem. (a) *Γ_q^1 has an irreducible module M_q', constructed by the methods of* [17], *of dimension $q^{m/2}$, the order of a Sylow group in Γ_q^1 (here m is the number of positive roots in $\sum$).* (b) *M_q' is equivalent to the restriction to Γ_q^1 of the Γ_q-module M_q' of* 11.3. (c) *All other irreducible Γ_q^1-modules have lower dimensions than M_q'.*

From 12.2 we know all irreducible Γ_q^1-modules (in fact all Γ_q-modules also by 7.4, 11.1 and 11.2) once we know those in $\mathfrak{M}'$. We consider the individual cases. For $\sum$ of type C_2, $p^{l/2} = 2$, so that the trivial 1-dimensional module and the 4-dimensional module M_2' of 11.3(b), on which Γ_q acts as the symplectic group, exhaust $\mathfrak{M}'$. For $\sum$ of type G_2, $p^{l/2} = 3$, we have in $\mathfrak{M}'$ the trivial module, the module $\mathfrak{g}_1$, of dimension 7, and the module M_3' of 11.3(b), of dimension $3^3 = 27$. Finally, for $\sum$ of type F_4, $p^{l/2} = 4$, there are the trivial module, the module $\mathfrak{g}_1$, of dimension 26, and the module M_2', of dimension $2^{12} = 4096$, leaving one module yet to be described.

References

[1] Brauer, R. and Nesbitt, C., On the modular representations of groups of finite order, U. of Toronto Studies (1937), 1-21.

[2] Brauer, R. and Nesbitt, C., On the modular characters of groups, Ann. of Math. **42** (1941), 556-590.

[3] Chevalley C., Sur certains groupes simples, Tôhoku Math. J. **7** (1955), 14-66.

[4] Curtis, C. W., Representations of Lie algebras of classical type with applications to linear groups, J. Math. and Mech. **9** (1960), 307-326.

[5] Curtis, C. W., On projective representations of certain finite groups, Proc. Amer. Math. Soc. **11** (1960), 852-860.

[6] Curtis, C. W., On the dimensions of the irreducible modules of Lie algebras of classical type, Trans. Amer. Math. Soc. **96** (1960), 135-142.

[7] Dickson, L. E., The abstract form (two papers), Quart. J. Math. **38** (1907), 141-158.

[8] Hertzig, D., Forms of algebraic groups, Proc. Amer. Math. Soc. **12** (1961), 657-660.

[9] Lang, S., Algebraic groups over finite fields, Amer. J. Math. **78** (1956), 555-563.

[10] Mark, C., Thesis, U. of Toronto.

[11] Ree, R., A family of simple groups associated with the simple Lie algebra of type (F_4), second paper: (G_2), Amer. J. Math. **83** (1961), 401-420, 432-462.

[12] Rosenlicht, M., Some rationality questions on algebraic groups, Ann. di Mat. **43** (1957), 25-50.

[13] Séminaire C. Chevalley, Classification des Groupes de Lie Algébriques (two volumes), Paris (1956-8).

[14] Séminaire "Sophus Lie," Paris (1954-5).

[15] Serre, J. P., Groupes algébriques et corps de classes, Hermann, Paris (1959).

[16] Steinberg, R., A geometric approach, Trans. Amer. Math. Soc. **71** (1951), 274-282.

[17] Steinberg, R., Prime power representations of finite linear groups II, Can. J. Math. **9** (1957), 347-351.

[18] Steinberg, R., Finite reflection groups, Trans. Amer. Math. Soc. **91** (1959), 493-504.

[19] Steinberg, R., Variations on a theme of Chevalley, Pacific J. Math. **9** (1959), 875-891.

[20] Steinberg, R., The simplicity of certain groups, Pacific J. Math. **10** (1960), 1039-1041.

[21] Steinberg, R., Générateurs, relations et revêtements de groupes algébriques, Colloque sur la théorie des groupes algébriques, Bruxelles (1961).

[22] Suzuki, M., On a class of doubly transitive groups, Ann. of Math. **75** (1962), 105-145.

[23] Tits, J., Sur les analogues algébriques des groupes semisimples complexes, Colloque d'algèbre supérieure, Bruxelles (1956), 261-289.

[24] Tits, J., Les "formes réelles" des groupes de type E_6, Séminaire Bourbaki 162, Paris (1958).

[25] Tits, J., Sur la trialité et certains groupes qui s'en déduisent, Paris Inst. Hautes Etudes Sci. Publ. Math. **2** (1959), 37-84.

[26] Weyl, H., Theorie der Darstellung kontinuierlicher halb-einfacher Gruppen durch lineare Transformationen III, Math. Zeit (1926), 377-395.

[27] Wong, W. J., Thesis, Harvard U.

Institute for Advanced Study, Princeton
University of California, Los Angeles

DIFFERENTIAL EQUATIONS INVARIANT UNDER FINITE REFLECTION GROUPS

BY

ROBERT STEINBERG

1. **Introduction and statements of results.** In this paper we study the characteristic functions (eigenfunctions) of those differential operators with constant coefficients that are invariant under finite linear groups, especially under finite reflection groups. Such operators and functions occur naturally in connection with harmonic analysis on semisimple Lie groups (see, e.g., [4]). Our main result (1.3 below) is a manifold of characterizations of finite reflection groups in terms of the above-mentioned characteristic functions.

To state this result, we introduce some notations which will be used throughout the paper. V is a vector space of finite dimension n over K, the complex field, V^* is the dual of V, S is the symmetric algebra on V, and S^* is the algebra of entire functions on V. For v in V, D_v is the operator on S^* defined by $(D_vF)(w) = \lim t^{-1}(F(w+tv)-F(w))$ ($t \in K$, $t \to 0$; $w \in V$; $F \in S^*$), and for s in S, D_s is then defined so as to make D an isomorphism. The space S^* (and also the space of polynomial functions on V) is in algebraic duality with S via the inner product $(s,F) = (D_sF)(0)$ ($s \in S$, $F \in S^*$). Any automorphism σ of V acts directly on S and contravariantly on S^* by the rule $(\sigma F)(v) = F(\sigma^{-1}v)$; one has $\sigma(D_sF) = D_{\sigma s}\sigma F$, so that the above inner product is preserved. Whenever a group G of automorphisms of V is being considered, I consists of the elements of S invariant under G, I_0 consists of the elements of I that vanish at 0, and S_0 is the ideal in S generated by I_0.

Assume now that G, of order g, is a finite group of automorphisms of V and that F is a nonzero element of S^* which is a characteristic function of all of the operators $D_i (i \in I)$: $D_iF = c_iF$. Then the map $i \to c_i$ is a homomorphism of the algebra I into K, which can be extended [5, p. 420] to a homomorphism of S into K; thus $c_i = i(L)$ for some L in V^* and all i in I. We are thus led to consider, for each L in V^*, the following system of differential equations, to be solved for F in S^*.

1.1 (Σ_L) $$D_iF = i(L)F \quad (i \in I).$$

We write Σ_L for this system and d_L for the dimension of the space of solutions of Σ_L. As a backdrop for our main theorem, we have the following preliminary results, of which at least (b′), (c), (d), and (e) are well known. Here,

Presented to the Society, February 22, 1962; received by the editors April 26, 1963.

and elsewhere, $\Sigma_0(G_L)$ denotes the system defined in terms of the subgroup G_L just as Σ_0 is defined in terms of G.

1.2 THEOREM. *Let G be a finite group of automorphisms of V, and for each L in V^* let G_L, of order g_L be the subgroup that fixes L.* (a) *If F_0 is a solution of $\Sigma_0(G_L)$, then $F_0 \exp L$ is a solution of Σ_L.* (b) *Each solution of Σ_L can be written $F = \sum_{\sigma \in G} \sigma(P_\sigma \exp L)$, each P_σ being a polynomial of degree at most $(g_L - 1)^n$.* (b′) *Each solution of Σ_0 is a polynomial of degree at most $(g-1)^n$.* (c) *The solutions of Σ_0 form the orthogonal complement of S_0 in S^*. Thus $d_0 = \dim S/S_0$.* (d) $g \leqq d_L < \infty$. (e) *If m is the number of elements in a minimal generating set for I_0, then $n \leqq m < \infty$.*

This brings us to our main theorem. We define a reflection to be an automorphism of V other than the identity which fixes a hyperplane pointwise and is of finite order (not necessarily two), and a reflection group to be a group generated by reflections.

1.3 THEOREM. *If G is a finite group of automorphisms of V, the following conditions are equivalent.* (a) *G is a reflection group.* (b) *There exists a polynomial P such that s is in S_0 if and only if $D_sP = 0$.* (c) *There exists a polynomial P such that F is a solution of Σ_0 if and only if $F = D_sP$ for some s in S.* (c′) *There exists a character ε from G into the multiplicative group of K such that, if $P^{(L)}$ denotes a polynomial of minimal degree among those which transform under G_L according to ε and are nonzero, and if $P_L = g^{-1} \sum_{L \in G} \varepsilon(\sigma)^{-1} \sigma(P^{(L)} \exp L)$, then F is a solution of Σ_L if and only if $F = D_sP_L$ for some s in S.* (c″) *For each L in V^* there exists a polynomial $P^{(L)}$ such that F_0 is a solution of $\Sigma_0(G_L)$ if and only if $F_0 = D_sP^{(L)}$ for some s in S.* (d) *d_L is independent of L.* (e) $d_0 = g$. (e′) $\dim S/S_0 = g$. (f) *If $i_0, i_1, \cdots, i_r$ in I are such that i_0 is not in the ideal in I generated by the others, and if $p_0, p_1, \cdots, p_r$ in S are homogeneous and such that $\sum p_a i_a = 0$, then p_0 is in S_0.* (g) *I is generated, as an algebra over K, by n of its elements.*

That every finite reflection group has properties (e′), (f) and (g) is due to Chevalley [1] (he considers only reflections of order two, but his methods apply equally well to reflections of any order), and that (g) implies (a) is due to Shephard and Todd [7]. To amplify 1.3 we introduce some more notation. H is the set of hyperplanes in which reflections of G take place, and for each h in H, R_h is a nonzero element of V^* which vanishes on h and $e(h)$ is the order of the (necessarily cyclic) group generated by the reflections in h that are in G.

1.4 THEOREM. *Let G be a finite reflection group on V. Let Π denote the product of all $R_h^{e(h)-1}$ as h varies over the reflecting hyperplanes for G, and for each L in V^* let $\Pi^{(L)}$ denote the corresponding product for the subgroup G_L that fixes L, and let $\Pi_L = g^{-1} \sum_{\sigma \in G} (\det \sigma^{-1}) \sigma(\Pi^{(L)} \exp L)$.* (a) *If P is a homogeneous* *polynomial for which* 1.3(c) *holds, then P is a constant multiple of* Π. (b) *If ε is such that* 1.3(c′) *holds, then* $\varepsilon = \det$. (c) *If $P^{(L)}$ and P_L are such that* 1.3(c′) *holds, they are constant multiples of* $\Pi^{(L)}$ *and* Π_L, *respectively.* (d) *To within multiplication by a constant,* Π_L *is the unique function which is a solution of* Σ_L *and transforms according to* det.

We remark that in case V is a Cartan subalgebra of a semisimple Lie algebra A over the complex field and G is the corresponding Weyl group the polynomial Π in 1.4 is just a constant multiple of the product of the positive roots (relative to some ordering of V) (see [4] where significant use is made of this polynomial), and, if L is identified with an element of V, then the value at L of any finite-dimensional irreducible character of A is just the ratio of two appropriately chosen values of Π_L (this is Weyl's formula; see, e.g., [6, pp. 255, 257] for the two most important cases: in the first, L is not on any reflecting hyperplane, while in the second, L is 0).

Clearly not every subgroup of a reflection group is a reflection group, but as a biproduct of 1.3 we prove:

1.5 THEOREM. *Let G be a reflection group on the space V and let U be a subspace* of V. *Then the subgroup of G that fixes U pointwise is also a reflection group.*

From this we have the following corollaries, important in the context of Lie algebras (see [6, p. 242] or [8, pp. 16–106]).

1.6 COROLLARY. *If v is a vector not on any of the reflecting hyperplanes of a reflection group G, the only element of G that fixes v is the identity.*

1.7 COROLLARY. *Let G be a reflection group, and assume that the linear functions R_h, defined just prior to* 1.4 *above, are such that for some vector v every $R_h(v)$ has a positive real part. Then the only element of G which permutes the R_h among themselves is the identity.*

As is well known, the solutions of a system of differential equations such as Σ_L enjoy certain mean value properties, in other words, satisfy certain difference equations [5, pp. 435–439]. The connection will be discussed in §8 below. Finally, to close this introduction, let us consider an example.

1.8 EXAMPLE. Let $x_1, x_2, \cdots, x_n$ be coordinates for V and let G be the symmetric group of degree n acting via permutations of the coordinates. Then G is generated by reflections in the hyperplanes $x_i - x_j = 0$ $(i < j)$, so that properties (b) and (c) of 1.3 hold with $P = \prod_{i<j}(x_i - x_j)$. This special case is due to E. Fischer and J. Schur [2], whose methods, however, are not applicable to the general case. That 1.3(e′) also holds here is a very old result. If we modify the example so that G consists of permutations of the coordinates combined with multiplications by arbitrary rth roots of 1, then P is to be replaced by $\prod(x_i^r - x_j^r)$.

The interested reader should have no trouble in constructing in these cases, for typical values of L, the function Π_L of 1.4 as well as the differential equations Σ_L and the corresponding difference equations Σ_L' considered in §8. For further examples, we refer to [7] where the classification of finite reflection groups is completed.

2. **General properties of Σ_L.** This section is devoted to the proof of 1.2. In the notation of (a), and with i in I, we have $D_i(F_0 \exp L) = (\exp L)D_jF_0$, with j in S defined by $j(L') = i(L'+L)$ $(L' \in V^*)$. Here j need not be invariant under G, but it is invariant under G_L. Thus $D_jF_0 = j(0)F_0 = i(L)F_0$, and (a) is proved.

2.1 Lemma. *If F is a solution of Σ_L, it is also a solution of the system Γ_L: $D_sF = 0$ $(s = \prod_{\sigma \in G}(v - (\sigma^{-1}L)(v)), v \in V)$.*

We have an identity $0 = \prod_{\sigma \in G}(v - \sigma v) = v^g + iv^{g-1} + \cdots$ with $i, \cdots$ in I_0. Thus $D_sF = (D_v^g + i(L)D_v^{g-1} + \cdots)F$, which, by Σ_L, is equal to $(D_v^g + D_iD_v^{g-1} + \cdots)F$, which is equal to D_0F, that is, to 0, by the identity above.

We now prove 1.2(b). By allowing v in the lemma to vary over a basis of V, we see that each solution of Σ_L can be written as the sum of a finite number of terms $P \exp L'$, with L' a linear function and P a polynomial. Consider a nonzero term of this form. Since $P \exp L'$ is a solution of Γ_L, P satisfies $D_tP = 0$ with $t = \prod(v - (\sigma^{-1}L)(v) + L'(v))$. If r (which depends on v) is the number of σ in G such that $(\sigma^{-1}L)(v) = L'(v)$, we may write $D_tP = 0$ as $D_v^rP = (c_1D_v + c_2D_v^2 + \cdots)$ D_v^rP $(c_j \in K)$, whence, since D_v^rP is a polynomial, $D_v^rP = 0$. Now L' must equal some $\sigma^{-1}L$ since otherwise we could find v so that r is 0 and then conclude $P = 0$. Consider the case $L' = L$. We choose a basis B for V such that for v in B and σ in G we have $(\sigma^{-1}L)(v) = L(v)$ only if σ is in G_L. For each v in B, the number r above is then g_L, so that the equations $D_v^rP = 0$ imply that the degree of P is at most $(g_L - 1)^n$. Thus we have (b), and as a special case, also (b').

For (c), we have the chain of equivalent statements: F is in the orthogonal complement of S_0; $(si, F) = 0$ for all s in S and i in I_0; $D_iF = 0$ for all i in I_0; F is a solution of Σ_0.

Let $s_1, s_2, \cdots, s_h$ be homogeneous elements of S which project into a basis for S/S_0. By induction on the degree, every element of S can be put in the form $\Sigma i_a s_a$ with i_a in I. By Galois theory [9, p. 156, Fundamental Theorem, part 4] applied to the quotient fields of S and I, we get $h \geqq g$, which, by (c), is the same as $d_0 \geqq g$. Writing $d_{0L} \geqq g_L$ for the corresponding inequality for the group G_L, we have, by (a), $d_L \geqq d_{0L}g/g_L \geqq g_Lg/g_L = g$. By (b), we have $d_L < \infty$, hence (d).

Finally, $n \leqq m$ in (e) because the two quotient fields mentioned above have the same degree of transcendence over K, namely n, and $m < \infty$ easily follows from $\dim S/S_0 < \infty$.

3. **Proof of Theorem 1.3, first part.** In this section we prove first that 1.3(a) implies 1.3(b) and that for P we may take the polynomial Π defined in 1.4, and

then we prove that 1.3(b) implies 1.3(c). For convenience, we call a function F skew if $\sigma F = (\det \sigma)F$ for every σ in G.

3.1 Lemma. (a) *The polynomial* Π *is skew.* (b) Π *divides every skew polynomial.*

Proof. If σ is a reflection in G and h is the corresponding hyperplane, then $\sigma R_h = (\det \sigma^{-1})R_h$ and $(\det \sigma)^{e(h)} = 1$. Thus $\sigma(R_h^{e(h)-1}) = (\det \sigma)R_h^{e(h)-1}$. If k is a reflecting hyperplane different from h, there is a smallest integer m such that $\sigma^m R_k = cR_k$ for some c in K. Since $\sigma^m R_k - R_k$ is a multiple of R_h, we have $c = 1$. Thus σ preserves the product of $R_k, \sigma R_k, \cdots, \sigma^{m-1}R_k$, whence the $R_l^{e(l)-1}$ for which $l \neq h$ may be arranged in cycles so that σ preserves the product of the terms in each cycle. Thus $\sigma\Pi = (\det \sigma)\Pi$, and since the reflections generate G, Π is skew. Now, assume that h and σ are as above and also that σ has order $e(h)$. Choose a basis $X, Y, \cdots$ of V^* so that $X = R_h$, $\sigma Y = Y$, $\sigma Z = Z, \cdots$. Then σ acts on a monomial $X^j Y^k \cdots$ via multiplication by $(\det \sigma)^{-j}$, a number which is equal to $\det \sigma$ only if $e(h)$ divides $j + 1$, hence only if j is at least $e(h) - 1$. Thus every skew polynomial is divisible by $X^{e(h)-1} = R_h^{e(h)-1}$, and so also by Π.

Assuming now that G is a finite reflection group, we prove 1.3(b) with $P = \Pi$. If s is in I_0, $D_s\Pi$ is skew and of lower degree than Π; thus it is 0 by 3.1(b), and the "only if" part of 1.3(b) holds. Conversely, let s in S be homogeneous and such that $D_s\Pi = 0$. We prove by downward induction on the degree of s that s is in S_0, the result being true for a sufficiently high degree because dim S/S_0 is finite, by 1.2(c,d). Let σ be a reflection in G. We may choose v in V so that $\sigma v = (\det \sigma)v$. Then $t = vs$ has higher degree than s and $D_t\Pi = 0$. Thus by the induction hypothesis vs is in S_0 and there is a relation $vs = \Sigma u_p i_p$ with u_p in S and i_p in I_0. On applying σ and then combining the two relations, we get $s - (\det \sigma)\sigma s = \Sigma((u_p - \sigma u_p)/v)i_p$, an element of S_0. Because G is a reflection group, we get $s \equiv (\det \sigma)\sigma s \bmod S_0$ for all σ in G, and then by averaging over G, $s \equiv s' \bmod S_0$, the element s' being such that $\sigma s' = (\det \sigma^{-1})s'$. Applying 3.1 to the group G^* dual to G, we may write $s' = \pi i$, with π defined for G^* just as Π is for G, and with i in I. If i has positive degree, s' is in S_0 and hence so is s. Assume then that i is constant, that Q is any homogeneous polynomial of the same degree as Π, and that R is the average of $(\det \sigma^{-1})\sigma Q$ under G. Then R is a multiple of Π by 3.1(b), so that $D_{s'}R = 0$ and $(s', R) = 0$. Thus $(s', Q) = g^{-1}\,\Sigma\sigma(s', Q) = g^{-1}\,\Sigma((\det \sigma^{-1})s', \sigma Q) = (s', R) = 0$, and since Q is arbitrary, $s' = 0$, so that s is in S_0 in this case also, and 1.3(b) is proved.

3.2 Corollary. *If G is a reflection group and Π is of degree N, then S_0 contains every homogeneous element of S of degree greater than N, and N is the smallest integer with this property.*

Now we deduce 1.3(c) from 1.3(b). Let DP denote the space of derivatives D_tP of P. We have the chain of equivalent statements: s is in the orthogonal complement of DP; $(s, D_tP) = 0$ for all t in S; $(t, D_sP) = 0$ for all t in S; $D_sP = 0$; s is in S_0. The last equivalence is by the assumption 1.3(b). Thus S_0 is the orthogonal complement of DP. Because DP is finite dimensional, it in turn is the orthogonal complement of S_0. Hence, by 1.2(c), it coincides with the space of solutions of Σ_0, which is 1.3(c).

4. **Continuation of the proof.** In this section we prove the equivalence of the conditions (c), (c′) and (c″) in 1.3.

4.1 LEMMA. (a) *If P satisfies* 1.3(c), *then P_0, the homogeneous part of highest degree of P, also does.* (b) *If P is homogeneous and satisfies* 1.3(c), *there exists a character ε on G such that $\sigma P = \varepsilon(\sigma)P$ for all σ in G.* (c) *If P and ε are as in* (b) *and Q in S^* satisfies $\sigma Q = \varepsilon(\sigma)Q$ $(\sigma \in G)$, then $Q = cP +$ higher terms $(c \in K)$.*

Proof. P_0 is a solution of Σ_0, whence $P_0 = D_sP$ $(s \in S)$, by 1.3(c), and clearly $s(0) \neq 0$. Hence the map $D_tP \to D_s(D_tP) = D_tP_0$ is an isomorphism of the space of solutions of Σ_0 into itself, and since the space is finite dimensional, the isomorphism is onto. Thus $P = D_tP_0$ for some t in S, and (a) follows. If P is as in (b), then σP satisfies Σ_0, so that $\sigma P = D_sP$ for some s in S. Since σP and P have the same degree, s may be taken constant, $s = \varepsilon(\sigma)$, and clearly ε is a character on G. In proving (c), we need only consider the case in which Q is a nonzero homogeneous polynomial which transforms according to ε and has minimal degree relative to these properties. Since for i in I_0, D_iQ has smaller degree than Q, it must be 0, so that Q is a solution of Σ_0, and $Q = D_sP$ for some s in S, by 1.3(c). For σ in G, $Q = D_{\sigma s}P$, so that by averaging over G we may take s in I. But then, because P satisfies Σ_0, $Q = s(0)P$, which implies (c).

4.2 LEMMA. *Let F satisfy Σ_L, and let $F = F_r + F_{r+1} + \cdots$ with F_j a homogeneous polynomial of degree j. Then* (a) *F_r satisfies Σ_0, and* (b) *if P satisfies* 1.3(c) *and r exceeds the degree of P, then $F = 0$.*

Proof. If i is homogeneous and in I_0, all terms of degree less than r in $D_iF = i(L)F$ must vanish, whence $D_iF_r = 0$, which is (a). Thus by 1.3(c) we have $F_r = D_sP$ $(s \in S)$ in (b). Hence F_r either vanishes or has degree at most that of P, which implies (b).

We come now to the proof that 1.3(c) implies 1.3(c′). For this we assume that P is homogeneous (see 4.1(a)), that ε is defined in terms of P as in 4.1(b), and that $P^{(L)}$ and P_L are defined in terms of ε according to 1.3(c′). If i is in S, homogeneous, of positive degree, and invariant under G_L, then $D_iP^{(L)}$ has smaller degree than $P^{(L)}$, hence is 0 because of the definition of $P^{(L)}$, whence $P^{(L)}$ is a solution of $\Sigma_0(G_L)$. By 1.2(a), then P_L and all of its derivatives are solutions of Σ_L. We prove the converse in a sharpened form.

4.3 LEMMA. *If $s_1, s_2, \cdots, s_h$ are homogeneous elements of S which project into a vector space basis in S/S_0, and if P_L is as above, then every solution of Σ_L has the form $F = D_sP_L$ with $s = \sum c_js_j$ $(c_j \in K)$.*

We prove this by downward induction on the number r that occurs in any representation of F in the form of 4.2, the result being true if r is large enough by 4.2(b). By 4.2(a) and 1.3(c) we may write $F_r = D_t P$ with t in S, and because P is homogeneous and is annihilated by D_s for every s in S_0, we may take t in the form $\Sigma a_j s_j$ ($a_j \in K$). Now $P_L \neq 0$ because distinct exponentials are linearly independent over the polynomials; so by 4.1(c) and 4.2(b) we may write $P_L = cP +$ higher terms with $c \neq 0$, c in K. But then $F - c^{-1} D_t P_L$ may be written in the form of 4.2 with $r + 1$ in place of r, whence the induction hypothesis may be applied to complete the proof of 4.3, and hence also the proof of 1.3(c′).

Assume now that 1.3(c′) holds. Because of the defining properties of $P^{(L)}$, we have already noted that $P^{(L)}$ is a solution of $\Sigma_0(G_L)$. Let F_0 be any solution of this system. By 1.2(a) and 1.3(c′), we may write $F_0 \exp L = D_t(P^{(L)} \exp L)$, with t in S. This implies that $F_0 = D_s P^{(L)}$, with s defined by $s(L') = t(L' + L)$ ($L' \in V^*$). Thus 1.3(c′) implies 1.3(c″).

Since 1.3(c) is just the case $L = 0$ of 1.3(c″), we have the equivalence of (c), (c′) and (c″) in 1.3.

5. **Completion of the proof.** Assuming now that 1.3(c) holds we prove that d_L is independent of L, and is in fact equal to dim S/S_0. For this it is enough to show that if $D_s P_L = 0$ with s as in 4.3, then $s = 0$. Assume that $s \neq 0$, that t denotes the sum of the terms of highest degree, say d, in the expression for s, and that N is the degree of P, assumed (see 4.1(a)) to be homogeneous. We have seen (three paragraphs back) that $P_L = cP +$ higher terms, with $c \neq 0$, c in K. Since the terms of degree $N-d$ in $D_s P_L$ are 0, we get $D_t P = 0$, then $(t, F) = 0$ for all solutions F of Σ_0 by 1.3(c), whence t is in S_0 by 1.2(c). From the way in which s and t have been chosen, this implies $t = 0$, a contradition. Thus $d_L = \dim S/S_0$, a number independent of L.

Next assume that 1.3(d) holds. If L is chosen so that G_L consists of the identity alone, then $d_L = g$ by 1.2(a) and 1.2(b). Thus $d_0 = g$ by 1.3(d), which is 1.3(e).

The equivalence of 1.3(e) and 1.3(e′) follows from 1.2(c). Assume 1.3(e′) holds. Let $s_1, s_2, \cdots, s_g$ be homogeneous elements of S which form a basis for a space complementary to S_0 in S. Every element of S can be written $s = \Sigma i_b s_b (i_b \in I)$, and by Galois theory [9, p. 156], the expression is unique. Write $p_a = \Sigma i_{ab} s_b$ (i_{ab} homogeneous and in I), so that $\Sigma\, \Sigma i_{ab} s_b i_a = 0$, whence $\Sigma i_{ab} i_a = 0$ for $b = 1, 2, \cdots, g$. If some i_{0b} is not in I_0, it is of degree 0 and we may take it to be 1. But then the bth equation above yields $i_0 + (i_{1b} i_1 + \cdots + i_{rb} i_r) = 0$, contradicting the original assumptions on i_0. Thus every i_{0b} is in I_0 and p_0 is in S_0, which is 1.3(f).

Now Chevalley [1] has proved that 1.3(f) implies 1.3(g), and Shephard and Todd [7] have proved that 1.3(g) implies 1.3(a). Thus the cycle is complete, and 1.3 has been established.

6. **Proof of 1.4.** Assume that G is a reflection group. We have proved in §3 that 1.3(c) is true with $P = \Pi$. If P' is another homogeneous polynomial for which

1.3(c) is true, then P' is a derivative of Π and of the same degree as Π, so is a constant multiple of Π, which proves 1.4(a). At the same time, this shows that the polynomial $P_0 = P^{(0)}$ of 1.3(c′) is a constant multiple of Π, whence, by 3.1(a), $\varepsilon = \det$, which is 1.4(b). Now by the equivalence of (a), (c) and (c″) in 1.3, G_L is a reflection group. Applying 3.1 to this group, we see that among the nonzero polynomials that transform under G_L according to det, the nonzero constant multiples of $\Pi^{(L)}$ are those of minimal degree, which implies 1.4(c). Now if P_L is a solution of Σ_L which transforms under G according to det, then $P_L = D_s \Pi_L$ with s in S, by (c), and by averaging over G we may take s in I. But then $P_L = s(L)\Pi_L$ because Π_L satisfies Σ_L, and this is (d).

As a consequence of 1.4(d), it may be observed that 1.3 and 1.4 are substantially true if the base field K is taken to be the real field, if L is taken in the complexification of V^* so that all $i(L)$ $(i \in I)$ are real, and only real solutions of Σ_L are considered. For then by 1.4(d) both the real and imaginary parts of Π_L are multiples of Π_L, and whichever of these is nonzero can take the place of Π_L in the development.

7. **Reflection subgroups.** It is enough to prove 1.5 in the case that $\dim U = 1$ since the general case then follows by induction. Thus, going over to the dual group, we need only prove that if L is in V^* then G_L is a reflection group. But, as has already been remarked, this is a consequence of the equivalence of (a), (c) and (c″) in 1.3.

In 1.6, the subgroup that fixes v is a reflection group by 1.5, but contains no reflections by assumption, hence consists of the identity, which proves 1.6.

Under the assumptions of 1.7, let σ in G permute the R_h among themselves, let k be the order of σ, and let $w = v + \sigma v + \cdots + \sigma^{k-1} v$. Since $R_h(\sigma^j v) = (\sigma^{-j} R_h)(v)$ and σ permutes the R_h, the real part of every $R_h(w)$ is positive, so that w is not on any reflecting hyperplane for G. Since $\sigma w = w$, we conclude that σ is the identity, by 1.6.

8. **Difference equations and mean values.** For v in V, we use T_v to denote the translation operator that acts on F in S^* via the rule $(T_v F)(v') = F(v' + v)$, and we call a finite sum $\Sigma c_v T_v$ $(c_v \in K)$ a difference operator. If G is a finite group of automorphisms of V, then in analogy with our previous development we are especially interested in the characteristic functions of those difference operators that are invariant under G. This leads us to the following system of difference equations, to be solved for F in S^*.

8.1 $$g^{-1} \sum_{\sigma \in G} T_{\sigma v} F = C(v) F \qquad (v \in V, C(v) \in K).$$

We remark that L. Flatto [3] has considered equations such as 8.1 with $C = 1$ and G the symmetry group of a tetrahedron or octahedron. Now if F satisfies Σ_L (see 1.1), then by Taylor's formula ($T_v = \exp D_v$), F satisfies 8.1 with C given by:

8.2 $$C = g^{-1} \sum_{\sigma \in G} \exp \sigma L.$$

Conversely, assume that the system of equations 8.1 has a nonzero solution F. (Here it is enough to assume that F is continuous on some open set in V, since then F is infinitely differentiable there (see [5, p. 438]), and this is all we need for the argument that follows. Similarly in our previous development it is enough to assume that the solution F of Σ_L is a distribution on some open set in V (see 2.1).) Setting $F_w(v) = g^{-1} \sum (T_{\sigma v}F)(w)$ for each w in V, we see that $(D_i F_w)(0) = (D_i F)(w)$ if i is in I, whence 8.1 yields $D_i F = (i, C)F$. But then (see the discussion before 1.1) we have $(i, C) = i(L)$ for some L in V^*, and F satisfies the corresponding system Σ_L. Here Σ_L is uniquely determined by C, and by our previous discussion C is given in terms of L by 8.2. We summarize:

8.3 *The system* 8.1 *has a nonzero solution if and only if C has the form* 8.2 *for some L in V^*. The system of difference equations Σ'_L composed of* 8.1 *and* 8.2 *is equivalent to the system of differential equations Σ_L of* 1.1.

From 8.3 we get at once:

8.4 THEOREM. *If Σ_L is replaced by Σ'_L in* 1.2, 1.3 *and* 1.4, *the results there remain true.*

Finally, let us remark a characteristic property of the function C of 8.2.

8.5 *The function C of* 8.2 *is a solution of Σ_L, it is invariant under G, and it satisfies $C(0) = 1$. It is uniquely determined by these properties.*

It is clear that C has the stated properties. Assume that C' does also. Let s be any element of S and i its average under G. Then $(s, C' - C) = (i, C' - C) = (D_i(C' - C))(0) = i(L)(C' - C)(0) = 0$, whence $C' - C = 0$.

BIBLIOGRAPHY

1. C. Chevalley, *Invariants of finite groups generated by reflections*, Amer. J. Math. **77** (1955), 778–782.
2. E. Fischer, *Über algebraische Modulsysteme und lineare homogene partielle Differentialgleichungen mit konstanten Koeffizienten*, J. Reine Angew. Math. **140** (1911), 48–81.
3. L. Flatto, *Classes of polynomials characterized by a mean value property*, Abstract 588-24, Notices Amer. Math. Soc. **9** (1962), 33.
4. Harish-Chandra, *Differential operators on a semi-simple Lie algebra*, Amer. J. Math. **79** (1957), 87–120.
5. S. Helgason, *Differential geometry and symmetric spaces*, Academic Press, New York, 1962.
6. N. Jacobson, *Lie algebras*, Interscience, New York, 1962.
7. G. C. Shephard and J. A. Todd, *Finite unitary reflection groups*, Canad. J. Math. 6 (1954), 274–304.
8. Séminaire "Sophus Lie", Ecole Normale Supérieure, Paris, 1955.
9. B. L. van der Waerden, *Modern algebra*, Vol. 1, Ungar, New York 1949.

UNIVERSITY OF CALIFORNIA,
LOS ANGELES, CALIFORNIA

REGULAR ELEMENTS OF SEMISIMPLE ALGEBRAIC GROUPS

by ROBERT STEINBERG

§ 1. Introduction and statement of results

We assume given an algebraically closed field K which is to serve as domain of definition and universal domain for each of the algebraic groups considered below; each such group will be identified with its group of elements (rational) over K. The basic definition is as follows. An element x of a semisimple (algebraic) group (or, more generally, of a connected reductive group) G of rank r is called *regular* if the centralizer of x in G has dimension r. It should be remarked that x is not assumed to be semisimple; thus our definition is different from that of [8, p. 7-03]. It should also be remarked that, since regular elements are easily shown to exist (see, e.g., 2.11 below) and since each element of G is contained in a (Borel) subgroup whose quotient over its commutator subgroup has dimension r, a regular element is one whose centralizer has the least possible dimension, or equivalently, whose conjugacy class has the greatest possible dimension.

In the first part of the present article we obtain various criteria for regularity, study the varieties of regular and irregular elements, and in the simply connected case construct a closed irreducible cross-section N of the set of regular conjugacy classes of G. Then assuming that G is (defined) over a perfect field k and contains a Borel subgroup over k we show that N (or in some exceptional cases a suitable analogue of N) can be constructed over k, and this leads us to the solution of a number of other problems of rationality. In more detail our principal results are as follows. Until 1.9 the group G is assumed to be semisimple.

1.1. *Theorem. — An element of* G *is regular if and only if the number of Borel subgroups containing it is finite.*

1.2. *Theorem. — The map* $x \to x_s$, *from* x *to its semisimple part, induces a bijection of the set of regular classes of* G *onto the set of semisimple classes. In other words:*

a) *every semi-simple element is the semisimple part of some regular element*;

b) *two regular elements are conjugate if and only if their semisimple parts are.*

The author would like to acknowledge the benefit of correspondence with T. A. Springer on these results (cf. 3.13, 4.7 *d)* below). The special case of *a)* which asserts the existence of regular unipotent elements (all of which are conjugate by *b)*) is proved in § 4. The other parts of 1.2 and 1.1, together with the fact that the number in 1.1, if finite, always divides the order of the Weyl group of G, are proved in § 3, where other characterizations of regularity may be found (see 3.2, 3.7, 3.11, 3.12 and 3.14). This material follows a preliminary section, § 2, in which we recall some basic facts about semisimple groups and some known characterizations of regular semisimple elements (see 2.11).

1.3. *Theorem.* — a) *The irregular elements of* G *form a closed set* Q.

b) *Each irreducible component of* Q *has codimension* 3 *in* G.

c) Q *is connected unless* G *is of rank* 1, *of characteristic not* 2, *and simply connected, in which case* Q *consists of* 2 *elements.*

This is proved in § 5 where it is also shown that the number of components of Q is closely related to the number of conjugacy classes of roots under the Weyl group. An immediate consequence of 1.3 is that the regular elements form a dense open subset of G.

It may be remarked here that 1.1 to 1.3 and appropriate versions of 1.4 to 1.6 which follow hold for connected reductive groups as well as for semisimple groups, the proofs of the extensions being essentially trivial.

In § 6 the structure of the algebra of class functions (those constant on conjugacy classes) is determined (see 6.1 and 6.9). In 6.11, 6.16, and 6.17 this is applied to the study of the closure of a regular class and to the determination of a natural structure of variety for the set of regular classes, the structure of affine r-space in case G is simply connected.

1.4. *Theorem.* — *Let* T *be a maximal torus in* G *and* $\{\alpha_i \mid 1 \leq i \leq r\}$ *a system of simple roots relative to* T. *For each* i *let* X_i *be the one-parameter unipotent subgroup normalized by* T *according to the root* α_i *and let* σ_i *be an element of the normalizer of* T *corresponding to the reflection relative to* α_i. *Let* $N = \prod_{i=1}^{r} (X_i \sigma_i) = X_1 \sigma_1 X_2 \sigma_2 \ldots X_r \sigma_r$. *If* G *is a simply connected group, then* N *is a cross-section of the collection of regular classes of* G.

In 7.4 an example of N is given: in case G is of type $SL(r+1)$ we obtain one of the classical normal forms under conjugacy. This special case suggests the problem of extending the normal form N from regular elements to arbitrary elements. In 7.1 it is shown that N is a closed irreducible subset of G, isomorphic as a variety to affine r-space V, and in 7.9 (this is the main lemma concerning N) that, if G is simply connected, and χ_i $(1 \leq i \leq r)$ denote the fundamental characters of G, then the map $x \rightarrow (\chi_1(x), \chi_2(x), \ldots, \chi_r(x))$ induces an isomorphism of N on V. Then in § 8 the proof of 1.4 is given and simultaneously the following important criterion for regularity is obtained.

1.5. *Theorem.* — *If* G *is simply connected, the element* x *is regular if and only if the differentials* $d\chi_i$ *are independent at* x.

At this point some words about recent work of B. Kostant are in order. In [3] and [4] he has proved, among other things, the analogues of our above discussed results that are obtained by replacing the semisimple group G by a semisimple Lie algebra L over the complex field (any algebraically closed field of characteristic 0 will serve as well) and the characters χ_i of G by the basic polynomial invariants u_i of L. The χ_i turn out to be considerably more tractable than the u_i. Thus the proofs for G with no restriction on the characteristic are simpler than those for L in characteristic 0. Assuming both G and L are in characteristic 0, substantial parts of 1.1, 1.2, and 1.3 can be derived from their analogues for L, but there does not seem to be any simple way of relating 1.4 and 1.5 to their analogues for L.

We now introduce a perfect subfield k of K, although it appears from recent results of A. Grothendieck on semisimple groups over arbitrary fields that the assumption of perfectness is unnecessary for most of what follows.

1.6. *Theorem. — Let* G *be over* k, *and assume either that* G *splits over* k *or that* G *contains a Borel subgroup over* k *but no component of type* A_n *(n even). Then the set* N *of* 1.4 *can be constructed over* k *(by appropriate choice of* T, σ_i, *etc.).*

Together with 1.4 this implies that if G is simply connected in 1.6 the natural map from the set of regular elements over k to the set of regular classes over k is surjective. For a group of type A_n (n even) we have a substitute (see 9.7) for 1.6 which enables us to show:

1.7. *Theorem. — Assume that* G *is simply connected and over* k *and that* G *contains a Borel subgroup over* k. *Then the natural map from the set of semisimple elements over* k *to the set of semisimple classes over* k *is surjective. In other words, each semisimple class over* k *contains an element over* k.

Theorems 1.6 and 1.7 are proved in § 9 where it is also shown (see 9.1 and 9.10) that the assumption that G contains a Borel subgroup over k is essential.

1.8. *Theorem. — Under the assumptions of* 1.7 *each element of the cohomology set* $H^1(k, G)$ *can be represented by a cocycle whose values are in a torus over* k.

In § 10 this result is deduced from 1.7 by a method of proof due to M. Kneser, who has also proved 1.7 in a number of special cases and has formulated the general case as a conjecture. In 9.9 and 10.1 it is shown that 1.7 and 1.8 hold for arbitrary simply connected, connected linear groups, not just for semisimple ones.

In § 10 it is indicated how Theorem 1.8 provides the final step in the proof of the following result, 1.9, the earlier steps being due to J.-P. Serre and T. A. Springer (see [12], [13] and [15]). We observe that G is no longer assumed to be semisimple, and recall [12, p. 56-57] that (cohomological) dim $k \leq 1$ means that every finite-dimensional division algebra over k is commutative.

1.9. *Theorem. — Let* k *be a perfect field. If* a) dim $k \leq 1$, *then* b) $H^1(k, G) = 0$ *for every connected linear group* G *over* k, *and* c) *every homogeneous space* S *over* k *for every connected linear group* G *over* k *contains a point over* k.

The two parts of 1.9 are the conjectures I and I′ of Serre [12]. Conversely *b)* implies *a)* by [12, p. 58], and is the special case of *c)* in which only principal homogeneous spaces are considered; thus *a)*, *b)* and *c)* are equivalent. They are also equivalent to: every connected linear group over k contains a Borel subgroup over k [15, p. 129].

After some consequences of 1.9, of which only the following (cf. 1.7) will be stated here, the paper comes to a close.

1.10. *Theorem. — Let k be a perfect field such that* $\dim k \leq 1$ *and* G *a connected linear group over k. Then every conjugacy class over k contains an element over k.*

After the remark that Kneser, using extensions of 1.8, has recently shown (cf. 1.9) that $H^1(k, G) = 0$ if k is a p-adic field and G a simply connected semisimple group over k, this introduction comes to a close.

§ 2. Some recollections

In this section we recall some known facts, including some characterizations 2.11 of regular semisimple elements, and establish some notations which are frequently used in the paper. If k is a field, k^* is its multiplicative group. The term " algebraic group " is often abbreviated to " group ". If G is a group, G_0 denotes its identity component. If x is an element of G, then G_x denotes the centralizer of x in G, and x_s and x_u denote the semisimple and unipotent parts of x when G is linear. Assume now that G is a semisimple group, that is, G is a connected linear group with no nontrivial connected solvable normal subgroup. We write r for the rank of G. Assume further that T is a maximal torus in G and that an ordering of the (discrete) character group of T has been chosen. We write Σ for the system of roots relative to T and X_α for the subgroup corresponding to the root α.

2.1. X_α *is unipotent and isomorphic (as an algebraic group) to the additive group (of* K*). If x_α is an isomorphism from* K *to* X_α, *then* $tx_\alpha(c)t^{-1} = x_\alpha(\alpha(t)c)$ *for all* α *and* c.

For the proof of 2.1 to 2.6 as well as the other standard facts about linear groups, the reader is referred to [8].

We write U (resp. U^-) for the group generated by those X_α for which α is positive (resp. negative), and B for the group generated by T and U.

2.2. a) U *is a maximal unipotent subgroup of* G, *and* B *is a Borel (maximal connected solvable) subgroup.*

b) *The natural maps from the Cartesian product* $\prod_{\alpha>0} X_\alpha$ *(fixed but arbitrary order of the factors) to* U *and from* $T \times U$ *to* B *are isomorphisms of varieties.*

In *b)* the X_α component of an element of U may change with the order, but not if α is simple.

2.3. *The natural map from* $U^- \times T \times U$ *to* G *is an isomorphism onto an open subvariety of* G.

We write W for the Weyl group of G, that is, the quotient of T in its normalizer. W acts on T, via conjugation, hence also on the character group of T and on Σ. For each w in W we write σ_w for an element of the normalizer of T which represents w.

2.4. a) *The elements* σ_w $(w \in W)$ *form a system of representatives of the double cosets of* G *relative to* B.

b) *Each element of* $B\sigma_w B$ *can be written uniquely* $u\sigma_w b$ *with u in* $U \cap \sigma_w U^- \sigma_w^{-1}$ *and b in* B.

The simple roots are denoted α_i $(1 \leq i \leq r)$. If $\alpha = \alpha_i$ we write X_i, x_i for X_α, x_α, and G_i for the group (semisimple of rank 1) generated by X_α and $X_{-\alpha}$. The reflection in W corresponding to α_i is denoted w_i. If $w = w_i$ we write σ_i in place of σ_w.

2.5. *The element σ_i can be chosen in G_i. If this is done, and $B_i = B \cap G_i = (T \cap G_i)X_i$, then G_i is the disjoint union of B_i and $X_i \sigma_i B_i$.*

The following may be taken as a definition of the term " simply connected ".

2.6. *The semisimple group G is simply connected if and only if there exists a basis $\{\omega_j\}$ of the dual (character group) of T such that $w_i \omega_j = \omega_j - \delta_{ij} \alpha_i$ (Kronecker delta, $1 \leq i, j \leq r$).*

An arbitrary connected linear group is simply connected if its quotient over its radical satisfies 2.6. If G is as in 2.6 we write χ_i for the ith fundamental character of G, that is, for the trace of the irreducible representation whose highest weight on T is ω_i.

2.7. *Let G be a semisimple group of rank r and x a semisimple element of G.*

a) *G_{x0} is a connected reductive group of rank r. In other words, $G_{x0} = G'T'$ with G' a semisimple group, T' a central torus in G_{x0}, the intersection $G' \cap T'$ finite, and* rank G' + rank $T' = r$. *Further G' and T' are uniquely determined as the commutator subgroup and the identity component of the centre of G_{x0}.*

b) *The unipotent elements of G_x are all in G'.*

Part *b)* follows from *a)* because G_{x0} contains the unipotent elements of G_x by [8, p. 6-15, Cor. 2]. For the proof of *a)* we may imbed x in a maximal torus T and use the above notation. If y in G_x is written $y = u\sigma_w b$ as in 2.4 then the uniqueness in 2.4 implies that u, σ_w and b are in G_x. By 2.1 and 2.2 we get:

2.8. *G_x is generated by T, those X_α for which $\alpha(x) = 1$, and those σ_w for which $wx = x$.*

Then G_{x0} is generated by T and the X_α alone because the group so generated is connected and of finite index in G_x (see [8, p. 3-01, Th. 1]). Let G' be the group generated by the X_α alone, and let T' be the identity component of the intersection of the kernels of the roots α such that $\alpha(x) = 1$. Then G' is semisimple by [8, p. 17-02, Th. 1], and the other assertions of *a)* are soon verified.

2.9. *Corollary.* — *In 2.7 every maximal torus containing x also contains T'.*

For in the above proof T was chosen as an arbitrary torus containing x.

2.10. *Remark.* — That G_x in 2.7 need not be connected, even if x is regular, is shown by the example: $G = PSL(2)$, $x = \text{diag}(i, -i)$, $i^2 = -1$. If G is simply connected, however, G_x is necessarily connected and in 2.8 the elements σ_w may be omitted. More generally, the group of fixed points of a semisimple automorphism of a semisimple group G is reductive, and if the automorphism fixes no nontrivial point of the fundamental group of G, it is connected. (The proofs of these statements are forthcoming.)

2.11. *Let G and x be as in 2.7. The following conditions are equivalent:*

a) *x is regular.*

b) *G_{x0} is a maximal torus in G.*

c) *x is contained in a unique maximal torus T in G.*

d) *G_x consists of semisimple elements.*

e) *If T is a maximal torus containing x then $\alpha(x) \neq 1$ for every root α relative to T.*

G_{x0} contains every torus which contains x. Thus *a)* and *b)* are equivalent and *b)* implies *c)*. If *c)* holds, G_x normalizes T, whence G_x/T is finite and $G_{x0}=T$, which is *b)*. By 2.7 *b)*, *b)* implies *d)*, which in turn, by 2.1, implies *e)*. Finally *e)* implies, by 2.8, that G_x/T is finite, whence *b)*.

2.12. *Lemma.* — *Let* $B'=T'U'$ *with* B' *a connected solvable group,* T' *a maximal torus, and* U' *the maximal unipotent subgroup. If* t *and* u *are elements of* T' *and* U', *there exists* u' *in* U' *such that* tu' *is conjugate to* tu *via an element of* U', *and* u' *commutes with* t.

For the semisimple part of tu is conjugate, under U', to an element of T' by [8, p. 6-07], an element which must be t itself because U' is normal in B'.

2.13. *Corollary.* — *In the semisimple group* G *assume that* t *is a regular element of* T *and* u *an arbitrary element of* U. *Then* tu *is a regular element, in fact is conjugate to* t.

By 2.12 we may assume that u commutes with t, in which case $u=1$ by 2.1 and 2.2 *b)*.

2.14. *The regular semisimple elements form a dense open set* S *in* G.

By 2.12, 2.13 and 2.11 (see *a)* and *e)*), $S\cap B$ is dense and open in B. Since the conjugates of B cover G by [8, p. 6-13, Th. 5], S is dense in G. Let A be the complement of $S\cap B$ in B, and let C be the closed set in $G/B\times G$ consisting of all pairs $(\bar{x},y)$ (here $\bar{x}$ denotes the coset xB) such that $x^{-1}yx\in A$. The first factor, G/B, is complete by [8, p. 6-09, Th. 4]. By a characteristic property of completeness, the projection on the second factor is closed. The complement, S, is thus open.

We will call an element of G *strongly regular* if its centralizer is a maximal torus. Such an element is regular and semisimple, the converse being true if G is simply connected by 2.10.

2.15. *The strongly regular elements form a dense open set in* G.

The strongly regular elements form a dense open set in T, characterized by $\alpha(t)\neq 1$ for all roots α, and $wt\neq t$ for all $w\neq 1$ in W. Thus the proof of 2.14 may be applied.

§ 3. Some characterizations of regular elements

Throughout this section and the next G denotes a semisimple group. Our aim is to prove 1.1 and 1.2 (of § 1). The case of unipotent elements will be considered first. The following critical result is proved in § 4.

3.1. *Theorem.* — *There exists in* G *a regular unipotent element.*

3.2. *Lemma.* — *There exists in* G *a unipotent element contained in only a finite number of Borel subgroups. Indeed let* x *be a unipotent element and* n *the number of Borel subgroups containing it. Then the following are equivalent:*

a) *n is finite.*

b) *n is* 1.

c) *If x is imbedded in a maximal unipotent subgroup* U *and the notation of* § 2 *is used, then for* $1\leq i\leq r$ *the* X_i *component of* x *is different from* 1.

Let T be a maximal torus which normalizes U, let $B=TU$, and let B′ be an arbitrary Borel subgroup. By the conjugacy theorem for Borel subgroups and 2.4 we have $B'=u\sigma_w B\sigma_w^{-1}u^{-1}$ with u and σ_w as in 2.4 *b)*. If *c)* holds and B′ contains x, then B contains $\sigma_w^{-1}u^{-1}xu\sigma_w$ and every X_i component of $u^{-1}xu$ is different from 1. Thus $w\alpha_i$ is positive for every simple root α_i and w is 1, whence $B'=B$ and *b)* holds. If *c)* fails, then for some i the Borel subgroups $u\sigma_i B\sigma_i^{-1}u^{-1}$ $(u\in X_i)$ all contain x, whence *a)* fails. Thus *a)*, *b)* and *c)* are equivalent. Since elements which satisfy *c)* exist in abundance, the first statement in 3.2 follows.

3.3. *Theorem. — For a unipotent element x of* G *the following are equivalent:*

a) *x is regular.*

b) *The number of Borel subgroups containing x is finite.*

Further the unipotent elements which satisfy a) *and* b) *form a single conjugacy class.*

Let y and z be arbitrary unipotent elements which satisfy *a)* and *b)*, respectively. Such elements exist by 3.1 and 3.2. We will prove all assertions of 3.3 together by showing that y is conjugate to z. By replacing y and z by conjugates we may assume they are both in the group U of § 2 and use the notations there. Let y_i and z_i denote the X_i components of y and z. By 3.2 every z_i is different from 1. We assert that every y_i is also different from 1. Assume the contrary, that $y_i=1$ for some i, and let U_i be the subgroup of elements of U whose X_i components are 1. Then y is in U_i, so that in the normalizer $P_i=G_iTU_i$ of U_i we have $\dim(P_i)_y=\dim P_i-\dim(\text{class of } y)\geq \dim P_i-\dim U_i=r+2$. This contradiction to the regularity of y proves our assertion. Hence by conjugating y by an element of T we may achieve the situation: $y_i=z_i$ for all i, or, in other words, zy^{-1} is in U′, the intersection of all U_i. Now the set $\{uyu^{-1}y^{-1}|u\in U\}$ is closed (by [7] every conjugacy class of U is closed). Its codimension in U is at most r because y is regular, whence its codimension in U′ is at most $r-(\dim U-\dim U')=0$. The set thus coincides with U′. For some u in U we therefore have $uyu^{-1}y^{-1}=zy^{-1}$, whence $uyu^{-1}=z$, and 3.3 is proved.

In the course of the argument the following result has been proved.

3.4. *Corollary. — If x is unipotent and irregular, then* $\dim G_x\geq r+2$.

If P_i is replaced by B in the above argument, the result is:

3.5. *Corollary. — If x is unipotent and irregular and* B *is any Borel subgroup containing x, then* $\dim B_x\geq r+1$.

3.6. *Lemma. — Let x be an element of* G, *and y and z its semisimple and unipotent parts. Let* $G_{y0}=G'T'$ *with* G′ *and* T′ *as in* 2.7, *and let r' be the rank of* G′. *Let* S *(resp.* S′*) be the set of Borel subgroups of* G *(resp.* G′*) containing x (resp. z):*

a) $\dim G_x=\dim G'_z+r-r'$.

b) *If* B *in* S *contains* B′ *in* S′ *then* $\dim B_x=\dim B'_z+r-r'$.

c) *Each element* B *of* S *contains a unique element of* S′, *namely,* $B\cap G'$.

d) *Each element of* S′ *is contained in at least one but at most a finite number of elements of* S.

We have $G_x = (G_y)_z$ by [8, p. 4-08]. Thus $\dim G_x = \dim G'_z + \dim T'$, whence *a)*. Part *b)* may be proved in the same way, once it is observed that $B_y = B'T'$. For B_y is solvable, connected by [8, p. 6-09], and contains the Borel subgroup $B'T'$ of G_y. Let B be in S. Let T be a maximal torus in B containing y, and let the roots relative to T be ordered so that B corresponds to the set of positive roots. The group G' is generated by those X_α for which $\alpha(y) = 1$, and the corresponding α form a root system Σ' for G' by [8, p. 17-02, Th. 1]. By 2.2 *a)* the groups $T \cap G'$ and X_α $(\alpha > 0, \alpha \in \Sigma')$ generate a Borel subgroup of G' which is easily seen to be none other than $B \cap G'$ (by 2.1 and 2.2 *b)*), whence *c)* follows. Let B' be in S'. Then a Borel subgroup B of G contains B' and is in S if and only if it contains $B'T'$. For if B contains x, it also contains y, then a maximal torus containing y by [8, p. 6-13], then T' by 2.9; while if B contains the Borel subgroup $B'T'$ of G_{y0}, it contains the central element y by [8, p. 6-15], thus also x. The number of possibilities for B above is at least 1 because $B'T'$ is a connected solvable group, but it is at most the order of the Weyl group of G because $B'T'$ contains a maximal torus of G (this last step is proved in [8, p. 9-05, Cor. 3], and also follows from 2.4).

3.7. *Corollary. — In* 3.6 *the element x is regular in* G *if and only if z is regular in* G′, *and the set* S *is finite if and only if* S′ *is.*

The first assertion follows from 3.6 *a)*, the second from *c)* and *d)*.

3.8. *Corollary. — In* 3.6 *the element x is regular in* G *if and only if the set* S *is finite.*

Observe that this is Theorem 1.1 of § 1. It follows from 3.7 and 3.3 (applied to z).

3.9. *Corollary. — The assertions* 3.4 *and* 3.5 *are true without the assumption that x is unipotent.*

For the first part we use 3.6 *a)*, for the second *b)* and *c)*.

3.10. *Conjecture. — For any x in* G *the number* $\dim G_x - r$ *is even.*

It would suffice to prove this when x is unipotent. The corresponding result for Lie algebras over the complex field is a simple consequence of the fact that the rank of a skew symmetric matrix is always even (see [4, p. 364, Prop. 15]).

3.11. *Corollary. — If x is an element of* G, *the following are equivalent.*

a) $\dim G_x = r$, *that is, x is regular.*

b) $\dim B_x = r$ *for every Borel subgroup* B *containing x.*

c) $\dim B_x = r$ *for some Borel subgroup* B *containing x.*

As we remarked in the first paragraph of § 1, $\dim B_x \geq r$. Thus *a)* implies *b)*. By 3.5 as extended in 3.9 we see that *c)* implies *a)*.

3.12. *Corollary. — In* 3.6 *let x be regular and n the number of Borel subgroups containing x.*

a) $n = |W|/|W'|$, *the ratio of the orders of the Weyl groups of* G *and* G′.

b) $n = 1$ *if and only if z is a regular unipotent element of* G *and y is an element of the centre.*

c) $n = |W|$ *if and only if x is a regular semisimple element of* G.

By 3.7, 3.2 and 3.3 the element z is regular and contained in a unique Borel subgroup B′ of G′. Let T be a maximal torus in B′T′. Then n is the number of Borel subgroups of G containing B′ and T. Now each of the $|W'|$ Borel subgroups of G′ normalized by T (these are just the conjugates of B′ under W′) is contained in the same number of Borel subgroups of G containing T, and each of the $|W|$ groups of the latter type contains a unique group of the former type by 3.6 *c)*. Thus *a)* follows. Then $n=1$ if and only if $|W'|=|W|$, that is, $G'=G$, which yields *b)*; and $n=|W|$ if and only if $|W'|=1$, that is, $G'=1$ and $G_{y0}=T'$, which by 2.11 (see *a)*, *b)* and *d)*) is equivalent to y regular and $x=y$, whence *c)*.

3.13. *Remark.* — Springer has shown that if x is regular in G then G_{x0} is commutative. Quite likely the converse is true (it is for type A_r). It would yield the following characterization of the regular elements, in the abstract group, G_{ab}, underlying G. The element x of G_{ab} is regular in G if and only if G_x contains a commutative subgroup of finite index. We have the following somewhat bulkier characterization.

3.14. *Corollary.* — *The element x of G_{ab} is regular if and only if it is contained in only a finite number of subgroups each of which is maximal solvable and without proper subgroups of finite index.*

For each such subgroup is closed and connected, hence a Borel subgroup. We remark that G_{ab} determines also the sets of semisimple and unipotent elements (hence also the decomposition $x=x_s x_u$), as well as the semisimplicity, rank, dimension, and base field (to within an isomorphism), all of which would be false if G were not semisimple. If G is simple, then G_{ab} determines the topology (the collection of closed sets) in G completely, which is not always the case if G is semisimple.

To close this section we now prove Theorem 1.2. Let y be semisimple in G, and $G_{y0}=G'T'$ as in 3.6. By 3.1 there exists in G′ a regular unipotent element z. Let $x=yz$. Then x is regular in G by 3.7 and $x_s=y$, whence *a)* holds. Let x and x' be regular elements of G. If x is conjugate to x', then clearly x_s is conjugate to x'_s. If x_s is conjugate to x'_s, we may assume $x_s=x'_s=y$, say. Then in G′ (as above) the elements x_u and x'_u are regular by 3.7, hence conjugate by 3.3, whence x and x' are conjugate.

§ 4. The existence of regular unipotent elements

This section is devoted to the proof of 3.1. Throughout G is a semisimple group, T a maximal torus in G, and the notations of § 2 are used. In addition V denotes a real totally ordered vector space of rank r which extends the dual of T and its given ordering.

4.1. *Lemma.* — *Let the simple roots α_i be so labelled that the first q are mutually orthogonal as are the last $r-q$. Let $w=w_1 w_2 \ldots w_r$.*

a) *The roots are permuted by w in r cycles.*

The space V *can be reordered so that*

b) *roots originally positive remain positive,*

and

c) *each cycle of roots under w contains exactly one relative maximum and one relative minimum.*

We observe that since the Dynkin graph has no circuits [9, p. 13-02] a labelling of the simple roots as above is always possible. In *c)* a root α is, for example, a maximum in its cycle under w if $\alpha > w\alpha$ and $\alpha > w^{-1}\alpha$ for the order on V. The proof of 4.1 depends on the following results proved in [16]. (These are not explicitly stated there, but see 3.2, 3.6, the proof of 4.2, and 6.3.)

4.2. *Lemma.* — *In* 4.1 *assume that* Σ *is indecomposable, that a positive definite inner product invariant under* W *is used in* V, *and that n denotes the order of w.*

a) *The roots of* Σ *are permuted by w in r cycles each of length n.*

If $\dim \Sigma > 1$, *there exists a plane* P *in* V *such that*

b) P *contains a vector v such that* $(v, \alpha) > 0$ *for every positive root* α,

and

c) *w fixes* P *and induces on* P *a rotation through the angle* $2\pi/n$.

For the proof of 4.1 we may assume that Σ is indecomposable, and, omitting a trivial case, that $\dim \Sigma > 1$. We choose P and v as in 4.2. Let α' denote the orthogonal projection on P of the root α. By 4.2 *b)* it is nonzero. Since by 4.2 *c)* the vectors $w^{-i}v$ $(1 \leq i \leq n)$ form the vertices a regular polygon, it can be arranged, by a slight change in v, that for each α these vectors make distinct angles with α'. It is then clear that there is one relative maximum and one relative minimum for the cycle of numbers $(w^{-i}v, \alpha')$. Since $(w^{-i}v, \alpha') = (w^{-i}v, \alpha) = (v, w^{i}\alpha)$, we can achieve *c)* by reordering V so that vectors v' for which $(v, v') > 0$ become positive. Then *a)* and *b)* also hold by 4.2 *a)* and 4.2 *b)*.

4.3. *Lemma.* — *Let* G *be simply connected, otherwise as above. Let* $\mathfrak{g}$ *be the Lie algebra of* G. *Let* $\mathfrak{t}$ *be the subalgebra corresponding to* T, *and* $\mathfrak{z}$ *the subalgebra of elements of* $\mathfrak{t}$ *which vanish at all roots on* T. *Let w be as in* 4.1. *Let x be an element of the double coset* $B\sigma_w B$, *and let* $\mathfrak{g}_x$ *denote the algebra of fixed points of x acting on* $\mathfrak{g}$ *via the adjoint representation. Then* $\dim \mathfrak{g}_x \leq \dim \mathfrak{z} + r$.

We identify $\mathfrak{g}$ with the tangent space to G at 1. Then by 2.3 we have a direct sum decomposition $\mathfrak{g} = \mathfrak{t} + \sum_{\alpha} K\mathfrak{x}_\alpha$ in which $K\mathfrak{x}_\alpha$ may be identified with the tangent space of X_α. We order the weights of the adjoint representation, that is, 0 and the roots, as in 4.1. By replacing x by a conjugate, we may assume $x = b\sigma_w$ $(b \in B)$.

1) *If* $\mathfrak{v}$ *in* $\mathfrak{g}$ *is a weight vector, then* $(1-x)\mathfrak{v} = \mathfrak{v} - c\sigma_w\mathfrak{v}$ + *terms (corresponding to weights) higher than (that of)* $\sigma_w\mathfrak{v}$ $(c \in K^*)$. This follows from 7.15 *d)* below, which holds for any rational representation of G.

2) *If the root* α *is not maximal in its cycle under w, then* $(1-x)\mathfrak{g}$ *contains a vector of the form* $c\mathfrak{x}_\alpha$ + *higher terms* $(c \in K^*)$. If $w\alpha > \alpha$ we apply 1) with $\mathfrak{v} = \mathfrak{x}_\alpha$, while if $w\alpha < \alpha$ we use $\mathfrak{v} = \sigma_w^{-1}\mathfrak{x}_\alpha$ instead.

3) *There exist* $r - \dim \mathfrak{z}$ *independent elements* $\mathfrak{t}_i$ *of* $\mathfrak{t}$ *such that for every i the space* $(1-x)\mathfrak{g}$ *contains a vector of the form* $\mathfrak{t}_i$ + *higher terms.* Because of 1), in which $c = 1$ if $\mathfrak{v}$ is in $\mathfrak{t}$, this follows from:

4) *The kernel of* $1 - \sigma_w$ *on* $\mathfrak{t}$ *is* $\mathfrak{z}$. Because the adjoint action of σ_w on $\mathfrak{t}$ stems from the action of w on T by conjugation, we may write w in place of σ_w, on $\mathfrak{t}$. Assume

$(1-w)\mathfrak{t}_0=0$ with $\mathfrak{t}_0$ in $\mathfrak{t}$. Then $(1-w_1)\mathfrak{t}_0=(1-w_2\ldots w_r)\mathfrak{t}_0$. If we evaluate the left side at the functions $\omega_2, \ldots, \omega_r$ of 2.6 or the right side at ω_1 then by 2.6 we always get 0, whence both sides are 0. By an obvious induction we get that $(1-w_i)\mathfrak{t}_0=0$ for all i, and on evaluation at ω_i, that $\mathfrak{t}_0(\alpha_i)=\mathfrak{t}_0((1-w_i)\omega_i)=0$. Thus $\mathfrak{t}_0$ is in $\mathfrak{z}$. One may reverse the steps to show that $\mathfrak{z}$ is contained in the kernel of $1-\sigma_w$, whence 4).

Lemma 4.3 is a consequence of 2) and 3).

4.4. *Remark.* — One can show that $\mathfrak{z}$ in 4.3 is the centre of $\mathfrak{g}$.

4.5. *Lemma.* — *Let the notation be as in* 4.1. *Let w_0 be the element of* W *which maps each positive root onto a negative one, and π the permutation defined by* $-w_0\alpha_i=\alpha_{\pi i}$ $(1\leq i\leq r)$. *Let σ_0 be an element of the normalizer of* T *which represents w_0. For each i let u_i be an element of $X_{\pi i}$ different from* 1 *and let* $x=u_1u_2\ldots u_r$. *Then* $\sigma_0 x\sigma_0^{-1}$ *is in* $B\sigma_w B$.

We have $\sigma_0 u_i\sigma_0^{-1}$ in $G_i - B$, hence in $B\sigma_i B$ by 2.5. Since

$$B\sigma_1\ldots\sigma_{i-1}B\sigma_i B=B\sigma_1\ldots\sigma_{i-1}X_i\sigma_i B=B\sigma_1\ldots\sigma_i B,$$

because w_i permutes the positive roots other than α_i by [8, p. 14-04, Cor. 3], and each root $w_1w_2\ldots w_{i-1}\alpha_i$ is positive (cf. 7.2 *a*)) we get 4.5.

4.6. *Theorem.* — *The element x of* 4.5 *is regular.*

By going to the simply connected covering group, we may assume that G is simply connected. For any subalgebra $\mathfrak{a}$ of $\mathfrak{g}$ we write $\mathfrak{a}_x$ for the subalgebra of elements fixed by x. Let $\mathfrak{b}$ and $\mathfrak{u}$ denote the subalgebras corresponding to B and U. By 4.3 and 4.5 we have $\dim\mathfrak{b}_x\leq\dim\mathfrak{g}_x\leq\dim\mathfrak{z}+r$. An infinitesimal analogue of 2.1 yields $x_\alpha(c)\mathfrak{t}_0=\mathfrak{t}_0+c'c\mathfrak{t}_0(\alpha)\mathfrak{x}_\alpha$ for all $\mathfrak{t}_0$ in $\mathfrak{t}$ and some c' in K, whence $\mathfrak{t}_x$ contains $\mathfrak{z}$, and $\dim\mathfrak{b}_x\geq\dim\mathfrak{z}+\dim\mathfrak{u}_x$. Combined with the previous inequality this yields $\dim\mathfrak{u}_x\leq r$, whence $\dim U_x\leq r$. From the form of x we see that B is the unique Borel subgroup containing x. Each element of G_x normalizes B, hence belongs to B by [8, p. 9-03, Th. 1], or else by 2.4. Now if ut $(t\in T, u\in U)$ is in B_x then, working in B modulo the commutator subgroup of U, and using the fact that each X_i component of x is different from 1, we get $\alpha_i(t)=1$ for all i, whence t is in the centre of G, a finite group. Hence $\dim G_x=\dim U_x\leq r$, as required.

4.7. *Remarks.* — *a)* The condition $\dim U_x=r$ on x in U is not enough to make x regular, as one sees by examples in a group of type A_2. The added condition that all X_i components are different from 1 is essential.

b) If the characteristic of K is 0, or, more generally, if $\dim\mathfrak{z}\leq 1$ in 4.3, we may conclude from 4.3 and 3.4 as extended in 3.9 that all elements of $B\sigma_w B$ are regular, and then (cf. 7.3) that all elements of N in 1.4 are regular. There is, however, an exception: $\dim\mathfrak{z}=2$ if G is of type D_r (r even) and of characteristic 2. It is nevertheless true that all elements of $B\sigma_w B$ are regular (cf. 8.8). By 4.5 this implies that if x is the regular element of 4.6 and t in T is arbitrary, then tx is regular. If u is an arbitrary regular element of U, however, tu need not be regular: consider in SL(3) the superdiagonal matrix with diagonal entries $-1, 1, -1$ and superdiagonal entries all 2. In contrast if t is regular and u is arbitrary, then tu is regular by 2.13.

c) In characteristic 0 one may, in the simply connected case, imbed the element x of 4.6 in a subgroup isomorphic to SL(2) and then use the theory of the representations of this latter group to prove that x is regular. This is the method of Kostant, worked out in [3] for Lie algebras over the complex field. In the general case, however, a regular unipotent element can not be imbedded in the group SL(2), or even in the $ax+b$ group: in characteristic $p \neq 0$, a unipotent element of either of these groups has order at most p, while in a group G of type A_r, for example, a regular unipotent element has order at least $r+1$, so that if $r+1>p$ the imbedding is impossible.

d) Springer has studied U_x (x as in 4.6) by a method depending on a knowledge of the structural constants of the Lie algebra of U. His methods yield a proof of the regularity of x only if

(*) p does not divide any coefficient in the highest root of any component of G,

but it yields also that U_x is connected if and only if (*) holds, a result which quite likely has cohomological applications, since (*) is necessary and quite close to sufficient for the existence of p-torsion in the simply connected compact Lie group of the same type as G (see [1]).

e) The group G of type B_2 and characteristic 2 yields the simplest example in which U_x is not connected (it has 2 pieces). In this group every sufficiently general element of the centre of U is an irregular unipotent element whose centralizer is unipotent. Hence not every unipotent element is the unipotent part of a regular element (cf. 1.2 *a)*).

§ 5. Irregular elements

Our aim is to prove 1.3. The assumptions of § 4 continue. We write T_i for the kernel of α_i on T, U_i for the group generated by all X_α for which $\alpha>0$ and $\alpha \neq \alpha_i$, B_i for T_iU_i $(1 \leq i \leq r)$. The latter is a departure from the notation of 2.5.

5.1. *Lemma. — An element of* G *is irregular if and only if it is conjugate to an element of some* B_i.

For the proof we may restrict attention to elements of the form $x=yz$ $(y \in T,\ z \in U \cap G_y)$ by 2.12. Let G′ be as in 3.6. The root system Σ' for G′ consists of all roots α such that $\alpha(y)=1$. It inherits an ordering from that of Σ. Assume first that x is in B_i. Then α_i is in Σ', and the X_i component of z is 1. Thus z is irregular in G′ by 3.2 and 3.3, whence x is irregular in G by 3.7. Assume now that x is irregular in G so that z is irregular in G′. If we write $z=\prod_\alpha u_\alpha$ $(u_\alpha \in X_\alpha,\ \alpha>0,\ \alpha \in \Sigma')$, we have $u_\alpha=1$ for some root α simple in Σ', by 3.2 and 3.3. We prove by induction on the height of α (this is $\sum_i n_i$ if $\alpha=\sum_i n_i\alpha_i$) that x may be replaced by a conjugate such that α above is simple in Σ. This conjugate will be in some B_i, and 5.1 will follow. We assume the height to be greater than 1. We have $(\alpha, \alpha_i)>0$ for some i, and α_i is not in Σ' since otherwise $\alpha-\alpha_i$ would be in Σ' in contradiction to the simplicity of α in Σ'. Thus $\sigma_i z \sigma_i^{-1}$ is in U. Since $w_i\alpha = \alpha - 2\alpha_i(\alpha, \alpha_i)/(\alpha_i, \alpha_i)$ has smaller height than α, we may apply our inductive assumption to $\sigma_i x \sigma_i^{-1}$ to complete the proof of the assertion and of 5.1.

5.2. *Lemma. — If B'_i is an irreducible component of B_i, the union of the conjugates of B'_i is closed, irreducible, and of codimension 3 in G.*

The normalizer P_i of B_i has the form $P_i = G_i B_i$ and is a parabolic subgroup of G, since it contains the Borel subgroup B. The number of components of T_i, hence of B_i, is either 1 or 2: if $\alpha_i = n\alpha'_i$ with α'_i a primitive character on T, then $(2\alpha'_i, \alpha_i)/(\alpha_i, \alpha_i)$ is an integer [8, p. 16-09, Cor. 1], whence $n = 1$ or 2. Thus P_i also normalizes B'_i, whence if easily follows that P_i is the normalizer of B'_i. Since G/P_i is complete (because P_i is parabolic) by [8, p. 6-09, Th. 4], it follows by a standard argument (cf. [8, p. 6-12] or 2.14 above) that the union of the conjugates of B'_i is closed and irreducible and of codimension in G at least $\dim(P_i/B'_i) = 3$, with equality if and only if there is an element contained in only a finite, nonzero number of conjugates of B'_i. Thus 5.2 follows from:

5.3. *Lemma. —* a) *There exists in $B'_i \cap T_i$ an element t such that $\alpha(t) \neq 1$ for every root $\alpha \neq \pm\alpha_i$.*

b) *If t is as in* a) *it is contained in only a finite number of conjugates of B'_i (or B_i).*

For *a)* we choose the notation so that $i = 1$. Then for some number $c_1 = \pm 1$, the set $B'_1 \cap T_1$ consists of all t for which $\alpha'_1(t) = c_1$. That values c_j may be assigned for $\alpha_j(t)$ $(2 \leq j \leq r)$ so that *a)* holds then follows by induction: having chosen $c_2, \ldots, c_j$ so that $\alpha(t) \neq 1$ if α is a combination of $\alpha_1, \alpha_2, \ldots, \alpha_j$ and $\alpha \neq \pm\alpha_1$, one has only a finite set of numbers to avoid in the choice of c_{j+1}. For *b)* let C be either B'_i or B_i, and let t be as in *a)*. Let yCy^{-1} be a conjugate of C containing t. Since B normalizes C we may take y in the form $u\sigma_w$ of 2.4. Writing $u^{-1}tu = tu'$, the inclusion $y^{-1}ty \in C$ yields

$$(*) \qquad \sigma_w^{-1} t \sigma_w . \sigma_w^{-1} u' \sigma_w \in C.$$

Since $\sigma_w^{-1} u \sigma_w$ is in U^-, so is $\sigma_w^{-1} u' \sigma_w$, whence $u' = 1$. Thus u commutes with t, hence it is in X_i because of the choice of t. By (*) we have $\sigma_w^{-1} t \sigma_w \in C$, hence $(w\alpha_i)(t) = 1$, and $w\alpha_i = \pm\alpha_i$. Thus $\sigma_w^{-1} u \sigma_w$ is in G_i and normalizes C, whence using $y = \sigma_w . \sigma_w^{-1} u \sigma_w$ we get $yCy^{-1} = \sigma_w C \sigma_w^{-1}$. The number in *b)* is thus finite and in fact equal to the number of elements of the Weyl group which fix α_i.

We now turn to the proof of Theorem 1.3. Parts *a)* and *b)* follow from 5.1 and 5.2. If $i \neq j$ the independence of α_i and α_j implies that each component of B_i meets each component of B_j. Thus by 5.2 the set Q is connected if $r > 1$. If $r = 1$, the irregular elements form the centre of G, whence *c)* follows.

5.4. *Corollary. — The set of regular elements is dense and open in* G.

This is clear.

5.5. *Corollary. — In the set of irregular elements the semisimple ones are dense.*

The set of elements of B_i of the form tu with t as in 5.3 *a)* and u in U_i is open in B_i, dense in B_i by 5.3 *a)*, and consists of semisimple elements: by 2.12 the last assertion need only be proved when u commutes with t and in that case $u = 1$ by 2.1 and 2.2 *b)*. By 5.1 this yields 5.5.

293

By combining 5.1, 5.5 and the considerations of 5.2 we may determine the number of components of Q. We state the result in the simplest case, omitting the proof, which is easy. We recall that G is an adjoint group if the roots generate the character group of T.

5.6. *Corollary.* — *If* G *is a simple adjoint group, the number of irreducible components of* Q *is just the number of conjugacy classes of roots under the Weyl group, except that when* G *is of type* C_r $(r \geq 2)$ *and of characteristic not* 2 *the number of components is* 3 *rather than* 2.

The method of the first part of the proof of 5.2 yields the following result, to be used in 6.11.

5.7. *Lemma.* — *The union of the conjugates of* U_i *is of codimension at least* $r+2$ *in* G.

§ 6. Class functions and the variety of regular classes

G, T, etc. are as before. By a function on G (or any variety over K) we mean a rational function with values in K. Each function is assumed to be given its maximum domain of definition. A function which is everywhere defined is called regular. A function f on G which satisfies the condition $f(x)=f(y)$ whenever x and y are conjugate points of definition of f is called a class function. As is easily seen, the domain of definition of a class function consists of complete conjugacy classes.

6.1. *Theorem.* — *Let* C[G] *denote the algebra (over* K*) of regular class functions on* G.

a) C[G] *is freely generated as a vector space over* K *by the irreducible characters of* G.

b) *If* G *is simply connected,* C[G] *is freely generated as a commutative algebra over* K *by the fundamental characters* χ_i $(1 \leq i \leq r)$ *of* G.

Let C[T/W] denote the algebra of regular functions on T invariant under W. Since two elements of T are conjugate in G if and only if they are conjugate under W (this follows easily from 2.4), there is a natural map β from C[G] to C[T/W].

6.2. *Lemma.* — *The map* β *is injective.*

For if f in C[G] is such that $\beta f=0$, then $f=0$ on the set of semisimple elements, a dense set in G by 2.14, e.g., whence $f=0$.

6.3. *Lemma.* — *If in* 6.1 *we replace* C[G] *by* C[T/W] *and the irreducible characters by their restrictions to* T, *the resulting statements are true.*

Let X, the character group of T, be endowed with a positive definite inner product invariant under W, and let D consist of the elements δ of X such that $(\delta, \alpha_i) \geq 0$ for all i. We wish to be able to add characters as functions on T. Thus we switch to a multiplicative notation for the group X. For each δ in D we write sym δ for the sum of the distinct images of δ under W. We write $\delta_1 < \delta_2$ if $\delta_1^{-1}\delta_2$ is a product of positive roots. Now the elements of X freely generate the vector space of regular functions on T [8, p. 4-05, Th. 2], and each element of X is conjugate under W to a unique element of D [8, p. 14-11, Prop. 6]. Thus the functions sym δ $(\delta \in D)$ freely generate C[T/W]. Now there is a 1—1 correspondence between the elements of D and the irreducible characters of G, say $\delta \leftrightarrow \chi_\delta$, such that one has $\chi_\delta|_T = \text{sym}\,\delta + \sum_{\delta'} c(\delta')\text{sym}\,\delta'$ $(\delta' < \delta, c(\delta') \in K)$

(see 7.15). Thus *a)* holds. Now if G is simply connected, the characters ω_i of 2.6 form a basis for D as a free commutative semigroup, and the corresponding irreducible characters on G are the χ_i. If $\delta = \prod_i \omega_i^{n(i)}$ is arbitrary in D, then on T we have $\chi_\delta = \prod_i \chi_i^{n(i)} + \sum_{\delta'} c(\delta')\chi_{\delta'}$ $(\delta' < \delta)$, whence by induction, the $\chi_i|_T$ generate the algebra C[T/W]. Using the above order one sees that the only polynomial in the $\chi_i|_T$ which is 0 is 0. Thus *b)* holds.

6.4. *Corollary.* — *The map* β *is surjective. Hence it is an isomorphism.*

The first statement follows from 6.3 *a)*, the second from 6.2.

Theorem 6.1 is now an immediate consequence of 6.3 and 6.4.

6.5. *Corollary.* — *For all* f *in* C[G] *and* x *in* G, *we have* $f(x) = f(x_s)$.

For this equation holds when f is a character on G.

6.6. *Corollary.* — *Assume that the elements* x *and* y *of* G *are both semisimple or both regular. Then the following conditions are equivalent.*

a) x *and* y *are conjugate.*

b) $f(x) = f(y)$ *for every* f *in* C[G].

c) $\chi(x) = \chi(y)$ *for every character* χ *on* G.

d) $\rho(x)$ *and* $\rho(y)$ *are conjugate for every representation* ρ *of* G.

If G *is simply connected,* c) *and* d) *need only hold for the fundamental characters and representations of* G.

Here *a)* implies *d)*, which implies *c)*, which implies *b)* by 6.1 *a)*; and the modified implications when G is simply connected also hold by 6.1 *b)*. To prove *b)* implies *a)* we may by 1.2 and 6.5 assume that x and y are semisimple, and then that they are in T and that $f(x) = f(y)$ for every f in C[T/W] by 6.4. Since W is a finite group of automorphisms of the variety T, it follows, among other things, by [10, p. 57, Prop. 18] that C[T/W] separates the orbits of T under W. Thus x and y are conjugate under W, and *a)* holds. This proves 6.6.

6.7. *Corollary.* — *If* x *is in* G, *the following are equivalent.*

a) x *is unipotent.*

b) *Either* b) *or* c) *of* 6.6, *or its modification when* G *is simply connected, holds with* $y = 1$.

Since x is unipotent if and only if $x_s = 1$, this follows from 6.5 and the equivalence of *a)*, *b)* and *c)* in 6.6.

6.8. *Corollary.* — *The set* S *of regular semisimple elements has codimension* 1 *in* G.

By 6.4 the function $\prod_\alpha (\alpha - 1)$ (α root) on T has an extension to an element f of C[G]. It is then a consequence of 2.11 (see *a)* and *e)*), 2.12, 6.5 and 2.13 that S is defined by $f \neq 0$, whence 6.8.

6.9. *Theorem.* — *Every element of* C(G), *the algebra of class functions on* G, *is the ratio of elements of* C[G].

Each element of C(G) is defined at semisimple elements of G by 2.14, hence at a dense open set in T, whence by the argument of the proof of 6.4, the natural map from C(G) to C(T/W) is an isomorphism. Now if f is in C(T/W), then $f=g/h$ with g and h regular on T, and because W is finite it can be arranged that h is in C[T/W], whence g is also, and 6.9 follows.

The class functions lead to a quotient structure on G which we now study. We say that the elements x and y of G are *in the same fibre* if $f(x)=f(y)$ for every regular class function f. We observe that if G is simply connected the fibres are the inverse images of points for the map p from G to affine r-space V defined thus:

6.10 $$p(x)=(\chi_1(x), \chi_2(x), \ldots, \chi_r(x)).$$

This is because of 6.1 *b)* and the surjectivity of p (see proof of 6.16). As the next result shows, the fibres are identical with the closures of the regular classes.

6.11. *Theorem. — Let* F *be a fibre.*

a) F *is a closed irreducible set of codimension* r *in* G.

b) F *is a union of classes of* G.

c) *The regular elements of* F *form a single class, which is open and has a complement of codimension at least* 2 *in* F.

d) *The semisimple elements of* F *form a single class, which is the unique closed class in* F *and the unique class of minimum dimension in* F, *and which is in the closure of every class in* F.

Clearly F is closed in G and a union of classes. By 1.2, 6.5 and 6.6 the fibre F contains a unique class R of regular elements and a unique class S of semisimple elements. Fix y in S and write $G_{y0}=G'T'$ as in 3.6. By 3.2 and 3.3 the regular unipotent elements are dense in U, hence also in the set of all unipotent elements. Applying this to G', and using 3.7, we see that among the elements x of F for which $x_s=y$ the regular ones, that is, the ones in R, are dense. Thus R is dense in F, which, being closed, is the closure of R. Since R is irreducible and of codimension r in G, the same is true of F. By 5.4 the class R is open in F. Applying 3.2, 3.3 and 5.7 to the group G' above, we see that the part of F—R for which $x_s=y$ has codimension at least $r+2$ in G_{y0}. Thus F—R itself has codimension at least $r+2$ in G, and at least 2 in F. It remains to prove that S is in the closure of every class in F, since the other parts of *d)* then follow, and by a shift to the group G' it suffices to prove this when $S=\{1\}$, that is, when F is the set of unipotent elements. Thus *d)* follows from:

6.12. *Lemma. — A nonempty closed subset* A *of* U *normalized by* T *contains the element* 1.

Let u in A be written $\prod_\alpha x_\alpha(c_\alpha)$ as in 2.2 *b)*. Let $n(\alpha)$ denote the height of α, and for each c in K let $u_c=\prod_\alpha x_\alpha(c^{n(\alpha)}c_\alpha)$. If $c\neq 0$, then u_c is conjugate to u via an element of T, whence it belongs to A. If f is a regular function on U vanishing on A, then $f(u_c)$ is a polynomial in c (by 2.2 *b)*) vanishing for $c\neq 0$, hence also for $c=0$. Thus u_0 is in A, which proves 6.12.

From 6.11 *d)* we get the known result.

6.13. *Corollary. — In a semisimple group a class is closed if and only if it is semisimple.*

More generally we have:

6.14. *Proposition.* — *In a connected linear group* G′ *each class which meets a Cartan subgroup is closed.*

Let B′ be a Borel subgroup of G′. Since G′/B′ is complete [8, p. 6-09, Th. 4], it is enough to prove 6.14 with B′ in place of G′. Let x be an element of a Cartan subgroup of B′. Then x centralizes some maximal torus T′ in B′ [8, p. 7-01, Th. 1], whence if B′=T′U′ as usual then the class of x in B′ is an orbit under U′ acting by conjugation on B′. Because U′ is unipotent it follows from [7] that this class is closed.

6.15. *Remarks.* — *a)* Almost all fibres in 6.11 consist of a single class which is regular, semisimple, and isomorphic to G/T. This follows from 2.15.

b) Almost all of the remaining fibres consist of exactly 2 classes R and S with $\dim \mathrm{R} = \dim \mathrm{S} + 2$.

c) It is natural to conjecture that every fibre is the union of a finite number of classes, or, equivalently, that the number of unipotent classes is finite. In characteristic 0 the finiteness follows from the corresponding result for Lie algebras [4, p. 359, Th. 1]. In characteristic $p \neq 0$ one may assume that G is over the field k of p elements and make the stronger conjecture that each unipotent class has a point over k, or equivalently, by 1.10, that each unipotent class is over k. The last result would follow from the plausible statement: if γ is an automorphism of K, the element $\prod_{\alpha>0} x_\alpha(c_\alpha)$ of U is conjugate to $\prod_\alpha x_\alpha(\gamma c_\alpha)$.

d) It should be observed that for a given type of group the number of unipotent classes can change with the characteristic. Thus for the group of type B_2 the number is 5 in characteristic 2 but only 4 otherwise.

e) The converse of 6.14 is false.

6.16. *Theorem.* — *Assume that* G *is simply connected and that* p *is the map* 6.10 *from* G *to affine r-space* V. *Then* G/p *exists as a variety, isomorphic to* V.

The points to be proved are 1), 2) and 3) below.

1) *p is regular and surjective.* Clearly p is regular. The algebra of regular functions on T is integral over the subalgebra fixed by W. Thus any homomorphism of the latter into K extends to one of the former [2, p. 420, Th. 5.5]. Applying this to the homomorphism for which $\chi_i|_\mathrm{T} \to c_i$ ($c_i \in \mathrm{K}, i \leq i \leq r$) (see 6.1 and 6.4), we get the existence of t in T such that $\chi_i(t) = c_i$ for all i, whence p is surjective.

2) *Let f be a function on* V *and x an element of* G. *Then f is defined at $p(x)$ if and only if $f \circ p$ is defined at x.* Write $f = g/h$, the ratio of relatively prime polynomials in the natural coordinates on V. Then the restrictions to T of $g \circ p$ and $h \circ p$, as linear combinations of characters on T, are also relatively prime: otherwise suitable powers of these functions would have a nontrivial common factor invariant under W, which by 6.1 and 6.4 would contradict the fact that g and h are relatively prime. If $h(p(x)) \neq 0$, then clearly f is defined at $p(x)$ and $f \circ p$ at x. Assume $h(p(x)) = 0$. Because g and h are relatively prime, f is not defined at $p(x)$. We may take x in B and write $x = tu$ with t in T and u in U. Let A be an open set in G containing x. Then $\mathrm{A}u^{-1} \cap \mathrm{T}$ is an open

subset of T containing t, and because $g\circ p$ and $h\circ p$ are relatively prime on T and $h(p(t))=h(p(x))=0$ by 2.12 and 6.5, it also contains a point t' at which $h\circ p=0$ and $g\circ p\neq 0$. Then A contains the point $t'u$ at which the same equations hold, at which $f\circ p$ is not defined. Since A is arbitrary, $f\circ p$ is not defined at x, whence 2). From this discussion we see that

(*) *the domain of definition of a class function on* G *consists of complete fibres relative to* p.

3) *Under the map* $f\to f\circ p$ *the field of functions on* V *is mapped (isomorphically) onto the field of functions on* G *constant on the fibres of* p. The latter field consists of class functions, so that 3) follows from 6.1 *b)* and 6.9.

We recall that the regular elements form an open subvariety G^r of G.

6.17. *Corollary.* — *If* G *is simply connected, the set of regular classes of* G *has a structure of variety, that of* V, *given by the restriction of* p *to* G^r.

This means that the restriction of p to G^r has as its fibres the regular classes of G, and that 1), 2) and 3) above hold with G^r in place of G. All of this is clear.

To close this section we describe the situation when G is not simply connected. The proofs, being similar to those above, are omitted. Let $\pi : G'\to G$ be the simply connected covering of G, and let F be the kernel of π. An element f of F acts on the ith fundamental representation of G' as a scalar $\omega_i(f)$. We define an action of F on V thus: $f.(c_i)=(\omega_i(f)c_i)$.

6.18. *Theorem.* — *Assume* G *semisimple but not necessarily simply connected. Then the set of regular classes of* G *has a structure of variety, isomorphic to that of the quotient variety* V/F.

§ 7. Structure of N

In this section G, N, etc. are as in 1.4. Our aim is to prove that N is isomorphic to affine r-space V, under the map p of 6.10 when G is simply connected.

7.1. *Theorem.* — *The set* N *of* 1.4 *is closed and irreducible in* G. *It is isomorphic as a variety to affine r-space* V *under the map* $(c_i)\to\prod_i (x_i(c_i)\sigma_i)$. *In particular, an element of* N *uniquely determines its components in the product that defines* N.

7.2. *Lemma.* — *Let* $\beta_i=w_1w_2\ldots w_{i-1}\alpha_i\ (1\leq i\leq r)$ *and* $w=w_1w_2\ldots w_r$.

a) *The roots* β_i *are positive, distinct and independent.*

b) *They form the set of positive roots which become negative under* w^{-1}.

c) *The sum of two* β*'s is never a root.*

Since β_i is α_i increased by a combination of roots $\alpha_j\ (j<i)$, we have *a)*. The roots $w^{-1}\beta_i=-w_rw_{r-1}\ldots w_{i+1}\alpha_i$ are all negative by *a)* applied with $\alpha_r, \ldots, \alpha_1$ in place of $\alpha_1, \ldots, \alpha_r$. Since w^{-1} is a product of r reflections corresponding to simple roots, no more than r positive roots can change sign under w^{-1} by [8, p. 14-04, Cor. 3], whence *b)*. If the sum of two β's were a root, this root would be a β by *b)*, which is impossible by *a)*.

7.3. *Lemma.* — *If* β_i *and* w *are as in* 7.2 *the product* $\prod_i X_{\beta_i}$ *in* U *is direct, and if* X_w *denotes this product and* $\sigma_w=\sigma_1\sigma_2\ldots\sigma_r$, *then* $N=X_w\sigma_w$.

The first part follows from *a)* and *c)* of 7.2, and the second from the equation $X_{\beta_i}=\sigma_1\ldots\sigma_{i-1}X_i\sigma_{i-1}^{-1}\ldots\sigma_1^{-1}$.

Consider now 7.1. By 2.2 *b)* the set $X_w\sigma_w$ is closed, irreducible, and isomorphic to V via the map $(c_i)\to\prod_i x_{\beta_i}(c_i)\sigma_w=\prod_i(x_i(a_ic_i)\sigma_i)$ (a_i fixed element of K^*), whence 7.1 follows.

7.4. *Examples of* N. — *a)* Assume $r=1$ and $G=SL(2, K)$. Here we may choose X_1 as the group of superdiagonal unipotent matrices and σ_1 as the matrix $\begin{pmatrix}0 & -1\\ 1 & 0\end{pmatrix}$. Then N consists of all matrices of the form $y(c)=\begin{pmatrix}c & -1\\ 1 & 0\end{pmatrix}$.

b) Assume $r>1$ and $G=SL(r+1, K)$. Here we may choose for $x_i(c)\sigma_i$ the matrix $1_{i-1}\overset{\circ}{+}y(c)\overset{\circ}{+}1_{r-i}$, with $y(c)$ as in *a)* and 1_j the identity matrix of rank j. Then the element $\prod_i(x_i(c_i)\sigma_i)$ of N has the entries $c_1, -c_2, \ldots, (-1)^{r-1}c_r, (-1)^r$ across the first row, 1 in all positions just below the main diagonal, and 0 elsewhere. We thus have one of the classical normal forms for a matrix which is regular in the sense that its minimal and characteristic polynomials are equal. We observe that the parameters c in this form are just the values of the characters χ_i at the element considered. A similar situation exists in the general case. The group X_w of 7.3 in the present case consists of all unipotent matrices which agree with the identity in all rows below the first.

Next we show (7.5 and 7.8 below) that N does not depend essentially on the choice of the σ_i and the labelling of the simple roots, or equivalently, the order of the factors in the product for N. The other choices necessary to define N, namely the maximal torus T and a corresponding system of simple roots, are immaterial because of well known conjugacy theorems.

7.5. *Lemma. — Let each* σ_i *be replaced by an element* σ_i' *equivalent to it* mod T, *and let* $N'=\prod_i(X_i\sigma_i')$. *Then there exist* t *and* t' *in* T *such that* $N'=t'N=tNt^{-1}$.

Because T normalizes each X_i and is itself normalized by each σ_i, the first equality holds. We may write $tNt^{-1}=tw(t^{-1})N$, with w as in 7.3. Thus the second equality follows from:

7.6. *Lemma. — If* w *is as in* 7.2, *the endomorphism* $1-w$ *of* T $(t\to tw(t^{-1}))$ *is surjective, or equivalently, its transpose* $1-w'$ *on the dual* X *of* T *is injective.*

Suppose $(1-w')x=0$ with x in X. Then $(1-w_1)x=(1-w_2\ldots w_r)x$. The left side being a multiple of α_1 and the right side a combination of $\alpha_2, \ldots, \alpha_r$, both sides are 0. Since x is fixed by w_1 it is orthogonal to α_1. Similarly it is orthogonal to $\alpha_2, \ldots, \alpha_r$, hence is 0. Thus $1-w'$ is injective.

7.7. *Remarks. — a)* The argument shows that the conclusion of 7.6 holds if w is the product of reflections corresponding to any r independent roots.

b) If G is simply connected, one can show by an argument like that in 4) of 4.3 that the kernel of $1-w$ on T is just the centre of G.

7.8. *Proposition. — For each i let y_i be an element of $X_i\sigma_i$. Then the products obtained by multiplying the y_i in the $r!$ possible orders are conjugate.*

This result is not used in the sequel. Consider the Dynkin graph in which the nodes are the simple roots and the relation is nonorthogonality. Since the graph has no circuits [9, p. 13-02], it is a purely combinatorial fact that any cyclic arrangement of the simple roots can be obtained from any other by a sequence of moves each consisting of the interchange of 2 roots adjacent in the arrangement and not related in the graph (see [16, Lemma 2.3]). Now if α_i and α_j are not related in the graph, that is, orthogonal, then G_i and G_j commute elementwise (because $\alpha_i \pm \alpha_j$ are not roots), so that in case y_i is in G_i for each i our result follows. In the general case, if one interchanges y_i and y_j in the above situation, a factor from T appears, but this can be eliminated by conjugation by a suitable element of T, whence 7.8 follows.

7.9. *Theorem. — Let G be simply connected and let p be the map 6.10 from G to affine r-space V. Then p maps N, as a variety, isomorphically onto V.*

As in § 6, D denotes the set of characters on T of the form $\omega = \sum_j n_j \omega_j$ ($n_j \geq 0$, ω_j as in 2.6). We write $n_j = n_j(\omega)$ in this situation.

7.10. *Definition.* — $\omega_j \prec \omega_i$ means that *a)* $i \neq j$, and *b)* there exists ω in D such that $\omega_i - \omega$ is a sum of positive roots and $n_j(\omega) > 0$.

7.11. *Lemma. — The relation $\prec$ of 7.10 is a relation of strict partial order.*

If $\omega_k \prec \omega_j$ and $\omega_j \prec \omega_i$, then $k \neq i$ since a sum of positive roots and nonzero elements of D can not be 0 unless it is vacuous. Thus 7.11 follows.

7.12. *Remark.* — For simple groups of type A_r, B_2, or D_4 the relation $\prec$ is vacuous; for the other simple groups it is nonvacuous.

7.13. *Lemma. — Assume that σ_i is in G_i, and let $T_i = G_i \cap T$. Then there exists a bijection β from T_i to $X_i - \{1\}$ such that $x = \beta t$ if and only if $(xt\sigma_i)^3 = 1$.*

The group G_i is isomorphic to SL(2) by [8, p. 23-02, Prop. 2]. Identifying T_i (resp. X_i) with the subgroup of diagonal (resp. unipotent superdiagonal) matrices of SL(2), we get 7.13 by a simple calculation.

7.14. *Lemma. — Assume that G is simply connected, and that σ_i is chosen in G_i for each i, in the definition of N. Let the isomorphisms $x_i : K \to X_i$ be so normalized that $x_i(-1) = \beta(1)$ if β is as in 7.13. Let ψ_i be the function on N defined by $\prod_j (x_j(c_j)\sigma_j) \to c_i$. Then there exist functions f_i and g_i ($1 \leq i \leq r$) such that:*

a) *f_i (resp. g_i) is a polynomial with integral coefficients in those ψ_j (resp. χ_j) such that $\omega_j \prec \omega_i$ (see 7.10).*

b) *On N we have $\chi_i = \psi_i + f_i$ and $\psi_i = \chi_i + g_i$.*

Let i be fixed and let V_i be the space of the ith fundamental representation of G. For each weight (character on T) ω, let V_ω be the subspace of vectors which transform according to ω. We recall, in the form of a lemma, the properties of irreducible representations needed for our proof.

7.15. *Lemma.* — a) $\sum_\omega V_\omega = V_i$, *the total space.*

b) *If* $\omega = \omega_i$, *the highest weight, then* $\dim V_\omega = 1$.

c) *If* $\omega_i - \omega$ *is not a sum of positive roots,* $V_\omega = 0$.

d) *If v is in* V_ω, *if* $1 \leq j \leq r$, *and if we set* $\omega(n) = \omega + n\alpha_j$ *for* $n \geq 1$, *then there exist vectors v_n in* $V_{\omega(n)}$ *such that* $x_j(c)v = v + \sum_n c^n v_n$ *for all c in* K.

The proofs may be found in [8, Exp. 15 and p. 21-01, Lemme 1].

Now let x be an element of N. We write $x = \prod_j y_j$ and $y_j = x_j(c_j)\sigma_j$, and proceed to calculate $\chi_i(x)$, in several steps:

1) *If v is in* V_ω *and* $\omega(n) = \omega + (n - n_j(\omega))\alpha_j$ *for* $n \geq 1$, *there exist vectors v_n in* $V_{\omega(n)}$ *such that* $y_j v = \sigma_j v + \sum_n \psi_j(x)^n v_n$. This follows from 7.15 *d)* because $\sigma_j v$ corresponds to the weight $w_j\omega = \omega - n_j(\omega)\alpha_j$.

2) *Let* π_ω *be the projection on* V_ω *determined by* 7.15 a). *Then* $\pi_\omega x \pi_\omega = \prod_j (\pi_\omega y_j \pi_\omega)$. This follows from 1) and the independence of the roots α_j.

3) $\chi_i(x) = \sum_\omega \operatorname{tr} \pi_\omega x \pi_\omega$. This follows from the orthogonal decomposition $1 = \sum_\omega \pi_\omega$, which holds by 7.15 *a)*.

4) *If* $\omega = \omega_i$, *the highest weight, then* $\operatorname{tr} \pi_\omega x \pi_\omega = \psi_i(x)$. Let v be a basis for V_ω (see 7.15 *b)*), and let $v' = -\sigma_i v$. Then $y_i = x_i(c_i)\sigma_i$ fixes the space V′ generated by v and v', by 7.15 *c)* and *d)*, and maps these vectors onto $-v' + ac_i v$ and bv $(a, b \in K)$, respectively. A simple calculation shows that $y_i^3 = 1$ on V′ if and only if $b = 1$ and $ac_i = -1$. Because of our normalization of x_i, this is true only if $c_i = -1$, so that $a = 1$. Thus $\pi_\omega y_i \pi_\omega v = c_i v$. If $j \neq i$, then $w_j\omega = \omega$ by 2.6, so that X_j and σ_j, and hence also the group G_j they generate, fix the line of v, and then v itself because G_j is equal to its commutator group. By 2) we conclude that $\pi_\omega x \pi_\omega v = c_i v$, whence 4) follows.

5) *If ω is in* D *and* $\omega \neq \omega_i$, *then* $\operatorname{tr} \pi_\omega x \pi_\omega$ *depends only on those* $\psi_j(x)$ *for which* $\omega_j \prec \omega_i$. We may assume $V_\omega \neq 0$. It follows from 1) and 2) that $\pi_\omega x \pi_\omega$ depends only on those $\psi_j(x)$ for which $n_j(\omega)$ is positive. Because $\omega_i - \omega$ is a sum of positive roots by 7.15 *c)*, this yields 5).

6) *If ω is not in* D, *then* $\pi_\omega x \pi_\omega = 0$. If j is such that $n_j(\omega) < 0$, then $\pi_\omega y_j \pi_\omega = 0$ by 1), whence 6) follows from 2).

7) *In terms of the* ψ_j *the function* χ_i *is a polynomial with integral coefficients.* That we have a polynomial follows from 1). The integrality follows from the fact, proved in [17] when the characteristic is not 0 and in [14] when the characteristic is 0, that there exists a basis of V_i relative to which each σ_j acts integrally and each $x_j(c_j)$ is a polynomial with integral coefficients.

To prove 7.14 now, we need only combine 3), 4), 5), 6) and 7) above to get the assertions concerning f_i and then solve the equations $\chi_i = \psi_i + f_i$ recursively for the ψ_i to get the assertions concerning g_i.

Now we can prove Theorem 7.9. By 7.5 we may assume σ_i is in G_i for each i. Then by 7.1 the functions ψ_i of 7.14 are affine coordinates on N, so that 7.9 follows from 7.14.

7.16. *Corollary.* — a) N *is a cross-section of the fibres of* p *in* 7.9.

b) *The corresponding retraction* q *from* G *to* N, *given by* $q(x)=\prod_i x_i(\chi_i(x)+g_i(x))\sigma_i$ *if the normalization of* 7.14 *is used, yields on* G *a quotient structure isomorphic to that for* p.

c) *The set* $s(\mathrm{N})$ *made up of the semisimple parts of the elements of* N *is a cross-section of the semisimple classes of* G.

The formula for q follows from 7.14, and the other parts of *a)* and *b)* from 7.9. Then *c)* follows from 6.11 *d)*. We observe that $s(\mathrm{N})$ is never closed or connected, only constructible.

§ 8. Proof of 1.4 and 1.5

It follows from 7.9 that if G is simply connected distinct elements of N lie in distinct conjugacy classes. Thus 1.4 and 1.5 are consequences of the following result.

8.1. *Theorem.* — *Let* G *be simply connected (and semisimple),* x *an element of* G, *and* N *as in* 1.4. *Then the following are equivalent.*

a) x *is regular.*

b) x *is conjugate to an element of* N.

c) *The differentials* $d\chi_i$ *are independent at* x.

First we prove some lemmas.

8.2. *Lemma.* — *Under the assumptions of* 8.1 *let* ψ_i *denote the restriction of* χ_i *to* T, *let* ω_0 *denote the product* $\prod_i \omega_i$ *of the fundamental weights, and let the function* f *on* T *be defined by* $\prod_i (d\psi_i)=f\prod_i(\omega_i^{-1}d\omega_i)$, *the products being exterior products of differential forms. Then* $f=\sum_w(\det w)w\omega_0=\omega_0\prod_\alpha(1-\alpha^{-1})$, *the sum over* w *in* W *and the product over the positive roots* α.

We will deduce this from $\psi_i=\operatorname{sym}\omega_i+\sum_\delta c_i(\delta)\operatorname{sym}\delta$ ($\delta\in\mathrm{D}$, $\delta<\omega_i$, $c_i(\delta)\in\mathrm{K}$, notation of 6.3). Replacing the c's by indeterminates, we may view the equations to be proved as formal identities with integral coefficients in the group algebra of the dual of T, thus need only prove them in characteristic 0. First f is skew: $wf=(\det w)^{-1}f$ for every w in W. We have $wd\psi_i=d\psi_i$, and if $w\omega_i=\prod_j\omega_j^{n(i,j)}$, then $w(\omega_i^{-1}d\omega_i)=\sum_j n(i,j)\omega_j^{-1}d\omega_j$, which, because $\prod_i\omega_i^{-1}d\omega_i\neq 0$, yields $f=wf.\det(n(i,j))=wf.\det w$. Because f is skew and the characteristic is 0, we have

$$(*)\qquad f=\sum_\delta c(\delta)\sum_w(\det w)w\delta \quad (\delta\in\mathrm{D},\ c(\delta)\in\mathrm{K}),$$

the inner sum being over W and the outer over D. From the expression for ψ_i, we have $d\psi_i=\omega_i(\omega_i^{-1}d\omega_i)+$ a combination of terms $\omega(\omega_j^{-1}d\omega_j)$, with ω lower (by a product of positive roots) than ω_i, whence $f=\omega_0+$lower terms. Thus in $(*)$ above $c(\omega_0)=1$ and $c(\delta)=0$ when δ is not lower than ω_0. If δ is lower than, and different from, ω_0, then δ is orthogonal to some α_i (if $\delta=\prod_i\omega_i^{n(i)}$, then some $n(i)$ is less than the corresponding object for ω_0, hence is 0), whence $\sum_w (\det w) w\delta = 0$. Thus (*) becomes $f = \sum_w (\det w) w\omega_0$. The final equality in 8.2 is a well known identity of Weyl [18, p. 386].

8.3. *Remark.* — $\prod_i (\omega_i^{-1} d\omega_i)$ above is, to within a constant factor, the unique differential r-form on T invariant under translations, that is, the "volume element" of T.

8.4. *Lemma.* — *Let* G' *denote the neighborhood* U^-TU *of* T (*see* 2.3), *and let* π *denote the natural projection from* G' *to* T. *For each* α *let* y_α *be the composition of the projection from* G' *to* X_α *and an isomorphism from* X_α *to* K.

a) *If* f *is a regular function on* G, *its restriction to* G' *is a combination of monomials in the functions* y_α *and* $\omega_i^{\pm 1} \circ \pi$.

b) *If* f *is also a class function and the combination is irredundant, then each monomial has a total degree in the* y_α*'s which is either* 0 *or at least* 2.

Here *a)* follows from 2.3. In *b)* no monomial could involve exactly one y_α (to the first degree), because then conjugation by t in T and use of 2.1 would yield $\alpha(t) = 1$ for all t in T, a contradiction.

8.5. *Lemma.* — *Let* ψ_i *be as in* 8.2 *and* π *as in* 8.4. *Then* $d\chi_i = d\psi_i \circ d\pi$ *at all points of* T.

Here the tangent space at t as an element of G is being identified with its tangent space as an element of G'. By 8.4 *b)* we have on G' an equation $\chi_i = \psi_i \circ \pi +$ terms of degree at least 2 in the y_α. Since each y_α is 0 on T, we have there $d\chi_i = d\psi_i \circ d\pi$.

8.6. *Lemma.* — *If* x *is semisimple,* a) *and* c) *of* 8.1 *are equivalent.*

We may take x in T. By 8.5 and the surjectivity of $d\pi$ (from the tangent space of x in G' to its tangent space in T), the $d\chi_i$ are independent at x if and only if the $d\psi_i$ are, and by 8.2 this is so if and only if $\alpha(x) \neq 1$ for every root α, that is, if and only if x is regular, by 2.11.

We can now prove 8.1. From 7.9 it follows that *b)* implies *c)*, and from 5.5 and 8.6 that *c)* implies *a)*. Now assume x is regular. By 7.9 there is a unique element y in both N and the fibre of p which contains x. Then y is regular because *b)* $\to$ *a)* has already been shown, whence x is conjugate to y by 6.11 *c)*. Thus *a)* implies *b)*, and 8.1 is proved.

Using the above methods one can also show:

8.7. *Theorem.* — *Without the assumption of simple connectedness in* 8.1, *conditions* a) *and* b) *are equivalent and are implied by*

c') *there exist* r *regular class functions on* G *whose differentials are independent at* x.

One can also show that the elements of N conjugate to a given one $\prod_i x_i(c_i)\sigma_i$ are those of the form $\prod_i x_i(\omega_i(f)c_i)\sigma_i$ $(f \in F)$, in the notation of the paragraph before 6.18.

8.8. *Remark.* — If $w = w_1 w_2 \ldots w_r$, all elements of the double coset $B\sigma_w B$ are regular, not just those of N. This depends on 7.3, 7.5 and the following result, whose proof is omitted.

8.9. *Proposition.* — *If* w *is as above, then the map from the Cartesian product of* $\sigma_w U^- \sigma_w^{-1} \cap U$ *and* $\sigma_w^{-1} U \sigma_w \cap U$ *to* U *given by* $(u_1, u_2) \to u_2^{-1} . u_1 . \sigma_w u_2 \sigma_w^{-1}$ *is bijective.*

§ 9. Rationality of N

Henceforth k denotes a perfect subfield of our universal field K, which for convenience is assumed to be an algebraic closure of k, and Γ denotes the Galois group of K over k. In this section G is a simply connected semisimple group. If G is (defined) over k, it is natural to ask whether N or a suitable analogue thereof can be constructed over k. As the following result shows, the answer is in general no.

9.1. *Theorem. — If* G *is over k, then a necessary condition for the existence over k of a cross-section* C *of the regular classes is the existence of a Borel subgroup over k.*

For the unique unipotent element of C is clearly over k, and so is the unique Borel subgroup that contains it (see 3.2 and 3.3).

As we now show, this necessary condition comes quite close to being sufficient. First we consider a more restrictive situation, that in which G splits over k, that is, is over k and contains a maximal torus which with all of its characters is over k.

9.2. *Theorem. — If* G *splits over k, then* N *in* 1.4 *(and hence also s*(N) *in* 7.16 c)) *can be constructed over k.*

Let G split relative to the maximal torus T. Since the simple root α_i is over k, so is X_i, and it remains to choose each σ_i over k. We start with an arbitrary choice for σ_i. Then the map $\gamma \rightarrow \sigma_i^{-1}\gamma(\sigma_i) = x_\gamma$ is a cocycle from Γ to a group isomorphic to K^*, namely, $G_i \cap T$. In other words:

9.3. a) $x_{\gamma\delta} = x_\gamma \gamma(x_\delta)$ *for all* γ *and* δ *in* Γ.

b) *There exists a subgroup* Γ_1 *of finite index in* Γ *such that* $x_\gamma = 1$ *if* γ *is in* Γ_1.

By a famous theorem of Hilbert (see, e.g., [11, p. 159]), this cocycle is trivial, that is, there exists t_i in T_i such that $x_\gamma = t_i \gamma(t_i^{-1})$ for all γ in Γ. Then $\sigma_i t_i$ is over k, as required.

9.4. *Theorem. — Assume that* G *is over k, and contains a Borel subgroup over k. Assume further that* G *contains no simple component of type* A_n *(n even). Then the set* N *of* 1.4 *can be constructed over k.*

Let B be a Borel subgroup over k. It contains a maximal torus T over k. If k is infinite, this follows from 2.14 and Rosenlicht's theorem [6, p. 44] that G_k is dense in G, while if k is finite with q elements and β is the qth power automorphism, one picks an arbitrary maximal torus T′, then x in B so that $x\beta(T')x^{-1} = T'$ (conjugacy theorem), then y in B so that $x = y^{-1}\beta(y)$ (Lang's theorem [5]), and then $T = yT'y^{-1}$. We order the roots so that B corresponds to the set of positive roots. Γ permutes the simple roots α_i in orbits. We order the α_i so that those in each orbit come together. If for each orbit we can construct over k the corresponding part of the product for N, then we can construct N over k. Thus we may (and shall) assume that there is a single orbit. Let Γ_1 be the stabilizer of α_1 in Γ, and k_1 the corresponding subfield of K. Then α_1 is over k_1, whence G_1 (the corresponding group of rank 1) is also, so that by 9.2 applied with G_1 in place of G the set $X_1\sigma_1$ can be constructed over k_1. Then Γ operates on this set to produce, in an unambiguous way, sets $X_i\sigma_i$ $(1 \leq i \leq r)$. But these sets commute

304

pairwise: the roots (in each orbit) are orthogonal because of the exclusion of the type A_n (n even). Their product is thus fixed by all of Γ, hence is over k, as required.

Observe that 9.2 and 9.4 yield 1.6.

9.5. *Corollary.* — *Under the assumptions of* 9.2 *or* 9.4 *the natural map (inclusion) from the set of regular elements over k to the set of regular classes over k is surjective. In other words, each regular class over k contains an element over k.*

Let C be a regular class over k. Then $C \cap N$ is over k by 9.2 or 9.4, and it consists of one element by 1.4, whence 9.5.

9.6. *Remark.* — For the group of type A_n (n even) we do not know whether there exists over k a global closed irreducible cross-section of the regular classes of G, or even of the fibres of the map p of 6.10 (which can be taken over k if V is suitably defined over k), although a study of the group of type A_2 casts some doubt on these possibilities. All that we can show, 9.7 *c)* below, is that there exists a local cross-section (covering a dense open set in V) with the above properties.

9.7. *Theorem.* — *Assume that* G *is over k, and contains a Borel subgroup over k. Assume that every simple component of* G *is of type* A_n *(n even). Then there exists in* G *a set* N′ *with the following properties.*

a) N′ *is a disjoint union of a finite number of closed irreducible subsets of* G.

b) N′ *is a cross-section of the fibres of* p *in* 6.10.

c) p *maps each component of* N′ *isomorphically onto a subvariety of* V, *and one component consisting of regular elements onto a dense open subvariety of* V.

d) $s(N')$ *is a cross-section of the semisimple classes of* G.

e) *Each component of* N′ *is over k.*

In order to continue our main development, we postpone the construction of N′ to the end of the section.

9.8. *Theorem.* — *If* G *(with or without components of type* A_n *(n even)) is over k and contains a Borel subgroup over k, the natural map from the set of semisimple elements over k to the set of semisimple classes over k is surjective.*

Observe that this is Theorem 1.7 of the introduction. As is easily seen, we may assume either that no components of G are of type A_n (n even) or that all are. In the first case we replace N by $s(N)$ and 1.4 by 7.16 *c)* in the proof of 9.5, while in the second case we use $s(N')$ and 9.7 *d)* instead.

9.9. *Remark.* — G need not be semisimple for the validity of 9.8. For let A be a connected linear group satisfying the other assumptions. If R is the unipotent radical, then A/R is a connected reductive group, hence the direct product of a torus and a simply connected semisimple group because A is simply connected, whence the result to be proved holds for A/R. A semisimple class of A over k thus contains an element x over k mod R. The map $\gamma \to x^{-1}\gamma(x)$ then defines a cocycle into R which is trivial because R is unipotent (see [12, Prop. 3.1.1]), whence 9.9.

Theorem 9.8 admits a converse.

9.10. *Theorem.* — *If G is over k and the map of 9.8 is surjective, then G contains a Borel subgroup over k.*

If k is finite, this follows from Lang's theorem (see the proof of 9.4), even without the assumption of surjectivity. Henceforth let k be infinite. Let F be the centre of G, n the order of F, h the height of the highest root, and c and c' elements of k^* such that $c=c'^n$ and c has order greater than $h+1$. Let T be a maximal torus over k (for the existence, see the proof of 9.4), and t' an element of T such that $\alpha_i(t')=c'$ for every α_i in some system of simple roots. Set $t=t'^n$, so that $\alpha_i(t)=c$.

1) *t is regular.* If α is a root of height m, then $\alpha(t)=c^m\neq 1$, whence 1). Since $c^m=c$ only if $m=1$ we also have:

2) *If α is a root such that $\alpha(t)=c$, then α is simple.*

3) *The class of t is over k.* Each element γ of the Galois group Γ acts as an automorphism on the root system, hence determines a unique element w_γ of the Weyl group such that $w_\gamma\circ\gamma$ permutes the simple roots. Since $\alpha_i(t')$ is independent of i and is in k, we have $\alpha_i((w_\gamma\circ\gamma)(t'))=((w_\gamma\circ\gamma)^{-1}(\alpha_i))(t')=\alpha_i(t')$, whence $(w_\gamma\circ\gamma)(t')=ft'$ for some f in F. Thus $(w_\gamma\circ\gamma)(t)=f^n t=t$, which yields 3).

4) *One can normalize the pair* T, *t above so that* 1) *and* 2) *hold and also t is over k.* By the surjectivity assumption in 9.10 there exists t'' over k and conjugate to t. Any inner automorphism which maps t to t'' maps T onto a maximal torus T'' which must be over k because it is the unique maximal torus containing t'' by 1) and 2.11, and also maps the simple system relative to T into one relative to T'' so that the equations $\alpha_i(t)=c$ are preserved. On replacing T, t by T'', t'', we get 4).

Now by 4) we have $(\gamma\alpha_i)(t)=(\gamma\alpha_i)(\gamma t)=\gamma(\alpha_i(t))=\gamma(c)=c$, whence $\gamma\alpha_i$ is simple by 2). Thus each γ preserves the set of positive roots, hence also the corresponding Borel subgroup, which is thus over k, as required.

It remains to construct the set N' of 9.7. If G is a group of type A_n (n even) in which T, etc. are given, the following notation is used. The simple roots are labelled $\alpha_1, \alpha_2, \ldots, \alpha_n$ from one end of the Dynkin graph to the other (see [8, p. 19-03]). We write $n=2m$, set $\alpha=\alpha_m+\alpha_{m+1}$, a root, let G_α denote the group of rank 1 generated by X_α and $X_{-\alpha}$, write T_α for $T\cap G_\alpha$, and σ_α for an element normalizing T according to the reflection relative to α. The group of automorphisms of the system of simple roots pairs α_i with α_{2m+1-i}, which is orthogonal to α_i unless $i=m$. Hence (see the proof of 9.4) only the part of N corresponding to α_m and α_{m+1} need be modified.

9.11. *Theorem.* — *Let G be as in 9.7. If G contains a single component, assume (in the above notation) that the choices σ_i and σ_α are normalized to be in G_i and G_α ($i\neq m, m+1$), that u_m and u_{m+1} are elements of X_m and X_{m+1} and different from 1, that N'' (resp. N''') is the product of $X_\alpha\sigma_\alpha$ (resp. $u_{m+1}u_m X_\alpha\sigma_\alpha T_\alpha$) and $\prod_j X_j\sigma_j$ ($j\neq m, m+1$), and that N' is the union of N'' and N'''. If G is a product of several components, assume that N' is constructed as a product accordingly. Then one has* a) *to* e) *of* 9.7.

We proceed to study N'' and N''' as we did N in § 7. The following observation will be useful.

9.12. *Lemma.* — a) *The sequence of roots* $S=\{\alpha_1, \ldots, \alpha_{m-1}, \alpha, \alpha_{m+2}, \ldots, \alpha_{2m}\}$ *yields a simple system of type* A_{2m-1}.

b) *If* G' *is the corresponding semisimple subgroup of* G, *then* N'' *as constructed in* G' *fulfills the rules of construction of* N *in* G.

The verification of *a)* is easy, while *b)* is obvious.

9.13. *Lemma.* — *The sets* N'' *and* N''' *are closed and irreducible in* G. *The natural maps from the Cartesian products* $X_\alpha \times \prod_j X_j$ *and* $X_\alpha \times T_\alpha \times \prod_j X_j$ *to* N'' *and* N''', *respectively, are isomorphisms of varieties. In particular each element of* N'' *or* N''' *uniquely determines its components.*

The assertions about N'' follow from 7.1 and 9.12. Those concerning N''' are proved similarly.

9.14. *Lemma.* — *If* u_m *and* u_{m+1} *in* 9.11 *are replaced by alternates* u'_m *and* u'_{m+1}, *then* N''' *is replaced by a conjugate, under* T.

We can find t in T to transform u_m and u_{m+1} into u'_m and u'_{m+1}, and, because only the values $\alpha_m(t)$ and $\alpha_{m+1}(t)$ are relevant (see 2.1), so that also $\alpha_j(t)=1$ if $j \neq m, m+1$; we are using the independence of the simple roots here. By conjugating N''' by t, we get 9.14.

9.15. *Lemma.* — *Let the functions* $\psi_i\ (i \neq m, m+1)$ *and* ψ_α *be defined on* N'' *as the functions* ψ_i *of* 7.14 *are defined on* N. *Further, set* $\chi_0=\chi_{2m+1}=1$ *and* $\psi_0=\psi_{2m+1}=1$. *Then on* N'' *on has*

a) $\chi_i=\psi_i+\psi_{i-1}$ *if* $1 \leq i \leq m-1$.

b) $\chi_i=\psi_i+\psi_{i+1}$ *if* $m+2 \leq i \leq 2m$.

c) $\chi_m=\psi_\alpha+\psi_{m-1}$.

d) $\chi_{m+1}=\psi_\alpha+\psi_{m+2}$.

1) *Let* ρ_i *be the* i^{th} *fundamental representation of* G *and* ρ'_i *that of* G' (*according to the sequence* S *in* 9.12). *Then the restriction of* ρ_i *to* G' *is isomorphic to the direct sum of* ρ'_i *and* ρ'_{i-1}. Here ρ'_0 is the trivial representation. We may identify G with SL(L) and G' with the subgroup SL(L')×SL(L''), if L' and L'' are vector spaces of rank $2m$ and 1 and L is their direct sum. Then ρ_i is realized by the action of G on the space $\wedge^i L$ of skew tensors of rank i over L. Combining this with the canonical decomposition $\wedge^i L=\wedge^i L' \oplus \wedge^{i-1} L' \otimes L''$, we get 1).

We will use the notation D, V_ω, π_ω, etc. of 7.14.

2) *If* G *in* 7.14 *is of type* A_r, *then one has*:

a) *The only weight* ω *in* D *such that* $V_\omega \neq 0$ *is* $\omega=\omega_i$.

b) *The function* f_i *is* 0.

Using the realization of ρ_i as in 1), we see that the transforms of V_{ω_i} under the Weyl group W generate V_i. Since D is a fundamental domain for the action of W, this proves *a)*. Referring to the proof of 7.14, the contribution to $\chi_i(x)$ coming from step 5) is 0, by *a)*, whence *b)* follows.

3) *Proof of* 9.15. — Writing 1) in terms of characters, $\chi_i = \chi'_i + \chi'_{i-1}$, and then using 9.12 and 7.14 as refined in 2 *b)* above, for the group G′, we get 9.15.

9.16. *Lemma.* — *Let* ψ_i *and* ψ_α *be as in* 9.15, *but on* N‴ *instead of* N″. *Let* u_m *and* u_{m+1} *be so chosen that the final stage of* ψ_α *(isomorphism from* X_α *to* K) *maps the commutator* (u_{m+1}, u_m) *onto* 1. *Let* φ_α *denote the composition of the projection* $N''' \to T_\alpha$ *and the evaluation* $t \to \alpha_m(t)$ *(or* $\alpha_{m+1}(t)$). *Then on* N‴ *one has* a) *and* b) *of* 9.15 *and also*

c) $\chi_m = \varphi_\alpha \psi_\alpha + \psi_{m-1}$,

d) $\chi_{m+1} = \varphi_\alpha + \varphi_\alpha \psi_\alpha + \psi_{m+2}$.

1) *Assume that* $1 \leq i \leq m$. *Then there exist exactly two weights* ω *such that* $(\omega, \beta) \geq 0$ *for all* β *in the sequence* S *of* 9.12, *and* $V_\omega \neq 0$. *For both*, $\dim V_\omega = 1$. *One is the highest weight* ω_i *and the other, say* ω'_i, *is orthogonal to all terms of* S *but the* $(i-1)^{\text{th}}$. The highest weights of the representations ρ'_i and ρ'_{i-1} in 1) of 9.15 satisfy the first two statements by 2 *a)* of 9.15 and 7.15 *b)*. Finally ω_i must correspond to ρ'_i rather than ρ'_{i-1} because ω_i is not orthogonal to the ith term of S.

Now let $x = y_\alpha \prod_j y_j = y_\alpha y$ be an element of N‴ with y_α in $u_{m+1} u_m X_\alpha \sigma_\alpha T_\alpha$ and y_j in $X_j \sigma_j$ $(j \neq m, m+1)$.

2) $\pi_\omega x \pi_\omega = \pi_\omega y_\alpha \pi_\omega \prod_j (\pi_\omega y_j \pi_\omega) = \pi_\omega y_\alpha \pi_\omega \cdot \pi_\omega y \pi_\omega$.

The proof is like that of 2) in the proof of 7.14.

3) $\chi_i(x) = \sum_\omega \operatorname{tr} \pi_\omega x \pi_\omega$ $(\omega = \omega_i, \omega'_i)$. This follows from 1) above, by a proof like that of 6) of 7.14.

4) *Proof of* a). — Since $1 \leq i \leq m-1$, both ω_i and ω'_i in 1) are orthogonal to α_m, α_{m+1} and α. Thus if $\omega = \omega_i$ or ω'_i and z is any element of the group generated by G_m and G_{m+1}, then $\pi_\omega z \pi_\omega = 1$ on V_ω, whence $\pi_\omega x \pi_\omega = \pi_\omega \sigma_\alpha y \pi_\omega$, and by a slight extension of 3) we get $\chi_i(x) = \chi_i(\sigma_\alpha y)$. Here $\sigma_\alpha y$ is in N″, so that 9.15 *a)* may be applied. The result is *a)*.

5) *Proof of* c). — Here $i = m$. If $\omega = \omega'_m$, then ω is orthogonal to α, whence $\pi_\omega x \pi_\omega = \pi_\omega \sigma_\alpha y \pi_\omega$ as in 4). Now applying 7.14 as refined in 2 *b)* of the proof of 9.15 to the representation ρ'_{m-1} of G′ (see step 1) of 9.15), we get

$$(*) \qquad \operatorname{tr} \pi_\omega x \pi_\omega = \psi_{m-1}(x).$$

Assume now that $\omega = \omega_n$. We write $y_\alpha = u_{m+1} u_m u_\alpha \sigma_\alpha t_\alpha$ as in 9.11, and normalize the choices σ_m and σ_{m+1} so that they are in G_m and G_{m+1} and $\sigma_\alpha = \sigma_{m+1} \sigma_m \sigma_{m+1}^{-1}$, and then write $y_\alpha = z_1 z_2 z_3 t_\alpha$ with $z_1 = u_{m+1} \sigma_{m+1}$, and $z_2 = \sigma_{m+1}^{-1} u_\alpha \sigma_\alpha \sigma_{m+1}$, and $z_3 = \sigma_{m+1}^{-1} \sigma_\alpha^{-1} u_m \sigma_\alpha$. Here z_1 and z_3 are in G_{m+1}, while z_2 is in G_m. The factor t_α acts on V_ω as the scalar $\alpha_m(t_\alpha) = \varphi_\alpha(x)$. Then because ω is orthogonal to α_{m+1} the factor z_3 may be suppressed. By the independence of α_m and α_{m+1} (see 7.15 *d)*) we may also suppress z_1. Thus $\pi_\omega x \pi_\omega = \varphi_\alpha(x) \pi_\omega z_2 \pi_\omega = \varphi_\alpha(x) \psi_\alpha(x)$ on V_ω, by 4) of 7.14. Combining this with (*) above, we get *c)*.

6) *Proof of* b) *and* d). — By applying to G an automorphism which fixes T and interchanges the roots α_i and α_{2n+1-i} $(1 \leq i \leq m)$, we get *b)* from *a)* and *d)* from *c)*, if we observe that in the latter case we must take the product of u_m and u_{m+1} in the opposite order, so that u_α in 5) above must be replaced by $(u_{m+1}, u_m)u_\alpha$, which because of the original assumption on this commutator yields the extra term φ_α.

9.17. *Remark.* — Observe that the extra term φ_α, which turns out to be just the term we need, owes its existence directly to the noncommutativity of X_m and X_{m+1}. This is only fair, since the present development does also.

9.18. *Corollary.* — $\sum_0^{n+1} (-1)^i \chi_i$ *is* 0 *on* N″ *and* $(-1)^{m+1}\varphi_\alpha$ *on* N‴.

If we use 9.15 and 9.16, then in the first case all terms cancel while in the second the one term remains.

One may also express 9.18 thus: if G is represented as $SL(n+1)$, the elements of N″ have 1 as a characteristic value, those of N‴ do not.

9.19. *Corollary.* — *Let* p *and* V *be as in* 6.10. *Let* f *be the function* $(c_1, \ldots, c_n) \rightarrow \sum_0^{n+1} (-1)^i c_i$ $(c_0 = c_{n+1} = 1)$, *and* V″ *and* V‴ *the subvarieties of* V *defined by* $f=0$ *and* $f \neq 0$, *respectively.*

a) p *maps* N″ *and* N‴ *isomorphically onto* V″ *and* V‴.

b) *All elements of* N‴ *are regular.*

The functions ψ_i $(i \neq m, m+1)$ and ψ_α may be used as coordinates on N″ by 9.12 and 7.1. So may the functions χ_i $(i \neq m)$, in terms of which the first set may be expressed by the recursive solution of *a)*, *b)* and *d)* of 9.15. The latter functions are the images under p of the canonical coordinates of V excluding the mth, which may be taken as coordinates on V″. Thus p maps N″ isomorphically onto V″. The proof for N‴ and V‴ is similar: first we normalize u_m and u_{m+1} as in 9.16, which is permissible by 9.4, and then in 9.16 we solve in turn for φ_α (see 9.18), ψ_i and $\varphi_\alpha\psi_\alpha$. The second isomorphism in *a)* implies that the differentials $d\chi_i$ are independent at all points of N‴, whence 1.5 implies *b)*.

9.20. *Remark.* — One can show that the regular elements of N″ are those for which $\sum_0^{n+1} (-1)^j j\chi_j \neq 0$.

Now we can prove 9.7 and 9.11. By 9.13 we have *a)*, and by 9.19 we have *b)* and *c)*, thus by *b)* also *d)*. The argument using k_1 and Γ_1 in the proof of 9.4 may be used to reduce the proof of *e)* to the case in which G consists of a single component. Proceeding as in the proof of 9.4 we are reduced to proving that the part of N″ and N‴ corresponding to the indices m, $m+1$, and α can be constructed over k. Since α is over k, so are T_α and X_α, and we can form $X_\alpha\sigma_\alpha$ over k by 9.3. Finally, by Hilbert's theorem [11, p. 159] and the k_1, Γ_1 reduction referred to above, we can choose u_m and u_{m+1} in 9.11 so that the class of $u_m u_{m+1}$ in $X_m X_{m+1} X_\alpha / X_\alpha$ is over k, whence *e)*.

§ 10. Some cohomological applications

The convention in § 9 concerning k and K continues.

First we prove 1.8. We recall that $H^1(k, G)$ consists of all cocycles from the Galois group Γ to the group G, that is, functions $\gamma \to x_\gamma$ which satisfy 9.3, modulo the equivalence relation, $(x_\gamma) \sim (x'_\gamma)$ if $x'_\gamma = a^{-1} x_\gamma \gamma(a)$ for some a in G and all γ in Γ. For the significance of this concept, as well as its basic properties, the reader is referred to [11, 12, 13]. We start with an arbitrary cocycle (x_γ) and wish to construct an equivalent one with values in a torus over k. Assume first that k is finite. Let q be the order of k, and β the qth power homomorphism. By Lang's theorem [5] there exists a in G such that $a^{-1} x_\beta \beta(a) = 1$. Since β and any subgroup Γ_1 of finite index generate Γ (in other words, the Galois group of any finite extension of k is generated by the restriction of β), it follows from 9.3 *b)* that $a^{-1} x_\gamma \gamma(a) = 1$ for all γ, whence $(x_\gamma) \sim (1)$. Assume now that k is infinite. We form $x(G)$, the group G twisted by the cocycle x (see, e.g., [13]). This is a group over k, isomorphic to G over K. If $x(G)$ is identified with G, then γ in Γ acts on $x(G)$ as $x(\gamma) = i(x_\gamma) \circ \gamma$; here $i(x_\gamma)$ denotes the inner automorphism by x_γ. By 2.15 and the Rosenlicht density theorem [6, p. 44] there exists in $x(G)$ an element y which is strongly regular and over k. Thus

$(*) i(x_\gamma)\gamma(y) = y$ for all γ in Γ.

Hence the conjugacy class of y in G is over k, whence by 1.7 it contains an element z over k. Writing $y = i(a)z$, with a in G, and substituting into $(*)$, we conclude that $a^{-1} x_\gamma \gamma(a)$ is in the centralizer of z, a torus because z is strongly regular, and over k because z is, whence 1.8.

10.1. *Corollary. — The assumption of semisimplicity in* 1.8 *can be dropped. In other words,* G *can be any simply connected, connected linear group with a Borel subgroup over k.*

By applying the semisimple case to G divided by its radical, we are reduced to the case in which G is solvable, which we henceforth assume. As in 9.4 we can find a Cartan subgroup C over k, and then the unique maximal torus T of C is over k and maximal also in G (see [8, p. 7-01 to p. 7-04]), whence we have over k the decomposition $G = UT$, with U the unique maximal unipotent subgroup. Now let $\gamma \to x_\gamma = u_\gamma t_\gamma$ be a cocycle. Then (t_γ) is also a cocycle, and (u_γ) is a cocycle in the group U twisted by (t_γ). Since U is unipotent, the last cocycle is trivial: $u_\gamma = a t_\gamma \gamma(a)^{-1} t_\gamma^{-1}$ for some a in U, by [12, Prop. 3.11]. Then $(x_\gamma) = (a t_\gamma \gamma(a)^{-1}) \sim (t_\gamma)$, whence 10.1 follows.

Next we consider 1.9. Assume that *a)* holds. By [12, Prop. 3.1.2] we have $H^1(k, G) = 0$ in case G is a torus, hence, by 1.8, also in case G is simply connected, semisimple, and contains a Borel subgroup over k, and then, by [12, Prop. 3.1.4], in case " simply connected " is replaced by " adjoint ". Now if G is an arbitrary semisimple adjoint group (over k, of course), there exists a group G_0 split over k and isomorphic to G over K, and the argument of [13, p. III-12] together with $H^1(k, G_0) = 0$ shows that G contains a Borel subgroup over k, whence $H^1(k, G) = 0$ by the result above. By [12, Prop. 3.1.4 Cor.] it now follows that *b)* holds in general. Now a result of

Springer [13, p. III-16, Th. 3] asserts that if $\dim k \leq 1$ and G and S are as in *c)*, then there exists a principal homogeneous space P and a G-map from P to S, all over k. By *b)*, P has a point over k, hence so does S, whence *c)*.

10.2. *Corollary. — Let k be a perfect field of* $\dim \leq 1$, *and* G *a connected linear group over k.*

a) G *contains a Borel subgroup over k.*

b) *Each conjugacy class over k contains an element over k.*

Observe that *b)* is the same as 1.10. Both results follow from 1.9. In the first case we take as the homogeneous space the variety of Borel subgroups, in the second case the conjugacy class under consideration.

10.3. *Corollary. — If k is as above and* G *is simply connected, the natural map from the set of semisimple classes of* G_k *to the set of semisimple classes of* G *over k is bijective.*

By 10.2 *a)* and 9.9 the map is surjective. To prove injectivity we must show that if x and y are semisimple elements of G_k which are conjugate in G they are also conjugate in G_k. We have $axa^{-1}=y$ with a in G. Then for γ in Γ we have $\gamma(a)x\gamma(a)^{-1}=y$, whence $a^{-1}\gamma(a)$ is in G_x. Now $\gamma \to a^{-1}\gamma(a)$ is a cocycle and G_x is connected (cf. 2.10), and over k because x is. Thus by 1.9 there exists b in G_x such that $b^{-1}a^{-1}\gamma(a)\gamma(b)=1$ for all γ. Thus ab is over k, and x and y are conjugate in G_k, under ab in fact, whence 10.3.

10.4. *Remarks. — a)* For regular classes 10.3 is false, since regular elements of G_k conjugate in G need not be conjugate in G_k.

b) For the split adjoint group of type A_r over any field k one can show, by the usual normal forms, that any elements of G_k, semisimple or not, are conjugate in G_k if they are conjugate in G. Does the same result hold for the other simple types, and is it enough to assume a Borel subgroup over k?

§ 11. Added in proof

M. Kneser has informed me that in 1.8 the assumption that G is simply connected can be dropped. If k is finite, the proof is as before (see § 10). If k is infinite, the key point is that the group $x(G)$ of the proof of 1.8 can be constructed even if (x_γ) is only a cocycle modulo the centre of G, so that if G is simply connected such a " cocycle " is equivalent to one with values in a torus over k. By applying this to the simply connected covering group of a group which is as in 1.8 but not simply connected, we get the improved version of 1.8. Proceeding then as in the proof of 10.1 we can drop the assumption of semisimplicity. The result is:

11.1. *Theorem. — Let k be a perfect field and* G *a connected linear group which is over k and contains a Borel subgroup over k. Then each element of* $H^1(k, G)$ *can be represented by a cocycle whose values are in a torus over k.*

Using 11.1 we now give a simplified proof of the implication *a)*→*b)* of 1.9. The assumption $\dim k \leq 1$ is used only in the proof, for which we refer the reader to [12, Prop. 3.1.2], that $H^1(k, G)=0$ if G is a torus over k, since we show:

11.2. *Theorem. — Let k be a perfect field and n a positive integer such that* $H^1(k, T)=0$ *for every torus* T *of rank n and over k. Then* $H^1(k, G)=0$ *for every connected linear group* G *of rank n and over k.*

By 11.1 and the assumption in 11.2 we have

(*) $H^1(k, G)=0$ if G in 11.2 contains a Borel subgroup over k.

In the general case let R be the radical of G and Z the centre of G/R. There exists a group G_0 (the split one, e.g.) which is over k and contains a Borel subgroup B over k, and an isomorphism φ over K of G_0 onto $(G/R)/Z$. Since G_0 is a centreless semisimple group, we have the split extension $\mathrm{Aut}\, G_0 = G_0E$, in which E is a finite group which fixes B (see [8, p. 17-07, Prop. 1]). For $\gamma\in\Gamma$, write $\varphi^{-1}\gamma(\varphi)=g_\gamma e_\gamma (g_\gamma\in G_0,\ e_\gamma\in E)$. Then (e_γ) is a cocycle and (g_γ) is a cocycle in the group G_0 twisted by (e_γ). In this group (g_γ) is equivalent to the trivial cocycle by (*) because B is over k. Thus $(g_\gamma e_\gamma)$ is equivalent to (e_γ) in $H^1(k, \mathrm{Aut}\, G_0)$, whence φ may be normalized so that $\varphi^{-1}\gamma(\varphi)=e_\gamma$. Then φB is a Borel subgroup over k in $(G/R)/Z$, and its inverse image is one in G, whence $H^1(k, G)=0$ by (*).

BIBLIOGRAPHY

[1] A. Borel, Sous-groupes commutatifs et torsion des groupes de Lie compacts connexes, *Tôhoku Math. J.*, 13 (1961), 216-240.

[2] S. Helgason, *Differential geometry and symmetric spaces*, Academic Press, New York (1962).

[3] B. Kostant, The principal three-dimensional subgroup and the Betti numbers of a complex simple Lie group, *Amer. J. Math.*, 81 (1959), 973-1032.

[4] —, Lie group representations on polynomial rings, *Amer. J. Math.*, 85 (1963), 327-404.

[5] S. Lang, Algebraic groups over finite fields, *Amer. J. Math.*, 78 (1956), 555-563.

[6] M. Rosenlicht, Some rationality questions on algebraic groups, *Ann. di Mat.*, 43 (1957), 25-50.

[7] —, On quotient varieties and the affine imbedding of certain homogeneous spaces, *Trans. Amer. Math. Soc.*, 101 (1961), 211-223.

[8] Séminaire C. Chevalley, *Classification des Groupes de Lie algébriques* (two volumes), Paris (1956-58).

[9] Séminaire « Sophus Lie », *Théorie des algèbres de Lie...*, Paris (1954-5).

[10] J.-P. Serre, *Groupes algébriques et corps de classes*, Hermann, Paris (1959).

[11] —, *Corps locaux*, Hermann, Paris (1962).

[12] —, Cohomologie galoisienne des groupes algébriques linéaires, *Colloque sur la théorie des groupes algébriques*, Bruxelles (1962), 53-68.

[13] —, *Cohomologie galoisienne*, Cours fait au Collège de France ((1962-3).

[14] D. A. Smith, *Dissertation*, Yale University (1963).

[15] T. A. Springer, Quelques résultats sur la cohomologie galoisienne, *Colloque sur la théorie des groupes algébriques*, Bruxelles (1962), 129-135.

[16] R. Steinberg, Finite reflection groups, *Trans. Amer. Math. Soc.*, 91 (1959), 493-504.

[17] —, Representations of algebraic groups, *Nagoya Math. J.*, 22 (1963), 33-56.

[18] H. Weyl, Theorie der Darstellung kontinuierlicher halb-einfacher Gruppen durch lineare Transformationen III, *Math. Zeit.* (1926), 377-395.

University of California, Los Angeles.

Reçu le 25-8-1964.

Proc. Internat. Conf. Theory of Groups, Austral. Nat. Univ. Canberra, August 1965, pp. 315–319.

On the Galois cohomology of linear algebraic groups

ROBERT STEINBERG

Our purpose is to discuss the theorem **A** below, some consequences, and some relatives. To do this we must start with some preliminaries.

Recall that a linear algebraic group G is a subgroup of some $GL_n(K)$ ($n \geq 1$, K algebraically closed field) which is the complete set of zeros of a set of polynomials in the matrix entries. The word algebraic will be omitted. For example, SL_n (the subgroup of elements of determinant 1), $O_n(K)$ (resp. $Sp_n(K)$) (the subgroup fixing a nonsingular symmetric (resp. skew) bilinear form), the subgroup of diagonal or superdiagonal elements of GL_n or SL_n, all are linear groups. An isomorphism of linear groups is required to be not only an isomorphism of the corresponding abstract groups but also birational (in terms of the matrix entries). We will use the Zariski topology, in which the closed subsets of G are the algebraic subsets (i.e., complete sets of zeros of polynomials in the matrix entries). In this topology all of the groups mentioned above are connected except for O_n which has two components, SO_n being the identity component. Recall that G is (defined) over the subfield k of K if the polynomial ideal which defines G has a basis of polynomials with coefficients in k. For example, $SL_n(K)$ is over any subfield of K while $O_n(K)$ is over any subfield which contains the coefficients of the bilinear form which defines the group.

Henceforth k will be a perfect field, K its algebraic closure, and Γ the Galois group of K over k. Let G be a linear algebraic group over k. Consider all maps $\gamma \to x_\gamma$ from Γ to G which

(1) *are cocycles:* $x_{\gamma\delta} = x_\gamma \gamma(x_\delta)$ $(\gamma, \delta \in \Gamma)$,

(2) *are continuous:* $x_\gamma = 1$ *for all* γ *in a subgroup of* Γ *corresponding to a finite extension of* k.

Here one uses the finite topology on Γ and the discrete topology on G. The group G acts on the continuous cocycles: $x_\gamma \cdot a = a^{-1}x_\gamma\gamma(a)$ ($a \in G$, $\gamma \in \Gamma$). Then, by definition, $H^1(k, G)$ is the set of orbits under this action. There is always at least one orbit, that containing the cocycle which is identically 1. If this is the only orbit, we write $H^1(k, G) = 0$.

There are several possible interpretations of H^1, but the one which is most important in the present context may be described as follows. Let S be some algebraic structure over k whose group of automorphisms over K is G. For example, S may be a quadratic form, or a central simple algebra, or a Lie algebra (G is the orthogonal group in the first case). Let S' be a second structure over k which is isomorphic over K to S. Let $\varphi: S \to S'$ be an isomorphism. Form

$$x_\gamma = \varphi^{-1}\gamma(\varphi) \quad (\gamma \in \Gamma, \gamma(\varphi) = \gamma \circ \varphi \circ \gamma^{-1}).$$

Then x is a cocycle, continuous because φ is over some finite extension of k. Here x depends on the choice of φ, but if a different isomorphism $\varphi \circ a$ ($a \in \operatorname{Aut} S = G$) is chosen then x is replaced by an equivalent cocycle. Thus it turns out that $H^1(k, G)$ classifies the structures S' over k which are isomorphic over K to S, modulo isomorphism over k.

Let us consider some examples. If $G = GL_1$, GL_n, SL_n, or Sp_n, then $H^1(k, G) = 0$. The first result is due to Hilbert (Theorem 90), the second to Speiser, the third follows easily from the first two, and the fourth is a consequence of the fact that any two nonsingular skew bilinear n-forms with coefficients in k are isomorphic over k. If $G = O_n$ then the field k plays a role: if k is the field of real (resp. rational) numbers, then the cardinality of $H^1(k, G)$ is $[n/2]+1$ (resp. infinite).

Using the notation $\dim k$ for the cohomological dimension of Γ (topologized as above), we can now state our central theorem.

A. *If k is a perfect field such that* (I) $\dim k \leq 1$, *then* (II) $H^1(k, G) = 0$ *for every connected linear group over k.*

Remarks. (a) Equivalent to $\dim k \leq 1$ is: there are no noncommutative finite dimensional division algebras over k. Some cases in which $\dim k \leq 1$ are as follows: if k is finite; if k is C_1 (i.e., for every n every homogeneous polynomial of degree n in $n+1$ variables over k has a nontrivial zero); if k is complete relative to a discrete valuation and has an algebraically closed residue class field.

(b) There is a converse to **A**: if k is a perfect field then (II) implies (I).

(c) In general the assumption that k is perfect cannot be dropped in **A**, but if it is assumed that G is semisimple (i.e., has no nontrivial connected solvable normal subgroup), it appears from recent work of Grothendieck (for which there is a simplified version due to Borel and Springer [7]), that it can be.

(d) Further details about the examples and cohomological concepts introduced above may be found in [5]. The proof of **A** in some special cases may be found in [4] and in general in [8].

We will now indicate, very briefly, some of the ideas involved in the proof. The case k finite follows almost immediately from the theorem of Lang [2] that if k is a finite field of q elements, if σ is the homomorphism which replaces all matrix entries of all elements of G by their qth powers, and if x is any element of G, then there exists a in G such that $x = a\sigma(a)^{-1}$ (if $\gamma \to x_\gamma$ is any continuous cocycle, one applies Lang's result with $x = x_\sigma$). In the general case we consider the result $\mathbf{A}^*$ obtained from **A** by adding the assumption:

(*) *G contains a Borel* (i.e., *a maximal connected solvable*) *subgroup over k.*

Assume $\mathbf{A}^*$ has been proved. Let G be as in **A**. Introducing a suitable group G^* which satisfies (*) and using $\mathbf{A}^*$ applied to G^*, one can make a favorable comparison of G and G^* to show that G also satisfies (*), whence $\mathbf{A}^*$ applied a second time, this time to G, yields **A** (see [8], p. 80). Thus we are reduced to proving $\mathbf{A}^*$. Recall that an algebraic group is called a torus if it is isomorphic (over K) to the direct product of a finite number of copies of GL_1. Now in case G is a torus, Serre ([4], Proposition 3.1.2) has proved $\mathbf{A}^*$. (This is the only place in our proof of **A** where the assumption $\dim k \leq 1$ is used.) Thus in the general case $\mathbf{A}^*$ follows from:

B. *If k is a perfect field and G is a connected linear group over k which satisfies* (*), *then every element of $H^1(k, G)$ can be reduced to a torus over k.*

The proof of **B** depends on the following result.

C. *Assume that G is as in* **B** *and also that G is semisimple and simply connected. Then every conjugacy class X of semisimple elements which is defined over k contains an element x over k.*

Recall that G is simply connected if every projective rational representation of G is a projection of a linear one, while an element of G is semisimple if it is diagonalizable (over K). (The groups SL_n and Sp_n are simply connected, while SO_n is doubly connected having as its simply connected covering the group Spin_n; all these groups are semisimple.) For the deduction of **B** from **C** see [8], p. 78. Finally, to prove **C** we construct a cross-section S, defined over k, of the collection of semisimple conjugacy classes ([8], 7.16, 9.4, 9.7); then the element $x = X \cap S$ fulfils the requirements of **C**. In this last step one must go rather deeply into the structure of linear groups; in the other steps one only skims the surface.

From the argument used to show that **A*** implies **A**, we see that the following result has also been proved.

D. *Let k be a perfect field. If* $\dim k \leq 1$, *then every connected linear group over k satisfies* (*).

A consequence of **D** is that the simple linear groups over k can be classified without much further work (see [4]), in terms of the corresponding classification over K; we recall that over K the simple groups are roughly in one-one correspondence with the simple Lie groups over the complex field (the classical groups together with the five exceptional groups) (see [3]). The converse of D turns out to be true ([6], p. 129).

We close by mentioning two other theorems related to the above results. These concern groups over fields of dim ≤ 2.

E. *If k is a p-adic field and G is a semisimple simply connected group over k, then $H^1(k, G) = 0$.*

F. *Same as* **E** *except that k is a complete field relative to a discrete valuation and the residue class field has* dim ≤ 1.

The first result is due to M. Kneser [1], whose proof uses **B** extensively. At the recently concluded 1965 A.M.S. Summer Institute in Algebraic Groups (see [9]), Bruhat, Tits, and Springer have indicated a proof of the more general result **F** by a simpler method which makes direct use of **A**.

References

[1] M. Kneser, Galois-Kohomologie halbeinfacher algebraischer Gruppen über p-adischen Körpern I, II, *Math. Z.* **88** (1965), 40–47; ibid. **89** (1965), 250–272.

[2] Serge Lang, Algebraic groups over finite fields, *Amer. J. Math.* **78** (1956), 555–563.

[3] Séminaire C. Chevalley, 1956–1958, *Classification des groupes de Lie algébriques, I, II*, Ecole Normale Supérieure, Paris, 1958.

[4] Jean-Pierre Serre, Cohomologie galoisienne des groupes algébriques linéaires, *Colloque sur la théorie des groupes algébriques, Bruxelles*, 1962, pp. 53–68; Gauthier-Villars, Paris, 1962

[5] Jean-Pierre Serre, *Cohomologie galoisienne*, Springer, Berlin–Göttingen–Heidelberg, 1964.

[6] T. A. Springer, Quelques rêsultats sur la cohomologie galoisienne, *Colloque sur la théorie des groupes algébriques, Bruxelles*, 1962, pp. 129–135; Gauthier-Villars, Paris, 1962.

[7] T. A. Springer and A. Borel, Rationality properties of linear algebraic groups, *Proc. Summer Inst. Algebraic Groups*, 1965, I-C; Amer. Math. Soc., Providence, R.I., 1965.

[8] R. Steinberg, Regular elements of semisimple, algebraic groups, *Inst. Hautes Études Sci. Publ. Math.* No. 25 (1965), 49–80.

[9] J. Tits, Simple groups over local fields, *Proc. Summer Inst. Algebraic Groups*, 1965, I-G; Amer. Math. Soc., Providence, R.I., 1965.

CLASSES OF ELEMENTS OF SEMISIMPLE ALGEBRAIC GROUPS

ROBERT STEINBERG

Given a semisimple algebraic group G, we ask: how are the elements of G partitioned into conjugacy classes and how do these classes fit together to form G? The answers to these questions not only throw light on the algebraic and topological structure of G, but are important for the representation theory and functional analysis of G; conversely, of course, we can expect the representations and functions on G to contribute to the answers to the questions. In what follows, we discuss some aspects of these questions, present some known results, and pose some unsolved problems.

To do this, we first recall some basic facts about semisimple algebraic groups. Our reference for all this is [7]. A linear algebraic group G is a subgroup of some $GL_n(K)$ ($n \geqslant 1$, K an algebraically closed field) which is the complete set of zeros of some set of polynomials over K in the n^2 matric entries. The Zariski topology, in which the closed sets are the algebraic subsets of G, is used. G is semisimple if it is connected and has no nontrivial connected solvable normal subgroup; it is then the product, possibly with amalgamation of centers, of some simple groups. The simple groups have been classified, in the Killing-Cartan tradition. Thus there are the classical groups (unimodular, symplectic, orthogonal, spin, etc.) and the five exceptional types. The rank of a semisimple group is the dimension of a maximal torus (a subgroup isomorphic to a product of GL_1's). We also need the notions of simple connectedness and Lie algebra as they apply to linear algebraic groups, but we omit the definitions. Henceforth K denotes a fixed algebraically closed field, p its characteristic, G a simply connected semisimple algebraic group over K, and r the rank of G. The simplest example is $G = SL_{r+1}(K)$. L denotes the Lie algebra of G. If x is an element of G, we write G_x (resp. L_x) for the centralizer of x in G (resp. L). We shorten conjugacy class to class, and dimension to dim.

Now we take up the problem of surveying the conjugacy classes of G and finding for them suitable representative elements. Recall that an element of G is semisimple if it is diagonalizable, and unipotent if its characteristic values are all 1, and that every element can be decomposed uniquely $x = x_s x_u$ as a product of commuting semisimple and unipotent elements. An element x will be called regular if $\dim G_x = r$, or, equivalently, if $\dim G_x$ is minimal [12, p. 49]; here x need not be semisimple and may in fact be unipotent. Various characterizations, hence alternate possible definitions, of regular elements may be found in [12]. The regular (resp. semisimple) elements form a dense open set with complement of codimension 3 (resp. 1) in G [12, 1.3 and 6.8]; thus most elements are regular (resp. semisimple). In [12, 1.4] we have constructed a closed irreducible cross-section, C, isomorphic to affine r-space, for the regular classes of G. Also, we have proved [12, 1.2]:

(1) *The map $x \to x_s$ sets up a 1-1 correspondence between the regular and the semisimple classes of G.*

Thus by choosing the elements of C and their semisimple parts we obtain a system of representatives for the regular classes and for the semisimple classes of G. If G is the group SL_n, the representatives of the regular classes chosen turn out to be in one of the Jordan canonical forms. Thus we are led to our first problem.

(2) Problem. *Determine canonical representatives, similar to those given by the Jordan normal forms in SL_n, for all of the classes of G, not just the regular ones.*

In proving the results described above, we are naturally led to consider the algebra of regular class functions, those regular functions on G that are constant on the conjugacy classes of G. This is a polynomial algebra generated by the characters $\chi_1, \chi_2, \ldots, \chi_r$ of the fundamental representations of G [12, 6.1]. (If $G = SL_{r+1}$, the χ's are just the coefficients of the characteristic polynomial with the first and last terms excluded.) First of all these functions give a very useful characterization of regular elements of G: x is regular if and only if the differentials of $\chi_1, \chi_2, \ldots, \chi_r$ are linearly independent at x [12, 1.5]. And secondly they are related to our classification problem by the following result [12, 6.17]:

(3) *The map $x \to (\chi_1(x), \chi_2(x), \ldots, \chi_r(x))$ on G induces a bijection of the regular classes, and of the semisimple classes, onto the points of affine r-space.*

(4) Problem. *Assume that x and y in G are such that for every rational representation (ρ, V) of G the elements $\rho(x)$ and $\rho(y)$ are conjugate in $GL(V)$. Prove that x and y are conjugate in G.*

By (3) this holds if x and y are both regular or both semisimple, and it can presumably de checked when G is a classical group, but we have not done this in all cases.

To continue our discussion, we will use the following result.

(5) *If y is a semisimple element of G, then G_y is a connected reductive group (i.e. the product of a semisimple group and a central torus).*

More generally one can show: The group of fixed points of every semisimple automorphism of G is connected. Such connectedness theorems are important in problems concerning conjugacy classes, as we shall see, and also in problems about cohomology [1, p. 224].

Assume now that x and x' in G have the same semisimple part y. Then x and x' are conjugate in G if and only if x_u and x'_u are conjugate in G_y, in fact, by (5), in the semisimple part of G_y. The latter group may not be simply connected, but this is immaterial for the study of unipotent elements. Thus the study of arbitrary classes is reduced to the study of unipotent classes, and these are the clases we now consider.

At present not much is known about the classification of the unipotent classes of G. Observe, though, that (1) implies that regular unipotent elements exist and that they in fact form a single conjugacy class.

(6) Problem. *Classify the unipotent classes and determine canonical representatives for them.*

If $p = 0$, the study of unipotent classes in G is equivalent to that of nilpotent classes in L, and a classification of these latter classes together with representative elements may be found in [3; 4]. The result, however, is in the form of a list, which, though finite, is very long, thus subject to error and inconvenient for applications. The main procedure in the construction of the list, that of imbedding each nilpotent element in an algebra of type sl_2 and then using the representation theory of this algebra, or the corresponding procedure with G in place of L, is not available if $p \neq 0$ [12, p. 60]. In studying the nilpotent classes of L (still for $p = 0$) one is naturally led to study the class H of G-harmonic polynomials on L: indeed these polynomials are in natural correspondence with the regular functions on the variety of all nilpotent elements of L, by results in [5]. To our knowledge, no one has succeeded in effectively relating H to the classification problem at hand.

(7) Problem. *Do this.*

(8) Problem. *Do the same for the unipotent classes of G (for p arbitrary), having first found an analogue for H.*

Now it follows from (5) that a unipotent element centralized by a semisimple element not in the center of G is contained in a proper semisimple subgroup of G. Thus an important special case of (6) is:

(9) Problem. *Same as (6) for the classes of unipotent elements x such that G_x is the product of the center of G and a unipotent subgroup.*

As is easily seen, the class of regular unipotent elements always satisfies this condition. If $p = 0$, then relatively few other classes do [3, Th. 10.6].

A more modest problem is as follows.

(10) Problem. *Prove that the number of unipotent classes of G is finite.*

Here only classes which satisfy (9) need be considered. Observe that a proof of (4) would yield the finiteness together with a bound on the number of classes. From the work of Kostant one obtains a rather lengthy proof of (10) together with the bound 3^r, in case $p = 0$. Richardson [6] has found a simple, short proof that works not only if $p = 0$, but, more generally, if p does not divide any coefficient of the highest root of any simple component of G, this root being expressed in terms of the simple ones. In the sequel such a value of p will be called good. Richardson's method depends eventually on the fact that an algebraic set has only a finite number of irreducible components; thus it yields no bound on the number of unipotent classes, but it does yield the following useful result.

(11) *If p is good and G has no simple component of type A_n, then L_x is the Lie algebra of G_x for every x in G.*

(12) Problem. *Extend (11) to all p and G, with the conclusion replaced by: L_x is the sum of the Lie algebra of G_x and the center of L.*

This is easy if x is semisimple and known if x is regular (cf. [12,4.3]).

Finally let us mention an old problem which seems to be closely related to (6).

(13) Problem. *Determine, within a general framework, the conjugacy classes of the Weyl group W of G, and assuming that W acts, as usual, on a real r-dimensional space V, relate them to the W-harmonic polynomials on V.*

Next we consider some results and problems about centralizers of elements of G. We have already stated, in (5) and (11) above, two such results. Springer [10] has proved that for any x in G the group G_x contains an Abelian subgroup of dimension r. It follows that if x is regular, then the identity component G_x^0 of G_x is Abelian.

(14) Problem. *If x is regular, prove that G_x is Abelian.*

(15) Problem. *Prove conversely that if G_x is Abelian, then x is regular.*

These results would yield an abstract characterization of the regular elements, but other such characterizations are known [12, 3.14]. Both problems may easily be reduced to the unipotent case. If x is a regular unipotent element, then, as already mentioned, G_x is the product of the center of G and a unipotent group U_x. Springer [11,4.11] has proved:

(16) *If p is good, then U_x is connected (hence Abelian).*

Thus (14) holds in this case. If p is bad, (16) is false, in fact $x \notin U_x^0$.

(17) Problem. *For p bad determine the structure of U_x/U_x^0 and whether U_x is Abelian.*

A solution to (17) would complete the proof of (14), and quite likely would also have cohomological applications (cf. [11, § 3]). We do not know whether (15) is true even if $p = 0$.

(18) Problem. *Prove that $\dim G_x - r$ is always even. In other word, the dimension of each conjugacy class is even.*

This has been proved in the following cases: if $p = 0$ [5, Prop. 15], if p is good (because of (11) the same type of proof as for $p = 0$ can be given), if x is semisimple (this is easy); and presumably can be checked when G is a classical group.

All of the above problems are special cases of one big final problem about centralizers.

(19) P r o b l e m. *For every element x of G determine the structure of G_x.*

Next we discuss briefly the individual classes, and their closures. Each class is, of course, an irreducible subvariety of G (because G is connected). A class is closed if and only if it is semisimple [12, 6.13]. More generally, the closure of a class consists of a number of classes of which exactly one is semisimple [12, 6.11]. Whether the number is always finite or not depends on (10). The most important case occurs when we start with the class U_0 of regular unipotent elements; then the closure U is just the variety of all unipotent elements of G. Although some results about U are known, e.g. U is a complete intersection specified by the equations $\chi_i(x) = \chi_i(1)$ $(1 \leqslant i \leqslant r)$ (the χ's are still the fundamental characters), and the codimension of $U - U_0$ in U is at least 2 [12, 6.11], we feel that U should be studied thoroughly.

(20) P r o b l e m. *Study U thoroughly.*

Of course this problem is related to (8).

Using the properties of U just mentioned and also (16), Springer has proved the following result.

(21) *If p is good, then the variety U (of unipotent elements of G) is isomorphic, as a G-space, to the variety N of nilpotent elements of L.*

If p is bad, this is false, because (16) is. If $p = 0$, Kostant [5] has obtained results about the structure and cohomology of N, in particular that N and any Cartan subalgebra of L intersect at 0 with multiplicity equal to the order of the Weyl group W.

(22) P r o b l e m. *Find a natural action of W on affine (dim G — r)-space so that the quotient variety is isomorphic to N.*

So far we have been considering G over an algebraically closed field K. From now on we shall assume that G is defined over a perfect field k which has K as an algebraic closure; thus the polynomials which define G as an algebraic group can be chosen so that their coefficients are in k. The problem is to study the classes of G_k, the group of elements of G which are defined over k, i.e. which have their coordinates in k. The main idea is to relate the classes of G_k to those of G with the help of the Galois theory. We will discuss mainly the problem of surveying, and finding canonical representatives for, the classes of G_k. Recalling our solution to this problem for the regular (and semisimple) classes of G, we naturally try to adapt our cross-section C of the regular classes to the present situation. Consider a regular class of G which meets G_k. This class, call it R, is necessarily defined (as a variety) over k. How can we ensure that the representative element $C \cap R$ of R is in G_k? The obvious way is to construct C so that it also is defined over k. For this to be possible, the unique unipotent element of C must be in G_k, and then the unique Borel (i.e. maximal connected solvable) subgroup of G containing this element [12, 3.2 and 3.3] must de defined over k; i.e. G must contain a Borel subgroup defined over k. This last condition is a natural one which arises in other classification problems, and, although a bit restrictive, holds for a significant class of groups, e.g. for the so-called Chevalley groups [2], of which one is the group SL_n. Conversely, if this condition holds, then it turns out that with a few exceptions (certain groups of type A_m (but not SL_{m+1}) are forbidden as simple components of G) C can be constructed over k [12, § 9]. Thus if we form $C \cap G_k$, we obtain a set of canonical representatives for those regular classes of G which meet G_k; but not necessarily for the regular classes of G_k since a single class of G may intersect G_k in several classes. In other words, elements of G_k conjugate in G may not be conjugate in G_k. Looking for an idea to remedy this situation, we consider the case $G = SL_n$. We have the Jordan normal forms for all the classes of $SL_n(K)$, not just the regular ones, and, though this fails in $SL_n(K)$, it does hold in $GL_n(k)$. Now the action of $GL_n(k)$ by conjugation is that of $PGL_n(k)$, i.e. of $\hat{G}_k$, if we let $\hat{G}$ denote the adjoint group of G, i.e. the quotient of G over its center, which is PSL_n in the present case. Thus we are led to the following problem.

(23) P r o b l e m. *Asume that G contains a Borel subgroup defined over k (and perhaps that G has no simple component of type A_m). (a) Prove that every class of G defined over k meets G_k, i.e. contains an element defined over k. (b) Prove that two elements of G_k which are conjugate in G are conjugate under $\hat{G}_k$.*

As we have seen, (a) holds for regular classes. It holds also for semisimple classes, without any restriction on the components of type A_m [12, 1.7].

(24) *Assume that G contains a Borel subgroup defined over k. Then every semisimple class of G defined over k meets G_k. (And conversely.)*

This result has a number of applications to classification problems [12, p. 51-2], especially if (cohomological) dim $k \leqslant 1$. We recall [8, p. 11-8] that dim $k \leqslant 1$ is equivalent to: every finite dimensional division algebra over k is commutative. This holds in several cases of interest in number theory, in particular if k is finite [8, p. II-10]. As a consequence of (24), we have [12, 1.9]:

(25) *If dim $k \leqslant 1$ and H is any connected linear group defined over k, then the Galois cohomology $H^1(k, H)$ is trivial.*

This result and an extension due to Springer [8, p. III-16] lead to the classification of semisimple groups defined over k, if dim $k \leqslant 1$ (we get [12, 10.2] that there is always a Borel subgroup defined over k), and answer most of the questions raised above. Concerning (23a), we get [12, 10.2]:

(26) *If k and H are as in* (25), *then every class of H defined over k meets H_k.*

Concerning (23b), we get [21, 10.3]:

(27) *If* $\dim k \leqslant 1$, *then two semisimple elements of G_k which are conjugate in G are conjugate in G_k.*

This result is false for arbitrary elements, e.g. for regular ones. To show a typical kind of argument, we give the proof of (27). Let y and z be semisimple elements of G_k conjugate in G; thus $aya^{-1} = z$ with a in G. Combining this equation with the one obtained by applying a general element γ of the Galois group of K over k, we see that $a^{-1}\gamma(a)$, call it x_γ, is in G_y. As we check at once, x satisfies the cocycle condition $x_{\gamma\delta} = x_\gamma \gamma(x_\delta)$, hence represents an element of $H^1(k, G_y)$, which is trivial, by (25), because G_y is connected, by (5). Thus $x_\gamma = b\gamma(b^{-1})$ for all γ and some b in G_y. We conclude that y and z are conjugate in G_k, by the element ab in fact. The same argument, with (16) in place of (5), shows that regular unipotent elements of $\hat{G}_k$ which are conjugate in $\hat{G}$ (the adjoint group) are conjugate in $\hat{G}_k$, if p is good (here $\dim k \leqslant 1$ is not necessary, but p good is).

By combining (26) and (27), we get a complete survey of the semisimple classes of G_k, if $\dim k \leqslant 1$; in particular, these classes correspond to the rational points of affine r-space, suitably defined over k (cf. [12, 10.3]). Thus, if k is a finite field of q elements, so that G_k is a simply connected version of one of the simple finite Chevalley groups or their twisted analogues, the number of such classes is q^r, a useful fact in the representation theory of these groups [11]. A related result, whose proof, unfortunately, does not use the same ideas, states that the number of unipotent elements of G_k is $q^{\dim G - r}$ [13].

Finally we conclude our paper with another problem.

(28) Problem. *If G is defined over k (any perfect field), prove that every unipotent class of G is defined over k.*

This would yield a proof of (10). In fact, assuming $p \neq 0$, as we may, choosing k as the field of p elements, applying (28) with G suitably defined, and then using (26), we would get (10) with the number bounded by $|G_k|$.

Dept. of Mathematics,
University of California, Los Angeles, USA

REFERENCES

[1] Borel A., Sous-groupes commutatifs et torsion des groupes de Lie compacts connexes, *Tôhoku Math. J.*, **13** (1961), 216-240.

[2] Chevalley C., Sur certains groupes simples, *Tôhoku Math. J.*, **7** (1955), 14-66.

[3] Дынкин Е. Б., Полупростые подалгебры полупростых алгебр Ли, *Матем. сб.*, **30** (72) (1952), 349-462.

[4] Kostant B., The principal three-dimensional subgroup and the Betti numbers of a complex simple Lie group, *Amer. J. Math.*, **81** (1959), 973-1032.
[5] Kostant B., Lie group representations on polynomial rings, *Amer. J. Math.*, **85** (1963), 327-404.
[6] Richardson R. W., Conjugacy classes in Lie algebras and algebraic groups, to appear.
[7] Séminaire C. Chevalley, Classification des Groupes de Lie Algébriques (two volumes), Paris (1956-8).
[8] Serre J.-P., Cohomologie Galoisienne, Lecture Notes, Springer-Verlag, Berlin.
[9] Springer T. A., Some arithmetical results on semisimple Lie algebras, Publ. Math. I.H.E.S., **30** (1966), 115-141.
[10] Springer T. A., A note on centralizers in semisimple groups, to appear.
[11] Steinberg R., Representations of algebraic groups, *Nagoya Math. J.*, **22** (1963), 33-56.
[12] Steinberg R., Regular elements of semisimple algebraic groups, Publ. Math. I.H.E.S., No. 25 (1965), 48-80.
[13] Steinberg R., Endomorpisms of linear algebraic groups, Memoirs Amer. Math. Soc., to appear.

О НЕКОТОРЫХ ПРОБЛЕМАХ БЕРНСАЙДОВСКОГО ТИПА

Е. С. ГОЛОД

Целью доклада является описание приема, который позволяет строить контрпримеры для некоторых проблем бернсайдовского типа [1] в случае «неограниченного» показателя. Как известно, в случае «ограниченного» показателя большинство из этих проблем имеет положительное решение [2, 3, 4], за исключением собственно проблемы Бернсайда о периодических группах [5]. Далее будет доказана следующая

Теорема. Пусть k — произвольное поле. Существует k-алгебра A (без единицы) с $d \geqslant 2$ образующими, которая бесконечномерна как векторное пространство над k, в которой всякая подалгебра с числом образующих $< d$ нильпотентна и которая такова, что $\bigcap_{n=1}^{\infty} A^n = (0)$.

Частными случаями этой теоремы являются результаты, полученные тем же методом в [6], а также следующие утверждения:

Следствие 1. Существует финитно-аппроксимируемая бесконечная p-группа с $d \geqslant 2$ образующими, в которой всякая подгруппа с числом образующих $< d$ конечна [1]).

[1]) Как доказал С. П. Струнков [7], если в группе с числом образующих $d \geqslant 3$ всякая собственная подгруппа конечна, то и сама группа также конечна.

Reprinted from:

MEMOIRS
of the
American Mathematical Society

Number 80

Endomorphisms of Linear Algebraic Groups

Robert Steinberg

American Mathematical Society
Providence, Rhode Island

0. Introduction

In this paper we are concerned with endomorphisms of linear algebraic groups and their fixed points. The development is divided roughly into three parts, one dealing with automorphisms, another with the case in which the set of fixed points is finite, and the third with prepatory material about reflection groups proved in abstract form. In more detail the main contents of the paper are as follows. Throughout the discussion G will denote a linear algebraic group. This will mean that an algebraically closed field K, fixed forever, has been prescribed, and that G is a subgroup of some $GL_n(K)$ and at the same time an algebraic set in the affine space determined by the n^2 matric coefficients. The Zariski topology, in which the closed sets are the algebraic sets, will be used. In addition σ will denote an endomorphism of G onto G, and G_σ its group of fixed points.

After some generalities (and definitions) about algebraic groups in §6, it is proved in §7 that σ always fixes a Borel subgroup of G, and also a maximal torus of that Borel subgroup in case σ is a semisimple automorphism. In §8 this is used to prove that in case G is simply connected and semisimple and σ is a semisimple automorphism then G_σ is connected; in particular, the centralizer of every semisimple element of G is connected. Also used in the proof is a (twisted version of a) result which states that if a reflection group acts effectively on an algebraic torus then the stabilizer of each point is again a reflection group provided the lattice of one-parameter subgroups is generated by elements in the directions of the reflections. This is proved in §4 for ordinary tori, and transferred in §5 to algebraic tori by a device due to T. A. Springer. A corollary of the connectedness theorem above, under the assumptions there, is that every element of G fixed by σ is contained in a Borel subgroup fixed by σ. In neither case can the assumption that G is simply connected by entirely omitted, but for the theorem it can be relaxed to: σ is fixed-point-free on the fundamental group of G, which in a sense is best possible (see 9.7), and for the corollary even further (9.14). These and related matters are discussed in §9 where the exact range of possibilities for the disconnectedness of G_σ in case G is not simply connected is determined (9.1). The results are formulated not for semisimple automorphisms, but, slightly more generally, for those that fix a Borel subgroup and a maximal torus thereof, and apply equally well to complex Lie groups (we take K to be the complex field) and to compact Lie groups (with connectedness taken in the usual sense), since Borel [2] has proved the basic connectedness theorem in that case.

In §10 we begin the study of the case in which G_σ is finite by proving the main (cohomological) tool: if G is connected and G_σ is finite, then $(1-\sigma)G = G$ (here and elsewhere $1-\sigma$ is defined by $(1-\sigma)x = x\sigma x^{-1}$). This extends a well known results of S. Lang [15]. We obtain as by-products that in case also σ is an

1991 *Mathematics Subject Classifications.* Primary 20-xx; Secondary 22-xx.

automorphism then G is necessarily solvable (19.12), a result proved by Winter [33] by a different method, and that there exists a dichotomy (10.13): if G is simple exactly one of the following holds: (a) σ is an automorphism; (b) G_σ is finite.

Assume henceforth that G is semisimple and G_σ finite. In §11 we obtain a (Bruhat) cellular decomposition for G_σ, identity the possibilities (see 11.6; each group turns out to be essentially a direct product of Chevalley groups [8] and their twisted analogues [16], [24], [29]), and present a new general formula for the order (11.16). The formula, a twisted version of Chevalley's, is in terms of the action of σ on the algebra of invariants under the Weyl group. It explains, for example, why, if k is a field of q elements, one can convert the order $q^{n(n-1)}\Pi(q^j - 1)$ $(2 \leq j \leq n)$ for $SL_n(k)$ into that for $SU_n(k)$ by simply replacing $q^j - 1$ by $q^j + 1$ whenever j is odd (11.20). Its proof depends on the evaluation of the series $\Sigma t^{N(w)}$, in which t is a variable, w runs over all elements of the Weyl group (relative to a suitable torus) fixed by σ, and $N(w)$ is the number of positive roots made negative by w. This is done in §2, by a method used by L. Solomon [22] in the untwisted case ($\sigma = 1$). The above series in the (untwisted) affine case (the reflections are then relative to affine hyperplanes) has arisen in the work of R. Bott [4] on the topology of compact Lie groups (he evaluates it as a by-product of deep topological considerations) and also in the work of N. Iwahori and H. Matsumoto [14] on semisimple groups over local fields. We have taken the opportunity in §3 to give an elementary derivation of Bott's formula, in a twisted form because of potential applications to noncompact Lie groups and to twisted versions of groups over local fields, by an extension of Solomon's method. In §12 the "topology" of G_σ (still supposed finite) is studied, the subgroup generated by the unipotent elements playing the role of the identity component. In particular it is shown (12.4) that if G is simply connected then G_σ is generated by its unipotent elements, hence is "connected". In §13 we have taken the opportunity to reformulate, in terms of G and σ, the principal results about the representations of G and G_σ proved in [26]. In particular, 13.1 (see also 13.2) combines in a beautiful way (we think) two of the main results there. The development in these three sections is to some extent expository presenting a unification and simplication of known results. Finally, in the last two sections we derive formulae for the number of semisimple classes, the number of maximal tori, and the number of unipotent elements (all characteristic values have to be 1) fixed by σ (see 14.8, 14.14, and 15.1). The first two formulae depend on the evaluation of certain averages over the Weyl group, and the third on the construction of a certain representation for G_σ (15.6). In case G (still semisimple) is defined over a field k of q elements, so that σ may be taken as the qth power map in the above considerations, the formulae yield, for the numbers of objects of the above types defined over k, $q^{\text{rank } G}$, $q^{\dim G - \text{rank } G}$, $q^{\dim G - \text{rank } G}$, respectively. Why these formulae should be pure powers of q and not involve terms of lower degree is not clear to the author, except in the first case since the semisimple classes of G correspond to the regular classes and the latter can, at least when G is simply connected, be given the structure of ordinary affine (rank G)-dimensional space (see [28] for all this). Quite likely the other formulae reflect interesting properties of the varieties of maximal tori and unipotent elements of G. A special case of the last result states that with k as above the number of unipotent elements of $SL_n(k)$ is $q^{n(n-1)}$, a fact already proved by several earlier authors (see [12] for a proof and references to earlier articles).

Throughout the work the notation G_σ and $1 - \sigma$ introduced above will be used. In addition, G_0 will denote the identity component of the group G, and G_x and i_x, repsectively, the centralizer of and the inner automorphism by the element x of G. Finally, $|S|$ will always denote the cardinality of the set S.

1. Preliminaries on reflection groups

Our purpose is to develop the basic properties of reflection groups in a form suitable for our later purposes. The results are for the most part not new (but see 1.26 and 1.29). Throughout the section V will denote a real inner-product space relative to a symmetric bilinear (but not necessarily nonsingular) form $(\ ,\ )$, Σ a set of vectors in V, Π a subset of Σ, and W a group of automorphisms of V such that:

1.1. *The elements of Σ are nonisotropic.*

1.2. *If α is in Σ than $-\alpha$ is also but no other multiple of α is.*

1.3 *The elements of Π are linearly independent.*

1.4 *Each element of Σ is a linear combination of the elements of Π such that all coefficients have the same sign.*

1.5 *For each $\alpha \in \Pi$, the corresponding reflection w_α, defined by $w_\alpha v = v - 2(v,\alpha)/(\alpha,\alpha) \cdot \alpha$, maps Σ onto itself.*

1.6 *W is generated by the reflections $w_\alpha (\alpha \in \Pi)$.*

The group W is our prime object of study. The systems Σ and Π are used mainly to facilitate this study. Abstractly, W can be any group with a set of involutory generators $(w_\alpha, w_\beta, \dots)$ for which the orders of the products $w_\alpha w_\beta$ are prescribed. The elements of Σ will be called roots, those with positive coefficients in 1.4 positive, and those in Π simple. The simple roots are characterized among the positive roots as those which are not positive linear combinations of others. The support of a vector $v = \Sigma c_\alpha \alpha$ $(\alpha \in \Pi)$ is the set of α such that $c_\alpha \neq 0$.

1.7. *For each $\alpha \in \Pi$, w_α permutes the positive roots other than α.*

Since w_α changes each vector by a multiple of α, this follows from 1.4.

1.8. *For $w \in W$ let $N(w)$ denote the number of positive roots made negative by w, and let $L(w)$ denote the minimal length of an expression $w = w_1 w_2 \cdots w_\ell$ ($w_i = w_{\alpha_i}, \alpha_i \in \Pi$). Then $N = L$.*

For the proof we use the following easy consequence of 1.7.

1.9. *If $\alpha \in \Pi$ and $w \in W$, then $N(ww_\alpha) - N(w)$ equals 1 or -1 according as $w\alpha$ is positive or negative.*

Now fix $w \in W$. By 1.9, $N(w) \leq L(w)$. Assume $N(w) < L(w)$ and $w = w_1 w_2 \cdots w_\ell$ as above with $\ell = L(w)$. By 1.9, $N(w_1 w_2 \cdots w_{j+1}) < N(w_1 w_2 \cdots w_j)$ and $w_1 w_2 \cdots w_j \alpha_{j+1} < 0$ for some $j < \ell$. Since $\alpha_{j+1} > 0$, we have $w_i w_{i+1} \cdots w_j \alpha_{j+1} < 0$ and $w_{i+1} \cdots w_j \alpha_{j+1} > 0$ for some $i \leq j$, whence $w_{i+1} \cdots w_j \alpha_{j+1} = \alpha_i$ by 1.7. But then $w_{i+1} \cdots w_{j+1} = w_i \cdots w_j$, and substituting the right side for the left in the expression for w and using $w_i^2 = 1$ we get a shorter expression. This contradiction proves 1.8.

1.10. *If $w \in W$ keeps all positive roots positive, or, equivalently, all simple roots positive, then $w = 1$.*

By 1.8 $L(w) = 0$, whence 1.10.

1.11 COROLLARY. *The restriction of W to the subspace of V generated by Π is faithful.*

1.12. *The following conditions are equivalent.* (a) *Σ is finite.* (b) *W is finite.* (c) *There exists $w_0 \in W$ such that $w_0\Pi = -\Pi$, or, equivalently, w_0 maps all positive roots onto negative ones. Further w_0 in* (c) *is unique.*

If Σ is finite, then W is finite by 1.5 and 1.11. Assume (b) holds. Choose w_0 to maximize the function N. Then by 1.9 $w_0\alpha$ is negative for every $\alpha \in \Pi$, hence for every positive root, whence (c) holds. If (c) holds, then the number of positive roots is $L(w_0)$ by 1.8, a finite number, whence (a). The uniqueness of w_0 follows from 1.10.

1.13. *Assume that (,) is positive semidefinite and either that Σ is finite or that all coefficients in 1.4 are integers. Then $\Sigma = W\Pi$.*

Assume $\beta \in \Sigma$. The first assumption and 1.1 imply $(\beta, \beta) > 0$, and then the other assumptions yield an easy proof that $\beta \in W\Pi$, by induction on the sum of the coefficients of β in 1.4, whence 1.13.

For $\pi \subseteq \Pi$ we write W_π for the group generated by all $w_\alpha (\alpha \in \pi)$.

1.14. *For* $\pi \subseteq \Pi$ *the conditions* 1.1 *to* 1.6 *hold with* Π, Σ, *and* W *replaced by* π, *the part of* Σ *supported by* π, *and* W_π.

This is easily verified.

1.15. *If* $\pi \subseteq \Pi$ *and* $w \varepsilon W_\pi$, *then* w *permutes the positive roots with support not in* π.

The proof is like that of 1.7.

1.16. *For* $\pi \subseteq \Pi$ *let* W_π *be as above and let* W'_π *be the set of* $w \varepsilon W$ *such that* $w\pi > 0$. (a) *Each* $w \varepsilon W$ *can be written uniquely* $w = w'w''$ *with* $w' \varepsilon W'_\pi$ *and* $w'' \varepsilon W_\pi$. (b) $N(w) = N(w') + N(w'')$ *in* (a).

Assume $w \varepsilon W$. Let w' minimize the function N over the elements of wW_π. Then $w'\alpha > 0$ for every $\alpha \varepsilon \pi$ by 1.9. Thus $w' \varepsilon W'_\pi$ and $w \varepsilon W'_\pi W_\pi$. Now assume $w = w'w'' = v'v''$ are two expressions as in (a). Then $w'w''v''^{-1}$ and w' are in W'_π. Since $w''v''^{-1}\pi$ is supported by π, it must be positive by what has just been said, hence equal to π. Thus $w'' = v''$ by 1.10 with W_π in place of W, and uniqueness holds in (a). Part (b) is an easy consequence of 1.15.

1.17. *Let* C *be the cone of all* $v \varepsilon V$ *such that* $(v, \alpha) \geq 0$ *for all* $\alpha \varepsilon \Pi$, *and let* V_0 *be the cone of all* $v \varepsilon V$ *such that* $(v, \alpha) < 0$ *for only a finite number of positive roots* α. *Then* C *is fundamental domain for* W *on* V_0.

We observe first that W really acts on V_0 because of 1.7. Assume $v \varepsilon V_0$. Choose $w \varepsilon W$ to minimize the number of positive roots α such that $(wv, \alpha) < 0$. Then $wv \varepsilon C$: if $(wv, \alpha) < 0$ for some $\alpha \varepsilon \Pi$, then because of 1.7 the element $w_\alpha w$ contradicts the choice of w. Thus each $v \varepsilon V_0$ is congruent to some $u \varepsilon C$. That u is unique is part (b) of the following result.

1.18. LEMMA. . *Assume* $u, u' \varepsilon C$, $w \varepsilon W$, *and* $wu = u'$. *Let* π *be the subset of* Π *orthogonal to* u. *Then* (a) $w \varepsilon W_\pi$, *and* (b) $u = u'$.

Write w in the form $w'w''$ of 1.16. Then $w''u = u$ because π is orthogonal to u, whence $w'u = u'$. To prove both parts of 1.18 we need only show that $w' = 1$. Assume $w' \neq 1$. Then $w'a < 0$ for some $a \varepsilon \Pi$, whence $0 \geq (u', w\alpha) = (u, \alpha) \geq 0$, $(u, \alpha) = 0$, $a \varepsilon \pi$, and $w'\alpha > 0$ because $w' \varepsilon W'_\pi$. This contradiction proves $w' = 1$, hence 1.18.

1.19. *For* $\pi \subseteq \Pi$ *let* C_π *be the part of* C *defined by* $(u, \alpha) = 0$ *for* $\alpha \varepsilon \pi$, $(u, \beta) > 0$ *for* $\beta \varepsilon \Pi - \pi$. *If* C_π *is nonempty, then the following are equal.* (a) W_π. (b) *The stabilizer of* C_π *in* W. (c) *The stabilizer of any point of* C_π.

The first group above contains the third by 1.18 with $u' = u$, the third contains the second because the stabilizer of C_π fixes it pointwise by 1.17, and the second contains the first because C_π is orthogonal to π. Hence the three groups are equal.

1.20. *If S is a finite subset of V_0, then the subgroup of W fixing S pointwise is a reflection group. In other words, every $w \in W$ which fixes S pointwise is a product of reflections of W which also do.*

Assume $u \in S$. By 1.17 we may assume $u \in C$. Then u belongs to some C_π and its stabilizer is W_π by 1.19. An obvious induction with S and W replaced by $S - \{u\}$ and W_π now yields 1.20.

1.21. **Remark**. This result as it applies to real finite reflection groups is often referred to as Chevalley's theorem, but it was already known by Cartan and Weyl. It is also true for complex groups [27].

1.22. Henceforth in this section, σ denotes an automorphism of V which fixes Π and Σ, hence normalizes W, and V_σ and W_σ denote the subsets of V and W fixed by σ.

1.23. *If $w \in W_\sigma$ and ρ is a σ-orbit of roots, then all elements of $w\rho$ have the same sign.*

Since σ preserves signs, this follows from the equation $w\sigma^n\alpha = \sigma^n w\alpha$ $(\alpha \in \rho)$.

1.24. *Notations.* Assume Π is finite and $\pi \subseteq \Pi$. (a) $W_{\pi\sigma}(t)$ denotes the formal power series $\Sigma t^{N(w)} = \Sigma t^{L(w)}$ summed over all $w \in W_\pi \cap W_\sigma$, with N and L as in 1.8; $W_\sigma(t)$ the same in case $\pi = \Pi$. (b) If σ fixes π, then $\varepsilon_\sigma(\pi) = (-1)^n$ with n the number of σ-orbits in π; if it does not, then $\varepsilon_\sigma(\pi) = 0$.

1.25. THEOREM. *Assume Π finite and the notation as in 1.24. Then $\Sigma\varepsilon_\sigma(\pi)W_\sigma(t)/W_{\pi\sigma}(t)$, summed over the subsets π of Π, equals t^m (m the number of positive roots) if w is finite, equals 0 if w is infinite.*

Assume $\pi \subset \Pi$ is fixed by σ. If w is in W_σ then w' and w'' in 1.16 also are, because of the uniqueness there. Thus by 1.16(b) we have $W_\sigma(t)/W_{\pi\sigma}(t) = \Sigma t^{N(w)}$, summed over all $w \in W_\sigma$ such that $w\pi > 0$. Therefore for a fixed $w \in W_\sigma$ the coefficient of $t^{N(w)}$ in the sum of 1.25 is $\Sigma\varepsilon_\sigma(\pi)$, summed over all π fixed by σ such that $w\pi > 0$, i.e., $(1-1)^n = 0^n$, if n is the number of σ-orbits of Π kept positive by w (see 1.24(b)). Hence the coefficient is 0 unless $n = 0$ when it is 1. Now by 1.12, if W if infinite then n is never 0, while if W is finite then $n = 0$ only if $w = w_0$ in which case $N(w)$ is the number of positive roots. Since w_0 is in W_σ (it equals $\sigma w_0 \sigma^{-1}$ by the uniqueness in 1.12), 1.25 follows.

1.26 COROLLARY. *The series $W_\sigma(t)$ above represents a rational function.*

If $\pi \subseteq \Pi$, then by 1.15 the function N on W_π computed in the context of 1.14 is the same as the restriction of N on W to W_π. Thus if also σ fixes π then $W_{\pi\sigma}(t)$ bears the same relationship to W_π as $W_\sigma(t)$ to W. By 1.25 then $\Sigma\varepsilon_\sigma(\pi)/W_{\pi\sigma}(t)$, summed over the proper subsets of Π, equals $(t^m - \varepsilon_\sigma(\Pi))/W_\sigma(t)$ or the same thing with t^m missing. Thus 1.26 follows by induction on $|\Pi|$.

1.27. COROLLARY. *If Π is finite, then $W(t) = \Sigma t^{L(w)}$ summed over all $w \varepsilon W$ represents a rational function.*

This is the case $\sigma = 1$ of 1.26.

Next we deduce from 1.25 a more general formula. Our proof differs from that given by Solomon [22] for the case W finite and $\sigma = 1$, and can be used in similar situations which arise later.

1.28. *Assume Π finite and that α and β are complementary σ-invariant subsets of Π. Then*

$$\Sigma_{\pi\supseteq\alpha}\varepsilon_\sigma(\pi-\alpha)/W_{\pi\sigma}(t) = \sum_{\substack{\pi\supseteq\beta \\ W_\pi \text{ finite}}} \varepsilon_\sigma(\pi-\beta)/W_{\pi\sigma}(t^{-1}).$$

First we observe that if W is finite, then $t^m/W_\sigma(t) = 1/W_\sigma(t^{-1})$. For the element w_0 of 1.12 is in W_σ, as noted above; hence there is a correspondence $w \leftrightarrow w' = w_0 w$ among the elements of W_σ such that $N(w) + N(w') = m$. Thus if $\pi \subseteq \Pi$ we get from 1.25: (*) $\Sigma\varepsilon_\sigma(\gamma)/W_{\gamma\sigma}(t)$, summed over $\gamma \subseteq \pi$, equals $1/W_{\pi\sigma}(t^{-1})$ if W_π is finite, 0 if W_π is infinite. We multiply this relation by $\varepsilon_\sigma(\pi-\beta)$ and sum over all $\pi \supseteq \beta$. On the right we get the right side of 1.28. On the left if we first sum $\varepsilon_\sigma(\pi-\beta)$ on $\pi \supseteq \beta\cup\gamma$ we get $\varepsilon_\sigma(\alpha)$ if $\beta\cup\gamma = \Pi$, i.e., if $\gamma \supseteq \alpha$, and 0 otherwise since if ρ is a σ-orbit outside $\beta\cup\gamma$ then the terms for which π contains ρ cancel those for which it doesn't. Thus the left becomes the left side of 1.28 as required.

If α is empty 1.28 becomes 1.25. If β is empty it becomes:

1.29. COROLLARY.

$$1/W_\sigma(t) = \sum_{\substack{\pi\subseteq\Pi \\ W_\pi \text{ finite}}} \varepsilon_\sigma(\pi)/W_{\pi\sigma}(t^{-1}).$$

In the next section we given an explicit formula for $W_\sigma(t)$ in case W is finite, 2.1; thus 1.29 gives one in general.

To close this section we will show how W_σ can be given the same structure as W. Let π be a σ-orbit of Π such that W_π is finite. Let V_π be the subspace of V generated by π. Then $V_\pi \cap V_\sigma$ consists of the real multiples of the sum of the elements of π, hence is one-dimensional. Among the elements of $\Sigma \cap V_\pi$, a finite set by 1.12, we select those for which the sum of the coefficients in 1.4 is maximal, then take their average. The result, an element of $V_\pi \cap V_\sigma$, will be denoted α_π.

1.30. *For each σ-orbit π of Π such that W_π is finite let α_π be as above and w_π the unique element of W_π such that $w_\pi \pi = -\pi$ (see 1.12).*
(a) *α_π is a nonisotropic element of V_σ.*
(b) *w_π is in W_σ and its restriction to V_σ is the reflection corresponding to α_π.*

The group generated by W_π and σ on V_π is finite, hence fixes a positive-definite symmetric bilinear form F. As is easily seen the action is irreducible. Thus $(\ ,\ )$ on V_π is a multiple of F, hence also is definite. Thus α_π is nonisotropic. As just before 1.26 we see that $w_\pi \in W_\sigma$. Now if $v \in V_\sigma$, then $w_\pi v - v$ is in $V_\sigma \cap V_\pi$, hence equals $c\alpha_\pi$ for some real number c. Since $w_\pi \alpha_\pi = -\alpha_\pi$, we have $c(\alpha_\pi, \alpha_\pi) = (w_\pi v, \alpha_\pi) - (v, \alpha_\pi) = (v, w_\pi \alpha_\pi) - (v, \alpha_\pi) = -2(v, \alpha_\pi)$, whence (b).

1.31. **Remark**. Because of (a) α_π may also be described as the most positive of the projections onto $V_\pi \cap V_\sigma$ of the various roots with support in π.

1.32. (a) *W_σ fixes V_σ and its restriction to V_σ is faithful.* (b) *If Π_σ is the set of α_π in 1.30 and $\Sigma_\sigma = W_\sigma \Pi_\sigma$, then the conditions 1.1 to 1.6 hold with V, Π, Σ, W replaced by $V_\sigma, \Pi_\sigma, \Sigma_\sigma, W_\sigma | V_\sigma$.*

Assume $w \in W_\sigma$, $w \neq 1$. Then $w\alpha < 0$ for some $\alpha \in \Pi$ by 1.10, whence $w_\rho < 0$ for every root ρ with support in π, the orbit of α, by 1.23. By 1.12 W_π is finite, and clearly $w\alpha_\pi \neq \alpha_\pi$. Thus $w \neq 1$ on V_σ, which proves (a). We have $N(ww_\pi) = N(w) - N(w_\pi)$ by 1.16. Thus by an obvious induction W_σ is generated by the various w_π. In the context of (b) the conditions 1.1 and 1.3 to 1.6 and the first part of 1.2 are easily verified. It remains to prove that if $w\alpha = c\alpha'$ ($c > 0$, $\alpha = \alpha_\pi$, $\alpha' = \alpha_{\pi'}$, π and π' as in 1.30) then $c = 1$. By 1.23 w maps the positive roots with support in π into those with support in π', and similarly for w^{-1} with π and π' interchanged. Thus the map is surjective, $w\alpha = \alpha'$, and $c = 1$.

1.33. Corollary. *Assume the conditions of 1.13 hold and W_π is finite for every σ-orbit π of Π. Then Σ_σ in 1.32 is the set obtained by averaging the elements of the various σ-orbits of Σ and then discarding those of the resulting vectors that are smaller multiples of others.*

2. The series in the finite case

Our aim in this section is to prove 2.1 below. This will be used in §11 to derive a uniform formula for the orders of the finite Chevalley groups and their twisted analogues. It will be assumed that W is a finite group generated by reflections acting on a real Euclidean space of finite dimension n, that Π and Σ have been chosen so that 1.1 to 1.6 hold (e.g., Σ can be chosen as the set of unit vectors orthogonal to the reflecting hyperplanes), and that σ is as in 1.22 and of finite order. The notations of §1 will be used. Chevalley [9] has shown that in the symmetric algebra S on V the subalgebra I of W-invariants is freely generated by n homogeneous (basic) elements of uniquely determined degrees.

2.1. THEOREM. *Let W and σ be as above. Assume that the basic invariants I_j have degrees $d(j)$ and are so chosen that each is a characteristic vector of $\sigma : \sigma I_j = \varepsilon_j I_j$, and let ε_{0j} be the characteristic values of σ acting on V ($j = 1, 2, \ldots, n$). Let $P_\sigma(t)$ denote the rational form $\Pi(1 - \varepsilon_j t^{d(j)})/(1 - \varepsilon_{0j} t)$, and as in 1.24 let $W_\sigma(t)$ denote the series $\Sigma t^{N(w)} (w \in W_\sigma)$. Then $W_\sigma(t) = P_\sigma(t)$.*

2.2 **Remarks**. (a) Since σ is of finite order, it acts completely reducibly on the invariants of a fixed degree. Thus a choice of the I_j as above is always possible. (b) It will be proved later that the ε_j form a permutation of the ε_{0j}. (c) If I^+ denotes the ideal of invariants without constant term and $J = I^+/I^{+^2}$, then $P_\sigma(t)$ may also be written $\det(1_J - (t\sigma)_J)/\det(1 - t\sigma)$. Here the denominator bears the same relation to the group of one element as the numerator does to W. Because of the tensor product decomposition $S = I \otimes H$, $P_\sigma(t)$ also equals the trace of $t\sigma$ on the space H of harmonic elements of S. (d) The proof to follow closely parallels a proof by Solomon [22] of the case $\sigma = 1$, when 2.1 becomes $\Sigma t^{N(w)} (w \in W) = \Pi(1 - t^{d(j)})/(1 - t)$. However, we have given a more elementary proof of the key lemma (2.3 below) which requires no results from algebraic topology.

The proof of 2.1 will be given in several steps. Since the group generated by W and σ is finite, we may assume our basic inner product is invariant under it. As is easily seen σ preserves the decomposition of V into the subspace generated by Π and its orthogonal complement. It follows that on restriction to this subspace neither $P_\sigma(t)$ nor $W_\sigma(t)$ changes. Thus we may assume throughout the rest of the proof that Π generates V.

2.3. *Assume $w \in W$. For $\pi \subseteq \Pi$ let C_π be as in 1.19 and let N_π denote the number of (simplicial) cones congruent to C_π under W and fixed by $w\sigma$. Then $\Sigma \varepsilon_\sigma(\pi) N_\pi = \det w$.*

Here and in similar situations which follow π runs over the σ-invariant parts of Π.

(1) *Let C' be congruent to C_π under W and fixed by $w\sigma$. Then π is invariant under σ.* Since σ normalizes W, we may assume $C' = C_\pi$, whence $w\sigma C_\pi = C_\pi$.

Since σC_π and C_π are parts of C, we get from 1.17 that $\sigma C_\pi = C_\pi$, whence $\sigma\pi = \pi$.

(2) *Let V' be the fixed-point-space for $w\sigma$ on V. Let K (resp. K') be the [conical simplicial] complex cut on V (resp. V') by the reflecting hyperplanes for W. Then the cells of K' are the intersections with V' of those cells of K which are fixed by $w\sigma$.* Let C' be a cell of K which intersects V'. Then any point in this intersection is fixed by $w\sigma$, whence C' is also. Conversely if C' is fixed by $\omega\sigma$, then so is its centroid, which thus is a point common to C' and V'.

(3) *Let C' and V' be as in* (1) *and* (2), *and let i be the number of σ-orbits of $\Pi - \pi$. Then* $\dim C' \cap V' = i$. Again we may assume $C' = C_\pi$. Then $w\sigma$ acts on C_π as σ by 1.19; hence it permutes the "vertices" of C_π in the same way as σ. These vertices correspond to the elements of $\Pi - \pi$ and each orbit contributes one dimension to $C_\pi \cap V'$, whence (3).

(4) *In the complex cut on a real k-dimensional vector space by a finite number of hyperplanes let N_i denote the number of cells of dimension i. Then* $\Sigma(-1)^i N_i = (-1)^k$. This follows from Euler's formula applied to the complex cut by the hyperplanes on the unit sphere centered at the origin, but it also follows directly by induction. In fact, if an extra hyperplane, say H, is added to the configuration, then each i-cell of the original complex which is cut into two parts by H has corresponding to it an $(i-1)$-cell in H, the one that separates these parts from each other, so that $\Sigma(-1)^i N_i$ remains unchanged.

We turn now to the proof of 2.3. We apply (4) to the configuration K' of (2). If we combine (1), (2) and (3) and observe that $(-1)^i = \varepsilon_\sigma(\Pi - \pi)$ in (3), we get $\Sigma\varepsilon_\sigma(\Pi - \pi)N_\pi$ on the left and $(-1)^k (k = \dim V')$ on the right. Since $w\sigma$ is orthogonal, its characteristic values other than ± 1 come in conjugate complex pairs. Thus $(-1)^k = (-1)^n \det w\sigma$, which equals $\varepsilon_\sigma(\Pi)\det w$ because $\det \sigma = (-1)^n \varepsilon_\sigma(\Pi)$. Thus if the original equation is divided by $\varepsilon_\sigma(\Pi)$, the result is 2.3.

2.4. **Remark**. One can also prove 2.3 by applying the Hopf trace formula (cf. [22], [18, p. 288]) to $w\sigma$ acting on the complex K of (2).

2.5. *For any complex character χ on $\langle W, \sigma\rangle$ (the group generated by W and σ) and any $\pi \subseteq \Pi$, let χ_π denote the character induced on $\langle W, \sigma\rangle$ by the restriction of χ to $\langle W_\pi, \sigma\rangle$. Then* $\Sigma\varepsilon_\sigma(\pi)\chi_\pi(w\sigma) = \chi(w\sigma)\det w$ $(w \in W)$.

Assume π is a σ-invariant part of Π. By 1.19 the stabilizer of C_π in W is W_π. Using as above (see (1)) that σ fixes the fundamental chamber C, we see that the stabilizer of C_π in $\langle W, \sigma\rangle$ is $\langle W_\pi, \sigma\rangle$. Thus the unit character on $\langle W_\pi, \sigma\rangle$ induces on $\langle W, \sigma\rangle$ the character whose value at x is the number of simplexes congruent to C_π under $\langle W, \sigma\rangle$, i.e., under W, and fixed by x. Thus if χ is the unit character 1, then 2.5 reduces to 2.3. If χ is arbitrary, then $\chi_\pi = 1_\pi\chi$ from the definitions, whence 2.5.

Let M be a (real finite-dimensional) $\langle W,\sigma\rangle$-module. We write $\hat{I}(M)$ for the subspace of skew-invariants under W (all $m \in M$ such that $wm = (\det w)m$), and for each σ-invariant part π of Π we write $I_\pi(M)$ for the subspace of invariants under W_π. These subspaces are all fixed by σ; thus we may speak of the trace of σ on them.

2.6. *For any $\langle W,\sigma\rangle$-module M,*

$$\sum \varepsilon_\sigma(\pi) tr(\sigma, I_\pi(M)) = tr(\sigma, \hat{I}(M)).$$

Let χ be the character of M. First we average $\chi_\pi(w\sigma)$ over $w \in W$. By the calculation of the proof of the Frobenius reciprocity theorem [11, p. 271] we get the average of $\chi(w\sigma)$ over W_π. Now since the average of w over W_π acts on M as the projection P_π on $I_\pi(M)$, we get $\chi(P_\pi\sigma)$, i.e., $tr(\sigma, I_\pi(M))$. Similarly the average of $\chi(w\sigma)\det w$ over W equals $tr(\sigma, \hat{I}(M))$. Substituting into 2.5 averaged over $w \in W$, we get 2.6.

2.7. *For each σ-invariant part π of Π let $d(\pi j)$, $\varepsilon_{\pi j}$, $P_{\pi\sigma}(t)$ be defined for W_π just as $d(j)$, ε_j, $P_\sigma(t)$ are for W. Then $\Sigma\varepsilon_\sigma(\pi)P_\sigma(t)/P_{\pi\sigma}(t) = t^m$, with m as in* 1.25.

Let $S = \Sigma S_k$ be the usual graded decomposition of the symmetric algebra on V. The identity to be proved may be written.

$$\sum \varepsilon_\sigma(\pi)\Pi(1-\varepsilon_{\pi j}t^{d(\pi j)})^{-1} = t^m\Pi(1-\varepsilon_j t^{d(j)})^{-1}. \tag{(*)}$$

Because of the definitions of the d's and ε's the coefficient of t^k on the left is $\Sigma\varepsilon_\sigma(\pi)tr(\sigma, I_\pi(S_k))$. Now as is well know $\hat{I}(S) = pI(S)$, p being the product of the positive roots (since α divides $(1-w_\alpha)q$ for any root α and any polynomial q). Clearly $dg\ p = m$ and $\sigma p = p$. Thus the coefficient of t^k on the right of (*) equals $tr(\sigma, \hat{I}(S_k))$, hence also the coefficient on the left by 2.6, whence 2.7.

Proof of 2.1.. This follows from 1.25 and 2.7 by induction on $|\Pi|$ (cf. the proof of 1.26).

2.9. COROLLARY. *In* 2.1 *the ε_j form a permutation of the ε_{0j}.*

This is easily reduced to the case that Π is orthogonally indecomposable and generates V, which we henceforth assume. Since $P_\sigma(t)$ takes on a finite nonzero value at $t = 1$, the value $|W_\sigma|$ by 2.1, the number 1 has the same multiplicity among the ε_j as among the ε_{0j}. If $\sigma^2 = 1$, then all ε's are ± 1; thus 2.9 holds in this case. Because of the classification of indecomposable reflection groups [10,

Ch. 11] the only other possibility is the triality case of D_4. Then two of the ε_{0j} are 1; hence so are two of the ε_j. The other two terms are, in both cases, the nontrivial cube roots of 1 since $\sigma^3 = 1$ and σ is a real transformation, whence 2.9.

2.10. **Remark**. A case by case check shows that the degrees of the basic invariants of $W_\sigma|V_\sigma$, viewed as a reflection group as in 1.32, are just those $d(j)$ for which $\varepsilon_j = 1$. we have no general explanation of this fact.

Finally, we present an extension of 2.5 in the spirit of 1.28.

2.11. *If α and β are complementary σ-invariant parts of* Π, *then* $\Sigma_{\pi\supseteq\alpha}\varepsilon_\sigma(\pi-\alpha)\chi_\pi(w\sigma) = (det\ w)\Sigma_{\pi\supseteq\beta}\varepsilon_\sigma(\pi-\beta)\chi_\pi(w\sigma)$.

We apply 2.5 with π, W_π and $\chi|W_\pi$ in place of Π, W and χ, and then induce from $\langle W_\pi, \sigma\rangle$ to $\langle W, \sigma\rangle$. The result is $\Sigma\varepsilon_\sigma(\gamma)\chi_\gamma(w\sigma) = (\det w)\chi_\pi(w\sigma)$, summed on $\gamma \subseteq \pi$. Then 2.11 follows from this just as 1.28 from (*) in the proof of that result.

2.12. COROLLARY. *If α and β are as above and the notations are as in 2.6, then*

$$\sum_{\pi\supseteq\alpha}\varepsilon_\sigma(\pi-\alpha)tr(\sigma, I_\pi(M)) = \sum_{\pi\supseteq\beta}\varepsilon_\sigma(\pi-\beta)tr(\sigma, \hat{I}_\pi(M)).$$

3. The affine case

Our main aim is to prove a twisted version of a theorem of Bott (see 3.8 and 3.10 below), the analogue of 2.1 for discrete groups generated by reflections in affine hyperplanes. In addition to the assumptions there, we will assume:

3.1. Π *(or Σ) generates V.*

3.2. $2(\alpha,\beta)/(\beta,\beta)$ *is an integer for all $\alpha, \beta \in \Sigma$.*

The first condition may be stated: W is *effective*, i.e., fixes no line of V. Because of 1.13 the second implies:

3.3. *The coefficients in* 1.4 *are all integers.*

Our main object of study is the group W' generated by the reflections in the affine hyperplanes $k + \alpha = 0$ (k integral, $\alpha \in \Sigma$). The collection $\{\lambda_\alpha = 2\alpha/(\alpha,\alpha)|\alpha \in \Sigma\}$ satisfies the same conditions as Σ, hence generates a complete (n-dimensional) lattice, which will be denoted L.

3.4. *The semidirect decomposition $W' = LW$ holds.*

Since W fixes L, LW is a group. Since (*) the reflection in $k + \alpha = 0$ is the product of a translation through $-k\lambda_\alpha$ and the reflection in $\alpha = 0$ and since the translation is the product of the two reflections, we get $W' \subseteq LW$ and $LW \subseteq W'$, whence 3.4.

3.5. **Remark**. It follows that W' is discrete and that V/W' has finite volume. Conversely, any group W' generated by affine reflections and having these properties can be realized in this way. First $W' = LW$ with L a complete lattice of translations and W a finite reflection group fixing a point which we take to be the origin [10, p. 191]. Our assertion then follows from:

3.6. *Let W be an effective finite reflection group and L a complete lattice fixed by W. For each direction orthogonal to a reflecting hyperplane for W, let the vector α be chosen so that $\lambda_\alpha = 2\alpha/(\alpha, \alpha)$ is the shortest translation of L in that direction and let Σ be the set of all such α.* (a) *The conditions* 1.1 *to* 1.6, 3.1 *and* 3.2 *hold for an appropriate choice of* Π.

(b) *The following conditions are equivalent.*

(1) *L is generated by the λ_α.*
(2) *LW is a reflection group.*
(3) *LW is the reflection group generated by the reflections in the hyperplanes $k + \alpha = 0$ (k integral, $\alpha \in \Sigma$).*

If Π is chosen as a simple system relative to some ordering of Σ (cf. the remark after 1.6), then all parts of (a) but 3.2 are verified at once, and 3.2 holds because W fixes L : $\lambda_\beta - w_\alpha\lambda_\beta$ is in L and in the line of α, hence is a multiple of λ_α, whence (λ_β, α) is integral by 1.5, which is 3.2. Let L_1 be the lattice generated by the λ_α. Then L_1W is the group described in (3) by (a) and 3.4. Thus (1) implies (3). But L_1W is also the largest reflection subgroup of LW : any reflection w in LW is in some hyperplane $k + \alpha = 0$ (k real, $\alpha \in \Sigma$), then $-k\lambda_\alpha \in L$ by (*) above, k is an integer, and $w \in L_1W$. Thus (2) implies (1).

Henceforth W' and L are as defined just before 3.5. We convert W' into a linear group so that our earlier results may be applied by making it act contragrediently on the space V' of inhomogeneous linear funcitons on V. We identify the subspace of homogeneous functions with V in the usual way, write elements of V' in the form $r + v$ (r real, $v \in V$), and extend the inner product of V to be trivial on the first component. Finally, let Σ' denote the set $\{k+\alpha | k \text{ real}, \alpha \in \Sigma\}$ and Π' the subset obtained by adjoining to Π the set $\{1 - \delta | \delta = \text{highest root for some irreducible component of } \Sigma\}$. Thus $\delta - \alpha$ is a positive combination in 1.4 for every root α in the component of δ; the existence of δ is proved in [7, p. 256].

3.7. *The reflection in the affine hyperplane $k + \alpha = 0$ of V acts on V' as the reflection in the hyperplane orthogonal to $k + \alpha$.*

In fact, the image of $r+v$ under the first reflection works out to $r-k(v,\lambda_\alpha)+v-(v,\lambda_\alpha)\alpha$ (recall $\lambda_\alpha = 2\alpha/(\alpha,\alpha)$), which may be written $r+v-(r+v,\lambda_{k+\alpha})(k+\alpha)$.

3.8. *The conditions* 1.1 *to* 1.6 *hold with* V', Σ', Π', W' *in place of* V, Σ, Π, W; *hence so do the other results of* §1.

Clearly 1.1 to 1.3 hold. Assume $k+\alpha \varepsilon \Sigma'$. Let δ be the highest root in the component of α. If $k > 0$, then 1.4, with integral coefficients in fact, follows from $k+\alpha = k(1-\delta)+(k-1)\delta+\delta+\alpha$, and similarly if $k < 0$, while if $k = 0$ it clearly holds. Finally 1.5 holds by 3.2, while if W'' is the group generated by all $w_\alpha(\alpha \varepsilon \Pi')$ then $W''\Pi' = \Sigma'$ by 1.13, so that W'' is equal to the group generated by all $w_\alpha(\alpha \varepsilon \Sigma')$, i.e., to W' by 3.7, and 1.6 holds.

Now we can state Bott's theorem [4].

3.8. Theorem. *Define* $W'(t)$ *for* W' *as* $W(t)$ *is defined for* W *in* 1.27 *(see also* 3.7 *and* 3.8*). Let* $d(j)$ $(j = 1, 2, \dots, n)$ *be the degrees of the basic invariants of* W *acting on* V. *Then* $W'(t) = \Pi(1-t^{d(j)})/(1-t)(1-t^{d(j)-1})$.

3.9. **Remarks**. (a) The denominator on the right causes no problems because each $d(j)-1$ is positive by 3.1. (b) The reflecting hyperplanes cut V into a number of simplicial chambers each of which is a fundamental domain for W' on V and one of which, denoted F, is defined by $\alpha \geq 0$ for $a \varepsilon \Pi'$, as we see by applying 1.17 to W' and V' and interpreting the result on V. (In the present case V_0 of 1.17 is the whole space by 3.7.) Because of 3.8 an element of Σ' is positive (resp. negative) if as a function on V it is positive (resp. negative) on F. Now $\Pi(1-t^{d(j)})/(1-t) = W(t)$ by 2.2(c). Thus by 1.15 with Π and π replaced by Π' and Π formula 3.8 is equivalent to $\Pi(1-t^{d(j)-1})^{-1} = \Sigma t^{N(w)}$, summed over those $w \varepsilon W'$ such that $w\Pi > 0$, i.e., $\Pi(w^{-1}F) > 0$. Here $w^{-1}F$ runs through the chambers contained in the cone $\Pi > 0$, and $N(w) = N(w^{-1})$ is, by the above remarks, just the number of reflecting hyperplanes separating $w^{-1}F$ from F. This is the form in which Bott originally gave his formula. His proof consists in interpreting both sides of the equation with t replaced by t^2 as the Poincaré series of the loop space of the simple compact Lie group corresponding to Σ, the infinite series arising from the Morse theory and the product from a comparison of the group with a product of spheres of dimensions $2d(j)-1$ $(1 \leq j \leq n)$. (c) The above formulas arise not only in Bott's work as just described but also in connection with certain cellular decompositions of linear algebraic groups defined over local fields [14]. Thus it can be expected that a twisted form of 3.8 will have applications to twisted versions of these groups, and we shall prove such a formula. The reader primarily interested in Bott's formula may take σ to be the identity in what follows.

Consider an affine-Euclidean automorphism σ of V which permutes the elements of Π'. It follows from 1.13 that σ fixes Σ', and then if σ_0 denotes the linear part of σ that σ_0 fixes Σ, hence normalizes W.

3.10. THEOREM. *Let $W, W', d(j)$ be as in 3.8 and σ, σ_0 as above. Let $\varepsilon_j, \varepsilon_{0j}$ be defined as in 2.1 but with σ_0 in place of σ. Let $Q_\sigma(t)$ denote the form $\Pi(1-\varepsilon_j t^{d(j)})/(1-\varepsilon_j t^{d(j)-1}) \cdot (1-\varepsilon_{0j}t)$ and $W'_\sigma(t)$ the series $\Sigma t^{N(w)} (w \varepsilon W'_\sigma)$. Then $W'_\sigma(t) = Q_\sigma(t)$.*

3.11. **Remarks**. (a) The ε_j need not form a permutation of the ε_{0j}, but they do so if σ is linear, i.e., fixes Π (see 2.2(b)). In this case (cf. 3.9(b)) 3.10 implies that $\Pi(1-\varepsilon_j t^{d(j)-1})^{-1} = \Sigma t^{N(w)}$, summed over those cells $w^{-1}F$ which are in the cone C and fixed by σ. (b) For each type of group W' (type $A_n, B_n, \ldots$) the $d(j)$ and the possibilities for σ are known. Once σ is given the ε_j and ε_{0j} are easy to work out, hence so is $Q_\sigma(t)$. Consider W' of type A_n. Here the $d(j)$ are $2, 3, \ldots, n+1$. If $n+1 = pq$ is a factorization and the elements of Π' are represented as usual as an extended Dynkin Diagram in the form of an $(n+1)$-cycle, let σ, for example, be the automorphism which moves each element p steps forward. Then the ε_{0j} consist of the qth roots of 1, with 1 counted $p-1$ times and the others p times, while the ε_j are all 1 since $\sigma_0 \varepsilon W$. Thus $Q_\sigma(t) = (1-t^{n+1})/(1-t^q)^p$. In case $p=1$, i.e., Π' itself is a σ-orbit, this reduces to 1, as it should since $W'_\sigma = \{1\}$ by 1.32. This is the only time that W'_σ is finite. In case $p = n+1$, i.e., σ is the identity, we get $(1-t^{n+1})/(1-t)^{n+1}$ as the value of the function of 3.10.

The proof of 3.10 also proceeds in several steps. Recall the decomposition $W' = WL$. Let T denote the torus V/L, and K the complex cut on T by the projections of the reflecting hyperplanes. The elements of W and σ act as simplicial mappings of K. For each proper part π of Π' let F_π be the "face" of F defined by $(\alpha = 0,\ \beta > 0 | \alpha \varepsilon \pi,\ \beta \varepsilon \Pi' - \pi)$ and T_π its projection on T. Since F is a fundamental domain for $W' = WL$ acting on V, its projection, i.e., the union of the T_π's, is faithful and yields a fundamental domain for W acting on T, and it further follows that each cell of K is congruent under W to a unique T_π.

3.12 *Assume $w \varepsilon W$. For each proper part π of Π' let N_π denote the number of cells of K congruent under W to T_π and fixed by $w\sigma$. Let E_i denote the ith exterior power of V. Then the following are equal:* (a) *$\Sigma\varepsilon_\sigma(\pi)N_\pi$ (π proper, σ-invariant),* (b) *$det(w-\sigma_0^{-1})$,* (c) *$-\varepsilon_\sigma(\Pi')\Sigma_{i=0}^n(-1)^i tr(w\sigma_0, E_i)$.*

We observe first that the last sum is $\det(1-w\sigma_0)$ and that (b) and (c) are equal because $-\varepsilon_\sigma(\Pi') = (-1)^n \det \sigma_0 = \det(-\sigma_0^{-1})$. Let T' be the fixed-point set of $w\sigma$ on T. Assume that T' is not empty. Then $w\sigma$ is conjugate to $w\sigma_0$ under a translation of T. Thus T' is a translate of the fixed-point set of $w\sigma_0$, i.e., of the kernel of $1 - w\sigma_0$ on T, and so consists of a finite number of translates, say d, of some subtorus of T. By going to V, we see that $\det(1 - w\sigma_0) \neq 0$ exactly when the dimension of T', call it k, is 0. This last condition holds even if T' is empty (in which case $k = -1$): if $\det(1 - w\sigma_0) \neq 0$, then $(1 - w\sigma_0)V = V$ so that $w\sigma$ is conjugate to $w\sigma_0$ by a translation and T' is not empty. Now we will use the following results.

3.13. *Assume that the complex cut on a k-dimensional torus by a finite number of translates of $(k-1)$-dimensional subtori is cellular, and that N_i denotes the number of i-dimensional cells. Then $\Sigma(-1)^i N_i = \delta_{0k}$.*

This follows from Euler's formula. A direct proof by induction also exists (cf. 2.3(4)) but will not be given here. We apply 3.13 to the complex cut by K on each component of T' and then add the results. As in the proof of 2.3 we get on the left $-\varepsilon_\sigma(\Pi')\Sigma\varepsilon_\sigma(\pi)N_\pi$, summed on the proper σ-invariant parts of Π'; on the right we get $d\delta_{0k}$. If $k \neq 0$, this is 0 and so is $\det(1 - w\sigma_0)\det(-\sigma_0^{-1})$ by the above, so that (a) and (b) of 3.12 are equal in this case. If $k = 0$, then T' is congruent to the kernel of $1 - w\sigma_0$, which consists of $|\det(1 - w\sigma_0)|$ points, represented in V by the lattice $(1 - w\sigma_0)^{-1}L$ taken mod L. Now $\det(1 - w\sigma_0)$ is positive: $\det(t - w\sigma_0)$ is positive for large positive values of t, hence for all $t > 1$ because $w\sigma_0$ is orthogonal. Thus the right side of our relation $d\delta_{0k}$ become $\det(1 - w\sigma_0)$, which proves the equality of (a) and (c) in this case and completes the proof of 3.12.

3.14. **Remark**. 3.12 also follows from the Hopf trace formula [18, p. 288] applied to $w\sigma$ acting on K. The alternating sum of the traces of $w\sigma$ yields on the chain groups $-\varepsilon_\sigma(\Pi')\Sigma\varepsilon_\sigma(\pi)N_\pi$ and on the homology groups $\Sigma(-1)^i tr(w\sigma_0, E_i)$ because of the canonical identification of E_i with the ith homology group of T (using the fact that T is a product of n circles).

If π is a proper part of Π', the group W'_π fixes a point of V (the equations $\alpha = 0$ for $\alpha \,\varepsilon\, \pi$ are consistent); hence in view of the decomposition $W' = WL$ it may be identified with a subgroup of W which we denote W_π (if $\pi \subseteq \Pi$ this reduces to the notation of §2). The groups W'_π and W_π act identically on T. Because of 1.19 and our present construction, this implies:

3.15. *If π is a proper part of Π', then W_π, the stabilizer of T_π in W, and the stabilizer of any point of T_π are all equal.*

Henceforth π denotes a σ-invariant proper part of Π', and the subscript T restriction to T. From the definitions,

3.16. *σ_T normalizes W_T and each $W_{\pi T}$.*

It follows that the group $\langle W_T, \sigma_T\rangle$ is finite.

3.17. *Let χ be a complex character on $\langle W_T, \sigma_T\rangle$ and χ_π the character on $\langle W_T, \sigma_T\rangle$ induced by the restriction of χ to $\langle W_{\pi T}, \sigma_T\rangle$. Then for $w \varepsilon W$ we have*

$$\Sigma\varepsilon_\sigma(\pi)\chi_\pi(w_T\sigma_T) = \chi(w_T\sigma_T)\det(w - \sigma_0^{-1}).$$

Because of 3.15 this follows from 3.12 just as 2.5 from 2.3.

Next we observe that there is a homomorphism φ of $\langle W_T, \sigma_T\rangle$ on $\langle W, \sigma_0\rangle$ such that $\varphi w_T = w$ for all $w \varepsilon W$ and $\varphi\sigma_T = \sigma_0$. For if we consider the natural homomorphisms of $\langle W', \sigma\rangle$ onto these two groups, the first by restriction to T, the second by extraction of linear parts, then the kernel of the first, L, is clearly contained in that of the second. This enables us to shift from a group on T to a linear group on V, and thus to complete the proof of 3.10.

3.18. *If M is a $\langle W, \sigma_0\rangle$-module and I, I_π are as in 2.6, then*

$$\Sigma\varepsilon_\sigma(\pi)tr(\sigma_0, I_\pi(M)) = -\varepsilon_\sigma(\Pi')\Sigma(-1)^i tr(\sigma_0, I(M \otimes E_i)).$$

In 3.17 we choose for χ the character of M converted into a $\langle W_T, \sigma_T\rangle$-module with the aid of the homomorphism φ above, write $\det(w - \sigma_0^{-1})$ in the form 3.12(c), and then average over $w \varepsilon W$. The result is 3.18.

3.19. *For π as above let $P_{\pi\sigma}(t)$ be as in 2.7 and $Q_\sigma(t)$ as in 3.10. Then $\Sigma\varepsilon_\sigma(\pi)/P_{\pi\sigma}(t) = -\varepsilon_\sigma(\Pi')/Q_\sigma(t)$.*

If the d's and ε's are as in 3.10 the identity to be proved may be written

$$((*)) \quad \sum \varepsilon_\sigma(\pi)\Pi(1 - \epsilon_{\pi j}t^{d(\pi j)})^{-1} = -\varepsilon_\sigma(\Pi')\Pi(1 - \epsilon_j t^{d(j)-1})/(1 - \varepsilon_j t^{d(j)}).$$

Here the products are over j from 1 to n. Now Solomon [21] has shown that if S is identified with the algebra of polynomials on V and $S \otimes E$ with the algebra of differential forms with polynomial coefficients then $I(S\otimes E)$ is an exterior algebra over $I(S)$ freely generated by the differentials of a basic set of generators of $I(S)$. This leads to the formal identity in t and u: $\Pi(1 + \varepsilon_j t^{d(j)-1}u)/(1 - \varepsilon_j t^{d(j)}) = \Sigma_{k,i} tr(\sigma_0, I(S_k \otimes E_i))t^k u^i$. Setting $u = -1$, we see that the coefficient of t^k on the right side of (*) is $-\varepsilon_\sigma(\Pi')\Sigma(-1)^i tr(\sigma_0, I(S_k \otimes E_i))$. As in 2.7 the coefficient on the left is $\Sigma\varepsilon_\sigma(\pi)tr(\sigma_0, I_\pi(S_k))$. By 3.18 with $M = S_k$ these two quantities are equal, whence (*) and 3.19.

3.20. Proof of 3.10. . We apply 1.25 with W' and Π' in place of W and Π. Isolating the term $\pi = \Pi'$ there and bearing in mind the present convention on π, we get (*) $\Sigma\varepsilon_\sigma(\pi)/W'_{\pi\sigma}(t) = -\varepsilon_\sigma(\Pi')/W'_\sigma(t)$. Now W'_π and σ fix a common point of V, e.g., the centroid of F_π. Thus $W'_{\pi\sigma}(t)$ as constructed from W'_π and σ is equal to the corresponding object constructed from W_π and σ_0, which by 2.1 applied to W_π is equal to $P_{\pi\sigma}(t)$. Comparing (*) and 3.19 we get $W'_\sigma(t) = Q_\sigma(t)$ as required.

3.21. **Remarks**. (a) One can extend 3.17 and 3.18 along the lines of 1.28. (b) Assume that σ in 3.10 is linear, so that it fixes both Π and Π' and we may form the two twisted systems $S = \{V_\sigma, \Pi_\sigma, \Sigma_\sigma, W_\sigma\}$ and $S' = \{V'_\sigma, \Pi'_\sigma, \Sigma'_\sigma, W'_\sigma\}$ as in 1.32. Then we claim that S satisfies the conditions 1.1 to 1.6, 3.1 and 3.2, and that S' is its affine extension. Since σ is linear, $\sigma(n+\alpha) = n+\sigma\alpha$ for $n+\alpha \in \Sigma'$, whence (*) $\Sigma'_\sigma = \mathbb{Z}+\Sigma_\sigma$ by 1.33 ($\mathbb{Z}$ = integers). Now if $\alpha, \beta \in \Sigma_\sigma$, then $-2(\beta,\alpha)/(\alpha,\alpha) + w_\alpha\beta = w_{1+\alpha}\beta \in \Sigma'_\sigma$ by (*), whence $2(\beta,\alpha)/(\alpha,\alpha)$ is an integer again by (*). Thus 3.2 holds for Σ_σ. The other points necessary to establish our claim are now easily verified. (c) There is a rather curious corollary to 3.12 or 3.17.

3.22. *Assume that σ is chosen so that $-1 \in W\sigma$.* (a) *If π is a proper σ-invariant part of Π', then $-1 \in W_\pi\sigma$ if and only if $-1 \in W_{\pi\sigma}|V_\sigma$.* (b) *If $n = \dim V$ and $k = \dim V_\sigma$ then $\Sigma|W/W_\pi| = 2^n$ and $\Sigma|W_\sigma/W_{\pi\sigma}| = 2^k$, both sums over those π for which* (a) *holds.*

Assume, for example, that W is of type E_6. Then σ is of order 2 and W_σ is of type F_4 as we see from Dynkin diagrams. The groups W_π of 3.22 are of types $E_6, A_5 \times A_1$ and D_5, and the corresponding groups $W_{\pi\sigma}$ of type $F_4, C_3 \times A_1$ and B_4. The equations read $1+36+27 = 2^6$ and $1+12+3 = 2^4$. If W is of type C_n, on the other hand, then the equations coincide and express 2^n as a sum of binomial coefficients in the usual way.

For the proof of 3.22 we observe first that σ is necessarily linear. We have $\sigma^2 = 1$ so that σ acts as -1 on the orthogonal complement of V_σ. Since also $-1 \in W_\pi\sigma$ if and only if $-1 \in W_{\pi\sigma}\sigma$ because -1 commutes with σ, we have (a). To prove (b) we consider -1 acting on T. The set of fixed points is finite and consists of $\det(1-(-1)) = 2^n$ points. These are just the centroids of the fixed cells in the complex K, for these centroids are clearly fixed and no cell could contain two fixed points since then the segment joining them would consist of an infinite number of fixed points. Now the collection of fixed cells consists of complete W-orbits and the orbit of T_π (as in 3.15) contains $|W/W_\pi|$ elements by 3.15. Thus $2^n = \Sigma|W/W_\pi|$ summed over the proper parts π of Π' such that -1 fixes T_π. Now if $-1 = w\sigma$, then -1 fixes T_π if and only if both W and σ do (cf. proof of 2.3), i.e., π is as in (a). Thus the first equation of (b) holds. Then so does the second since we may replace W by $W_\sigma|V_\sigma$ and σ by 1 and use 3.21(b).

4. Reflection groups, lattices and tori

The lemma 3.15 has the following consequence.

4.1. *The stabilizer in W of each point of T is a reflection group.*

Since each point of T is congruent under W to a point of some T_π, this is clear.

The purpose of this section is to extract the essence of this result and to provide a converse. The focus is on tori and reflection groups thereon. (Each reflection is relative to some subtorus of codimension one, a Euclidean metric being assumed).

4.2. *Let T be a torus, $p : V \to T$ the universal covering of T, and L the (lattice) kernel of p, so that $T = V/L$. Let W be a finite reflection group acting effectively on T and L_0 the sublattice of L generated by the translations orthogonal to the reflecting hyperplanes. Assume* (a) $L = L_0$. *Then* (b) *the stabilizer in W of each point t of T is a reflection subgroup.*

This follows from 3.6 (see (1) and (3)) and 4.1.

4.3. **Remarks**. (a) It follows from 3.6 that (a) is equivalent to: the inverse image of W in the affine group of V, viz. LW, is a reflection group. (b) If W is of type A_1, E_8, F_4 or G_2, the condition (a) always holds. (c) if W is not effective, (a) has to be replaced by (*) L/L_0 is torsion free. Even in this form 4.2 fails if in (b) we take more than one point (cf. 1.20) [although it does hold in some cases, e.g., if Σ (see 3.6) is of type A_n or C_n]. The reason is that (*) may fail to hold when W is replaced by a reflection subgroup.

4.4. *In* 4.2 *drop the assumption that* (a) *holds. Let W_t be the group in* (b) *and W_{t0} the subgroup generated by the reflections in W_t.* (a) *W_t/W_{t0} is isomorphic to a subgroup of L/L_0.* (b) *Every subgroup of L/L_0 can be realized in* (a).

Observe that L/L_0 is Abelian and isomorphic to LW/L_0W, the quotient of the complete inverse image of W by its largest reflection subgroup (3.6 again). Clearly (a) implies 4.2 and (b) its converse. The proof depends on the following result.

4.5. *Let A be an abstract group, C a central subgroup of A; let B denote the group A/C and π the natural map from A to B. Let σ be an automorphism of A which fixes C. Then we have an exact sequence*

$$A_\sigma \xrightarrow{\pi} B_\sigma \xrightarrow{\delta} ((1-\sigma)A \cap C)/(1-\sigma)C \longrightarrow 1.$$

Here $\delta = \rho \circ (1-\sigma) \circ \pi^{-1}$, ρ being the natural map from C to $C/(1-\sigma)C$.

The relevant points, that δ is well defined, that δ is a homomorphism, and that the sequence is exact, are easily verified.

Returning to 4.4, we let $T^0 = V/L_0$ and write W^0 for W acting on T^0. We apply 4.5 with $A = T^0W^0$ (semidirect product), $B = TW$, $\pi : A \to B$ the natural map, and σ the inner automorphism of A effected by an element t^0 of T^0 such that $\pi t^0 = t$. To do this we must show that the kernel C of π is central, which amounts to: $(1-w)L \subseteq L_0$ because of the natural identification of C and L/L_0. If w is a reflection this follows from the definition of L_0; if w is arbitrary, we write $w = w'w''$ with each of w' and w'' a product of a smaller number of reflections and then use the equation $1 - w = 1 - w' + w'(1 - w'')$. Now $A_\sigma = T^0W^0_{t^0}$, $B_\sigma = TW_t$, $\pi T^0 = T$, and σ acts trivially on C. Thus 4.5 yields: $W_t/\pi W^0_{t^0}$ is isomorphic to a subgroup of L/L_0. Now $W^0_{t^0}$ is a reflection group by 4.2; hence $\pi W^0_{t^0}$ is also, is in fact the largest reflection subgroup W_{t0} of W_t: if w is a reflection in W_t, it fixes t, so that if v is chosen in V above t then $(1-w)v$ is in L, hence in L_0 by the definition of L_0, whence w^0 fixes t^0 and $w \,\varepsilon\, \pi W^0_{t^0}$. From this (a) follows. The subgroup in (a) is just $(1-\sigma)W^0 \cap C$. Because of the definitions, (b) follows from:

4.6. *Assume as in* 4.2 *that* $L = L_0$. *Let* C *denote the group of fixed points of* W *on* T, *and* C_0 *any subgroup of* C. *Then there exists* $t \,\varepsilon\, T$ *so that*

$$\{(1-w)t | w \,\varepsilon\, W\} \cap C = C_0.$$

For $c \,\varepsilon\, C$ choose $v_c \,\varepsilon\, p^{-1}c$. As is well known:

4.7. *The condition* $c \,\varepsilon\, C$ *is equivalent to:* (v_c, α) *is an integer for every* $\alpha \,\varepsilon\, \Sigma$ *(see* 3.6*).*

For the last condition is equivalent to: $(1 - w_\alpha)v_c$ is a multiple of λ_α by 1.5. Now if F is as in 3.9(b) then $wv_c \,\varepsilon\, F$ for some $w \,\varepsilon\, W'$ (see 3.8). Since $pwv_c = wpv_c = wc = c$, we may assume $v_c \,\varepsilon\, F$, a vertex of F by 4.7. In particular, C is finite. Going back to T we see that the elements of C_0 are the vertices of one of the cells T_π of 3.15. Let t be its centroid. For a fixed $c \,\varepsilon\, C$ the equation $(1-w)t = c$ has a solution in W exactly when $t = wt + c$ does, i.e., when $T_\pi = wT_\pi + c$ does. Since C_0, the group of vertices of T_π, is central, this last condition can hold only if $C_0 = C_0 + c$, i.e., $c \,\varepsilon\, C_0$. Conversely, if $c \,\varepsilon\, C_0$ then translation by c is a simplicial mapping of the complex K of §3, so that $T_\pi - c = wT_{\pi'}$ for some $w \,\varepsilon\, W$ and $\pi' \subseteq \Pi'$, then $\pi' = \pi$ as we see by checking vertices, and the above condition holds. Thus 4.6 follows.

Now for t as above we have $|C_0|^2 t = 0$. Thus:

4.8. *In* 4.6 *we may choose* t *to be of finite order.*

4.9. *Assume T, W, etc. are as in* 4.2 *and that* (a) *holds. Let σ be an automorphism of T fixing a fundamental cell for W, hence normalizing W. Then W_σ acts as an effective reflection group on $T_{\sigma 0}$ and on $T/(1-\sigma)T$, and in both cases* 4.2(a) *holds.*

Lifting σ to a linear automorphism of V and using 1.32, we see that $W_\sigma|V_\sigma$ is an effective reflection group. Now pV_σ is a closed connected subgroup of T_σ for which the quotient is discrete, by 4.5 with $A = V$, $B + T$, $C + L$, $\pi = p$; hence it is equal to $T_{\sigma 0}$. Thus $p : V_\sigma \to T_{\sigma 0}$ is the universal covering. The kernel L_σ consists of all translations $\lambda = \Sigma n_\alpha \lambda_\alpha$ ($\alpha \varepsilon \Pi, \lambda_\alpha$ as in 3.6, n_α integral, $n_{\sigma\alpha} = n_\alpha$), hence is generated by the sums of the λ_α's over the various σ-orbits on Π, i.e., by multiples of the elements of Π_σ (see 1.32). Thus 4.9 holds in this case. Now consider the composite natural map $q : V \to T \to T/(1-\sigma)T$. Since $(1-\sigma)V$ is in the kernel and V_σ is its orthogonal complement (since W and σ generate a finite group we may assume σ is orthogonal), the restriction $q : V_\sigma \to T/(1-\sigma)T$ is onto. The kernel, $V_\sigma \cap (1-\sigma)V + L$, is just the orthogonal projection of L on V_σ, hence is a lattice generated by the projections of the basic translations λ_α ($\alpha \varepsilon \Pi$), so that (cf. 1.30) 4.9 holds in this case also.

4.10. *Let everything be as in* 4.9 *and $t \varepsilon T$. Then those $w \varepsilon W_\sigma$ for which $(1-w)t \varepsilon (1-\sigma)T$ form a reflection subgroup.*

If we project t onto $\bar{t} \varepsilon T/(1-\sigma)T$, the condition becomes $w\bar{t} = \bar{t}$, so that 4.10 follows from 4.2 and 4.9.

In the same way we can extend 4.6.

4.11. *Let $T, W, L = L_0$ be as in 4.2, C as in 4.6, σ as in 4.9, and C_0 a subgroup of C containing $(1-\sigma)C$. Then there exists $t \varepsilon T$ such that $\{(1-w)t|w \varepsilon W_\sigma\}(1-\sigma)T \cap C = C_0$.*

Theorem 4.6 holds with $T/(1-\sigma)T$ and W_σ in place of T and W, by 4.9. Thus there exists $t \varepsilon T$ such that the left side of the desired equation equals $C_0((1-\sigma)T \cap C)$. Thus 4.11 follows from:

4.12. *If T, C and σ are as above, then $(1-\sigma)T \cap C = (1-\sigma)C$.*

The character group X of T may be identified with the set of linear functions on V which are integral on L. As such it is generated by the dual basis to $\{2\alpha/(\alpha,\alpha)|\alpha \varepsilon \Pi\}$, i.e. by $\{\omega_\alpha|(2\omega_\alpha,\beta)/(\beta,\beta) = \delta_{\alpha\beta}; \alpha,\beta \varepsilon \Pi\}$, and contains Σ and Π (see 3.6), via the usual identification of V and its dual. If $t \varepsilon T$, we have the equivalent conditions: $t \varepsilon C; w_\alpha t = t$ for all $\alpha \varepsilon \Pi$, by 1.6; $\omega_\beta((1-w_\alpha)t) = 0$ for all $\alpha, \beta \varepsilon \Pi$; $\alpha(t) = 0$ for all $\alpha \varepsilon \Pi$, by 1.5 and the definition of ω_β. Now assume $t \varepsilon T$ such that $(1-\sigma)t \varepsilon C$. Then $\alpha((1-\sigma)t) = 0$ for all $\alpha \varepsilon \Pi$, i.e.,

$\alpha(t)$ is constant on each orbit of σ on Π. Now orbit representatives project onto linearly independent elements of V_σ (the projections have disjoint supports in Π), hence restrict to linearly independent characters on $T_{\sigma 0}$ (see also the proof of 4.9). By the elementary divisor theorem [31, II, p. 107] applied to the lattice generated by these characters and the lattice of all characters on $T_{\sigma 0}$, there exists $t_0 \varepsilon T_\sigma$ such that $\alpha(t_0) = \alpha(t)$ for every orbit representative α, i.e. for every $\alpha \varepsilon \Pi$ since $t_0 \varepsilon T_\sigma$. Thus $tt_0^{-1} \varepsilon C$ and $(1-\sigma)t \varepsilon (1-\sigma)C$, whence 4.12 follows.

5. The transition to algebraic tori

Our purpose is to carry over the results of the preceding section to algebraic tori. Let K be an algebraically closed field and p its characteristic exponent (see §6). An algebraic torus shall mean an algebraic group isomorphic to the product of a finite number of copies of the multiplicative group K^*. In what follows T^a will be an algebraic torus, L its (lattice) group of one-parameter subgroups, i.e., homomorphisms of K^* into T^a, V the extension of L to a real vector space, and T the torus V/L. The group X of characters of T^a, i.e., homomorphisms of T^a into K^*, is a lattice in $\mathbb{Z}$-duality with L (see [19, p. 405]), by extension in $\mathbb{R}$-duality with V, and finally in $\mathbb{R}/\mathbb{Z}$-duality with T. Observe that X is in duality with both T^a (algebraically) and T (topologically).

5.1. (a) *For each $t^a \varepsilon T^a$ there exists $t \varepsilon T$ such that if $\chi \varepsilon X$ then $\chi(t^a) = 0$ if and only if $\chi(t) = 0$.* (b) *Every $t \varepsilon T$ of finite order prime to p can be realized in* (a).

This basic lemma and its proof are due to T. A. Springer. Let X_0 be the annihilator of t^a in X and T_0 the annihilator of X_0 in T. The group X/X_0 is a finitely-generated Abelian group, hence is isomorphic to a direct product of a lattice and a finite group. Since it is also isomorphic to a subgroup of K^* (to the group of values $X(t^a)$), the finite group is cyclic [31, I, p. 112]. Thus T_0, the dual of X/X_0, is the product of a torus and a cyclic group, hence it has a generator, say t, as a closed subgroup of T (see [13, p. 136]). The annihilator of t, i.e., of T_0, in X is X_0 by topological duality, so that (a) holds. If t is as in (b), we can reverse the procedure and define t^a as a generator of the annihilator in T^a of the annihilator X_0 of t in X. The assumptions on t imply that X/X_0 is finite of order prime to p, so that X_0 is the annihilator of t^a and (b) follows.

5.2. **Remarks**. (a) The various assertions about annihilators and duality become transparent once compatible bases of X and X_0 are chosen [31, II, p. 107]. (b) In case K is not the algebraic closure of a finite field, one need only assume in 5.1(b) that the closed subgroup generated by t taken modulo its identity component has order prime to p.

Each automorphism, in fact endomorphism, of T^a acts also on L, on V and on $T = V/L$. A group W of automorphisms of T^a will be called a reflection group if it is so on V, and similarly for an automorphism fixing a fundamental cell for W.

5.3. THEOREM. *The results* 4.2, 4.6, 4.10, 4.11, 4.12 *hold if T is replaced by an algebraic torus T^a and if L and V are defined as above.*

Let the required results be labeled 4.2′, 4.6′, etc. The proof depends on the following extension of 5.1.

5.4. *If t^a and t are as in* 5.1, *if α is any endomorphism of T^a, and if $\chi \in X$, then $\chi(\alpha(t^a)) = 0$ if and only if $\chi(\alpha(t)) = 0$.*

This follows from 5.1 with χ replaced by $\chi \circ \alpha$.

Consider now 4.2′. Assume $t^a \in T^a$. Choose t as in 5.1(a). For $w \in W$, the condition $wt^a = t^a$ may be written $(w-1)t^a = 0$, hence is equivalent to $wt = t$ by 5.1. Thus 4.2′ follows from 4.2. Now let C^a be defined for T^a as C is for T (see 4.6), and let C_0^a be any subgroup of C^a. Let X_0 be the annihilator of C_0^a in X and C_0 the annihilator of X_0 in T. Since C_0^a is of finite order prime to p (it is a subgroup of some K^{*n}), so is X/X_0, and then also C_0. (In fact the three groups are isomorphic). Since $(1-w)C_0^a = 0$ for all $w \in W$, it follows that $(1-w^{-1})X \subseteq X_0$ and then that $(1-w)C_0 = 0$, so that C_0 is a subgroup of C. Choose $t \in T$ as in 4.6. By 4.8 we may assume t is of finite order, and since the order of C_0 is prime to p we may replace t by an appropriate t^{p^k} and assume further that the order of t is prime to p. Now choose t^a as in 5.1(b). For $w \in W$ the condition $(1-w)t \in C$ is equivalent to $(1-w')(1-w)t = 0$ for all $w' \in W$, hence to $(1-w)t^a \in C^a$ by 5.4. Thus, again by 5.4, if we write C_1 for $\{(1-w)t | w \in W\} \cap C$ and C_1^a for the corresponding group in T^a, then C_1 and C_1^a have the same annihilator in X, which is in fact X_0 since $C_1 = C_0$, the annihilator of X_0. Thus X_0 is the annihilator in X of both C_0^a and C_1^a. Being finite, each of the latter groups is conversely the annihilator of X_0 in T^a. Hence they are equal, which yields 4.6′. In the same way we can deduce 4.10′, 4.11′, and 4.12′ from 4.10, 4.11 and 4.12. Alternatively, we can parallel the proofs given in §4. We omit the details.

5.5. **Exercise**. Assume T^a and W as above, that $L = L_0$, and that Σ in 3.6 is identified with a subset of X in the natural way. If σ is an automorphism of T^a which fixes Σ and some $t^a \in T^a$ such that $\alpha(t^a) \neq 1$ for every $\alpha \in \Sigma$, then σ fixes a fundamental chamber for W. (Hint: use 5.4.)

6. Structure of algebraic groups

In this section we recall the main properties of linear algebraic groups and introduce some notations which will be used henceforth. The basic source of this material is [19]. K will be an algebraically closed field, over which all algebraic groups will be taken. K^* will be the multiplicative group of K and p its characteristic exponent (which is the characteristic of K or 1 according as the characteristic differs from 0 or not). Let G be a connected (linear algebraic) group. Then G has a unique maximal connected solvable normal subgroup R, called the radical of G. If G/R is given the structure of an algebraic group, it has a trivial radical, i.e., is semisimple. The semisimple groups turn out to be products, with some amalgamation of (finite) centers, of simple groups, which have been classified by Chevalley [19] in the Killing-Cartan tradition (types $A_n, B_n, \ldots, G_2$). Assume now that G is a semisimple group in which a maximal torus T has been chosen. X will always denote the (discrete) character group of T, written additively most (but not all) of the time, Σ the subset of roots, and X_α the one-parameter unipotent subgroup corresponding to the root α. We have:

6.1. *X_α is unipotent and isomorphic (as an algebraic group) to the additive group (of K). If x_α is an isomorphism from K to X_α, then $tx_\alpha(k)t^{-1} = x_\alpha(a(t)k)$ for all t and k.*

N will denote the normalizer of T and $W = N/T$ the Weyl group. As is known W acts faithfully on T as a reflection group in the sense of §5 with Σ as a corresponding root system, so that the notations w_α, etc. may be used. Assume now that an ordering of Σ has also been given. Then U (resp. U^-) will denote the group generated by those X_α for which α is positive (resp. negative), and B the group generated by T and U.

6.2. (a) *U is a maximal unipotent subgroup of G, and B is a Borel (maximal connected solvable) subgroup of G.* (b) *The natural maps from the Cartesian product $\Pi X_\alpha(\alpha > 0)$ (arbitrary order of the factors) to U and from $T \times U$ to B are isomorphisms of varieties.* (c) *The natural map from $U^- \times T \times U$ to G is an isomorphism onto an open subvariety of G.*

6.3. *Let $\{n_w | w \in W\}$ be a system of representatives for W, i.e., for N/T, in N.* (a) *$\{n_w\}$ is also a system of representatives for the double cosets of G relative to B.* (b) *If we set $U_w = U \cap n_w^{-1} U^- n_w$, then each element of Bn_wB can be written uniquely $u_1 n u_2$ with $u_1 \in U$, $n \in n_w T$, $u_2 \in U_w$.*

In terms of a W-invariant inner product on X and its extension to a real vector space V, we have the clearly equivalent conditions, each of which may be taken as a definition:

6.4. *The semisimple group G is simply connected if one of the following holds.*

(a) *If L is the group of one-parameter subgroups of T and L_0 is the subgroup generated by $\{2\alpha/(\alpha,\alpha)|\alpha \varepsilon \Sigma\}$, then $L = L_0$* (b) *There exists a basis $\{\omega_\alpha|\alpha$ simple root$\}$ of X so that $(2\omega_\alpha,\beta)/(\beta,\beta) = \delta_{\alpha\beta}$ or, equivalently, so that $w_\alpha\omega_\beta = \omega_\beta - \delta_{\alpha\beta}\alpha$ for all simple roots α and β.*

6.5. **Reductive groups**. A connected group G is called reductive if its radical is a torus, or, equivalently (see [3, p. 63]) if it can be decomposed $G = G'T'$ with G' a semisimple group and T' a central torus. In this decomposition T' is uniquely determined as the identity component of the center of G, and G' as the derived group of G or else as the largest semisimple subgroup.

7. Endomorphisms of algebraic groups

Our purpose is to prove 7.2 and 7.5 below. Let G be a linear algebraic group and σ an endomorphism of G. We want to study G_σ. Now G, σG, $\sigma^2 G, \ldots$ is a decreasing sequence of closed subgroups [19, p. 304], hence must level off: $\sigma^n G = \sigma^{n+1} G$ for some n. Since all these groups contain G_σ, we may as well assume, after replacing G by $\sigma^n G$, that σ maps G onto G. Here are some elementary results which hold in this case.

7.1. *Let G be a linear algebraic group and σ an endomorphism of G onto G.* (a) *Ker σ is finite and contained in the center of G_0. If the center of G_0 is finite, in particular if G_0 is semisimple, then ker $\sigma = 1$, so that σ is an automorphism of the abstract group underlying G.* (b) *If H is a subgroup of G, then* $\dim \sigma H = \dim H$. *If H is connected and $\sigma H \subseteq H$, then $\sigma H = H$. If H is normal in G, then so is σH.* (c) *If R is the radical of G, then $\sigma R = R$.*

Ker σ has dimension 0 by [19, p. 304], hence is finite. Now ker σ is contained in G_0, since otherwise σG would have fewer connected components than G. Since ker σ is a discrete normal subgroup of the connected group G_0, it must be central. Since σ is onto, it follows that for every positive integer n the map σ : ker $\sigma^n \to$ ker σ^{n-1} is onto, so that $|\text{ker } \sigma^n| = |\text{ker } \sigma|^n$. Since ker σ^n is contained in the center of G_0, it follows that if the latter group is finite then $|\text{ker } \sigma| = 1$, which proves (a). Since ker σ is finite $\dim \sigma H = \dim H$ (loc. cit.). Ths implies that if also $\sigma H \subseteq H$ then σH is of finite index in H, hence is equal to H in case H is connected. If H is normal and x is arbitrary in G, we may write $x = \sigma y$ and $x(\sigma H)x^{-1} = (\sigma y)(\sigma H)(\sigma y^{-1}) = \sigma H$, so that σH is also normal. Since σR is a solvable connected subgroup which is normal in G by a part of (b), we have $\sigma R \subseteq R$, and equality holds by another part of (b).

Our main aim is the proof of the following result.

7.2. THEOREM. *Let G be a linear algebraic group and σ an endomorphism of G onto G, then σ fixes a Borel subgroup of G.*

This follows from:

7.3. LEMMA. *Let G be a connected linear algebraic group, σ an endomorphism of G onto G, and B a Borel subgroup fixed by σ. Then the map $\alpha : G \times B \to G$ defined by $\alpha(x, b) = xb\sigma x^{-1}$ is surjective.*

If σ is the identity, this reduces to the fact that every element of G is conjugate to some element of B. Assuming the lemma for the moment, let G and σ be as in 7.2 and let B be a Borel subgroup of G. Then σB is also a Borel subgroup by 7.1, so that (*) $y \cdot \sigma B \cdot y^{-1} = B$ for some y in G [16, p. 609]. Applying 7.3 to $i_y\sigma$ we get $y^{-1} = xby \cdot \sigma x^{-1} \cdot y^{-1}$ for some $x \,\varepsilon\, G$ and $b \,\varepsilon B$, whence $y = b^{-1}x^{-1}\sigma x$. Substituting into (*), we see that σ fixes the Borel subgroup xBx^{-1}, which proves 7.2, mod 7.3.

Since the radical of G is a normal subgroup contained in B, we need only prove 7.3 when G is semisimple, which we henceforth assume. We first show that the image of α is closed. The subset of $G \times G$ consisting of all (x, y) such that $x^{-1}y\sigma x \,\varepsilon\, B$ is closed and consists of complete cosets relative to the subgroup $B \times 1$; hence its projection in $G/B \times G$, call it S, is closed. Since G/B is complete [19, p.609], the projection of S on the second factor, i.e., the image of α, is also closed. Next we show that at some point the differential $d\alpha$ is onto. Let $\mathbf{g}$ be the Lie algebra of G and $\mathbf{b}$ that of B. We identify $\mathbf{g}$ with the tangent space to G at 1. A simple calculation shows that $d\alpha$ followed by translation to 1 in G maps the tangent space to $(1, b)$ in $G \times B$ onto $\mathbf{b} + (1 - d(i_b\sigma))\mathbf{g}$. We can choose b so that $i_b\sigma$ fixes a maximal torus T in B (by the conjugacy theorem [19, p. 609]). The negative roots are then permuted in orbits, so that by modifying b by a sufficiently general element of T we can assume that $i_b\sigma$ fixes no element in the Lie algebra $\mathbf{u}^-$ of U^- (see §6). Thus the above map covers $\mathbf{b} + \mathbf{u}^- = \mathbf{g}$, as required. By the first part the image of α is closed, by the second it contains a nonempty open subset of G, which is dense because G is connected; hence it is all of G, which is 7.3.

7.4 COROLLARY. *If G and σ are as in 7.2, then any connected solvable subgroup of G_σ is contained in a Borel subgroup of G fixed by σ.*

The subgroup acts by conjugation on the variety of Borel subgroups fixed by σ, which is nonempty by 7.2 and complete by [19, p. 609]. Thus by Borel's theorem [19, p. 514] there is a fixed point, whence 7.4.

Next we present a related result which will be used later. An automorphism of an algebraic group G will be called semisimple if it can be achieved by conjugation by a semisimple element of some algebraic group containing G. If G is semisimple, then its group of outer automorphisms is finite so that every automorphism σ can be realized by conjugation in a larger group, and the condition that σ is semisimple is equivalent to each of: (1) σ acts semisimply (i.e. diagonally) on the Lie algebra of G; (2) σ^k is a semisimple inner automorphism for some integer k prime to p.

7.5. THEOREM. *Every semisimple automorphism σ of an algebraic group G fixes a Borel subgroup and a maximal torus thereof.*

Because of 7.2 this follows from:

7.6. *If G in 7.5 is solvable, then σ fixes a maximal torus.*

We may assume G connected, write $G = TU$ so that T is a maximal torus and U is the normal subgroup consisting of all unipotent elements of G [19, p. 606], and extend G so that σ is realized by conjugation by the semisimple element s. Since sTs^{-1} and T are maximal tori in G, we have $usTsu^{-1} = T$ for some $u \in U$ (by conjugacy), and on replacing us by its semisimple part we may assume it to be semisimple. To prove that σ fixes a maximal torus, i.e., that i_s does, it is thus enough to show (*) if us is semisimple it is conjugate to s. By an obvious induction on the length of the derived series of U, we can assume that U is commutative. In this case $u = u_1^{-1}u_2su_1s^{-1}$ with $u_1 \in U$ and $u_2 \in U_s$ [19, p. 412], so that us is conjugate to u_2s. The last element being semisimple we must have $u_2 = 1$, which proves (*) and 7.6.

7.7. **Remark**. This result may be viewed as an extension to disconnected groups of the fact that every semisimple element of a connected algebraic group is contained in a maximal torus. It has also been proved by Winter [33] by a different method.

8. A basic theorem on connectedness

This section is devoted to the proof of the follow result.

8.1. THEOREM. *Let G be a simply connected semisimple algebraic group and σ a semisimple automorphism of G. Then G_σ is a connected reductive group.*

Because of 7.5 this is a consequence of the following result, which is slightly more general because σ need not be semisimple.

8.2. THEOREM. *Let G be as in 8.1 and σ an automorphism of G which fixes a Borel subgroup B and a maximal torus T of B. Then G_σ is a connected reductive group.*

Since the simple components of G are all simply connected (see, e.g., 6.4 and the discussion at the start of §6) and σ permutes them, we may assume that G itself is simple. We use the notations of §6 associated with B and T. Since σ fixes T it also acts on N/T, i.e., on W, on X, and on Σ, permuting the positive roots since it fixes B, and for each root α we have $\sigma x_{\sigma\alpha}(k) = x_\alpha(c_{\sigma\alpha}k)$ for all $k \varepsilon K$ and some $c_\alpha \varepsilon K^*$.

(1) G_σ *is generated* by U_σ *and* N_σ. This follows from the uniqueness in 6.3.

(2) U_σ *is connected.* For each positive root α we form the set S_α of all (positive) roots whose projections on V_σ (see 1.22) are proportional to that of α. Using 1.32(b), the fact that σ preserves the set of positive roots, and the classification of simple root systems [20, p. 1308], we conclude:

(2′) S_α is of one of the following types: (a) an orbit $\{a, \sigma\alpha, \dots\}$ of roots no two of which add up to a root; (b) a set of the form α, $\sigma\alpha$, $\beta = \alpha + \sigma\alpha$, which occurs only if Σ is of type A_{2n}.

(2″) In both cases above $\{\alpha, \sigma\alpha, \dots\}$ is part of a simple system of roots fixed by σ.

(2‴) The corresponding reflection in W_σ (see 1.30) is $w_\alpha w_{\sigma\alpha} \cdots$ in the first case and w_β in the second.

¿From 6.2(b) it follows that if we form the parts of U_σ supported by the various S_α's and take their product we get U_σ. Thus (2) follows from:

(2⁗) *The part of* U_σ *supported by each* S_α *is connected. It is different from* 1 *if and only if in case* (2′a) $c_\alpha c_{\sigma\alpha} \cdots = 1$ *and in case* (2′b) $c_\alpha c_{\sigma\alpha} = \pm 1$, *and then it is a one-parameter group.* In case (a) the elements $x_\alpha(k_\alpha)x_{\sigma\alpha}(k_{\sigma\alpha}) \cdots$ fixed by σ are determined by the equations $k_\gamma = c_{\sigma\gamma}k_{\sigma\gamma} (\gamma = \alpha, \sigma\alpha, \dots)$, which have a nonzero solution for $k_\alpha, k_{\sigma\alpha}, \dots$ only if $c_\alpha c_{\sigma\alpha} \cdots = 1$, in which case k_α may be chosen arbitrarily and the other k's solved for. In case (b) we adjust the homomorphism x_β so that the commutator relation $(x_\alpha(k), x_{\sigma\alpha}(\ell)) = x_\beta(k\ell)$ holds. Because of the equation $c_\alpha c_{\sigma\alpha} = -c_\beta$ which follows from this relation, the conditions for $x_\alpha(k_\alpha)x_{\sigma\alpha}(k_{\sigma\alpha})x_\beta(k_\beta)$ to be fixed by σ are: $k_\alpha = c_{\sigma\alpha}k_{\sigma\alpha}$, $k_{\sigma\alpha} = c_\alpha k_\alpha$, $k_\alpha k_{\sigma\alpha} + (1 + c_\alpha c_{\sigma\alpha})k_\beta = 0$. If $c_\alpha c_{\sigma\alpha} \neq \pm 1$, then all k's must be 0. If $c_\alpha c_{\sigma\alpha} = 1$ and $p \neq 2$ then k_α may be chosen arbitrarily and the other k's solved for. Finally, if $c_\alpha c_{\sigma\alpha} = -1$ (here $p = 2$ is allowed), then k_β becomes arbitrary and the other k's 0, whence (2⁗).

(3) T_σ *is connected.* This is because σ permutes the elements of the basis $\{\omega_\alpha\}$ of X.

(4) *If* w *is a reflection of* W_σ *such that the corresponding set* S_α *of* (2′) *supports some* $u \varepsilon U_\sigma - \{1\}$ *(see 2⁗), then* w *is represented in* N *by an element of* $G_{\sigma 0}$, *in fact by an element of* $U_\sigma^- U_\sigma U_\sigma^-$. By 6.3 with U^- and U interchanged, $u = u_1 n u_2$ with $u_1, u_2 \varepsilon U_\sigma^-$, and by the uniqueness there, all comonents u_1, n, u_2 are in G_σ. The element of W represented by n is different from 1, is in the group generated by $w_\alpha, w_{\sigma\alpha}, \dots$, and is fixed by σ, hence is w, as required.

(5) *Choose* $t \varepsilon T$ *so that* $\alpha(t) = c_{\sigma\alpha}$ *for every simple root* α. *Then* N_σ *is generated by* T_σ *and representatives in* N *of those* $w \varepsilon W_\sigma$ *such that* $(1-w)t \varepsilon (1-\sigma)T$. First each $n \varepsilon N_\sigma$ represents a $w \varepsilon W_\sigma$. Second if we define ρ by $\sigma = i_t\rho$, so that $W_\sigma = W_\rho$, then each $w \varepsilon W_\sigma$ is represented by an element

of N_ρ. By 1.32 it is enough to prove this when w has the form w_π, i.e., when $\alpha, \sigma\alpha, \ldots$ of (2′) are all simple. But then our assertion follows form (2) and (4) with σ replaced by ρ and the c's by 1. Now if $n \varepsilon N_\rho$ represents $w \varepsilon W$, then $t_1 n \varepsilon N_\sigma$ for some $t_1 \varepsilon T$ if and only if $(\sigma - 1)n \varepsilon (1-\sigma)T$, which works out to $(1-w)t \varepsilon (1-\sigma)T$ and proves (5).

(6) *Every $w \varepsilon W_\sigma$ which is represented in N_σ is a product of reflections (in the sense of* 1.32*) which also are.* The results of §5 are applicable to the torus T. If the lattice L of one-parameter subgroups of T and its dual X are embedded in a real vector space in the usual way, then the basis of L dual to the basis $\{\omega_\alpha\}$ of X is just $\{2\alpha/(\alpha,\alpha)\}$ (see 6.4), so that $L_0 = L$. Thus (6) follows from (5) and 4.10′ (see 5.3).

(7) *If a reflection $w \varepsilon W_\sigma$ is represented in N_σ, then for the corresponding set S_α of* (2′) *equation* (a) *or* (b) *of* (2′′′′) *holds.* We use (5) with B replaced by the Borel subgroup containing T and corresponding to a simple system of roots chosen as in (2′′). We apply $\omega_\alpha + \omega_{\sigma\alpha} + \cdots$ to the condition $(1-w)t \varepsilon (1-\sigma)T$. By 6.4 and (2′′′) we get $(\alpha + \sigma\alpha + \cdots)(t) = 1$ and $(\alpha + \sigma\alpha)(t^2) = 1$ in cases (a) and (b) of (2′), which because of the definition of t are equivalent to (a) and (b) of (2′′′′), as required.

Now we can prove 8.2. Let G_1 be the group generated by T_σ, U_σ and U_σ^-. By (6), (7), (2′′′′) and (4) we get $N_\sigma \subseteq G_1$, so that $G_\sigma = G_1$ by (1). Then G_σ is connected by (2) and (3). Let R be the unipotent radical of G_σ. By 7.4 there exists a Borel subgroup of G containing R and fixed by σ which we may take to be the group B above. Thus $R \subseteq U_\sigma$. It follows from (2′′′′) and (4) that the only subgroup of U_σ normalized by N_σ is 1. Thus $R = 1$, G_σ is reductive, and 8.2 is proved.

8.3. **Remarks**. (a) It should be clear that U_σ and N_σ form the makings of a Bruhat decomposition 6.3 of G_σ, and that the corresponding root system may be identified with a subsystem of the system Σ_σ of 1.32. (b) If every $w \varepsilon W_\sigma$ is represented in N_σ, for example if $\sigma = \rho$ with ρ as in (5), then G_σ is even semisimple since T_σ and Σ_σ have the same rank. (c) In case σ is inner the situation becomes transparent, even when G is not simply connected. If $\sigma = i_x$ with $x \varepsilon T$, then it follows from 6.3(b) that G_x is generated by T, those X_α for which $\alpha(x) = 1$, and those n_w for which $wx = x$, so that G_{x0} is generated by T and the X_α alone, hence is reductive (see [19, p. 17-02]). If W_x (resp. W_{x0}) is the subgroup (resp. reflection subgroup) of W generated by all w (resp. w_α) such that $wx = x$ (resp. $\alpha(x) = 1$), it follows that G_x/G_{x0} is isomorphic to W_x/W_{x0}. The situation is similar to that of 4.4, but not exactly so since it can happen that $w_\alpha x = x$ even though $\alpha(x) \neq 1$ (see 8.7(b) below). If G is simply connected this possibility is obviated as in (7) above which becomes very simple when σ is inner, so that $W_x = W_{x0}$ by 4.2′ and G_x is connected.

8.4. COROLLARY. *Every element of G fixed by σ is contained in a Borel subgroup fixed by σ.*

Since G_σ is connected, each of its elements is contained in a Borel subgroup, so that 8.4 follows from 7.4.

8.5. COROLLARY. *The centralizer of a semisimple element of G is connected.*

This is the special case of 8.2 discussed in 8.3(c).

8.6. COROLLARY. *Every two commuting semisimple elements of G are contained in a maximal torus.*

If x and y are the elements, we need only choose a maximal torus of G_x which contains y.

8.7. **Remarks** (a) "Two" can not be replaced by "three" in 8.6 or "*a*" by "two" in 8.5 since the semisimple part of the centralizer in 8.5 need not be simply connected (cf. 9.1 below). (b) The extent to which simpleconnectedness is necessary for the above results is studied in the next section; that it can not be entirely dropped is shown by the following example: $G = PSL_2(K)$, $x = \operatorname{diag}(i, -i)$, $i^2 = -1$, $\sigma = i_x$. Here G_σ consists of the monomial matrices, so that 8.2, 8.5 and 8.6 all fail, and since the only Borel subgroups fixed by σ are the superdiagonal and subdiagonal subgroups 8.4 does also.

Finally, to close this section we sketch an alternate proof of the connectivity in 8.1 which in part parallels a proof of Borel for compact Lie groups [2, p. 225], and which brings up other points of interest. Clearly it will be enough to prove:

8.8. *Assume G and σ are as in* 8.1. *If $x \varepsilon G_\sigma$, then $x \varepsilon G_{\sigma 0}$ in case either* (a) *x is semisimple or* (b) *x is unipotent.*

Part (a) follows at once from:

8.9. *If* (a) *holds, then x is contained in a Borel subgroup B and a maximal torus T of B, both fixed by σ.*

For T_σ is connected (see (3) above). In 8.9 assume first that x is regular. Thus x is contained in a unique maximal torus T, clearly fixed by σ, and $\alpha(x) \neq 1$ for every root α (see, e.g., [28, p. 53]). Then $L = L_0$ as in 6.4, whence σ fixes a positive system of roots by 5.5, hence also the corresponding Borel subgroup (see 6.2(a)). Since G_{x0} is reductive (even if G is not simply connected; see 8.3(c)), and the set of fixed points of an automorphism of a nontrivial semsimple group is always infinite (see 10.5 below), the general case of 8.9 may be reduced to the first case as in [2, p. 225]. We omit the details.

For the proof of 8.8(b) the assumptions can be considerably weakened.

8.10. *The conclusion of* 8.8(b) *holds if G is any connected linear algebraic group (and σ is semisimple).*

Since this is proved for solvable groups in [19, p. 602], it follows from:

8.11. *In* 8.10 *every unipotent element u fixed by σ is contained in a Borel subgroup fixed by σ.*

By 7.2 the automorphism $i_u\sigma$ fixes a Borel subgroup B of G. Its unipotent and semisimple parts i_u and σ do also, whence 8.11.

8.12. **Remark**. The case of 8.10 in which σ is inner is proved in [19, p. 615], but the proof just give is simpler.

9. Extensions of 8.2.

It will be convenient to formulate our results as in 8.2 in terms of automorphisms (of semisimple groups) which fix a Borel subgroup and a maximal torus thereof, which will be called quasisemisimple (abbreviated quass). We observe that every semisimple automorphism is quass, by 7.5, that a quass automorphism is semisimple if and only if its class in Aut $G/\text{Int } G$ has order prime to p, and that every class in Aut $G/\text{Int } G$ contains a quass element. The main result is as follows.

9.1. THEOREM. *Let G be a semisimple algebraic group and σ a quass automorphism of G. Let $\pi : G' \to G$ be the universal covering, F the kernel of π, i.e., the fundamental group of G, and C' the center of G'.* (a) *$G_\sigma/G_{\sigma 0}$ is isomorphic to a subgroup of $F/(1-\sigma)F$ which contains $((1-\sigma)C' \cap F)/(1-\sigma)F$.* (b) *As σ varies over its class in Aut $G/$Int G, every such subgroup is achieved.*

In the statement and proof of this theorem, the following facts about the universal covering are used. G' is a simply connected semisimple group and π is an isogeny (i.e. a surjective homomorphism with a finite kernel, a part of the center of G' (cf. 7.1)). Every (quass) automorphism σ of G can be lifted uniquely to one of G', also denoted σ (this is proved in 9.16 below). The action of σ on F and C' depends only on the class of Aut $G/\text{Int } G$ to which σ belongs.

9.2. LEMMA. (a) $\pi G'_\sigma = G_{\sigma 0}$. (b) *$G_\sigma/G_{\sigma 0}$ is isomorphic to* $((1-\sigma)G' \cap F)/(1-\sigma)F$.

We apply 4.5 with A, B, and C replaced by G', G, and F. By the exactness in the third position it is enough to prove (a). Now G'_σ is connected by 8.2. Hence so is $\pi G'_\sigma$. Thus $\pi G'_\sigma \subseteq G_{\sigma 0}$. Also $\pi G'_\sigma = \ker \delta$ is a closed subgroup of G_σ, of finite index since F is finite. Thus $\pi G'_\sigma \supseteq G_{\sigma 0}$ and 9.2 follows.

Part (a) of 9.1 follows at once from 9.2(b).

9.3. COROLLARY. *G_σ is connected if and only if $(1-\sigma)G' \cap F = (1-\sigma)F$.*

9.4. COROLLLARY. *$G_{\sigma 0}$ is a reductive group, and $G_\sigma/G_{\sigma 0}$ is Abelian and consists of semisimple elements.*

The first result follows from 9.2(b), the second from 8.2 and 9.2(b).

To prove 9.1(b) we use the notations $B', T', \ldots$ for G' just as we used $B, T, \ldots$ in the proof of 8.2.

9.5. LEMMA. $(1-\sigma)G' \cap F = \{(1-w)t' | w \,\varepsilon\, W'_\sigma\}(1-\sigma)T' \cap F$.

Assume $f = (1-\sigma)x$ with $f \,\varepsilon\, F$, $x \,\varepsilon\, G'$. Then $f\sigma x = x$. If $x = u_1 n u_2$ as in 6.3 with n representing $w \,\varepsilon\, W'$, then the uniqueness implies that w, u_1 and u_2 are all fixed by σ, so that $f = (1-\sigma)n$. Proceeding as in (5) of the proof of 8.2, we get $(1-\sigma)n \,\varepsilon\, \{(1-w)t' | w \,\varepsilon\, W'_\sigma\}(1-\sigma)T' \cap F$, whence 9.5.

Now we can prove 9.1(b). Let $T = \pi T'$, a maximal torus of G. The elements of $i_T\sigma$ are all quass. On lifting them to G' and combining 9.2, 9.5, and part 4.11 of 5.3 (which is applicable because W' acts trivially on the center of G', in particular on F), we get 9.1(b).

9.6. **Remark**. It follows from 4.8 and 5.1(b) that in 9.1(b) we can get by with automorphisms of finite order.

9.7. COROLLARY. *Let G be a semisimple algebraic group and F its fundamental group. Then the following conditions on a class S of Aut G/Int G are equivalent.* (a) *The action on F of some (hence of every) $\sigma \,\varepsilon\, S$ is fixed-point-free.* (b) *For every quass $\sigma \,\varepsilon\, S$ the group G_σ is connected.*

Since F is finite, (a) is equivalent to $(1-\sigma)F = F$, hence to (b) by 9.1(a).

As an illustration of how (a) $\to$ (b) goes beyond 8.2 which it clearly generalizes, we have:

9.8. *If G is the adjoint group of type A_{2n}, E_6 or D_4 and σ is any quass outer automorphism in the first two cases or quass triality automorphism in the third case, then G_σ is connected.*

9.9. COROLLARY. *For a semisimple algebraic group G the following conditions are equivalent.* (a) *The fundamental group F is trivial.* (b) *The centralizer of every semisimple element of G is connected.* (c) *Every two commuting semisimple elements of G are contained in a maximal torus.*

Here (a) and (b) are equivalent by 9.7 with $S =$ Int G, and (b) implies (c) as in 8.6. If (c) holds and x is semisimple, then every semisimple element of G_x is contained in G_{x0} by (c) and every unipotent one by 9.4 (or by 8.10), whence (b).

9.10. **Caution**. If F is trivial G need not be simply connected, although it is so in case $p = 1$. In the general case inseparability can occur [19, Exp. 18].

9.11. **Remark**. The implication (a) $\to$ (c) can be generalized as follows. If A is an Abelian subgroup of semisimple elements such that A/A_0 is generated by two elements at least one of which has order prime to that of F, then A is contained in a maximal torus of G. This easily follows from: if σ is a quass automorphism of G of finite order q prime to that of F, then G_σ is connected. To see this assume $y \in G_\sigma$ and choose $y' \in \pi^{-1}y$, so that $\sigma y' = fy'$ with $f \in F$, whence $f^q = 1$, $f = 1$, $y' \in G'_\sigma$ and $y \in G_{\sigma 0}$ by 9.2 (cf. [2, p. 227]).

Next we observe that the lower limit in 9.1 comes directly from the center of G; more precisely:

9.12. Corollary. *If C is the center of G in* 9.1, *then the following groups are naturally isomorphic.* (a) $((1-\sigma)C' \cap F)/(1-\sigma)F$. (b) $C_\sigma/\pi C'_\sigma$.
(c) $C_\sigma G_{\sigma 0}/G_{\sigma 0}$.

Here (a) and (b) are isomorphic by 4.5 applied with C', C, F in place of A, B, C, while (b) and (c) are because $C_\sigma \cap G_{\sigma 0} = \pi C'_\sigma$ by 9.2(a).
We observe that the group in (b) measures the amount by which $\pi : C'_\sigma \to C_\sigma$ fails to be surjective. The equality of (b) and (c) may be stated: the natural map $C_\sigma/\pi C'_\sigma \to G_\sigma/G_{\sigma 0}$ is injective. More precisely one can show that if B and T are as usual then the maps $C_\sigma/\pi C'_\sigma \to T_\sigma/T_{\sigma 0} \to B_\sigma/B_{\sigma 0}$ are bijective.

9.13. Corollary. *If G and S are as in* 9.7 *and C is the center of G, the following conditions on S are equivalent.* (a) $(1-\sigma)C' \cap F = (1-\sigma)F$ *for some, hence for every,* $\sigma \in S$. (b) *The natural map* $\pi : C'_\sigma \to C_\sigma$ *is surjective.*
(c) G_σ *is connected for some quass* $\sigma \in S$.

Here (a) and (c) are equivalent by 9.1, while (a) and (b) are by 9.12.
We observe that the conditions of 9.13 always hold when G is an adjoint group since then $F = C'$.

9.14. Corollary. *If everything is as in* 9.13, *then the following conditions on S are equivalent.* (a) $(1-\sigma)C' \supseteq F$ *for some, hence for every* $\sigma \in S$.
(b) *For every quass* $\sigma \in S$, $G_\sigma = C_\sigma G_{\sigma 0}$. (c) *For every quass* $\sigma \in S$, $G_\sigma/G_{\sigma 0}$ *is isomorphic to* $F/(1-\sigma)F$. (d) *For all quass* $\sigma \in S$, $G_\sigma/G_{\sigma 0}$ *has the same value, in the sense of isomorphism.* (e) *For every quass* $\sigma \in S$, *every element x of G fixed by σ is contained in a Borel subgroup of G fixed by σ.*

The equivalent of (a), (b), (c) and (d) follows from 9.1 and 9.12. Assume that (a) holds, and let σ and x be as in (e). Choose $x' \in \pi^{-1}x$. Then $(1-\sigma)x' \in F$, whence by (a), $(1-\sigma)x' = (1-\sigma)c$, i.e., $x'c^{-1} \in G'_\sigma$ for some $c \in C'$. By 8.4 there exists a Borel subgroup of G' fixed by σ and containing $x'c^{-1}$, hence containing

x' since c is in the center [19, p. 609]. Projecting this group to G, we get (e). Now assume (e). Choose $\sigma \varepsilon S$, σ quass, so that $(1-\sigma)G' \supseteq F$ (see 9.1(b), 9.2(b)). Assume $f \varepsilon F$. Then $f = (1-\sigma)x'$ for some $x' \varepsilon G'$, and since (e) holds x' is contained in a Borel subgroup B' of G' fixed by σ. Applying 5.3 and 4.12 to B' modulo its normal subgroup of unipotent elements, we conclude that $f \varepsilon (1-\sigma)C'$. Thus (a) holds, and 9.14 is proved. Here we have used the fact that if $c \varepsilon T'$ is fixed by every $w \varepsilon W'$, then $c \varepsilon C'$. This follows from the simple connectedness of G': if for every simple root α we apply ω_α to the equation $w_\alpha c = c$, we get $\alpha(c) = 1$, whence our assertion.

As an example to illustrate 9.12 and 9.14 we have:

9.15. *If $p \neq 2$, $G = SO_{2n}$, and σ is any automorphism (necessarily outer) effected by a semisimple element of O_{2n} not in SO_{2n}, then G_σ has two components and* 1 *and* -1 *always lie in different components.*

As a final example, typical of how some of the above results can be extended to groups which are not semisimple, we have: if G is a connected linear algebraic group whose semisimple part is simply connected, then the conclusion of 8.5 holds. The reader might wish to prove this for himself.

Finally we prove the fact used in 9.1, that every automorphism of G can be lifted to G'. This follows from:

9.16. *Let G be a semisimple algebraic group and $\pi : G' \to G$ its universal covering. Let $\sigma : G \to G$ be an isogeny. Then there exists a unique isogeny $\sigma' : G' \to G'$ such that $\pi\sigma' = \sigma\pi$.*

By modifying σ by an inner automorphism, we may assume σ fixes a maximal torus G. Set $T' = \pi^{-1}T$. Let Σ, Σ' and X, X' be the corresponding root systems and character groups. Let γ be the automorphism of the real vector space generated by X' defined by $\gamma\pi^* = \pi^*\sigma^*$ (here σ^* denotes the transpose of the restriction of σ to T, and similarly for π^*). By [19, p. 1809] the existence of σ' in 9.16 is equivalent to the following two properties of γ. (1) γ maps every root in Σ' onto an integral multiple of another. (2) $\gamma X' \subseteq X'$. Since π is the universal covering, we have (*) $\pi^*\Sigma = \Sigma'$. Since σ is an isogeny, for each $\alpha \varepsilon \Sigma$ there exists $\beta \varepsilon \Sigma$ and a positive integer n such that $\sigma^*\alpha = n\beta$ [19, p. 1804]. If we apply π^* and use (*) and the definition of γ we get (1). Pick a basis B of X' compatible with the decomposition of Σ' into its simple components. We must show $\gamma B \subseteq X'$, i.e., if $\rho \varepsilon \gamma B$ and $\alpha \varepsilon \Sigma'$ then $2(\rho,\alpha)/(\alpha,\alpha)$ is an integer (see 6.4). We may assume ρ and α in the same component since otherwise $(\rho,\alpha) = 0$. We have (3) $\rho = \gamma\omega$ and $q\alpha = \gamma\beta$ with $\omega \varepsilon B$ and $\alpha \varepsilon \Sigma'$ in the same component and q a positive integer by (1). Since γ normalizes W', the function $x, y \to (\gamma x, \gamma y)$ on $X' \times X'$ is a symmetric bilinear form invariant under W', hence is a multiple of $x, y \to (x, y)$ on each simple component. Using this and (3), we get $2(\rho, \alpha)/(\alpha, \alpha) = q2(\gamma\omega, \gamma\beta)/(\gamma\beta, \gamma\beta) = q2(\omega, \beta)/(\beta, \beta)$, an integer, which proves (2) and the existence of σ'. If 9.16 also holds with σ'' in place of σ', then $\pi\sigma' = \pi\sigma''$ on G', so that $\sigma' - \sigma''$ is a homomorphism of G' into its center. Since G' is connected, or else since G' equals its derived group, this homomorphism is trivial, which proves the uniqueness in 9.16.

9.17. **Remarks**. (a) From the above discussion it should be clear that for each G there is a unique covering (isogeny) $\pi : G' \to G$ which is universal subject to (*) above. The fact that G' is simply connected is proved in [19, p. 2301]. (b) Theorem 8.1 holds for complex semisimple Lie groups with connectedness taken in the usual sense as we see by taking K to be the complex field there, and also for compact semisimple Lie groups as has been shown by Borel and others [2]. (Of course all elements and automorphisms are semisimple in this case.) Since the results of this section are deduced from 8.2 and formal properties of root systems, they remain equally valid in these cases. (In the compact case "Borel subgroup ..." in 9.14(e) has to be replaced by "maximal torus for which there is a fundamental chamber fixed by σ.")

10. Automorphisms versus finiteness

Henceforth we are concerned with the case in which G_σ in §7 is finite. In this section we study the general situation and develop the main tools. Our immediate aim is:

10.1 THEOREM. *Let G be a connected linear algebraic group and σ an endomorphism of G onto G. If G_σ is finite, then $(1 - \sigma)G = G$.*

In the various steps of the proof we do not assume G_σ to be finite unless we explicitly say so.

10.2 *G_σ is finite if and only if $(1 - \sigma)G$ contains a dense open part of G.*

This is because $\dim(1 - \sigma)G = \dim G - \dim G_\sigma$.

10.3. *If H is a normal subgroup of G and G_σ is finite, then so is $(G/H)_\sigma$.*

By 10.2.

10.4. *If G is solvable, then* 10.1 *holds.*

We assume $G \neq 1$, let H be the second last term in the derived series of G, and use induction on the length. Assume $x \in G$. By the induction assumption applied to G/H (see 10.3) there exists $y \in G$ so that $y^{-1}x\sigma y \in H$. Since H is Abelian, $1 - \sigma$ acts as a homomorphism there; thus $(1 - \sigma)H$ is closed [19, p.

304] hence equal to H by 10.2. Thus there exists $z \varepsilon H$ so that $y^{-1}x\sigma y = z\sigma z^{-1}$, i.e., $x = (1-\sigma)(yz)$, whence 10.4.

10.5. *If G is semisimple, then G_σ is finite if and only if the differential $d\sigma$ (at the identity) is nilpotent.*

Assume G_σ finite. Let B be a Borel subgroup fixed by σ (see 7.2). For each $b \varepsilon B$ we have $i_b\sigma$ conjugate to σ by 10.4. Thus we may assume that σ fixes a maximal torus T in B. Then σ permutes the positive roots up to factors $p^m (m \geq 0)$ [19, p. 1804]; and on each orbit of roots a factor > 1 must occur, since otherwise the sum of the roots of that orbit would be fixed by σ, the map $1 - \sigma^* : X \to X$ would not be injective, and the map $1 - \sigma : T \to T$ would not be surjective in contradiction to 10.4. It follows that some $\sigma^n (n > 0)$ multiplies each root by a factor $p^m (m > 0)$, whence (e.g. by 6.2) $d\sigma^n = 0$ and $d\sigma$ is nilpotent. We see incidentally that G_σ can be finite only if $p > 1$. Assume $d\sigma$ nilpotent. Then $1 - d\sigma$, the differential of $1 - \sigma$, is surjective at the identity, so that $(1-\sigma)G$ covers an open part of G and G_σ is finite by 10.2.

10.6. *If G is semisimple, G_σ is finite, and either $\tau = i_x\sigma$ for some x in G or $\tau = \sigma^n (n > 0)$, then G_τ is finite.*

For in both cases the nilpotence of $d\sigma$ implies that of $d\tau$.

10.7. *If G is semisimple, then* 10.1 *holds.*

Assume $x \varepsilon G$. Let $\tau = i_x\sigma$. By 10.6 and 10.2 applied twice, $(1-\sigma)G$ and $(1-\tau)G \cdot x$ overlap: $(1-\sigma)y = (1-\tau)z \cdot x$ for some $y, z \varepsilon G$. Then $x = (1-\sigma)(z^{-1}y)$, whence 10.7.

Now we can prove 10.1. Assume $x \varepsilon G$. Let R be the radical of G. By 10.3 and 10.7 there exists $y \varepsilon G$ so that $y^{-1}x\sigma y \varepsilon R$, and then by 10.4, $z \varepsilon R$ so that $y^{-1}x\sigma y = (1-\sigma)z$, whence $x \varepsilon (1-\sigma)G$.

10.8. **Remark**: The above proof simplifies considerably in an important special case, in which G is defined over a finite field of q elements and σ is the Frobenius map (which replaces each matric entry of G by its qth power). Then $d\sigma = 0$ even if G is not semisimple so that we need only combine the argument of 10.7 and the last few lines of 10.5. This proof is a variant of one due to Lang [15]. A shorter proof of 10.1, which does not bring up the interesting points 10.5 and 10.6, is as follows.

First prove 10.4 as before. Then assume $x \varepsilon G$. By 7.2 there is a Borel subgroup B fixed by σ and by 7.3 elements y and b of G and B such that $x = yb\sigma y^{-1}$. By 10.4 there exists $c \varepsilon B$ such that $b = c\sigma c^{-1}$. Then $x = (1-\sigma)(yc)$ as required.

10.9. Corollary. *If G_σ is finite, $x \varepsilon G$, and $\tau = i_x\sigma$, then τ is conjugate to σ, hence G_τ is isomorphic to G_σ.*

For if $x = (1-\sigma)y$, then $\tau = i_y\sigma i_y^{-1}$.

10.10. Corollary. *If G_σ is finite, then σ fixes a Borel subgroup and a maximal torus thereof. Further any two such couples are conjugate under an element of G_σ.*

The first statement is by 10.9 and the usual conjugacy theorems. Assume B, B' are the Borel subgroups in the second statement. Then $xBx^{-1} = B'$ for some $x \varepsilon G$, and $x^{-1}\sigma x$ normalizes B since σ fixes B and B', hence belongs to B, hence has the form $b\sigma b^{-1}$ for some $b \varepsilon B$ by 10.1, whence B is conjugate to B' by the element xb of G_σ. The maximal tori can now be treated similarly.

10.11. Corollary. *Assume G_σ is finite and A and B are subgroups of G fixed by σ such that $A \supset B$ and B is connected.* (a) $(1-\sigma)B = B$. (b) *The natural map $A_\sigma \to (A/B)_\sigma$ is surjective.*

Here (a) follows from 10.1. Assume $aB \varepsilon (A/B)_\sigma$ (with $a \varepsilon A$). Then $a^{-1}\sigma a \varepsilon B$. Thus $a^{-1}\sigma a = b\sigma b^{-1}$ with $b \varepsilon B$ by (a). Since $ab \varepsilon A_\sigma$ we have (b).

10.12. Corollary. *If σ is an automorphism in* 10.1 *and G_σ is finite, then G is solvable.*

Let R be the radical of G. We use 10.1 with G replaced by G/R, which is permissible by 10.3. Since $d\sigma$ is then a nilpotent automorphism, we conclude that G/R is trivial, as required.

Remark. A similar but different proof has been found, independently, by Winter [33].

10.13. Corollary. *For an endomorphisms σ of a simple group there exists the dichotomy:* (a) *σ is an automorphism,* (b) *G_σ is finite.*

We are excluding the trivial group from the list of simple groups. We may assume σ is surjective since otherwise σ is trivial. If σ is an automorphism, then G_σ is infinite by 10.12. Assume now that G_σ is infinite. By 10.9 we may replace σ by any $i_x\sigma$, hence assume that σ fixes a Borel subgroup and a maximal torus thereof. If σ^n is now chosen, as in the proof of 10.5, to multiply each root by a factor p^m, the factors are all equal since the root system is irreducible, all equal to 1 since otherwise $d\sigma$ would be nilpotent in contradiction to 10.5. It follows [19, p. 1809] that σ is an automorphism, whence 10.13.

10.14. **Remark**. This result justifies to some extent the dichotomy of this paper in which only endomorphisms which satisfy (a) or (b) are studied.

10.15. COROLLARY. *Let σ and τ be endomorphisms of the simple group G. If G_σ is finite, then so are $G_{\sigma\tau}$ and $G_{\tau\sigma}$.*

If σ is not an automorphism, then neither are $\sigma\tau$ or $\tau\sigma$; thus 10.5 follows from 10.13.

10.16. **Caution**. This result is false for semisimple groups even if G_τ is also finite (cf. 10.6).

11. The classical finite simple groups

Our objective is to study G_σ in case G is semisimple and σ as usual is an endomorphism of G onto G such that G_σ is finite. In this section we identify the possibilities for G_σ, obtain a cellular decomposition, and obtain a new formula for the order (see 11.6, 11.1, and 11.16 below). The development is to a certain extent expository, providing a unification and simplification of known results, since each group turns out to be essentially a product of Chevalley groups and their twisted analogues. If G and σ are as above, we know that ker σ is trivial by 7.1 so that σ is an automorphism of the abstract group underlying G, and that $p > 1$ by 10.5. As is permissible by 10.10 we choose B and T in §6 to be fixed by σ; then U, U^-, N and W (i.e. N/T) also are. The transpose of the restriction of σ to T, extended to the real vector space V generated by X, will be denoted σ^*.

11.1. THEOREM. *Let G and σ be as above.* (a) *Each $w \in W_\sigma$ is represented by some $n_w \in N_\sigma$.* (b) *For each $w \in W_\sigma$ the group U_w of 6.3 is fixed by σ.* (c) *If n_w is as in* (a), *than 6.3 holds with $G_\sigma, W_\sigma, N_\sigma$, etc. in place of G, W, N, etc.*

Here (a) follows from 10.11 with A and B replaced by N and T, while (b) follows from the definition of U_w. Now if $x \in G_\sigma$ is written $x = u_1 n u_2$ as in 6.3, so that also $x = \sigma u_1 \cdot \sigma n \cdot \sigma n_2$, we have $w \in W_\sigma$ by 6.3(a) and then $u_1, n, u_2 \in G_\sigma$ by 6.3(b), whence 11.1.

11.2 *There exists a permutation ρ of the roots and for each root α a power $q(\alpha)$ of p such that :* (a) *ρ permutes the positive roots.* (b) *$\sigma^*\rho\alpha = q(\alpha)\alpha$ for every root α.* (c) *$\sigma x_\alpha(k) = x_{\rho\alpha}(c_\alpha k^{q(\alpha)})$ for some $c_\alpha \in K^*$ and all $k \in K$.*

The groups $X_\alpha (\alpha > 0)$ are the minimal subgroups of U that are normalized by T [19, p. 1305], hence they are permuted by σ, and similarly for negative roots. If we define ρ by $\sigma X_\alpha = X_{\rho\alpha}$, then (a) holds, and then by [19, p. 1804] so do (b) and (c) (which are proved by setting $\sigma x_\alpha(1) = x_{\rho\alpha}(c_\alpha)$ and then applying σ to the equation 6.1 with $k = 1$).

We observe that a different choice of B, T or the isomorphisms x_α will not change ρ or q in any essential way.

11.3. **Remark**. It follow from 11.2(a) that W_σ in 11.1 can be given the structure of a reflection group as in 1.32. The fact that σ^* permutes the roots only up to positive multiples is of no importance there.

The condition for G_σ to be finite discussed in the proof of 10.5 reads:

11.4. *If α runs over any ρ-orbit of roots, then the product $\Pi q(\alpha) > 1$.*

For the identification of the groups G_σ we need:

11.5. *In any irreducible component Σ_1 of Σ, q is constant on roots of a given length, and if q is not constant on all of Σ_1 and α and β are long and short roots, respectively, then $(\alpha,\alpha)/(\beta,\beta) = p$ and $q(\beta)/q(\alpha) = p$.*

This is a variant of [19, p. 1806]. The proof is as follows. Let α and β be roots in Σ_1. Assume first they have the same length. Then $\beta = w\alpha$ for some $w \varepsilon W$. Since σ^* clearly normalizes the action of W on V, $\sigma^{*-1} w \sigma^* = w_1$ for some $w_1 \varepsilon W$. This equation and 11.2(b) imply that $q(\alpha)\sigma^* \rho w\alpha = q(\alpha) q(w\alpha) w\alpha = q(w\alpha) w \sigma^* \rho\alpha = q(w\alpha)\sigma^* w_1 \rho\alpha$. Since $\rho w\alpha$ and $w_1\rho\alpha$ are roots and $q(\alpha)$ and $q(w\alpha)$ are positive numbers, $q(\alpha) = q(w\alpha) = q(\beta)$. Now assume α long, β short, and $q(\alpha) \neq q(\beta)$. We may also assume $(\alpha,\beta) > 0$. In this case (*) $\langle\beta,\alpha\rangle = 1$ and $\langle\alpha,\beta\rangle = (\alpha,\alpha)/(\beta,\beta)$, in the notation $\langle\beta,\alpha\rangle = 2(\beta,\alpha)/(\alpha,\alpha)$: for the product of the positive integers $\langle\beta,\alpha\rangle$ and $\langle\alpha,\beta\rangle$ is less than 4 by Schwarz's inequality. By 11.2(b), $\sigma^* w_{\rho\alpha} = w_\alpha \sigma^*$. Applying both sides to $\rho\beta$ and using 1.5 and 11.2(b), we get $\langle\rho\beta,\rho\alpha\rangle q(\alpha) = \langle\beta,\alpha\rangle q(\beta)$. Since $\langle\beta,\alpha\rangle = 1$, $\langle\rho\beta,\rho\alpha\rangle = 1, 2,$ or 3 and $q(\beta)/q(\alpha)$ is a power of p different from 1, we conclude that $q(\beta)/q(\alpha) = p$ and $\langle\rho\beta,\rho\alpha\rangle = p$. Then $\langle\rho\alpha,\rho\beta\rangle = 1$ so that the equation just used with α and β interchanged yields $\langle\alpha,\beta\rangle = p$, i.e., $(\alpha,\alpha)/(\beta,\beta) = p$ by (*). We see incidentally that in this situation ρ maps long roots onto short ones and vice versa.

11.6. **Identification of G_σ**. We consider the possibilities when G is simple, i.e., Σ is irreducible. We know in any case that ρ permutes the positive roots, hence the simple ones, and preserves orthogonality of roots since σ^* normalizes W and orthogonal roots correspond to commuting reflections. (a) Assume $q(\alpha)$ is constant. Then ρ preserves root lengths, e.g., by the remark following the proof of 11.5, hence induces an automorphism of the Dynkin diagram. The cases are: (1) ρ is the identity, (2) ρ is not the identity and either $\rho^2 = 1$ and G is of one of the types $A_n (n \geq 2)$, $D_n (n \geq 4)$, E_6, or else $\rho^3 = 1$ and G is of type D_4. Consider case (1). The isomorphisms x_α can be adjusted so that each $c_\alpha = 1$ in 11.2 (write $c_\alpha = c_\alpha'^{1-q}$ and then replace k by $c'_\alpha k$), and then $x_\alpha(k)$ is fixed by σ exactly when $k \varepsilon F_q$, the field of q elements. It is clear that G_σ is just a Chevalley group parametrized by F_q [8] (in Chevalley's treatment G is an adjoint group, but this makes little difference (cf. 12.8)). Similarly in case (2)

G_σ is one of the twisted analogues of the Chevalley groups considered in [24]. (b) Assume $q(\alpha)$ is not constant. Then ρ interchanges long and short roots, so that $\rho^2 = 1$ and G is of type C_2, F_4 or G_2. By 11.5 the corresponding values of p are 2, 2 or 3, and if $q(\alpha) = p^a$ for a long root, then $q(\beta) = p^{a+1}$ for a short root, so that G_σ can be parametrized by a field of p^{2a+1} elements. Thus we recognize G_σ in the first case as a Suzuki group [29] and in each of the other cases as a so called Ree group [16]. We thus see that versions (cf. 12.8 below) of most of the know finite simple groups can be realized in the form G_σ. The simplicity, incidentally, is most easily proved from 11.1 and 11.3 by the method of Tits [30]. We observe that in case (a) G_σ can be realized as the group of rational points of G for an appropriate definition of G over a finite field (of q elements, σ being the Frobenius q^{th} power map), while in case (b) it can not. If G is not simple, then G_σ is essentially a product of groups of the type just discussed (see §6). Combining this with 10.11(b) we can therefore assert:

11.7. *If G is a connected linear algebraic group and σ is an endomorphism of G such that G_σ is finite, then each composition factor of G_σ is either cyclic or else one of the groups discussed in* 11.6.

Next we will develop a formula for the order of G_σ.

11.8. *If U_1 is a subgroup of U which is normalized by T and fixed by σ, then $|U_{1\sigma}| = \Pi q(\alpha)$, the product being taken over those roots α for which $X_\alpha \subseteq U_1$.*

Let Σ_1 be the set of roots α for which $X_\alpha \subseteq U_1$. We proceed by induction on $|\Sigma_1|$. Let Σ_2 be the set of roots such that $X_\alpha \subseteq U_2$, the derived group of U_1, and $\Sigma_3 = \Sigma_1 - \Sigma_2$. Then [19, p. 1305] U_1 (resp. U_2) is generated by the X_α it contains and U_1/U_2 is canonically isomorphic to the direct product of the X_α for $\alpha \varepsilon \Sigma_3$; and all of these groups are fixed by σ. By 10.11 with U_1, U_2 in place of A, B we have $|U_{1\sigma}| = |U_{2\sigma}||(U_1/U_2)_\sigma|$. Now $|U_{2\sigma}| = \Pi q(\alpha)(\alpha \varepsilon \Sigma_2)$: if Σ_1 is empty this is clear, and if not it comes from the induction assumption. Thus it remains to show that $|(U_1/U_2)_\sigma| = \Pi q(\alpha)(\alpha \varepsilon \Sigma_3)$. Now in the representation of U_1/U_2 as a direct product of the $X_\alpha(\alpha \varepsilon \Sigma_3)$, σ acts on the factors according to the formula 11.2. Breaking up Σ into orbits under the action of ρ and changing the notation slightly, we are reduced to showing that the number of solutions of the system $c_1 k_1^{q(1)} = k_2,\ c_2 k_2^{q(2)} = k_3, \dots, c_n k_n^{q(n)} = k_1$ is $\Pi q(j)$, with the $c_j \varepsilon K^*$ and the $q(j)$ positive integral powers of p with a product > 1 (by 11.4). Now we may use the first $n-1$ equations to express $k_2, \dots, k_n$ in terms of k_1 and then reduce the last equation to the form $ck_1^q = k_1$ with $c \varepsilon K^*$ and $q = \Pi q(j)$. The last equation has q solutions. Hence so does the system.

11.9. *Let Q denote the product of all $q(\alpha)$ for which $\alpha > 0$, and for each*

$w \varepsilon W$ let Q_w denote the product of all $q(\alpha)$ for which $\alpha > 0$ and $w\alpha < 0$.
(a) $|U_\sigma| = Q$. (b) *For each $w \varepsilon W_\sigma$, $|U_{w\sigma}| = Q_w$.*

Because of 11.1(b) this follows from 11.8.

11.10. *For each ρ-orbit π of simple roots let $Q_\pi = \Pi q(\alpha)(\alpha \ \varepsilon \ \pi)$. Then $|T_\sigma| = |det(\sigma^* - 1)| = \Pi(Q_\pi - 1)$, the product over the various orbits.*

By 11.2(b), $|\det(\sigma^* - 1)| = \Pi(Q_\pi - 1)$, a number prime to p. By the duality between X and T this number is also the order of $\ker(\sigma - 1)$ on T, which is $|T_\sigma|$.

11.11 *If Q, Q_w and Q_π are as in 11.9 and 11.10, then $|G_\sigma| = Q \cdot \Pi(Q_\pi - 1) \cdot \Sigma Q_w(w \ \varepsilon \ W_\sigma)$.*

This follows from 11.1, 11.9 and 11.10.

11.12. *U_σ is a maximal unipotent (i.e., a p-Sylow) subgroup of G_σ.*

For Q, the order of U_σ, is the largest power of p dividing $|G_\sigma|$, because each Q_π and $Q_w(w \neq 1)$ is divisible by p by 11.4.

11.13. COROLLARY. *If $\tau = \sigma^n$ and Q_τ and Q_σ are the orders of p-Sylow subgroups of G_τ and G_σ, then $Q_\tau = Q_\sigma^n$.*

This follows from 11.9 and 11.12.

The formula 11.11 is only useful for crude purposes. To get our final formula we will have to assume that G is σ-simple, i.e. σ permutes the simple components of G in a single orbit. This presents no serious loss in generality since we may recover the general case by taking products.

11.14. *If G is σ-simple, then the following are true.* (a) *ρ permutes the irreducible components of Σ in a single orbit.* (b) *$\sigma^* = q\sigma_1$ with q a positive number and σ_1 a transformation of (finite) order equal to that of ρ.*

Since the irreducible components of Σ correspond to the simple components of G [19, p. 1713], we have (a). Let r be the order of ρ. By 11.2(b), $\sigma^{*r}\alpha = f(\alpha)\alpha$ with $f(\alpha)$ positive for each root α. To prove (b) we must show f is constant. If α and β are roots which are linearly independent and not orthogonal, there is a root of the form $\alpha + c\beta (c = \pm 1)$. Then $f(\alpha + c\beta)(\alpha + c\beta) = f(\alpha)\alpha + cf(\beta)\beta$ since σ^* is linear, and $f(\alpha) = f(\alpha + c\beta) = f(\beta)$ since α and β are linearly independent. Thus f is constant on each component of Σ. Now evaluating $\sigma^{*r+1}\rho\alpha$ in two different ways $(r + 1 = 1 + r)$, we get $q(\alpha)f(\alpha)\alpha = q(\alpha)f(\rho\alpha)\alpha$, whence $f(\alpha) = f(\rho\alpha)$. Thus f is constant by (a).

11.15. **Remark**. In case (a) of 11.6 the number of q above is just the common value of all $q(\alpha)$ and $\sigma_1 = \rho$ (on the roots), while in case (b) q is their average value $(p^a p^{a+1})^{1/2}$ and $\sigma_1 \neq \rho$.

We are now in a position to apply the considerations of §2 with σ_1 of 11.14 in place of σ of 2.1, since σ_1 fixes a positive chamber and permutes the collection of unit vectors in the directions of the roots (this scaling down is immaterial since 2.1 is a result about W rather than Σ).

11.16. THEOREM. *Assume that G is σ-simple and that q and σ_1 are as in 11.14. Let I_j be basic invariants for W acting on V, so chosen that each is a characteristic vector for $\sigma_1 : \sigma_1 I_j = \varepsilon_j I_j$ with ε_j a root of 1, and $d(j)$ the degree of I_j $(j = 1, 2, \dots, n)$. Let $N = \Sigma(d(j) - 1)$ be the number of positive roots. Then $|G_\sigma| = q^N \Pi(q^{d(j)} - \varepsilon_j)$.*

We get the formula for N, which is of course well known, by comparing degrees in 2.1. The rest of the proof will be given in several steps.

11.17. *Let a denote the number of roots such that $X_\alpha \subseteq U_1$ in* 11.8, *let N denote the total number of positive roots, and for each $w \, \varepsilon \, W$ let $N(w)$ denote the number of roots α such that $\alpha > 0$ and $w\alpha < 0$.* (a) $|U_{1\sigma}| = q^a$ *in* 11.8. (b) $|U_\sigma| = q^N$ *and* $|U_{w\sigma}| = q^{N(w)}$ *in* 11.9.

If we form the product $\Pi q(\alpha)$ over a ρ-orbit of r elements, we get q^r by 11.2(b), 11.14(b) and the fact that σ_1 has the same order as ρ. This yields (a), hence also (b).

11.18. *If Q_w is as in* 11.11, *then* $\Sigma Q_w(w \, \varepsilon \, W_\sigma) = \Pi(q^{d(j)} - \varepsilon_j)/|det(\sigma^* - 1)|$.

We have $\Sigma Q_w = \Sigma q^{N(w)}$ by 11.17(b). We now apply 2.1 with σ replaced by σ_1 and t by q. We get the right side of 11.18 in view of the equation $\det(\sigma^* - 1) = \det(q\sigma_1 - 1) = \Pi(q\varepsilon_{0j} - 1)$, if the ε_{0j} denote the characteristic values of σ_1.

If we now combine 11.10, 11.11, 11.17 and 11.18 we get 11.16.

11.19. **Remarks**. (a) The first factor in the formula for $|G_\sigma|$ represents the action of σ^* on the Jacobian of the I's, since the latter is a scalar multiple of the product of the positive roots, and the second factor may be written $|\det(\sigma_J^* - 1_J)|$ with J as in 2.2(c). In this form the formula is valid even if G is not σ-simple as we see by taking products. The same is true of the formula written as $|G_\sigma| = Q^2 |\det \sigma^*| \det(1_J - \sigma_J^{*-1})$ with Q as in 11.9. (b) If G is σ-simple and simply connected, then G_σ modulo its center is simple if we exclude a few cases (cf. 11.6 and 12.8). Thus the order of the simple group of the family becomes $|G_\sigma|/|C_\sigma|$ if C denotes the center of G.

11.20. *The individual cases of* 11.16. We will discuss the formula for $|G_\sigma|$ in the cases of 11.6, leaving the discussion of the values of $|C_\sigma|$ to the interested reader. Recall that in 11.6(a) $\sigma_1 = \rho$. If σ_1 is the identity, then the ε_j are all 1 and 11.16 becomes Chevalley's formula for the orders of his groups (see [8] where the $d(j)$ for the various types and the orders of the centers are also given (cf. 12.8)). Now assume in case (a) that $\sigma_1 \neq 1$, $\sigma_1^2 = 1$ and $-\sigma_1 \varepsilon W$. Then σ_1 acts on I_j just as -1 does, so that $\varepsilon_j = (-1)^{d(j)}$. Thus we get the formula for $|G_\sigma|$ from Chevalley's by simply replacing the term $q^{d(j)} - 1$ by $q^{d(j)} + 1$ whenever $d(j)$ is odd. For example when G is of type E_6 the $d(j)$ are 2,5,6,8,9,12 so that only 5 and 9 have to be treated as above. This explains the formula in this case, proved in [24, p. 888] by a tricky calculation, and also explains the alternation in the other cases: G of type $A_n (n \geq 2)$ (and G_σ the unitary group), and G of type $D_n(n$ odd) (and G_σ a type of orthogonal group). Next assume in (a) that $\sigma_1 \neq 1$, $\sigma_1^2 = 1$ and $-\sigma_1 \not\varepsilon W$. Then G must be of type D_n (n even), but the following discussion is valid even if n is not even. Relative to a suitable basis $v_1, v_2, \ldots, v_n$ of V the I_j may be taken as Πv_i and the first $n-1$ elementary symmetric polynomials in the v_i^2, and σ_1 as the change in sign of v_n [20, p. 1308]. Thus $|G_\sigma| = q^d(q^n + 1)\Pi(q^{2i} - 1)$, the product on i from 1 to $n - 1$, and $d = n(n-1)$, in agreement with [1]. Finally assume in (a) that $\sigma_1 \neq 1$, $\sigma_1^3 = 1$, so that G_σ is a triality form of D_4. The $d(j)$ are $2,4,4,6$ (see the previous case), and the ε_j are $1, \varepsilon, \varepsilon^2, 1$ with $\varepsilon \neq 1$, $\varepsilon^3 = 1$: this is true to within a permutation by 2.9, and ε and ε^2 have to occur in the same dimension since σ_1 is a real transformation. Thus $|G_\sigma| = q^{12}(q^2-1)(q^6-1)(q^4-\varepsilon)(q^4-\varepsilon^2) = q^{12}(q^2-1)(q^6-1)(q^8+q^4+1)$, which explains the factor $q^8 + q^4 + 1$ in [24, p. 888]. Now consider case (b). If G is of type C_2 (resp. G_2), then $d(j) = 2$, h with $h = 4, 6$ (resp.). The quadratic invariant is always fixed by σ_1 (since W and σ_1 generate a finite group), thus the other one changes sign by 2.9. Thus $|G_\sigma| = q^h(q^2 - 1)(q^h + 1)$ (cf. [16, p. 432], [29]). Finally if G is of type F_4 we have $d(j) = 2, 6, 8, 12$, and we claim that $\varepsilon_j = 1, -1, 1, -1$, respectively. We know by 2.9 that exactly two of the ε_j are 1 and as above that ε_1 is one of them. We consider the form $\Sigma\alpha^8 + \Sigma(\sqrt{2}\beta)^8$, the first sum on the long roots, the second on the short ones; clearly it is fixed by W and σ_1 (see 11.2(b), 11.5, and 11.14(b)). It is not a multiple of the quadratic invariant: to see this choose a basis v_1, v_2, v_3, v_4 for V so that the roots are obtained from $v_1 + v_2$ (short), $2v_1$ (long), and $v_1 + v_2 + v_3 + v_4$ (long) by arbitrary permutations and sign changes, and then compute the terms of the form that involve only v_1 and v_2. Thus it may be taken as I_3, whence $\varepsilon_3 = 1$. It follows that $|G_\sigma| = q^{24}(q^2 - 1)(q^6 + 1)(q^8 - 1)q^{12} + 1)$ (cf. [16, p. 401]).

12. Topology of G_σ

The group G_σ continues as in §11. In this section we consider the "connectedness" and "simply-connectedness" properties of G_σ. The role of identity component is played by $G_{\sigma u}$, the group generated by the unipotent elements of G_σ.

12.1 **Remarks**. (a) The situation is parallel to the "continuous" case in which σ is an automorphism since in that case if each $w \varepsilon W_\sigma$ is represented in N_σ (cf. 11.1 and 8.3(b)), then $G_{\sigma 0}$ turns out to be not only reductive but semisimple, hence is generated by its unipotent elements. (b) In most (but not all) cases $G_{\sigma u}$ turns out to be the derived group of G_σ.

First we extend 11.1 to $G_{\sigma u}$.

12.2. THEOREM. (a) *In* 11.1 *each* n_w *can be chosen in* $G_{\sigma u}$, *in fact in the group generated by* U_σ *and* U_σ^-. (b) *If this is done, then* 11.1 *is true with* $G_{\sigma u}$ *and* $N_\sigma \cap G_{\sigma u}$ *in place of* G_σ *and* N_σ.

Let π be a σ-orbit of simple roots, S the set of all positive roots with support in π, and U_1 the group generated by all X_α for which $\alpha \varepsilon S$. By 11.8 there exists $u \varepsilon U_{1\sigma}$, $u \neq 1$. As in step (4) of the proof of 8.2 this implies that the group generated by U_σ and U_σ^- contains a representative for w_π (see 1.30). Now the elements w_π for the various ρ-orbits generate W_σ (see 1.32 and 11.3). This proves (a), hence also (b).

12.3. COROLLARY. (a) G_σ *is generated by* U_σ, U_σ^- *and* T_σ; *and* $G_{\sigma u}$ *by* U_σ *and* U_σ^- *alone.* (b) $G_\sigma/G_{\sigma u}$ *is isomorphic to* $T_\sigma/(T_\sigma \cap G_{\sigma u})$, *hence is Abelian.*

Here (a) follows from 12.2(a) and (b) from the fact that T_σ normalizes U_σ and U_σ^-.

12.4. THEOREM. *If* G *is simply connected, then* $G_\sigma = G_{\sigma u}$, *i.e.,* G_σ *is generated by its unipotent elements.*

In other words in view of our analogy, G_σ is "connected" (cf. 8.1). By 12.3 we must show: (*) $T_\sigma \subseteq G_{\sigma u}$. For each simple root α let G_α be the group generated by X_α and $X_{-\alpha}$. As is known G_α is semisimple and $T_\alpha = T \cap G_\alpha$ is a maximal torus of it [19, p. 1702], of dimension 1. Let $\lambda_\alpha : K^* \to T_\alpha$ be an isomorphism. Then $\lambda_\alpha \varepsilon L$, the group of one-parameter subgroups of T and $w_\alpha \lambda_\alpha = -\lambda_\alpha$. It follows from 6.4 that the λ_α are up to sign the basic generators of L. Thus T is the direct product of the T_α. Breaking the set of simple roots up into ρ-orbits, we see that it is enough to prove (*) when there is a single orbit. Now if G_1 is a simple component of G, if r is the number of components, and if $\tau = \sigma^r$, then G_σ is isomorphic to $(G_1)_\tau$. Thus it may further be assumed that G is simple, that Σ is irreducible. By now the only possibilities are: G is of type A_1, A_2, C_2 or G_2, and the corresponding group G_σ is SL_2, SU_3, C_2' (Suzuki group), or G_2' (Ree group) (see 11.6). In the second case (*) can be proved by simple calculations. We exclude this case henceforth. Then the following holds:

(**) the element w of W_σ other than 1 is just -1; in other words $w\alpha = -\alpha$ for every root α. Assume now $t \in T_\sigma$. Choose n_w as in 12.2 and $t_1 \in T$ so that $t_1^2 = t$. By (**) $n_w t_1 n_w^{-1} = t_1^{-1}$, whence $t_1 n_w t_1^{-1} = t n_w$. If $p \neq 2$, then n_w is semisimple because $n_w^2 \in T$. Thus n_w and tn_w are semisimple elements of G_σ which conjugate in G. They are thus also conjugate in G_σ (the argument, taken from [26, p. 39], is given in 12.5 below). The last statement also holds if $p = 2$ since then $|T_\sigma|$ is odd by 11.10 and t_1 may be chosen in T_σ. Since $n_w \in G_{\sigma u}$, so is tn_w, and then also t, as required.

In the course of the proof we have used:

12.5. *If x and y are semisimple elements of G_σ which are conjugate in G, assumed to be simply connected, then they are conjugate in G_σ.*

Assume $zxz^{-1} = y$ with $z \in G$. Then $\sigma z \cdot x \cdot \sigma z^{-1} = y$, so that $z^{-1}\sigma z \in G_x$. By 8.1 G_x is connected. Thus by 10.1 there exists $u \in G_x$ such that $z^{-1}\sigma z = u\sigma u^{-1}$. Then $zu \in G_\sigma$, and $(zu)x(zu)^{-1} = y$, as required.

The reader may have noticed that in this last step of the proof of 12.4 its continuous analogue 8.1 has been used.

12.6. Corollary. *If the assumption that G is simply connected is dropped, $\pi : G' \to G$ is the universal covering, and F is the kernel of π, then the following are true.* (a) *$\pi G'_\sigma = G_{\sigma u}$.* (b) *$G_\sigma/G_{\sigma u}$ is isomorphic to $F/(1-\sigma)F$.* (c) *Dually $G_{\sigma u}$ is isomorphic to G'_σ/F_σ.*

We apply 4.5 with G', G, F in place of A, B, C. By 12.4 G'_σ is generated by its unipotent elements. Thus $\pi G'_\sigma \subseteq G_{\sigma u}$. Since F is semisimple, so is $G_\sigma/\ker \delta$, whence $G_{\sigma u} \subseteq \ker \delta$. Since $\pi G_\sigma = \ker \delta$, by exactness we get (a) which implies (c). By 10.1 $(1-\sigma)G' \supseteq F$, whence (b).

12.7. Corollary. *π induces a bijection of the unipotent elements of G'_σ onto those of G_σ.*

This is clear.

12.8. **Remarks**. (a) If G is a simple adjoint group then with a few exceptions $G_{\sigma u}$ is simple (cf. 11.6), and most of the known finite simple groups are obtained in this way. It follows from 12.6(a) that these groups may also be obtained by starting with G simply connected and then dividing G_σ by its center. (b) In analogy with the continuous case $\pi : G'_\sigma \to G_{\sigma u}$ is a universal covering in the following sense: with a few exceptions, every projective representation of $G_{\sigma u}$ can be lifted uniquely to a linear representation of G'_σ (see [25] for a part of the proof).

13. About representations

In this section we present a reformulation and unification of the principal results of [26]. We recall that an irreducible rational representation of a simply connected group is characterized by its highest weight, a character on T which is a nonnegative integral combination of the ω_α of 6.4 [19, Exp. 15, 16].

13.1. THEOREM. *Let G be a simply connected semisimple algebraic group and σ an endomorphism of G onto G such that G_σ is finite. Let $q(\alpha)$ be as in 11.2, and let $\mathcal{R}$ denote the set of irreducible rational representations for which the highest weight $\lambda = \Sigma\lambda(\alpha)\omega_\alpha$ satisfies $0 \le \lambda(\alpha) \le q(\alpha) - 1$, so that $|\mathcal{R}| = |\det \sigma^*| = \Pi q(\alpha)$ (α simple). Then the collection $\Pi_{i=0}^{\infty} R_i \circ \sigma^i$ (tensor product, $R_i \,\varepsilon\, \mathcal{R}$, most R_i trivial) is a complete set of irreducible rational representations of G, each counted exactly once.*

13.2. **Remarks**. (a) In case all $q(\alpha) = p$ and $\rho = 1$ in 13.1 (so that G may be thought of as being defined and "split" over the prime field with σ the pth power map), this reduces to [26, Th. 1.1]. (b) However, 13.1 also contains a refinement of (a) proved in [26], namely, that in case G is simple and two different root lengths occur and are related to p as in 11.5, then each of the basic representations of (a) splits as a tensor product of two others, for one of which λ vanishes on all long (simple) roots, for the other on all short ones. This can be deduced from 13.1 as follows. Let G' be the simply connected group whose root system Σ' is the dual of that, Σ, of G (abstractly Σ' is got from Σ by the inversion $\alpha \to \alpha' = 2\alpha/(\alpha, \alpha)$, hence interchanges long and short roots). Let σ be an endomorphism of $G \times G'$ (interchanging the factors) such that $\sigma^*\alpha = p\alpha'$ and $\sigma^*\alpha' = \alpha$ (resp. $\sigma^*\alpha = \alpha'$ and $\sigma^*\alpha' = p\alpha$) for every long (resp. short) root α of Σ (thus σ^2 is the pth power map). The existence of σ, by no means simple, is proved in [19]. If we now apply 13.1 with G replaced by $G \times G'$ and σ as just described, and use the canonical decomposition of an irreducible representation of $G \times G'$ as a tensor product of representations of G and G', we easily get our assertion. (c) Conversely, 13.1 is a formal consequence of the special case (a) as refined in (b). The proof, a bookkeeping job, consists in breaking each $R_i \circ \sigma^i$ into the basic components described in (a) and (b) and then reassembling the pieces. (d) Among the elements in $\mathcal{R}$ there is one which is especially interesting, namely, the one for which the highest weight λ takes on its highest value: $\lambda = \Sigma(q(\alpha)-1)\omega_\alpha$, which may be written $(\sigma^*-1)\omega$ with $\omega = \Sigma\omega_\alpha$, the famous half the sum of the positive roots. In this case the representation is very much the same as in characteristic 0; in particular the degree is, as in characteristic 0, $\Pi q(\alpha)(\alpha > 0)$ (cf. [26, Cor. 8.3]). Every other representation in $\mathcal{R}$ has a smaller degree as we see from a comparison with the characteristic 0 case where Weyl's formula [32, p. 389] may be used. (e) If G is not simply connected, then 13.1 holds for projective representations of G (which correspond to linear representations of the universal covering).

13.3. THEOREM. *Let G, σ and $\mathcal{R}$ be as in* 13.1. (a) *The elements of $\mathcal{R}$ remain distinct and irreducible on restriction to G_σ.* (b) *A complete set of irreducible representations of G_σ (over K) is obtained in this way.*

13.4. **Remarks**. (a) If G is not simply connected, then 13.3 holds for projective representations of $G_{\sigma u}$ (cf. 12.8(b)). (b) By writing G as a product of its σ-simple components we may reduce 13.3 to the case in which G is simple which is proved in [26]. Once (a) has been proved, (b) follows, by a theorem of Brauer and Nesbitt [5, p. 14], from:

13.5. THEOREM. *If G, σ and $\mathcal{R}$ are as in* 13.1, *then the number of semisimple conjugacy classes of G_σ is $|\mathcal{R}|$.*

This is proved in [26, 3.9]. Another proof, which yields a more general result, will be given in the next section.

As a final related result we state without proof:

13.6. THEOREM. *In* 13.3 *the representation algebra of G_σ over K has a definition in terms of generators and relations (y_α (α simple): $y_\alpha^{q(\alpha)} = y_{\rho\alpha}$) with ρ as in* 11.2.

14. Semisimple classes and maximal tori

In this section we determine the number of semisimple classes and the number of maximal tori of G fixed by σ (with G and σ as in §11) (see 14.8, 14.14 below), and obtain 13.5 as a consequence. Although the two numbers depend eventually on the evaluation of averages over the Weyl group of reciprocal sets of numbers (see 14.4 and 14.6), both turn out to be powers of p (in fact products of $q(\alpha)$'s (see 11.2)), and their product is just the principal term in the formula 11.16 for the order of $|G_\sigma|$. We start our development with some preliminary material about reflection groups.

14.1. LEMMA. *Let V be a real finite-dimensional Euclidean space and E_i the component of degree i of its exterior algebra E. If W is a (not necessarily finite) reflection group which acts effectively on V, then it also acts effectively on each E_i.*

We proceed by induction on n, the dimension of V. Let $v_1, v_2, \dots, v_n$ be a basis of V so chosen that the reflections $w_1, w_2, \dots, w_n$ in the corresponding (orthogonal) hyperplanes belong to W. Let V' be the subspace generated by $v_1, v_2, \dots, v_{n-1}$, let v be a nonzero vector orthogonal to V', and let W' be the restrictions to V' of the group generated by $w_1, w_2, \dots, w_{n-1}$. We have the decomposition $E_i = F_1 + F_2$ with $F_1 = E_i(V')$ and $F_2 = E_{i-1}(V') \wedge v$. Now let $x = x_1 + x_2$ ($x_1 \varepsilon F_1$, $x_2 \varepsilon F_2$) be fixed by W. Then x_1 is fixed by W', hence is 0 by the induction hypothesis applied to W'. Similarly $x_2 = 0$ unless $i = 1$, in which case $x = cv$ for some number c. Then $w_n x = x$ implies that either $c = 0$ or $w_n v = v$. The last equation is not possible because v_n is not orthogonal to v. Thus $x = 0$ as required.

Within the framework of the preceding proof one can also prove the two following results, which will not be used here.

14.2. *If W in 14.1 is not effective on V, then $E_W = E(V_W)$.*

14.3. *If W in 14.1 is irreducible on V, then it is irreducible on each E_i ($i = 1, 2, \ldots, n$) and the corresponding representations are all inequivalent.*

14.4. THEOREM. *Assume in 14.1 that W is finite besides being effective. Let γ be any endomorphism of V. Then $|W|^{-1}\Sigma_{w \varepsilon W}\, det(1 - \gamma w) = 1$.*

Let $v_1, v_2, \ldots, v_n$ be a basis of V. Then $\det(1 - \gamma w)v_1 \wedge \cdots \wedge v_n = (1 - \gamma w)v_1 \wedge \cdots \wedge (1 - \gamma w)v_n$, which can be written as a sum of 2^n terms of the form $\pm y_1 \wedge \cdots \wedge y_i \wedge \gamma w y_{i+1} \wedge \cdots \wedge \gamma w y_n$, the y's forming a permutation of the v's. If the part of this term involving w is averaged over W, then by 14.1 the result is 0 unless $i = n$, i.e., unless the term involves w vacuously. Thus the average of $\det(1 - \gamma w)$ is 1 as required.

14.5. **Remarks**. (a) Similarly one can prove that if β is a second endomorphism then the average value of $\det(\beta - \gamma w)$ is $\det \beta$. (b) The reader might wish to consider the case in which γ is in matric form and W consists of the matrices $\mathrm{diag}(\pm 1, \pm 1, \ldots)$. (c) The preceding results 14.1 to 14.4 are also true if V is a complex unitary space and a reflection is defined to be any automorphism whose fixed-point set is a hyperplane.

Parallel to 14.4, but involving reciprocals, we have:

14.6. THEOREM. *Assume that W is finite but not necessarily effective. Let γ be an invertible linear transformation which normalizes W and is such that every $1 - \gamma w$, and also $1_J - \gamma_J$ (J as in 2.2(c)), is invertible. Then*

$$|W|^{-1} \sum_{w \varepsilon W} det(1 - \gamma w)^{-1} = det(1_J - \gamma_J)^{-1}.$$

Since γ normalizes W, it acts on J, hence the right side of 14.6 makes sense. We introduce a small parameter t and prove 14.6 with γ replaced by $t\gamma$. Because of the identity $\det(1 - \gamma w t)^{-1} = \Sigma tr(\gamma w,\ S_k)t^k$ ($S = \Sigma\, S_k$ is the usual grading of the symmetric algebra on V), which becomes clear once γw has been put in superdiagonal form, the coefficient of t^k on the left of 14.6 is the average of $tr(\gamma w,\ S_k)$, which is $tr(\gamma,\ I(S_k))$ since the average of w on S_k is the projection on $I(S_k)$. Let the homogeneous basic W-invariants $I_1, I_2, \ldots$ of degrees $d(1), d(2), \ldots$ be so chosen that γ acts superdiagonally relative to them: $\gamma I_j = e_j I_j +$ lower terms. Then since the I's are basic, $tr(\gamma,\ I(S_k))$ is $\Sigma e_1^{p(1)} e_2^{p(2)} \cdots$ summed over all sequences $p(1), p(2), \ldots$ such that $p(1)d(1) + p(2)d(2) + \cdots = k$, thus is also the coefficient of t^k in $\Pi(1 - e_j t^{d(j)})^{-1}$, i.e., in $\det(1_J - \gamma_J)^{-1}$, which proves 14.6.

We will also use the following simple combinatorial principle.

14.7. *Let S be a set, W a finite group acting on S, and σ a map of S into S which respects the equivalence relation S/W.*
Then $|(S/W)_\sigma| = |W|^{-1}\Sigma_{w\varepsilon W}|ker(\sigma - w)|$.

Here $\ker(\sigma - w)$ denote the set of solutions of $\sigma x = wx$. Let S_0 be the set of all $x \varepsilon S$ such that $\sigma x = wx$ for some $w \varepsilon W$. Since S_0 is clearly W-stable, each $x \varepsilon S_0$ has $|W/W_x|$ conjugates, all in S_0. Thus $|(S/W)_\sigma|$ equals $|W|^{-1}\Sigma|W_x|$ summed on $x \varepsilon S_0$, hence also $|W|^{-1}\times$ (the number of couples x, w such that $\sigma x = wx$), which is the right side of 14.7.

This concludes the preliminaries.

14.8. THEOREM. *Assume as before that G is a semisimple algebraic group, σ an endomorphism of G onto G such that G_σ is finite, and σ^* and $q(\alpha)$ are as in* 11.2. (a) *The number of semisimple conjugacy classes of G fixed by σ is $|det\ \sigma^*| = \Pi q(\alpha)$ (α simple).* (b) *If also G is simply connected, then the number of semisimple conjugacy of G_σ is the same.*

We first prove:

14.9. *In* 14.8 *each of the numbers $det(\sigma^* - w)$* (a) *has the same sign as $det\ \sigma^*$, and* (b) *is prime to p.*

Let $\tau = w^{-1}\sigma^*$. Because the number of roots is finite, there exists a positive integer n such that $\tau^n\alpha = f(\alpha)\alpha$ with $f(\alpha) > 0$ for all roots α. Since σ^* normalizes W, we have $\tau^n = w_1\sigma^{*n}$ with $w_1 \varepsilon W$. Then $w_1 = 1$ by 1.10, so that each $f(\alpha)$ is p to a positive power by 11.4. If in the identity

$$\det(\tau - c)\det(\tau^{n-1} + c\tau^{n-2} + \cdots + c^{n-1}) = \Pi(f(\alpha) - c^n)$$

(α simple) we now set $c = 1$ and then let c decrease to 0, we see that $\det(\tau - 1)$ is prime to p and has the same sign as $\det\ \tau$, whence 14.9.

We consider now 14.8(a). Every semisimple element of G is conjugate to an element of T (chosen as at the beginning of §11), and two elements of T are conjugate in G if and only if they are conjugate under W, as easily follows from the uniqueness in 6.3. Thus the number sought in (a) is just $|(T/W)_\sigma|$. By 14.7

this equals $|W|^{-1}\Sigma|\ker(\sigma - w)|$. By 14.9(b) and the usual duality $|\ker(\sigma - w)| = |\det(\sigma^* - w)|$, which because of 14.9(a) may be written $|\det \sigma^*| \det(1 - \sigma^{*-1}w)$. Substituting this into the sum and then using 14.4 with $\gamma = \sigma^{*-1}$, we get $|\det \sigma^*|$ as required. To prove 14.8(b) we will show that the natural map from the classes described in (b) to those described in (a) is bijective. It is injective by 12.5 (the simpleconnectedness is used here). It is also surjective, since we have more generally:

14.10. *If G is any connected linear group and σ is as usual, then every class of G fixed by σ contains an element fixed by σ.*

Let C be the class and $x \varepsilon C$ so that $y\sigma x y^{-1} = x$ for some $y \varepsilon G$. By 10.1 there exists $z \varepsilon G$ so that $y = (1-\sigma)z$. Then $z^{-1}xz \varepsilon C \cap G_\sigma$ as required.

14.11. COROLLARY. *Let G be a semisimple algebraic group defined over a finite field k of q elements. Let r be the rank of G.* (a) *The number of semisimple conjugacy classes of G defined over k is q^r.* (b) *The number of semisimple conjugacy classes of G_k is also q^r in case G is simply connected.*

We need only take σ in 14.8 to the the qth power map.

We observe that the formulae of 14.11 also hold if G in 14.8 is σ-simple and q is defined as in 11.14.

By essentially the same calculation as in the proof of 14.8 one can show:

14.12. *If G is a semisimple group of rank r and if n is a multiple of p such that $|n| > 1$, then the number of semisimple classes fixed by the map $x \to x^n$ is $|n|^r$.*

Here the group can be an algebraic group or a complex or compact Lie group (in which case $p = 1$). As it applies to Lie groups this result bears a superficial resemblance to H. Hopf's theorem that the topological degree of the map in 14.12 is n^r, even if G is not semisimple [13].

As a combinatorial corollary of the preceding considerations (e.g. of 14.12), we have:

14.13. *If the symmetric group S_{r+1} acts naturally on the sequences of complex numbers $(c_1, c_2, \dots, c_{r+1})$ $(\Pi c_i = 1)$, then the number of classes fixed by the map $c_i \to c_i^n$ $(|n| > 1)$ is $|n|^r$.*

Now we consider maximal tori fixed by σ.

14.14. THEOREM. *Assume that G and σ are as in 14.8(a) and that Q denotes the order of a maximal unipotent (i.e. a p-Sylow) subgroup of G_σ, so that $Q = \Pi q(\alpha)$, the product taken over the positive roots. Then the number of maximal tori of G fixed by σ is Q^2.*

14.15. **Remark**. In the next section we will show this is also the number of unipotent elements fixed by σ. We know of no way of relating these facts.

Since T is fixed by σ and N is the normalizer of T, the number sought in 14.14 is $|(G/N)_\sigma|$. If we consider G/T instead with W acting from the right ($w \cdot xT = xTn_w^{-1}$), then this is that same as $|((G/T)/W)_\sigma|$, which by 14.7 may be written (*) $|W|^{-1}\Sigma|\ker_{G/T}(\sigma - w)|$. Fix $w \varepsilon W$, write n for n_w, choose $g \varepsilon G$ so that $n^{-1} = (1-\sigma)g$ (by 10.1), and set $\tau = i_n^{-1}\sigma$. A direct calculation shows that left multiplication by g maps $\ker_{G/T}(\sigma - w)$ onto $(G/T)_\tau$, so that the two sets have the same size. Now τ is conjugate to σ, under i_g in fact, so that $|(G/T)_\tau| = |G_\tau|/|T_\tau|$ by 10.11; and $|G_\tau| = |G_\sigma|$, and $|T_\tau| = |\ker_T(w^{-1}\sigma - 1)| = |\ker_T(\sigma - w)| = |\det \sigma^*|^{-1}\det(1-\sigma^{*-1}w)^{-1}$ by 14.9 as in the proof of 14.8. Substituting into (*) we get $|G_\sigma||\det \sigma^*|^{-1}|W|^{-1}\Sigma \det(1-\sigma^{*-1}w)^{-1}$. If we use 14.6 with $\gamma = \sigma^{*-1}$ and then 11.19(a), we get Q^2 as required.

14.16. Corollary. *Let G be a connected linear algebraic group defined over a finite field k of q elements. Let n be the dimension of G and s that of a Cartan subgroup. Then the number of maximal tori (or Cartan subgroups) defined over k is q^{n-s}.*

The Cartan subgroups are the centralizers of the maximal tori [19, p. 701], hence are in $1-1$ correspondence with them, and identical with them in the semisimple case. In this case 14.16 follows from 14.14 applied to the qth power map σ since then $Q^2 = q^{2N}$ (see 11.17) with $2N$ = total number of roots = $n - s$. In the general case let R be radical of G and let $G_1 = G/R$. Let C_1 be a Cartan subgroup of G_1 fixed by σ (i.e. defined over k) and S its inverse image in G. Then S contains a Cartan subgroup C of G [19, p. 705], and this may be chosen to be fixed by σ (by the conjugacy theorem and 10.9). Since S is solvable, C is its own normalizer in S [19, p. 604]. Hence the number of Cartan subgroups of S fixed by σ is $|(S/C)_\sigma|$. Let S_u and C_u be the unipotent parts of S and C. Since $S = S_uC$ (because C contains a maximal torus of S) and $C_u = S_u \cap C$, we may identify S/C with S_u/C_u. Thus the preceding number becomes $|(S_u/C_u)_\sigma|$, which equals $|S_{u\sigma}|/|C_{u\sigma}|$ by 10.11. Now if A is a connected unipotent group defined over k, then $|A_\sigma| = q^{\dim A}$: by Rosenlicht [17] A possesses a normal series defined over k such that each quotient is isomorphic to (the additive group of) K, so that by 10.11 we may assume $A = K$, and then σ has the form $\sigma k = ck^q (c \varepsilon K^*)$, whence our assertion. It follows that the number of Cartan subgroups of G which are in S, i.e., which map onto C_1, is q^k with $k = \dim S_u - \dim C_u = \dim S - \dim C = \dim R + \dim C_1 - \dim C$, which reduces 14.16 to the semisimple case and thus proves it.

14.17 Remark. The number $n - s$ in 14.16 is just the dimension of the variety of Cartan subgroups of G.

15. Unipotent elements

This, our final section, is devoted to the proof of the following result.

15.1. THEOREM. *Assume that G is a semisimple algebraic group and σ an endomorphism of G onto G such that G_σ is finite. Then the number of unipotent elements of G_σ is the square of the number of elements in a maximal unipotent subgroup.*

The last number is, of course, Q in our usual notation. Since an element is unipotent if and only if it is a p-element (its order is some power of p), this can be reformulated:

15.2. *The number of p-elements of G_σ is the square of the number of elements of a p-Sylow subgroup.*

As an interesting consequence, we have:

15.3. COROLLARY. *Let G be a connected linear algebraic group defined over a finite field k of q elements. Let n be the dimension of G and r that of a maximal torus (i.e., r is the rank of G). Then the number of unipotent elements of G defined over k, i.e., of G_k, is q^{n-r}.*

We observe that $n - r$ is just the dimension of the variety of all unipotent elements of G, and that the same formula holds if G in 15.1 is σ-simple and q is defined as in 11.14.

The deduction of 15.3 from 15.1 is analogous to that of 14.16 from 14.14, hence will be left to the reader.

The proof of 15.1 depends on the following two results.

15.4. *If G and σ are as in* 15.1 *and x is a semisimple element of G_σ, there exists a semsimple subgroup G' of G_x which is fixed by σ and contains all of the unipotent elements of G_x.*

15.5. THEOREM. *Assume that G and σ are as in* 15.1 *and also that G is simply connected. Then there exists a complex irreducible character χ on G_σ with the following property. If x is any semisimple element of G_σ and $Q(x)$ is the order of a maximal unipotent (p-Sylow) subgroup of $G_{\sigma x}$, then $\chi(x) = \pm Q(x)$.*

Here $G_{\sigma x} = G_\sigma \cap G_x$.

Let us deduce 15.1 from 15.4 and 15.5, by induction on $\dim G$. By 12.7 and 9.16, we may assume that G is simply connected. The degree $\chi(1)$ of χ in 15.5 is equal to $Q(1)$, the order of a p-Sylow subgroup of G_σ. By a theorem of Brauer and Nesbitt [6], $\chi(x) = 0$ unless x is semisimple. Thus by 15.5 and the orthogonality relations for characters, $|G_\sigma| = \Sigma Q(x)^2$ (x semisimple). Let $P(x)$ denote the number of unipotent elements of $G_{\sigma x}$. Each element of G_σ can be written uniquely xu with x and u commuting semisimple and unipotent elements respectively [19, p. 408], and for a fixed x there are $P(x)$ possibilities for u. Thus $|G_\sigma| = \Sigma P(x)$ (x semisimple). Combined with the preceding equation this yields (*) $\Sigma P(x) = \Sigma Q(x)^2$ (x semisimple). If x is not in the center of G_σ, then $\dim G_x < \dim G$. The induction hypothesis may be applied by 15.4 to yield $P(x) = Q(x)^2$. We may thus cancel in (*) all terms that correspond to values of x not in the center of G_σ. If we then divide by the order of the center, we get $P(1) = Q(1)^2$, which is 15.1 (mod 15.4 and 15.5).

Next we consider 15.4. By 9.4 the group G_{x0} is reductive and contains all of the unipotent elements of G_x; hence so does its semisimple component G' (see 6.5). Since x is fixed by σ so is G', whence 15.4.

It remains to prove 15.5 which is not as simple. First we construct the needed representation. For $w \in W_\sigma$ let $\varepsilon(w)$ denote the determinant of the restriction of w to V_σ (see 1.32). In the group algebra of G_σ over the complex field let $e = \Sigma\, \varepsilon(w)n_w \cdot \Sigma b$, the first sum over W_σ, the second over B_σ, let E be the left ideal generated by e, and let R be the natural representation of G_σ on E by left multiplication.

15.6. THEOREM. *Let the notations be as above.* (a). *The dimension of E is $Q = |U_\sigma|$. The elements ue ($u \in U_\sigma$) form a (linear) basis of E.* (b) *The representation R is integral relative to the basis in* (a). (c) *It is absolutely irreducible and it remains so when reduced mod p relative to the basis in* (a).

This is proved in [23] on the basis of certain axioms (1) to (14) which, because of the properties of the decomposition 11.1 developed in §12, are at once seen to hold in the present case. We omit the detailed verification.

We will show, in several steps, that the character χ of the representation R of 15.6 has the property in 15.5.

(1) *If $x \in T_\sigma$, then $\chi(x) = Q(x)$.* We have $xue = xux^{-1}e$ since $xe = e$. Thus x permutes the basis elements ue, and its character by 15.6 is $\chi(x) = |U_{\sigma x}|$. Now U_x is a maximal unipotent subgroup of G_x (by [19, p. 1702]), and both are fixed by σ. By 11.12 (applied to the semisimple component of G_x) $U_{x\sigma}$ is a maximal unipotent subgroup of $G_{x\sigma}$. Thus $|U_{x\sigma}| = Q(x)$, whence (1).

(2) *For each semisimple $x \in G_\sigma$ there exists some $\tau = \sigma^n$ (n a positive integer) such that x is conjugate in G_τ to some $t \in T_\tau$.* Since x is semisimple it is conjugate to some $t \in T$. Since t is of finite order and T has only finitely many elements of a given finite order, there exists a positive integer n such that $\sigma^n t = t$. We set $\sigma^n = \tau$; then $t \in T_\tau$. Now x and t are semisimple elements of G_τ which are conjugate in G. Since G is simply connected, they are also conjugate in G_τ, by 12.5, whence (2).

(3) *Let* $\tau = \sigma^n$ *(n positive). Write* R_σ *for the representation in* 15.6 *and* R_τ *for the corresponding one in terms of* τ. *Then the restriction of* R_τ *to* G_σ *is the* n^{th} *tensor power of* R_σ. The proof is a bit long and is postponed to the end of this section.

(4) *Deduction of* 15.5. Let x be a semisimple element of G_σ, and let n, τ and t be as in (2). We have $\chi_\sigma(x)^n = \chi_\tau(x) = \chi_\tau(t) = Q_\tau(t) = Q_\tau(x) = Q_\sigma(x)^n$. The first equality is by (3), the second and fourth by the conjugacy between x and t, the third by (1) applied with τ in place of σ, and the fifth by 11.13 applied with the group G' of 15.4 in place of G. Since $\chi_\sigma(x)$ and $Q_\sigma(x)$ are integers, they must agree up to sign, which yields 15.5.

It remains to prove (3). Recall that G is simply connected.

(5) *Let* $\lambda = (\sigma^* - 1)\omega$ *as in* 13.2(d) *and* P_λ *the corresponding rational irreducible representation of* G. *Let* $\bar{R}_\sigma$ *be the reduction mod p of* R_σ (see 15.6(c)). *Then* P_λ *and* $\bar{R}_\sigma$ *are equivalent on* G_σ. By 13.2(d) and 13.3 any irreducible representation of G_σ whose degree is $\Pi q(\alpha)$ $(\alpha > 0)$, i.e., $|U_\sigma|$, must be equivalent to P_λ. By 15.6(c), $\bar{R}_\sigma$ is such a representation, whence (5).

(6) If $\tau = \sigma^n$, *the restriction of* $\bar{R}_\tau$ to G_σ *is the* n^{th} *tensor power* of $\bar{R}_\sigma$. Let μ and P_μ be defined as in (5) but with τ in place of σ. Thus

$$\mu = (\tau^* - 1)\omega = (\sigma^{*n} - 1)\omega = \lambda + \sigma^*\lambda + \cdots + \sigma^{*n-1}\lambda.$$

The terms on the right are the highest weights of the n representations $P_\lambda \circ \sigma^i$ $(i = 0, 1, \ldots, n-1)$. The tensor product $\Pi P_\lambda \circ \sigma^i$ contains P_μ as a component by the above equation for μ, hence is equivalent to it by 13.1 (or else by a comparison of degrees). On restricting to G_σ where σ acts trivially and using (5) for σ and for τ we get (6).

Proof of (3). We will show that $\chi_\sigma(x)^n = \chi_\tau(x)$ for every $x \in G_\sigma$. As already noted both numbers are zero unless x is semisimple. Assume then that x is semisimple, and let m be its order. Let M (resp. $\bar{M}$) be the group of mth roots of 1 in a field containing the characteristic values of $R_\sigma(x)$ (resp. $\bar{R}_\sigma(x)$). Because p is prime to m, there exists an isomorphism Θ of M onto $\bar{M}$. Because also $R_\sigma(x)$ is integral (see 15.6(b)) it follows that if $s_1, s_2, \ldots$ are the characteristic values of $R_\sigma(x)$, each written according to its multiplicity, then $\Theta(s_1), \Theta(s_2), \ldots$ are those of $\bar{R}_\sigma(x)$. The same remarks apply to the characteristic values $t_1, t_2, \ldots$ of $R_\tau(x)$. Consider the equation $\Sigma\Theta(t_j) = (\Sigma\Theta(s_i))^n$. By (6) the terms on the left form a permutation of the terms obtained on the right by formal expansion. Since Θ is an isomorphism, we get $\Sigma t_j = (\Sigma s_i)^n$, i.e., $\chi_\tau(x) = \chi_\sigma(x)^n$, whence (3). The proof of 15.1 is now complete.

15.7. **Remarks**. (a) Our proof of (3) by reduction mod p is by no means elementary, although it is quite natural. Perhaps some reader can replace it by a simple proof based directly on the construction of the representation R. (b) In case all $q(\alpha)$ above are equal, say to q, then the number $Q(x)$ of 15.5 may be written $q^{d(x)}$, with $d(x)$ the dimension of a maximal unipotent subgroup of G_x. Observe that $\lambda = (q-1)\omega$ in (5) in this case. There exists an analogue in characteristic 0. Assume that G is semisimple and simply connected, that $p = 1$, that n is a positive integer, and that x belongs to one of the classes described in 14.12. Then in the irreducible representation of G whose highest weight is $(n-1)\omega$ the character of x is $\pm n^{d(x)}$. This and a corresponding result for compact Lie groups may be deduced from Weyl's formula [32, p. 389]. (c) As an exercise the reader is asked to prove in 15.1 that the number of semisimple elements of G_σ is a multiple of $\det(\sigma_J^* - 1_J)$ (see 11.19(a)).

References

1. E. Artin, *Orders of classical simple groups*, Comm. Pure Appl. Math. **8** (1955), 455.
2. A. Borel, *Sous-groupes commutatifs et torsion des groupes de Lie compacts connexes*, Tôhoku Math. J. **13** (1961), 216–240.
3. A. Borel and J. Tits, *Groupes réductifs*, I.H.E.S. Publ. Math. **27** (1965), 55–151.
4. R. Bott, *An application of the Morse theory to the toplogy of Lie groups*, Bull. Soc. Math. France **84** (1956), 251–282.
5. R. Brauer and C. Nesbitt, *On the modular representations of groups of finite order*, U. of Toronto Studies (1937), 1–21.
6. ______, *On the modular characters of groups*, Ann. of Math **42** (1941), 556–590.
7. E. Cartan, *Complément au mémoire "sur la géométrie des groupes simples"*, Annali Mat. **5** (1928), 253–260.
8. C. Chevalley, *Sur certains groupes simples*, Tôhoku Math. J. **7** (1955), 14–66.
9. ______, *Invariants of finite groups generated by reflections*, Amer. J. Math. **77** (1955), 778–782.
10. H. S. M. Coxeter, *Regular polytopes*, 2nd ed., Macmillan, New York, 1963.
11. C. W. Curtis and I. Reiner, *Representation theory* ..., Interscience, New York, 1962.
12. M. Gerstenhaber, *On the number of nilpotent matrices with coefficients in a finite field.*, Ill. J. Math. **5** (1961), 330–333.
13. H. Hopf, *Über den Rang geschlossenen Liescher Gruppen*, Comm. Math. Helv. **13** (1940), 119–143.

14. N. Iwahori and H. Matsumoto, *On some Bruhat decomposition* ..., I.H.E.S. Publ. Math. **25** (1965), 5–48.
15. S. Lang, *Algebraic groups over finite fields*, Amer. J. Math. **78** (1956), 555–563.
16. R. Ree, *A family of simple groups* ... (two papers), Amer. J. Math. **83** (1961), 401–520, 432–462.
17. M. Rosenlicht, *Some rationality questions on algebraic groups*, Annali Mat. **43** (1957), 25–50.
18. H. Seifert and W. Threlfall, *Lehrbuch der Topologie*, Chelsea, New York, 1947.
19. Séminaire C. Chevalley, *Classification des groupes de Lie algébriques*, (two volumes), Paris, 1956–8.
20. Séminaire "Sophus Lie", *Théorie des algèbres de Lie et Topologies des groupes de Lie*, Paris, 1955.
21. L. Solomon, *Invariants of finite reflection groups*, Nagoya Math. J. **22** (1963), 57–64.
22. ______, *The orders of the finite Chevalley groups*, J. Algebra **3** (1966), 376–393.
23. R. Steinberg, *Prime power representations of finite linear groups.* II, Can. J. Math **9** (1957), 347–351.
24. ______, *Variations of a theme of Chevalley*, Pacific J. Math. **9** (1959), 875-891.
25. ______, *Générateurs, relations et revêtements de groupes algébriques*, Colloque sur la théorie des groupes algébriques, Bruxelles, 1962, pp. 113–127.
26. ______, *Representations of algebraic groups*, Nagoya Math. J. **22** (1963), 33–56.
27. ______, *Differential equations invariant under finite reflection groups*, Trans. Amer. Math. Soc. **112** (1964), 392–400.
28. ______, *Regular elements of semisimple algebraic groups*, I.H.E.S. Publ. Math. **25** (1965), 49–80.
29. M. Suzuki, *On a class of doubly transitive groups*, Ann. of Math. **75** (1962), 105–145.
30. J. Tits, *Algebraic and abstract simple groups*, Annals of Math. **80** (1964), 313–329.
31. B. L. Van der Waerden, *Modern algebra*, (two volumes), Ungar, New York, 1950.
32. H. Weyl, *Theorie der Darstellung*..., Math Zeit. **24** (1926), 377–395.
33. D. Winter, *On automorphisms of algebraic groups*, Bull. Amer. Math. Soc. **42** (1966), 706–708.

Department of Mathematics, University of California, Los Angeles, California 90095-1555

E-mail address: rst@math.ucla.edu

Reprinted from ILLINOIS JOURNAL OF MATHEMATICS
Vol. 13, No. 1, March 1969
Printed in U.S.A.

ALGEBRAIC GROUPS AND FINITE GROUPS

BY
ROBERT STEINBERG

Our object is to indicate how large classes of finite simple groups, specifically those introduced by Chevalley [5], Suzuki [21], Ree [14], and us [16], can be studied profitably with the aid of the theory of linear algebraic groups. We shall refer to these groups as groups of Chevalley type, the first-mentioned as untwisted, the rest as twisted.

First we recall some facts about linear algebraic groups, all taken from [7]. Assume given an algebraically closed field k. A (linear) algebraic group is a subgroup of some $GL_n(k)$ which is at the same time an algebraic set (i.e. the complete set of solutions of a set of polynomials) in the space determined by the n^2 matric coefficients. The Zariski topology, in which the closed sets are the algebraic sets, is used. An algebraic group is said to be simple if it is connected, has only discrete, i.e. finite, nontrivial normal (algebraic) subgroups, and is not Abelian. The simple algebraic groups have been classified by Chevalley [7] in the Killing-Cartan tradition. Thus there are the classical types A_n, B_n, C_n, D_n and the five exceptional types E_6, E_7, E_8, F_4 and G_2. As examples we may mention the groups PSL_n or SL_n (of type A), SO_n or Spin_n (of type B or D depending on the parity of n), Sp_n (of type C), and the group of automorphisms of the Cayley algebra (of type G_2).

The connection between simple algebraic groups and simple finite groups comes from the fact that many of the latter, all of those mentioned in the first sentence above, arise as fixed-point groups of endomorphisms of the former.

The basic tool for studying the latter with the aid of the theory of algebraic groups is the following extension of a result of Lang [13].

(A) Let G be a connected linear algebraic group and σ an (algebraic) endomorphism of G onto G such that G_σ, the group of fixed points, is finite. Then the map $\varphi : G \to G$ defined by $\varphi x = x\sigma x^{-1}$ is surjective.

This is proved in [20]. Here we will sketch a proof in a special case, in which $\sigma x = x^{(q)}$, the result of replacing each entry of x by its qth power, it being assumed that this operation maps G onto itself. Here q is a power of p, the characteristic of k, and is assumed to be greater than 1. From the rules of differentiation, the differential at $y = 1$ of the map $y \to y\sigma y^{-1}$ is the same as that of $y \to y$, hence is surjective. It follows that the map covers an open set in G. The same holds for the map $z \to zg\sigma z^{-1}$ with g an arbitrary element of G. Since G is connected, it is irreducible as an algebraic set [7, Exp. 3]. Thus the preceding open sets intersect: $y\sigma y^{-1} = zg\sigma z^{-1}$ for some y, z. Then $g = \varphi x$ with $x = z^{-1}y$, which proves (A), in this special case.

An immediate consequence of (A) is that if i_g is any inner automorphism, then $i_g\sigma$ is conjugate to σ: if $g = x\sigma x^{-1}$, then $i_g\sigma = i_x\sigma i_x^{-1}$. For the study of

G_σ this means that σ may be altered by an arbitrary inner automorphism and thus be brought to a form in which G_σ can be analyzed exactly. In particular, we get [20]:

(B) If G and σ are as in (A), then every composition factor of G_σ is either cyclic or of Chevalley type.

As a second application of (A) let us consider the conjugacy classes of G_σ. In an ideal situation we would have:

(C) (1) Every class of G fixed by σ contains an element fixed by σ.
(2) Every such class meets G_σ in a single class.

Here (1) is true. Let C be the class. If $y \in C$, then $g\sigma y g^{-1} = y$ for some $g \in G$, whence, writing $g = x\sigma x^{-1}$ by (A), we conclude that σ fixes $x^{-1}yx \in C$. On the other hand (2) is in general false. Assume however that C is such that

(*) G_x is connected for each $x \in C$.

Then (2) holds: if $x, y \in C_\sigma$ and $g \in G$ are such that $gxg^{-1} = y$, then $\sigma g.x.\sigma g^{-1} = y$, whence $g^{-1}\sigma g \in G_x$, $g^{-1}\sigma g = h\sigma h^{-1}$ with $h \in G_x$ by (*) and (A) applied to G_x, and x is conjugate to y under $gh \in G_\sigma$, which yields (2).

The condition (*), hence also (2), holds in a special case, important for representations in characteristic p, viz. when C is semisimple, i.e. consists of diagonalizable elements, and the group G is simple and simply connected (we will not define this term, but remark that SL_n has to be taken rather than PSL_n, and Spin_n rather than SO_n) (see [7, Exp. 23]), the connection with representations coming from the fact that in this case p must be nonzero and the diagonalizable elements of G_σ are those of order prime to p. It leads to a complete survey of the semisimple classes of the finite simple groups of Chevalley type (see [19]). For arbitrary classes, no such survey has as yet been given, except for the classical groups and one or two other types.

One of the most important structural properties of a simple (or more generally connected) algebraic group G is its decomposition into double cosets relative to a Borel (i.e. maximal connected solvable) subgroup, and this carries over to G_σ (as in (A)), basically by (A), where it becomes the decomposition into double cosets relative to the normalizer B of a p-Sylow subgroup P (p is the characteristic of k). In SL_n, for example, the essence of the decomposition is that a system of representatives of the monomial subgroup modulo the diagonal subgroup is also a system of representatives of the double cosets relative to the superdiagonal subgroup. Tits [22] has abstracted the essence of this (Bruhat) decomposition in his theory of B-N pairs. His methods yield a very simple proof of simplicity, i.e. of:

(D) If G and σ are as in (A) and G is simple and simply connected, then G_σ is simple over its center, with a finite number of exceptions which can be listed.

The simple finite groups obtained this way are of course just the groups of Chevalley type. Henceforth G and σ will be as in (D).

The above decomposition leads to a rather uniform determination of the automorphisms of our groups (cf. [17]), as follows. Let α be an automorphism. By Sylow's theorem the group B above may be taken to be fixed by α, and so may some conjugate of P which intersects B trivially, as may easily be proved. By now the situation is quite rigid and the possibilities can be analyzed. In SL_2, for example, the subdiagonal and superdiagonal subgroups are both fixed. Here one knows that α is the conjugation by a diagonal element of GL_2 composed with an automorphism of the base field, and this is the case to which much of the discussion can be reduced. The Schreier conjecture, that the group of outer automorphisms of a simple finite group is always solvable, is verified in every case.

We may also determine the possible isomorphisms among our groups, by extending a method of Artin [3], as follows. The abstract group G_σ usually determines p, the characteristic of k, as the prime making the largest contribution to its order, thus also the p-Sylow subgroup P, its normalizer B, and the numbers of elements in the double cosets relative to B. It rarely happens that these numbers are the same for two choices of G_σ, so that the possible isomorphisms can be severely limited and then analyzed. In some exceptional cases p does not make the largest contribution to the order of G_σ and further argument is necessary.

The Bruhat decomposition leads to a simple presentation of G_σ in terms of generators and relations, at least if G_σ is not twisted, which can be used to study its Schur multiplier and thus the connection between its projective and linear representations (see [18]). The result is that SL_n, Sp_n, Spin_n, E_8, F_4, G_2, and the simply connected versions of E_6 and E_7 all have trivial Schur multipliers, if a finite number of cases are excluded (see [18]; the true exceptions and their exact nature has not been completely worked out yet). Without exception the Schur multiplier is a p-group, since in our presentation each relation either is confined to a p-group or else expresses the conjugacy of two p-elements; hence it causes no trouble for representations in characteristic p. A number of the twisted cases have been treated by Grover [12], and the Suzuki groups by Alperin and Gorenstein [2]. There remain the groups SU_{2n+1} and the Ree groups.

The above ideas are also important in the representation theory of G_σ as far as it has till now been developed.

In characteristic p the theory is in pretty good shape (see [19]). The irreducible representations have been classified, in terms of "highest weights", certain characters on the group B above, and a tensor product theorem, expressing all of them in terms of a few, p^r (r = rank G = dimension of a maximal diagonalizable subgroup), has been proved. Contributors here have been Brauer and Nesbitt [4], Chevalley [7, Exp. 20], and Wong [24]. The information, however, is incomplete, and it is not even known what the degrees of the representations are, except in a few scattered cases. Each representation can be realized in the decomposition of the reduction mod p of a corresponding representation of an algebraic group of characteristic 0, where essentially all is known, by Weyl's character formula [23, p. 389], but the exact nature of the degeneracy that occurs in this reduction is not known. Curtis [8], using the theory of B-N pairs, has presented a more elementary version of the classification, by a construction in the group algebra itself. His approach, however, does not yield the tensor product theorem mentioned above, or, e.g., the degrees.

For representations over the complex field, in contrast, the theory is in poor shape. Aside from the groups GL_n, whose characters have been determined by Green [11], only a few cases, of small dimensions, have been treated. Gelfand and Graev [10] have announced the following theorem.

(E) Assume that G_σ is untwisted. Let r be a one-dimensional representation of P (a p-Sylow subgroup, as above) such that
(**) r has a nontrivial restriction to each one-parameter subgroup of P corresponding to a simple root.
Then the induced representation R of G_σ is multiplicity-free.

This is proved in [10] for SL_n, and in a parallel fashion in [25] for all the groups, however with an oversight, since the proof uses the fact that, in the notation of [25], the group U_2 is the derived group of U, which fails in some (a finite number of) cases. A version of (E) is undoubtedly true for the twisted groups as well. It seems to us that the determination of the irreducible components of R would be a major step in the construction of a representation theory for G_σ, and that if the same were done without the condition (**) on r then the result would be decisive. A second theorem announced in [10], and proved there for the groups SL_n, states that the components thus obtained contain a complete set of irreducible representations of G_σ, i.e., by Frobenius reciprocity, the restriction to P of every representation of G_σ contains a one-dimensional representation. This is in general false. As M. Kneser has observed, the representation of the group of type C_2 over a field of 3 elements onto the group of elements of determinant 1 in the Weyl group of type E_6 in its usual representation as a reflection group provides a counter-example. It would be useful to know for which groups the theorem holds, for which it fails.

If we induce from larger subgroups than P we can expect the decomposition into irreducible components to be easier. In this connection we would like to mention two results.

(*F*) Let $B = PT$ be a semidirect product decomposition of the normalizer B of P, r a one-dimensional representation of B which is trivial on P, and R the induced representation of G_σ. If the restriction of r to T is not fixed by any non-identity element of the Weyl group W (of G_σ relative to T), then R is irreducible.

This follows from a theorem of Mackey (see [9, p. 51]). At the other extreme we have:

(G) If r in (F) is the trivial representation, then the multiplicities of the irreducible components of R are just the degrees of the irreducible representations of W. Or, equivalently, the algebra of complex-valued functions on G_σ which are invariant under left and right translations by the elements of B is isomorphic to the group algebra of W.

This is proved in [1, p. 81].

We will close our discussion by presenting a formula for the orders of the groups G_σ. Let T be a maximal torus of G (a torus is a group isomorphic to a product of GL_1's, e.g. the diagonal subgroup of SL_n), taken to be fixed by σ, which is permissible by (A), and of dimension r (the rank). Then X, the character group of T, is isomorphic to Z^r. The Weyl group $W = N(T)/T$ (of G, not of G_σ) acts on T, hence on X, on the latter as a reflection group (see [7, Exp. 11]). By a result of Chevalley [6], the algebra of invariants under W in the symmetric algebra on X extended to the reals is generated by r homogeneous elements $I_1, I_2, \cdots, I_r$ of uniquely determined degrees $d_1, d_2, \cdots, d_r$. σ acts on T, hence on X, and there may be written $q\sigma_1$ with q positive and σ_1 of finite order. σ_1 normalizes W, hence acts on the invariants, and may be assumed to reproduce the I's with scalar factors $\varepsilon_1, \varepsilon_2, \cdots, \varepsilon_r$. Then $|G_\sigma| = q^d \prod (q^{d_i} - \varepsilon_i)$, with $d = \sum (d_i - 1)$. This can be proved by a method of Solomon (see [15] and [20]). The groups $SL_n(q)$ and $SU_n(q^2)$, for example, can be obtained from $SL_n(k)$ by taking σ in the first case the qth power operation, in the second case its composition with the inverse transpose. Taking T to be the diagonal subgroup, we get σ_1 to be 1 (the identity) in the first case, -1 in the second. Now -1 acts as 1 on a homogeneous invariant of even degree, as -1 on one of odd degree. Thus we can convert the formula for the order of $SL_n(q)$ into that of $SU_n(q^2)$ by simply replacing the factor $q^j - 1$ by $q^j + 1$ whenever j is odd. It is suspected that a similar replacement (basically $q \to -q$) will convert the formulas for the degrees of the irreducible representations, in fact all of the entries in the character table, of the first group into those of the second.

References

1. *Algebraic groups and discontinuous groups*, Amer. Math. Soc. Symposia in Pure Mathematics, Vol. IX.
2. J. Alperin and D. Gorenstein, *The multipliers of certain simple groups*, Proc. Amer. Math. Soc., vol. 17 (1966), pp. 515–519.
3. E. Artin, *Orders of classical simple groups*, Comm. Pure Appl. Math., vol. 8 (1955), pp. 455–472.
4. R. Brauer and C. Nesbitt, *On the modular characters of groups*, Ann. of Math., vol. 42 (1941), pp. 556–590.
5. C. Chevalley, *Sur certains groupes simples*, Tôhoku Math, J., vol. 7 (1955), pp. 14–66.
6. ———, *Invariants of finite groups generated by reflections*, Amer. J. Math., vol. 77 (1955), pp. 778–782.

7. ———, *Classification des groupes de Lie algébriques* (two volumes), Séminaire, Paris, 1956–58.
8. C. W. Curtis, *Irreducible representations of finite groups of Lie type*, J. Reine Angew Math., vol. 219 (1965), pp. 180–199.
9. W. Feit, *Characters of finite groups*, W. A. Benjamin, New York, 1967.
10. I. M. Gelfand and M. I. Graev, *Construction of irreducible representations . . .*, Soviet Math. Dokl., vol. 147 (1962), pp. 529–532.
11. J. A. Green, *The characters of finite linear groups*, Trans. Amer. Math. Soc., vol. 80 (1955), pp. 402–447.
12. J. Grover, *Coverings of groups of Chevalley type*, Ph.D. thesis, U.C.L.A., 1966.
13. S. Lang, *Algebraic groups over finite fields*, Amer. J. Math., vol. 78 (1956), pp. 555–563.
14. R. Ree, *A family of simple groups . . .* (two papers), Amer. J. Math., vol. 83 (1961), pp. 401–420, 432–462.
15. L. Solomon, *The orders of the finite Chevalley groups*, J. Algebra, vol. 3 (1966), pp. 376–393.
16. R. Steinberg, *Variations on a theme of Chevalley*, Pacific J. Math., vol. 9 (1959), pp. 875–891.
17. ———, *Automorphisms of finite linear groups*, Canad. J. Math., vol. 12 (1960), pp. 606–615.
18. ———, *Générateurs, relations et revêtements de groupes algébriques*, Colloque sur la Théorie des groupes algébriques, Brussels, 1962, pp. 113–127.
19. ———, *Representations of algebraic groups*, Nagoya Math. J., vol. 22 (1963), pp. 33–56.
20. ———, *Endomorphisms of linear algebraic groups*, Mem. Amer. Math. Soc., no. 80 (1968).
21. M. Suzuki, *On a class of doubly transitive groups*, Ann. of Math., vol. 75 (1962), pp. 105–145.
22. J. Tits, *Algebraic and abstract simple groups*, Ann. of Math., vol. 80 (1964), pp. 313–329.
23. H. Weyl, *Theorie der Darstellung . . .*, Math. Zeitschr., vol. 24 (1926), pp. 377–395.
24. W. Wong, Ph.D. thesis, Harvard Univ., 1961.
25. T. Yokonuma, *Sur le commutant d'une représentation . . .*, C. R. Acad. Sci. Paris Sér. A, vol. 264 (1967), pp. 433–436.

University of California
Los Angeles, California
Yale University
New Haven, Connecticut

E. CONJUGACY CLASSES

T. A. Springer and R. Steinberg

In this part conjugacy classes of elements of groups that are for the most part semisimple will be considered. We shall treat only the absolute case (in which the groups take their values in an algebraically closed field) and the finite case (in which the groups are the finite Chevalley groups and their twisted analogues). Our purpose is to present a (by no means complete) survey of known results together with some new results, and to focus attention on a number of basic unsolved problems. For standard facts about algebraic groups, many of which will be used without specific reference, the reader may consult [10] or [2], and for facts about reflection groups and root systems [7].

CHAPTER I. BASIC RESULTS AND BACKGROUND

§1. Some recollections

We recall some basic facts about linear algebraic groups and introduce some notations to be used in connection with them. Let G be a connected semisimple algebraic group (semisimple means that the radical, the maximal connected solvable normal subgroup, is trivial) in which a maximal torus T (a group isomorphic to a product of GL_1's) has been chosen. X will always denote the (discrete) character group of T, Σ the subset of roots, and U_a the one-parameter unipotent subgroup corresponding to the root a. Recall that an element is unipotent if its eigenvalues are all 1. Identifying G with its group G(k) of rational points over some algebraically closed field of definition, we have:

1.1. U_a *is unipotent and isomorphic (as an algebraic group) to the additive group* G_a *(of* k). *If* x_a *is an isomorphism from* k *to* U_a, *then* $tx_a(c)t^{-1} = x_a(a(t)c)$ *for all* $t \in T$, $a \in \Sigma$ *and* $c \in k$.

N will denote the normalizer of T and W = N/T the Weyl group. As is known, W acts faithfully on T, hence also on X, as a reflection group with Σ a corresponding root system. We write w_a for the reflection corresponding to a. Assume now that an ordering of Σ has also been given. Then U (resp. U^-) will denote the group generated by those U_a for which $a > o$ (resp. $a < o$), and B the group generated by T and U.

1.2. (a) U *is a maximal connected unipotent subgroup of* G, *and* B *is a maximal connected solvable subgroup (i.e., a Borel subgroup) of* G.

(b) *The natural maps from the Cartesian product* $\prod_{a>o} U_a$ *(fixed, but arbitrary, order of factors) to* U *and from* $T \times U$ *to* B are isomorphisms of varieties.

For $w \in W$, n_w will always denote a representative for w in N (recall $W = N/T$). We have:

1.3. (a) $\{n_w | w \in W\}$ is not only a system of representatives for N/T (or, equivalently, for $T\backslash N/T$), but also for $B\backslash G/B$.

(b) If we set $U_w = U \cap n_w^{-1} U^- n_w$ (or, equivalently, $U_w = \prod U_a$, the product over all a such that $a > o$ and $wa < o$), then each element of Bn_wB can be written uniquely unv with $u \in U$, $n \in n_w T$ and $v \in U_w$.

Further, $(\ ,\ )$ will denote a fixed W-invariant, positive definite, bilinear form on X.

1.4. Reductive groups. A connected linear algebraic group is called reductive if its radical is a torus, necessarily central, or, equivalently, (see [6, Prop. 2.2]) if it can be decomposed $G = G'T'$ with a G' a semisimple group and T' a central torus. In this decomposition T' is uniquely determined as the identity component of the center of G, and G' as the derived group of G or else as the largest semisimple subgroup. It easily follows that 1.1, 1.2 and 1.3 hold when G is a reductive group.

1.5. EXAMPLES. The group SL_n is semisimple. The standard choice for T is the diagonal subgroup. For each couple (i, j) of distinct integers in the range from 1 to n there is a root $a = a(i, j)$ defined by $\operatorname{diag}(t_1, t_2, \ldots, t_n) \longrightarrow t_i t_j^{-1}$, the correspondence group U_a consisting of those matrices that agree with the identity outside the (i, j) position. N consists of the monomial matrices and $W = N/T$ may be identitied with the symmetric group S_n acting on T via permutation of the coordinates, with w_a corresponding to the transposition (ij). The $a(i, j)$ for which $i < j$ may be taken as positive. Then B, U, U^- consist, respectively, of those elements of SL_n that are superdiagonal, unipotent superdiagonal, unipotent subdiagonal. The group $G = GL_n$ is reductive. If it is written $G = G'T'$ as in 1.4, then $G' = SL_n$ and T' consists of the scalar multiples of the identity.

1.6. Further notations. Given an algebraic group G, we write G^o for its identity component and $\underline{g}$ for its Lie algebra. Given an element or subset S of G then $Z_G(S)$ denotes the centralizer of S in G. More generally, if G acts on a set M and S is an element or subset of M then $Z_G(S)$ is the pointwise stabilizer of S in G. Finally, ${}^xS$ stands for xSx^{-1}.

§2. The σ-setup

One of our objects is to study the finite Chevalley groups and their twisted analogues. Their connection with algebraic groups is provided by the following fact: each of them can be realized as the group of fixed points, G_σ, of an endomorphism σ (in the sense of algebraic groups) of a connected (in fact semisimple) linear algebraic group G onto itself.

2.1. EXAMPLES. Let G be a semisimple group defined over a finite field k_o of q elements. The Frobenius map $c \longmapsto c^q$ on the base field extends to an endomorphism of G onto itself. The group G_σ consists of all elements whose coordinates satisfy $c = c^q$, i.e., whose coordinates lie in k_o; hence G_σ is a finite group. The Chevalley groups, introduced by Chevalley in his famous paper [9] and studied at great length in [26] are of this type (the action of σ on each U_a being given by $\sigma x_a(c) = x_a(c^q)$ and on T by $\sigma t = t^q$). The groups $SU_n(k_o)$, $SO^-_{2n}(k_o)$ (the second orthogonal group) and the twisted forms of D_4 and E_6 that arise from the extra symmetries of the underlying root systems are also of this form, but now σ

permutes the groups U_a and acts on X and on T accordingly (see §1). Now let G be a simple algebraic group of type C_2 over a field of characteristic $p = 2$. Let n be a positive integer, and set $r = 2^n$, $s = 2^{n+1}$. Then as is known (see, e.g., [26, §11]) there exists an endomorphism σ such that if a and b are the simple roots with a long, then $\sigma x_a(c) = x_b(c^r)$ and $\sigma x_b(c) = x_a(c^s)$ with similar equations for $-a$, $-b$. The groups G_σ in this case are just the Suzuki groups. A similar construction for the type F_4 with $p = 2$ and type G_2 with $p = 3$ yields the Ree groups. (See, e.g., [26, §11].)

The basic tool for carrying out our study is the following extension of a theorem of Lang.

2.2. THEOREM. Let G be a connected linear algebraic group and σ an endomorphism of G onto G such that G_σ is finite. Then the map $f : x \longrightarrow x\sigma(x)^{-1}$ of G into G is surjective.

This is proved in [27, §10]. In view of its importance we shall sketch a proof. Consider first the case in which G is defined over a finite field of q elements and σ is the Frobenius endomorphism. Computing the differential of f at the identity element of G, we get $df = 1 - d\sigma$. But $d\sigma$ is 0: if x is any function on G, then $dx^q = qx^{q-1}dx = 0$ since the characteristic divides q. Thus df is an isomorphism at the identity. By an analogue of the implicit function theorem [2, p. 75] it follows that the image of f contains a (Zariski) open part of G. If a is a fixed element of G then similarly the map $x \longrightarrow xa\sigma(x)^{-1}$ contains an open part of G. These two open sets intersect since G is connected and hence irreducible as an algebraic set. Thus there exist y, z in G such that $y\sigma(y)^{-1} = za\sigma(z)^{-1}$. Then $a = f(x)$ with $x = z^{-1}y$, as required. In the general case 2.2 may be reduced to the case in which G is semi-simple. What emerges then is that $d\sigma$ is nilpotent due to the finiteness of G_σ so that the earlier argument may be used to complete the proof.

2.3. COROLLARY. *If* a *is an arbitrary element of* G *(as in* 2.2), *then the map* $x \longrightarrow xa\sigma(x)^{-1}$ *is surjective.*

Assume $g \in G$. Choose y and b so that $g = y\sigma(y)^{-1}$ and $a = b\sigma(b)^{-1}$. Then $g = xa\sigma(x)^{-1}$ with $x = yb^{-1}$, as required.

2.4. REMARK. In case G is defined over a finite field k_o and σ is the Frobenius endomorphism the conclusion of 2.2 may be stated: $H^1(k_o, G) = 0$. More generally one knows that $H^1(k_o, G) = 0$ if k_o is any perfect field of cohomological dimension ≤ 1 [25, §10]. Thus the inferences which we shall draw from 2.2 may be to a large extent drawn in this case also.

We start with a preparatory lemma.

2.5. *Let* σ *be an endomorphism of a linear algebraic group* G *onto itself.*

(a) Ker σ *has dimension* 0, *i.e., is finite.*

(b) *Let* A *be a subgroup (algebraic) of* G. *If* $\sigma A \supseteq A$, *then* $\sigma A = A$. *If* $\sigma A \subseteq A$, *then* $\sigma A = A$ *if and only if* $\ker \sigma \cap A \subseteq A^o$; *in particular, this holds if* A *is connected.*

(c) *Let* C *be the center of* G^o *(the identity component of* G). *Then* $\ker \sigma \subseteq C^o$. *In particular, if* C *is finite, e.g., if* G^o *is semi-simple, then* $\ker \sigma = 1$ *so that* σ *is an automorphism of the abstract group underlying* G.

The group $\ker \sigma$ has dimension $\dim G - \dim \sigma G = 0$, hence is finite. Thus $\dim \sigma A = \dim A$ for any subgroup A, so that both inclusions in (b) represent finite extensions of closed subgroups. Since σA can not have more connected components than A, the first statement in (b) follows. Assume $\sigma A \subseteq A$. Then $\sigma A^o \subseteq A^o$ and equality must hold since both groups are connected. Thus $\sigma A = A$ if and only if $\sigma(A/A^o) = A/A^o$, which is equivalent to σ being injective on A/A^o since this group is finite, i.e., to $\ker \sigma \cap A \subseteq A^o$, which proves the second statement in (b). It is clear that in the situation just discussed σ and σ^{-1} permute the cosets of A^o in A. Taking $A = G$, we see that $\sigma^{-1}C \subseteq G^o$. We claim in fact that $\sigma^{-1}C \subseteq C$. Assume $a \in \sigma^{-1}C$, so that $\sigma a \in C$ and $aga^{-1}g^{-1} \in \ker \sigma$ for all $g \in G^o$. The map $g \longrightarrow aga^{-1}g^{-1}$ from the connected set G^o to the discrete set $\ker \sigma$ (by (a)) must be constant, of value 1 clearly, so that $a \in C$, whence our claim. We conclude from $\sigma^{-1}C \subseteq C$ that $\ker \sigma \subseteq C$ and that $\sigma C \supseteq C$, i.e., $\sigma C = C$ by the first part of (b). By the second part $\ker \sigma = \ker \sigma \cap C \subseteq C^o$, whence (c).

Assume now that A is a group on which an endomorphism σ acts. We shall let $H^1(\sigma, A)$ denote A modulo the equivalence relation: $a \sim b$ if $a = cb\sigma(c)^{-1}$ for some $c \in A$.

2.6. Assume G and σ are as in 2.2, and let A be a (closed) subgroup of G fixed by σ. Then the natural map from $H^1(\sigma, A)$ to $H^1(\sigma, A/A^o)$ is bijective.

It is clearly surjective. Let $a, b \in A$ represent the same element of $H^1(\sigma, A/A^o)$. Thus $a \equiv cb\sigma(c)^{-1} \bmod A^o$ for some $c \in A$, or, after replacing b by an element equivalent to it in G, $a \equiv b \bmod A^o$. Write $b = g^{-1}\sigma(g)$ with $g \in G$, as in 2.2. Then σ fixes the connected group gA^og^{-1}, so that if we apply 2.2 in this situation, which is permissible by 2.5(b), we see that $gab^{-1}g^{-1} = gcg^{-1}\sigma(gcg^{-1})^{-1}$ for some $c \in A^o$. This simplifies to $a = cb\sigma(c)^{-1}$, which proves the injectivity.

2.7. Assume G and σ are as in 2.2, and let M be a nonempty left homogeneous space for G on which σ acts. (Thus G permutes *the elements of* M *transitively and* $\sigma(gm) = \sigma(g)\sigma(m)$ *for all* $g \in G$, $m \in M$).

(a) M *contains a point fixed by* σ.

(b) *Fix* $m_o \in M_\sigma$ *and set* $A = Z_G(m_o)$. *Assume that* A *is a closed subgroup of* G. *Then the elements of the orbit space* $G_\sigma \backslash M_\sigma$ *are in one-one correspondence with those of* $H^1(\sigma, A)$, *or, equivalently, with those of* $H^1(\sigma, A/A^o)$.

Choose $m \in M$ and $g \in G$ so that $g\sigma(m) = m$, which is possible since M is nonempty and homogeneous. If $g = x\sigma(x)^{-1}$ as in 2.2, then σ fixes $x^{-1}m$, which proves (a). Next start with an element of $G_\sigma \backslash M_\sigma$. Represent it by some $m \in M_\sigma$ and then choose $g \in G$ so that $gm_o = m$. Applying σ, we get $\sigma g.m_o = m$, so that $g^{-1}\sigma(g)$ is in A and hence represents an element, say h, of $H^1(\sigma, A)$. It is immediate that h is independent of the choice of m and g above, so that a map from $G_\sigma \backslash M_\sigma$ to $H^1(\sigma, A)$ has been defined. This map is injective, since if $m' \in M_\sigma$ and g' and h' are defined accordingly as above then $h = h'$ implies that $g^{-1}\sigma(g) = ag'^{-1}\sigma(g')\sigma(a)^{-1}$ for some $a \in A$, whence $gag'^{-1} \in G_\sigma$, which, because of $gag'^{-1}m' = gam_o = gm_o = m$, shows that m and m' represent the same element of $G_\sigma \backslash M_\sigma$. It is surjective since if $a \in A$ is arbitrary we may, by 2.2, write $a = g^{-1}\sigma(g)$ with $g \in G$ and then verify that $gm_o \in G_\sigma$. The final equivalence in 2.7 follows from 2.6.

2.8. COROLLARY. (a) *If* A *in* 2.7 *is connected, then* M_σ *consists of a single* G_σ *orbit. More generally, this last condition holds if and only if the map* $x \longrightarrow x\sigma(x)^{-1}$ *on* A, *or, equivalently on* A/A^o, *is surjective.*

(b) *If* σ *is trivial on* A/A^o, *then the orbits correspond to the conjugacy classes of* A/A^o, *and in case also* A/A^o *is Abelian, to its elements.*

This is clear.

2.9. *Assume* G *and* σ *are as in* 2.2. *Then* σ *fixes a Borel subgroup* B *and a maximal torus* T *contained in it. Any two such couples are conjugate by an element of* G_σ.

G acts transitively, by conjugation, on the set M of Borel subgroups, and σ acts compatibly. By 2.7(a) there exists a Borel subgroup fixed by σ, and by 2.7(b) any two such are conjugate by an element of G_σ since the normalizer of a Borel subgroup B is B itself, hence connected. The maximal tori may now be treated similarly, but working in B instead of in G.

2.10. "Classification Theorem." Let G be simple, σ as in 2.2, B and T as in 2.9, and then the other notations as in §1. It easily follows that σ permutes the U_a's, also those that are positive, and also those that are simple, and that it acts on T accordingly. The possibilities for G_σ are thus severely limited and can be analyzed. They are in fact just the groups listed in 2.1. For further details see [27, §11].

2.11. *Assume as in* 2.2 *and that* A *and* B *are subgroups of* G *fixed by* σ *such that* $A \supset B$ *and* B *is connected. Then the natural map (inclusion)* $A_\sigma \longrightarrow (A/B)_\sigma$ *is surjective.*

Assume $a \in A$ is such that $aB \in (A/B)_\sigma$. Then aB is a homogeneous space for B acting by multiplication on the right, hence has a point fixed by σ by 2.7(a).

2.12. REMARK. If G in 2.2 is semisimple and B and T are as in 2.9, then each element of $W_\sigma = (N/T)_\sigma$ is represented in N_σ by 2.11. From this a Bruhat decomposition (see 1.3) for G_σ may be inferred from that of G, with B, U, U^-, W replaced by B_σ, U_σ, U^-_σ, W_σ, respectively.

§3. Generalities about conjugacy classes

In this section G is a linear algebraic group, and σ, when it is introduced, is assumed to be as in 2.5.

3.1. We recall the basic facts about the decomposition $x = x_s x_u = x_u x_s$ of an element x of G into its semisimple (i.e., diagonalizable) and unipotent parts. The components x_s and x_u are uniquely determined by x and may be expressed as polynomials in x. Thus $Z(x) = Z(x_s) \cap Z(x_u)$. Further the decomposition is preserved by any homomorphism, in particular by σ. Thus if $x \in G_\sigma$, then $x_s, x_u \in G_\sigma$. In case G is semisimple and nontrivial and σ acts as in 2.2 it follows from the analysis referred to in 2.10 that the characteristic of the base field must be nonzero, say p. Thus in this case for an element of G_σ "unipotent" is equivalent to "p-element" (of order a power of p), and "semisimple" to "p'-element" (of order prime to p).

3.2. *Let σ be an endomorphism of a linear algebraic group G and S an element or subset of G^o fixed by σ. If A is either the centralizer or the normalizer of S in G then $\sigma A = A$.*

It is straightforward that $\sigma A \subseteq A$, while $\ker \sigma \subseteq C^o \subseteq A^o$ by 2.5(c), so that 3.2 follows from the last part of 2.5(b).

3.3. EXERCISE. Show that $S \subseteq G^o$ in 3.2 can not be dropped.

The general connection between the conjugacy classes of G and those of G_σ is as follows.

3.4. *Assume G and σ are as in 2.2 and that C is a class of G fixed by σ.*

(a) *C contains an element fixed by σ.*

(b) *If x is such an element, then the classes of G_σ into which*

$C \cap G_\sigma$ <u>splits are in one-one correspondence with the elements of</u> $H^1(\sigma, Z(x)/Z(x)^o)$.

(c) <u>If</u> $Z(x)$ <u>is connected in</u> (b) <u>then no splitting takes place. In other words, two elements of</u> $C \cap G_\sigma$ <u>which are conjugate in</u> G <u>are also conjugate in</u> G_σ.

This follows from 2.7 and 2.8(a) applied to G acting on C by conjugation.

3.5. EXAMPLES. (a) $G = GL_n$. In this case all centralizers are connected, as will be seen later. It follows from 3.4(c) that if k is a finite field, then two elements of $GL_n(k)$, or of $U_n(k)$, are conjugate there if and only if they are conjugate in $GL_n(\bar{k})$. In the first case we take σ to be the Frobenius endomorphism of $GL_n(\bar{k})$, in the second case, its composition with the inverse transpose.

(b) $G = SL_n$. Here (and in the other classical groups) the situation is less favorable. Assume, for example, that $n = 2$ and that the characteristic is not 2. Let $x(c) = \begin{bmatrix} 1 & c \\ 0 & 1 \end{bmatrix}$. Then $Z(x(1)) = \{\pm x(c) \mid c \in \bar{k}\}$, hence consists of two components. The class of $x(1)$ splits accordingly into two classes of $SL_2(k)$ represented by $x(1)$ and $x(c)$ with c a nonsquare.

3.6. We discuss briefly the situation in the Lie algebra $\underline{g}$ of G, assuming as we may $G \subset$ some GL_n and $\underline{g} \subset \underline{g\ell}_n$ (for the basic facts about Lie algebras of algebraic groups see [2, p. 114]). If $X \in \underline{g}$, then X has the decomposition $X = X_s + X_n$ into semisimple and nilpotent parts to which substantially all the statements of 3.1 apply (see [2, p. 150]). If σ is the Frobenius endomorphism, so that $\sigma G = G$ and $\sigma\underline{g} \subseteq \underline{g}$, then the obvious analogues of 3.2 (S is now a subset of $\underline{g}$) and 3.4 (dealing with the splitting under $Ad(G_\sigma)$ of the classes of $\underline{g}$ under $Ad(G)$ fixed by σ) are true. Further all centralizers of elements of $\underline{g\ell}_n$ in GL_n are connected. For if

$X \in \underline{g\ell}_n$ then $X + C \in GL_n$ for some scalar C, whence our assertion follows from 3.5(a).

§4. Bad primes and others

In this section we discuss some primes which play a role in what follows. Throughout Σ will denote an irreducible root system. For $a \in \Sigma$, set $a^* = 2a/(a, a)$, the coroot of a. Then $\Sigma^* = \{a^* \mid a \in \Sigma\}$ is also a root system, the dual of Σ.

4.1. DEFINITIONS. A prime p is bad (for Σ or for any reductive group with Σ its root system) if $L(\Sigma)/L(\Sigma_1)$ has p-torsion for some (integrally) closed subsystem Σ_1 of Σ. Here $L(\Sigma)$ denotes the lattice generated by Σ. The prime p is a torsion prime if $L(\Sigma^*)/L(\Sigma_1^*)$ has p-torsion for some closed subsystem Σ_1 of Σ.

4.2. Caution. Since Σ_1^* need not be closed in Σ^*, the torsion primes of Σ and the bad primes of Σ^* need not be the same.

4.3. Let $\{a_1, a_2, \ldots, a_r\}$ be a simple system of roots and $h = \Sigma m_i a_i$ the corresponding highest root. Then the following conditions on a prime p are equivalent.

(a) p is bad (see 4.1).

(b) p equals some m_i.

(c) p divides some m_i.

(d) $p \leq$ some m_i.

For the various root systems, the bad primes are as follows:

(e) For type A_r: none.

(f) For types B_r, C_r, D_r: 2.

(g) For types E_6, E_7, F_4, G_2: 2, 3.

(h) For type E_8: 2, 3, 5.

4.4. Let $h^* = \Sigma m_i^* a_i^*$ be the coroot of the highest root expressed in *terms of the coroots of the simple roots. Then the following conditions on a prime* p *are equivalent.*

(a) p *is a torsion prime (see* 4.1).

(b, c, d) *Same as* 4.3(b, c, d) *with* m_i^* *in place of* m_i.

The torsion primes are as listed in 4.3(e, f, g, h) *except that the cases* $C_r(r \geq 2) : 2$ *and* $G_2 : 3$ *should be removed.*

Let us consider 4.3, for example. The highest roots of the various root systems are, of course, well known (see, e.g., [7]), so that the equivalence of (e), (f), (g), (h) and the equivalence of (b), (c), (d) can be verified. Assume that (b) holds, say $p = m_1$. Let Σ_1 be the root system generated by $h, a_2, \ldots, a_r$. Then $L(\Sigma)/L(\Sigma_1)$ is of order p so that (a) holds. To continue we need the following result, proved in [4].

4.5. *The maximal closed subsystems of* Σ *are the following:*

(a) $\langle a_1, a_2, \ldots, \hat{a}_i, \ldots, a_r \rangle$ *with* $m_i = 1$,

(b) $\langle h, a_2, \ldots, \hat{a}_i, \ldots, a_r \rangle$ *with* m_i *prime.*

Now let Σ_1 be a closed subsystem of Σ such that $L(\Sigma)/L(\Sigma_1)$ has p-torsion. We may assume that Σ_1 is maximal by the equivalence of (b) and (d) and the following fact: max m_i diminishes if we replace Σ by any closed irreducible subsystem Σ' (pick compatible orderings, express the highest root of Σ' in terms of the simple roots of Σ', and then express these in terms of the simple roots of Σ). In the cases (a) and (b) of 4.5 we have no torsion and torsion of order m_i, respectively, so that (b) must hold with $p = m_i$. Thus 4.3(a) implies 4.3(b), which completes the proof of 4.3.

4.6. REMARK. As already observed, for the various types $h = \Sigma m_i a_i$ is well known. $h^* = \Sigma m_i^* a_i^*$ can then be found as follows, because of the definitions. Let c be the square length ratio of a long root to a short root (or 1 if there are no short roots). Then

$m_i^* = m_i$ if a_i is long and $m_i^* = m_i/c$ if a_i is short.

4.7. EXERCISE. Prove that if Σ_1 is a subsystem of Σ then the torsion primes (and the bad primes) of Σ_1 are among those of Σ. Because of 4.2 this is not entirely trivial.

Now let $\underline{L}$ be a complex simple Lie algebra whose root system is Σ. Relative to a Cartan decomposition $\underline{L} = \underline{H} + \sum_{a\in\Sigma} \mathbb{C}X_a$ let $\{X_a, H_i = H_{a_i} | a \in \Sigma, i = 1, 2, \ldots, r\}$ be a Chevalley basis for $\underline{L}$ and let $\underline{L}_{\mathbb{Z}}$ be the corresponding algebra over $\mathbb{Z}$. We wish to calculate the discriminant of the Killing form of $\underline{L}_{\mathbb{Z}}$. For this we shall need various formulas in $\underline{L}_{\mathbb{Z}}$ for which we refer the reader to [9].

4.8. *Let* h^* *be as in* 4.4, $m^* = \Sigma m_i^* + 1$, *and* f *the determinant of the Cartan matrix* $((a_i, a_j^*))$. *Let* $\underline{L}_{\mathbb{Z}}$ *be as above,* d *its dimension, and* n *the number of positive roots. If more than one root length occurs let* c *be as in* 4.6 *and* n_1 *(resp.* n_2*) the number of short (resp. short simple) roots. Let* δ *be the discriminant of the Killing form on* $\underline{L}_{\mathbb{Z}}$.

(a) $\delta = (-1)^n (2m^*)^d c^{n_1+n_2} f$.

(b) *If* $\underline{L}_{\mathbb{Z}}$ *is replaced by* $L'_{\mathbb{Z}}$, *obtained by replacing* $\{H_i\}$ *by* $\{H'_j\} \subseteq \underline{H}$ *defined by* $a_i(H'_j) = \delta_{ij}$, *then* f *should be replaced by* $1/f$ *in the expression for* δ.

The proof will be given in several steps.

(1) *If* a *is a long root, then*

$$(H_a, H_a) = \sum_{b\in\Sigma} (b, a^*)^2 = 4m^* .$$

Since H_a commutes with every H_i and $[H_a, X_b] = (b, a^*)X_b$, the first equality holds. In what follows, each unspecified summation is to be taken over the positive roots b. We have, for any i,

$(\Sigma b, a_i^*) = 2$ since w_{a_i} permutes the positive roots other than a_i, hence maps Σb onto $\Sigma b - 2a_i$, on the other hand maps it onto $\Sigma b - (\Sigma b, a_i^*)a_i$ by the formula for a reflection. Combined with the equation $h^* = \Sigma m_i^* a_i^*$ this yields $\Sigma(b, h^*) = 2\Sigma m_i^*$. Each term on the left is 0 or 1 except for the term $b = h$ which is 2. Hence replacing each term on the left by its square increases the sum by exactly 2, so that $\Sigma(b, h^*)^2 = 2\Sigma m_i^* + 2 = 2m^*$. Doubling this equation yields the second equality of (1) when a is h, hence when a is any long root.

(2) <u>If</u> b <u>is a short root, then</u> $(H_b, H_b) = 4cm^*$. Recall that H_b corresponds to $\frac{2}{(b, b)}b$ in the isometry from $\underline{H}$ to its dual given by the Killing form, if b is any root. Hence if a is a long root, then $(H_b, H_b)/(H_a, H_a) = (a, a)/(b, b) = c$, so that (2) follows from (1).

(3) <u>If</u> a <u>is any root, then</u> $(X_a, X_{-a}) = \frac{1}{2}(H_a, H_a)$. Working in the universal enveloping algebra, if we multiply $H_a = X_aX_{-a} - X_{-a}X_a$ on the left by H_a, and $2X_a = H_aX_a - X_aH_a$ on the right by X_{-a}, and then subtract the results, we see that $H_a^2 - 2X_aX_{-a}$ is the commutator of X_a and H_aX_{-a}, so that its trace is 0 in any representation, not just in the adjoint representation.

(4) Now we can calculate the discriminant δ. Since $(H_i, X_a) = 0$ and $(X_a, X_b) = 0$ unless $a = -b$, the discriminantal matrix, and hence also δ, decomposes according to the sum $\underline{L}_{\mathbb{Z}} = \underline{H}_{\mathbb{Z}} + \sum_{a>0} (\mathbb{Z}X_a + \mathbb{Z}X_{-a})$. The contribution of $\underline{H}$ to δ is $\det(H_i, H_j)$, which may be written (multiply the ith row by $2/(H_i, H_i)$) as $\prod_i (H_i, H_i)/2.f$. The contribution of $\mathbb{Z}X_a + \mathbb{Z}X_{-a}$ is $-(X_a, X_{-a})(X_{-a}, X_a) = -(H_a, H_a)^2/4$ by (3). If we multiply all these contributions and then fill in the values of the (H_a, H_a)'s as given by (1) and (2), then after some simplification we get (a). If $\{H_i\}$ is replaced by $\{H_j'\}$, then because of the equations $H_i = \sum_j a_j(H_i)H_j' = \sum_j (a_j, a_i^*)H_j'$ whose determinant is f, the

contribution of $\underline{H}_{\mathbb{Z}}$ above has to be divided by f^2, which proves (b).

4.9. COROLLARY. *If* Σ *is exceptional (of type* E, F *or* G) *and* δ *is as in* 4.8(a) *or* (b), *then the primes that divide* δ *are just the bad primes.*

This is proved by verification. Consider, for example, the type F_4. Here $h = 2a_1 + 3a_2 + 4a_3 + 2a_4$ with the first two roots long and the last two short, and $c = 2$. Thus $h^* = 2a_1^* + 3a_2^* + 2a_3^* + a_4^*$ by 4.6, so that $m^* = 9$. The primes that divide δ are thus 2 and 3, in accordance with 4.3(g).

We conclude with yet another set of primes.

4.10. *The primes that divide the order of the Weyl group corresponding to* Σ *are as follows.*

(a) *For type* A_r: *those* $\leq r + 1$.

(b) *For types* B_r, C_r, D_r: *those* $\leq r$.

(c) *For types* G_2, F_4: 2, 3.

(d) *For type* E_6: 2, 3, 5.

(e) *For types* E_7, E_8: 2, 3, 5, 7.

The proof is by verification.

§5. A finiteness theorem

Our object is the following result of Richardson [18].

5.1. THEOREM. *Let* G *be a (connected) reductive group,* $G \subset GL_n$. *Suppose* (*) *there exists a subspace* $\underline{m}$ *of* $\underline{g\ell}_n$ *such that*

(1) $\underline{g\ell}_n = \underline{g} \oplus \underline{m}$ *and*

(2) $\underline{m}$ *is stable under* Ad(G).

Then every conjugacy class of GL_n *meets* G *in finitely many classes of* G.

For a point v of a variety V we write $T(V)_v$ for the tangent space to V at v. Set $G_1 = GL_n$ and let C_1 be a conjugacy class of G_1. Let Z be an irreducible component of $C_1 \cap G$. Since there are finitely many possibilities for Z it will be enough to show that Z consists of a single class of elements of G. Let C be a class of G contained in Z and g an element of C. Consider the map f from G_1 to $C_1 g^{-1}$ defined by $f(x) = xgx^{-1}g^{-1}$. Clearly f fixes e, the unit element of G.

LEMMA. $(df)_e \colon \underline{g}_1 \longrightarrow T(C_1 g^{-1})_e$ <u>is surjective</u>.

We have $\dim T(C_1 g^{-1})_e = \dim G_1 - \dim Z_{G_1}(g)$. Thus we must prove that $\ker (df)_e$ and $Z_{G_1}(g)$ have the same dimension. The former is an associative algebra, consisting of all $X \in \underline{g\ell}_n$ such that $gXg^{-1} = X$. The latter consists of the invertible elements of this algebra, which form an open part, hence it has the same dimension.

Consider now the following cycle of inclusions.

$$T(Zg^{-1})_e \subset T(C_1 g^{-1})_e \cap T(G)_e = (1-\mathrm{Ad}(g))\underline{g}_1 \cap \underline{g}$$
$$= (1-\mathrm{Ad}(g))\underline{g} \subset T(Cg^{-1})_e \subset T(Zg^{-1})_e \ .$$

Here the first inclusion holds because $Zg^{-1} \subset C_1 g^{-1} \cap G$, the second because, by the lemma, $T(C_1 g^{-1})_e = (df)_e(\underline{g}_1) = (1-\mathrm{Ad}(g))\underline{g}_1$, the third because, by the assumption (*), $(1-\mathrm{Ad}(g))\underline{g}_1 = (1-\mathrm{Ad}(g))\underline{g} \oplus (1-\mathrm{Ad}(g))\underline{m}$ and $(1-\mathrm{Ad}(g))\underline{g}_1 \cap \underline{g} = (1-\mathrm{Ad}(g))\underline{g}$, the fourth because $(1-\mathrm{Ad}(g))\underline{g} = (df)_e(\underline{g})$, and the fifth because $C \subset Z$. It follows that all terms of the cycle are equal, in particular that $T(C)_g = T(Z)_g$. Thus C contains an open part of Z. This applies to any class C of G contained in Z. Since Z is irreducible, there can be only one such class, as required.

5.2. COROLLARY. <u>If</u> G <u>is as in</u> 5.1 <u>and</u> $g \in G$, <u>then</u> $Z_{\underline{g}}(g)$ <u>is the Lie algebra of</u> $Z_G(g)$.

By the cycle of inclusions in the above proof $(1-\mathrm{Ad}(g))\underline{g}$ and C have the same dimension. Since $\dim Z_{\underline{g}}(g) = \dim \ker (1-\mathrm{Ad}(g)) = \dim \underline{g} - \dim (1-\mathrm{Ad}(g))\underline{g}$ and $\dim Z_G(g) = \dim G - \dim C$, so do $Z_{\underline{g}}(g)$ and $Z_G(g)$. Since the former contains the Lie algebra of the latter, we have 5.2.

5.3. LEMMA. *Let* G *be a linear algebraic group and* p *its characteristic. Then the condition* (*) *of* 5.1, *in fact the condition* (**) *the trace form* $T(X, Y) = \mathrm{tr}\, XY$ *is nondegenerate on* $\underline{g}$, *holds for some faithful representation of* G *or a group isogenous to* G *in each of the following cases.*

(a) $p = 0$ *and* G *is simple.*

(b) p *is not a bad prime (see* 4.1 *and* 4.3) *and* G *is simple and not of type* A_n.

(c) $G = GL_n$.

It is enough to prove (**) since then (*) holds with $\underline{m}$ the orthogonal complement (relative to T) of $\underline{g}$ in $\underline{g\ell}_n$. If G is of type B, C, or D in (b), then $p \neq 2$ by 4.3(f) and in the usual representation of G as a classical group $\underline{g}$ consists of the elements of $\underline{g\ell}_n$ skew with respect to some nondegenerate bilinear form on the base space. Since the spaces of skew and symmetric elements are mutually orthogonal (check this), the nondegeneracy of T on $\underline{g}$ follows from that on $\underline{g\ell}_n$, in the present case. If G is exceptional in (b) or $p = 0$ as in (a), we replace G by Ad(G), acting by automorphisms of $\underline{g}$, and its Lie algebra by $\mathrm{ad}(\underline{g})$. It is known that $\underline{g}$ may be obtained by extension of scalars from some algebra between the algebras $\underline{L}_{\mathbb{Z}}$ and $\underline{L}_{\mathbb{Z}}'$ of 4.8. Thus by 4.9 the Killing form on $\underline{g}$ is nondegenerate. If G is as in (c) the situation is clear.

5.4. THEOREM. Let G be a (connected) reductive group. Assume that the characteristic p is good (i.e. not bad for any simple component of G). Then the number of unipotent conjugacy classes of G is finite.

By 1.4 we may assume that G is semisimple. Since G is isogenous to a product of simple groups, and since the number of unipotent classes does not change under an isogeny (because the kernel consists of central semisimple elements), we may assume that G is simple. If G is of type A_n, the number of unipotent classes is finite by the Jordan normal form in SL_{n+1}, in fact equal to $p(n+1)$, the number of partitions of $n+1$. Combining this fact with 5.1 and 5.3(b) we conclude that 5.4 also holds when G is not of type A_n, as required.

5.5. QUESTION. Do 5.1 and 5.4 hold without the assumptions (*) and "p is good" respectively?

5.6. Substantially the same proofs work in the Lie algebra $\underline{g}$ of G to yield in 5.1 every class of $\underline{g\ell}_n$ meets $\underline{g}$ in finitely many classes of $\underline{g}$, in 5.2 if $X \in \underline{g}$ then $Z_{\underline{g}}(X)$ is the Lie algebra of $Z_G(X)$, in 5.4 the number of nilpotent classes of $\underline{g}$ is finite. The last result has a very useful consequence.

5.7. Let G be a nontrivial (connected) reductive group of good characteristic and X a nonzero nilpotent element of $\underline{g}$. Then there exists a torus S in G and a nontrivial character a on S such that $Ad(s)X = a(s)X$ for all $s \in S$.

Let $N(X) = \{x \in G \mid Ad(x)X = a(x)X \text{ for some } a(x) \in k^*\}$. Since the number of nilpotent classes of $\underline{g}$ is finite, there are infinitely many nonzero multiples of X conjugate to X; in fact all are since the image of $a : N(X) \longrightarrow k^*$ is necessarily closed. Hence $N(X)/Z(X)$ is a torus and S may be taken to be any maximal torus of $N(X)$.

CHAPTER II. SEMISIMPLE ELEMENTS

§1. Maximal tori

We recall that every semisimple element of a connected group is contained in a maximal torus and every torus, of course, consists of semisimple elements.

1.1. *Let* G *be a connected (linear algebraic) group and* σ *an endomorphism of* G *onto* G *such that* G_σ *is finite. Then every semisimple element of* G *fixed by* σ *is contained in a maximal torus fixed by* σ.

Let x be such an element. Since x is contained in a maximal torus, the group $Z(x)^o$ contains x and has the same rank as G. This group contains a maximal torus fixed by σ by I.2.9 and this torus contains x because x is central.

In the absolute case all maximal tori are conjugate. In the finite case the situation is as follows.

1.2. *Let* G *be reductive and* σ *as in* 1.1.

(a) G *contains a maximal torus fixed by* σ.

(b) *If* T *is such and* $W = N/T$ *is its Weyl group, then the classes of maximal tori fixed by* σ *under conjugation by* G_σ *are in one-one correspondence with the elements of* $H^1(\sigma, W)$.

(c) *If* σ *fixes each element of* W, *i.e., commutes with the action of* W *on* T, *then the classes in* (b) *correspond to the conjugacy classes of* W.

We apply I.2.7 with G acting by conjugation on the set of maximal tori. We need only observe that N is the stabilizer of T and that $N^o = T$.

1.3. *Twisting*. Let T be as in 1.2, $w \in W$, and correspondingly

$n_w \in N$. Write $n_w = g^{-1}\sigma(g)$. Then referring to the proof of 2.7 we see that σ fixes $T' = {}^gT$ and that every torus fixed by σ can be obtained in this way, by "twisting" by some $w \in W$. If we identify T' with T according to the isomorphism Int(g) we see that the original action σ has to be replaced by $w \circ \sigma$. Replacing w by an element cohomologous to it in $H^1(\sigma, W)$ amounts to replacing $w \circ \sigma$ by something conjugate to it under W.

1.4. LEMMA. Let G, σ, T be as in 1.2 and let σ^* denote the action of σ on X, the dual of T, and also on $X_{\mathbb{R}}$, its real extension.

(a) There exists a permutation π of the roots relative to T and powers q_a of the characteristic such that $\sigma^* a = q_a \pi a$ for every root a. Further $\prod q_a > 1$ for every orbit.

(b) Assume further that G is simple. Then $\sigma^* = q\tau$ with $q > 0$ and τ an isometry of $X_{\mathbb{R}}$. If θ is any π-orbit, then $\prod_{a\in\theta} q_a = q^{|\theta|}$. Further q depends only on σ, not on T. Finally $q > 1$.

The first assertion is standard (see [10, p. 18-06]). If m is the order of π then σ^{*m} maps each $a \in \Sigma$ onto a positive multiple of itself, hence must be a constant on each irreducible component of Σ. From this the first and second statements of (b) follow. The third follows from 1.3 and the fact that each w is an isometry. Now assume $q = 1$. Then every $q_a = 1$ so that σ^* permutes the roots, and by picking T appropriately (see 1.3) we may assume it also permutes the positive ones. But then σ^* fixes the sum of the positive roots, a nonzero character on T. Thus T_σ is infinite, a contradiction. Hence $q > 1$ in (b) and $\prod q_a > 1$ in (a), as required.

1.5. EXAMPLES.

(a) The Chevalley groups and their twisted analogues other

than the Suzuki and Ree groups (see I.2.1). Here the q_a are all equal, equal, in fact, to the number of elements q of the base field. More generally, when T is any torus defined over a field of q elements and σ is the Frobenius automorphism, then $\sigma^* = q\tau$ with τ an automorphism.

(b) Consider the Suzuki groups. Here $\sigma^* a = 2^{n+1} b$ and $\sigma^* b = 2^n a$ in the notation of I, 2.1. Thus $q = 2^{n+1/2}$ and τ interchanges $a/\sqrt{2}$ and $\sqrt{2}b$, hence is the reflection in the line bisecting the angle between a and b. The situation for the Ree groups is similar.

1.6. REMARK. There is a simple formula for the number of maximal tori fixed by σ. We state the result when G is simple. If q is as in 1.4(b) and m is the number of roots, i.e., the dimension of the variety of maximal tori, then the number of maximal tori fixed by σ is q^m. This is worked out in [27, 14.16].

Now we can determine the structure of T_σ.

1.7. <u>Assume that</u> G <u>is semisimple and that</u> T, X, σ^* <u>are as in</u> 1.4.

(a) T_σ <u>is in duality with, hence is isomorphic to,</u> $X/(\sigma^* - 1)X$. <u>Its order is</u> $|\det(\sigma^* - 1)|$.

(b) <u>If also</u> G <u>is simple,</u> $\sigma^* = q\tau$ <u>as in</u> 1.4(b), <u>and</u> f_τ <u>is the characteristic polynomial of</u> τ, <u>then the order of</u> T_σ <u>is</u> $|f_\tau(q)|$.

On taking σ^* mod p (the characteristic) relative to X, we see from 1.4(a) that σ^* becomes nilpotent. Hence $\sigma^* - 1$ is injective on X and $X/(\sigma^* - 1)X$ has finite order prime to p. Then picking compatible bases for X and $(\sigma^* - 1)X$ (elementary divisor theorem), we see that this order is $|\det(\sigma^* - 1)|$ and, since $|X/(\sigma^* - 1)X|$ is prime to p, that $(\sigma^* - 1)X$ is the annihilator of its annihilator in T. The annihilator of $(\sigma^* - 1)X$ in T, however, is just T_σ: if $t \in T$, and t annihilates $(\sigma - 1)X$, then $\chi(\sigma t) = \chi(t)$ for all $\chi \in X$, so that $\sigma t = t$ and $t \in T_\sigma$; and conversely. Thus (a) holds, and by 1.4(b) so does (b).

Before giving some examples, we observe:

1.8. <u>If</u> G, σ, T <u>are as in</u> 1.2 <u>and no root relative to</u> T <u>vanishes on</u> T_σ, <u>then</u> N_σ <u>is the normalizer of</u> T_σ <u>in</u> G_σ <u>and</u> N_σ/T_σ <u>is isomorphic to</u> W_σ.

Assume that x normalizes T_σ. Write $x = unv$ as in I.1.3(b). We have ${}^{nv}T_\sigma = {}^{u^{-1}}T_\sigma$. By I.1.1 and I.1.3(b) the left side is in B^- and the right side in B, hence both sides in T. From I.1.1 it follows that u centralizes T_σ, and since no root vanishes on T_σ that $u = 1$. Similarly $v = 1$. Thus $x \in N$ and the first statement holds. The second does also, by I.2.11.

1.9. REMARK. The condition on T_σ in 1.8 is equivalent to: $(\sigma^* - 1)X$ contains no root; thus it certainly holds most of the time. It can fail, however, e.g., if T is a split torus over the field of two elements, in which case $T_\sigma = \{1\}$. It would, perhaps, be worthwhile to work out the exact exceptions. The condition is also equivalent to T being the unique maximal torus containing T_σ. It should be remarked that in the representation theory one considers $N_{G_\sigma}(T)/T_\sigma$, which is always isomorphic to W_σ.

1.10. EXAMPLES.

(a) The Chevalley groups. Here we can take T to be a split maximal torus, i.e., diagonalizable over the base field, so that $\sigma t = t^q$ for all $t \in T$. Thus σ^* is multiplication by q and τ is the identity. By 1.2(c) the classes of maximal tori fixed by σ correspond to the conjugacy classes of the Weyl group. If T' is another maximal torus fixed by σ, obtained by twisting T by $w \in W$ (see 1.3), then $\sigma^* = qw$ on $X(T')$, and all $w \in W$ are realizable in this situation. By 1.7 the characteristic polynomial of w at q yields the order of

T'_σ and the corresponding matrix is a relation matrix. Further if $W = W(T')$ then W_σ is just $Z_W(w)$, the centralizer of w in W. Carter, in his section of these notes, presents a classification of the conjugacy classes of the various Weyl groups, thus completing the classification of the maximal tori (of the Chevalley groups). He also gives the corresponding characteristic polynomials and tables of $|Z_W(w)|$ for the exceptional groups.

(b) SL_n. Here the Weyl group is the symmetric group S_n, so that the classes of maximal tori correspond to the partitions of n (the class of an element of S_n is determined by its cycle structure). As an exercise we suggest the following: if w is an n-cycle and T is a maximal torus twisted according to w, then T_σ is cyclic of order $(q^n-1)/(q-1)$ and $N(T_\sigma)/T_\sigma$ is cyclic of order n.

(c) SU_n. Here we may take σ on SL_n as Frobenius combined with inverse transpose. If T is the group of diagonal matrices, σ^* acts on X as $-q$. Thus the classes of maximal tori correspond to those of SL_n, under the correspondence $-qw \sim qw$ of the actions of σ^*, hence also to the partitions of n. To convert the formula for the order of a torus of SL_n to the corresponding one of SU_n we simply replace q by $-q$. A similar connection exists between the untwisted and twisted groups of type E_6, and those of type D_{2n+1}, i.e., the orthogonal groups $SO_{4n+2}(q)$ and $SO^-_{4n+2}(q)$ of indices $2n+1$ and $2n$, respectively.

(d) Suzuki groups. Let T be a torus as in 1.5(b). By 1.3 the classes of maximal tori correspond to the classes of elements $w\sigma$, i.e., of elements $\tau' = w\tau$, under conjugation by W. From the explicit representation of W as a dihedral group of order 8 on $X_{\mathbb{R}}$, we see that there are three such classes represented by τ' the reflection τ of 1.5(b), the rotation through 45° and the rotation through 135°. The corresponding characteristic polynomials are

q^2-1, $q^2-\sqrt{2}q+1$, and $q^2+\sqrt{2}q+1$. By 1.7 if we take q as in 1.5(b) we get the corresponding values of $|T_\sigma|$. (Despite appearances these values are all integers.) These groups are all cyclic. Consider the second case, for example. Choosing a basis of X consisting of b and $a+b$ (we are assuming that G is adjoint, that the roots generate X), we see by 1.7 that

$\begin{bmatrix} q/\sqrt{2}-1 & q/\sqrt{2} \\ q/\sqrt{2} & -q/\sqrt{2}-1 \end{bmatrix}$ is a relation matrix for T_σ, from which it easily follows that T_σ is cyclic. By 1.8 we see that $N(T_\sigma)/T_\sigma$ in the above cases is cyclic of order 2, 4, 4, respectively.

(e) GL_n and U_n. Since GL_n is not semisimple, the preceding results don't quite apply. However, by the remark at the end of 1.5(a) we can adapt 1.7 and get substantially the same results for GL_n and U_n as for SL_n and SU_n.

(f) Remaining cases. By the classification of the simple groups G_σ given in I.2.1 and I.2.10 the only groups yet to be discussed are the twisted group of type D_{2n} and the Ree groups of type G_2 and F_4. The classification of maximal tori is in the first case a straightforward exercise which will be left to the interested reader and in the second case very easy (see the paragraph before 5.20), leaving the third case yet to be done (in this connection see 5.21 below).

1.11. DEFINITION. If G, σ, T are as in 1.2 then T is minisotropic if it is contained in no proper parabolic subgroup of G fixed by σ. (A parabolic subgroup is one containing a Borel subgroup.)

This concept is important in the representation theory of G_σ.

1.12. <u>If</u> G, σ, T <u>are as in</u> 1.2, <u>then</u> T <u>is minisotropic if and only if</u> σ^* <u>fixes no ray of</u> $L(\Sigma)_{\mathbb{R}}$, <u>or, equivalently, in case</u> $\sigma^* = q\tau$ <u>as in</u> 1.4(b) <u>or</u> 1.5(a), τ <u>acting on</u> $L(\Sigma)_{\mathbb{R}}$ <u>does not have</u> 1 <u>as an</u>

eigenvalue.

Assume that T is contained in a proper parabolic subgroup P fixed by σ. Let U_o be the unipotent radical of P (see [6, §4]). Thus U_o is a product of certain groups U_a ($a \in S$) and all a may be taken positive. σ^* permutes the elements of S up to positive multiples (see 1.4(a)). If θ is any orbit then positive numbers c_a ($a \in \theta$) can be found so that σ^* fixes the ray through $\Sigma c_a a$. (Check this.) Conversely, assume that σ^* fixes some ray of $L(\Sigma)_{\mathbb{R}}$, through v, say. Let $\Sigma_o = \{a \in \Sigma | (v, a) \geq 0\}$. Then Σ_o is a proper parabolic subsystem of Σ (it is parabolic because either $a \in \Sigma_o$ or $-a \in \Sigma_o$ for each $a \in \Sigma$, and proper since otherwise v would be orthogonal to Σ, hence equal to 0). The corresponding parabolic subgroup (generated by T and the U_a such that $a \in \Sigma_o$) is clearly fixed by σ, so that T is not minisotropic. The last assertion of 1.12 is clear.

1.13. REMARK. Assume that G in 1.12 is defined over a finite field k and σ is the Frobenius endomorphism. It then follows that if G' is the semisimple component of G (see I.1.4) then T is minisotropic if and only if $T \cap G'$ is anisotropic, i.e., has no nontrivial character defined over k.

1.14. EXAMPLES. If G is a Chevalley group and T is a maximal torus "twisted by $w \in W$" (see 1.10(a)), then T is minisotropic if and only if w is fixed-point-free. There is always at least one class of compact tori, corresponding to $w = \prod w_a$, the product over a simple set of roots (the "Coxeter class"), but there may be more: for example, $w = -1$ always corresponds to one. In SL_n (or in GL_n) there is a single class (this is the only simple group with this property), corresponding to the class of n-cycles of S_n (see 1.10(b)). In SU_n (or U_n), on the other hand, there are many, corresponding to the classes of elements of S_n that are products of cycles of odd length (see 1.10(c)). This explains to some extent why, despite the similarities between the two groups, the representation theory of U_n seems to be more difficult (not yet done) than that of GL_n (done). In the Suzuki groups, the first class listed in 1.10(d) is not minisotropic, the second and third are.

1.15. REMARK. For the representation theory it would be interesting to know for which minisotropic maximal tori the subgroup T_σ contains a regular element, one contained in a unique maximal torus.

§2. Simply connected groups and adjoint groups

Let G be a connected (linear algebraic) group. We consider couples $(\pi, \tilde{G})$ such that $\tilde{G}$ is a connected group and π is an isogeny (surjective homomorphism with finite kernel) of $\tilde{G}$ into G which maps a maximal connected unipotent subgroup of $\tilde{G}$ isomorphically (in the sense of algebraic groups) onto one of G (in characteristic 0 the last condition is automatic; in characteristic $p \neq 0$ it rules out unwanted inseparability). If there exists such a couple that dominates all others then it is unique in an obvious sense and we call it the universal covering of G. If (id, G) is the universal covering of G we say that G is simply connected. The following facts are known (see [10]).

2.1. *Let* G *be a semisimple group.*

(a) *There exists a universal covering* $(\pi, \tilde{G})$ *of* G, *and* $\tilde{G}$ *is simply connected (clearly).*

(b) *Let* T *be a maximal torus of* G, $\{a_i\}$ *a simple system of roots, and* $\{a_i^*\}$ *as in* I.4. *Then* G *is simply connected if and only if there exists a basis* $\{\lambda_j\}$ *of* $X(T)$ *such that* $(\lambda_j, a_i^*) = \delta_{ij}$, *or, equivalently,* $w_i\lambda_j = \lambda_j - \delta_{ij}a_i$.

The equivalence is by the formula for a reflection.

There is another way of stating this condition. Let L be the lattice of one-parameter subgroups (homomorphisms of the one-dimensional torus into T). L is in $\mathbb{Z}$-duality with X (see [10, p. 9-06]). For $a \in \Sigma$, $(X, a^*) \subseteq \mathbb{Z}$. Thus a^* may be identified with an element of L (in fact, a^* is just the restriction to the diagonal matrices of any homomorphism of SL_2 into G which maps the superdiagonal and subdiagonal unipotent matrices isomorphically onto U_a and U_{-a}). The following is then clear.

2.2. <u>The condition for simple connectedness in</u> 2.1(b) <u>is equivalent to</u>: Σ^* <u>generates the lattice</u> L <u>of one-parameter subgroups of</u> T.

We recall that, dually:

2.3. G in 2.1 is said to be adjoint if Σ generates X.

2.4. EXAMPLES. Consider SL_n as in I.1.5. Let $t = \mathrm{diag}(t_1, t_2, \ldots, t_n)$. We may take $a_i : t \longrightarrow t_i t_{i+1}^{-1}$. Define λ_j by $t \longrightarrow t_1 t_2 \ldots t_j$. Then since w_i interchanges t_i and t_{i+1}, it modifies λ_i by $t \longrightarrow t_i^{-1} t_{i+1}$, i.e., by $-a_i$ in additive notation, and if $j \neq i$ it fixes λ_j since then $\{t_1, t_2, \ldots, t_j\}$ contains both t_i and t_{i+1} or neither of them. Thus SL_n is simply connected. In the same way one can verify that Sp_n is also. On the other hand, SO_n is not, $\pi : Spin_n \longrightarrow SO_n$ being its universal covering. The groups PSL_n, PSp_n and PSO_n are all adjoint groups. The groups of type E_8, F_4 and G_2 are automatically both simply connected and adjoint.

§3. <u>Semisimple elements</u>

We consider first the absolute case.

3.1. <u>Assume that</u> G <u>is reductive and that</u> T <u>is a maximal torus of</u> G. <u>Then the semisimple conjugacy classes of</u> G <u>are in natural correspondence</u>

with the elements of T/W.

For every semisimple element of G is conjugate to an element of T, while two elements of T which are conjugate in G are conjugate under W: if $xt_1 = t_2x$ with $t_1, t_2 \in T$ and $x = unv$ as in I.1.3(b), then by the uniqueness there $nt_1 = t_2n$.

3.2. *Assume that* G *is semisimple and that* T *and* W *are as in* 3.1.

(a) T/W *is isomorphic to the algebraic variety whose coordinate algebra is* $C(G)$, *the algebra of regular class functions on* G.

(b) $C(G)$ *is freely generated as a linear space by the irreducible characters of* G.

(c) *In case* G *is also simply connected* $C(G)$ *is freely generated as a commutative algebra by the fundamental characters* $\chi_1, \chi_2, \ldots, \chi_r$. *Thus* T/W *is isomorphic to affine* r-*space* A_r *under the map* $t \longrightarrow (\chi_1(t), \chi_2(t), \ldots, \chi_r(t))$.

Here χ_j is the character of the irreducible representation whose highest weight is λ_j (see 2.1(b)). This (3.2) is proved in [25, §6]. We remark that in case $G = SL_n$ 3.2(c) amounts to the fundamental theorem on symmetric polynomials since the χ_i's are then just the elementary symmetric polynomials of the eigenvalues.

3.3. COROLLARY. *Let* G *be semisimple and* x *and* y *semisimple elements of* G. *Then the following conditions are equivalent:*

(a) x *and* y *are conjugate.*

(b) $\chi(x) = \chi(y)$ *for every irreducible character* χ *of* G, *or, equivalently, for every class function.*

(c) $\rho(x)$ *and* $\rho(y)$ *are conjugate in* $GL(V)$ *for every irreducible representation* (ρ, V) *of* G. *If* G *is simply connected,* (b) *and* (c) *need only hold for the fundamental characters and representations.*

Each of the three conditions on x and y is implied by the following by 3.1 and 3.2.

3.4. PROBLEM. Are the first and third conditions of 3.3 equivalent if x and y are not semisimple?

3.5. COROLLARY. *Let G be semisimple and x in G. Then x is unipotent if and only if $\chi(x) = \chi(1)$ for every irreducible character χ, or in case G is simply connected $\chi_j(x) = \chi_j(1)$ for every j.*

We have $\chi(x) = \chi(x_s)$, as we see by putting $\rho(x)$ in its Jordan normal form, and x is unipotent if and only if $x_s = 1$. Thus 3.5 follows from 3.3.

3.6. COROLLARY. *In a semisimple group a conjugacy class is closed if and only if it is semisimple.*

Let C be the class and y an element of C. Assume y semisimple. Then C is specified as the set of elements x of G such that x has the same minimal polynomial as y in some faithful linear realization (this polynomial has no multiple roots, hence defines a semisimple set) and $\chi(x) = \chi(y)$ for every character χ by 3.3, hence is a closed set. The converse follows from:

3.7. *If G is semisimple then the closure of any class contains with each of its elements its semisimple part.*

Let C and y be as before. We may imbed y in B as in I.1.2 so that $y_s \in T$ and $y_u \in U$. Write $y_u = \prod x_a(c_a)$ as in I.1.2(b). Let $n(a)$ denote the height of a, and for each scalar c let $y_u(c) = \prod x_a(c^{n(a)}c_a)$. If $c \neq 0$ then $y_s y_u(c)$ is conjugate to y via an element of T, hence belongs to C. If f is a regular function on G vanishing on C then $f(y_s y_u(c))$ is a polynomial in c vanishing for $c \neq 0$, hence also for $c = 0$. Thus C contains $y_s y_u(0) = y_s$, as required.

3.8. REMARK. More generally one can show that if G is a connected group then any class meeting a Cartan subgroup is closed [25, 6.14]. If we apply this to the semidirect product formed by a semisimple group G acting on its Lie algebra $\underline{g}$, we soon see that a class in $\underline{g}$ (under the action of G) is closed if and only if it is semisimple.

To continue we need the following important result, which will be discussed in the next section.

3.9. In a semisimple simply connected group the centralizer of every semisimple element is a connected reductive group.

3.10. COROLLARY. Assume that G is semisimple and simply connected and that σ is as in 1.1. Then the natural map (inclusion) yields a bijection from the semisimple classes of G_σ to the semisimple classes of G fixed by σ.

This follows from 3.9 and I.3.4, parts (a) and (c).

3.11. COROLLARY. If G and σ are as in 3.10 and T is a maximal torus fixed by σ, then the semisimple classes of G_σ are in natural correspondence with the elements of $(T/W)_\sigma$ and with the elements of $(A_r)_\sigma$ (see 3.2(c)).

By 3.10, 3.1 and 3.2(c).

The first part of 3.11 may be restated: $t \in T$ is conjugate to an element of G_σ if and only if $w \circ \sigma$ fixes t for some $w \in W$; thus it may be viewed as a classification of the semisimple elements of G_σ according to the types of tori fixed by σ in which they lie.

3.12. COROLLARY. Let G, σ, T be as in 3.11 and $x \in T_\sigma$. Then the classes of maximal tori fixed by σ represented by tori containing x are those obtained by twisting T by some $w \in Z_W(x)$ (see 1.3).

Let $T' = {}^gT$ $(g \in G)$ be fixed by σ. By 3.1 we may assume that $x = {}^gx$, i.e., $g \in Z_G(x)$. Then $g^{-1}\sigma(g) \in N \cap Z_G(x)$ so that the twisting is by some $w \in Z_W(x)$. Conversely, if $w \in Z_W(x)$, we can write $n_w = g^{-1}\sigma(g)$ with $g \in Z_G(x)$ by I.2.2 applied to $Z_G(x)$ which is connected by 3.9, and then work backwards.

3.13. COROLLARY. Let G, σ, T be as in 3.11.

(a) The number of semisimple classes of G_σ is $|\det \sigma^*|$.

(b) If T is contained in a Borel subgroup fixed by σ, this number is $\prod_{a \text{ simple}} q_a$ with q_a as in 1.4(a).

(c) If G is simple the number is q^r with q as in 1.4(b).

By 1.3 we may assume in (a) that T is contained in a Borel subgroup fixed by σ. Choosing in $X_{\mathbb{R}}$ the basis consisting of a set of simple roots we see that (a) and (b) are equivalent and clearly (a) implies (c). Now σ acts on the coordinate χ_a (write $\chi_i = \chi_a$ if $a_i = a$) of A_r (see 3.2(c)) thus: $\sigma^*\chi_a = \chi_a \circ \sigma = \chi_{\pi a}^{q_a}$ by 1.4(a). The number of points of A_r fixed by σ is thus $\prod q_a$. By 3.11 this yields (b).

3.14. REMARKS. In the case that σ is the Frobenius endomorphism, what we have just proved also follows from: every vector space defined over a field k has a basis defined over k [19, p. 159, Prop. 3]. It is also possible to prove 3.13 by evaluating $|(T/W)_\sigma|$ combinatorially [27, 14.8].

3.15. EXAMPLES. In SL_n the semisimple classes correspond to the possible characteristic polynomials, since the coefficients of the latter excluding the first and last are just the χ_i's. Since these coefficients are just the elementary symmetric polynomials in the eigenvalues, 3.2(c) amounts to the fundamental theorem on symmetric polynomials. There are q^{n-1} semisimple classes in $SL_n(k)$, $|k| = q$, since each coefficient can take on any value in k. In $SU_n(k)$ the number of classes is the same, but the coefficients must satisfy $\chi_{n-i} = \chi_i^q$ instead. In the Suzuki group of 1.5(b) the number of semisimple classes is 2^{2n+1}.

Most of the preceding results have natural analogues in $\underline{g}$ which we now discuss.

3.16. (See 3.1.) *If G is reductive, then the semisimple conjugacy classes of $\underline{g}$ (conjugacy under Ad(G)) are in natural correspondence with the elements of $\underline{t}/W$.*

The proof is like that of 3.1.

3.17. (See 3.2(a).) *If G is semisimple and adjoint, then $\underline{t}/W$ is isomorphic to the algebraic variety whose coordinate algebra is $C(\underline{g})$ (the Ad(G)-invariant polynomials on $\underline{g}$).*

Since $\underline{t}/W$ has $C(\underline{t})^W$ as its coordinate algebra this amounts to proving:

3.17'. *In 3.17 the natural map f (induced by restriction) from $C(\underline{g})$ to $C(\underline{t})^W$ is an isomorphism.*

In char $\neq 0$ this result is new, while in char 0 no decent proof appears in the literature. Hence we shall sketch a proof.

(1) Define a morphism φ from $G/T \times \underline{t}$ to $\underline{g}$ by $\varphi(gT, X) = \mathrm{Ad}(g)X$. The Weyl group W acts on $G/T \times \underline{t}$ by $w.(gT, X) = (gn_w^{-1}T, \mathrm{Ad}(n_w)X)$ and $\varphi \circ w = \varphi$. Let $x = (T, X)$ with X regular in $\underline{t}$, i.e., such that $(da)(X) \neq 0$ for all roots a on T; such an X exists because G is adjoint. It is easily checked that $d\varphi$ is surjective at x. Hence φ is separable. Since also $\varphi^{-1}(\varphi x)$ has exactly $|W|$ points, standard facts from algebraic geometry (see, e.g., [10], in particular Exp. 5 and pages 5-07 and 5-08) imply that the quotient variety $V = W \backslash (G/T \times \underline{t})$ is birationally equivalent to $\underline{g}$.

(2) If F is a class function (perhaps not regular) on $\underline{g}$, then F is defined at a point X if and only if it is defined at a conjugate of X_s in $\underline{t}$, and furthermore $F(X) = F(X_s)$. This can be proved like the corresponding result in G (see [25, p. 65, last paragraph]). It follows that the map f of 3.17' is injective.

(3) Assume $F \in C(\underline{t})^W$. Define F_1 on $G/T \times \underline{t}$ by $F_1(gT, X) = F(X)$. Then F_1 is a W-invariant regular function on $G/T \times \underline{t}$, hence a regular function on V. By (1) there is a rational function H on $\underline{g}$, corresponding to F_1 via φ, which is $Ad(G)$-invariant and which agrees with F at its points of definition in $\underline{t}$. It remains, for the surjectivity of f in 3.17', to prove that H is in fact a polynomial on $\underline{g}$. Write $H = H_1/H_2$ as the ratio of relatively prime polynomials. Since $Ad(G)$ equals its own derived group, it has no nontrivial characters into k^*, so that H_1 and H_2 are both $Ad(G)$-invariant. We have $H_1(X) = F(X)H_2(X)$ for $X \in \underline{t}$. Thus $H_2(X) = 0$ implies $H_1(X) = 0$, for all $X \in \underline{g}$ by (2). This can only be if H_2 is constant, so that H is a polynomial as required.

From what has been said, proving the analogue of 3.2(c) comes down to showing that $C(\underline{t})^W$ is a polynomial algebra (on $r = \dim t$ generators). In char. 0 this is a well known result of Chevalley, while in char. $\neq 0$ it can be checked in some of the classical and low-dimensional cases. It holds in particular when G is of type A_2 in char. 3. In this case, however, when G is simply connected the degrees of the basic generators of $C(\underline{t})^W$ are 2 and 3, while when G is adjoint they are 1 and 6. Thus these degrees may change under isogeny.

3.18. PROBLEM. Assume that G is semisimple (and perhaps also adjoint). Prove that $C(\underline{t})^W$ is a polynomial algebra.

Concerning 3.9 we mention the following result whose proof will appear elsewhere.

3.19. *Assume that* G *is semisimple and simply connected. If the characteristic of* G *is not a torsion prime (see* I.4.1), *then* $Z_G(X)$ *is connected (and reductive) for every semisimple element* X *of* $\mathfrak{g}$, *and conversely*.

3.20. EXERCISE. Formulate and prove analogues in $\mathfrak{g}$ of the following results in G: the equivalence of 3.3(a) and 3.3(c), 3.5, 3.6 (see 3.8), and 3.7. Do the same for 3.10 through 3.13, restricting the characteristic as in 3.19.

§4. *The connectedness theorem* 3.9

This is proved in [27, 8.1], in a more general form. In view of the importance of 3.9 and a lemma 4.2 that comes up in the proof we shall sketch a proof here.

4.1. *Assume* G *semisimple,* T *a maximal torus, and* S *a subset of* T. *Let* Σ_1 *be the system of roots vanishing on* S *and* W_1 *the centralizer of* S *in* W.

(a) $Z_G(S)$ *is generated by* T, *the* U_a *such that* $a \in \Sigma_1$, *and the* n_w *such that* $w \in W_1$.

(b) $Z_G(S)^o$ *is generated by* T *and the* U_a's, *hence is reductive with* Σ_1 *as its root system.*

By the uniqueness in I.1.3(b) and I.1.2(b) and the equation in I.1.1 we get (a) without difficulty (cf. the proof of 3.1). The subgroup generated by T and the U_a's is connected, clearly, and of finite index in $Z_G(S)$ since it is normalized by the n_w's, so that it must equal $Z_G(S)^o$. It is then standard that $Z_G(S)^o$ is reductive with Σ_1 as its root system [10, Exp. 17].

We observe next that if a is a root such that $w_a \in W_1$, then $a \in \Sigma_1$ and $n_{w_a} \in Z_G(S)^o$, if G is simply connected: we may take a simple, $a = a_i$ as in 2.1(b), then apply λ_i to $t = w_{a_i}(t)$ to get

$a_i(t) = 1$, and then choose $n_{w_a} \in \langle U_a, U_{-a} \rangle \subseteq Z_G(S)^o$. Applying this and 4.1 to the case that S has a single element, and using the condition 2.2 for simple-connectedness, we see that the proof of 3.9 (all that remains is the connectedness) will be completed by:

4.2. THEOREM. *Let* T *be an algebraic torus*, W *a finite reflection group on* T *(i.e.*, W *acts on* X(T) *as a finite group generated by reflections)*, L *the lattice of one-parameter subgroups of* T, *and* L^o *the sublattice generated by the elements in the directions in which reflections take place. Assume that* $L = L^o$. *Then* $Z_W(t)$ *is a reflection subgroup for every* $t \in T$.

4.2(a). *Reduction to the case of an ordinary torus (product of circles).* Let X be the dual of T and $\hat{T}$ the topological dual of X. We have $X \cong \mathbb{Z}^r$ so that $\hat{T}$ is an ordinary torus. The group W acts on T, hence on X and $\hat{T}$, and the condition $L = L^o$ carries over to $\hat{T}$ since L is the $\mathbb{Z}$-dual of X. Thus the desired reduction follows from:

(*) There exists $\hat{t} \in \hat{T}$ such that $Z_W(t) = Z_W(\hat{t})$. Let A be the annihilator of t in X and B the annihilator of A in $\hat{T}$. The group X/A is finitely generated, hence is the product of a lattice and a finite group. Since it is isomorphic to a subgroup of k^* (k is some algebraically closed field of definition for T and t), to the set of values X(t) in fact, the finite group is cyclic. Thus B, the topological dual of X/A, is the product of a torus and a cyclic group, and hence possesses a topological generator $\hat{t}$. Now if $w \in W$, then $(1-w)X$ vanishes on t if and only if it does on $\hat{t}$, i.e., w fixes t if and only if it fixes $\hat{t}$, which proves (*).

4.2(b). *Proof of* 4.2 *when* T *is an ordinary torus.* Let $p : V \longrightarrow T$ be the universal covering. Then the one-parameter subgroups correspond to the paths $(0\ v)$ $(v \in \ker p)$ so that L may be identified with $\ker p$. Let R be the set of minimal elements of L in the directions in which reflections take place.

(1) If $r, s \in R$ and $(r, s) < 0$, then $r+s = 0$ or $r+s \in R$.

If $r, s \in R$, then $(s, r^*)r = (1-w_r)s \in L$, whence $(s, r^*) \in \mathbb{Z}$. It follows that R is a root system and (1) holds. Now given $t \in T$ choose $v \in p^{-1}t$ so that the length $|v|$ is minimal. Let $w(t) = t$, so that $wv-v \in L$. Since R generates L by assumption we may write

(2) $wv-v = \sum_1^n r_i$ with $r_i \in R$. Choose this expression to be minimal. Then (1) implies

(3) $(r_i, r_j) \geq 0$ for all i, j. We have

$$\begin{aligned} n|v|^2 &\leq \sum_1^n |v + r_i|^2 \\ &\leq \sum_1^n |v + r_i|^2 + 2 \sum_{i<j} (r_i, r_j) \\ &= (n-1)|v|^2 + |v + \Sigma r_i|^2 \\ &= (n-1)|v|^2 + |wv|^2 = n|v|^2 . \end{aligned}$$

The first inequality is by the choice of v, the second by (3). Clearly both must be equalities. Thus (4) $|v| = |v + r_i|$ for every i, and (5) $(r_i, r_j) = 0$ for $i < j$. If we write $w_i = w_{r_i}$, then (4) together with the formula for a reflection yields (6) $w_i v = v + r_i$. Thus $\prod w_i \cdot v = v + \Sigma r_i = wv$ by (5) and (6), whence $\prod w_i^{-1} wv = v$. Since $w_i \in Z_W(t)$ by (6), we may assume $wv = v$ at the start. But then it is standard (since we now are dealing with a reflection group on a Euclidean space) that w is a product of reflections of W that fix v, so that we are done.

4.3. EXAMPLES. If $x = \operatorname{diag}(i, -i) \in PSL_2$, then $Z(x)$ consists of the monomial matrices, hence has two connected components. If we go to SL_2, then $Z(x)$ becomes connected by 3.9, but then the centralizer of a unipotent element becomes disconnected (see I. 3. 5(b)). The final remedy is to go to GL_2 (or in higher dimensions GL_n) where all centralizers are connected. This process of making the center of the simply connected group connected by extending it to a central torus, however, will not work for the other types of groups. In a group of type G_2 in characteristic 0, for example, it may be verified that if $x = x_a(1)x_{a+3b}(1)$ (with a, b simple and the notation as in I. 1. 1), then $Z(x)$ is a connected unipotent group extended by the symmetric group S_3.

4.4. COROLLARY (Extension of 3.9). _Assume_ G _semisimple, but perhaps not simply connected. Let_ $\pi : \tilde{G} \longrightarrow G$ _be the universal covering and_ $F = \ker \pi$ _(the fundamental group of_ G). _Then for any semisimple_ $x \in G$, $Z(x)/Z(x)^o$ _is isomorphic to a subgroup of_ F, _hence is Abelian and consists of semisimple elements._

Define $\varphi : Z(x) \longrightarrow F$ thus. For $g \in Z(x)$ choose $\tilde{g} \in \pi^{-1}g$ and $\tilde{x} \in \pi^{-1}x$ and then set $\varphi g = (\tilde{g}, \tilde{x})$ (commutator). Since F is central φ is well-defined and is a homomorphism. We have $Z(x)^o \subset \ker \varphi$ since F is finite, $\ker \varphi \subset \pi Z(\tilde{x})$ by the definition of φ, and $\pi Z(\tilde{x}) \subset Z(x)^o$ since $Z(\tilde{x})$ is connected by 3.9. Thus $\ker \varphi = Z(x)^o$ and 4.4 follows.

4.5. REMARK. It can be shown that in 4.4 all subgroups of F are realizable. Since $F = 1$ is equivalent, up to an inseparable isogeny, to G being simply connected, this provides a converse to 3.9 [27, 9.1].

4.6. COROLLARY. _Assume as in_ 4.4 _and that_ x _is of finite order prime to_ $|F|$. _Then_ $Z(x)$ _is connected._

We proceed as in the proof of 4.4. Since $(\tilde{g}, \tilde{x})$ is central it follows that $(\tilde{g}, \tilde{x})^n = (\tilde{g}, \tilde{x}^n)$ for every n, so that $(\tilde{g}, \tilde{x})$ has order prime to $|F|$. Since it belongs to F it must equal 1. Thus $Z(x) = \ker \varphi = Z(x)^o$, as required.

As a consequence of 4.6 we see that the bijection of 3.10 carries over to the nonsimply connected case provided that we stick to elements of order prime to $|F|$. In SO_n, for example, we are all right if we stick to elements of odd order.

4.7. REMARKS on the main lemma 4.2 of the proof of 3.9.

(a) If we drop the assumption $L = L^o$, then in analogy with 4.4, we may show that if $Z_W(t)^o$ is the subgroup of $Z_W(t)$ generated by its reflections then $Z_W(t)/Z_W(t)^o$ is isomorphic to a subgroup of $\operatorname{tors}(L/L^o)$.

(b) The transfer lemma 4.2(a) (see (*)), and the proof given, do not depend on the condition $L = L^o$. Further if t has finite order, then $\hat{t}$ may be chosen to have the same order. This result (with the condition $L = L^o$ still dropped) may be extended as follows. (**) Let S be a finite W-stable subgroup of T. Then there exists a subgroup $\hat{S}$ of $\hat{T}$ and a W-isomorphism φ of S onto $\hat{S}$. To see this we let $\hat{S}$ be the annihilator in $\hat{T}$ of the annihilator of S in X and define φ thus: if the group of $|S|^{th}$ roots of 1 in k^* is identified with that in $\mathbb{C}$ by some (unnatural) isomorphism, then $\varphi(s)$ is the element of S such that $\chi(\varphi(s)) = \chi(s)$ for all $\chi \in X$. The proof is then clear. If we assume that (T, W) (resp. $(\hat{T}, W)$) has been abstracted from a semisimple algebraic group G (resp. semisimple compact Lie group $\hat{G}$), then (**) enables us to transfer from G (or from G_σ in 1.1) to $\hat{G}$ certain problems about semisimple elements. The point is that in $\hat{G}$ the situation is much more favorable, especially if we go to the simply connected covering (a process compatible with the transfer since $Z_T(W)$ is the center of G and similarly for $\hat{G}$), since then there is a very convenient fundamental simplex for $\hat{T}/W$ available [4]. See, in particular, Iwahori's article in this volume.

§5. Several semisimple elements

Throughout this section G is a semisimple group. As a consequence of 3.9 we have:

5.1. *If G is simply connected, then any two commuting semisimple elements of G are contained in a maximal torus.*

If t and x are the elements, then any maximal torus of $Z_G(t)$ containing x will do.

5.2. REMARKS. Conversely the conclusion implies that G is simply connected up to inseparable isogeny (see 4.5). For if $F \neq 1$, then t above can be chosen so that $Z(t)$ is not connected (see 4.5) and then x any semisimple element of $Z(t) - Z(t)^o$ (see 4.4). The conclusion also fails if "two" is replaced by "three," the point being that even though G is simply connected the semisimple component of $Z_G(t)$ above need not be. To clarify this point, we observe the following:

5.3. *Let T be a maximal torus in G, Σ_1 a closed (under integral combinations) subsystem of Σ, and G_1 the corresponding semisimple subgroup of G (generated by all U_a such that $a \in \Sigma_1$). Assume that G is simply connected. Then the following conditions on G_1 are equivalent.*

(a) *G_1 is simply connected.*

(b) *$L(\Sigma^*)/L(\Sigma_1^*)$ has no torsion.*

(c) *Any long root which is rationally dependent on Σ_1 is contained in Σ_1.*

Here "long" is with respect to the irreducible component containing the root. Let $T_1 = T \cap G_1$, a maximal torus of G_1. For any $a \in L(\Sigma^*)$ such that $na \in L(\Sigma_1^*)$ for some n, we have $\operatorname{im} a = \operatorname{im} na \subseteq T_1$.

Thus $L(\Sigma_1^*)$ is the complete group of one-parameter subgroups into T_1 if and only if it has no torsion in the corresponding group for T. Thus (a) and (b) are equivalent by 2.2. Assume (a). Let a be a long root rationally dependent on Σ_1. Then a^* represents a one-parameter group into T_1, thus by (a) and 2.2 can be written $a^* = \Sigma a_i^*$ $(a_i \in \Sigma_1)$. Then $a = \Sigma |a|^2/|a_i|^2 . a_i$ with each coefficient integral since a is long, so that $a \in \Sigma_1$ and (c) holds. Finally assume (c). Let Σ_2 be the rational closure of Σ_1 in Σ. We claim that $L(\Sigma_1^*) = L(\Sigma_2^*)$ (but not $\Sigma_1 = \Sigma_2$). Assume $a \in \Sigma_2$. If a is long, then $a^* \in L(\Sigma_1^*)$ by (c), while if a is short the same conclusion holds since Σ_2^*, like any root system, is generated by its short roots: the roots a^* such that a is long. Now any simple system of Σ_2^* may be extended to one of Σ^*: simply extend the given ordering on Σ_2^* so that the positive elements of Σ_2^* are all less than the other positive elements of Σ^*. Thus $L(\Sigma_2^*)$, i.e., $L(\Sigma_1^*)$, has no torsion in $L(\Sigma^*)$, (b) holds, and 5.3 is proved.

5.4. COROLLARY. G_1 <u>is simply connected in each of the following cases.</u>

(a) Σ_1 <u>contains all of the long roots of</u> Σ.

(b) G_1 <u>is the semisimple component of a parabolic subgroup of</u> G.

For 5.3(c) clearly holds in (a) and also in (b) since then Σ_1 consists of the roots that are rational (or, equivalently, integral) combinations of a subset of some system of simple roots.

5.5. EXAMPLES. The inclusion $F_4 \supset D_4$, $F_4 \supset B_4$, $G_2 \supset A_2$ are all of simply connected groups, while $F_4 \supset C_4$, $G_2 \supset A_1^2$, $E_8 \supset D_8$ are not.

Returning now to our main development, we have:

5.6. THEOREM. <u>Assume that</u> G <u>is simply connected.</u> <u>Let</u> t <u>be a semisimple element such that for some</u> n <u>not divisible by any torsion</u>

prime (see I. 4. 1 *and* I. 4. 4) t^n *is in the center of* G. *Then the semisimple component of* $Z_G(t)$ *is simply connected.*

The proof of this result will be given elsewhere.

5.7. REMARK. By using the argument of the proof of 4.4 we may replace the assumption of simpleconnectedness by: n is prime to $|F|$; and similarly for the consequences of 5.6 to follow.

5.8. THEOREM. *Assume* G *simply connected,* A *a subgroup of commuting semisimple elements,* $A = A^o . \prod_i A_i$ *with* A^o *connected and* A_i *cyclic of order, say,* n_i. *Let* ρ *be the number of* n_i's *not prime to all the torsion primes (see* I. 4. 4).

(a) *If* $\rho = 0$, *then* $Z_G(A)$ *is a connected reductive group whose semisimple component is simply connected.*

(b) *If* $\rho \leq 1$, *then* $Z_G(A)$ *is a connected reductive group.*

(c) *If* $\rho \leq 2$, *then* A *is contained in a maximal torus of* G.

We observe first that $Z_G(A^o)$ satisfies (a). Since A^o is a torus we can put it in some maximal torus T and then analyze $Z_G(A^o)$ as in 4.1, the connectedness then being a general fact [10, p. 6-14, Cor. 2], and the torsion in 5.3(b) being ruled out by the fact that A^o is divisible. Or else we can argue as follows. A^o is generated as a closed group by its set, say S, of elements of finite order divisible by no torsion prime [10, p. 4-07, Prop. 2]. Hence $Z_G(A^o) = \bigcap_{s \in S} Z_G(s)$. On the right we can restrict s to a finite set by the descending chain condition on closed subgroups, thus in effect reduce 5.8 to the case $A^o = 1$. Let a_1 be a generator of A_1. Then $Z_G(a_1)$ satisfies the conclusion of (a). By induction on the number of factors A_i we conclude that (a) holds, the inductive step being applicable to $Z_G(a_1)$ by I. 4. 7. In (b) we write $A = A_1 A'$ with A' the product of the A_i's for which n_i is favorable. By 3.9 applied to $a_1 \in Z_G(A')$, which is permissible by (a), $Z_G(A) = Z_G(a_1) \cap Z_G(A')$ is a connected reductive group, as required. In (c) we assume the decomposition $A = A_1 A'$ so that A' satisfies (b). Then any maximal torus of $Z_G(A')$ which contains a_1 fulfills the requirements of (c).

To apply 5.8 to the case of finite groups we need:

5.9. LEMMA. Let G and σ be as in 1.1 and A a subset of G fixed by σ and contained in a maximal torus. Then A is contained in a maximal torus fixed by σ.

The proof is like that of 1.1.

5.10. COROLLARY. Assume as in 5.8 and also that σ is as in 1.1 and $A \subseteq G_\sigma$.

(a) If $\rho \leq 2$, then the torus of 5.8(c) may be taken to be fixed by σ.

(b) If $\rho \leq 1$ and $A' \subseteq G_\sigma$ is conjugate to A in G, ${}^gA = A'$, then it is so in G_σ, ${}^hA = A'$, and in such a way that ${}^ha = {}^ga$ for all $a \in A$.

To get (a) we apply 5.9. To get (b) we list the elements $a_1, a_2, \ldots$ of A, consider the action of G by conjugation on $G \times G \times \ldots$ restricted to the orbit through $a_1 \times a_2 \times \ldots$ and then combine 5.8(b) and I.2.2.

5.11. EXERCISE. If G is not simply connected, then 5.8(b), 5.8(c) and 5.10 continue to hold if we decrease the allowed values of ρ by 1.

5.12. EXAMPLES. In SL_n and Sp_n over an algebraically closed field a set of commuting semisimple elements can always be diagonalized by 5.8(c) and I.4.4(e) and (f), a well-known result. Over a finite field 5.10(a) and 5.10(b) provide refinements. In SO_n

(any form) we are all right in 5.8 and 5.10 provided A is a group of odd order by I.4.4(b) (the nonsimpleconnectedness causing no trouble since $|F| = 1$ or 2 (see 5.7)). In 5.8(c) and 5.10(a) we can even accommodate a cyclic factor of even order (two if we go to $Spin_n$).

5.13. REMARK. The torsion primes are essential in each part of 5.8 as one can show by examples. In fact if p is any torsion prime and G is of characteristic not p, then there exists a subgroup A isomorphic to $(Z/pZ)^{\rho+1}$ for which the conclusion fails. If G is not simply connected and $p \mid |F|$, then $(Z/pZ)^{\rho}$ will do. Such examples are given in [1] (for compact groups but the transfer to algebraic groups is clear).

In 5.8 although we can not always put A in a maximal torus at least we can put it in the normalizer of one, as we shall now show. To compensate for the weaker conclusion we can weaken the assumption that A is Abelian.

5.14. DEFINITION. A subgroup A of an algebraic group is supersolvable if there exists a series $A = A_0 \supset A_1 \supset \dots \supset A_n = 1$ such that each A_i is normal in A and each A_i/A_{i+1} is cyclic of finite order or else is a torus.

5.15. EXAMPLES. Every nilpotent, in particular every Abelian, subgroup (algebraic of course) is supersolvable. The group S_3 is supersolvable but not nilpotent.

We recall that since G is semisimple its group of automorphisms is a finite extension of its group of inner automorphisms (corresponding to the symmetries of the Dynkin diagram), hence may be given the structure of algebraic group.

5.16. THEOREM. *Let* A *be a supersolvable group of semisimple automorphism of* G.

(a) A _stabilizes some maximal torus_ T.

(b) _If_ σ _is as in_ 1.1 _and_ σ _fixes each term of some supersolvable series for_ A _(see_ 5.14), _then_ T _can be chosen to be fixed by_ σ.

Theorem 5.16(a) for compact Lie groups is due to Borel and Serre [3]. We use induction on dim G. Let $G_1 = Z_G(A_{n-1})^o$, a smaller group if $A_{n-1} \neq 1$ as we clearly may assume. We claim that (1) G_1 is reductive, and (2) each maximal torus of G_1 is contained in a unique one of G. If A_{n-1} is a torus, then, being connected, it acts by the inner automorphisms of some torus of G, which can always be included in a maximal torus, so that we get (1) by 4.1, and we also get (3) A_{n-1} stabilizes a maximal torus T and a Borel subgroup of G containing it. If A_{n-1} is cyclic, then (1) and (3) also hold, as is shown in [27]. But (3) implies (2): Let T_0 be the identity component of the group of all $t \in T$ such that $a_1(t) = a_2(t) = \ldots = a_r(t)$ (the a_i's being a system of simple roots on T), a one-dimensional torus. Set $a_0 = a_1 | T_0$. Since A_{n-1} permutes the a_i's, it fixes T_0 and a_0 and hence acts trivially on T_0, so that $T_0 \subseteq G_1$. Choose $t \in T_0$ so that $a_0(t)^n \neq 1$ whenever $1 \leq n \leq$ max. root height. Then $a(t) = a_0^{\text{ht } a}(t) \neq 1$ for every root a. It follows from 4.1 that $T = Z_G(t)^o$ so that T is the unique maximal torus of G containing t, hence containing any maximal torus of G_1 containing t, whence (2). By (1) we may apply our induction hypothesis to A acting on the semisimple component of G_1 to get a maximal torus of G_1 fixed by A, and by (2) get a corresponding one in G, which proves (a). To get (b) we simply choose the torus in G_1 to be fixed by σ, which is possible by I.2.9.

5.17. COROLLARY. _Let_ A _be a supersolvable subgroup of semisimple elements of_ G. _Then_ A _normalizes some maximal torus,_ *which may be taken to be fixed by* σ *in case* σ *acts as in* 5.16(b).

We apply 5.16 to A acting by conjugation on G.

5.18. COROLLARY (Blichfeldt). *Every finite nilpotent subgroup of* $SL_n(C)$ *(or of* $GL_n(C)$*) is equivalent to a monomial group.*

For the monomial matrices form the normalizer of the maximal torus consisting of the diagonal matrices.

Thus 5.16 may be viewed as a considerable extension of Blichfeldt's theorem, which continues to hold, for example, if $SL_n(C)$ is extended by its outer automorphism "inverse transpose."

5.19. COROLLARY. *Let* σ *act on* G *(still semisimple) as in* 1.1. *Let* p *be a prime, prime to the characteristic of* G, *and* S_p *a Sylow* p-*subgroup of* G_σ.

(a) S_p *normalizes a maximal torus fixed by* σ.

(b) *If in addition* p *is prime to* $|W|$ *(see* I.4.10), *then* S_p *is contained in a maximal torus fixed by* σ, *hence is Abelian.*

Since every p-group is nilpotent, hence supersolvable, 5.17 implies (a) which in turn implies (b).

To illustrate 5.19 we shall discuss the semisimple Sylow subgroups S_p of the groups $G_2(q)$ and their twisted (Ree) analogues. We have the factorization $|G_2(q)| = q^6(q^2-1)(q^6-1) = q^6(q-1)^2(q+1)^2(q^2-q+1)(q^2+q+1)$, in which, as may easily be verified, only the primes 2 and 3 can divide as many as two factors. If $p \neq 2, 3$, then p divides exactly one of the last four factors. Each of these factors is the characteristic polynomial of some element of W (identity, -identity, rotation through 60°, rotation through 120°, respectively), hence by 1.7(b) is $|T_\sigma|$ for some maximal torus fixed by σ (the Frobenius map in the present case). Thus $S_p \subseteq T_\sigma$, S_p is

Abelian in each case, in accordance with 5.19(b) and the fact that p is prime to $|W|$, i.e., to 12. The exact structure of T_σ, hence of S_p, can be worked out from 1.7(a); it is the product of two cyclic groups of the same order in the first two cases, a cyclic group in the last two. The groups $N(S_p)/S_p$, by 1.8 and 1.10(a), work out to be dihedral of order 12 in the first two cases, cyclic of order 6 in the last two. If $p = 2$ or 3, then the torus must be one of the first two types, whichever one will absorb the largest power of p (first type if $p = 2$ and $q \equiv 1 \bmod 4$, etc.), so that the toral part T_p of S_p is again a product of two cyclic groups of the same order; and S_p/T_p is isomorphic to the p-part of W: a (2, 2) group if $p = 2$ and a (3) group if $p = 3$. A discussion of how S_p/T_p and T_p are composed to form S_p requires further details which will not be given here. For the twisted group of type $^2G_2$, the appropriate factorization is $q^6(q^2-1)(q^6+1) = q^6(q^2-1)(q^2+1)(q^2-\sqrt{3}q+1)(q^2+\sqrt{3}q+1)$ (recall that $q^2 = 3^{2n+1}$ in the present case). The last factors on the right correspond to the four types of maximal tori fixed by σ, so that we may continue as before, the discussion being simpler now since fewer cases have to be considered. Every odd S_p turns out to be cyclic with $N(S_p)/S_p$ cyclic of order 6, while S_2 is dihedral of order 8.

5.20. EXERCISE. Fill in the details of the discussion just given. Do the same for the Suzuki groups $^2C_2(q)$ $(q^2 = 2^{2n+1})$ using the equation $q^4(q^2-1)(q^4+1) = q^4(q^2-1)(q^2-\sqrt{2}q+1)(q^2+\sqrt{2}q+1)$ (cf. 1.10(d)).

5.21. REMARK. For the other type of Ree group, $^2F_4(q)(q^2 = 2^{2n+1})$, the discussion has not yet been carried out. We would like to record here the following factorization which we think is relevant.

$$q^8-q^4+1 = (q^4-\sqrt{2}q^3+q^2-\sqrt{2}q+1)(q^4+\sqrt{2}q^3+q^2+\sqrt{2}q+1) \quad .$$

If σ acts on a maximal torus contained in a Borel subgroup also fixed by σ, then the factors on the right are just the characteristic polynomials of σ^* and $-\sigma^*$. (Recall that σ^* denotes the action of σ on $X(T)$.)

CHAPTER III. REGULAR ELEMENTS AND UNIPOTENT ELEMENTS

§1. Regular elements

G continues to be a semisimple group. As we have seen in the last chapter our knowledge of semisimple elements and classes of G (or of G_σ) is in reasonably good shape. Now if $x \in G$ is arbitrary, then $Z_G(x)$ is just the centralizer of x_u in $Z_G(x_s)$ (see I.3.1), a reductive group containing x_u in its identity component by II.4.4. Therefore the study of arbitrary classes can to a large extent be reduced to the unipotent case. Here our results are fragmentary (for general semisimple groups; for some types of groups there are no problems; e.g., in SL_n the unipotent classes correspond to the partitions of n via the Jordan normal form). It turns out, however, that there is a particular class of unipotent elements which dominates all others. The basic definition, in which x need not be unipotent, is as follows.

1.1. DEFINITION. An element x of G (or of any reductive group) is regular if $\dim Z_G(x)$ is minimal, or, equivalently, dim (class of x) is maximal.

This concept is developed extensively in [25] and [21] so that our development here will be sketchy.

1.2. <u>The minimum</u> m <u>in</u> 1.1 <u>is just</u> r, <u>the rank of</u> G.

If x is a sufficiently general element of a maximal torus T, then $Z_G(x) = T$, so that $m \leq r$. If x is arbitrary, we put it in a Borel subgroup B. Since B has dimension r over its commutator subgroup, the class of x in B has codimension at least r, so that $m \geq r$.

The inequality $m \geq r$ can be improved as follows.

1.3. $Z_G(x)$ _always contains an Abelian subgroup of dimension_ r.

This holds if x is semisimple (put x in a maximal torus), and in the general case follows from the fact that the semisimple elements are dense (see [22]).

1.4. COROLLARY. _If_ x _is regular_, $Z(x)^o$ _is Abelian_.

1.5. EXAMPLE. The following assertions about elements of SL_n are not difficult to prove. Prove them.

(a) An element is regular and semisimple if and only if its eigenvalues are all distinct.

(b) A unipotent element is regular if and only if its Jordan normal form consists of a single block.

(c) Various intermediate cases can occur. Give an example. (See 1.6.)

(d) The following conditions are equivalent.

(1) x is regular.

(2) The minimum and the characteristic polynomials of x are equal.

(3) Every element of Z(x) is a polynomial in x.

(4) The underlying space k^n is a cyclic x-module.

The following result, whose proof is immediate, focuses our study of regular elements onto the unipotent case.

1.6. _If_ $x \in G$, _then_ x _is regular in_ G _if and only if_ x_u _is regular in_ $Z_G(x_s)^o$.

If x is semisimple, then $x_u = 1$, and 1 is regular in $Z_G(x_s)^o$ only if this group is a torus. Thus (see II.4.1):

1.7. COROLLARY. _The following conditions on a semisimple element_ x _of_ G _are equivalent_.

(a) x _is regular_.

(b) $Z(x)^o$ *is a torus.*

(c) x *is contained in a unique maximal torus.*

(d) $Z(x)$ *consists of semisimple elements.*

(e) $a(x) \neq 1$ *for every root* a *relative to some, or every, maximal torus containing* x.

For regular unipotent elements the situation is as follows.

1.8. THEOREM. (a) *The set* V *of unipotent elements of* G *is a closed irreducible subvariety of codimension* r.

(b) *The regular elements of* V *form a single class. It is open in* V *and its complement has codimension at least* 2.

The argument for (a) is taken from [10, p. 6-12]. Let B and U be as in I.1 and $X = \{gB, x) \in G/B \times G \mid g^{-1}xg \in U\}$. Clearly X is closed. Since G/B is complete, the projection of X on the second factor is also closed, i.e. V is closed. Alternately V is closed because it is defined by the equations $\chi(s) = \chi(1)$ by II.3.5. The set X is irreducible since its inverse image in $G \times G$ is the image of the irreducible variety $G \times U$ under the map $(g, u) \longrightarrow (g, gug^{-1})$; hence V is also irreducible. Consider the projection of X into the first factor. Each fibre is isomorphic to U (the fibre above gB is gUg^{-1}). Hence $\dim X = \dim G/B + \dim U = \dim G - r$. Projecting into the second factor we get $\dim V \leq \dim G - r$ with equality if and only if some fibre is finite, i.e., if and only if for some $x \in U$ the set $\{gB \mid g^{-1}xg \in U\}$ is finite. Since B is its own normalizer and also the normalizer of U, the last condition works out to: the number of conjugates of B, i.e., of Borel subgroups, containing x is finite. This number can in fact be made 1 by choosing $x = \prod_{a>0} x_a(c_a)$ as in I.1 so that $c_a \neq 0$ for every simple root a, which completes the proof of (a). The result just observed is a special instance of the following result [23, 1.5]: the fibres above (for projection on the second factor) are all connected, i.e., the Borel subgroups containing a unipotent element x form a connected set, or yet again, the set of fixed points for x acting on G/B by left multiplication is connected. Alternately one can prove that V is irreducible and has codimension r by showing that the regular unipotent elements form a single class whose closure is V, as is done in [25]. The crucial point in (b) is the existence of regular unipotent elements. This is proved in [25, §4]. It follows from 1.8(a) and I.5.4 in case the characteristic of G is good (not bad) (see I.4.3) since then V is the union of a finite number of classes and must have the same dimension as one of them. Another proof which works in the same case is given in [21] and then extended to the exceptional cases, with a good deal of work, in [15]. Now every conjugacy class is open in its closure. Hence every class of regular unipotent elements is open in V. Since V is irreducible, there can be only one such class. The final point in 1.8 is proved in [25, 5.7].

1.9. COROLLARY. *The map $x \longrightarrow x_s$ on G induces a bijection of the regular classes onto the semisimple classes.*

Assume that x and y are regular elements such that x_s and y_s are conjugate, ${}^g x_s = y_s$. Then ${}^g x_u$ and y_u are regular unipotent elements of $Z_G(y_s)^o$ by 1.6, hence are conjugate there, ${}^{hg}x_u = y_u$ with $h \in Z_G(y_s)^o$ by 1.8(b). Then ${}^{hg}x = y$ so that x and y are conjugate.

1.10. COROLLARY. *Let V' be a fibre of the map f that assigns to each element of G the collection of values of the algebra of class functions there.*

(ab) *The conclusions of 1.8(a) and 1.8(b) hold with V' in*

<u>place of</u> V.

(c) <u>The semisimple elements of</u> V' <u>form a single class</u> S, <u>and</u> V' <u>consists of the</u> $x \in G$ <u>such that</u> $x_s \in S$.

(d) S <u>is closed, in the closure of every class contained in</u> V', <u>and of minimum dimension among the classes of</u> V', <u>and is characterized by any of these properties</u>.

We observe that (ab) is an extension of 1.8 by II.3.5. The regular elements of V' form a single class by 1.6 and 1.8(b). The other parts of (ab) are proved as in 1.8. Then 1.9 together with $\chi(x) = \chi(x_s)$ implies (c). The class S is closed by II.3.6, in the closure of every class of V' by (c) and II.3.7, and hence of minimum dimension among the classes of V'. The equivalence of these three conditions is clear.

We remark that the map f puts on the set of regular classes a structure of affine variety, that of ordinary affine r-space in case G is simply connected [25, 6.17].

1.11. <u>The set</u> G^r <u>of all regular elements of</u> G <u>is open and each irreducible component of the complement has codimension</u> 3.

This is proved in [25, §5]. The result on the codimension combined with the normality of G implies that every regular function on G^r can be extended to one on G, a result that would fail if we were to stick to regular semisimple elements since then the codimension would be 1. (The favorable condition is codimension ≥ 2, which also holds in 1.8(b)).

We continue with some other characterizations of regular elements.

1.12. <u>An element</u> x <u>of</u> G <u>is regular if and only if the number of Borel subgroups containing it is finite</u>.

This is proved in [25, 3.8], where it is shown further that the number is $|W(G)/W(Z_G(x_s)^o)|$ (see II.4.1), hence is 1 if x is unipotent.

This last fact is in accordance with:

1.13. *Let* x *be a unipotent element imbedded in a maximal connected unipotent subgroup* U *as in* I.1: $x = \prod_{a>0} x_a(c_a)$. *Then* x *is regular if and only if* $c_a \neq 0$ *for every simple root* a.

This is proved in [25, §3]. In SL_n the second condition is just 1.5(b).

1.14. *Assume as in* 1.13.

(a) $Z_G(x) = Z(G) \cdot Z_U(x)$.

(b) *If the characteristic* p *of* G *is good* (I.4.3) *then* $Z_U(x)$ *is connected.*

(c) *If* G *is adjoint in* (b), *then* $Z_G(x)$ *is connected.*

(d) *If* p *is bad, then* $Z_U(x)$ *is not connected since in fact* $x \notin Z_U(x)^o$.

This ((a) follows from 1.13 and I.1.3) is proved in [21].

As an example of 1.14(d) we consider the group C_2 with $p = 2$. If a and b are the simple roots and $x = x_a(1)x_b(1)$, then $Z_U(x)^o = U_{a+b}U_{a+2b}$ and $Z_U(x) = \langle x, Z_U(x)^o \rangle$.

It turns out that the last condition always holds.

1.15. *In* 1.14(d) $Z_U(x)$ *is generated by* x *and* $Z_U(x)^o$. *Hence* $Z_U(x)/Z_U(x)^o$ *is cyclic.*

This is proved in [15], where the order of $Z_U(x)/Z_U(x)^o$ is also determined. If G is simple, then this order is 4 in case $p = 2$ and the type is E_7 or E_8 and it is p in all other cases.

1.16. COROLLARY. *If* x *is a regular unipotent element of* G, then $Z_G(x)$ is Abelian.

$Z_U(x)^o$ is Abelian by 1.4, then $Z_U(x)$ is by 1.15, and finally $Z_G(x)$ is by 1.14(a).

With some care we can extend this to arbitrary regular elements.

1.17. In 1.16 $Z_G(x)$ is Abelian even if x is not unipotent, provided that G is simply connected.

$Z_G(x_s)$ is connected by 3.9 and it contains x_u as a regular element by 1.6. By 1.16 applied to $Z_G(x_s)$ (or rather its semisimple component) and x_u in place of G and x, the centralizer of x_u in $Z_G(x_s)$, in other words $Z_G(x)$, is Abelian.

1.18. PROBLEM. Conversely, if $Z_G(x)$ is Abelian does it follow that x is regular? If so, we would have a very pleasing characterization of regular elements.

We turn now to the finite group G_σ.

1.19. Let σ be as in II. 1. 1.

(a) G_σ contains regular unipotent elements.

(b) If x is one of them and U is the unique maximal connected unipotent subgroup containing x (see 1.13), then the number of classes of such elements is $|Z(G)/(1-\sigma)Z(G)| \cdot |Z_U(x)/Z_U(x)^o|$.

(c) There is a single such class if and only if σ is fixed-point-free on $Z(G)$ and the characteristic is good.

Part (a) follows from I. 3. 4(a). The number in (b) is $|H^1(\sigma, Z(x)/Z(x)^o)|$ by I. 3. 4(b). This works out to the expression in (b) by 1.14(a) and 1.15. This expression is 1 if and only if both factors are, so that the finiteness of $Z(G)$ and 1.14(d) imply (c).

Finally, some formulas.

1.20. Let σ be as in II. 1. 1 and G simple.

(a) <u>The number of regular unipotent elements of</u> G_σ <u>is</u> $|G_\sigma|/q^r$, <u>with</u> q <u>as in</u> II. 2. 4(b).

(b) <u>The total number of unipotent elements of</u> G_σ <u>is</u> $q^{\dim G - r} = q^{\dim V}$, <u>with</u> V <u>as in</u> 1.8, <u>the variety of unipotent elements</u>.

Part (a) will be left as an exercise. (Count the number in any Borel subgroup fixed by σ using 1.13 and then use the fact that each is contained in a unique Borel subgroup, again by 1.13). Part (b), which appears to be considerably more difficult, is proved in [27, §15].

The fact that the number in (b) is just what it would be if V were ordinary affine space suggests some nice geometric property of V, perhaps that it can be cut up into pieces in some reasonable way and the pieces reassembled to form affine space. This turns out to be the case in the group SL_2. V is then isomorphic to the variety of nilpotent elements (subtract the identity), hence in terms of coordinates to the variety $x^2 = yz$. (Check this.) The projection in the z-direction now maps this variety with the generator $y = 0$ missing isomorphically onto the xy-plane with the line $y = 0$ missing.

In a later section regular nilpotent elements of $\underline{g}$ will be discussed.

§2. A normal form for regular elements

G continues to be semisimple. We shall discuss an extension, to arbitrary semisimple simply connected groups, of the Jordan normal form for regular elements of SL_n. This form is, by 1.5(d)(4),

$$2.1. \quad \begin{bmatrix} & & & & 1 \\ -1 & & & & c_1 \\ & -1 & & & c_2 \\ & & \cdot & & \cdot \\ & & & \cdot & \cdot \\ & & & \cdot & \cdot \\ & & & -1 & c_{n-1} \end{bmatrix}$$

The c's here are, up to sign, the interior coefficients of the characteristic polynomial. As the c's vary each possible characteristic polynomial, hence each regular class, is achieved exactly once, so that we have in 2.1 a cross-section of the regular classes of SL_n, isomorphic to affine r-space $(n = r+1)$. To attain our extension of 2.1 we rewrite it:

$$2.2. \quad \begin{bmatrix} & & & & & 1 \\ -1 & & & & & \\ & -1 & & & & \\ & & \cdot & & & \\ & & & \cdot & & \\ & & & & \cdot & \\ & & & & -1 & \end{bmatrix} \cdot \begin{bmatrix} 1 & & & & -c_1 \\ & 1 & & & -c_2 \\ & & \cdot & & \\ & & & \cdot & \\ & & & \cdot & -c_{n-1} \\ & & & & 1 \end{bmatrix}$$

With the situation as in I.1.5, the first factor of 2.2 normalizes T and may be written n_w with $w \in W$ being represented by the n-cycle $(123\ldots n)$ in S_n. Writing this as $(12)(23)\ldots(n-1\ n)$, we see that w is just the product of the simple reflections. The second factor of 2.2, as the c's vary, runs over an r-dimensional Abelian group, which is in fact just the group U_w of I.1.3(b) since the positions occupied by the c's are those that w maps from above the diagonal to below the diagonal.

2.3. THEOREM. _Assume_ G _simply connected (and semisimple). Let_ T _be a maximal torus and then the other notations as in_ I.1. _Let_ w _be the product of the simple reflections (in an arbitrary order),_ n_w _a representative of_ w _in_ N, _and_ U_w _as in_ I.1.3(b).

Set $C = n_w U_w$. *Then* C *is a cross-section of the collection of regular classes of* G.

2.4. COROLLARY. C_s *(the set of semisimple parts of the elements of* C*) is a cross-section of the collection of semisimple classes.*

The corollary follows from 1.9.

The proof of 2.3 is given in [25, §8]. We shall not discuss it here except to mention two important auxiliary results.

2.5. *Assume* G *and* C *as in* 2.3. *Let* $\{\chi_i\}$ *be the fundamental characters and* f *the map* $x \longrightarrow (\chi_1(x), \chi_2(x), \ldots, \chi_r(x))$ *from* G *to affine* r-*space* A_r.

(a) *The map* $f : C \longrightarrow A_r$ *is an isomorphism of algebraic varieties.*

(b) $x \in G$ *is regular if and only if* $(df)_x$ *is surjective, in other words, if* $d\chi_1, \ldots, d\chi_r$ *are linearly independent at* x.

In SL_n, for example, c_i in 2.1 is, up to sign, the ith coefficient of the characteristic polynomial, hence the ith elementary symmetric polynomial in the eigenvalues, hence the trace of the ith exterior power, which is just χ_i, so that (a) is clear. In (b) if we stick to the torus T of diagonal elements and write $x = \mathrm{diag}(x_1, \ldots, x_n)$, then the Jacobian determinant of the χ's relative to appropriate coordinates of T works out to the Vandermondian $\prod_{i<j} (x_i - x_j)$, which is nonzero if and only if the eigenvalues x_i are distinct, i.e., no root vanishes at x, in accordance with 1.5(a) and 1.7(e). In the general case the situation is similar.

In SL_n we have normal forms for nonregular elements also, consisting of several blocks as in 2.1.

2.6. PROBLEM. Extend the normal form of 2.3 to nonregular elements.

We can now strengthen our hold on the fibres of 1.10, in

particular on the variety of unipotent elements, in the simply connected case.

2.7. Assume as in 2.5. Let V' be as in 1.10, any fibre of the map f (see II.3.2(c)): $V' = f^{-1}(c_1, c_2, \ldots, c_r)$.

(a) The regular elements of V' are all simple.

(b) V' is nonsingular in codimension 1.

(c) The ideal of V', in the algebra of regular functions on G, is generated by $\{\chi_i - c_i\}$, so that the latter ideal is prime for all $\{c_i\}$, and V' is a complete intersection.

(d) V' is normal.

The simple elements of V' (of any variety) form a non-empty open set, which must intersect the class of regular elements since the latter is open and V' is irreducible. Thus some, hence by homogeneity every, regular element is simple. Then (b) follows from (a) and 1.8(b). Let x be a regular element of V'. The condition 2.5(b) is just the one that $\{\chi_i - c_i\}$ be extendable to a system of local coordinates at x in G. This implies the various assertions of (c) except for the primeness which comes from the irreducibility of V'. Then (d) follows from (b) and the last assertion of (c) [12, p. 100, 5.8.6].

2.8 REMARK. It also follows that in (a), conversely, the simple elements are all regular.

2.9. LEMMA. Assume as in 2.3, and also that G is simple, that σ acts as in II.1.1, excluding the cases in which G_σ is SU_{2n+1} or a Suzuki or Ree group. Then the cross-section C of 2.3 can be taken to be fixed by σ (by appropriate choice of T, n_w, etc.).

Choose B and T as in I.2.9 to be fixed by σ and the other notations as in I.1. Then σ permutes the simple roots as in II.1.4(a), and each orbit consists of mutually orthogonal roots, as may be checked in the cases not excluded by our assumptions. Since reflections corresponding to orthogonal roots commute, it is clear that σ will fix $w = w_1 w_2 \dots w_r$, the product of the simple reflections, if the first few w_i's are those in the first orbit, the next few in the second orbit, and so on. By I.2.11 we may choose $n_w \in G_\sigma$, and it is then clear that σ fixes $n_w U_w$.

2.10. REMARK. If G_σ is SU_{2n+1}, there is an appropriate analogue of C, consisting however of two pieces [25, Th. 9.7], but for the Suzuki and Ree groups no such analogue has yet been found.

2.11. COROLLARY. Under the assumptions of 2.9 the set C_σ is a cross-section of the regular classes of G fixed by σ, and $C_{\sigma s}$ is a cross-section of the semisimple classes of G_σ.

If a regular class A is fixed by σ, then by 2.9 so is $A \cap C$, a point by 2.3. This is the first statement, which by 1.9 implies that $C_{\sigma s}$ is a cross-section of the semisimple classes fixed by σ, hence also of the semisimple classes of G_σ by II.3.10.

2.12. EXAMPLE. The class A of n_w of 2.3 in G, which turns out to be independent of the choices made to define n_w, is an interesting one. The corresponding class of w in W is just the "Coxeter class." The order of n_w is finite, equal to that of w in fact, hence independent of the base field, and A is characterized as the unique class of regular elements of minimal order. The class A can vary with the base field from being semisimple (most of the time) to being unipotent (e.g., SL_4 in characteristic 2), but always remains regular.

§3. Unipotent elements (relation with nilpotent elements)

G is semisimple over an algebraically closed field k of characteristic p. In this section we will discuss a connection between unipotent elements in G and nilpotent elements in its Lie algebra $\underline{g}$, in good characteristics (as in I.5.4). We begin by stating Lie algebra analogues of some results of §1.

3.1. DEFINITION. A nilpotent element $X \in \underline{g}$ is called regular if the dimension of its centralizer $Z_G(X)$ equals the rank r of G.

3.2. EXERCISE. State and prove the analogues of 1.5(b) and 1.5(d) for regular nilpotent elements of $\underline{s\ell}_n$.

3.3. THEOREM. (a) The set V of nilpotent elements of $\underline{g}$ is a closed irreducible subvariety of the affine space $\underline{g}$, of codimension r.

(b) If the characteristic p is good, the regular elements of V form a single class. It is open in V and its complement has codimension at least 2.

The proof of (a) is similar to that of 1.8(a) (see [21]). Again the crucial point in (b) is the existence of regular nilpotents in $\underline{g}$. If p is good, this follows from 3.3(a) and I.5.6 (another proof is indicated in [21], 5.9). The proof of [25] for G does not carry over to $\underline{g}$. The final point of (b) is proved as in [25].

Let B = T.U be a Borel subgroup of G, let $\underline{u}$ be the Lie algebra of U.

3.4. Let $\underline{X} = \{(gB, X) \in G/B \times \underline{g} \mid \mathrm{Ad}(g)^{-1} X \in \underline{u}\}$, let π be the canonical projection of $\underline{X}$ onto $\underline{g}$. Then $\pi(\underline{X}) = \underline{V}$ and the fibers of π are connected.

That $\pi(\underline{X}) = V$ means that every nilpotent in $\underline{g}$ is conjugate to an element of $\underline{u}$ (which follows from [2, 14.17]. For the connectedness statement see [23, 2.5].

3.5. *Let* $X_a \in \underline{u}$ *be a tangent vector to the unipotent subgroup* U_a *of* U *(notation of* I.1). *Then if* $X \in \underline{u}$ *is a regular nilpotent element, we have* $X = \sum_{a>0} c_a X_a$ *with* $c_a \neq 0$ *for all simple* a. *Conversely, if* p *is good, such an element is regular.*

This is proved in [21].

3.6. PROBLEM. Prove that regular nilpotent elements exist in all characteristics.

By 3.5 one must show that $\sum_{a \text{ simple}} X_a$ is regular.

3.7. *Let* p *be good, let* $X \in \underline{u}$ *be a regular nilpotent element.*

(a) $Z_G(X) = Z_U(X).Z(G)$.

(b) $Z_U(X)$ *is connected.*

(c) *If* G *is adjoint, then* $Z_G(X)$ *is connected.*

The proof is similar to that of 1.14.

3.8. *Let* G *be defined over a finite subfield of* k, *let* σ *be the corresponding Frobenius endomorphism. Suppose* p *good.*

(a) $\underline{g}_\sigma$ *contains a regular nilpotent element.*

(b) *If* G *is adjoint, these form a single orbit of* G_σ *in* $\underline{g}_\sigma$.

The proof is like that of 1.19.

3.9. REMARKS. (a) A regular nilpotent element of $\underline{g}$ is *not* a regular element in the "classical" sense (cf. [2, p. 286], for example).

(b) In characteristic 0, Kostant has proved Lie algebra analogues of 2.3, 2.5 and 2.7 (see [14]). In characteristic $p > 0$ no such results seem to be known. We mention explicitly one question.

3.10. QUESTION. Is $\underline{V}$ normal?

3.11. *Let* G *be adjoint, suppose* p *is good. Let* x *be a regular*

<u>unipotent element in</u> G. <u>There exists a regular nilpotent</u> X <u>in</u> <u>g</u> <u>with</u> $Z_G(X) = Z_G(x)$.

By 1.8(b) and 1.13 we may take $x = \prod_{a \text{ simple}} x_a(1)$. One then checks (using [21]) that there is an $X = \sum_{a>0} c_a X_a$ in $\underline{u}$ with $c_a \neq 0$ for all simple a, such that $Ad(x)X = X$. By 3.5, X is regular. X is in the Lie algebra of $Z_G(x)$ (I.5.1). By 1.14, 1.16 we know that $Z_G(x)$ is connected abelian, whence $Z_G(x) \subset Z_G(X)$. But since $Z_G(X)$ is connected (3.7(c)), of the same dimension as $Z_G(x)$, the two groups must be equal.

We can now state the main theorem of this section.

3.12. THEOREM. (a) <u>Suppose that</u> G <u>is simply connected and that</u> p <u>is good</u>. <u>There exists a morphism</u> $f : V \longrightarrow \underline{V}$, <u>which induces a homeomorphism of topological spaces and which commutes with the actions of</u> G <u>on</u> V <u>and</u> $\underline{V}$.

(b) <u>If, moreover,</u> G <u>is defined over a finite subfield of</u> k, f <u>can be taken to commute with the Frobenius endomorphism</u> σ <u>in</u> V <u>and</u> $\underline{V}$.

This is proved in [23], in the more general setting of an arbitrary groundfield. We will make a few remarks about the idea of the proof. Suppose G simple and not of type A_r (this is a case which is easily dealt with). We can then pass to the adjoint group and use 3.11. We conclude that the open subvarieties U and $\underline{U}$ of V resp. $\underline{V}$, formed by the regular elements, are isomorphic as varieties on which G acts.

Since V-U has dimension ≥ 2 (1.8(b)) and since V is normal by 2.7, we see that there is a morphism $f : V \longrightarrow \underline{V}$, commuting with G and inducing the isomorphism $U \longrightarrow \underline{U}$. If $\underline{V}$ were normal, we could go to the other way round and obtain an isomorphism $V \xrightarrow{\sim} \underline{V}$. But since the normality of $\underline{V}$ is not known, a different argument is needed. It turns out that by using the connectedness result of 3.4 and that mentioned in the proof of 1.8, together with Grothendieck's Stein factorization of a proper morphism, one can deduce 3.12 (which is somewhat weaker than an isomorphism of V and $\underline{V}$).

3.13. *If f is as in 3.12(a) we have $f(e) = 0$.*

For 0 is the only nilpotent in $\underline{g}$ whose centralizer is all of G (check this).

3.14. EXAMPLES. (a) $G = SL_n$. In this case we may take for f the map $x \longmapsto x-1$ (in the matrix algebra M_n). The same map also works for GL_n.

(b) $G = SO_n$ and $p \neq 2$. We may take now the well-known Cayley parametrization map $x \longmapsto (1-x)(1+x)^{-1}$. Similarly in the symplectic groups. So the f of 3.12 may be viewed as a generalized Cayley parametrization.

We next discuss applications of 3.12.

3.15. *Suppose that p is good. Let S be a subset of G. Then the unipotent elements of $Z_G(S)$ are contained in $Z_G(S)^o$.*

Let $(\pi, \tilde{G})$ be the universal covering of G (II.2.1). π induces a homeomorphism of the unipotent variety of $\tilde{G}$ onto that of G, similarly for the nilpotent varieties. One concludes that it suffices to prove 3.15 for a simply connected G. In that case, let f be as in 3.12. Let x be unipotent in $Z_G(S)$. Put $X = f(x)$. For any $t \in k$, let $x_t = f^{-1}(tX)$. The set $\{x_t\}$ is a closed, connected set in V (being the image under the homeomorphism f^{-1} of the affine line kX in $\underline{V}$), which is contained in $Z_G(S)$ (since f commutes with G). It contains x and by 3.13 also e. Hence $x \in Z_G(S)^o$.

3.16. REMARKS. (a) 3.15 is a counterpart to the well-known result that (in any connected linear algebraic group G) a unipotent element, which commutes with a semisimple element x, lies in $Z(x)^o$ (see [2], 11.12). For G semisimple, this is proved in II.4.4. Observe that by 1.14(d) the restriction on p is essential.

(b) 1.14(d) also shows that 3.12(a) cannot hold in bad characteristics: if it did one could deduce 3.15.

To continue, we need a result about semisimple elements. Suppose G is simple. Let T be a maximal torus in G, let Σ be the root system of G with respect to T, let $\{a_1, \ldots, a_r\}$ be a system of simple roots and $h = \Sigma m_i a_i$ the corresponding highest root.

With these notations we have (in all characteristics):

3.17. *Suppose* G *adjoint and simple. Let* S *be a set of unipotent elements in* G. *If* x *is a semisimple element in* $Z_G(S)$, *then there exists an* i *such that* $x^{m_i} \in Z_G(S)^o$.

Put $G_1 = Z_G(x)^o$. By the result recalled in 3.16(a) we have $S \subset G_1$. Let T be a maximal torus of G containing x, then $T \subset Z_G(x)^o$, whence also $x \in Z_G(x)^o$. The center C of G_1 is contained in $Z_G(S)$, and $x \in C$. If $x^t \in C^o$, then clearly $x^t \in Z_G(S)^o$. Consequently we are finished if we show that the order of any element of C/C^o divides some m_i. Σ and Σ_1 being the root systems of G resp. G_1 with respect to T, the character group of C/C^o is isomorphic to the torsion part of $L(\Sigma)$ $L(\Sigma_1)$ (notation of I.4.1), because G is adjoint. Hence 3.17 will follow from the following lemma.

3.18. LEMMA. *Let* Σ *be irreducible. The order of a torsion element of* $L(\Sigma)/L(\Sigma_1)$ *divides some* m_i.

To prove this it suffices to deal with the case that $L(\Sigma)$ and $L(\Sigma_1)$ have the same rank. This is similar to the reduction made in I.4.5 to the case of a maximal closed subsystem Σ_1 of Σ, and comes from the fact that the coefficients of the highest root of a closed subsystem of Σ divide the m_i (this one deduces from the fact that any positive integer smaller than one of the m_i divides some m_j, as a case by case check shows).

Now recall [4] that if Σ' is a maximal closed subsystem of Σ, there exists an m_i (call it m_1) which is a prime, such that $\Sigma' = \langle h, a_2, \ldots, a_r \rangle$. Also, $\{-h, a_2, \ldots, a_r\}$ is a simple system of roots of Σ'. Put $a_1' = -h$, $a_i' = a_i$ $(i \geq 2)$. If h' is the corresponding highest root of Σ', then $h' = a_1' + \sum_{i \geq 2} m_i' a_i'$, where $m_i' \leq m_i$ (as one sees by expressing everything in the a_i). Now the closed subsystem Σ_1 can be found up to isomorphism from Σ by repeating the previous construction for an irreducible subsystem of Σ' and continuing in this manner. Since this procedure does not affect a_1', which will remain a simple root with coefficient 1 in the highest roots, we have $a_1' \in L(\Sigma_1)$. Also, if $m_i = 1$ for some $i > 1$, we have $m_i' = 1$. Using these facts it is not difficult to prove 3.18, checking through the possible cases. The possible maximal subsystems are listed in [4]. Details are left to the reader.

3.19. <u>Assume</u> G <u>is simple</u>. <u>Let</u> S <u>be a set of unipotent elements in</u> G. <u>The order of</u> $Z_G(S)/Z_G(S)^o$ <u>is divisible only by the primes which are bad for</u> G <u>and those dividing the order of the fundamental group of</u> Ad G.

Let $\overline{G}$ be the adjoint group of G and $\pi : G \longrightarrow \overline{G}$ the canonical isogeny. $A = \ker \pi$ is a finite central subgroup of G. Since replacing S by its (Zariski) closure does not change $Z_G(S)$ we may assume S to be closed. Put $\overline{S} = \pi(S)$. Let $\overline{x} \in Z_{\overline{G}}(\overline{S})$, take $x \in \pi^{-1}(\overline{x})$. Then for $s \in S$ we have $xsx^{-1} = f(s)s$, with a continuous map $f : S \longrightarrow A$. Since f must be constant on the irreducible components of S, we conclude that $\pi Z_{\overline{G}}(S)$ has finite index in $Z_{\overline{G}}(\overline{S})$. Hence $\pi Z_G(S)^o = Z_{\overline{G}}(\overline{S})^o$. Moreover, $\pi^{-1} Z_{\overline{G}}(\overline{S})^o = Z_G(S).A$.

If the characteristic p is good, the preceding remarks together with 3.15 and 3.17 establish 3.19. If p is bad, 3.17 can still be used.

3.20. COROLLARY. The conclusion of 3.19 also holds with S a connected semisimple subgroup of G.

Let in that case S' denote the set of unipotent elements of S. S' generates S, hence $Z_G(S') = Z_G(S)$. Apply 3.19 to S'.

3.21. EXAMPLE (of application of 3.17). Let $G = SO_n$ or $G = Sp_n$, in characteristic $\neq 2$. Let S be a set of unipotent elements in G. Then $Z_G(S)/Z_G(S)^o$ is an abelian group of type $(2, 2, \ldots, 2)$.

The proof is left as an exercise. The case when S consists of a single element will also be discussed in Chap. IV.

There is another particular case which deserves mention.

3.22. Let $G = GL_n$. For any subset S of G, the centralizer $Z_G(S)$ is connected.

Imbed G in the algebra A of $n \times n$-matrices. Let B be the subalgebra of A centralizing S, then $Z_G(S)$ is the group B^* of invertible elements of B. Let $b \in B^*$. To prove that b is in the identity component of B^*, it suffices to prove that $k[b]^*$ is connected. But $k[b]$ is a direct sum of local Artin rings, for each of which the connectedness of the group of nonsingular elements is immediate. This implies the connectedness of $k[b]^*$.

3.23. REMARK. The argument proves that the group of invertible elements of any associative algebra with identity is connected.

We now turn to another application of 3.12. Before stating the next result we recall that a symplectic structure on a nonsingular algebraic variety V is an exterior 2-form ω on V, of rank equal to $\dim V$ in all points of V. It follows that $\dim V$ must be even if V carries a symplectic structure.

3.24. Assume either (i) $G = GL_n$ or (ii) G is simple, not of type A_r and p is good for G. Let $x \in G$ be unipotent. There exists a G-invariant symplectic structure on the homogeneous space $G/Z_G(x)$.

Using 3.12 in case (ii) and 3.14(a) in case (i) one sees that it suffices to prove the corresponding assertion for $V = G/Z_G(X)$, X denoting a nilpotent element in $\underline{g}$. Let v_o denote the image in V of the neutral element. Let $\underline{z}$ be the Lie algebra centralizer of X in $\underline{g}$. By I.5.6, $\underline{z}$ is the Lie algebra of $Z_G(X)$. It follows that the tangent space T of V in v_o can be identified with $\underline{g}/\underline{z}$. Let $F(\ , \ )$ be a nondegenerate symmetric bilinear form on $\underline{g} \times \underline{g}$, such that the linear transformations $\mathrm{ad}Y$ $(Y \in \underline{g})$ are skew for F (see I.5.3). Let $Y, Z \in T$ be cosets of $Y_1, Z_1 \in \underline{g}$. Then $\omega(Y, Z) = F([XY_1], Z_1)$ defines a nondegenerate skew symmetric form on $T \times T$. We then obtain the desired symplectic structure by translation.

3.25. Assume G semisimple and p good. For any $x \in G$ we have that $\dim G - \dim Z_G(x)$ is even.

By II.4.1 this is true if x is semisimple. Using the facts recalled in I.3.1 one sees that it suffices to prove 3.25 for x unipotent, in which case it is a consequence of 3.24 (check this).

3.26. Same assumptions as in 3.24. Let $2d = \dim G/Z_G(x)$. There exists a differential form of degree $2d$ on $G/Z_G(x)$, which is G-invariant and everywhere nonzero.

Let ω be the symplectic structure of 3.23, then ω^d has the desired properties.

3.27. REMARKS. (a) If in 3.24 G is defined over a subfield ℓ of k, one can take ω to be rational ℓ.

(b) Let, moreover, ℓ be a locally compact field. Then the group $G(\ell)$ of ℓ-rational points of G is also locally compact and

3.26 implies then that for any $x \in G(\ell)$ the centralizer of x in $G(\ell)$ is a unimodular locally compact group (which in characteristic 0 is a result of Harish-Chandra).

Our final application of 3.12 is to Lie algebras over finite fields.

3.28. Let G be semisimple, simply connected, defined over a finite field k with q elements. Assume p to be good. Then the number of nilpotent elements in $\underline{g}$ which are rational over k is $q^{\dim G - \operatorname{rank} G}$. This follows by 3.12 and 1.20(b).

3.29. QUESTION. Is 3.28 true in all characteristics?

§4. Classification of nilpotent elements

G is either an adjoint simple algebraic group or $G = GL_n$, over the algebraically closed field k. The characteristic p of k is assumed to be good (not bad, I.4.1) for G. In this section we shall review some results of Dynkin's paper [11], viz. those which bear upon the classification of unipotent conjugacy classes in G or, what amounts to the same by 3.12, the classification of nilpotent conjugacy classes of $\underline{g}$. In [11] these problems are not treated directly, but one finds there a method for dealing with the problem of classifying 3-dimensional simple subalgebras of $\underline{g}$ in characteristic 0, which is equivalent to classifying nilpotents of $\underline{g}$ (in char. 0). Dynkin's method hinges on the theorem of Jacobson-Morozov, which is true in characteristic $p > 0$ only under restrictive assumptions. Because of this the methods give only incomplete results; further investigation will be necessary to improve them.

For some particular groups better results are indeed available. The case of symplectic and orthogonal groups will be discussed in Chap. IV (for a treatment of G_2 in arbitrary characteristics see R. Jeurissen, Thesis, Utrecht, 1969-70).

As before, T denotes a maximal torus in G, Σ the set of roots of G with respect to T and $\{a_1, \ldots, a_r\}$ a set of simple roots of Σ. For $a = \Sigma n_i a_i$, we denote by $h(a) = \Sigma n_i$ the height of a. There is a unique root in Σ with maximal height m (recall that G is simple). U_a denotes the one-parameter unipotent group defined by $a \in \Sigma$ and $X_a \in \underline{g}$ a nonzero tangent vector to U_a.

We first state some known results.

4.1. <u>There exists a faithful rational representation</u> (ρ, V) <u>of</u> G <u>or a group isogenous to</u> G <u>such that the symmetric bilinear form on</u> $\underline{g} \times \underline{g}$ <u>defined by</u> $F(X, Y) = T_r(d\rho(X). d\rho(Y))$ <u>is nondegenerate</u>. F <u>is invariant under all</u> Ad(g) $(g \in G)$ <u>and the</u> ad(X) <u>are skew for</u> F.

($d\rho$ denotes the differential of ρ at e, which is a homomorphism of $\underline{g}$ into the endomorphisms of V).

4.1 follows from I. 5. 3.

4.2. <u>For all nilpotent</u> $A \in \underline{g}$ <u>we have</u> $(\text{ad } A)^{2m+1} = 0$.

For $i \neq 0$, let $\underline{g}_i$ be the subspace of $\underline{g}$ spanned by the X_a with $h(a) = i$. Put $\underline{g}_o = \underline{t}$, the Lie algebra of T. Then $[\underline{g}_i, \underline{g}_j] \subset \underline{g}_{i+j}$ and the $\underline{g}_i$ define a grading of $\underline{g}$. A is conjugate to an element of $\underline{u} = \sum_{i>0} \underline{g}_i$ ([2], 14.17), so that we may take $A \in \underline{u}$ in which case the assertion is obvious.

4.3. <u>Let</u> $A \neq 0$ <u>be a nilpotent in</u> $\underline{g}$, <u>suppose</u> $(\text{ad } A)^n = 0$.

(i) <u>If</u> $p = 0$ <u>or</u> $n < p$ <u>there exists</u> B, H <u>in</u> $\underline{g}$ <u>such that</u> $[H, A] = 2A$, $[H, B] = -2B$, $[A, B] = H$. <u>Let</u> v <u>be the subalgebra spanned by</u> A, B, H.

(ii) <u>If</u> $p = 0$ <u>or</u> $2n < p$ <u>the restriction to</u> $\underline{v}$ <u>of the adjoint representation of</u> $\underline{g}$ <u>is a direct sum of irreducible representations of</u> $\underline{v}$. <u>If</u> $p > 0$ <u>these can be obtained by reduction</u> mod p <u>from irreducible representations of the</u> 3-<u>dimensional Lie algebra over</u> $\mathbb{Z}$ <u>with the same structure as</u> $\underline{v}$.

(i) is the theorem of Jacobson-Morozov. For a proof see [8]. (ii) is clear in characteristic 0, for $p > 0$ see loc. cit.

4.4. Let (π, V) be an irreducible representation of $\underline{v}$ as in 4.3(ii). It is well-known that there exists a basis $(e_i)_{0\leq i<d}$ of V such that

$$\pi(A)e_i = e_{i+1} \quad (0 \leq i < d-1), \quad \pi(A)e_{d-1} = 0 \quad ,$$

$$\pi(B)e_i = i(d-i)e_{i-1}, \quad \pi(H)e_i = (2i+1-d)e_i \quad .$$

The representation is determined by its dimension d.

4.5. In the sequel we suppose (unless the contrary is stated) that $p = 0$ or $p \geq 4m+3$. Let $A \neq 0$ be nilpotent in $\underline{g}$. 4.3 is applicable, let $\underline{v}$ be as in 4.3. We conclude that $\underline{g}$ is a direct sum

$$g = \bigoplus_{i=1}^{s} \underline{v}_i \quad ,$$

of $\underline{v}$-stable subspaces $\underline{v}_i$ in each of which $\underline{v}$ acts irreducibly as described in 4.3(ii). Put $d_i = \dim \underline{v}_i$. There exists $E_i \in \underline{v}_i$ such that the

$$(\mathrm{ad}\ A)^j E_i \quad (1 \leq i \leq s, \ 0 \leq j < d_i)$$

form a basis of $\underline{g}$. Also

$$(4.6) \qquad [H, (\mathrm{ad}\ A)^j E_i] = (2j+1-d_i)(\mathrm{ad}\ A^j E_i) \quad .$$

For $i \in \mathbb{Z}$ denote by $\underline{g}(i)$ the subspace of $\underline{g}$ formed by the $X \in \underline{g}$ with $[H, X] = iX$. Then

$$(4.7) \qquad [g(i), g(j)] \subset \underline{g}(i+j), \ \underline{g}(i) = 0 \ \text{if} \ i \geq 2h+1 \quad .$$

$\underline{g} = \bigoplus g(i)$ defines a structure of graded Lie algebra on $\underline{g}$.

4.8. Let ρ be the homomorphism $\underset{=}{G}_m \longrightarrow GL(\underline{g})$ defined by

$$\rho(x)X = x^i X \text{ if } X \in \underline{g}(i) \ .$$

By (4.7) we conclude that the $\rho(x)$ are automorphisms of $\underline{g}$, so that $\rho(\underset{=}{G}_m)$ is a subgroup of the identity component of the automorphism group of $\underline{g}$, which is G (by [24], 4.2). Hence there exists a one-parameter multiplicative subgroup λ of G, i.e., a homomorphism $\lambda : \underset{=}{G}_m \longrightarrow G$ such that $\rho = \mathrm{Ad} \circ \lambda$. We denote by S the 1-dimensional subtorus $\lambda(\underset{=}{G}_m)$ of G. We order the character group of S corresponding to the natural order of the character group $\underset{=}{G}_m$. We call λ (resp. S) a one-parameter subgroup of G (resp. a 1-dimensional subtorus) adapted to A.

4.9. kH <u>is the Lie algebra of</u> S.

It is immediate that there exists H' in the Lie algebra $\underline{s}$ of S such that $[H', X] = iX$ for $X \in \underline{g}(i)$. It follows that $H-H'$ is in the center of $\underline{g}$, which is $\{0\}$ (as follows from [24], 2.6).

4.10. <u>Assumptions and notations as before. Let</u> $Z = Z_G(A)$ <u>be the centralizer of</u> A <u>in</u> $\underline{g}$, <u>let</u> R <u>be its unipotent radical.</u>

(i) S <u>normalizes</u> Z. <u>The weights of</u> S <u>in</u> $\underline{z}$ <u>are</u> ≥ 0, <u>the weights of</u> S <u>in</u> $\underline{r}$ <u>are</u> > 0.

(ii) <u>Let</u> C <u>be the centralizer of</u> S <u>in</u> Z^o. C <u>is a connected reductive group and</u> Z <u>is the semi-direct product of</u> C <u>and</u> R.

Since $A \in \underline{g}(2)$, we have that $\mathrm{Ad}(S)A \subset kA$, which implies that S normalizes Z. By I.5.6 we know that $\underline{z}$ is the centralizer of A in $\underline{g}$. It then follows that $\underline{z}$ is spanned by the elements

$$X_i = (\mathrm{ad}\ A)^{d_i - 1} E_i \ (1 \leq i \leq s) \ .$$

That the weights of S in $\underline{z}$ are ≥ 0 is now a consequence of (4.6). The Lie algebra $\underline{c}$ of C is the set of $X \in \underline{z}$ fixed by $\mathrm{Ad}(S)$ ([2], p. 229, corollary) and is consequently spanned by the X_i with $d_i = 1$. Let $\underline{r}_1$ be the subalgebra of $\underline{z}$ spanned by the X_i with $d_i > 1$. Clearly $\underline{z} = \underline{c} + \underline{r}_1$ and $\underline{r}_1 = \underline{z} \cap \operatorname{ad} A(\underline{g})$. Let F be a symmetric bilinear form as in 4.1. Since the ad X are skew for F, the orthogonal $\underline{z}^{\perp}$ of $\underline{z}$ with respect to F is ad $A(\underline{g})$, whence $\underline{r}_1 = \underline{z} \cap \underline{z}^{\perp}$. It follows that the restriction of F to $\underline{c} \times \underline{c}$ must be nondegenerate, which can only be if C is reductive ([7], prop. 6, p. 81). On the other hand, we have $F(\underline{r}, \underline{z}) = 0$ (same reference), whence $\underline{r} \subset \underline{r}_1$. Since Z^{o}/R is a reductive group and since S acts on its Lie algebra with non-negative weights, S must act trivially on Z^{o}/R. This implies that $\underline{r}_1 \subset \underline{r}$, whence $\underline{r} = \underline{r}_1$. This proves (i).

As to (ii), we have already seen that C is reductive. It is connected by ([10], 6-14, Th. 6). By ([2], p. 229, corollary), $C \cap R = \{e\}$. Since $\underline{g} = \underline{c} + \underline{r}$, it then follows that $G = C.R$.

4.11. *Let* B_1, H_1 *be elements in* $\underline{g}$ *such that* $[H_1, A] = 2A$, $[H_1, B_1] = -2B_1$, $[A, B_1] = H_1$. *Then there exists* $x \in R$ *such that* $H_1 = \operatorname{Ad}(x)H$, $B_1 = \operatorname{Ad}(x)B$.

Let $\underline{w}$ be the subalgebra spanned by A, B_1, H_1, let S_1 be defined via $\underline{w}$ like S.

Denote by $N = \{x \in G \mid \operatorname{Ad}(x)A \in kA\}$ the normalizer of A in G. If T is a maximal torus of Z, then ST is one in N. From the conjugacy of maximal tori of N one obtains, using 4.10, that after conjugation with an element of R, we may assume that $S_1 \subset ST$. By 4.9 it follows that then $H_1 = H + Z$, where $Z \in \underline{c}$. Then $Z = [A, B-B_1] \in \operatorname{ad}(A)\underline{g}$. But $\underline{c} \cap \operatorname{ad} A(\underline{g}) = 0$ (proof of 4.10), whence $Z = 0$, $H = H_1$. Then $[A, B-B_1] = 0$ and $B-B_1 \in \underline{z} \cap \underline{g}(-2) = \{0\}$ (by 4.10(i)).

4.12. COROLLARY. *Let* λ *and* λ_1 *be two one-parameter subgroups of* G *adapted to* A. *There exists* $x \in R$ *such that* $\lambda_1 = \operatorname{Int}(x) \circ \lambda$.

4.13. EXERCISE. Assumptions of 4.5. Prove that there is a one to one correspondence between conjugacy classes of nilpotents of $\underline{g}$ and conjugacy classes of 3-dimensional simple subalgebras of $\underline{g}$ (for $p = 0$ this is proved in [11] and [13]).

Let $M = Z_G(S)$, the centralizer of S in G. Then $\mathrm{Ad}(M)\underline{g}(i) \subset \underline{g}(i)$ and $\underline{m} = \underline{g}(0)$ ([2], p. 229, corollary).

4.14. (i) $\mathrm{ad}\, A : \underline{m} \longrightarrow \underline{g}(2)$ <u>is a surjective linear transformation.</u>

(ii) <u>The morphism</u> $x \longmapsto \mathrm{Ad}(x)A$ <u>of</u> M <u>to</u> $\underline{g}(2)$ <u>is dominant and separable. In particular,</u> M <u>has an open orbit in</u> $\underline{g}(2)$.

$\underline{g}(2)$ is spanned by the $(\mathrm{ad}\, A)^j E_i$ with $d_i = 2j-1$ (notations of 4.5), hence $\underline{g}(2) \subset \mathrm{ad}\, A(\underline{g})$.

This implies (i). The map of (i) is the tangent map of that in (ii), in the neutral element of M. (ii) then follows from standard facts in algebraic geometry ([2], 17.3, p. 75).

4.15. <u>Let</u> λ <u>be a one-parameter subgroup of</u> G, <u>adapted to some nilpotent element</u> A. <u>Then</u> A <u>lies in the open orbit of</u> M <u>in</u> $\underline{g}(2)$, <u>hence the conjugacy class of</u> A <u>is uniquely determined by</u> λ (<u>observe that the grading of</u> $\underline{g}$ <u>and</u> S <u>are determined by</u> λ).

4.15 follows from 4.14(ii). It is a result due to Kostant ([13], lemma 4.2C, p. 990).

We keep the previous notations. Let P be the parabolic subgroup of G defined by λ. The Lie algebra of P is

$$\underline{p} = \bigoplus_{i \geq 0} \underline{g}(i) ,$$

that of its unipotent radical U is

$$\underline{u} = \bigoplus_{i > 0} \underline{g}(i) .$$

With $M = Z_G(S)$ as before, we have $P = M.U$, a semidirect product. The next statement is a reformulation of 4.10(i).

4.16. We have $Z \subset P$, $Z \cap U = R$, $Z^o \cap M = C$.

4.17. P is uniquely determined by A.

This follows from 4.12 and 4.16.

4.18. REMARK. 4.17 shows that we can "attach" to any nilpotent in $\underline{g}$ a parabolic subgroup of G. By 3.12 it then follows that a similar result is true for unipotents in G. The assumptions of 4.5 on p, which we had to make, are unnatural and it seems likely that the result is true without any assumption on p.

According to Tits, the tool for these matters is the flag complex of G, discussed in ([17], p. 55-63). The fixed point conjecture of loc. cit., p. 64 (line 16), should be used.

Let $\underline{u}_i = \bigoplus_{j \geq i} \underline{g}(j)$. This is a subalgebra of $\underline{u}$ for $i \geq 1$, which is stable under Ad(P). In fact, using ([5], Th. 9.8) one sees that in the present situation, $\underline{u}_i$ is the Lie algebra of a normal subgroup U_i of P $(i \geq 0)$. We have $A \in \underline{u}_2$. By 4.12, $\underline{u}_i$ is independent of the choice of λ.

4.19. (i) ad $A : \underline{p} \longrightarrow \underline{u}_2$ is a surjective linear transformation.

(ii) The morphism $x \longmapsto \mathrm{Ad}(x)A$ of P to $\underline{u}_2$ is dominant and separable. In particular, P has an open orbit in $\underline{u}_2$.

This is again a result of Kostant's ([13], Th. 4.3, p. 991). The proof is like that of 4.14.

We say that A is even if $\underline{g}(i) = 0$ for i odd. By 4.12 this is independent of the choice of λ.

4.20. Same assumptions and notations. Let A be even.

(i) The P-orbit of A in $\underline{u}$ is open.

(ii) dim Z = dim M.

(iii) A is contained in finitely many conjugates of $\underline{u}$.

If A is even, we have $\underline{u}_2 = \underline{u}$. Hence (i) follows from 4.19(ii).

By (i) and 4.16, dim P - dim Z = dim $\underline{u}$ = dim U. Since dim P = dim M + dim U, (ii) follows.

Let $F = \bigcup_{x \in G} \mathrm{Ad}(x)\underline{u}$. By a Lie algebra analogue of ([10], 6-11, lemme 2) this is a closed, irreducible subset of $\underline{g}$, of dimension $\leq$ dim G - dim M. Since the orbit O = Ad(G)A has dimension equal to dim G - dim M by (ii), and since $O \subset F$, we have dim F = dim G - dim M. Hence there is a nonempty open subset O' of F consisting of elements of $\underline{g}$ which are contained in only finitely many conjugates of $\underline{u}$. F being irreducible, we have $O \cap O' \neq \emptyset$, from which we conclude that $A \in O'$. This proves (iii).

There is a particular sort of even nilpotents, which will be of importance.

4.21. DEFINITION. Let G be semisimple (p arbitrary). A unipotent element $x \in G$ (resp. a nilpotent $X \in \underline{g}$) is called semi-regular if any semisimple element of $Z_G(x)$ (resp. $Z_G(X)$) lies in the center of G.

4.22. *If p is good and G is adjoint, then the centralizer in G of a semi-regular unipotent element of G (resp. a semi-regular nilpotent of* $\underline{g}$*) is connected.*

This follows from 3.12 and 3.15.

4.23. *Assumptions of 4.5, notations as before; assume moreover that G is not a simple or reductive group of type* A_r*. Let* $A \in \underline{g}$ *be semi-regular.*

(i) *A is even.*

(ii) $x \longmapsto \mathrm{Ad}(x)A$ *defines an isomorphism of M onto an open subset of* $\underline{g}(2)$.

Let λ be a one-parameter subgroup adapted to A, put $s = \lambda(-1)$. Then $s^2 = e$ and Ad(s)A = A. Since $p \neq 2$, s is semisimple. The semi-regularity of A implies that s = e. Hence Ad(s) = id, which can only be if $\underline{g}(i) = 0$ for i odd. This proves (i).

With the notation of 4.10, it follows from the semi-regularity of A that $C = \{e\}$. By 4.16 and 4.22 we see that $Z \cap M = \{e\}$. (ii) is then a consequence of 4.14(ii).

4.24. We now return to the classification of nilpotents of $\underline{g}$. The assumptions and notations remain the same.

Let T be a maximal torus of M, this is also a maximal torus of G. Let Σ be the root system of G with respect to T. Order Σ such that the one-parameter groups U_a with $a > 0$ are all in $\underline{p}$. Let Δ be the set of simple roots for this order. It is known that there is a subset Δ_1 of Δ such that the following holds:

$U_a \subset P$ (resp. U) if and only if $a = \sum_{b \in \Delta} n(b)b$ with $n(b) \geq 0$ for all $b \in \Delta_1$ (resp. $n(b) > 0$ for some $b \in \Delta_1$). For $a \in \Sigma$ define the integer $h_1(a)$ by: $X_a \in \underline{g}(h_1(a))$. Then $h_1(r+s) = h_1(r) + h_1(s)$, $h_1(a) \geq 0$ if and only $U_a \subset P$, $h_1(a) > 0$ if and only if $U_a \subset U$.

From the preceding description and the definition of the grading of $\underline{g}$ it follows that $h_1(a)$ can be found as follows: there is a subset Δ_2 of Δ_1 such that

$$h_1\Big(\sum_{b \in \Delta} n(b)b\Big) = \sum_{b \in \Delta_2} n(b) + 2 \sum_{b \in \Delta_1 - \Delta_2} n(b) .$$

We have $\Delta_2 = \emptyset$ if A is even (in particular in the situation of 4.23).

We now define the <u>Dynkin diagram</u> $D_G(A)$ <u>or</u> $D(A)$ <u>of</u> A as the Dynkin diagram of Σ, with numbers attached to the nodes as follows: to the nodes corresponding to elements of $\Delta - \Delta_1$, Δ_2, $\Delta_1 - \Delta_2$ we attach 0, 1, 2, respectively ([11], no. 27, p. 164).

4.25. (i) $D(A)$ <u>is uniquely determined by</u> A.

(ii) $D(A) = D(A')$ <u>if and only</u> A <u>and</u> A' <u>are conjugate</u>.

(i) follows from 4.12 and the conjugacy of maximal tori of M.

(ii) follows from (i) and 4.15.

It should be pointed out that, $D(A)$ being given, 4.15 provides a way of finding a representative of the conjugacy class of A.

4.26. EXERCISES. (i) The Dynkin diagram of a regular nilpotent of $\underline{g}$ has the number 2 in all nodes.

(ii) The Dynkin diagram of the nilpotent $\begin{pmatrix} 010 \\ 000 \\ 000 \end{pmatrix}$ of $\underline{s\ell}_3$ is

$\overset{1}{\circ}\!\!-\!\!\!-\!\!\!-\!\!\overset{1}{\circ}$.

4.27. A natural question which arises now is that of describing the possible Dynkin diagrams of nilpotents. For G of exceptional type (E_6, E_7, E_8, F_4, G_2) these can be found in ([11], p. 177-185). For G of classical type see Chap. IV.

Dynkin's method to derive the Dynkin diagrams of the exceptional groups is of a general nature and will now be described. If $A \in \underline{g}$ is nilpotent but not semi-regular (4.21), let x be a non-central semi-simple element of G centralizing A.

Then by ([2], p. 229, cor.) A is in the Lie algebra of $Z_G(x)^o$, which is a proper connected reductive subgroup of G, of the same rank as G. Continuing in this manner, one sees that for any nonzero nilpotent $A \in \underline{g}$ there exists a reductive subgroup H of G, with rank equal to rank G such that $A \in \underline{h}$ and that A is semi-regular in the Lie algebra of the semi-simple part of H. The following problems now arise:

(i) determine the semi-regular nilpotents,

(ii) determine the possible H,

(iii) let A_1, A_2 be semi-regular nilpotents in $\underline{h}_1$, $\underline{h}_2$. When are A_1, A_2 conjugate in G?

Notice that this method works in any characteristic, and works as well for the classification of unipotents in G.

A partial solution to problem (i) is given in the next result.

4.28. THEOREM. Suppose G is simple adjoint; let p be as in 4.5.

(i) If G is not of type D_r $(r \geq 4)$ or E_r $(r = 6, 7, 8)$, then a semi-regular nilpotent in g is regular.

(ii) In type D_r $(r \geq 4)$ there are $[\frac{r-2}{2}]$ non-regular classes of semi-regular nilpotents. Their Dynkin diagrams are

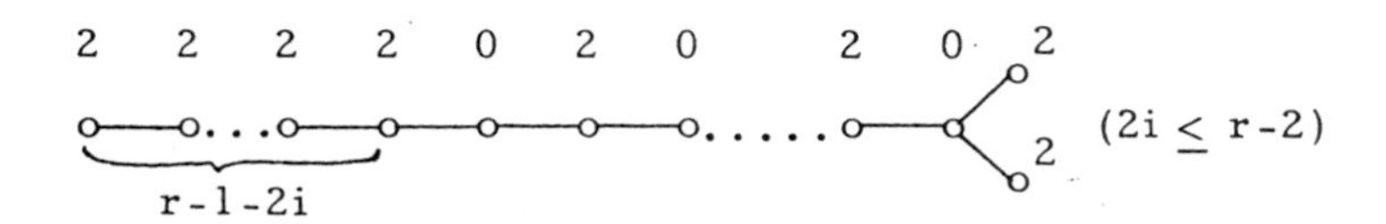

(iii) In type E_6 there is one class of non-regular semi-regular nilpotents, in E_7 and E_8 there are two of them. Their Dynkin diagrams are

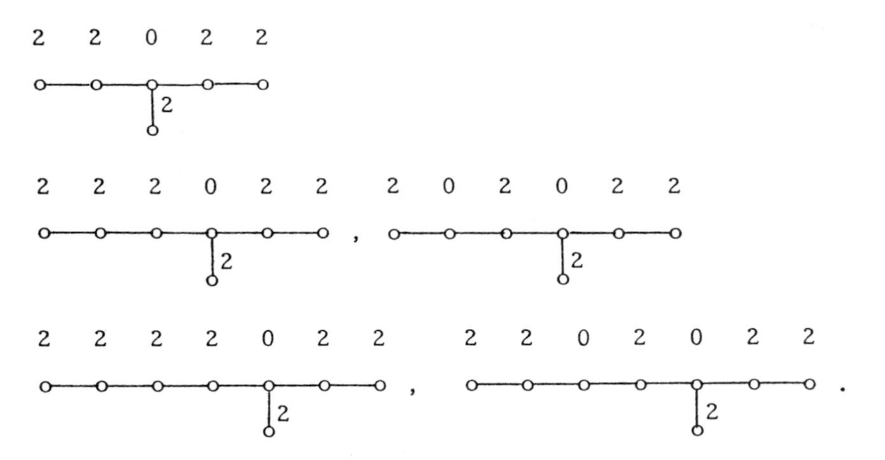

In characteristic 0, this is proved in ([11], Th. 9.2, p. 169 and Th. 9.3, p. 170).

We shall prove 4.28 for the classical types in Chap. IV, even for $p \neq 2$.

The proof of 4.28 for the exceptional types, given in ([11], nos. 33, 34, p. 187-191) works under our restrictions on p. We have the following consequence of 4.28.

4.29. THEOREM. Assumptions and notations of 4.24. Let A be a nonzero nilpotent in $\underline{g}$. There exists $t \leq$ rank G positive roots r_i $(1 \leq i \leq t)$ such that A is conjugate to $\sum_{i=1}^{t} X_{r_i}$.

The argument of 4.27 shows that it is sufficient to prove this in the case of a semi-regular A. In that case one can make an explicit check, based on 4.28 and 4.15. We omit the details.

4.30. PROBLEM. Prove that for any unipotent $x \in G$ there exists $s \leq$ rank G positive roots r_i such that x is conjugate to $x_{r_1}(1)x_{r_2}(1) \ldots x_{r_s}(1)$. (The notation is that of I.1.) This does not appear to be a straightforward consequence of 4.29.

4.31. REMARK. Let W be the Weyl group of the root system Σ of G with respect to T. Let $w_r \in W$ be the reflection defined by $r \in \Sigma$. If A is a semi-regular nilpotent in $\underline{g}$, and A is conjugate to $\sum_{i=1}^{t} X_{r_i}$ as in 4.29, it might be true that the conjugacy class in W of $C = w_{r_1} w_{r_2} \ldots w_{r_t}$ depends only on the conjugacy class of A. This is so if A is regular: C is then a "Coxeter element" of W (see 2.12), and there is some experimental evidence in the other cases. See also Part G.

4.32. It is not difficult to deal with the problems (ii) and (iii) of 4.27 (at least theoretically). The maximal $H \neq G$ have been determined by Borel-de Siebenthal and Dynkin and from these one can derive the others (see [11], table 9, p. 147 and table 11, p. 149). To solve (iii) it suffices, by 4.25(ii), to solve the following, more general problem:

(iv) Let H be a reductive subgroup of the same rank as G, let A be a nilpotent of $\underline{h}$. Find the Dynkin diagram $D_G(A)$ of A as a nilpotent in $\underline{g}$ from $D_H(A)$.

Let H_1 be the semisimple part of H. Let T be a maximal torus of H, $T_1 = H_1 \cap T$ is then one in H_1. Let $\Sigma(\Sigma_1)$ be the root system of G with respect to T (resp. of H_1 with respect to T_1). We may identify Σ_1 with a closed subsystem of Σ. The reader may now convince himself that, Σ, Σ_1 and $D_H(A)$ being given, $D_G(A)$ is completely determined. Hence (iv) can be solved, the solution requiring only computations in root systems. From this one also concludes

4.33. THEOREM. _Let_ $p \geq 4m+3$. _The number of unipotent classes of_ G _and the number of nilpotent classes of_ $\mathfrak{g}$ _equals the number of nilpotent classes of the complex simple Lie algebra of the same type as_ G.

4.34. PROBLEM. Is 4.33 true under the weaker assumption that p is good?

CHAPTER IV. CONJUGACY CLASSES IN CLASSICAL GROUPS

In the case of classical groups, the known results on conjugacy classes and centralizers go beyond those obtainable from the general theory of the preceding chapters. In the present chapter we give a brief discussion of some of these results.

Of the literature on the subject we mention the paper [28] of G. E. Wall, which contains more general results than those exposed here. He also deals with the characteristic 2 cases which we exclude. The method followed here is somewhat different from that of [28], it was given in [20] for the symplectic case.

Some material, which is relevant in connection with the previous chapters (e.g., relating to connectedness of centralizers and to the Dynkin classification) has been included.

§1. GL_n, SL_n

Contrary to our previous conventions, k denotes in this chapter a non-necessarily algebraically closed field. $\bar{k}$ is an algebraic closure of k.

Let $V = k^n$ be the canonical n-dimensional vector space over k. We denote in this section by G the algebraic group GL_n $(= GL_n(\bar{k}))$. G is defined over k, let $G(k) (= GL_n(k))$ be its group of k-rational points. If X is a linear transformation of V we denote by $A(X)$ (or A) the k-algebra of linear transformations of V generated by X. A is isomorphic to $k[T]/f\ k[T]$, where f is the minimum polynomial of X. V considered as an $A(X)$-module is denoted by $V(X)$.

1.1. <u>Let</u> X <u>and</u> X' <u>be linear transformations of</u> V.

(i) <u>If there exists</u> $Y \in G(k)$ <u>such that</u> $X' = YXY^{-1}$, <u>there is an isomorphism</u> $f : A(X) \longrightarrow A(X')$ <u>with</u> $f(X) = X'$;

(ii) <u>If</u> f <u>is as in</u> (i), <u>there is a bijection of the set of all</u> Y *onto the set of* k-*isomorphisms* $g : V(X) \longrightarrow V(X')$ *with* $g(ax) = f(a)g(x)$ *for* $a \in A(X)$, $x \in V(X)$.

1.2. *Let* X *be a linear transformation of* V. *The centralizer of* X *in* $G(k)$ *is isomorphic to the group of all* $A(X)$-*automorphisms of* $V(X)$.

The proofs of 1.1 and 1.2 are immediate.

1.3. REMARK. Let H be the group scheme "general linear group of $V(X)$" over $A(X)$. One can show that the centralizer of X in G, as a group scheme over k, is isomorphic to $\prod_{A(X)/k} H$, where $\prod_{A(X)/k}$ is the functor "restriction of the base ring from $A(X)$ to k."

1.4. We now restrict ourselves to the case of a nilpotent X. Let $X^d = 0$, $X^{d-1} \neq 0$. Then $A = A(X) \simeq k[T]/T^d k[T]$. For $h \leq d$, let $M(h)$ denote the A-module $X^{d-h}A$. The theory of Jordan normal forms shows:

1.5. (i) $V(X)$ *is isomorphic to a direct sum* $\bigoplus_{i=1}^{s} M(d_i)$;

(ii) *The* d_i *are unique up to permutation.*

1.6. By 1.5 there exist e_i $(1 \leq i \leq s)$ in V and integers $d_i > 0$ such that $X^{d_i} e_i = 0$ and that $X^j e_i$ $(0 \leq j < d_i,\ 1 \leq i \leq s)$ is a k-basis of V. Moreover, if f_i $(1 \leq i \leq s)$ has the same properties (for the same d_i), there exists $Y \in G(k)$ centralizing X with $Ye_i = f_i$.

Define a k-homomorphism λ of the multiplicative group G_m into G $(= GL_n)$ by

$$\lambda(x)X^j e_i = x^{1-d_i+2j}(X^j e_i) \quad (x \in \overline{k}^*) \ .$$

$S = \lambda(G_m)$ is a 1-dimensional k-torus in SL_n.

Let Z be the centralizer of X in G, let $\underline{z}$ be the centralizer of X in the Lie algebra $\underline{g\ell}_n$, i.e., the set of linear transformations of

$V \otimes_k \bar{k}$ centralizing X. $\underline{z}$ is an associative algebra, and Z is its group of invertible elements. It follows that $\dim Z = \dim \underline{z}$, which implies that the algebraic group Z is defined over k, by ([2], Prop. 6.7, p. 180). Let Y be a linear transformation in Z or $\underline{z}$. If

$$Ye_i = \sum_{j=1}^{s} \sum_{h=0}^{d_j-1} a_{ijh} X^h e_j ,$$

we must have

$$X^{d_i+h} e_j = 0 \text{ if } a_{ijh} \neq 0 ,$$

whence

$$Ye_i = \sum_{j=1}^{s} \sum_{\max(0, d_j-d_i) \leq h < d_j} a_{ijh} X^h e_j .$$

Hence, λ being as above,

$$\lambda(x) Y \lambda(x)^{-1} = \sum_{j=1}^{s} \sum_{\max(0, d_j-d_i) \leq h < d_j} a_{ijh} x^{2h+d_i-d_j} X^h e_j .$$

Order the characters of S such that the order corresponds, via λ, to the canonical order of the characters of G_m.

1.7. (i) S *normalizes* Z *and* $\underline{z}$. *The weights of* S *in* $\underline{z}$ *are* ≥ 0.

(ii) *Let* R *be the unipotent radical of* Z. *Its Lie algebra* $\underline{r}$ *is the subalgebra of* $\underline{z}$ *spanned by the weight vectors of* S *corresponding to strictly positive weights*. R *is defined over* k.

(iii) *Let* C *be the centralizer of* S *in* Z. C *is a connected reductive algebraic group, which is defined over* k. Z *is the semi-direct product of* C *and* R.

That S normalizes Z and $\underline{z}$ is clear. (i) then follows from the formulas given in 1.6. The proofs of (ii) and (iii) are similar to those of III. 4. 10(i), (ii) and are omitted. That R and C are defined over k is proved by using ([2], Prop. 9.3, p. 230).

1.8. COROLLARY. For $j \geq 1$, let r_j be the number of d_i equal to j.

(i) There is a k-isomorphism $\phi: C \longrightarrow \prod_{i \geq 1} GL_{r_i}$.

(ii) $\dim R = \sum_i (r_i + r_{i+1} + \ldots)^2 - \sum_i r_i^2$.

These statements follow from 1.7 by using 1.6.

1.9. EXERCISE. Let k be a finite field. Prove the formula for the order of the centralizer of a unipotent element in $G(k)$, given in Part D, 2.2.

1.10. EXERCISE. (i) Show that ϕ in 1.8(ii) is such that $\det(\phi^{-1}(x_1, x_2, \ldots)) = \prod_{i \geq 1} (\det x_i)^i$.

(ii) With the previous notations, let $Z_1 = Z \cap SL_n$, let Z_1^o be the identity component of Z_1. Deduce from (i) that Z_1/Z_1^o is a cyclic group of order equal to the greatest common divisor of the $d_i (1 \leq i \leq s)$.

(iii) Let k be a finite field. Using (ii) and I. 3.4, discuss the splitting of the conjugacy class in SL_n of a unipotent element of $SL_n(k)$.

1.11. Let now k be algebraically closed, of characteristic p. The notations remain as before. We want to discuss the connection with the theory of III. 4. The nilpotent linear transformation X is clearly an element of the Lie algebra $\underline{sl}_n$ of SL_n.

1.12. (i) Let $p \geq n$ or $p = 0$. There exist linear transformations Y, H of V, with trace 0, such that $[H, X] = 2X$, $[H, Y] = -2Y$, $[X, Y] = H$.

(ii) H can be chosen such that kH is the Lie algebra of S.

Using Jordan normal forms, it suffices to prove (i) for the case that V has a basis $\{e, Xe, \ldots, X^{n-1}e\}$. The formula given in III. 4.4 shows how to define Y and H. (ii) then also holds.

1.13. It follows from 1.12 that, under the restrictions on p of III.4.5 (which could be relaxed a little here, because of 1.12(i)), $\lambda(S)$ is a 1-parameter subgroup (resp. a 1-dimensional torus) in SL_n adapted to X, in the sense of III.4.8.

Let $(v_i)_{1\leq i\leq n}$ be the basis $(X^j e_i)$ of V of 1.6, suitably ordered. Let T be the maximal torus in SL_n consisting of the linear transformations in SL_n which fix all lines kv_i $(1 \leq i \leq n)$. Using the explicit description of the root system of SL_n with respect to T ([10], 20-02) one obtains the following recipe for determining the Dynkin diagram of the nilpotent X (III.4.24).

If $v_h = X^j e_i$, put $t_h = 1-d_i+2j$. Assume the numbering of the v_h has been chosen such that $t_1 \leq t_2 \leq \dots \leq t_n$. Then the Dynkin diagram of X is

$$\overset{t_2-t_1}{\circ}\!\!-\!\!\!-\!\!\!-\!\!\overset{t_3-t_2}{\circ} \;\cdots\cdots\; \overset{t_{n-1}-t_{n-2}}{\circ}\!\!-\!\!\!-\!\!\!-\!\!\overset{t_n-t_{n-1}}{\circ}$$

(It follows readily that the integers $t_i - t_{i-1}$ are 0, 1 or 2.)

1.14. EXERCISES. (i) Prove the recipe of 1.13.

(ii) Let the r_i be as in 1.8. In the sequence $(t_1, t_2, \dots, t_n)$, the integer $\pm 2i$ $(i > 0)$ occurs $\sum_{2j+1>2i} r_{2j+1}$ times and the integer $\pm(2i+1)(i > 0)$ occurs $\sum_{2j>2i+1} r_{2j}$ times.

(iii) The Dynkin diagram of X is invariant under the non-trivial automorphism of the Dynkin diagram of type A_{n-1}.

§2. Unitary, orthogonal, symplectic groups

k and V are as in §1.

2.1. Let σ_0 be an automorphism of k with $\sigma_0^2 = \mathrm{id}$. Let $\langle , \rangle$ be a nondegenerate σ_0-sesquilinear form on $V \times V$. We assume that

(a) $\langle x, y\rangle = \varepsilon\sigma_0\langle y, x\rangle$, $\varepsilon^2 = 1$;

(b) if $\sigma_0 = \text{id.}$, then k is of characteristic $\neq 2$.

There are three cases:

(1) $\sigma_0 \neq \text{id.}$, we may then assume that $\varepsilon = 1$ (unitary case),

(2) $\sigma_0 = \text{id.}$, $\varepsilon = 1$ (orthogonal case).

(3) $\sigma_0 = \text{id.}$, $\varepsilon = -1$ (symplectic case).

We denote by G the algebraic subgroup of GL_n which leaves $\langle , \rangle$ invariant. G is defined over k.

The group G(k) of k-rational points of G is a unitary, orthogonal or symplectic group.

Let $\underline{g}$ be the Lie algebra of G. Its algebra $\underline{g}(k)$ of k-rational elements is the Lie algebra of the linear transformations X of V such that

$$\langle Xx, y\rangle + \langle x, Xy\rangle = 0 .$$

2.2. Let $X \in G(k)$, let $A = A(X)$ be as in §1. σ_0 extends to an automorphism σ of A such that $\sigma X = X^{-1}$ and that

$$\langle ax, y\rangle = \langle x, \sigma(a)y\rangle \quad (x, y \in V,\ a \in A(X)) . \tag{2.3}$$

If $f = \sum_{i=0}^{d} a_i T^i$ is a polynomial in $k[T]$ of degree d, we write

$$\sigma f = \sum_{i=0}^{d} (\sigma_0 a_i) T^{d-i} .$$

If f is the minimum polynomial of X, then σf is a multiple of f. Since $A \cong k[T]/fk[T]$, it follows that A is a direct sum of σ-stable subalgebras which are either of the form

$$B = k[T]/g^d k[T] \oplus k[T]/(\sigma g)^d k[T] ,$$

where g is irreducible and σg is not a multiple of g (σ permuting the two summands), or of the form

$$C = k[T]/g^d k[T] \quad ,$$

with g irreducible, σg a multiple of g.

2.4. (i) <u>There exists a</u> k-<u>linear function</u> ℓ <u>on</u> A <u>such that the symmetric bilinear function</u> $(a, b) \longmapsto \ell(ab)$ <u>on</u> $A \times A$ <u>is nondegenerate</u>. <u>For such an</u> ℓ <u>there exists</u> $\alpha \in A$ <u>with</u> $\ell(\sigma a) = \ell(\alpha a)$.

(ii) <u>For</u> B <u>of</u> 2.3 <u>we may take</u> ℓ <u>such that</u> $\alpha = 1$. <u>For</u> C <u>this is also true, unless</u> $\sigma_0 = \text{id.}$, $g(T) = T \pm 1$. <u>In that case we may assume that</u> $\alpha = (-1)^{d-1}$.

To prove 2.4, it suffices to consider the case that $A = B$ or $A = C$. The second assertion of (i) is a consequence of the first one. In the case of C, any ℓ such that ℓ is nonzero on $g^{d-1}k[T] \bmod g^d k[T]$ satisfies the requirement of (i). A little computation shows that, unless we are in the case excluded in (ii), there is a multiple of $g^{d-1} \bmod g^d k[T]$ which is not of the form $c - \sigma c$ $(c \in C)$, from which (ii) follows for that case.

The excluded case is easily dealt with.

The proof for B is similar (but simpler).

2.5. As in §1, let $V(X)$ denote V, considered as an A-module. It follows from 2.4 that there exists a σ-sesquilinear form $F(X)$ on $V(X)$ such that for all $a \in A(X)$ we have, ℓ being as in 2.4(i),

$$\langle ax, y \rangle = \ell(aF(x)(x, y)) \quad . \tag{2.6}$$

We then have by 2.4(i)

$$F(X)(x, y) = \varepsilon\alpha\sigma(F(X)(y, x)) \quad ,$$

ε being as in 2.1 and α as in 2.4(i).

Let $X, X' \in G(k)$. Assume that there is an isomorphism $f : A(X) \longrightarrow A(X')$ with $f(X) = X'$ - which is necessary for conjugacy of X and X' in $GL_n(k)$ (1.1(i)). Putting $\ell' = \ell \circ f$, ℓ' is a k-linear function on $A(X')$ with the properties of 2.4(i). Using ℓ' we define the sesquilinear form $F(X')$ on $V(X')$.

The next results play the same roles for $G(k)$ as 1.1 and 1.2 do for $GL_n(k)$.

2.7. _There is a bijection of the set of all_ $Y \in G(k)$ _such that_ $X' = YXY^{-1}$ _onto the set of_ k-_isomorphisms_ $g : V(X) \longrightarrow V(X')$ _with_ $g(ax) = f(a)g(x)$, $F(X')(gx, gy) = f(F(X)(x, y))$ $(x, y \in V)$.

2.8. _The centralizer of_ X _in_ $G(k)$ _is isomorphic to the group of all_ $A(X)$-_automorphisms_ g _of_ $V(X)$ _such that_ $F(gx, gy) = F(x, y)$ $(x, y \in V)$.

2.9. REMARK. Similar to 1.3, the centralizer of X in G, as a group scheme over k, is isomorphic to $\prod_{A(X)/k} H$, where H is the "unitary" group scheme over $A(X)$, defined by F.

2.10. The preceding results have counterparts in $\underline{g}$. If $X \in \underline{g}(k)$, one can extend σ_0 to an automorphism σ of $A(X)$ by defining $\sigma X = -X$. Then (2.3) is again verified. If $f = \sum_{i=0}^{d} a_i T^i \in k[T]$, define

$$\sigma f = \sum_{i=0}^{d} a_i (-T)^i .$$

Then the assertions of the last lines of 2.2 remain true. The same holds for 2.4, with the modification that "$g(T) = T \pm 1$" in 2.4(ii) is replaced by "$g(T) = T$."

2.5, 2.7 and 2.8 now carry over without difficulty.

2.11. As 2.7 shows, in order to investigate conjugacy in $G(k)$ or $\underline{g}(k)$ we have to know about equivalence of forms like $F(X)$. This will be discussed now.

We write A, V for A(X), V(X). Let F be an A-linear σ-sesquilinear form on $V \times V$ such that

$$F(x, y) = \eta\sigma(F(y, x)) ,$$

where $\eta = \pm 1$. We call F an η-hermitian form on $V \times V$. F is non-degenerate if $F(x, y) = 0$ for all $y \in V$ implies that $x = 0$. The F(X) of 2.[illegible] is nondegenerate (because $<, >$ is). Let F and F' be two nondegenerate η-hermitian forms on $V \times V$. It suffices to consider the cases $A = B$ and $A = C$ of 2.2. If $A = B$ one easily sees that F and F' are always equivalent. So we may assume that $A = k[T]/g^d k[T]$, where g is irreducible and σg is a multiple of g. Let $\pi = g \bmod g^d k[T]$. One checks that by suitable normalization of g it may be assumed that $\sigma\pi = \pi$ unless we are in the exceptional case of 2.4(ii). In that case we may assume that $\sigma\pi + \pi \in \pi^2 A$. Let $m = A/\pi A$. This is a finite extension of k. The automorphism σ of A induces an automorphism of m, also denoted by σ, which extends σ_0. For $0 \leq i \leq d$ let $V_i = \{x \in V \mid \pi^i x = 0\}$, $W_i = V_i/V_{i-1} + V_i \cap \pi V$. Then W_i is a vector space over m. If $x \in V_i$, we have $\pi^i F(x, y) = 0$, whence $F(x, y) \in \pi^{d-i} A$. If $x, y \in V_i$, put $F(x, y) = \pi^{d-i} a$ and $h_i(x, y) = a \bmod \pi A$.

2.12. (i) <u>h_i is a σ-sesquilinear form on $W_i \times W_i$ and $h_i(x, y) = \eta\sigma h_i(y, x)$ if $\sigma\pi = \pi$, $h_i(x, y) = (-1)^{d-i}\eta\sigma h_i(y, x)$ otherwise;</u>

(ii) <u>h_i is nondegenerate.</u>

(i) is easily proved. (ii) follows from the nondegeneracy of F.

Let h_i' have the same meaning for F' as h_i for F.

2.13. THEOREM. F <u>and</u> F' <u>are equivalent if and only if</u> h_i <u>and</u> h_i' <u>are equivalent for</u> $0 \leq i < d$.

2.13 is proved in [20]. The proof is an essentially straightforward induction on d and is omitted. For related results see ([28], Th. 2.4.1, p. 26) and ([16], §3).

2.14. From 2.7 and 2.13 one obtains invariants for the conjugacy classes of $G(k)$ and $\underline{g}(k)$. We only deal with the case that $X \in G(k)$ (resp. $\underline{g}(k)$) has minimum polynomial g^d, with g irreducible and σg a multiple of g. The notations are as in 2.11. Let $M(h)$ denote the A-module $\pi^{d-h}A$ $(0 < h \leq d)$. The theory of elementary divisors shows that the A-module V is isomorphic to a direct sum $\bigoplus_{i=1}^{s} M(d_i)$. Let r_i be the number of d_j equal to i, then $\dim W_i = r_i$ $(1 \leq i \leq d)$. We assume the r_i to be fixed, i.e. the elementary divisors of X are given. In the cases of 2.1 one then finds the following results.

Unitary case: Invariants for the conjugacy class of X are the equivalence classes of the (skew-) hermitian forms h_i on W_i $(1 \leq i \leq d)$.

Orthogonal and symplectic case, $g(T) \neq T \pm 1$ (resp. $g(T) \neq T$): Same situation.

Orthogonal case, $g(T) = T \pm 1(T)$: The h_i are symmetric bilinear for odd i, their equivalence classes are the nontrivial invariants of the conjugacy class of X, the h_i are skew-symmetric bilinear and nondegenerate for even i, hence r_i is even for even i.

Symplectic case, $g(T) = T \pm 1(T)$: Same with even and odd interchanged.

The question which equivalence classes of h_i can occur depends on the initial form $\langle , \rangle$ and can be studied by using (2.6)

2.15. EXERCISES. (i) Show that in the symplectic case all equivalence classes of h_i do occur.

(ii) k algebraically closed, orthogonal and symplectic case. Show that $X, X' \in G(k)$ are conjugate in $G(k)$ if and only if they are in $GL_n(k)$.

(iii) Same assumptions. Let $X \in GL_n(k)$ be unipotent, let r_j be as above. X is conjugate to an element of $G(k)$ if and only r_j is even for even j in the orthogonal case resp. for odd j in the symplectic case.

2.16. From now on suppose that X is unipotent or nilpotent. Then $A = k[\pi]$ and $\sigma\pi = \pi$ (if $\sigma_0 \neq$ id.), $\sigma\pi + \pi \in \pi^2 A$ (if $\sigma_0 =$ id.). As ℓ of 2.4 we now take the linear function defined by

$$\ell(a_0 + a_1\pi + \ldots + a_{d-1}\pi^{d-1}) = a_{d-1} .$$

The following auxiliary result is easily proved.

2.17. (i) _If_ $a \in A$, $\sigma a = -a$, _then there exists_ $b \in A$ _such that_ $a \in k\,(b\sigma b)$.

(ii) $\sigma_0 =$ id. _If_ $a \in A$, $\sigma a = -a$, _then there exists_ $b \in A$ _such that_ $a \in k\,(\pi b \sigma b)$.

The next result is useful for finding explicit "normal forms." F is a nondegenerate η-hermitian form as in 2.11.

2.18. (a) $\sigma_0 \neq$ id. _There exist elements_ e_i $(1 \leq i \leq s)$ _in_ V _such that_ (i) V _is the direct sum of the_ A-_modules_ Ae_i, (ii) $F(e_i, e_j) = 0$ _if_ $i \neq j$, $F(e_i, e_i) \in k^*$.

(b) $\sigma_0 =$ id. _There exist elements_ e_i $(1 \leq i \leq s)$, f_j, g_j $(1 \leq j \leq t)$ _in_ V _such that_ (i) V _is the direct sum of the_ A-_modules_ Ae_i, Af_j, Ag_h, (ii) $F(e_i, e_j) = 0$ if $i \neq j$, $F(e_i, e_i) \in k^*$, $F(e_i, f_j) = F(e_i, g_j) = 0$, $F(f_i, g_j) = \delta_{ij}$, $F(f_i, f_j) = F(g_i, g_j) = 0$.

In case (a) we have an orthogonal basis for V, in case (b) a mixture of an orthogonal and a symplectic basis. We will only sketch the proof for the more complicated case (b).

If $F(x, x)$ is non-invertible for all $x \in V$, then $F(x, y) + F(y, x) \in \pi A$ for $x, y \in V$. The non-degeneracy of F then shows that we must have $\eta = -1$. Hence if $\eta = 1$, there exists $e_1 \in V$ such that $F(e_1, e_1)$ is invertible. By 2.17(i) we may assume that $F(e_1, e_1) \in k^*$. One can then write V as the direct sum of Ae_1 and its orthogonal complement V_1. If $\eta = -1$, one shows that one can split off a submodule $Af_1 + Ag_1$, where f_1 and g_1 have the properties of (ii). Induction on the length on V then proves the assertion.

2.18 implies a normal form for X, using 2.5. We state the result only for a nilpotent $X \in \underline{g}(k)$. The proof is straightforward.

2.19. _Let_ $X \in \underline{g}(k)$ _be nilpotent._

(a) $\sigma_0 \neq \mathrm{id}$. _There exist_ e_i $(1 \leq i \leq s)$ _and integers_ $d_i > 0$ _such that_ (i) $X^{d_i} e_i = 0$, _the_ $X^a e_i$ $(0 \leq a \leq d_i, 1 \leq i \leq s)$ _form a_ k-_basis for_ V, (ii) _there exist_ $\alpha_i \in k^*$ _such that_

$$\langle X^a e_i, X^b e_j \rangle = 0 \text{ if } i \neq j \text{ or } a + b \neq d_i - 1 ,$$
$$\langle X^a e_i, X^{d_i - a - 1} e_i \rangle = (-1)^a \alpha_i .$$

(b) $\sigma_0 = \mathrm{id}$. _There exist_ e_i, f_j, g_j $(1 \leq i \leq s, 1 \leq j \leq t)$ _and integers_ $d_i, \delta_j > 0$ _such that_ (i) $X^{d_i} e_i = X^{\delta_j} f_j = X^{\delta_j} g_j = 0$, _the_ $X^a e_h, X^b f_i, X^c g_j$ $(0 \leq a < d_k, 0 \leq b < \delta_i, 0 \leq c < \delta_j, 1 \leq h \leq s, 1 \leq i, j \leq t)$ _form a_ k-_basis of_ V, (ii) _the value of_ $\langle , \rangle$ _on a pair of these basis vectors is_ 0, _except the following ones_

$$\langle X^a e_i, X^{d_i - a - 1} e_i \rangle = (-1)^a \alpha_i, \text{ where } \alpha_i \in k^* ,$$
$$\langle X^a f_j, X^{\delta_j - a - 1} g_j \rangle = \varepsilon \langle X^{\delta_j - a - 1} g_j, X^a f_j \rangle = (-1)^a .$$

2.20. COROLLARY. _In case_ (b) _we have_ $(-1)^{d_i} = (-1)^{\delta_j - 1} = -\varepsilon$.
For $\alpha_i = \langle e_i, X^{d_i - 1} e_i \rangle = \varepsilon \langle X^{d_i - 1} e_i, e_i \rangle = \varepsilon (-1)^{d_i - 1} \alpha_i$, whence $(-1)^{d_i} = -\varepsilon$. Then use induction on d.

2.21. EXERCISE. Deduce from 2.20 the facts stated in 2.14 about the parity of the r_i in the orthogonal and symplectic case.

2.22. We next study centralizers of unipotent elements of $G(k)$ or nilpotent elements of $\underline{g}(k)$. It suffices to consider the nilpotent case: one can use a Cayley transform $X \longmapsto (\alpha - X)(\alpha + X)^{-1}$ (with $\alpha \in k$, $\sigma_0 \alpha = \alpha^{-1}$) to pass from unipotents to nilpotents (one can also argue directly). Let $X \in \underline{g}(k)$. We use the notations of 2.19. Define a k-homomorphism $\lambda : G_m \longrightarrow G$ by

$$\lambda(x) X^a e_i = x^{1-d_i+2a} X^a e_i \quad \text{(in cases (a) and (b))}$$

$$\left.\begin{aligned} \lambda(x) X^b f_j &= x^{1-\delta_j+2b} X^b f_j \\ \lambda(x) X^b g_j &= x^{1-\delta_j+2b} X^b g_j \end{aligned}\right\} \quad \text{(in case (b))} .$$

That $\lambda(x) \in G$ follows from 2.19. $S = \lambda(G_m)$ is a 1-dimensional k-torus in the identity component G^o of G. Let Z be the centralizer of X in G, let $\underline{z}$ be the centralizer of X in $\underline{g}$. We have $\dim Z = \dim \underline{z}$ (as a consequence of I. 5. 6), so that again Z is defined over k. Order the characters of S such that the order corresponds, via λ, to the canonical order of the characters of G_m.

2.23. (i) S normalizes Z and $\underline{z}$. The weights of S in $\underline{z}$ are ≥ 0.

(ii) Let R be the unipotent radical of Z. Its Lie algebra $\underline{r}$ is the subalgebra of $\underline{z}$ spanned by the weight vectors of S corresponding to strictly positive weights. R is defined over k.

(iii) Let C be the centralizer of S in Z. C is a connected reductive algebraic group, which is defined over k. Z is the semi-direct product of C and R.

This is exactly as 1.7. The proof of (i) and (ii) is like that of their counterparts in 1.7, similarly (iii) with Z and C replaced by their identity components Z^o, C^o. That (iii) also holds for Z and

C follows by an explicit check, using a basis with the properties of 2.19.

2.24. As before, let r_j denote the number of d_i (resp. the number of d_h and δ_i) which are equal to j. We know that r_j is even for even j in the orthogonal case and for odd j in the symplectic case (2.14). Observe that $\sum_{j=1}^{d} jr_j = n$. Let h_i be as in 2.11 (observe that now $m = k$). C is as in 2.23.

2.25. C(k) <u>is isomorphic to</u>

$$\prod_{i=1}^{d} U(h_i, k) \text{ in the unitary case,}$$

$$\prod_{\substack{i=1 \\ i \text{ even}}}^{d} Sp_{r_i}(k) \times \prod_{\substack{i=1 \\ i \text{ odd}}}^{d} O(h_i, k) \text{ in the orthogonal case,}$$

$$\prod_{\substack{i=1 \\ i \text{ odd}}}^{d} Sp_{r_i}(k) \times \prod_{\substack{i=1 \\ i \text{ even}}}^{d} O(h_i, k) \text{ in the symplectic case.}$$

This follows by using the basis of 2.19, observing that h_i of 2.11 which are not skew symmetric are equivalent to suitable diagonal forms $\mathrm{diag}(\alpha_{h(1)}, \ldots, \alpha_{h(r)})$.

We can now deal with the connectedness of Z. Let Z^o be the identity component of Z.

2.26. (i) Z <u>is connected in the unitary case.</u>

(ii) <u>In the orthogonal (symplectic) case</u> Z/Z^o <u>is a product of cyclic groups of order</u> 2, <u>where</u> a <u>is the number of odd (resp. even)</u> i <u>such that</u> $r_i > 0$.

Since a unitary group G is a form of GL_n, (i) follows already from III, 3.22. (ii) is a consequence of 2.25. It is a refinement of III, 3.21.

2.27. Orthogonal case. (i) Let G^o be the identity component of G (a special orthogonal group), let $Z_1 = Z \cap G^o$. Then Z_1/Z_1^o is a product of max (0, a-1) cyclic groups of order 2, where a is the number of odd i such that $r_i > 0$.

(ii) $Z \subset G^o$ if and only if $r_i = 0$ for odd i. In that case we have $n \equiv 0 \pmod 4$.

(i) and the first assertion of (ii) follow from 2.25 and 1.10(i). The second assertion of (ii) is then a consequence of the facts mentioned in 2.24.

2.28. Orthogonal and symplectic case. We have

$$\dim R = \frac{1}{2}\sum_i (r_i + r_{i+1} + \dots)^2 - \frac{1}{2}\sum_i r_i^2 - \frac{1}{2}\varepsilon \sum_{i \text{ even}} r_i .$$

Using 2.18(b) one can determine dim $\underline{z}$, which equals dim Z. 2.28 then follows by using 2.25. See also ([28], p. 33).

2.29. EXERCISE. Using the results of this section, determine the unipotent conjugacy classes and their centralizers in the finite groups $Sp_4(k)$ (char $k \neq 2$). (The result is given in: B. Srinivasan, Trans. Amer. Math. Soc. 131 (1968), p. 489).

From now on let k be algebraically closed. We first state III. 4.28 for classical groups.

2.30. Let H be a simple algebraic group over k of type A_r, B_r, C_r, D_r. Suppose the characteristic p of k is good. Let X be a semi-regular nilpotent element in the Lie algebra $\underline{h}$ of H.

(i) If H is of type A_r, B_r, C_r we have that X is a regular nilpotent.

(ii) If H is of type D_r there are $[\frac{r-2}{2}]$ nonregular classes of semi-regular nilpotents in $\underline{h}$.

This is an easy consequence of 1.8, 2.25, 2.27.

2.31. COROLLARY. Let $H = SO_{2r}$ be of type D_r, let X be semi-regular nilpotent in $\underline{h}$. There exists an integer ℓ with $0 \leq 2\ell \leq r-2$ such that X is conjugate to a regular nilpotent element of the subalgebra

$\underline{so}_{2\ell+1} \oplus \underline{so}_{2(r-\ell)-1}$ of $\underline{so}_{2r}$.

This follows by using the normal form of 2.19, together with 2.25 and 2.27.

2.32. Let $G = Sp_n$ or $G = O_n$. As in 1.12 and 1.13 it follows that $\lambda(S)$ of 2.24 is a 1-parameter subgroup (resp. a 1-dimensional torus) in G^o which is adapted to the nilpotent X, provided that the characteristic p of k is subject to the restrictions of III.4.5, which we will assume now.

We can now determine the Dynkin diagram of X. We will only give the results. Proofs are left as exercises.

(a) $G = Sp_n$. Let $(v_i)_{1\leq i\leq n}$ be the basis of V formed by the $X^a e_h$, $X^b f_i$, $X^c g_j$ of 2.19, suitably ordered. If $v_\ell = X^a e_h$, $X^b f_i$, $X^c g_j$, put $t_\ell = 1-d_h+2a$, $1-\delta_i+2b$, $1-\delta_j+2c$, respectively. Assume the numbering of the v_ℓ has been chosen such that $t_1 \geq t_2 \geq \ldots \geq t_n$. We have $n = 2r$, and $t_1, \ldots, t_r$ are ≥ 0. G is of type C_r. The Dynkin diagram of X is

$$\overset{t_1-t_2}{\circ}\!\!-\!\!-\!\!-\overset{t_2-t_3}{\circ}\ \cdots\ \overset{t_{r-1}-t_r}{\circ}\Longleftarrow\overset{2t_r}{\circ}$$

(b) $G = O_n$, $G^o = SO_n$. Let $r = [\frac{n}{2}]$. With the same notations we have again that $t_1, \ldots, t_r$ are ≥ 0. If $n = 2r+1$ (G of type B_r) the Dynkin diagram is

$$\overset{t_1-t_2}{\circ}\!\!-\!\!-\!\!-\overset{t_2-t_3}{\circ}\ \cdots\ \overset{t_{r-1}-t_r}{\circ}\Longrightarrow\overset{t_r}{\circ}$$

If $n = 2r$ (G^o of type D_r) the Dynkin diagram is

$$\begin{array}{l} t_1-t_2 \quad t_2-t_3 \quad t_{r-2}-t_{r-1} \quad t_{r-1}-t_r \\ \circ\!\!-\!\!-\!\!-\circ \quad \cdots \quad \circ \!<\! \begin{array}{l}\circ\ t_{r-1}-t_r\\ \circ\ t_{r-1}+t_r\end{array}\end{array}$$

or

$$\begin{array}{l} t_1-t_2 \quad t_2-t_3 \quad t_{r-2}-t_{r-1} \\ \circ\!\!-\!\!-\!\!-\circ \quad \cdots \quad \circ \!<\! \begin{array}{l}\circ\ t_{r-1}+t_r\\ \circ\ t_{r-1}-t_r\end{array}\end{array}$$

(Observe that a class in O_n may split in SO_n!)

2.33. EXERCISES. (i) If G^o is of type D_r, r odd, the Dynkin diagram of X is invariant under the automorphisms of the Dynkin diagram of type D_r. If G^o is of type D_r, r even, this is no longer true (use 2.27(ii)).

(ii) Discuss the case that G^o is of type D_4.

(iii) Prove the assertion about Dynkin diagrams in III. 4.28(ii).

1. A. Borel, Sous-groupes commutatifs et torsion des groupes de Lie compacts connexes, Tôhoku Math. J., vol. 13 (1961), 216-240.

2. ———, Linear algebraic groups, W. A. Benjamin, Inc., New York (1969).

3. ——— and J.-P. Serre, Sur certains sous-groupes des groupes de Lie compacts, Comm. Math. Helv. 27 (1953), 128-139.

4. ——— and J. de Siebenthal, Les sous-groupes fermés de rang maximum des groupes de Lie clos, Comm. Math. Helv. 23 (1949), 200-221.

5. ——— and T. A. Springer, Rationality properties of linear algebraic groups II, Tôhoku Math. J., vol. 20 (1968), 443-497.

6. ——— and J. Tits, Groupes réductifs, Publ. Math. I.H.E.S., no. 27 (1965), 55-151.

7. N. Bourbaki, Groupes et algèbres de Lie, chap. I, IV, V, VI.

8. F. Bruhat, Sur une classe de sous-groupes compacts maximaux des groupes de Chevalley sur un corps p-adique, Publ. Math. I.H.E.S. no. 23 (1964), 46-74.

9. C. Chevalley, Sur certains groupes simples, Tôhoku Math. J., vol. 7 (1955), 14-66.

10. ———, Séminaire sur la classification des groupes de Lie algébriques, 2 vol., Paris (1958).

11. E. B. Dynkin, Semisimple subalgebras of semisimple Lie algebras Am. Math. Soc. Transl. Ser. 2, 6 (1957), 111-245 (= Mat. Sbornik N.S. 30 (1952), 349-462).

12. A. Grothendieck and J. Dieudonné, Eléments de géometrie algébrique IV, Publ. Math. I.H.E.S. no. 24 (1965).

13. B. Kostant, The principal three-dimensional subgroup and the Betti numbers of a complex simple Lie group, Amer. J. Math. 81 (1959), 973-1032.

14. ———, Lie group representations in polynomial rings, ibid., 85 (1963), 327-404.

15. B. Lou, The centralizer of a regular unipotent element in a semi-simple algebraic group, Bull. A.M.S., vol. 74 (1968), 1144-1146.

16. J. Milnor, On isometries of inner product spaces, Inv. math., vol. 8 (1969), 83-97.

17. D. Mumford, Geometric invariant theory, Erg. Math. Bd. 34, Springer-Verlag, 1965.

18. R. Richardson, Conjugacy classes in Lie algebras and algebraic groups, Ann. of Math., vol. 86 (1967), 1-15.

19. J.-P. Serre, Corps locaux, Hermann, Paris (1962).

20. T. A. Springer, Over symplectische transformaties, Thesis, University of Leiden, 1951.

21. ———————, Some arithmetic results on semi-simple Lie algebras, Publ. Math. I.H.E.S., no. 30 (1966), 115-141.

22. ———————, A note on centralizers in semisimple groups, Indag. Math., vol. 28 (1966), 75-77.

23. ———————, The unipotent variety of a semisimple group, Algebraic Geometry (papers presented at the Bombay Colloquium, 1968), 373-391, Tata Institute, 1969.

24. R. Steinberg, Automorphisms of classical Lie algebras, Pac. J. Math., vol. 11 (1961), 1119-1129.

25. ———————, Regular elements of semisimple algebraic groups, Publ. Math. I.H.E.S., no. 25 (1965), 49-80.

26. ———————, Lectures on Chevalley groups, Yale Univ. Lecture Notes (1967-68).

27. ———————, Endomorphisms of linear algebraic groups, A.M.S. Memoirs no. 80 (1968).

28. G. E. Wall, On the conjugacy classes in the unitary, symplectic and orthogonal groups, J. Austr. Math. Soc., vol. 3 (1963), 1-62.

Additional results since the first edition of these notes.

1. Proofs for the results II.3.19 and II.5.6 have now appeared (see [St_2]).
2. The problems posed in I.5.5, III.1.18 and III.3.6 have all been answered in the affirmative (in [L_1], [Ku] and [K] respectively), in the second case under the assumption that the characteristic is good. Further, in the third case the results of 3.5 and 3.7 are all true. It follows that 3.11 fails when p is bad. The answer to II.3.18 is negative: consider the adjoint group of type A_4 in characteristic 5.
3. As regards the classification of nilpotent and unipotent elements discussed in III.4, another approach is now available (see [B-C]). The results of [L_2] lead to a very different kind of classification of unipotent elements.
4. Many interesting results related to the unipotent variety and its desingularization have been obtained. In this desingularization, first given in [23], the fibre above each element consists of the Borel subgroups containing that element. (The variety X of III.3.4 is just the Lie algebra version of this desingularization.) For results about the fibres see, besides the original paper [23], the references [St_1], [St_3] and [Sp] (which is quite comprehensive and contains further references). For the connection with Kleinian (also called rational) singularities, see [Br], [St_1] and [S] (a comprehensive treatment). Finally, it has been shown that the Weyl group operates in the cohomology groups of the fibres. This leads to an a priori parametrization of the irreducible representations of Weyl groups using unipotent conjugacy classes, see [Spr_1], [Spr_2].

Additional References

[B-C] P. Bala - R.W. Carter, Classes of unipotent elements in simple algebraic groups, Math. Proc. Camb. Phil. Soc. 79 (1976), 401-425 and 80 (1976), 1-18.

[Br] E. Brieskorn, Singular elements of semisimple algebraic groups, Proc. Intern. Congress Math. at Nice (1970), vol. 2, 279-284.

[K] S. Keny, Centralizers of nilpotent elements in the Lie algebra of a simple algebraic group, to appear.

[Ku] J. Kurtzke, Centralizers of irregular elements in reductive algebraic groups, Pacific J. Math. 104 (1983), 133-154.

[L_1] G. Lusztig, On the finiteness of the number of unipotent classes, Inv. Math. 34 (1976), 201-213.

[L_2] ————, Intersection complexes on a reductive group, Inv. Math. 75 (1984), 205-272.

[S] P. Slodowy, Simple singularities and simple algebraic groups, Lecture Notes in Math. No. 815 (1980), Springer Verlag.

[Sp] N. Spaltenstein, Classes unipotentes et sous-groupes de Borel, Lecture Notes in Math. No. 946 (1980), Springer Verlag.

[Spr_1] T.A. Springer, Trigonometric sums, Green functions of finite groups and representations of Weyl groups, Inv. Math. 36 (1976), 173-207.

[Spr_2] ————, Quelques applications de la cohomologie d'intersection, Sém. Bourbaki exp. 589, Astérisque 92-93 (1982), 249-273.

[St_2] R. Steinberg, Torsion in reductive groups, Advances in Math. 15 (1975), 63-92.

[St_1] ————, Conjugacy classes in algebraic groups, Lecture Notes in Math. No. 366 (1974), Springer Verlag.

[St_3] ————, On the desingularization of the unipotent variety, Inv. Math. 36 (1976), 209-224.

Séminaire BOURBAKI
25e année, 1972/73, n° 435

435-01
Juin 1973

ABSTRACT HOMOMORPHISMS OF SIMPLE ALGEBRAIC GROUPS

[after A. BOREL and J. TITS]

by Robert STEINBERG

§ 1. Introduction

Let G be a simple algebraic group (i.e. affine, connected, and having only finite, hence central, nontrivial normal subgroups) defined over an algebraically closed field k, and let $G(k)$ denote the group of rational points of G. According to Chevalley [6, exp. 7] :

1.1. A subgroup of G (identified with $G(k)$) is a Cartan subgroup if and only if it is maximal nilpotent and has every subgroup of finite index of finite index in its normalizer.

Similarly the structure of $G(k)$ as an abstract group determines the Borel subgroups, the maximal unipotent subgroups, ..., even the field k, up to isomorphism, and eventually almost completely determines the structure of G as an algebraic group. The end result (proved in § 2 below) can be stated thus :

1.2. THEOREM.- Let G, k be as above with G simply connected and let G' be an algebraic group over an algebraically closed field k'. Let $\alpha : G(k) \to G'(k')$ be an isomorphism of groups. Then there exists an isomorphism of fields $\varphi : k \to k'$ and a k'-isogeny $\beta : {}^{\varphi}G \to G'$ such that $\alpha = \beta \circ \varphi^{o}$ on $G(k)$.

Here ${}^{\varphi}G$ is the group over k' obtained by transfer of base field and $\varphi^{o} : G \to {}^{\varphi}G$ the corresponding map. (If G is given as a matrix group defined over k, then ${}^{\varphi}G$ is got by applying φ to the equations over k defining G and φ^{o} by applying φ to the matrix entries.)

Further a statement concerning the uniqueness of φ and β, especially simple when G is simply connected, can be given (see 2.3).

This theorem classifies, not only the possible structures of algebraic group for $G(k)$, but also the various groups of this type up to abstract isomorphism, thus also their abstract automorphisms. Further it does so in a precise way since the isogenies (i.e. rational homomorphisms, surjective with finite kernel) among the various simple groups are quite well known [6, Exp. 18].

The purpose of the article [4], of which we are giving an account here, is to prove a vast generalization of this result in which the class of groups is considerably extended and in which homomorphisms, not just isomorphisms, are considered, and then to apply this result to diverse situations concerning isomorphisms, automorphisms, continuity of homomorphisms, representations (homomorphisms into some $GL(V)$), etc..

First let us state this result. Let k be an infinite field and G a simple algebraic group defined over k and of positive k-rank, i.e. "isotropic". (Thus G contains nontrivial k-split tori and many rational unipotent elements.) Let H be a subgroup of $G(k)$ containing G^+, the group generated by the rational points of the unipotent radicals of the rational parabolic subgroups of G. (If k is perfect G^+ is the group generated by the rational unipotent elements.) Finally let G' be a simple algebraic group over an infinite field k' and $\alpha : H \to G'(k')$ a homomorphism. Then the generalization mentioned above is :

1.3. THEOREM ((A) of [4]).- Let everything be as just stated. Assume that G is simply connected or G' adjoint and that $\alpha(G^+)$ is dense in G' (in the Zariski topology). Then there exists a homomorphism $\varphi : k \to k'$, a k'-isogeny $\beta : {}^{\varphi}G \to G'$ with $d\beta \neq 0$ (called <u>special</u> for this reason), and a homomorphism $\gamma : H \to$ center of $G'(k')$, all three unique, such that $\alpha(h) = \gamma(h)\beta(\varphi^{o}(h))$ for all $h \in H$.

1.4. <u>Remarks</u>.- (a) G^+ is always dense in G. If P, P^- are opposed proper parabolic subgroups of G, then their unipotent radicals U, U^- generate G, as is easily seen. Taking P, P^- to be defined over k as we may since G is isotropic [3], we see that $\langle U(k), U^-(k)\rangle$, hence also G^+, is dense in G.

(b) It is conjectured that γ above is always trivial. This is certainly so in case G' is adjoint since then the center of $G'(k')$ is trivial. Assume that G is simply connected. Then there is a long standing conjecture, proved in most cases (for G split, quasi-split, a classical group,...), that $G(k) = G^+$ (and hence that the only choice for H in 1.3 is G^+). Now G^+ always equals its own derived group [4 : 6.4], hence has only trivial homomorphisms into Abelian groups. Thus this conjecture would imply the present one.

(c) Even if G is not simply connected, then the groups G^+ and $G(k)$ can be determined quite explicitly in most cases, and hence also the possibilities for H.

(d) An easy consequence of 1.3 is that if G and k are as there and k' is an algebraically closed field in which k has no imbedding, e.g. a field of a different characteristic, then a homomorphism from G^+ to any k'-group G' is necessarily trivial.

The proof of 1.3 and some related results will be discussed in § 3 and § 4.

In view of 1.4 (a) the theorem applies in case H' is a group constructed in the same way as H and $\alpha(H) = H'$. It therefore yields a classification of the groups of this type, and of their automorphisms. For the groups PSL_n over infinite fields, for example, the result can be formulated geometrically as follows, in view of the fundamental theorem of projective geometry : every isomorphism between two groups of this class is effected by a collineation of the underlying projective spaces and every automorphism by a collineation or a correlation. This is the original result of this type and was first proved in a classical memoir [11] of Schreier and van der Waerden who proved also that over finite fields the only exceptions are $PSL_2(\mathbb{F}_7) \cong PSL_3(\mathbb{F}_2)$ and $PSL_2(\mathbb{F}_4) \cong PSL_2(\mathbb{F}_5)$. Later it was extended by Dieudonné, Hua, Rickert, O'Meara and others (see [8, 10, 17] for further references) to include many of the other classical groups ((projective) symplectic, unitary, orthogonal, spin,...), on a case by case basis. Theorem 1.3 unifies these results as they apply to isotropic groups (for unitary and orthogonal groups, the defining form must have positive Witt index) and at the same time extends them to the exceptional groups. The earlier proofs, however, were decidedly more elementary.

Let us remark also that substantially the same results hold over finite fields, as was shown in a number of cases by various of the authors mentioned above and in the general case by the present author [13]. The proofs, indicated in § 2, are, at least in the split case, identical with those in the algebraically closed case from a certain point on.

Now let k, k' above be nondiscrete locally compact topological fields with

k' not isomorphic to $\mathbf{C}$. Then one can show (easily if k , k' are real) that every homomorphism $\varphi : k \to k'$ is necessarily a topological isomorphism of k onto a closed subfield of k' , hence is continuous [4, § 2.3]. It follows, in this case, that if $G(k)$ and $G'(k')$ in 1.3 are viewed as topological groups in the natural way, then α must be continuous. In [4] this result is extended to semisimple groups and it is shown that the assumption of isotropicity is not needed. It then includes the result of E. Cartan [5] and van der Waerden [19] that every homomorphism of a compact connected semisimple Lie group into a compact Lie group is continuous, and that of Freudenthal [9] that every isomorphism of a connected Lie group with absolutely simple Lie algebra onto a Lie group is continuous. These results are discussed further in § 5.

Finally let us consider (abstract) representations.

1.5. THEOREM ((B) of [4]).- Let k , G , G^+ , H be as in 1.3, k' an algebraically closed field, and $\rho : H \to PGL_n(k')$ $(n \geq 2)$ a projective representation which is irreducible on G^+ . Then there exist irreducible rational projective representations π_i of G and distinct homomorphisms $\varphi_i : k \to k'$ $(1 \leq i \leq m < \infty)$ such that ρ is the restriction to H of the tensor product of the representation ${}^{\varphi_i}\pi_i \circ \varphi_i^o$.

This result, conjectured by the present author in case k is algebraically closed in [14], together with a statement of uniqueness, will be proved in § 6 below. Since the only continuous homomorphisms of the complex field into itself are the identity and ordinary complex conjugation the above result overlaps the classical result that every irreducible differentiable complex representation of a connected complex Lie group is the tensor product of a holomorphic representation and an antiholomorphic one (see, e.g., [12, p. 22-12]).

§ 2. Algebraically closed fields and finite fields

We start with 1.2 since its proof, which is quite simple, will serve as a model for that of 1.3, which is not. Let everything be as in 1.2 and identify G with $G(k)$. For the standard facts about affine algebraic groups, many to be used without explicit reference, we cite [2, 3, 6].

2.1. In G one has the following abstract characterizations.

(a) The maximal tori : as in 1.1.

(b) The Borel subgroups : those that are maximal solvable and without proper subgroups of finite index.

(c) The maximal connected unipotent subgroups : the derived subgroups of the Borel subgroups.

(d) car k : if p is a prime then car $k = p$ if and only if there is no p-torsion in any maximal torus.

Here G need not be simple, only connected reductive (and nontrivial in the case of (d)). The proof of (a) is given in [6, Exp. 7]. If B satisfies the properties listed in (b), then it is closed by the maximality, connected since it has no subgroups of finite index, hence Borel by the maximality. Conversely, let B be Borel. Then B is maximal solvable [6, p. 9-05]. Write $B = TU$ (semidirect) with T a torus and U the subgroup of unipotent elements of B. Let B' be a subgroup of finite index. Since T is divisible it has no proper subgroup of finite index. Hence $B' \supseteq T$. Since G is reductive, U is the product of one-parameter subgroups each normalized by T according to a nontrivial character (root). If t is a regular element of T (at which no root vanishes) it follows that the map $u \to t^{-1}u^{-1}tu$ on U is surjective (in fact, bijective). Thus if u' is arbitrary, then tu' is conjugate to t, hence semisimple, hence contained in a maximal torus of B, hence contained in B' by the earlier argument. Thus $B' \supseteq U$, $B' \supseteq B$, $B' = B$, which yields (b). Since U above is maximal unipotent connected, (c) follows, for example, from the surjectivity above. Finally since a torus is diagonalizable (d) is clear.

2.2. Proof of 1.2. We observe first that G' is simple. For since B above does not have a proper subgroup of finite index neither does G, which is generated by its Borel subgroups, hence neither does G', which is thus connected. And since G does not have nontrivial infinite normal subgroups, neither does G'. In G let B, B^- be opposite Borel subgroups and U, U^- their unipotent radicals, so that $B \cap B^- = T$ is a maximal torus and one has $B = TU$, $B^- = TU^-$. It follows from 2.1 that the groups $B' = \alpha(B)$, $B^{-\prime} = \alpha(B^-)$, etc., have the same properties in G', and that car k = car k'. Further α matches up the normalizers of T and T', hence also the corresponding Weyl groups. Let a be a simple root for G relative to T, U_a the corresponding one-parameter subgroup of U, and n_a an element of the normalizer of T representing the reflection corresponding to a. Since the union of B and another B, B double coset can be a group only if the coset has the form Bn_aB (a simple) and since $U_a = U \cap {}^{n_a}U^-$, it follows that $\alpha(U_a) = U'_{a'}$ and $\alpha(n_a) = n_{a'}$ for some simple root a' for G'. Since ${}^{n_a}U_a = U_{-a}$ and similarly for a' one obtains from α an isomorphism from $\langle U_a, U_{-a}\rangle$ to $\langle U'_{a'}, U'_{-a'}\rangle$ which may be viewed as an isomorphism from $SL_2(k)$ to $SL_2(k')$ or to $PSL_2(k')$ (see, e.g., [6] and recall that G is simply connected) preserving superdiagonal, subdiagonal, and diagonal elements. Let φ and χ on k and k^* be defined by

$\alpha(1 + xE_{12}) = 1 + \varphi(x)E'_{12}$ and $\alpha(\operatorname{diag}(y,y^{-1})) = \operatorname{diag}(\chi(y), \chi(y)^{-1})$, with E_{12}, E'_{12} the appropriate matrix units. Here χ is well-defined even if the second group is $PSL_2(k')$ since then $SL_2(k)$ has no center, hence car k = 2 and car k' = 2. Here one can normalize the identification of $\langle U_a, U_{-a}\rangle$ with $SL_2(k)$ so that $\varphi(1) = 1$ (or one can do this for a'). We claim that then φ is an isomorphism of fields and that $\chi = \varphi/k^*$. One verifies that for given $y \neq 0$ the product $(1 + yE_{12})(1 + zE_{21})(1 + yE_{12})$ normalizes the diagonal subgroup only if $z = -y^{-1}$. Let $w(y)$ denote this product when $z = -y^{-1}$. It follows that $\alpha(w(y)) = w'(\varphi(y))$. Since also $\operatorname{diag}(y,y^{-1}) = w(y)w(1)^{-1}$ and $\varphi(1) = 1$ it follows that $\chi = \varphi/k^*$. Finally since φ is additive and χ is multiplicative, φ is an isomorphism as asserted. It depends on a. Now any root r is simple relative to some ordering of the roots. We write φ_r for the corresponding isomorphism. Also we let $u_r : k \to U_r$ denote a parametrization. Let a and b be simple roots with $a \neq b$ and $(a,b) \neq 0$. Then one has a commutator relation

$(u_a(x), u_b(y)) = \prod_{i,j>0} u_{ia+jb}(C_{abij}\, x^i y^j)$ with C_{abij} fixed in k and $C_{ab11} \neq 0$. On applying α and comparing with the corresponding relation in G' we get $\varphi_{a+b} = \varphi_a^m$ and $\varphi_{a+b} = \varphi_b^n$ with $(a+b)' = ma' + nb'$. Since φ_{a+b}, φ_a and φ_b are isomorphisms it follows that m and n are powers of the characteristic exponent p of k. Since G is simple and its root system irreducible it follows that there exists an isomorphism $\varphi : k \to k'$ and integers m_r, all nonnegative, some equal to 0, such that $\varphi_r = Fr^{m_r} \circ \varphi$ (Fr = Frobenius), first for all simple roots r and then for all roots as we see by making the Weyl group act. We then set $\beta = (\varphi^o)^{-1} \circ \alpha : {}^{\varphi}G \to G'$. Identifying ${}^{\varphi}G$ with G according to φ^o we have a normalization in which $\varphi = id$. Then β is a morphism on each U_r and on each $T_r = \langle U_r, U_{-r}\rangle \cap T$, by the above. Hence β is a morphism on U^-TU since this set is naturally isomorphic to the Cartesian product of all of the groups U_r (r a root) and all of the groups T_a (a simple), arranged in some order. Finally since this set is open in G and β is a homomorphism of groups β is a morphism on G, which proves 1.2.

2.3. Uniqueness. One can choose φ and β in 1.2 so that β is special. Then they are unique. More precisely, if φ and β are so chosen and if $\bar{\varphi}$ and $\bar{\beta}$ satisfy the conclusions of 1.2 then there exists $m \geq 0$ such that $\varphi = Fr^m \circ \bar{\varphi}$ and $\bar{\beta} = \beta \circ (Fr^m)^o$.

Proof. Choose r so that $m_r = 0$ above. Then $\beta : {}^{\varphi}U_r \to U'_r$ is an algebraic isomorphism. Thus $d\beta \neq 0$ and β as constructed above is special. For the other assertions we replace G by ${}^{\bar{\varphi}}G$, thus normalize to the case $\bar{\varphi} = id$. Then on U_r, imbedded as k in SL_2 as above, we have $\bar{\beta} = \beta \circ \varphi$. Since $\bar{\beta}$ is a morphism and β an algebraic isomorphism, it follows that φ is a morphism, hence is of the form Fr^m $(m \geq 0)$. Then $\bar{\beta} = \alpha = \beta \circ (Fr^m)^o$. Hence if $\bar{\beta}$ is also special, $d\bar{\beta} \neq 0$, then $m = 0$, $\beta = \bar{\beta}$, and $\varphi = id = \bar{\varphi}$.

2.4. A slight extension. If we assume that α is surjective instead of bijective in 1.2 and 2.3 then the conclusions still hold.

Proof. $\ker\alpha$ is central in G since G is simple so that $G/\ker\alpha$ is also a simple algebraic group. Applying 2.1 to this group we see that the properties of

B , T ,... are preserved by α , which is all that is needed for the rest of the proof.

2.5. Special isogenies and central isogenies. We recall some facts about isogenies of connected semisimple algebraic groups. An isogeny $\pi : G \to G'$ is central if $\ker d\pi$ is central, or, equivalently, if π is an algebraic isomorphism on unipotent subgroups, or, again, if, when restricted to corresponding maximal tori, π^* maps one root system onto the other. Every central isogeny is special and conversely for simple groups a special isogeny can be noncentral only in the exceptional case that G is of type B_n , C_n , F_4 , G_2 and car k = 2 , 2 , 2 , 3 , resp. . The central isogenies are those that figure in the definition of universal covering, hence of simpleconnectedness. If G is simply connected and F the quotient of the weight lattice by the root lattice then the central isogenies $G \to \cdot$ are in correspondence with the subgroups of F , and also with the subgroups of Z(G) in case (car k, $|F|$) = 1 . For all this see [4, § 3 ; 6, Exp. 18].

2.6. The simple connectedness of G . This assumption can not be entirely dropped in 1.2 since then α need not be a morphism even if it is a morphism on each $\langle U_r , U_{-r}\rangle$ and each such group is isomorphic to SL_2 , as is shown by the example π (nat) : $SL_4 \to PSL_4$, car k = 2 , $\alpha = \pi^{-1}$. However such examples are the only ones possible :

2.7. Corollary.- In 1.2 as extended in 2.4 drop the assumption that G is simply connected. Then $\alpha = \beta \circ \varphi^o \circ \gamma$ with φ and β as in 1.2 and γ the inverse of purely inseparable central isogeny. If G is simply connected or G' adjoint, then γ may be omitted.

Proof. Let $\pi : \widetilde{G} \to G$ be the universal covering of G . By 2.4 one has $\alpha \circ \pi = \widetilde{\beta} \circ \varphi^o$ with $\varphi^o : \widetilde{G} \to {}^{\varphi}\widetilde{G}$. By replacing $\widetilde{G}$ by ${}^{\varphi}\widetilde{G}$ and G by ${}^{\varphi}G$ we may assume that φ = id . Factor π thus : $\widetilde{G} \xrightarrow{\pi_s} G_1 \xrightarrow{\pi_i} G$ with π_s separable and π_i purely inseparable, both central. Then π_s is a quotient map and $\widetilde{\beta}$ is constant on its fibres. Hence there is a (unique) morphism $\beta : G_1 \to G'$ such that $\widetilde{\beta} = \beta \circ \pi_s$. Then $\alpha = \beta \circ \gamma$ with $\gamma = \pi_i^{-1}$, as required. Further if $d\widetilde{\beta} \neq 0$ then $d\beta \neq 0$ so that β can be chosen to be special. Finally, if G' is adjoint then $\tilde{\beta}$ factors through π itself so that α is a morphism and γ may be omitted. We have used here (and elsewhere) a theorem of Chevalley [6, p. 18-07] which gives conditions under which one isogeny emanating from a (connected) semi-simple group can be factored through another. These conditions are seen to be verified here since G' is adjoint and π is central.

The discussion of uniqueness here, which is very easy, will be omitted.

2.8. Automorphisms. Let G be simple (and k still algebraically closed) and α an (abstract) automorphism of G. Then there exist φ, β, γ as in 2.7 (with $k' = k$ and $G' = G$) such that $\alpha = \gamma^{-1} \circ \beta \circ \varphi^{o} \circ \gamma$. Here γ is necessary only if $\operatorname{car} k = 2$ and G is of type D_{2n} corresponding to a semispinorial representation (thus not simply connected and not adjoint).

Proof. By 2.7 we may suppose that G is not simply connected and not adjoint. Thus we are not in the exceptional cases of 2.5 and every special isogeny is central. Let $\pi : \tilde{G} \to G$ be the universal covering. By 2.4 applied to $\alpha \circ \pi : \tilde{G} \to G$ we have $\alpha \circ \pi = \tilde{\beta} \circ \varphi^{o}$ with φ an automorphism of k and $\tilde{\beta} : {}^{\varphi}\tilde{G} \to G$ a central isogeny. Since ${}^{\varphi}\tilde{G}$ is simply connected, $\tilde{\beta}$ is thus a universal covering, thus equivalent to π. Thus $\operatorname{dg} \tilde{\beta} = \operatorname{dg} \pi = \operatorname{dg} {}^{\varphi}\pi$. Let G, hence also $\tilde{G}$, etc., not be of type D_{2n}. Then the group F of 2.5 is cyclic [6]. Thus there is at most one subgroup of a given index, thus at most one central isogeny ${}^{\varphi}G \to \cdot$ of a given degree. Thus ${}^{\varphi}\pi$ and $\tilde{\beta}$ are equivalent and there exists an algebraic isomorphism $\beta : {}^{\varphi}G \to G$ such that $\tilde{\beta} = \beta \circ {}^{\varphi}\pi$. Then $\alpha \circ \pi = \beta \circ {}^{\varphi}\pi \circ \varphi^{o} = \beta \circ \varphi^{o} \circ \pi$. Since π is surjective, $\alpha = \beta \circ \varphi^{o}$. Now let G be of type D_{2n}. In any case $\tilde{\beta} = \alpha \circ \varphi^{o^{-1}} \circ {}^{\varphi}\pi$. Thus $\ker \tilde{\beta} = \ker {}^{\varphi}\pi$. If $\operatorname{car} k \neq 2$ then again $\tilde{\beta}$ is equivalent to ${}^{\varphi}\pi$ for now the central isogenies coming from ${}^{\varphi}G$ correspond to the central subgroups of ${}^{\varphi}G$ since F is now a $(2,2)$ group, of order prime to $\operatorname{car} k$. Finally if $\operatorname{car} k = 2$ then π is purely inseparable, hence an abstract isomorphism. If we apply what has been proved to ${}^{\pi^{-1}}\alpha$ on $\tilde{G}$ we get 2.8 with $\gamma = \pi^{-1}$, which ends the proof.

Conversely, if G is of this exceptional type then γ may be needed : let $\tilde{\alpha}$ be an algebraic automorphism of $\tilde{G}$ which maps the semispinorial representation defining G onto another one, and then $\alpha = {}^{\pi}\tilde{\alpha}$.

2.9. Split groups and quasisplit groups. For split groups the proofs of 1.2 and the later results work equally well over arbitrary fields, with $G(k)$, $G'(k')$ replaced by G^+ , G'^+ , once it is known that α must preserve the properties of B , T , U ,... . For quasisplit groups (those having a rational Borel subgroup) only a little more work is needed (see [13] where automorphisms are considered). For infinite fields, this preservation will be shown in § 3 below in a very general setting.

2.10. Finite groups. For finite fields the preservation can be proved in most cases as follows. First G is quasisplit by a theorem of Lang, so that rational B and U exist. One then shows that except for a few cases of small rank car k is that prime which makes the largest contribution to the order of G^+ . This involves an exhaustive analysis of the group orders $|G^+|$ for the various types of simple groups and finite fields (see [1]). Thus G^+ determines $p = \text{car } k$, hence also, up to conjugacy, the Sylow p-subgroup $U(k)$, and finally $B(k)$, the normalizer of $U(k)$. The excluded cases then involve further considerations, and eventually one gets the result as stated for algebraically closed fields with a few exceptions, e.g. $PSL_2(\mathbf{F}_7) \cong SL_3(\mathbf{F}_2)$,... . If we are interested only in automorphisms then this development can be dispensed with since we have the preservation a priori (see [13]).

§ 3. Proof of 1.3

We give only several indications, supposing first G' adjoint. If X is a subgroup of G , let us write X' for $\overline{\alpha(X \cap H)}$ (closure in G').

3.1. (cf. 7.1 of [4]) Let S be a k-subtorus of G and U a connected unipotent subgroup normalized by S such that $Z_G(S) \cap U = \{1\}$. Suppose that $H \cap S$ is dense in S and that $H \supseteq U(k)$. Then U' is a connected unipotent k'-subgroup of G' .

Proof. The set A of $s \in S$ such that $Z(s) \cap U = \{1\}$ is dense open in U , and for each $s \in A \cap H$ the map $u \to (s,u)$ on U is a k-isomorphism of varieties, thus (*) it maps $U(k)$ onto itself [2 : 9.3]. SU is solvable, thus $L = (SU)'$ is also. Let us choose E of finite index in $S \cap H$ such that E' is connected, hence contained in L^o. Since $S \cap H$ is dense in S, there exists $s \in A \cap E$. By (*), $\alpha(U(k)) = (\alpha(s), \alpha(U(k))) \subset (L^o, L) \subset L^o$ and then $\alpha(U(k)) \subset DL^o$, the derived group of L^o. But $DL \subset U'$ since $D(SU) \subset U$. Thus $V = DL^o$, a connected unipotent group by the theorem of Lie-Kolchin.

3.2. One has $\mathrm{car}\ k = \mathrm{car}\ k'$.

Proof. Since G is isotropic, one can realize the situation in 3.1 with S and U nontrivial [3]. Then U' is nontrivial, for G^+ is generated by $U(k)$ as a normal subgroup of itself [16] and $\alpha(G^+)$ is dense in G'. If $\mathrm{car}\ k = 0$ then $G(k)$ is divisible, hence $\alpha(G(k))$ is also, hence $\mathrm{car}\ k' = 0$. If $\mathrm{car}\ k = p \neq 0$, then $U(k)$ has only p-elements, thus $\alpha(U(k))$ has also, and $\mathrm{car}\ k' = p$.

3.3. (7.2 of [4]) Let P, P^- be opposite parabolic k-subgroups of G, U, U^- their unipotent radicals, $Z = P \cap P^-$.

(a) U' and $U^{-\prime}$ are connected unipotent k'-groups and $U^{-\prime}Z'U'$ is dense open in G'.

(b) P', $P^{-\prime}$ are opposed parabolic k'-subgroups of G', U', $U^{-\prime}$ their unipotent radicals and $Z' = P' \cap P^{-\prime}$.

Proof. (a) There exists a split torus S normalizing U and such that $Z_G(S) \cap U = \{1\}$, and it can be imbedded in a split semisimple k-subgroup of G [3]. It follows that $(S \cap G^+)$ is dense in S, and U' and $U^{-\prime}$ are connected unipotent by 3.1. Let $\Omega' = U^{-\prime}Z'U'$. Now G is the union of a finite number of translates of U^-ZU by elements of G^+ [4 : 6.11] ; thus H and $U^-(k)(Z \cap H)U(k)$ have the same property. Thus $G' = H'$ is the union of a finite number of translates of Ω'. Thus Ω' contains a nonempty open subset of G, thus is itself open since it is a double coset : every double coset AcB is an orbit for the action $g \to agb^{-1}$ of $A \times B$ on G, thus is locally closed.

(b) Let T' be a maximal torus of Z'^o and V^-, V opposed maximal connected unipotent subgroups of Z'^o normalized by T' and such that $V^-.T'.V$ contains a non empty open subset of Z'^o. By (a) and the density of $\alpha(H)$ in G', $U^{-\prime}.V^-.T'.V.U'$ contains (thus is) a nonempty open subset of G', and $U^{-\prime}.V^-$, $V.U'$ are connected unipotent groups normalized by T'. As G' is simple $U^{-\prime}.V^-.T'$ and $T'.V.U'$ are opposed Borel subgroups of G'. From this the assertions of (b) follow without trouble.

In (b) G' can be reductive, and in (a) arbitrary.

Now let S_m be a maximal k-split torus of G, a_m the maximal root on S_m relative to some ordering, $\check{a}_m$ the coroot of a_m, and $S = \check{a}_m$ (Mult) the corresponding one-dimensional torus. Let $Z = Z_G(S)$ and U (resp. U^-) = the connected unipotent subgroup of G corresponding to the positive (resp. negative) weights on S relative to some ordering. Then Z is connected reductive and $P = ZU$, $P^- = ZU^-$ are opposite parabolic subgroups. Further all of these groups are k-groups. For all of this see [3]. This is the set-up in which 3.2 and 3.3 will be used.

3.4. Definition of φ. One uses the action of S on U in much the same way as in 2.2 where the group SL_2 was considered (there S = diagonal subgroup, U = superdiagonal unipotent subgroup), the multiplicative structure of k being embodied in S and the additive structure in U. But now the situation is more complicated since U is not just the group Add. However, U *is* the extension of one vector space by another such that S acts according to a character a on one and according to $2a$ on the other [4, § 8]. From this and a suitable "preservation theorem", refining 3.2 and 3.3, and a good deal of further work, one can construct a homomorphism $\varphi : k \to k'$ such that if G is replaced by ${}^{\varphi}G$ then α becomes on $U(k)$ the restriction of a special morphism $\beta_U : U \to U'$, and similarly for U^-.

3.5. Completion of proof. The group Z acts on U (*via* a rational representation, in fact, if one of the above vector spaces is trivial), as does Z' on U', and the last action is faithful since G' is adjoint. From this one gets a morphism β_Z on Z whose restriction to $Z \cap H$ agrees with α. Since the map $U^- \times Z \times U \to U^-ZU = \Omega$ is an isomorphisms of varieties we deduce a morphism β_Ω on Ω which agrees with α on $\Omega \cap H$. Finally we define β on G thus. Write $x \in G$ as gy with $g \in G^+$, $y \in \Omega$, and then set $\beta(x) = \alpha(g)\beta_\Omega(y)$. This defines $\beta(x)$ uniquely since for fixed $g \in G^+$, $\beta_\Omega(gx) = \alpha(g)\beta_\Omega(x)$ for $x \in \Omega \cap g^{-1}\Omega$ for this holds on the dense set $\Omega \cap g^{-1}\Omega \cap H$. Since β is a morphism of varieties on Ω, it is so on each $g\Omega$, hence also on G. But β is also a homomorphism on the dense subgroup H since $\alpha = \beta|_H$, clearly. Thus β

is a morphism of groups. Further β is special since β_U is.

3.6. G <u>simply connected</u>. Consider this case now. Let $\pi : G' \to \mathrm{Ad}\, G'$ be the natural map. Applying the case just proved to $\pi \circ \alpha$ we get φ and β as before such that $\pi \circ \alpha = \beta_1 \circ \varphi^0$ (on H). Since G, hence ${}^{\varphi}G$, is simply connected and π is central there exists an isogeny $\beta : {}^{\varphi}G \to G'$ such that $\beta_1 = \pi \circ \beta$, thus $\pi \circ \alpha = \pi \circ \beta \circ \varphi^0$. Since π is central, α and $\beta \circ \varphi^0$ agree on H up to a map μ into the center of G', and clearly μ is a homomorphism.

3.7. <u>Uniqueness</u>. This can be proved as in 2.3.

3.8. <u>An example</u>. Consider π (nat) : $SL_3(R) \to PSL_3(R)$, $\gamma = \pi^{-1}$. This shows that, even if car k = 0, if G is not simply connected and G' is not adjoint then 1.3 may fail (cf. 2.7).

3.9. <u>A complement</u>. Suppose that G', H' are like G, H in 1.3 and that $\alpha(H) = H'$. Then $\varphi : k \to k'$ in 1.3 is an isomorphism, not just a homomorphism.

This is proved in [4 : 8.11].

3.10. <u>Automorphisms</u>. The result is like that in 2.8 except that now a homomorphism $\mu : H \to Z(G) \cap H$ must be included and in the exceptional case of type D_{2n} of 2.8 one does not know (but one supposes) that car k = 2 since H may not contain Z(G) (as in the example of 3.8 ; it does so if G is split, quasisplit,...). The proof is similar.

§ 4. Extension and reformulation

We wish to extend 1.3 to the case where G' is reductive. This can not always be done : Let $\alpha : SL_n(C) \to SL_{2n}(R)$ be the map obtained by replacing each complex coordinate by two real coordinates. Clearly there is no homomorphism $\varphi : C \to R$. The image group is semisimple, not simple (as an algebraic group). This process which produces from a group defined over C (SL_n in this case) a corresponding group defined over R (the image) is called restriction of scalars and works whenever we have a finite dimensional field (or even algebra) k' separable over a given field k and a group G defined over k'. We write $R_{k'/k}\, G$ for the resulting group over k. There exists a natural isomorphism

$R^{o}_{k'/k} : G(k') \to (R_{k'/k}G)(k)$ (for this see [16 : I, § 1, 6.6]).

4.1. THEOREM (8.16 of [4]).- Assume as in 1.3 except that G' is reductive. Let G'_i $(1 \le i \le m)$ be the normal k'-subgroups of G' that are k'-simple (perhaps not absolutely simple). Then there exist finite separable extensions k_i $(1 \le i \le m)$ of k', field homomorphisms $\varphi_i : k \to k_i$, and a special k'-isogeny $\beta : \prod_{i=1}^{m} R_{k_i/k'}({}^{\varphi_i}G) \to G'$ and a homomorphism $\mu : H \to Z(G')(k')$ such that $\beta(R_{k_i/k'}({}^{\varphi_i}G)) = G'_i$ and $\alpha(h) = \mu(h).\beta(\prod_{i=1}^{m} R^{o}_{k_i/k'}(\varphi_i^{o}(h)))$ for all $h \in H$.

We give the proof in case G' is adjoint.

Then G' is the direct product of the G'_i 's, and $G'_i = R_{k_i/k'}G''_i$ with k_i/k' finite separable and G''_i absolutely simple. Let $\pi_i : G' \to G'_i$ be the natural projection. One then applies 1.3 to each of the maps $(R^{o}_{k_i/k'})^{-1} \circ \pi_i \circ \alpha : G \to G''_i$ and collects the results to get 4.1.

4.2. Uniqueness. We consider only the case : k' algebraically closed. Then 4.1 simplifies since the G'_i themselves are absolutely simple, each $k_i = k'$, and each R and each $R^{o} = id$. Then the possibility of making β special and the resulting uniqueness easily follow from that of 1.3. We see further that $\varphi_i = Fr^{m} \circ \varphi_j$ $(i \ne j)$ could never occur since then the image of ${}^{\varphi_i}G \times {}^{\varphi_j}G$ in $G'_i \times G'_j$ would be the graph of a morphism $G'_j \to G'_i$ and thus not dense.

4.3. A reformulation of 4.1. Under the hypotheses of 4.1 there exists a finite dimensional separable commutative k'-algebra L, a homomorphism $\varphi : k \to L$, a k'-isogeny $\beta : R_{L/k'}{}^{\varphi}G \to G'$ and a homomorphism $\mu : H \to$ center $G'(k')$ such that $\alpha(h) = \mu(h).\beta(R^{o}_{L/k'}(\varphi^{o}(h)))$ for all $h \in H$.

Proof. In 4.1 let $L = \dot{\Sigma}\, k'_i$, $\varphi = \dot{\Sigma}\, \varphi_i$,... .

4.4. A conjecture. The result 4.3 remains true if G' is arbitrary, with, perhaps, some mild changes (like dropping the separability).

The authors of [4] indicate that they have proved this in a number of cases and expect to return to it later. It holds, for example, if G is split, simply connected, semisimple and k is infinite and not nonperfect of car 2.

§ 5. Continuity of homomorphisms

There are many results in [4] on this subject. We discuss here only one or two of them related to the development given so far. We now assume that k is given a nondiscrete locally compact topology which makes it into a topological field, hence $G(k)$ into a Lie group, and similarly for k' and G', and further that G is semisimple and G' reductive.

5.1. DEFINITION.- Given a connected normal k-subgroup G_1 of G, we say that $G_1(k)$ is a complex factor of $G(k)$ if either (1) $k \cong \mathbf{C}$ or (2) $k \cong \mathbf{R}$ and G_1 is isogenous to a group of the form $R_{\bar{k}/k} G_2$.

5.2. THEOREM (9.8, 9.13 (ii) of [4]).- Let G, G', k, k' be as above. Suppose that G possesses no nontrivial normal k-anisotropic factor and that G' possesses no nontrivial complex factor. Let H and α be as in 1.3 ($G(k) \supset H \supset G^+$, $\alpha(H)$ Zariski-dense in G'). Then α is continuous. In particular $k \not\cong \mathbf{C}$. Further each surjective homomorphism of $G(k)$ onto $G'(k')$ is continuous and each such isomorphism is a topological isomorphism.

If G is simple then the first statement follows from 4.1 and the fact, mentioned above, that if $k' \neq \mathbf{C}$ then every homomorphism $\varphi : k \to k'$ is continuous. The general case follows from this case by a series of simple reductions. The last statement then follows since φ is then necessarily surjective by 3.9.

As just seen, the assumption of isotropicity on G has been used to deduce 5.2 from 4.1, but as shown in [4] it is, in fact, not needed here and in many other results. For example :

5.3. THEOREM (9.13 (i) of [4]). In 5.2 replace the assumption of isotropicity on G by : the universal covering of G is separable (which holds if G is simply connected or if car $k = 0$, and in many other cases). Then the last conclusion there holds.

The proof of 5.3 is based on a line of reasoning (due to van der Waerden [19]) not in the spirit of the above development and will be omitted.

In closing, let us mention that in particular this result implies the result of Freudenthal in the introduction, extended to other types of groups and fields.

5.4. Added remark. One of the authors of [4] has proved, over R, a very general theorem [18, § 4] of the type we have been discussing. It implies that 4.4 holds if $k = k' = R$ and G is simply connected as an algebraic group (but not necessarily semisimple) and equal to its own derived group. The two basic cases are $G = SL_2(R)$ and $G = Spin_3(R)$. From these the general case is deduced.

§ 6. Irreducible representations

We recall that a projective (resp. linear) representation of a group is a homomorphism into some [finite-dimensional] $PGL(V)$ (resp. $GL(V)$). We shall identify isomorphic representations. The principal result in [4] in this area is the following result and its refinement in 6.4.

6.1. THEOREM (10.3 of [4]).– Let k, G, H be as in 1.2, and let k' be an algebraically closed field. Let $\rho : H \to PGL_n(k')$ $(n \geq 2)$ be a projective representation irreducible on G^+. Then there exist homomorphisms $\varphi_i : k \to k'$, finite in number, and irreducible rational projective representations π_i of ${}^{\varphi_i}G$ such that, on H, ρ is equivalent to the tensor product of the $\pi_i \circ \varphi_i^o$.

Let $G' = \overline{\rho(G^+)}$. By 3.1 this group is connected, thus by the lemma of Schur it is also reductive, semisimple, adjoint. Let $\{G_i'\}$ be its simple factors. The identity representation of $G' = \Pi\, G_i'$ is irreducible, thus can be written as a tensor product $\Pi\, \lambda_i$ with λ_i an irreducible rational projective representation of G_i'. By 4.1 there exist homomorphisms $\varphi_i : k \to k'$ and special isogenies $\beta_i : {}^{\varphi_i}G \to G_i'$ such that $\rho = \Pi(\lambda_i \circ \beta_i \circ \varphi_i^o)$ on G^+. If ρ' denotes this product and $h \in H$, then $\rho'(h)\rho(h)^{-1}$ centralizes $\rho(G^+)$ (G^+ is normal in H) and is thus equal to 1 by Schur's lemma. This gives 6.1 with $\pi_i = \lambda_i \circ \beta_i$.

6.2. Refinement and uniqueness. In 6.1 uniqueness does not hold if $\mathrm{car}\, k = p \neq 0$, for, if β is an irreducible rational projective representation then ${}^{Fr}\beta \circ Fr^o$ is one also. The situation is, in fact, more complicated than this, for one has the following result [14, th. 6.1].

6.3. THEOREM.- Assume car $k = p \neq 0$. Let $M(G)$ denote the set of irreducible rational projective representations of G for which the dominant weight is a linear combination of the fundamental weights with all coefficients between 0 and $p - 1$. (Up to isomorphism, there are p^ℓ such, ℓ = rank G.) Then every irreducible rational projective representation of G is isomorphic to a finite tensor product of the form $\prod_i \pi_i \circ Fr^i$ with $\pi_i \in M(Fr^i_G)$, uniquely up to trivial factors.

For simplicity we shall write $M(G)$ in this situation. If car $k = 0$, then $M(G)$ is defined as above with no restriction on coefficients.

6.4. THEOREM.- In 6.1 it can be arranged that the φ_i are distinct and that each π_i is nontrivial and in $M({}^{\varphi_i}G)$. Then the decomposition is unique. Conversely, if the φ_i and π_i are such, then the resulting product is irreducible.

Proof. The first statement follows easily from 6.1, 6.3 and the last remark of 4.2. For the uniqueness, we put together in blocks the terms of the product for which the corresponding φ_i's differ only by a power of Fr. We get a coarser factorization $\Pi = \Pi^1 \Pi^2 \ldots$ with $\Pi^j = \pi^j \circ \varphi^j$ and $\pi^j = \Pi_i \pi^j_i \circ Fr^i$ with $\pi^j_i \in M({}^{\varphi^j}G)$. Now $\Pi^j(G)$ is the image of ${}^{\varphi^j}G$ under π^j, hence is simple, equal to one of the G'_i. It follows that the $\Pi^j(G)$ form a permutation of the G'_i. The uniqueness of the φ^j and π^j follows from 4.2, and then the uniqueness of the π^j_i from 6.3. The final statement is proved in [14, th. 5.1].

6.5. Linear representations. The preceding results extend to linear representations if one assumes that G is simply connected and adds a homomorphism into the center of $GL_n(k')$. The proof is rather easy.

We thus see that the theory of abstract irreducible representations of H is very much like that of rational ones, e.g. in case k is algebraically closed so that $H = G$ $(= G(k))$: Let B be a Borel subgroup. Then B fixes a unique line of V and acts on it according to some character, the "highest weight", which conversely determines ρ uniquely.

We close with a reformulation of the conjecture 4.4 in terms of representations. We recall that a function $f : H \to k'$ is called a representative function if the space generated by its translates over k' (left or right) is finite dimensional. As one sees these functions are the matrix coefficients of the finite-dimensional representations of H over k'. They form a k'-algebra. Let L be a finite-dimensional commutative k'-algebra and $\varphi : k \to L$ a homomorphism. Now if f is a polynomial function on G defined over k, and g is a k'-linear function from L to k' one sees easily that $g \circ \varphi \circ f : H \to k'$ is a representative function. For example, if $d : k \to k$ is a derivation, then $d \circ f$ is of this form, with L the algebra of dual numbers over k and φ the map $x \to x + dx \cdot \epsilon$. This case is related to a number of examples given in [4 : 8.18 (b), 9.15 (a)] to show the pathology that can occur if various assumptions are omitted.

6.6. Reformulation of 4.4. Under the assumptions of 4.4 every representative function on H is a polynomial in functions of the above form.

BIBLIOGRAPHY

[1] E. ARTIN - Orders of classical simple groups, Comm. Pure Appl. Math., 8(1955), 455-472.

[2] A. BOREL - Linear algebraic groups, Benjamin, New York, 1969.

[3] A. BOREL et J. TITS - Groupes réductifs, Publ. Math., I.H.E.S., 27 (1965), 55-151 ; 41 (1972), 253-276.

[4] A. BOREL et J. TITS - Homomorphismes "abstraits" de groupes algébriques simples, Ann. of Math., 97(1973), 499-571.

[5] E. CARTAN - Sur les représentations linéaires des groupes clos, Comm. Math. Helv., 2 (1930), 269-283.

[6] C. CHEVALLEY - Classification des groupes de Lie algébriques, Notes from Inst. H. Poincaré, 2 volumes, Paris (1956-58).

[7] M. DEMAZURE et P. GABRIEL - Groupes algébriques, T. I. Masson, Paris (1970).

[8] J. DIEUDONNÉ - La géométrie des groupes classiques, Second Edition, Springer Verlag, Berlin (1963).

[9] H. FREUDENTHAL - Die Topologie der Lieschen Gruppen ..., Ann. of Math., 42 (1941), 1051-1074 ; 47 (1946), 829-830.

[10] O. T. O'MEARA - The automorphisms of the orthogonal groups..., Amer. J. Math., 90 (1968), 1260-1306.

[11] O. SCHREIER und B. L. VAN DER WAERDEN - Die Automorphismen der projectiven Gruppen, Abh. Math. Sem. Hamburg, 6 (1928), 303-322.

[12] Séminaire "Sophus Lie" - Notes from Inst. H. Poincaré, Paris (1955).

[13] R. STEINBERG - Automorphisms of finite linear groups, Canad. J. Math., 12 (1960), 606-615.

[14] R. STEINBERG - Representations of algebraic groups, Nagoya Math. J., 22 (1963), 33-56.

[15] R. STEINBERG - Lectures on Chevalley groups, Yale University lecture notes, (1967).

[16] J. TITS - Algebraic and abstract simple groups, Ann. of Math., 80 (1964), 313-329.

[17] J. TITS - Homomorphismes et automorphismes "abstraits" de groupes algébriques et arithmétiques, Int. Math. Congress, Nice, 2 (1970), 349-355.

[18] J. TITS - Homomorphismes "abstraits" de groupes de Lie, to appear.

[19] B. L. VAN DER WAERDEN - Stetigkeitssätze für halb-einfache Liesche Gruppen, Math. Zeit. 36 (1933), 780-786.

Reprinted from ADVANCES IN MATHEMATICS
All Rights Reserved by Academic Press, New York and London

Vol. 15, No. 1, January 1975
Printed in Belgium

Torsion in Reductive Groups

ROBERT STEINBERG

Department of Mathematics, University of California,
Los Angeles, California 90024

INTRODUCTION

In [3, p. E-12], the notion of torsion primes for semisimple algebraic groups was defined (see 2.1 below for a somewhat different definition, for reductive groups), and there it was stated (on p. E-41) without proof that:

0.1 THEOREM. *Let G be a simply connected semisimple algebraic group and t a semisimple element such that $t^n \in Z(G)$, the center of G, for some n divisible by no torsion prime. Then the semisimple component of $Z_G(t)$ is simply connected.*

Our first objective here is to supply a proof of this result and to indicate, as is done in [3], its connection with the problem of extending to several elements the theorem [12, 8.1] that with G as above, $Z_G(t)$ is connected for every semisimple element t. This is done in §2 where various extensions and converses are also considered. This follows preliminary material in §1 where simple proofs of some results of de Siebenthal have been included.

Our second objective is, with the aid of these results, to supply a proof of the following theorem also stated in [3] (on p. E-35).

0.2 THEOREM. *If G is as in* 0.1 *and the characteristic of the base field is not a torsion prime* (*as in* 1.3 *below*), *then $Z_G(H)$ is connected* (*and reductive*) *for every semisimple $H \in \mathfrak{g}$, the Lie algebra of G.*

This and related matters, some also mentioned in [3], are discussed in §3.

Our results were obtained in 1963 at the time of the first writing of [12]. We omitted their proofs from [12] to avoid digressions and from [3] to keep the length down. The starting point, occurring in

63

[12, 1.20], is the theorem that in a finite reflection group acting on real vector space, the centralizer of any collection of points is again a reflection group. In a final section we indicate the extent to which this result holds when the real field is replaced by an arbitrary Abelian group.

The considerations of §2 lead naturally to connections among: the torsion primes, the coefficients of the coroot of the highest root, and the imbeddability of elementary Abelian p-groups in tori. These connections were obtained in [1] for compact Lie groups (which is not an essentially different case), however, with many ad hoc verifications using the classification, as was lamented by Borel himself with the aid of a quotation from G. B. Shaw. For this reason in the present development we have avoided proofs by classification like the plague, even when such proofs could be accomplished "avec un coup d'œil."

1. The Geometry of the Highest Root

Throughout this work Σ will denote a root system in the classical sense, W its Weyl group, (,) a positive definite inner product invariant under W, $\Sigma^* = \{\alpha^* = 2\alpha/(\alpha, \alpha), \alpha \in \Sigma\}$ the dual system, and V the real space extending $L(\Sigma^*)$, the lattice generated by Σ^*. The elements of Σ should be thought of primarily as functions on L and its extensions such as V. We write $\{\alpha_i, 1 \leqslant i \leqslant r\}$ for a basis (simple system), $-\alpha_0 = \sum_{i=1}^{r} n_i\alpha_i$ for the corresponding highest root (in case Σ is irreducible), and $-\alpha_0^* = \sum n_i^*\alpha_i^*$ for its coroot. We have

1.1 (a) $n_i^* = n_i(\alpha_i, \alpha_i)/(\alpha_0, \alpha_0)$.

(b) If α is a long root (i.e., as long as α_0) and $\alpha^* = \sum m_i^*\alpha_i^*$, then $m_i^* \leqslant n_i^*$ for all i.

Here (a) is clear and (b) then follows.

We set $n_0^* = 1$ so that

1.2
$$\sum_{i=0}^{r} n_i^*\alpha_i^* = 0.$$

1.3 Definition. A prime p is a *torsion prime* for a root system Σ if $L(\Sigma^*)/L(\Sigma_1^*)$ has p-torsion for some closed subsystem Σ_1 of Σ.

We mean closed with respect to the taking of integral combinations, or, equivalently, negatives and sums, whenever these operations lead to roots. From the definition we get at once:

1.4. If Σ_1 is a closed subsystem of Σ, then the torsion primes for Σ_1 are among those for Σ.

Presently we shall show that p is a torsion prime just when it equals some $n_i{}^*$ for some irreducible component of Σ.

1.5 Lemma of the String. *Let Σ be an irreducible root system and $\alpha_0, \alpha_1, \ldots$ the extended system of simple roots so labeled that $\alpha_0, \alpha_1, \ldots, \alpha_q$ is a minimal string connecting (i.e., $(\alpha_i, \alpha_{i+1}) \neq 0$ for all i) α_0 to a root for which $n^* = \max n_i{}^*$ is achieved.*

(a) *Each α_i $(0 \leqslant i \leqslant q)$ is a long root and $n_i{}^* = i + 1$, so that in particular $n^* = q + 1$.*

(b) *If $n^* > 1$, then the string is a simple string connected to the other simple roots only at α_q.*

(c) *If $\{\omega_i \mid 1 \leqslant i \leqslant r\}$ denote the fundamental weights defined by $(\omega_i, \alpha_j{}^*) = \delta_{ij}$, then for $1 \leqslant i \leqslant q$, ω_i is a sum of long roots, in fact is equal to $-\sum_{j=0}^{i-1} (i - j)\alpha_j$.*

(d) *If v_i in V is defined by $\alpha_j(v_i) = 0$ for $j \neq 0, i$, and $\alpha_0(v_i) = -1$, or, equivalently, $\alpha_i(v_i) = 1/n_i$, then for $1 \leqslant i \leqslant q$ we have $n_i v_i \in L(\Sigma^*)$.*

Proof. If $n^* = 1$, $q = 0$, then the only assertion being made is that α_0 is a long root, which is, of course, well-known (see, e.g., [6, p. 165, Proposition 25]). Assume henceforth that $n^* > 1$. We form the inner product of 1.2 with $\alpha_0, \alpha_1, \ldots$ in turn and then discard a number of terms of the form $(\alpha_i, \alpha_j{}^*)n_j{}^*$, $i \neq j$, all $\leqslant 0$. Thus we get

1.6
$$(\alpha_0, \alpha_0{}^*)\, n_0{}^* + (\alpha_0, \alpha_1{}^*)\, n_1{}^* \geqslant 0,$$
$$(\alpha_i, \alpha_{i-1}^*)\, n_{i-1}^* + (\alpha_i, \alpha_i^*)\, n_i{}^* + (\alpha_i, \alpha_{i+1}^*)\, n_{i+1}^* \geqslant 0 \qquad (1 \leqslant i < q).$$

Equality holds only if all the discarded terms are 0. We have $(\alpha_i, \alpha_i{}^*) = 2$ and

1.7
$$(\alpha_0, \alpha_1{}^*) \leqslant -1,$$
$$(\alpha_i, \alpha_{i+1}^*) \leqslant -1 \qquad (1 \leqslant i < q).$$

Substituting this into 1.6 we get

1.8
$$2n_0{}^* - n_1{}^* \geqslant 0,$$
$$-n_{i-1}^* + 2n_i{}^* - n_{i+1}^* \geqslant 0 \qquad (1 \leqslant i < q).$$

On adding we get $n_0^* + n_{q-1}^* - n_q^* \geqslant 0$, hence $n_q^* \leqslant n_{q-1}^* + 1$ since $n_0^* = 1$. But $n_q^* \geqslant n_{q-1}^* + 1$ from the definition of the string. Thus equality holds here and also in all of the above inequalities. Thus all roots of the string have the same length by the equality in 1.7, and all $n_i^* = i + 1$ by induction and the equality in 1.8, which proves (a). By the equality in 1.6, α_0 is connected only to α_1, α_i only to α_{i-1} and α_{i+1} for $1 \leqslant i < q$, which proves (b). By (a) and (b), the last expression in (c) is a sum of long roots and its inner product with α_k^* is, for $1 \leqslant k < i$ equal to $-(i - k + 1) + 2(i - k) - (i - k - 1) = 0$, for $k = i$ equal to 1, and for $k > i$ equal to 0; hence it equals ω_i, as asserted. In (d), $n_i v_i$ and ω_i are both orthogonal to all α_j $(j \neq i)$, hence are in the same line; $n_i v_i = (2/(\alpha_i, \alpha_i))\omega_i$ as we see by taking inner products with α_i. By (c), $\omega_i = \sum \beta_j$, a sum of long roots. Since α_i is also long by (a), we have $n_i v_i = \sum \beta_j^* \in L(\sum^*)$, as required.

1.9 *Remark.* The argument used to prove (a) and (b) can be carried one step further to show that if $n^* > 1$, then α_q is a branch point.

1.10 Corollary. (a) *The coefficients n_i^* form a connected string of integers starting with* 1 *or with* 2 *and going up.*

(b) *For a prime p, the following conditions are equivalent.*

(1) *$p \leqslant n^*$, the largest n_i^*.*

(2) *$p =$ some n_i^*.*

(3) *$p \mid$ some n_i^*.*

Here (a) follows from 1.5(a), and (b) follows from (a).

1.11 *Remark.* In the same way one may prove an analogous result about $-\alpha_0$ itself and its coefficients n_i and one may add to (b) the condition (4) p is a coefficient of some root. This is because every positive root may be written as a sum of simple roots so that every partial sum is a root.

1.12 Theorem. *Let Σ be a root system and p a prime. Then the following conditions are equivalent.*

(a) *p is a torsion prime for Σ (see* 1.3*).*

(b) *p satisfies any of the equivalent conditions of* 1.10(b) *for some irreducible component of Σ.*

(c) *$L(\Sigma^*)/L(\Sigma_1{}^*)$ has order p for some maximal closed subsystem Σ_1 of Σ.*

1.13 COROLLARY. *For Σ irreducible of type A_r, C_r, B_r ($r \geqslant 3$), D_r, E_6, E_7, E_8, F_4, G_2, the torsion primes are those $\leqslant n^* = 1, 1, 2, 2, 3, 4, 6, 3, 2$, respectively. If Σ is reducible, its torsion primes are those of its various components.*

This follows from 1.10, 1.12, and a list of highest roots (see, e.g., [6, pp. 200–221]).

To prove 1.12, we may as well assume that Σ is irreducible. If (b) holds, we set $i = p - 1$ in 1.5(a), so that $n_i{}^* = p$, and also $n_i = p$ since α_i is long. If Σ_1 consists of all roots $\sum m_j\alpha_j$ for which $p \mid m_i$, it readily follows that Σ_1 fulfils the requirements of (c). Clearly (c) implies (a). For the proof that (a) implies (b) and for other purposes, we recall some known facts. We assume Σ irreducible and the other notation as above.

1.14. Let S be the simplex in V defined by $\alpha_i \geqslant 0$ ($1 \leqslant i \leqslant r$), $\alpha_0 \geqslant -1$. Then S is a fundamental domain for W extended by the translations of $L(\Sigma^*)$ acting on V. Hence S projects faithfully into the torus $T = V/L(\Sigma^*)$ and there becomes a fundamental domain for W.

For the proof see either [6, p. 75] or [12, p. 29].

1.15. Let v_i be the vertex of S in V defined as in 1.5(d). Then the roots integral at v_i (or, equivalently, those vanishing at v_i if v_i is projected into T, so that the roots become characters on T) form a closed subsystem Σ_i of rank r. It consists of all roots $\alpha = \sum_{j=1}^{r} m_j\alpha_j$ such that $m_i = 0, \pm n_i$ and has $\{\alpha_j \mid j \neq 0, i\} \cup \{\alpha_0\}$ as a basis.

The last point is proved thus: Let α be as given. If $\alpha > 0$ and $m_i = 0$, then $\alpha = \sum_{j \neq i} m_j\alpha_j$ with each $m_j \geqslant 0$, while if $m_i = -n_i$, then $\alpha = \alpha_0 + \sum_{j \neq i} (m_j + n_j)\alpha_j$, with each $m_j + n_j \geqslant 0$ since $-\alpha_0$ is the highest root.

1.16. The subsystem Σ_i of 1.15 is maximal if and only if n_i is prime. Every maximal subsystem of rank r is, up to conjugacy under W, equal to such a Σ_i.

This is proved in [5], as follows. If n_i is not prime and $p \mid n_i$, then the roots $\sum m_j\alpha_j$ for which $p \mid m_i$ form a larger subsystem because of 1.11. Now let Σ' be any maximal subsystem of rank r. Since $L(\Sigma') \subsetneqq L(\Sigma)$, there exists a point v at which Σ' is integral and Σ is not, and this point may be taken in S by 1.14, at a vertex v_i since Σ' has rank r. Then $\Sigma' \subseteq \Sigma_i$ (as in 1.15) and by maximality equality holds, as required.

1.17. If $n_i = 1$, then $\{\alpha_j \mid j \neq i, 0\}$ is a basis of a maximal subsystem, and every maximal subsystem of rank $<r$ is obtained this way.

This is easily verified.

1.18. If $n_i = 1$, in other words, if all roots are integral at v_i, then $\{\alpha_j \mid j \neq 0, i\} \cup \{\alpha_0\}$ is a basis for Σ itself and $-\alpha_i$ is the corresponding highest root.

For $-\alpha_i$ in terms of this basis has the same sum of coefficients as $-\alpha_0$ in terms of the original basis.

Resuming the proof of 1.12, we prove next:

1.19. *Let Σ be irreducible and Σ' a closed irreducible subsystem. Then $n^*(\Sigma') \leqslant n^*(\Sigma)$.*

Let Σ'' be the rational closure of Σ' in Σ. Then every simple system of Σ'' can be extended to one of Σ; for example, by extending it to an arbitrary basis in the usual sense (maximal linearly independent) of Σ and then using the ordering in which the last nonzero coefficient relative to this basis counts. By an obvious induction, this reduces the proof of 1.19 to two cases: (1) that of 1.16; (2) that of 1.17, but with n_i perhaps different from 1. Let $-\alpha_0$, $-\alpha_0'$ be the highest roots of Σ, Σ' with respect to compatible orderings. We express $-\alpha_0' = \sum n_i'\alpha_i'$ in terms of the simple roots of Σ' and these in turn $\alpha_i' = \sum m_{ij}\alpha_j$ in terms of the simple roots of Σ, so that $-\alpha_0' = \sum l_j\alpha_j$ with $l_j = \sum n_i'm_{ij}$. Now every short root must have a short simple root in its support. It follows that $\max n_i'$ (α_i' short) $\leqslant \max l_j \leqslant \max n_j$ (α_j short), and similarly for long roots. Thus if $-\alpha_0'$ is long (as $-\alpha_0$ always is), then the same inequalities hold on the coefficients of $-\alpha_0'^*$ and $-\alpha_0^*$, whence 1.19 holds. This covers case (1) above and leaves the special case of (2) in which two root lengths occur and $-\alpha_0'$ is short. Then α_i must be long, the other simple roots short. Since Σ is indecomposable and $-\alpha_0'$ is a strictly positive combination of the α_j ($j \neq i$), we have $(-\alpha_0', \alpha_i) < 0$. Set $\alpha = \alpha_i - (\alpha_i, -\alpha_0'^*)(-\alpha_0')$, a long root since it equals $w_\beta\alpha_i$ with $\beta = -\alpha_0'$. The coefficients of $\alpha^* = \alpha_i^* + (\alpha_i^*, \alpha_0')(-\alpha_0'^*)$ then dominate those of $-\alpha_0'^*$ and are in turn dominated by those of $-\alpha_0^*$ by 1.1(b), whence 1.19.

Now let p be a torsion prime in 1.12(a). To prove (b), that $p \leqslant n^*$, in view of 1.19, again by induction we are reduced to the two cases just considered. Now in (2) there is no torsion; so (1) must hold. Then by 1.15, $L(\Sigma^*)/L(\Sigma'^*)$ is cyclic of order n_i^*. Thus $p \mid n_i^*$. We have completed the proof of 1.12.

We continue with a technical lemma needed later. We recall that if the notation is as above, then the *central elements* of T (when it is put in as a maximal torus of a semisimple group) are those at which all roots vanish (or, in V above, are integral), or, equivalently, belong to $Z_T(W)$, or yet again (in case Σ is irreducible) those represented in the fundamental simplex S by the origin and the vertices v_i with $n_i = 1$ as in 1.15. These equivalences are all classical [6] and furthermore rather easy to prove.

1.20. LEMMA. *Let Σ be irreducible, p a prime, and t a central element of order p in T, represented in V by the first vertex, v_1, of S.*

(a) *There exist $u \in T$, $w \in W$ such that*

(1) $(1 - w)u = t$.

(2) $u^p \in \langle t \rangle$, *even* $= 1$ *in case* $p \neq 2$.

(3) $w^p = 1$.

(b) *If $p = 2$, then w in* (a) *may be chosen so that if $\beta_1, \beta_2, \ldots$ are the positive roots made negative by w then $1/2 \sum \beta_i^*$ is an integral multiple of v_1.*

Proof. We represent $\langle t \rangle = C$, say, by the corresponding set of p vertices of the fundamental simplex S, and choose u as the centroid of the corresponding face. Then u^p is the product of the elements of C, which is t if $p = 2$ and 1 if p is odd since then the nontrivial elements of C cancel in reciprocal pairs, so that (a2) holds. Now $S - t$ is also one of the standard fundamental cells (cut from T by the equations $\alpha = 0$ for all $\alpha \in \Sigma$) for the action of W on T, hence has the form wS for some $w \in W$. Let $\sigma = w^{-1} \circ \rho_{-t}$, translation by $-t$, so that $\sigma S = S$. Since W acts trivially on C, $\sigma C = C - t = C$, σ fixes the corresponding face of S and hence also its centroid u, so that $w^{-1}(u - t) = u$, $t = (1 - w)u$, and (a2) holds. Then (a3) also holds for $\sigma^p S = S$ implies $w^{-p}S = S$ since t has order p, and then w^{-p} is 1 since it fixes the set of simple roots. Now we claim that (*) w^{-1} as just chosen makes negative just those positive roots with α_1 in their supports. This is known [6, p. 176, Proposition 6] and proved thus. ρ_{-t} maps t and the inequalities defining S there to 0 and those defining wS there. Hence $\{\alpha_0, \alpha_2, \alpha_3 \cdots\}$ is a simple system for the chamber containing wS, and w^{-1} maps it onto $\{\alpha_1, \alpha_2, \ldots\}$ and also matches up the corresponding lowest roots α_1 and α_0 (see 1.18). Hence w^{-1} keeps positive those positive roots with support in $\{\alpha_2, \alpha_3, \ldots\}$, i.e., with support not containing α_1. Now let α be positive with α_1 in its support. Write $\alpha = \alpha_1 + \beta$ with supp $\beta \leqslant \{\alpha_2, \alpha_3, \ldots\}$. Since w^{-1} supp β = supp $w^{-1}\beta$ by the above, we have on taking heights that $h(w^{-1}\alpha) = h(w^{-1}\alpha_1) + h(w^{-1}\beta) = h(\alpha_0) + h(\beta) = h(\alpha_0) + h(\alpha) - 1 < 0$ since α_0 is the lowest root. Thus $w^{-1}\alpha < 0$ and (*) holds. Consider now (b), in which $p = 2$ and $w^{-1} = w$. The roots as in (*) are permuted by every w_j ($j \neq 1$), whence their sum is kept fixed. Hence $1/2 \sum \beta_i{}^*$ is orthogonal to every α_j ($j \neq 1$) and so must be a real multiple of v_1, say cv_1. Then $2cv_1 \in L(\Sigma^*)$. Since the order of $v_1 \bmod L(\Sigma^*)$ is 2, we get $2c \in 2\mathbb{Z}$, $c \in \mathbb{Z}$, whence (b).

We close this section by proving some results of de Siebenthal [10], obtained by him by case-by-case verification. Like him, we shall not need them later, but we give proofs since we are set up for them and the only other general proofs in the literature are unnecessarily complicated [4, §4].

1.21. *Let Σ, as above, be a root system and Σ' a proper subsystem of the same rank ordered in some way. Then $s^* = \sum \alpha^*$ ($\alpha \in \Sigma'$, $\alpha > 0$) is singular relative to Σ, i.e., orthogonal to some root.*

1.22 *Remarks.* (a) For α a simple root of Σ', $w_\alpha s^* = s^* - 2\alpha$ since w_α permutes the positive roots of Σ' other than α. Hence $\alpha(s^*) = 2$, independent of α, by the formula for a reflection. Hence the line determined by s^* is just "the diagonal" of the given basis of Σ', where all simple roots are equal. Thus 1.21 may be reformulated to say that this diagonal in V is singular. The corresponding result in T is true since s^* can be interpreted as a one-parameter group into T whose image is just the identity component of the diagonal there; looked at this way, it is seen to be true even if T is an algebraic torus. (b) We do not require Σ' to be (integrally) closed (as did the earlier authors) only to satisfy $w_\alpha\Sigma' = \Sigma'$ for all $\alpha \in \Sigma'$. This extra bit of generality actually simplifies our development. For example, Σ' might be the set of short roots of an irreducible root system Σ with two different root lengths and hence not be closed.

Proof of 1.21. Assume not. Then we have an ordering of Σ in which the positive roots are those for which $(s^*, \alpha) > 0$, compatible with the given ordering on Σ' by 1.22 above. Label the simple roots $\alpha_1, \alpha_2, \ldots, \alpha_r$ of Σ so that $\alpha_1, \alpha_2, \ldots, \alpha_q$ are those lying in Σ'. Let α be some other simple root of Σ', $\alpha = \sum m_i \alpha_i$. Then $\sum m_i(s^*, \alpha_i) = 2$. By the choice of α, at least two terms occur on the left. Hence exactly two do and $m_i = 1$, $(s^*, \alpha_i) = 1$ for both of them, so that $\alpha = \alpha_i + \alpha_j$, say, with α_i, α_j not in Σ' and adjacent in the graph of simple roots of Σ. Now these simple roots have a graph which is a forest (no circuits), hence have at most $(r - q - 1)$ adjacent pairs (in fact, exactly $r - q - p$ if p is the number of trees). Hence there are at most $(r - q - 1) + q = r - 1$ roots simple relative to Σ', which must therefore have smaller rank than Σ, a contradiction.

1.23 *Remark.* Even without the assumption of equal rank, s^* in 1.21 is likely to be singular. In fact, by further argument it can be shown that the only case is which s^* is regular when Σ is irreducible is: $\sum$ of type A_r (r even), Σ' any subsystem of rank $r - 1$.

1.24 COROLLARY. *In* 1.21 *above,* $s = \sum \alpha$ $(\alpha > 0,\ \alpha \in \Sigma')$ *is singular relative to* Σ.

To get this we apply 1.21 with Σ, Σ', s^* replaced by Σ^*, Σ'^*, s.

1.25 COROLLARY. *Assume* Σ *irreducible with* $-\alpha_0 = \sum n_i \alpha_i$ *the highest root and* n_1 *prime. Let* v *be the point of* V *at which* $\alpha_0, \alpha_2, \alpha_3, \ldots$ *are all* 1. *Then* α_1 *is also integral at* v, *which is thus central.*

Proof. Let $\Sigma' = \Sigma_1$ as in 1.15, with $\alpha_0, \alpha_2, \alpha_3, \ldots$ as its basis and s^* defined accordingly as in 1.21. Then by 1.21 and 1.22, s^* is a multiple of v and $\alpha(v) = 0$ for some root $\alpha = a_1\alpha_1 + a_2\alpha_2 + \cdots$ which may be taken positive. We have $0 < a_1 < n_1$ since $\alpha \notin \Sigma'$ clearly. Hence

$$(*) \qquad a_1\alpha_1(v) = -a_2\alpha_2(v) - a_3\alpha_3(v) \cdots \in \mathbb{Z}.$$

But also by the choice of v

$$(**) \qquad n_1\alpha_1(v) = -\alpha_0(v) - n_2\alpha_2(v) - \cdots = -1 - n_2 - n_3 - \cdots \in \mathbb{Z}.$$

Since n_1 is prime, a_1 and n_1 are relatively prime, so that by (*) and (**) $\alpha_1(v) \in \mathbb{Z}$, whence 1.25.

1.26 COROLLARY. *If* $-\alpha_0 = \sum n_i \alpha_i$ *as above and* n_1 *is prime, then* $n_1 \mid h = 1 + \sum n_i$.

This follows from 1.25 and equation (**).

The last two results and their proofs come from [4, §4].

2. Torsion in Reductive Groups

Let G be a (connected) reductive algebraic group over an algebraically closed field k (or else a connected compact Lie group), $G = G_1T_1$ with G_1 the semisimple component, T_1 the radical, a central torus. We write $F = F(G)$ for the fundamental group of G_1. A reductive subgroup of G will be called *regular* if it contains a maximal torus of G. Its root system may thus (and will) be identified with a subsystem of that of G.

2.1 Definition. A prime p is a *torsion prime* for a reductive group G if $F(G')$ has p-torsion for some regular reductive subgroup G' of G whose root system is integrally closed in that of G.

2.2 *Remarks.* (a) Since every semisimple group is imbeddable in some SL_n, some condition on the allowable subgroups is needed. (b) In the context of compact Lie groups, 2.1 is equivalent to: $H^1(G')$ has p-torsion for some regular subgroup G'. This turns out to be equivalent to: $H^*(G)$ has p-torsion, but at the moment only by a long series of case-by-case considerations (see [1]). (c) The condition on root systems requires further explanation. This will be given below (in 2.9 especially).

Because of the definitions we have:

2.3. Let $G = G_1T_1$ be as above and G' a regular reductive subgroup.

(a) G and G_1 have the same torsion primes.

(b) The torsion primes of G' are among those of G.

To go further we introduce the data consisting of T, a maximal torus (of G), X its character group, L its lattice of one-parameter subgroups, in natural $\mathbb{Z}$-duality with X, and Σ the root system. We have Σ imbedded in X and Σ^* in L.

2.4. *We have* $F = F(G) = \operatorname{tors} L/L(\Sigma^*)$.

If G is semisimple, this may be taken as the definition of F. (If we write F_s for the separable part of F, of order prime to char k, and F_i for the inseparable part, of order a power of char k, so that $F = F_sF_i$, then F_s is isomorphic to the kernel of the universal covering $\pi: G' \to G$, while $F_i \neq \{1\}$ signifies that $\ker d\pi \neq 0$, i.e., that π is not separable.) In particular, $L = L(\Sigma^*)$ is the condition for simple connectedness. If G is arbitrary, T can be written as the direct product of a maximal torus of G_1 and another torus, so that 2.4 still holds.

2.5 Lemma. *If G is reductive, then its torsion primes are those of its root system Σ (see 1.3) together with those of its fundamental group F, i.e., those of $L/L(\Sigma^*)$.*

Proof. We may assume G semisimple. Clearly the torsion primes of F are torsion for G. So are those of Σ for if p is one of them then by [8, Exp. 17] we may choose G' as the subgroup corresponding to $\Sigma' = \Sigma_{p-1}$ (see 1.15 and 1.5(a)) to get p-torsion in $L(\Sigma^*)/L(\Sigma'^*)$, hence also in $L/L(\Sigma'^*) = F(G')$. Conversely, let p be a torsion prime. Then $L/L(\Sigma'^*)$ has p-torsion for some $\Sigma' = \Sigma(G')$ with G' as in 2.1, so that either $L(\Sigma^*)/L(\Sigma'^*)$ does, i.e., Σ does, or else $L/L(\Sigma^*)$ does, i.e., F does, as required.

2.6 Corollary. *If G_1 is simply connected, then its torsion primes are those of its root system.*

2.7 Corollary. *If G is simple, then it can have p-torsion beyond that of its root system only in the cases: type A_r, $p \mid (r+1)$; type C_r, $p = 2$.*

Proof. The possibilities for F in the various cases are, of course, well known.

2.8 Corollary. *Each torsion prime for G divides the order of the Weyl group (but not conversely).*

Proof. We may assume G simple, adjoint, of rank r, say. By a formula of Weyl $|W| = f \cdot r! \cdot \Pi n_i$ with $f = |F|$ and $\sum n_i \alpha_i$ the highest root. Since n_i^* divides n_i, the corollary follows from 1.12.

2.9 Lemma. *Let G and Σ be as above.*

(a) *Every closed subsystem of Σ supports a regular reductive subgroup of G.*

(b) *If Σ' is the root system of a regular reductive subgroup and Σ'' its (integral) closure in Σ, then $\Sigma' \neq \Sigma''$ only if* char $k = p \neq 0$ *and p is the square-length ratio of two elements in the same component of Σ''. In that case, $L(\Sigma''^*)/L(\Sigma'^*)$ is an elementary p-group.*

2.10 *Remark.* This shows that the extra condition on root systems of 2.1 is needed only in the rather exceptional circumstances of (b). It has no bearing in the present section since G has no semisimple elements of order p, but does so in the next section where elements of the Lie algebra are considered.

Proof. Here (a) is standard [8, Exp. 17]. In (b) we may assume that $\Sigma'' = \Sigma$, that $\Sigma \neq \Sigma'$, and that Σ is irreducible so that there are at most two root lengths. Hence there exist $\alpha, \beta \in \Sigma'$ such that $\alpha + \beta \in \Sigma - \Sigma'$. Then (*) $(\alpha, \beta) \geqslant 0$ since otherwise $\alpha + \beta$ would equal either $w_\alpha \beta$ or $w_\beta \alpha$ and hence be in Σ'. Then by (*) $\alpha + \beta$ is longer than α and β so that (**) $\alpha + \beta$ is long, α and β short. Further, $\alpha + 2\beta$ is longer than $\alpha + \beta$ so that it cannot be a root. If q is the smallest nonnegative integer such that $\beta - q\alpha$ is not a root, then by symmetry $\beta - q\alpha = w_\alpha(\beta + 2\alpha)$, so that $q = (\beta, \alpha^*) + 2 = |\alpha + \beta|^2/|\alpha|^2$ by (**). Now if U_α, U_β are the one-parameter unipotent subgroups corresponding to α, β, then (U_α, U_β) (commutator) $\subseteq \Pi\, U_{i\alpha+j\beta}$ $(i, j \geqslant 1)$ and $U_{\alpha+\beta}$ is present on the right unless q above is 0 in k. Hence $|\alpha + \beta|^2/|\alpha|^2 = q = 0$ in k. However, $|\alpha + \beta|^2/|\alpha|^2 = 2$ or 3, a prime. Hence $q = p$. Now if $\{\alpha_i\}$ is a basis for Σ' and $\alpha = \sum m_i\alpha_i$ $(m_i \in \mathbb{Z})$ is any root of Σ, then $\alpha^* = \sum m_i^*\alpha_i^*$ with $m_i^* = m_i|\alpha_i|^2/|\alpha|^2 \in q^{-1}\mathbb{Z} = p^{-1}\mathbb{Z}$. Thus $L(\Sigma^*)/L(\Sigma'^*)$ is an elementary p-group, which proves (b).

We continue with a final preliminary result.

2.11. Let G' be a regular reductive subgroup of G and Σ' its root system. Then the following conditions are equivalent.

(a) The natural map $F(G') \to F(G)$ is injective.

(b) $L(\Sigma^*)/L(\Sigma'^*)$ has no torsion.

(c) Every long root of Σ rationally dependent on Σ' is in Σ'.

The equivalence of (a) and (b) follows rather easily from 2.4. Now assume (b). Let α be a root as in (c). Then by (b) $\alpha^* = \sum \beta_i^*$ $(\beta_i \in \Sigma')$. We observe now as in the proof of 2.9 that (*) if $\beta, \gamma \in \Sigma'$, $(\beta, \gamma) < 0$, then $\beta + \gamma \in \Sigma'$. Hence if the above expression for α^* is minimal, then $(\beta_i, \beta_j) \geqslant 0$ for all i, j. Then since α is long, hence α^* short, there can be only one summand, so that $\alpha = \beta_1 \in \Sigma'$, which is (c). Now if Σ'' is the rational closure of Σ' in Σ, then $L(\Sigma^*)/L(\Sigma''^*)$ has no torsion (since a basis of Σ'' can be extended to one of Σ). Thus in proving that (c) implies (b) we may assume that $\Sigma'' = \Sigma$ and then must show that $L(\Sigma'^*) = L(\Sigma^*)$ (but not that $\Sigma' = \Sigma$: consider, e.g., $\Sigma' = \{$long roots$\}$ in a two root length system). If α is a long root in Σ, then α^* is short and in Σ'^* by assumption. Since Σ^*, like any root system, is generated by its short roots, we get $L(\Sigma'^*) = L(\Sigma^*)$, as required.

2.12 Definition. If any of the equivalent conditions of 2.11 hold, we shall say that G' is *simply connected in* G.

2.13 *Remarks.* (a) The relation just defined is transitive. (b) If G (or rather its semisimple component) is simply connected and G' is simply connected in G, then G' is simply connected. (c) If (b) of 2.11 fails, we still have a homomorphism $F(G') \to F(G)$, the kernel being tors $L(\Sigma^*)/L(\Sigma'^*)$. (d) If Σ' above is rationally closed in Σ, then G' is simply connected in G (e.g., a Levi component of a parabolic subgroup is such). (e) If G' is simply connected in G, then Σ' is (integrally) closed in Σ. For (*) above implies that any short root integrally dependent on Σ' is in Σ'.

We turn now to one of our main concerns, centralizers of semisimple elements. First we recall some facts from [12, §7].

2.14 Lemma. *Let G be reductive, T a maximal torus, t an element or subset of T, Σ' the (closed) subsystem of roots vanishing at t, W' its Weyl group, and W'' the centralizer of t in W.*

(a) *$Z_G(t)$ is generated by T, those U_α such that $\alpha \in \Sigma'$, and those n_w such that $w \in W''$.*

(b) *$Z_G(t)^0$ is generated by T and the U_α's alone. It is (regular) reductive with Σ' as its root system.*

(c) *W' is normal in W'' and $Z_G(t)/Z_G(t)^0$ is isomorphic to W''/W'.*

We are using the notation of [12] for algebraic groups, except that U_α, not X_α, denotes the unipotent group corresponding to α. Here (a) comes from the Bruhat lemma [12, 6.3] and then everything else from the easily proved fact that W'' fixes Σ'. (If G were a compact Lie group instead, then a corresponding result would hold with $\langle U_\alpha, U_{-\alpha}\rangle$ replaced by an analogous compact group, SL_2 by SU_2 most of the time.)

We observe that $Z_G(t)^0$ fulfils the conditions of 2.1.

As a consequence of 2.14 we have:

2.15 Theorem. *If G is simply connected and t a single semisimple element, then $Z_G(t)$ is connected.*

The point is that $W'' = W'$ in 2.14(c), a geometric property of reflection groups acting suitably on tori, which is proved in [12, §5] and also in [3, pp. 36–37] in a more direct way. For compact Lie groups, a different proof may be found in [1, p. 225].

From 2.15 one can deduce:

2.16 COROLLARY. *Assume G reductive but perhaps not simply connected, t as in* 2.14.

(a) *$Z_G(t)/Z_G(t)^0$ is isomorphic to a subgroup of $F_s(G)$. Every subgroup is attainable.*

(b) *If $t^n \in Z(G)$, then $y^n = 1$ for every $y \in Z_G(t)/Z_G(t)^0$.*

(c) *In* (b), *if n is prime to $|F_s|$, then $Z_G(t)$ is connected.*

Here (a) and (b) (with the assumption $t^n = 1$) follow from [12, 9.1 and argument of 9.11] for semisimple groups, but the transition to reductive groups is immediate. Then (c) follows from (b). Observe that (a) provides both an extension and a converse for 2.15. In particular, $Z_G(t)$ is always connected if the universal covering of the semisimple component of G is purely inseparable.

What happens if we consider several semisimple elements?

2.17 LEMMA. *Assume in* 2.14 *that t is a subtorus of T. Then $Z_G(t)$ is connected and it is simply connected in G (see* 2.12).

Proof. The connectivity is a standard fact [8, Exp. 6, Theorem 6]. Since t is now a divisible group, Σ' is rationally closed in Σ and the second assertion holds by 2.13(d).

2.18 COROLLARY. *Let A be a solvable (not necessarily closed) subgroup of semisimple elements of G (still reductive).*

(a) *$Z_G(A)^0$ is reductive.*

(b) *$Z_G(A)/Z_G(A)^0$ is solvable, consists of semisimple elements, and its torsion primes (those that divide its order) are among those of $A/A^0 \cdot (A \cap Z(G))$.*

(c) *If A/A^0 is nilpotent, then so is $Z_G(A)/Z_G(A)^0$.*

Proof. A^0 is connected and solvable, hence contained in a Borel subgroup, hence in a torus [8, Exp. 6, Lemma 1], so that $\bar{A}^0$ is a torus. In view of 2.17 and the fact that $\bar{A}$ splits over $\bar{A}^0$ (which is divisible), we may replace G by $Z_G(A^0)$ and A by A/A^0 and thus assume that A is finite. In that case we shall prove a somewhat stronger statement.

2.19. *In* 2.18 *let A be replaced by a finite solvable group of semisimple automorphisms of G, $Z_G(A)$ by G_A (the group of fixed points), and $A/A^0 \cdot (A \cap Z(G))$ by A. Then the conclusions there hold.*

This reduces to 2.18 in case the automorphisms are all inner. Assume first that $A = \langle\sigma\rangle$, a cyclic group, of order m say. Then by [12, 8.1 and proof of 9.1] (a) holds if G is semisimple, hence also if G is reductive, and (b) holds if G is semisimple. Consider (b) in case $G = T$, a torus. Replacing T by $T/T_\sigma{}^0$, we may assume T_σ $(= \ker(1-\sigma))$ finite, hence $(1-\sigma)$ surjective on T, hence injective on $X = X(T)$. Then T_σ is in duality with the subgroup of $X/(1-\sigma)X$ consisting of the elements of order not divisible by char k. Now $X/(1-\sigma)X$ has order $\det_X(1-\sigma)$, and σ has on X a characteristic polynomial which is a product of cyclotomic polynomials $\varphi_d(t)$ $(d \mid m, d > 1)$. Thus $\det(1-\sigma)$ is a product of $\varphi_d(1)$'s. However, $\varphi_d(1)$ is p if d is a power of a prime p, is 1 otherwise. Thus if p is torsion for $T_\sigma = T_A$, it is torsion for some d, hence for A, and (b) holds. Now let G be reductive, $G = ST$, with S semisimple and T the radical, a torus. Set $H = S \times T$, $\pi\colon H \to G$ the natural map, and $F = \ker \pi$, a finite central subgroup. We use the exact sequence of cohomology [9, p. 133, Proposition 1] $H_A \to^{\pi} G_A \to^{\delta} H^1(A, F)$. Here $\pi H_A{}^0 = G_A{}^0$ and $G_A{}^0 \subseteq \ker\delta$ since $H^1(A, F)$ is finite. Thus H_A may be replaced by $H_A/H_A{}^0$ and G_A by $G_A/G_A{}^0$. Now $H_A/H_A{}^0$ satisfies (b) by what has been proved for semisimple groups and for tori, and $H^1(A, F)$ does also [9, p. 138, Corollary 1]. Hence so does $G_A/G_A{}^0$. Thus (b) holds in case A is cyclic. Consider now the general case. Let A' be a proper nontrivial normal subgroup of A. We have $G_A \supset G_A \cap G^0_{A'} \supset G_A{}^0$. The first quotient is isomorphic to a subgroup of $G_{A'}/G^0_{A'}$ and the second equals $K_{A''}/K^0_{A''}$ with $K = G^0_{A'}$ and $A'' = A/A'$. Thus (b) follows by induction, and so does (a) since $G_A{}^0 = (K_{A''})^0$, clearly. If A is nilpotent and p a prime, we choose A' to be the complement of a Sylow p-subgroup of A. Then from what has been proved, the first quotient above is a p'-group, the second a p-group, so that $G_A/G_A{}^0$ has a normal Sylow p-subgroup. Since p is arbitrary, $G_A/G_A{}^0$ is nilpotent, which proves (c).

2.20 *Example.* If A/A^0 is Abelian in 2.18(c), then $Z_G(A)/Z_G(A)^0$ need not be Abelian. Let G be adjoint of type D_4 in char $\neq 2$ and A the subgroup of T defined by $\alpha_1 = \alpha_2 = \alpha_3 = \pm 1$, $\alpha_4 = \pm 1$, in terms of a root basis for which α_4 is at the center of the Dynkin diagram. The following may be verified. A is a (2, 2) group consisting of the identity and three involutions conjugate to one another. No root vanishes on A. Thus $Z_G(A)/Z_G(A)^0$ is isomorphic to $Z_W(A)$ by 2.14(c). Let R be the root lattice, R' the sublattice vanishing on A. Then A is dual to R/R' and $Z_W(A) = Z_W(R/R')$. To find the latter group, write $\alpha_1 = x_1 - x_2$,

$\alpha_4 = x_2 - x_3$, $\alpha_2 = x_3 - x_4$, $\alpha_3 = x_3 + x_4$ in terms of basis of $\mathbb{Z}^4$. Then R is the sublattice in which the sum of the coordinates is even, R' the one in which all coordinates have the same parity. W acts on the x's via sign changes, even in number, combined with arbitrary permutations. All such sign changes lie in $Z_W(R/R')$, but only the permutations of the Klein (2, 2) group do. The group $Z_W(R/R')$ is not Abelian since, for example, the permutation (13)(24) does not fix the change of sign of the first two coordinates. $Z_W(A)$ is in fact the "metaplectic group" of the (2, 2) group.

We come back now to the torsion primes.

2.21 THEOREM. *Let G be reductive, t a semisimple element, and n an integer such that $t^n \in Z(G)$. Then the torsion primes of $L(\Sigma^*)/L(\Sigma'^*)$ (with Σ, Σ' as in* 2.14) *all divide n.*

2.22 COROLLARY. *If no torsion prime for Σ divides n, then $Z_G(t)^0$ is simply connected in G.*

This follows from 1.3 and 2.21. In view of 2.13(b), it implies Theorem 0.1 of the introduction.

Proof of 2.21. We may assume G semisimple, then simply connected by going to the covering group, then simple since G is at this stage a product of simple groups. In that case we prove a sharper result:

2.23. *If G in* 2.21 *is simple, then* $|\operatorname{tors} L(\Sigma^*)/L(\Sigma'^*)|$ *divides n.*

Proof. By 2.9(a) we may assume that Σ itself is the rational closure of Σ' in Σ. We consider the compact torus $T^c = \mathbb{R} \otimes L/L$ (L is the lattice of one-parameter subgroups of T) for which L and X have the same interpretations as they do for T. Since G is simply connected, $L = L(\Sigma^*)$, so that T^c is just the torus labeled T in 1.14. By [12, 5.1] there exists $t^c \in T^c$ such that the same characters vanish at t^c and at t. Since all roots vanish at t^n and $\alpha(t^n) = 0$ is equivalent to $(n\alpha)(t) = 0$, it follows that all roots vanish at $(t^c)^n$. Now t^c is equivalent to a point of the fundamental domain S of 1.14, to a vertex since Σ' and Σ have the same rank, to a vertex other than 0 since otherwise $\Sigma' = \Sigma$ and we are done, hence to some v_i as in 1.15; and then $\Sigma' = \Sigma_i$. Now $\alpha(mv) = 0$ in T is equivalent to $\alpha(mv) \in \mathbb{Z}$ in the covering space V above. Hence $n_i v_i$ is the smallest multiple of v_i at which all roots vanish. Hence n is a multiple of n_i, which in turn is a multiple of $n_i{}^*$ since

$n_i/n_i^* = (\alpha_0, \alpha_0)/(\alpha_i, \alpha_i)$, an integer since α_0 is a long root. Since $L(\Sigma^*)/L(\Sigma_i^*)$ has order n_i^* by 1.15, we are done.

A number of the results so far may be put together as follows.

2.24 Theorem. *Let G be reductive, A a subgroup of T, X^0 the annihilator of A in X, and p a prime. If X/X^0 has no p-torsion or if $A/A^0 \cdot (A \cap Z(G))$ has none or if G has none, then $Z_G(A)/Z_G(A)^0$ and $L(\Sigma^*)/L(\Sigma'^*)$ have none.*

Proof. As is easily seen, the first assumption implies the second. Thus there remain four results to be proved. We label them 11, 12, 21, 22. Then 11 follows from 2.18(b) and 22 from 1.3 and 2.5. For 21 and 12 we may, as earlier, assume that A is finite. Then 21 and 12 follow from 2.5, 2.16(a), and 2.21 in case A is cyclic, hence in general by an obvious induction.

2.25 Corollary. *Let G be reductive and A a commutative subgroup of semisimple elements. Write $A/A^0 \cdot (A \cap Z(G))$ as a product of, say a cyclic subgroups, and let exactly b of these have torsion in common with G.*

(a) *If $b \leqslant 1$, then A is contained in a torus.*

(b) *If $b = 0$, then in addition $Z_G(A)$ is connected and simply connected in G.*

(c) *If G is simply connected, then the values of b in* (a) *and the first part of* (b) *may be increased by* 1.

Proof. By 2.17, we may assume A finite and then eliminate the cyclic subgroups having no torsion in common with G, thus assume that $a = b$. Then (b) is obvious and (a) also since every semisimple element is contained in some torus. If G is simply connected and C is one of the cyclic subgroups still remaining, then we may apply (a) and (b) as already proved with $Z_G(C)$ in place of G, by 2.15, to get (c).

2.26 *Examples.* (a) In SL_n or Sp_n there are no torsion primes. Thus every commuting set A of semisimple elements can be put in a torus and has a connected centralizer, a classical result. (b) In SO_n the only torsion prime is 2, by 2.5. Thus the conclusions of (a) hold if A/A^0 is of odd order, and in any case $Z_G(A)/Z_G(A)^0$ is a 2-group by 2.24.

In the last example, the diagonal elements of order 2, which cannot be imbedded in any torus, show that the assumption there is essential. Such examples, consisting of elementary p-groups, exist whenever p is a torsion prime, as we shall now show. More specifically, we shall prove the following two theorems.

2.27 THEOREM. *Let G be reductive and p a prime different from* char k. *Then the following conditions are equivalent.*

(a) $p \nmid |F|$ (*F is the fundamental group of G*).

(b) *$Z_G(t)$ is connected for every element t of order p or, equivalently, for every* rank 1 *elementary p-subgroup.*

(c) *Every* rank 2 *elementary p-subgroup is contained in a torus.*

2.28 THEOREM. *Let G and p be as in* 2.27. *Then the following conditions are equivalent.*

(a) *p is not a torsion prime for G.*

(b) *$Z_G(P)$ is connected for every* rank $\leqslant 2$ *elementary p-subgroup P.*

(c) *$Z_G(P)$ is connected for every elementary p-subgroup P.*

(d) *Every* rank $\leqslant$ 3 *elementary p-subgroup P is contained in a torus.*

(e) *Every elementary p-subgroup P is contained in a torus.*

Proof. In 2.27, (a) implies (b) by 2.16(a) or 2.24, while if (b) holds and t_1, t_2 generate P as in (c), then any maximal torus of $Z_G(t_1)$ containing t_2 will do in (c), leaving only "(c) implies (a)" to be proved. Consider now 2.28. Here (a) implies (c) (and (e)) by 2.16(a,b) or 2.24, while (c) implies (b) and (e) implies (d) trivially. Consider now (c_i), (e_i) obtained from (c), (e) by sticking to subgroups of rank i. Then (c_i) implies (e_{i+1}) for every i as in the proof that (b) implies (c) in 2.27, which is just the case $i = 1$. Thus (c) implies (e) and (b) implies (d), and only "(d) implies (a)" remains here. To prove the remaining assertions, we may assume G semisimple. We use the following two lemmas.

2.29 LEMMA. *Let G be simply connected and p a torsion prime for G other than* char k. *Then there exists an element t of order p such that*

(a) *$Z_G(t)$ is semisimple.*

(b) *Both the center and the fundamental group of $Z_G(t)$ have elements of order p.*

2.30 LEMMA. *Let G be as in* 2.29, *p any prime, and t an element of order p of the center of G. Then there exist elements u, v of G such that*

(a) $(u, v) = t$.

(b) $u^p, v^p \in \langle t \rangle$, *even* $= 1$ *in case p is odd.*

Assume these lemmas for a moment. Suppose in 2.27 that (a) fails. We must show that (c) also fails. Let $\pi\colon G' \to G$ be the universal covering. Choose $t \in \ker \pi$ of order p and then u and v as in 2.30. Then $(\pi u)^p = 1$, $(\pi v)^p = 1$, and πu, πv cannot be put into the same torus T of G since then $\pi^{-1}T$ would be a torus in G', hence an Abelian group, containing u, v, a contradiction since $(u, v) = t \neq 1$. Thus (c) fails and 2.27 is completely proved. Now assume that (a) fails in 2.28. If $p \nmid |F|$, we choose t as in 2.29, while if $p \mid |F|$, we set $t = 1$. Thus in both cases $Z_G(t) = G_1$, say, is a semisimple group for which 2.27(a) fails. Thus 2.27(c) also fails and there is a rank-2 elementary p-subgroup P of G_1 not contained in any torus of G_1. Then $\langle P, t\rangle$ has rank $\leqslant 3$ and is not contained in any torus of G, for any such torus would have to centralize t and thus be contained in G_1. Thus (d) fails. In other words, (d) implies (a) in 2.28 and that theorem is also proved, mod 2.29 and 2.30.

2.31 *Remark.* We may replace the inequality in (b) of 2.28 by an equality since P above always has rank 2, and in case $p \nmid |F|$ do the same in (d) by 2.27.

It remains to prove 2.29 and 2.30. We recall the basic transfer situation of [12, 5.1]. Let $G, T, X, L, W, \ldots$ be as at the beginning of this section. Let $V = \mathbb{R} \otimes_{\mathbb{Z}} L$ and $T^c = V/L$, a compact torus. Then X may be viewed as the character group of T^c.

2.32 Lemma. *Let ψ be a fixed (unnatural) isomorphism of* tors k^* *into* $\mathbb{R}/\mathbb{Z}$ *and φ its extension from* tors $T =$ tors $k^* \otimes L$ *to* tors $T^c =$ tors $\mathbb{R}/\mathbb{Z} \otimes L$. *Then φ is a W-isomorphism and for any subset S of* tors T *the annihilators of S and $\varphi(S)$ in X are equal.*

Proof. The first point holds because the two actions of W are extensions of that on L, and the second because φ is injective.

Proof of 2.29 *and* 2.30. We have $L = L(\Sigma^*)$ since G is simply connected, so that the torus T^c of 2.32 is just the torus of 1.14, labeled T there. In 2.29 we choose $t \in T$ so that $\varphi(t) \in T^c$ is just the ith vertex of S, $i = p - 1$, as in 1.5(d). Then $Z_G(t)$ is connected reductive by 2.15, and the roots vanishing at t form the system Σ_i of 1.15, of the same rank as Σ, so that $Z_G(t)$ is semisimple. It contains t, of order p by 1.5(a,d) and 2.32, in its center, and has $L(\Sigma^*)/L(\Sigma_i{}^*)$, i.e., $\mathbb{Z}/p\mathbb{Z}$, as its fundamental group. Thus 2.29 is proved. Now let t be as in 2.30. By 2.32, $\varphi(t)$ has order p and is in "the center" of T^c. We choose $\varphi(u)$, w accordingly as in 1.20 and then shift back to $u, w \in T$, W via 2.32. The equations (a1, a2) of 1.20 continue to hold, and it remains to show that w can be represented in $N(T)$ by a v for which 2.30(b) holds. For each simple root α, we choose n_α to represent w_α in $N(T)$ and to be in the corresponding rank-1 subgroup of G (isomorphic to SL_2 in the present case, to SU_2 if G were a compact Lie group). Let $w = w_1 w_2 \cdots w_s$ be a minimal expression as a product of simple reflections. Provisionally, we set $v = n_1 n_2 \cdots n_s$, the corresponding product of n_α's.

(1) If α and β are distinct simple roots, then $n_\alpha n_\beta n_\alpha \cdots = n_\beta n_\alpha n_\beta \cdots$ (ord $w_\alpha w_\beta$ terms on each side). For by grouping the terms correctly one can show that the ratio of the two sides is in Im α^* and also in Im β^*, hence is 1 since $\{\alpha^*, \beta^*\}$ is part of a basis of L (cf. [11, Lemma 56(a)]).

(2) The value of v is independent of the minimal expression chosen for w. As is known, any minimal expression for w can be transformed into any other as a consequence of the relations in (1) with w's in place of n's. Hence (2) follows from (1).

(3) $n_\alpha^2 = \alpha^*(-1)$, an element of order 2, for every simple α. For it is known that there exists a homomorphism of SL_2 into G which maps the off-diagonal matrix $(0, 1; -1, 0)$ onto n_α and $\operatorname{diag}(a, a^{-1})$ onto $\alpha^*(a)$ for all $a \in k^*$.

(4) In the group generated by the chosen n_α's those elements that lie in T all have order 1 or 2. Let $n = n_1 n_2 \cdots n_q$ be one such. Then correspondingly $w_1 w_2 \cdots w_q = 1$. Now any relation in W is a consequence of those mentioned in (2) and the relations $w_\alpha^2 = 1$. It follows from (3), e.g., by induction on q, that n can be written as a product of $\alpha^*(-1)$'s ($\alpha \in \Sigma$), whence (4).

We return to the proof of 2.30. Since $w^p = 1$, v^p is in T, hence has order 1 or 2 by (4). If p is odd and v^p has order 2, we replace v by $v' = v^{p+1}$ and then have $v'^p = (v^p)^{p+1} = 1$, i.e., 2.30(b) since $p + 1$ is even. Assume now that $p = 2$. Then $w = w^{-1}$ so that $v^2 = n_1 n_2 \cdots n_s \cdot n_s \cdots n_2 n_1$ by (2). By (3) this may be simplified from the center outward to yield $v^2 = \Pi \beta_i^*(-1)$ with the product over all $\beta_i = w_1 w_2 \cdots w_{i-1} \gamma_i$ with γ_i denoting the simple root corresponding to w_i $(1 \leqslant i \leqslant s)$, i.e., over all positive roots made negative by w [6, p. 158, Corollary 2]. Now each $\alpha_i^*(-1)$ (α_i simple) is characterized in T or in T^c by the equations $\omega_j(\alpha_i^*(-1)) = (-1)^{\delta_{ij}}$ with $\{\omega_j\}$ as in 1.5(c). Hence by 2.32 it will be enough to show that $\varphi(v^2) = \Pi \beta_i^*(-1)$ (in T^c) is a multiple of $\varphi(t)$. In the covering space V of T^c in which

$\beta_i^*(-1)$ may be represented by $1/2\beta_i^*$, this amounts to showing that $1/2 \sum \beta_i^*$ is a multiple of v_1 representing t in V. Since this has been done in 1.20(b), we are done with 2.30, hence also with 2.27 and 2.28.

2.33 *Remark.* It seems likely to us that the replacement $v \to v^{p+1}$ above is unnecessary and that u and v as chosen originally are in fact conjugate.

There is one more item to be discussed in this section.

2.34 COROLLARY. *Let G be reductive, G_1 and G_2 reductive subgroups containing the same maximal torus T and corresponding to integrally closed subsystems of roots (which is automatic most of the time by* 2.9(b)). *Then $(G_1 \cap G_2)^0$ is reductive and $(G_1 \cap G_2)/(G_1 \cap G_2)^0$ is nilpotent and its torsion is contained in that of Σ.*

2.35 LEMMA. *Let $\Sigma_1, \Sigma_2, W_1, W_2$ be the root systems, Weyl groups of G_1, G_2. The $G_1 \cap G_2$ is generated by T, those U_α such that $\alpha \in \Sigma_1 \cap \Sigma_2$, and those n_w such that $w \in W_1 \cap W_2$; and $(G_1 \cap G_2)^0$ by T and the U_α's alone. The Weyl group W_3 of $\Sigma_1 \cap \Sigma_2$ is normal in $W_1 \cap W_2$ and $(G_1 \cap G_2)/(G_1 \cap G_2)^0$ is isomorphic to $(W_1 \cap W_2)/W_3$.*

Proof. An element of $G_1 \cap G_2$ has three Bruhat decompositions, one in G, one in G_1, one in G_2, which must be identical. From this, 2.35 readily follows.

Proof of 2.34. By the lemma, $(G_1 \cap G_2)^0$ is reductive. Since W_1, W_2, W_3 depend only on Σ_1, Σ_2, we may be 2.9 switch to any group with Σ as its root system, thus assume that (*) G is semisimple, simply connected and of char 0. Set $A_1 = Z_T(\Sigma_1)$. Then $\Sigma_1 = Z_\Sigma(A_1)$ since Σ_1 is integrally closed. Hence $G_1 = Z_G(A_1)^0$ by 2.14, and similarly $G_2 = Z_G(A_2)^0$. We have $Z_G(A_1A_2) \supset G_1 \cap G_2 \supset (G_1 \cap G_2)^0 = Z_G(A_1A_2)^0$, the last equality by 2.14(b) and 2.35. We conclude by applying 2.18, 2.24, and 2.6 to the outside terms.

2.36 *Complements.* (a) Conversely, every torsion prime p for Σ can be realized in 2.34. For assuming as we may that (*) above holds, we may choose u, v of order p in T so that $Z_G(u, v)$ is disconnected by 2.28 and then set $G_1 = Z_G(u)$ and $G_2 = Z_G(v)$. (b) If Σ_1, Σ_2 are rationally closed in Σ, then $G_1 \cap G_2$ is connected. For in this case we may take $A_1 = Z_T(\Sigma_1)^0$ and similarly for A_2 and then use the fact that A_1A_2, a torus, has a connected centralizer. (c) As an example, we see that if G is of type A_n or C_n in 2.34, then $G_1 \cap G_2$ is always connected.

A final remark: The results and proofs of this section hold equally well for (connected) compact Lie groups, subject to minor modifications that have been indicated from time to time (cf. [1]).

3. The Infinitesimal Case

In this section we carry over our earlier results, especially 2.27 and 2.28 (see 3.13 and 3.14 below), to semisimple elements of $\mathfrak{g}$, the Lie algebra of G.

3.1 The Lie algebra of a torus. Let T be an algebraic torus over k with L and X as before. Then T may be identified with $k^* \otimes_{\mathbb{Z}} L$. This is clear if the rank is 1 since $k^* \otimes_{\mathbb{Z}} \mathbb{Z} = k^*$ and then if the rank is arbitrary as we see by taking direct products. Explicitly, $\sum c_i \otimes \lambda_i \sim \Pi \lambda_i(c_i)$ if the elements of L are considered to be one-parameter subgroups. The Lie algebra $\mathfrak{t}$ of T then becomes $k \otimes_{\mathbb{Z}} L$ since that of k^* is k. For each χ in X, there is then a linear function on $\mathfrak{t}$, the differential of χ, also to be denoted χ, which sends $\sum c_i \otimes \lambda_i$ to $\sum c_i(\lambda_i, \chi)$. Thus $\mathfrak{t}$ comes with a natural $\mathbb{Z}$-structure, hence with a natural structure of variety over k_0, the prime field. As is easily seen, H in $\mathfrak{t}$ is in $\mathfrak{t}(k_0) = k_0 \otimes L$ if and only if $X(H) \subseteq k_0$.

3.2 *Example.* Assume that T above is a maximal torus of a simply connected semisimple algebraic group, so that the simple coroots $\{\alpha_i^*\}$ form a basis for L and their images $\{1 \otimes \alpha_i^*\}$ one for the k_0-structure of $\mathfrak{t}$. If $\{\omega_j\}$ is the dual basis of X consisting of the fundamental weights, then $\omega_j(1 \otimes \alpha_i^*) = \delta_{ji}$. Thus $1 \otimes \alpha_i^*$ is just that element of $\mathfrak{t}$ which in the classical theory is denoted H_{α_i}, and similarly for every root α.

We recall that a subalgebra of the Lie algebra of an algebraic group is said to be *algebraic* if it is the Lie algebra of an algebraic subgroup.

3.3 Lemma. *If* T, $\mathfrak{t}$, *etc. are as above and* $\mathfrak{t}_1$ *is a subalgebra of* $\mathfrak{t}$, *then the following are equivalent.*

(a) $\mathfrak{t}_1$ *is an algebraic subalgebra of* $\mathfrak{t}$.

(b) $\mathfrak{t}_1 = k \otimes L_1$ *for some sublattice* L_1 *of* L *such that* L/L_1 *has no torsion.*

(c) $\mathfrak{t}_1$ *is defined over* k_0, *the prime field.*

(d) $\mathfrak{t}_1$ *has a basis of elements of* $\mathfrak{t}(k_0)$.

Proof. If $\mathfrak{t}_1$ in (a) is the Lie algebra of the subtorus T_1, then the corresponding lattice L_1 satisfies (b) since if $\lambda \in L$ and $n\lambda \in L_1$, then $\operatorname{Im} \lambda = \operatorname{Im} n\lambda \subseteq T_1$, whence $\lambda \in L_1$. Thus (a) implies (b). If (b) holds, then a basis of L_1 can be extended to one of L. It then follows, from uniqueness of expression in terms of a basis, that $T_1 = k^* \otimes L_1$ is a subtorus of T with Lie algebra $\mathfrak{t}_1$. Thus (b) implies (a). If $\{\lambda_i\}$ is a basis of L_1 as in (b), then $\{1 \otimes \lambda_i\}$ is one for $\mathfrak{t}_1$. Thus (b) implies (d). That (d) is equivalent to (c) is a standard (elementary) fact from Galois theory which holds for arbitrary fields and vector spaces. Finally, assume $\mathfrak{t}_1$ has a basis as in (d). If $\sum c_i \otimes \lambda_i \in k_0 \otimes L$ is in the basis, it may be written $c \otimes \lambda$ since the c_i's are all rational numbers and may be taken to a common denominator. Thus the basis may be taken in the form $\{1 \otimes \lambda_i\}$. The λ_i's are not uniquely determined, only mod pX ($p = \operatorname{char} k$), so we have to make a choice which we do. We then set $L_1 = \sum \mathbb{Q}\lambda_i \cap L$. Then $k \otimes L_1$ contains $\mathfrak{t}_1$ and on the other hand does not have a larger dimension, both by our construction. Thus the spaces are equal, (d) implies (b), and we are done.

3.4 Corollary. *The intersection of any family of algebraic subalgebras of* $\mathfrak{t}$ *is algebraic.* (*By the equivalence of* (a) *and* (c) (*or* (a) *and* (b)).

3.5 *Remark.* T_1 above is not in general uniquely determined by $\mathfrak{t}_1$ if $\operatorname{char} k = p \neq 0$, for, by adding arbitrary elements of pL to a basis for L_1, we may change L_1, hence also T_1, without changing $\mathfrak{t}_1 = k \otimes L_1$.

3.6 Lemma. *Let* T, $\mathfrak{t}$, *etc. be as above and* $H \in \mathfrak{t}$.

(a) *There exists a unique smallest algebraic subalgebra* $\mathfrak{t}_1$ *of* $\mathfrak{t}$ *containing* H.

(b) *Let* $H = \sum_{i=1}^{s} c_i \otimes \lambda_i \in k \otimes L$. *Then the following conditions are equivalent.*

(1) s *is minimal.*

(2) $\{1 \otimes \lambda_i\}$ *is a basis of* $\mathfrak{t}_1$ (*see* (a)).

(3) $\{c_i\}$ *and* $\{1 \otimes \lambda_i\}$ *are both linearly independent over* k_0.

(c) *If* $\chi \in X$, *then* $\chi(H) = 0$ *if and only if* $\chi(\mathfrak{t}_1) = 0$. *If* $\sigma \in \operatorname{Aut}(T)$, *then* σ *fixes* H *if and only if it fixes every point of* $\mathfrak{t}_1$. *In both cases* $\mathfrak{t}_1$ *may be replaced by* $\mathfrak{t}_1(k_0)$.

Proof. (a) This follows from 3.4.

(b) Clearly $\mathfrak{t}_1 \subseteq \langle\{1 \otimes \lambda_i\}\rangle$. Thus $\min s = \dim \mathfrak{t}_1$ and the minimum occurs exactly when $\{1 \otimes \lambda_i\}$ is a basis of $\mathfrak{t}_1$. The equivalence of (1) and (3) is a standard fact from elementary linear algebra.

(c) Write H minimally as in (b). If $\chi(H) = 0$, then $\sum c_i\chi(1 \otimes \lambda_i) = 0$, whence every $\chi(1 \otimes \lambda_i) = 0$ by (b3), and $\chi(\mathfrak{t}_1) = 0$ by (b2). The reverse conclusion is clear. If σ is as in (c), then $\sigma H = H$ if and only if $((1 - \sigma)X)(H) = 0$ and similarly with $1 \otimes \lambda_i$ in place of H. Thus the second part of (c) follows from the first, and the third from 3.3.

3.7 Lemma. *Let G be reductive, T a maximal torus, H an element or subset of* $\mathfrak{t}$, *and Σ', W', W'' as in* 2.14 *with H in place of t. Then the conclusions* (a), (b), *and* (c) *modified accordingly hold.*

Proof. For each root α, let x_α: $k \to U_\alpha$ be a parametrization. For $H \in \mathfrak{t}$, we have the equation (*) $x_\alpha(c)H = H - c\alpha(H)X_\alpha$ with X_α a suitable nonzero tangent vector to U_α (the image under dx_α of the standard tangent vector to k^*, in fact), got by differentiating $x_\alpha(c)\,tx_\alpha(c)^{-1} = x_\alpha(c(1 - \alpha(t)))t$ in G. It follows that the parts of G listed in (a) are all in $Z_G(H)$. Conversely, let $x = bn_wu$ (normal form of [12, 6.3]) be in $Z_G(H)$. We also have (**) $x_\alpha(c)X_\beta = \sum c_ic^iX_{\beta+i\alpha}$ for suitable c_i's ($i \geqslant 0$). Hence $uH = H + \sum c_\alpha X_\alpha$ ($c_\alpha \in k$, $\alpha > 0$, $w\alpha < 0$), so that $n_wuH = n_wH + V$ ($V \in \mathfrak{u}^-$). Since also $n_wuH = b^{-1}H = H + U_1$ ($U_1 \in \mathfrak{u}$), we get $U_1 = 0$ so that b fixes H, and $V = 0$ so that u fixes H, and finally $n_wH = H$ so that w also fixes H. Now if $u = \Pi\, x_\alpha(c_\alpha)$ ($\alpha > 0$), then since u fixes H, it follows from (*) and (**) that $\alpha(H) = 0$ for every α in the support of u, by induction on the height of α. Thus (a) holds when H is a single element, hence, by induction, also when H is a set of several elements. The proofs of (b) and (c) are substantially as in 2.14 and will be omitted.

3.8 Corollary. *Let G be reductive and H a semisimple element of* $\mathfrak{g}$. *Then there is a unique minimal* (*one contained in all others*) *algebraic Lie subalgebra* $\mathfrak{t}_1$ *of* $\mathfrak{g}$ *containing H. Further $Z_G(H) = Z_G(\mathfrak{t}_1)$.*

3.9 Definition. By the *rank* of H we shall mean the dimension of $\mathfrak{t}_1$.

Proof. We use the fact that (in any algebraic group) every such H is in the Lie algebra of some maximal torus. This is proved like the corresponding result in the group (cf. [8, Exp. 6, Theorem 5]). Let T be such a torus, and let $\mathfrak{t}_1$ be as in 3.6. Then $Z_G(H) = Z_G(\mathfrak{t}_1)$ by 3.6(c) and 3.7. Let $\mathfrak{a}$ be any minimal algebraic subalgebra of $\mathfrak{g}$ containing H. We must show that $\mathfrak{a} = \mathfrak{t}$. Let A be a subgroup of G corresponding to $\mathfrak{a}$. Then H is in the Lie algebra of some torus of A. By minimality, A itself must be a torus, hence contained in some maximal torus T' of G. Now T and T' are maximal tori of $Z_G(H)^0$, i.e., of $Z_G(\mathfrak{t}_1)^0$, hence are conjugate there: ${}^xT' = T$. Then ${}^x\mathfrak{a} \subseteq \mathfrak{t}$, so that ${}^x\mathfrak{a} \supseteq \mathfrak{t}_1$ by 3.6, whence $\mathfrak{a} \supseteq {}^{x^{-1}} \mathfrak{t}_1 = \mathfrak{t}_1$ and $\mathfrak{a} = \mathfrak{t}_1$ by minimality, as required.

3.10 *Remark.* It is not known to the author whether 3.8 is true without the assumptions on G and H, or, more generally, whether the intersection of two algebraic subalgebras is always an algebraic subalgebra (if char $k \neq 0$).

3.11 COROLLARY. *In* char 0, $Z_G(H)$ *in* 3.7 *is connected for every subset* H *of* $\mathfrak{t}$.

Proof. Assume $w \in W''$. By a theorem of Chevalley [12, 1.21], w is a product of reflections each in W''. Let w_α be any of them. Then $(1 - w_\alpha)H = 0$ so that $\chi(1 - w_\alpha)H = 0$ for every $\chi \in X$, i.e., $(\chi, \alpha^*)\, \alpha(H) = 0$ by the formula for a reflection. Set $\chi = \alpha$, $(\chi, \alpha^*) = 2$. Then $2\alpha(H) = 0$, $\alpha(H) = 0$. Thus $\alpha \in \Sigma'$, $w_\alpha \in W'$, $w \in W'$. Hence $W'' = W'$ and $Z_G(H)$ is connected by 3.7(c).

3.12 *Remark.* A different proof in case $k = \mathbb{C}$ is given in [8, Lemma 5]. The proof just given presents two obstructions in char $p \neq 0$. The first is the extension of Chevalley's theorem. This turns out to be all right (see 4.6 below) as long as p is not a torsion prime for G. The second, involved in the step from $(1 - w_\alpha)H = 0$ to $\alpha(H) = 0$, is not so serious since we need only the existence of some $\chi \in X$ such that $(\chi, \alpha^*) = 1$, which we always have in case each component of type C_r is simply connected, hence if G itself is so.

Turning now to the case char $k \neq 0$, we shall prove the following analogues of 2.27 and 2.28.

3.13 THEOREM. *Let G be a reductive group and p a prime equal to* char k. *Then the following are equivalent.*

(a) $p \nmid F(G)$ (*see* §2); *i.e., the universal covering of the semisimple component of G is separable.*

(b) *$Z_G(H)$ is connected for every* rank-1 *semisimple element H of* $\mathfrak{g}$ (*see* 3.9).

3.14 THEOREM. *Let G and p be as in* 3.13. *Then the following are equivalent.*

(a) *p is not a torsion prime for G.*

(b) *$Z_G(H)$ is connected for every* rank $\leqslant 2$ *semisimple element H of* $\mathfrak{g}$.

(c) *$Z_G(H)$ is connected for every semisimple element H of* $\mathfrak{g}$.

(d) *$Z_G(H)$ is connected for every commutative set of semisimple elements of* $\mathfrak{g}$.

3.15 *Remarks.* (a) 3.11 and 3.14 yield Theorem 0.2 of the introduction.

(b) Conditions such as 2.28(d) have no place here since every set as in (d) can be put in the Lie algebra of some torus. We can prove this by starting with a single element H, proving the analogue of 3.7 with $Z_{\mathfrak{g}}(H)$ in place of $Z_G(H)$, and then proceeding by induction.

The proof of 3.13 and 3.14 depend on another transfer lemma. Let G and p be as above and let G' be of the same type as G but over an algebraically closed field k' of char $\neq p$. If T' is a maximal torus of G' and L', X', etc. are defined accordingly, we require the existence of an isometry from L onto L' matching up X with X', Σ with Σ', and W with W'. We may construct G', for example, by starting with the direct product of a simply connected semisimple group G'' and a torus T''', both over k', with $\Sigma''*$ matched up with $\Sigma*$ (hence L'' with $L(\Sigma*)$) and L''' with the orthogonal complement of $L(\Sigma*)$ in L, and then dividing out an appropriate finite subgroup of the center (which may be partly infinitesimal (see [2, §17])).

3.16 LEMMA. *Let ψ be a fixed isomorphism of the additive group of k_0 onto the group of p-th roots of* 1 *of k' and φ its extension from $\mathfrak{t}(k_0)$ onto the group $T'(p)$ of p-th roots of* 1 *of T'* (*cf.* 3.1). *Then the conclusions of* 2.32 *with the obvious substitutions hold.*

Proof. Like that of 2.32.

Proof of Theorems 3.13 *and* 3.14. Let T be a maximal torus of G and G', T', φ as in 3.16 so that φ maps $\mathfrak{t}(k_0)$ onto $T'(p)$ isomorphically. Let H, $H' = \varphi(H)$ be a corresponding pair of elements. Let W', W'' be defined for H as in 3.7, and W''', W'''' accordingly for H' as in 2.14. Then $W' = W'''$ and $W'' = W''''$ by 3.16, so that $Z_G(H)$ and $Z_{G'}(H')$ are isomorphic over their identity components by 3.7 and 2.14, hence are both connected or both disconnected. Since every rank-1 semisimple element of $\mathfrak{g}$ is conjugate to an element of $\mathfrak{t}(k_0)$ and every order p element of G' to an element of $T'(p)$, it follows that the conditions 3.13(b) for G and 2.27(b) for G' are equivalent. But the last condition is equivalent to 2.27(a) (for G'), which in turn is equivalent to 3.13(a) since $F(G) \sim L/L(\Sigma^*) \sim F(G')$ and char $k = p$, char $k' \neq p$. Thus (a) and (b) of 3.13 are equivalent and that theorem is proved. In the last three parts of 3.14, H may be replaced by a subspace of some $\mathfrak{t}(k_0)$, of dimension $\leqslant 2$ in part (b), without changing $Z_G(H)$, by 3.6 and 3.8. Hence (c) and (d) are equivalent and 3.14 can be deduced from 2.28 just as 3.13 has been deduced from 2.27, with the aid of 3.16 and a little care. We see further that $\leqslant$ may be replaced by $=$ in 3.14(b) in case $p \nmid |F|$, as in 2.28(b).

4. The Abstract Essence

The reader may have observed that in §3 once past 3.7 the discussion was no longer concerned with $\mathfrak{g}$ per se, only with the action of W on $\mathfrak{t}$, and similarly in §2. We wish to extract the geometric essence of those discussions in the form of extensions of Chevalley's theorem mentioned in 3.11. Thus to start with we assume given only a lattice L, its dual X, a finite reflection group W acting on L and X compatibly, and any Abelian group A. We write ${}^AL$ for $A \otimes_{\mathbb{Z}} L$ and wish to study $Z_W(H)$ when H is an element or subgroup of ${}^AL$; in §2 we did this for $A = k^*$ or $\mathbb{R}/\mathbb{Z}$, in §3 for $A = k$. The root system Σ is not given *a priori* so that we are free to choose it so as to facilitate our study. First we define Σ^* (as in [12, 3.6]) to consist of the minimal elements of L in the various directions in which the reflections of W take place. For each α^* in Σ^* there then exists α in $\mathbb{Q} \otimes X$ such that $\alpha(\alpha^*) = 2$ and $w_{\alpha^*}\lambda = \lambda - \alpha(\lambda)\alpha^*$ for all λ in L, and these α's form Σ. Here every $\alpha(\lambda)$ is in $\mathbb{Z}$ since α^* is primitive in L, so that each α is in fact in X. It is easily verified that Σ^* and Σ are root systems, the integrality coming from the above equation with $\lambda = \beta^* \in \Sigma^*$, so that we are in the situation of §2 and §3. Our choice of Σ^* also yields:

4.1 Lemma. *If H is a subset of ${}^AL$ and α a root, then $\alpha(H) = 0$ if and only if $w_\alpha \in Z_W(H)$.*

Proof. The last condition, $(1 - w_\alpha)H = 0$, i.e., $\alpha(H)\alpha^* = 0$ by the formula for a reflection, is equivalent to $\alpha(H) = 0$ since α^* is primitive.

The results we have in mind may now be stated, with $L, X, \ldots, A$ as we have just introduced them. By 2.14 and 3.7, the analogue in W of a subgroup of G that is connected is a subgroup that is generated by reflections. Therefore, for each subgroup Y of W we write Y^0 for its largest reflection subgroup, generated by the reflections that it contains.

4.2 THEOREM. *Let H be an element of ${}^{A}L$. Assume that every finite subgroup of A is cycle, or, equivalently, that* tors $X(H)$ *is cyclic.*

(a) *$Z_W(H)/Z_W(H)^0$ is isomorphic to a subgroup of $F =$* tors $L/L(\Sigma^*)$, *and if A contains an element of order n, then every subgroup of F of order n is realizable.*

(b) *If $H^n \in Z$, then $y^n = 1$ for all $y \in Z_W(H)/Z_W(H)^0$.*

Here Z denotes "the center" $Z_{A_L}(W)$.

4.3 COROLLARY. *$Z_W(H)$ is a reflection group in* (a) *in case $F = 0$, i.e. (W, L) is "simply connected," in* (b) *in case n is prime to $|F|$.*

This is clear.

4.4 *Remarks.* In theorems such as this the only elements of A that come into play are those of $X(H)$ so that the assumptions made are effectively equivalent. The condition on A is that it is imbeddable in some k^* (in $\mathbb{R}/\mathbb{Z}$ if A is small enough), so that in view of 2.14 and the fact that the data $L, X, \ldots, k$ are realizable in some reductive algebraic group, the above results may be extracted from 2.16. Alternatively, the earlier proofs may be transferred to the present context.

4.5 THEOREM. *Let H be any subgroup of ${}^{A}L$. Then $Z_W(H)/Z_W(H)^0$ is nilpotent. Let X^0 be the annihilator of H in X and p a prime. If X/X^0 has no p-torsion or if (L, W) has none, then $Z_W(H)/Z_W(H)^0$ and $L(\Sigma^*)/L(\Sigma'^*)$ have none.*

Here $\Sigma' = Z_\Sigma(H)$ and the torsion primes for (L, W) are meant to be those of 2.5.

4.6 COROLLARY. *If* tors X/X^0 *is relatively prime to* tors (L, W) (*resp.*

to tors Σ), *then* $Z_W(H)$ *is a reflection group* (*resp.* $Z_W(H)^0$ *is simply connected in* W).

In view of 4.5 this is clear.

4.7 *Remarks.* If A itself, or equivalently $X(H)$ by 4.4, is free from the torsion of (L, W) (resp. of Σ) in 4.6, then X/X^0 will always be torsion free (and similarly in 4.5). In particular, if A is $\mathbb{Z}$ or $\mathbb{R}$, hence has no torsion at all, we get Chevalley's theorem itself, made more precise by the condition in brackets. Another case, related to 2.17, in which the same strong conclusions hold, is that in which A is assumed to be divisible in 4.6, p-divisible in 4.5.

Proof of 4.5. This may be deduced from 2.24 as 4.2 from 2.16 once it is observed that ${}^AL$ may be replaced by ${}^{\mathbb{C}}L$ and H by the annihilator of X^0 in ${}^{\mathbb{C}}L$.

Now given an element H of order p, we may define its p-rank as the rank of the elementary p-group X/X^0, or, equivalently, of $X(H)$, and similarly if H is a subgroup. The obvious analogues of 3.13 and 3.14 are then true. Their formulations and proofs will be left to the reader.

Finally a word about duality. Let A^* be the character group of A into some Abelian group, so that ${}^{A^*}X$ is that of ${}^AL$. Then if we are interested in stabilizers in W of sets of characters, for example, for subgroups of T in §2, we may apply the preceding results with the roles of L and X, of Σ^* and Σ, and of A and A^* interchanged.

References

1. A. Borel, Sous-groupes commutatifs et torsion des groupes de Lie compacts connexes, *Tôhoku Math. J.* **13** (1961), 216–240.
2. A. Borel, "Linear Algebraic Groups," Benjamin, New York, 1969.
3. A. Borel *et al.*, "Seminar on Algebraic Groups and Related Finite Groups, Lecture Notes in Mathematics", 131, Part E, Springer, Berlin, 1970.
4. A. Borel and F. Hirzebruch, Characteristic classes and homogeneous spaces **III**, *Amer. J. Math.* **82** (1960), 491–504.
5. A. Borel and J. de Siebenthal, Les sous-groupes fermés de rang maximum des groupes de Lie clos, *Comm. Math. Helv.* **23** (1949), 200–221.
6. N. Bourbaki, "Groupes et Algèbres de Lie," Chapters 4–6, Hermann, Paris, 1968.
7. B. Kostant, Lie group representations on polynomial rings, *Amer. J. Math.* **85** (1963), 327–404.

8. Séminaire C. Chevalley: "Sur la Classification des Groupes de Lie Algébriques," 2 Vol., Inst. H. Poincaré, Paris, 1956–58.
9. J.-P. Serre, "Corps Locaux," Hermann, Paris, 1968.
10. J. de Siebenthal, Sur les sous-groupes fermés connexes des groupes de Lie clos, *Comm. Math. Helv.* **25** (1951), 210–256.
11. R. Steinberg, "Lectures on Chevalley Groups," Yale University Math. Dept., 1968.
12. R. Steinberg, Endomorphisms of linear algebraic groups, *Amer. Math. Soc. Mem.* No. 80 (1968).

Printed by the St Catherine Press Ltd., Tempelhof 37, Bruges, Belgium.

Topology Vol. 14, pp. 173–177. Pergamon Press, 1975. Printed in Great Britain

ON A THEOREM OF PITTIE

ROBERT STEINBERG

(*Received* 1 *October* 1974)

§1. INTRODUCTION

HARSH V. PITTIE[3] has proved the following result:

THEOREM 1.1. *Let G be a connected compact Lie group with $\pi_1 G$ free and G' a (closed) connected subgroup of maximal rank. Then $R(G')$ is free (as a module) over $R(G)$ (by restriction).*

Here $R(G)$ denotes the complex representation ring of G. For the bearing of (1.1) on the K-theory of G the reader may consult [3]. Pittie's proof actually omits a few cases, which can however be checked out by hand. Here we present an elementary proof which yields an explicit basis for $R(G')$ over $R(G)$ (see (2.2) and (2.3(a)) below) and then a converse after suitably weakening the assumption on $\pi_1 G$.

THEOREM 1.2. *Let G be a connected compact Lie group and S its semisimple component. Then the following conditions are equivalent.*

(a) *$R(G')$ is free over $R(G)$ for every connected subgroup G' of maximal rank.*

(b) *$R(T)$ is free over $R(G)$ for some maximal torus T.*

(c) *$R(G)$ is the tensor product of a polynomial algebra and a Laurent algebra.*

(d) *$R(S)$ is a polynomial algebra.*

(e) *S is a direct product of simple groups, each simply connected or of type SO_{2r+1}.*

Since $\pi_1 G$ is free if and only if S is simply connected, because G is the product of S and a central torus, the equivalence of (a) and (e) provides the just-mentioned extension and converse of (1.1).

As a result of our development we also obtain:

THEOREM 1.3. *Theorems* 1.1 *and* 1.2 *are true for linear algebraic groups over algebraically closed fields (instead of compact Lie groups) and their rational representations.*

§2. PROOF OF (1.1)

We may, and shall, assume that G is semisimple, hence simply connected since $\pi_1 G$ is free, as is indicated in [3]. Let T be a maximal torus of G', hence also of G, and W' and W the corresponding Weyl groups, and X the character group (lattice) of T. As is known (see [1]), $R(G)$ may be identified with $\mathbf{Z}[X]^W$ via restriction to T, even if G is not semisimple, and similarly for $R(G')$. To prove (1.1), therefore, we need only produce a free basis for $\mathbf{Z}[X]^{W'}$ over $\mathbf{Z}[X]^W$. This puts us in the realm of weights, roots and reflection groups, for which we use [2] as a general reference. Let $\Sigma \subseteq X$ be the root system of G relative to T, Σ^+ the set of positive roots and Π the corresponding basis of simple roots relative to some, fixed, ordering. The condition that G be simply connected is:

2.1. The fundamental weights $\{\lambda_a\}$, defined by $(\lambda_a, b^*) = \delta_{ab}$ $(a, b \in \Pi)$ with $b^* = 2b/(b, b)$, form a basis for X.

We generalize our problem slightly by allowing W' to be any reflection subgroup of W. Let Σ' be the corresponding root system, consisting of the roots orthogonal to the reflecting hyperplanes for W', and W'' the subset of W keeping Σ'^+ positive. Finally, for $v \in W''$ let λ_v denote the product in X of those λ_a for which $a \in \Pi$ and $v^{-1}a < 0$, and $e_v = \Sigma x^{-1} v^{-1} \lambda_v \in \mathbf{Z}[X]$, the sum over $x \in W'(v)\backslash W'$ with $W'(v)$ denoting the stabilizer of $v^{-1}\lambda_v$ in W'.

THEOREM 2.2. *Assume G simply connected and the other notations as above. Then $\mathbf{Z}[X]^{W'}$ is free over $\mathbf{Z}[X]^W$ with $\{e_v | v \in W''\}$ as a basis.*

Remarks 2.3. (a) Observe that each $v^{-1}\lambda_v$ is dominant for Σ'. For $(v^{-1}\lambda_v, a) = (\lambda_v, va) \geq 0$ since $va > 0$ for all $a \in \Sigma'^+$. It follows from (2.2) and the above discussion that (1.1) holds with a basis consisting of those irreducible representations of G' for which the highest weights are $\{v^{-1}\lambda_v \,|\, v \in W''\}$. It also follows that the rank is $|W|/|W'|$, in (1.1) or in (2.2), either by Galois theory or by (2.5(a)) below. (b) In the principal case in which $W' = \{1\}$, in which G' is a torus in (1.1), we get $\mathbf{Z}[X]$ free over $\mathbf{Z}[X]^W$ with $\{w^{-1}\lambda_w | w \in W\}$ as a basis.

MAIN LEMMA 2.4. *Let $\{e_v\}$ be as above and $\{f_v\}$ ($v \in W''$) any collection of elements of $\mathbf{Z}[X]^{W'}$. Set $D = \det ue_v$, $E = \det uf_v$ ($u, v \in W''$).*

(a) *$D \neq 0$.*

(b) *D divides E and the ratio is in $\mathbf{Z}[X]^W$.*

Granted this lemma, we may prove (2.2) as follows. If $f \in \mathbf{Z}[X]^{W'}$, then the system $\Sigma a_v ue_v = uf$ has a unique solution for $a_v \in \mathbf{Z}[X]^W$, hence the equation $\Sigma a_v e_v = f$ does also, whence (2.2).

It remains to prove (2.4).

LEMMA 2.5. *Let everything be as above.*

(a) *W'' is a system of representatives for W/W'.*

(b) *If Σ' has a basis consisting of a subset of Π, then $\ell(ux) = \ell(u) + \ell(x)$ for $u \in W''$, $x \in W'$.*

Here $\ell(w)$ denotes the number of positive roots made negative by w. Fix $w \in W$. Then $w^{-1}\Sigma^+ \cap \Sigma'$ and Σ'^+ are two positive systems for Σ', hence (*) they are congruent under a unique $x \in W'$. Then $u = wx^{-1} \in W''$ and $w = ux \in W'' \cdot W'$. Conversely, if w has this form, we may work backwards to conclude that x satisfies (*), hence is uniquely determined. This proves (a). The number of roots in Σ'^+ made negative by ux as in (b) is $\ell(x)$ since x fixes Σ' and u fixes the signs of the roots in Σ', while the number in $\Sigma^+ - \Sigma'^+$ is $\ell(u)$ since x fixes this set, whence (b).

LEMMA 2.6. *Assume as before and that $w \in W$ keeps $\Sigma^+ - \Sigma'^+$ positive. Then $w \in W'$, in fact w is in the subgroup generated by the simple reflections that W' contains.*

Assume w as given, $w \neq 1$. Then $wa < 0$ for some simple root a, so that $\ell(ww_a) < \ell(w)$. By our assumption $a \in \Sigma'^+$, so that w_a preserves $\Sigma^+ - \Sigma'^+$ and hence ww_a keeps it positive. By induction on $\ell(w)$ we conclude that ww_a is in the above subgroup, whence w is also.

LEMMA 2.7. *For $v \in W''$ we have $vW(v) \subseteq W''W'(v)$.*

Recall that $W(v)$, for example, denotes the stabilizer of $v^{-1}\lambda_v$ in W. As is known, this is a reflection group. Let $\Sigma(v)$ be the corresponding system of roots, those orthogonal to v. We have $v\Sigma'(v) = v\Sigma' \cap v\Sigma(v)$. Hence $v(\Sigma'^+ - \Sigma'^+(v))$ is disjoint from $(v\Sigma(v))^+$, and it is positive since $v \in W''$. Now if $w \in W(v)$, then $vwv^{-1} \in {}^vW(v)$, the group corresponding to the root system $v\Sigma(v)$, which is the subset of Σ orthogonal to λ_v and hence is like Σ' in (2.5(b)) since λ_v is dominant. By the above disjointness, $vwv^{-1} . v(\Sigma'^+ - \Sigma'^+(v)) > 0$. If we write $vw = ux$ as in (2.5(b)), this yields $x(\Sigma'^+ - \Sigma'^+(v)) > 0$ since u fixes signs on Σ'. Thus $x \in W'(v)$ by (2.6) with Σ, Σ' there replaced by Σ', $\Sigma'(v)$ here, whence (2.7).

LEMMA 2.8. *For each root a let n_a denote the number of pairs in W/W' interchanged by left multiplication by w_a.*

(a) *n_a is constant on W-conjugacy classes of roots.*

(b) *If a is simple then n_a is the number of v's in W'' such that $v^{-1}a < 0$.*

If a and b are conjugate, then so are w_a and w_b, hence also their left multiplications on W/W', whence (a). In (b) let w_a fix vW'. Then $v^{-1}w_a v \in W'$, whence $v^{-1}a \in \Sigma'$ and $v^{-1}a > 0$ since $v \in W''$. Now assume w_a does not fix vW', i.e. $v^{-1}a \notin \Sigma'$. Then $v\Sigma'^+$ is disjoint from a and positive, whence $w_a v\Sigma'^+$ is also positive and $w_a v \in W''$. Now just one of $v^{-1}a$, $(w_a v)^{-1}a$ is negative. Thus $v^{-1}a < 0$ for exactly n_a choices of $v \in W''$.

LEMMA 2.9. *If D is as in (2.4) and n_a as in (2.8) then D has $\Pi\lambda_a^{n_a}$ ($a \in \Pi$) as its unique highest term and $\pm\Pi\lambda_a^{-n_a}$ as its unique lowest term.*

This is relative to the usual partial order in which $\lambda > \mu$ denotes that $\lambda\mu^{-1}$ is a product of positive roots. Let A denote the matrix (ue_v). Recall that $ue_v = \Sigma ux^{-1}v^{-1}\lambda_v$, summed over $x \in W'(v)\backslash W'$. Consider the vth column of A. We have $\lambda_v \geq ux^{-1}v^{-1}\lambda_v$ for all terms there. We claim that equality can hold on or above the diagonal only for the term with $u = v$ and $x \in W'(v)$, if we order the rows so that u is above u' whenever $\ell(u) < \ell(u')$. Assume equality. Then $v^{-1}ux^{-1} \in W(v)$ by definition, so that $ux^{-1} \in W''W'(v)$ by (2.7), and $x \in W'(v)$ by (2.5(a)), so that $uv^{-1}\lambda_v = \lambda_v$. Thus uv^{-1}, hence also vu^{-1}, is in the group generated by the reflections for the simple roots orthogonal to λ_v, which are those kept positive by v^{-1} by the definitions. Applying (2.5(b)) to this situation we get $\ell(u^{-1}) = \ell(v^{-1}) + \ell(vu^{-1})$. On or above the diagonal where $\ell(u) \leq \ell(v)$ this can hold only if $\ell(vu^{-1}) = 0$, whence $u = v$ and our claim. It follows that $D = \det A$ has $\Pi\lambda_v$ as its unique highest term. Now λ_a ($a \in \Pi$) makes a contribution to λ_v just when $v^{-1}a<0$. Thus by (2.8(b)) the highest term is as in (2.9). Now each $w\in W$ permutes the rows of A by (2.5(a)) and the invariance of e_v under W', hence fixes D up to sign. It follows that there is a unique lowest term, $\pm w_0\Pi\lambda_a{}^{n_a}$, with w_0 the element of W that makes all positive roots negative. Now if $b=-w_0a$ then $n_a=n_b$ by (2.8(a)). Thus the lowest term is as in (2.9), as required.

Consider now (2.4). We show that $D_1=\Pi(a^{1/2}-a^{-1/2})^{n_a}(a\in\Sigma^+)$ divides E and that $D_1=D$. Assume $a\in\Sigma^+$. As noted earlier there are n_a pairs of rows of (uf_v) which are interchanged by w_a. If we subtract row w_au from row u for such a pair then all entries of the result are divisible by $a-1$ since $w_a\lambda=\lambda a^n$, $n=-(\lambda,a^*)$, for $\lambda\in X$. Thus $(a-1)^{n_a}$ divides E, and since $\mathbf{Z}[X]$ is a u.f.d., so do $\Pi(a-1)^{n_a}$ and D_1. In particular D_1 divides D. To prove $D_1=D$ we need only show that the highest and lowest terms match up, i.e. by (2.9), that $\Sigma n_a a\ (a>0)=\Sigma n_a\lambda_a\ (a\in\Pi)$, with the operation of X now written as addition. If s denotes the left side and b a simple root then w_b maps b on $-b$ and permutes the other positive roots. Thus $(1-w_b)s=2n_bb$, and $(s,b^*)=2n_b$ by the formula for a reflection, so that s equals the right side by (2.1). Finally, each $w\in W$ acts on the rows of (ue_v) and (uf_v) just as it does on W/W', hence fixes E/D. This proves (2.4), hence also (2.2) and (1.1).

§3. PROOF OF (1.2) AND (1.3)

In this section G is a simply connected group, T is a maximal torus, and the other notations of §2 are used. Further r_a $(a\in\Pi)$ denotes the irreducible representation of G with highest weight λ_a, so that $R(G)$ is a polynomial algebra in the r_a's. If z is in the center of G, then $r_a(z)=\lambda_a(z)$. id. Thus there is a natural action of z on $R(G)$ and $\mathbf{Z}[X]$ with their scalars extended from $\mathbf{Z}$ to $\mathbf{C}$ such that $zr_a=\lambda_a(z)r_a$ and $z\lambda_a=\lambda_a(z)\lambda_a$ for all a. Observe that z fixes roots and commutes with W. We call z a pseudoreflection if it is one on $\Sigma\mathbf{C}\,r_a$ or $\Sigma\mathbf{C}\,\lambda_a$, i.e. if $\lambda_a(z)=1$ for every a but one.

MAIN LEMMA 3.1. *Let G be simply connected and Z a subgroup of the center of G. Then the following conditions are equivalent.*

(a) *$R(G')^Z$ is free over $R(G)^Z$ for every connected subgroup G' of maximal rank.*

(b) *$R(T)^Z$ is free over $R(G)^Z$ for some maximal torus T.*

(c) *$R(G)^Z$ is a polynomial algebra over $\mathbf{Z}$.*

(d) *Z is a direct product of the centers of a number of the simple components of G of type* Spin_{2r+1}.

(e) *Z is generated by pseudoreflections.*

(f) *$R(G)^Z$ has a generating set of the form $\{r_a{}^{m_a}\mid a\in\Pi\}$.*

(g) *X^Z has a basis of the form $\{m_a\lambda_a\}$.*

(h) *(X^Z,W,Σ_1) is the data for a simply connected group for some choice of an abstract root system $\Sigma_1\subseteq X^Z$.*

Consider now (1.2) in which, as noted earlier, G may be assumed semisimple. Since every semisimple group may be written G/Z with G and Z as in (3.1), and since $R(G)^Z, R(T)^Z,\dots$ have the same significance for G/Z as $R(G), R(T),\dots$ have for G, Theorem 1.2 follows from the equivalence of (a), (b), (c) and (d) of (3.1).

We first prove the equivalence of the last four parts of (3.1), which have been added mainly for convenience. If (e) holds and Z acts as a product of cyclic groups, the one on r_a being of order m_a, say, then $R(G)^Z=\mathbf{Z}[r_a\text{'s}]^Z=\mathbf{Z}[r_a{}^{m_a}\text{'s}]$, whence (f). Conversely, if this equation holds then Z is a subgroup of the above product, is the whole product in fact since otherwise some nontrivial character $\Pi\lambda_a{}^{d_a}$ $(0\le d_a<m_a)$ would vanish on Z and $\Pi r_a{}^{d_a}$ would contradict the last equation, whence (e). Since $\Pi r_a{}^{d_a}$ is in $R(G)^Z$ if and only if $\Sigma\, d_a\lambda_a\in X^Z$ (additive notation here), (f) and (g) are equivalent. Observe that in (h) the elements of Σ_1 are multiples of those of Σ since their directions are determined by the reflections of W. If (g) holds then $m_aa=(1-w_a)m_a\lambda_a\in X^Z$ and $(m_aa,(m_bb)^*)$ is always integral since $\{(m_bb)^*\}$ is a basis for the dual of X^Z. It readily follows that $\{m_aa\mid a\in\Pi\}$ is a basis for a root system Σ_1 for which (h) holds. Conversely, if (h) holds and $\{m_aa\}$ is a basis for Σ_1, then $\{m_a\lambda_a\}$ is the corresponding basis of X^Z (see 2.1), whence (g).

Next we prove that (e) $\Rightarrow$ (a) $\Rightarrow$ (b) $\Rightarrow$ (c) $\Rightarrow$ (e). If (e) holds, so does (h) and then also (a) by (1.1) which in (2.2) has been reduced to a theorem about (X,W,Σ). Clearly (a) implies (b). Assume (b). Then $A=\mathbf{C}R(T)^Z$ is free, hence also integral, over $B=\mathbf{C}R(G)^Z$. Localize B at the point q of Spec B where all r_a's are 0 and A at a point p of Spec A above q. The first condition makes sense since each r_a has some power in B. We now invoke a result of Auslander–Buchsbaum–Serre.

3.2 A (commutative, Noetherian) local ring R is regular if and only if its cohomological dimension $d(R)$ is finite.

Now let $\{x_1, x_2, \ldots, x_r\}$ with $r = |\Pi|$, the rank of G, be a basis for X^Z, imbedded in A in the natural way, and $x_i(p) = c_i$. Then A_p is the tensor product of r algebras, the ith equal to $\mathbf{C}[x_1, x_i^{-1}]$ localized at $x_i = c_i$, hence is regular, whence $(d(A_p) < \infty$ by (3.2). Since A_p is free over B_q (with basis any basis for $R(T)^Z$ over $R(G)^Z$), $d(B_q) \leq d(A_p)$, so that B_q is regular by (3.2). Thus $\dim_{\mathbf{C}} m/m^2 = r$, if m denotes the maximal ideal of B_q. The monomials $\Pi r_a^{d_a}$ that lie in m form a multiplicative semigroup. Let C be its minimal generating set, consisting of those elements that are not products of others. Clearly B is a basis for m/m^2 over $\mathbf{C}$ so that $|B| = r$. However for each $a \in \Pi$ some $r_a^{m_a}$ lies in B. Thus B consists of the $r_a^{m_a}$'s, and $R(G)^Z$ is a polynomial algebra on the $r_a^{m_a}$'s, whence (f) and also (c). Now assume (c). The free generating set for $R(G)^Z$ may be taken in the ideal m just considered. Then the proof just given shows that (f) holds, hence also (e).

It remains only to prove the equivalence of (d) and (e). For this we may assume that G is simple since if $z \in Z$ is a pseudoreflection it acts nontrivially on just one r_a, hence belongs to some simple component of G. Let V be the universal covering space for T, a real Euclidean space, and for convenience take the character values $\lambda(v)$ to be in $\mathbf{R}/\mathbf{Z}$ rather than in the complex numbers of norm 1. Then there is the famous fundamental simplex $S: \{v \in V \mid a(v) \geq 0\ (a \in \Pi), h(v) \leq 1\}$. Here $h = \Sigma\, h_a a$ is the highest root. This sum is to be taken over Π and similarly for the sums on $a, b, \ldots$ that follow. The center of G is represented in S by 0 and the vertices z_a of S corresponding to a's for which $h_a = 1$. For any such we have

$$b(z_a) = \delta_{ba} \qquad (b \in \Pi). \tag{3.3}$$

For a again arbitrary write

$$\lambda_a = \Sigma\, n_{ab} b \qquad (n_{ab} \in \mathbf{Q}). \tag{3.4}$$

We claim that for the dual root system Σ^*, in which a is replaced by $2a/(a, a)$ and similarly for λ_a, the corresponding equation reads

$$\lambda_a^* = \Sigma\, n_{ba} b^*. \tag{3.5}$$

For substituting the definitions into (3.4) we get (3.5) with the coefficient of b^* equal to $n_{ab}(b, b)/(a, a)$. But $(\lambda_a, \lambda_c) = n_{ac}(c, c)/2$ by (3.4) and (2.1), whence $n_{ac}(c, c) = n_{ca}(a, a)$ by symmetry. The coefficient of b^* thus becomes n_{ba}, whence (3.5). Now assume that $z_a \in Z$ acts as a pseudoreflection on $\mathbf{C}R(G) = \mathbf{C}[r_a\text{'s}]$. Then $\lambda_c(z_a)$ is integral with just one exception, say for $c = b$. But $\lambda_c(z_a) = n_{ca}$ by (3.4). Thus (by (3.5)) $n_{ba} b^*$, hence also some submultiple of b^*, is a weight. This implies that Σ^* is of type C_r and b^* is the unique long simple root, as is well known and proved thus: in any other case there is a simple root c^* such that $(b^*, c^{**}) = -1$, so that b^* is primitive as a weight. Then Σ is of type B_r (and $G = \mathrm{Spin}_{2r+1}$), and a is the long root at the end of the Dynkin diagram and $\{1, z_a\}$ is the center of G since this a is the only simple root for which $h_a = 1$. Conversely, if Σ and a are as just mentioned it can be verified that λ_a^* in (3.5) has exactly one nonintegral coefficient so that z_a is a pseudoreflection. Thus (d) and (e) are equivalent, and (3.1) is completely proved.

Remarks 3.6. (a) For a proof of the equivalence of (b), (c) and (e) in a more general setting see [4], from which our proof that (b) implies (c) is taken. We could avoid the other heavy commutative algebra used there because of the simple action of Z in our case. (b) The geometric essence of the equivalence of (c) and (e) in its general form is that, in an algebraic or analytic variety acted on by a finite group Z of order not divisible by the characteristic a nonsingular point p remains nonsingular in the quotient space if and only if Z^p acting on the tangent space at p is generated by pseudoreflections.

Finally, we consider (1.3). Since every irreducible representation of G is trivial on the unipotent radical of G, we may assume G reductive. Then we may reduce (1.3) to properties of abstract root systems and reflection groups, as we reduce (1.1) to (2.2), properties which have been proved above.

REFERENCES

1. J. F. ADAMS: *Lectures on Lie Groups.* Benjamin, New York (1969).
2. N. BOURBAKI: *Groupes et algèbres de Lie,* Chapters IV, V and VI. Hermann, Paris (1968).
3. H. V. PITTIE: Homogeneous vector bundles on homogeneous spaces, *Topology* **11** (1972) 199–203.
4. J.-P. SERRE: *Colloque d'algèbre,* no. 8. Ecole Normale Supérieure de Jeunes Filles, Paris (1967).

University of California, Los Angeles

Inventiones math. 36, 209 – 224 (1976)

Inventiones mathematicae

On the Desingularization of the Unipotent Variety

Robert Steinberg (Los Angeles)

To Jean-Pierre Serre

1. Introduction

Springer [13] has proved the following result.

Theorem 1.1. *Let G be a (connected) semisimple algebraic group with separable universal covering, V the variety of unipotent elements of G, B a Borel subgroup of G, and W the subset of all (u, gB) in $V \times G/B$ such that $g^{-1}vg \in B$. Then π: $W \to V$ is a desingularization of V.*

Here the nonsingular elements of V are just the regular elements, all conjugate, studied in some detail in [16]. Next, in case G is simple, come the subregular elements, dense among the singular elements of V and forming a single class of codimension 2 along which V has its generic singularity. This turns out to be a Kleinian singularity, an isolated rational double point on an algebraic surface, the one with the same label as that of G: for example, the surface singularity $xy = z^{r+1}$, commonly labeled A_r, occurs as the generic singularity for V in case G is of type A_r. Furthermore, with some restrictions on char k, the characteristic of the base field, the universal deformation of the surface singularity can be naturally achieved within the corresponding group G. In view of these remarkable facts (and others), conjectured by Grothendieck and proved by Brieskorn and Tits (mostly in unpublished notes; see, however, [5] and the last part of [17]) attention is focused on the deeper singularities of V, hence on all of the fibres of π. Observe that the fibre above u is $(G/B)_u$, the variety of "flags" fixed by u, or, equivalently, the variety of Borel subgroups containing u. In this article we obtain some results about the dimensions of these fibres and the numbers of irreducible components, relating the latter to elements of the Weyl group W.

The basic idea, introduced in [17], is that given two components Y, Z of $(G/B)_u$ there exists a unique w in the Weyl group such that $g_1 B$ and $g_2 B$ are in the attitude w, that is, $g_1^{-1} g_2 \in BwB$, for a dense open set of $(g_1 B, g_2 B)$ in $Y \times Z$. This enables us to get a sort of classification of pairs of components in terms of elements of W and of components in terms of involutions and to prove that (*) $\dim (G/B)_u \leq 1/2(\dim G_u - r)$. Here and elsewhere G_u is the centralizer of u and r

is the rank of G. All of this is done in Section 3, much of it for parabolics more general than Borels. In Section 4 we establish equality in (*) (also conjectured by Grothendieck), under restrictions which are probably not necessary, via a criterion proved in Section 3 and explicit construction of components. Here $\dim G_u$ can be considered to be known, for the classical groups in terms of data specifying the conjugacy class of u (see, e.g., [15, Ch. IV]), for the exceptional groups in terms of tables given by Elkington [7]. In Section 5 we consider the group $G=SL_n$ where more explicit results can be obtained. In particular, we show that the number of components of $(G/B)_u$ equals the degree of an appropriate complex irreducible representation of the Weyl group, S_n in this case, and then offer a second proof to illustrate an alternate approach. In Section 6 we prove 1.1 and then indicate how the previous results can be extended to classes that are not unipotent. The reason for doing the first thing is that 1.1 is in [13] erroneously stated (and "proved") for adjoint groups (cf. 6.1 below); in [17] the proof is correct but there is an unnecessary assumption.

Since completing this paper, I have received a preprint of a paper written by Springer [14]. This contains (among many other results) a proof of the inequality for $\dim(G/B)_u$ above (a proof now several years old) and a connection between the étale cohomology of $(G/B)_u$ and the representations of W, in a sense dual to our connection between the components of $(G/B)_u$ and the elements of W. The focus is entirely different there, as are the methods, involving as they do a theory of trigonometric sums on finite Lie algebras, the theory of reduction mod p, and the étale cohomology.

As general references for the theory of algebraic groups we cite [3, 4], for the unipotent variety [13, 15, 16].

2. Notation and Other Preliminaries

Throughout this work G is a semisimple (algebraic) group, B a fixed Borel (subgroup), T a maximal torus of B, $\Sigma(\Sigma^+)$ the corresponding system of roots (positive roots), W the Weyl group, and U_α the one-dimensional unipotent subgroup corresponding to the root α. If H is a (closed) subgroup containing T, then $\Sigma_H \subseteq \Sigma$ and $W_H \subseteq W$ are the corresponding objects for H. As is known $T \cdot \prod U_\alpha$ ($\alpha \in \Sigma_H$, or, equivalently, $U_\alpha \subseteq H$) is a dense open part of H, the direct product of its factors as an algebraic variety. From this it follows that $L(H)$, the Lie algebra of H, is the direct sum of $L(T)$ and the $L(U_\alpha)$'s.

Lemma 2.1. *Two connected subgroups H, K of G are transversal if they have a maximal torus in common, in particular if both are parabolic. Then HK/K is isomorphic to $H/(H \cap K)$.*

Proof. We must show that $L(H \cap K) = L(H) \cap L(K)$. This follows from the remarks about $L(H)$ above. The second part then follows since it is the essence of the Bruhat lemma that any two Borels have a maximal torus in common [17, p. 72]. For the first part to hold G can be reductive, for the second part arbitrary. The third part then follows since the natural map from $H/(H \cap K)$ to HK/K is a separable bijective map of homogeneous H-spaces, hence defines a quotient for $H/(H \cap K)$ (see [3, p. 180]).

Besides the above, P and Q always denote parabolics, taken to contain B (that is, to be standard) whenever it is convenient, U_P the unipotent radical of P (written U in case $P=B$) and M_P its Levi component containing T, and u a unipotent element of G. Finally we write, for example, $n(u)$ in place of n_u on occasion to avoid multiple indices.

Proposition 2.2. *The number of unipotent conjugacy classes of G is finite.*

We are indebted to G. Lusztig (letter, Jan. 1976) for this result, long expected to be true, but previously proved, by Richardson [10], only with some (mild) restrictions on char k. It follows that the number of singularities and fibres to be studied in 1.1 is finite.

We continue with some remarks on the fibres of 1.1, more generally, on the sets $(G/P)_u$. They are closed in G/P, hence are projective varieties. Further they are connected, as will soon be seen.

Lemma 2.3. *Let P and Q be parabolic with $P\subseteq Q$. Then the natural map from $(G/P)_u$ to $(G/Q)_u$ is surjective.*

Proof. If u fixes gQ, then $g^{-1}ug\in Q$, hence $q^{-1}g^{-1}ugq\in B\subseteq P$ for some $q\in Q$, and u fixes gqP, even if u is not unipotent.

Corollary 2.4. *$(G/P)_u$ above is connected.*

Proof. Following Tits we can make this more precise as follows. For each simple root α let P_α be the corresponding rank 1 parabolic containing B. For α not in the support of P and $g\in G$ call $gP_\alpha P/P$ a line of type α. It is a projective line since it is isomorphic by 2.1 to $P_\alpha/(P_\alpha\cap B)=P_\alpha/B$. Then (*) $(G/P)_u$ is arc connected, by lines of the various types α. This follows from 2.3 and Tits' result that this is true when P is Borel (see [10, Section 1]).

Corollary 2.5. $\dim(G/P)_u$ *is a decreasing function of P. More generally, so is the number of components of dimension $\geqq n$ for each fixed n.*

Remarks 2.6. (a) If u is regular, then $(G/B)_u$ is a point. If u is subregular and G is simple then $(G/B)_u$ is a "Dynkin curve", a finite union of lines, one of each type α if all roots have the same length, whose intersection pattern is that of the Dynkin diagram of G (see [17, Section 3.10]). This is exactly the singular set for the minimal desingularization of the corresponding Kleinian surface singularity mentioned in the introduction. (b) If $G=SL(V)$, then G/B is the ordinary flag variety on V. Each simple root corresponds to a number i between 0 and $\dim V$, and a line of type i is obtained by keeping fixed all components but the i-th of a given flag $V_0\subset V_1\subset V_2\subset\cdots\subset V_n$ and varying it subject to the constraint $V_{i-1}\subset V_i\subset V_{i+1}$. (c) If $P=B$ there is just one line of each type α through a given point, but if $P\neq B$ this is not the case. For example, if $G=SL(V)$ and P is the stabilizer of an i-dimensional subspace of V then only lines of type i occur in G/P and each such is the set of i-dimensional subspaces subject to the above constraint with V_{i-1} and V_{i+1} fixed, which in case $i=1$ reduces to the set of points of an ordinary projective line in $\mathbb{P}(V)$.

3. Generic Attitudes

Let C be a unipotent conjugacy class and P and Q (standard) parabolic subgroups. We wish to study $S=S(C,P,Q)\subseteq C\times G/P\times G/Q$ consisting of all (v,gP,hQ) such that v fixes gP and hQ, i.e., such that $g^{-1}vg\in P$ and $h^{-1}vh\in Q$. In this way we shall get information about the components of $(G/P)_v$ and $(G/Q)_v$, their dimensions and their numbers, and a connection with certain elements of $W_P\backslash W/W_Q$.

Proposition 3.1. *Let $S\subseteq C\times G/P\times G/Q$ be as above and $u\in C$.*

(a) *S is closed and its dimension is* $\dim C+\dim(G/P)_u+\dim(G/Q)_u$.

(b) *There is a one-to-one correspondence between the (irreducible) components of S and the G_u-components of $(G/P\times G/Q)_u$, or, equivalently, the orbits of G_u on the set of pairs of components of $(G/P)_u$ and $(G/Q)_u$, given thus. If X is one of the latter, then $Y={}^G(u,X)$ is one of the former.*

Proof. Since all fibres of S over C are isomorphic to $(G/P\times G/Q)_u$, we have (a). Since the Y's in (b) cover S irredundantly and are finite in number, it remains to prove that each of them is closed. Let $X'=G_uX$. It is the union of the components equivalent to X under G_u, hence is closed. Thus the set in $G/G_u\times G/P\times G/Q$ of all (gG_u,g_1P,g_2Q) with $(g^{-1}g_1P,g^{-1}g_2Q)\in X'$, which is well-defined because X' is invariant under G_u, is also closed. Hence so is its image in $C\times G/P\times G/Q$ under the map $(gG_v,g_1P,g_2Q)\to({}^gv,g_1P,g_2Q)$ for this is a map of homogeneous $G\times G\times G$ spaces, hence is open, and being bijective it is also closed. That is, Y is closed.

Remarks 3.2. A connected group must fix every component of a variety on which it acts. Thus orbits in (b) are G_u/G_u^0 orbits and their number in case G_u is connected is just the product of the numbers of components of $(G/P)_u$ and $(G/Q)_u$.

Proposition 3.3. *For $w\in W_P\backslash W/W_Q$ let S_w denote the subset of all (u,g_1P,g_2Q) in S with g_1P,g_2Q in the attitude w, i.e., of the form gP,gwQ for some $g\in G$.* (a) *$S=\bigcup S_w$ is a disjoint union of locally closed subsets.* (b) *S_w is isomorphic to the set of all points of $C\times G/(P\cap{}^wQ)$ such that the first coordinate fixes the second, under the map $(u,gP,gwQ)\to(u,g(P\cap{}^wQ))$.* (c) $\dim S_w\leqq\dim G-r$ *with equality if and only if $C\cap P\cap{}^wQ$ is dense in the variety V_w of unipotent elements of $P\cap{}^wQ$. In that case S_w is a dense open part of one of the irreducible components of S, and it is nonsingular if also $P\cap{}^wQ$ is solvable, in particular if P or Q is Borel.* (d) $\dim S\leqq\dim G-r$ *with equality if and only if there exists some w as in* (c). *Further the number of irreducible components of S of this dimension is the number of such w's.*

Proof. We decompose $G/P\times G/Q$ into its orbits under G. These are represented by the points (P,wQ) with $w\in W_P\backslash W/W_Q$ since the latter is a system of representatives for $P\backslash G/Q$ (see [4]). This yields (a). The natural map from the orbit through (P,wQ) to $G/(P\cap{}^wQ)$ is a bijective map of homogeneous G-spaces which is separable since P and ${}^wQ$ are parabolic and hence transversal. Hence it is an isomorphism and (b) follows. The fibres of S_w over the second factor are all isomorphic to $C\cap V_w$. Thus $\dim S_w=\dim G/(P\cap{}^wQ)+\dim(C\cap V_w)$. Since $\dim(P\cap{}^wQ)=r+\dim V_w$, which holds in any connected group of rank r, this yields $\dim S_w = \dim G - r - (\dim V_w - \dim(C \cap V_w)) \leqq \dim G - r$, with equality if and only if $C \cap V_w$ is dense in V_w. Here $P \cap {}^wQ$ is connected, and hence V_w is irreducible, because the intersection of Borel subgroups is always connected. From this inequality and (a) the inequality in (d) follows. Now assume equality in (c). Since V_w is irreducible, so is $C \cap V_w$ and hence also $S_w = {}^G(C \cap V_w, P \cap {}^wQ)$. Since also S_w is locally closed and of dimension that of S, it is open in one of the components of S; and the last part of (d) also follows. Finally, if $P \cap {}^wQ$ is solvable then V_w is a subgroup. Thus the open part $C \cap V_w$ is nonsingular and then so is S_w because of the homogeneous action of G on its base space $G/(P \cap {}^wQ)$.

Remark 3.4. Let Y and Z be irreducible subsets of G/P and G/Q. Then there is a unique orbit ${}^G(P, wQ)$ containing a dense open part of $Y \times Z$. The corresponding element $w \in W_P \backslash W / W_Q$, or any of its representatives in W, will be called the attitude of Y to Z.

Theorem 3.5. *For u unipotent and P and Q parabolic as above we have:* (a) $\dim(G/P)_u \leqq 1/2(\dim G_u - r) = d_u$, *say.* (b) *Consider the components Y of $(G/P)_u$ and Z of $(G/Q)_u$ of dimension d_u. The set of G_u-orbits of such pairs (Y, Z) is in one-to-one correspondence with the set of w's of $W_P \backslash W / W_Q$ such that $C(u)$ cuts the unipotent variety of $P \cap {}^wQ$ densely. In this correspondence, w is just the attitude of Y to Z.*

Proof. Observe that d_u above is independent of P and Q. From 3.1 (a), 3.3 (d) and the equation $\dim C = \dim G - \dim G_u$ we get $\dim(G/P)_u + \dim(G/Q)_u \leqq \dim G_u - r = 2d_u$, with equality as in 3.3 (d). With $Q = P$, this yields (a). Then equality holds above if and only if both $(G/P)_u$ and $(G/Q)_u$ have dimension d_u. Thus (b) follows from 3.1 (b), 3.3 (c) and the definition of S_w.

Remarks 3.6. (a) Equality holds in 3.5 (a) in case P is Borel under additional conditions to be discussed in the next section, but in accordance with 2.5 it tends to fail if P is "large" compared to u. It fails, for example, if P is not Borel and $C = \{1\}$ or, at another extreme, if $P = G$ and C is not regular. (b) As mentioned before the unipotent elements of $P \cap {}^wQ$ form an irreducible set. Thus by 2.2 there exists, for each w, exactly one class C for which 3.5 (b) holds. In other words, surprisingly (to us), for u unipotent the attitude of a pair of components of $(G/P)_u$ and $(G/Q)_u$ of dimension d_u uniquely determines the class of u in G and the G_u-orbit of the pair of components. (c) Henceforth the notation d_u of 3.5(a) will be used.

Corollary 3.7. *Let m_u* (resp. m'_u) *denote the number of components of $(G/P)_u$* (resp. $(G/Q)_u$) *of dimension d_u and n_u the number of orbits of G_u, or, equivalently, of $A_u = G_u/G_u^0$ on the set of pairs of such components* (a) $\sum n_u = |W_P \backslash W / W_Q|$, *the sum over a system of representatives for the unipotent classes.* (b) $\sum |A_u|^{-1} m_u m'_u \leqq |W_P \backslash W / W_Q| \leqq \sum m_u m'_u$, *with equality in case all G_u's are connected.*

Proof. Here (a) follows from 2.2 and 3.5 (b), and then (b) from the obvious inequalities $|A_u|^{-1} m_u m'_u \leqq n_u \leqq m_u m'_u$.

Remarks 3.8. (a) $|W_P \backslash W / W_Q| = \sum l_\rho l'_\rho$, summed over the complex irreducible representations ρ of W with l_ρ the multiplicity of ρ in the representation of W induced by the trivial representation of W_P and l'_ρ the same for W_Q; for both numbers equal the intertwining number of the two induced representations. (b) For $a \in A_u$ let i_a (resp. i'_a) denote the number of the above components of $(G/P)_u$ (resp. $(G/Q)_u$) fixed by a. Then $n_u = |A_u|^{-1} \sum i_a i'_a$. For in any finite action the number of orbits is the average number of fixed points. In this form and with P and Q both Borel 3.7 (a) is due to Springer [14]. (c) Here is a simple example in which components are being moved by G_u. Take G of type B_2 in char $k \neq 2$ and $P=Q=B$. There are four unipotent classes, for three of which $(G/B)_u$ is irreducible, so that $n_u = m_u = 1$. The fourth is the subregular class (see [17, Section 3.10]). Here there is a line of type β meeting two lines of type α which are interchanged by G_u. Here α and β are the long and short simple roots. Thus there are 9 pairs of components but only 5 G_u-orbits, corresponding to the $8-3$ missing elements of W. In characteristic 2 the situation is better since then the two lines of type α fuse and another unipotent class appears, and G_u stabilizes every component so that $n_u = m_u^2$ in all cases.

Corollary 3.9. *Let Y, Z, w be as in* 3.5 (b), *and let A be the transporter of u into $P \cap {}^wQ$.* (a) *$A^{-1}P$ covers a dense open part of $G_u Y$.* (b) *If w' is like w, then it leads to the same first component $G_u Y$ as w if and only if ${}^P(C \cap P \cap {}^wQ)$ covers a dense open part of $C \cap P \cap {}^{w'}Q$.* (c) *PwQ/Q contains a dense open part of Z or one of its G_u-translates for a dense open set of u in $C \cap P \cap {}^wQ$.* (d) *If P has a dense orbit on $C \cap P \cap {}^wQ$, then G_u^0 has one on Y, and if $P \cap {}^wQ$ has one on $C \cap P \cap {}^wQ$, then G_u^0 has one on $Y \times Z$. In particular, if $Q=P$ and $Z=Y$ in the last case, then G_u^0 is generically doubly transitive on Y.*

Proof. The fibre over u in S_w covers a dense open part of $G_u(Y \times Z)$ and it may be identified with $A^{-1}(P \cap {}^wQ)$ in $G/(P \cap {}^wQ)$ by 3.3(b). Projection on the first factor yields (a). Choosing a generic fibre of this projection, normalized so that (P, wQ) is on this fibre and hence $u \in P \cap {}^wQ$, we get (c). In (b) let w and w' lead to the same first component. Then $B^{-1}P = B'^{-1}P$ is dense open in $G_u Y$ for open parts B, B' of A, A' which can be assumed invariant under $P \cap {}^wQ$ on the left and G_u on the right since A, A' are so, and such that $B'^{-1}(P \cap {}^{w'}Q)$ is dense in $A'^{-1}(P \cap {}^{w'}Q)$, the component of $S_{w'}$ over u, which is permissible by (a). Calculating dim ${}^G({}^{B'}u, P \cap {}^{w'}Q)$ in two ways as in the proof of 3.3, we see that ${}^{B'}u$ is dense in $C \cap P \cap {}^{w'}Q$. Let $v' = {}^{b'}u$ with $b' \in B'$ arbitrary. Choose $b \in B$ so that $b^{-1}P = b'^{-1}P$ and set ${}^bu = v$. Then $b'b^{-1}$ is in P and it transforms v into v', which proves half of (b). By working backward we get the other half. Let C' be the dense orbit in the first part of (d) and A' the transporter of u into C'. Then as before $A'^{-1}P$ is dense in $G_u Y$. But if $a \in A'$, then $A' \subseteq PaG_u$. Thus $A'^{-1}P = G_u a^{-1} P$. Thus G_u has a dense orbit on $G_u Y$ and G_u^0 has one on its component Y. The second part of (d) is proved in the same way.

Remarks 3.10. (a) Parts (a) to (c) lead in principle to a classification of the components of $(G/Q)_u$ of dimension d_u, if such exist, in terms of their attitudes to a fixed such component of a suitably chosen P, e.g., $P=B$ or $P=Q$, given thus in case G_u is connected. First find w so that 3.5 (b) holds (for $P=Q=B$ see Section 4 below) and then all w''s equivalent to w as in 3.9 (b). Then for u generic in $P \cap {}^wQ$, the sought components lie, generically, one in each of the spaces $Pw'Q/Q$. (b) The second condition of (d) holds, for example, in case $P=Q=B$ and $w = w_1 w_0$ with w_1

the product of the simple reflections (or any subset thereof) and w_0 the element of W that makes every positive root negative. For then $P \cap {}^wQ = T \cdot U_w$, which has as its support the roots $\alpha > 0$ such that $w_1^{-1}\alpha < 0$, a linearly independent set as is known (see [16, Section 7.2]), so that T has a dense orbit on U_w. We observe that the class C involved depends on the order in which the simple roots are taken. For another example, assume $Q = P$ arbitrary, $w = -1 \in W$, and C the class cutting the unipotent variety of $P \cap {}^wP$ (now reductive) densely.

Corollary 3.11. *If u and w are as in* 3.3 (c), *then $d_u \leqq l(w)$, the length of w.*

Proof. $\dim BwQ/Q \leqq \dim BwB/B$ (since $Q \supseteq B) = l(w)$. Thus 3.11 follows from 3.9 (c) in case $P = B$. If $P \neq B$ then there is a class $C(u')$ which intersects the set of unipotent elements of $B \cap {}^wQ$ in a dense part, and $d_{u'} \leqq l(w)$ by the case just proved. Since $C(u')$ is in the closure of $C(u)$ it easily follows that $d_u \leqq d_{u'}$, whence our result.

In case $Q = P$ in 3.5(b) the self-inverse elements of $W_P \backslash W / W_P$ play a special role, for if Y is a component of $(G/P)_u$ of dimension d_u and w is the attitude of Y to Y, then clearly w represents a self-inverse double coset. More generally this is so for a self-inverse pair of such components, a pair (Y, Z) equivalent to (Z, Y) under G_u. Thus we get:

Corollary 3.12. (a) *The set of G_u-orbits of self-inverse pairs of components of $(G/P)_u$ of dimension d_u is in one-to-one correspondence with the set of self-inverse elements of $W_P \backslash W / W_P$ such that* 3.5(b) *holds with $C = C(u)$ and $Q = P$.* (b) *If s_u is the number of orbits in* (a) *then $\sum s_u$, the sum as in* 3.7(a), *equals the number of self-inverse elements of $W_P \backslash W / W_P$.*

Remark 3.13. If G_u is connected in (a), then s_u is just the number of components of $(G/P)_u$ of dimension d_u, and similarly in (b).

As in 3.7 the second item in (b) has an interesting expression as a sum over the representation classes of W, as we shall now show.

Proposition 3.14. *The number $\sum s_u$ of* 3.12 (b) *also equals the sum of the multiplicities of the complex irreducible representations of W in the representation of W induced from the trivial representation of W_P.*

Proof. Since the irreducible representations of W are all real (in fact rational, by [2]), this follows from:

Lemma 3.15. *Let A be a finite group, B a subgroup, ρ the representation of A induced by the trivial representation of B and $\rho = \sum n_i \rho_i$ its decomposition into irreducible components. Let ε_i be* 1 *if ρ_i is real, -1 if ρ_i is not real but its character χ_i is real,* 0 *otherwise. Then $\sum \varepsilon_i n_i$ equals the number of self-inverse elements of $B \backslash A / B$.*

Proof. We adapt the method of Frobenius and Schur [8] who proved the case $B = \{1\}$. According to them $\varepsilon_i = |A|^{-1} \sum \chi_i(x^2)$ summed over A, as are the following sums. Thus if χ is the character of ρ then $\sum \varepsilon_i n_i = |A|^{-1} \sum \chi(x^2) = |A|^{-1} |B|^{-1} \sum \delta({}^yx^2, B)$ (using the definition of $\rho) = |B|^{-1} \sum \delta(x^2, B)$. Here $\delta(a, B)$ is 1 if $a \in B$, and 0 if not. To complete the proof we show that if $D = BaB$ is any double coset then the equation $x^2 \in B$ has $|B|$ solutions in D if D is self-inverse, none if not. If there is a solution then clearly D is self-inverse, while if D is self-inverse: $a^{-1}=b_1ab_2$ then ab_1 is a solution. Now assume D self-inverse with $a^2\in B$. Then $x=b_1ab_2$ is a solution if and only if $b_2b_1\in B\cap {}^aB$, yielding $|B|\,|B\cap {}^aB|$ pairs (b_1, b_2). Since each x has $|B\cap {}^aB|$ representations as b_1ab_2, we are done.

Corollary 3.16. *If P is Borel and s_u is as in 3.12 (b) then $\sum s_u$ equals the number of involutions of W (including 1), which in turn equals the sum of the degrees of the irreducible representations of W.*

Proof. The induced representation of 3.14 is the regular representation of W, in which each irreducible representation has a multiplicity equal to its degree.

Examples 3.17. Assume P Borel in all cases. (a) If $C(u)$ is dense in the unipotent radical of some parabolic subgroup Q, then $Y=Q/B$ is admissible above (see 4.1 below) and the (involutary) element of W corresponding to (Y, Y) in 3.12 (a) is w_{0P}, the element of W_P of maximum length. In particular if $Q=B$ then u is regular and $w=1$, while if $Q=G$ then $u=1$ and $w=w_0$. (b) The two lines of type α in 3.8 (b) provide an example of a self-inverse pair (Y, Z) with $Y\neq Z$. As will be seen below, such an example can not occur in SL_n.

4. Construction of Components

In this section we prove that equality holds in 3.5 (a) if P is Borel, under extra conditions on G, and we construct some components of dimension d_u. We start with a special case.

Proposition 4.1. *Let $C=C(u)$ be as before. Assume that there exists a parabolic subgroup P such that $C\cap U_P$ is dense in U_P. Then $\dim(G/B)_u=d_u$. In fact if $u\in C\cap U_P$ then P/B is a component of dimension d_u.*

Proof. Let w_{0P} be the element of W_P of maximal length, so that $U\cap {}^wU=U_P$. By 3.11, $d_u\leqq l(w_{0P})=\dim P/B\leqq\dim(G/B)_u\leqq d_u$, whence our result.

Corollary 4.2. *If P is parabolic then P has a dense orbit on U_P. In fact $C\cap U_P$ in 4.1 is such. Further $G_u^0\subseteq P$ if $u\in C\cap U_P$, and G_u is transitive on the conjugates of P containing u in their unipotent radicals, which are thus finite in number.*

Proof. Since G_u^0P/B is irreducible and contains the component P/B of $(G/B)_u$, we have $G_u^0P=P$, $G_u^0\subseteq P$. Thus $\dim C_P(u)=\dim P-\dim(G_u\cap P)=\dim P-\dim G_u=\dim P-r-2\dim P/B$ (by 4.1) $=\dim U_P$ since P/U_P is reductive of rank r. Since this holds for any u, we get $C_P(u)=C\cap P$. Finally if $u\in {}^gU_P$ then $g^{-1}ug\in C\cap U_P$, whence $p^{-1}g^{-1}ugp=u$ for some $p\in P$ by what has just been proved, so that ${}^gU_P={}^{gp}U_P$ with $gp\in G_u$ as stated.

Remark 4.3. This result is due to Richardson [11]. Our proof is somewhat different.

We continue with another construction of components using the elements of M_P rather than those of U_P.

Proposition 4.4. *Let M be the Levi component of a parabolic subgroup of G, R a parabolic subgroup of M, u a unipotent element of M, and A/R a component of $(M/R)_u$ of dimension* $1/2\,(\dim M_u - r)$. *Choose any parabolic subgroup Q of G with $M = M_Q$ and set $P = RU_Q$, a parabolic subgroup of G. Then $G_u^0 AP/P$ is (an open part of) a component of $(G/P)_u$ of dimension* $1/2\,(\dim G_u - r) = d_u$. *Further G_u^0 has a dense orbit on this component if M_u^0 has one on A/R.*

Proof. Since u stabilizes the big cell decomposition $U_Q^- M U_Q$, in which $U_Q^- M$ denotes a parabolic subgroup opposite to Q, we have $\dim G_u = \dim M_u + \dim (U_Q^-)_u + \dim (U_Q)_u$. Combining this with $\dim M_u = r + 2 \dim A/R$ and $A/R \simeq ARU_Q/RU_Q = AP/P$, and using the definition of d_u we get $\dim (U_Q^-)_u + \dim (U_Q)_u + 2 \dim AP/P = 2 d_u$. However, $\dim G_u^0 AP/P \geqq \dim (U_Q^-)_u + \dim AP/P$ since $G_u^0 AP$ contains $(U_Q^-)_u^0 AP$, a transversal product since $P \subseteq Q$, and similarly with U_Q^- replaced by U_Q and P by $P^- = RU_Q^-$. Thus $\dim G_u^0 AP/P + \dim G_u^0 AP^-/P^- \geqq 2 d_u$. But each term on the left $\leqq d_u$ by 3.5 (a). Thus each equals d_u (and equality holds everywhere else above). The last statement is clear.

Corollary 4.5. *Let M be as above and u a regular element of M. Let P be any parabolic subgroup with $M_P \subseteq M$,* e.g. $P = B$. *Then* $\dim (G/P)_u = d_u$. *Further $(G/P)_u$ contains a component of dimension d_u on which G_u^0 acts with a dense orbit,* viz., $G_u^0 P/P$.

Proof. As may easily be seen the parabolic subgroups P that arise in 4.4 are just those for which $M_P \subseteq M$. Now since u is regular in M, $\dim M_u = r$. Thus A may be (must be) taken to be R in 4.4, and then 4.5 follows.

To get our main result we shall have to assume that G is classical or else that the characteristic is 0 or sufficiently large (as in [15, Ch. III]), that G is "very good".

Theorem 4.6. *Assume that G is very good and that u is any unipotent element. Then* $\dim (G/B)_u = 1/2\,(\dim G_u - r) = d_u$.

Proof. We recall the classification of unipotent elements as given by Bala and Carter [1] and based on earlier work by Dynkin [6], Kostant [9], Springer [15], and others. We call a parabolic subgroup P distinguished if $\dim P/U_P = \dim U_P/U_P'$. For example Borels are always distinguished and in SL_n the converse holds. In type D_r or E_r the group P_α is distinguished if α is the simple root at the branch point of the Dynkin diagram. The result of Bala-Carter is:

4.7. *Assume that the characteristic is very good. Then for each unipotent conjugacy class C of G there exists a unique G-conjugacy class of pairs (M, A) such that M is the Levi component of some parabolic subgroup of G and A is a distinguished parabolic subgroup of M such that $C \cap U_A$ is dense in U_A.*

Further for the exceptional groups they give the possibilities for M and A in terms of tables, based in part on earlier tables of Dynkin [6].

For the classical groups (of any characteristic) 4.7 can be proved directly (as is done, e.g., for SL_n in Section 5 below). We assume this has been done. Now to prove 4.6 we can use 3.5 (a) with $P = Q = B$ and $w = w_{0A} w_{0M} w_0$ or else the construction of 4.4 with R a Borel subgroup of A, which is permissible by 4.1, and which works even if A is not distinguished in M.

Remarks 4.8. (a) In case G is very good 4.7 is equivalent to a corresponding statement about a nilpotent class $C(X)$ of $L(G)$, which by induction is easily reduced to the case in which X is not fixed by any nontrivial torus of G. The critical point of the proof of 4.7 is that in this case if X is imbedded in a simple three-dimensional subalgebra with standard basis $\{X, Y, H\}$ then the eigenvalues of ad H are all even. (See, e.g., [15, Ch. III, 4.20] for how this implies 4.7.) Unfortunately, this result in [1] (and also in [6]) occurs only toward the end of the classification after many case-by-case considerations. An a priori proof would be most welcome. A global form of the result, which may be true even if G is not very good, is:

4.9. *Let G be a semisimple group and G_1 a semisimple subgroup of rank* 1 *not centralized by any nontrivial torus. If G is of adjoint type then so is G_1.*

(b) As mentioned in the introduction for each class $C(u)$ of 4.6 the dimension of G_u (and a good deal more about its structure) is known, hence also d_u. Unfortunately our knowledge of G_u/G_u^0 is rather spotty.

5. The Group SL_n

This is the most favorable of all cases. In the first place,

Lemma 5.1. *G_u stabilizes every component of $(G/P)_u$, for every P and u.*

For by going to GL_n we do not change things essentially, and there all centralizers are connected, being open parts of vector spaces, viz. the corresponding centralizers in End_n.

It follows that the G_u-orbits of 3.5 (b) are all trivial, so that $n_u = m_u m_u'$ in 3.7 and the components of $(G/P)_u$ in 3.12 (a) are themselves in correspondence with the self-inverse elements of $W_P \backslash W/W_P$.

In the second place, the unipotent classes of G, the conjugacy classes of $W = S_n$ and its representations, and the families of associated parabolic subgroups are all classified by the same set, the set of partitions of n, i.e., decreasing sequences $\lambda = (\lambda_1, \lambda_2, \ldots)$ of non-negative integers with sum n, the first via the Jordan normal form and the fourth thus: if T is taken to be diagonal and B superdiagonal, then $M(\bar{\lambda}) = \prod GL(\lambda_i) \cap SL(n)$ is the Levi component of a standard parabolic $P(\bar{\lambda})$ and $U_{P(\bar{\lambda})}$ is supported by the superdiagonal positions not used by $M(\bar{\lambda})$. The bar here indicates that the parts of λ need not be taken in decreasing order. We recall also that each partition λ has a diagram consisting of the lattice points (i,j) with $1 \leqq j \leqq \lambda_i$, and a dual μ whose diagram is the transpose of that of λ. We can verify Theorem 4.6 here by observing that $C(\lambda)$, the class corresponding to λ, is the regular class of $M(\lambda)$ and is dense in every $U_{P(\bar{\mu})}$, and using 4.5 (which yields components of dimension d_u on which $G_{u(\lambda)}$ is dense) or 4.1. From the known value of dim $G_{u(\lambda)}$ (see, e.g., [15, Ch. IV]), the value of d_λ here works out to $\sum_{i<j} \min(\lambda_i, \lambda_j)$, or in terms of the dual, $1/2 \sum (\mu_i^2 - \mu_i)$.

Since the sums on the right and on the left of 3.7 (see 3.8 (a)) and 3.14 can now be taken over the same set, the set of partitions of n, one may expect (or hope) to have term-by-term equality and this turns out to be the case as we shall now show.

We use the standard partial order for partitions in which $\lambda \geqq \mu$ if and only if $\lambda_1 \geqq \mu_1, \lambda_1+\lambda_2 \geqq \mu_1+\mu_2, \ldots$. As is known and easy to prove this is just the condition for $C(\lambda)$ to have $C(\mu)$ in its closure. Dual to this we have the following result of Frobenius.

Lemma 5.2. *Let σ_λ denote the complex representation of W induced by the trivial representation of W_λ. Then the irreducible representations of W can be labeled ρ_λ so that $\sigma_\lambda = \rho_\lambda +$ higher terms, for all λ.*

For this and other facts about the representations of S_n we cite [12], where other references may be found.

Lemma 5.3. *For all partitions λ, μ we have:*

(a) $\dim (G/B)_{u(\lambda)} = d(\lambda)$.

(b) $\dim (G/P(\mu))_{u(\lambda)} = d(\lambda)$ if $\lambda = \mu$,
$< d(\lambda)$ if $\lambda \ngeqq \mu$.

Proof. That (a) holds has been observed above. By 3.5 (b) with P, Q replaced by $B, P(\mu)$ equality holds in (b) if and only if $C(\lambda)$ cuts some $U \cap {}^wP(\mu)$ densely. This holds if $\lambda = \mu$, for then $w = w_{0M} w_0$ with $M = M(\mu)$ yields $U_{w_{0M}}$ for this intersection. In general we may assume $w \sum_M^+ > 0$. Then ${}^{w^{-1}}U \cap P(\mu)$ contains $U_{w_{0M}}$, whence $C(\lambda)$ must contain $C(\mu)$ in its closure if equality is to hold in (b). As remarked above this is so if and only if $\lambda \geqq \mu$, whence (b).

Theorem 5.4. *If λ and μ are any partitions then the number of components of $(G/P(\mu))_{u(\lambda)}$ of dimension $d(\lambda)$ equals the multiplicity of ρ_λ in σ_μ.*

Proof. Let $m(\lambda, \mu)$ be the first number, $n(\lambda, \mu)$ the second. We prove their equality in several steps.

(1) $m(\lambda, \lambda) > 0$; $m(\lambda, \mu) = 0$ if $\lambda \ngeqq \mu$, and similarly for n in place of m. This follows from 5.3 and 5.2.

(2) $\sum_\lambda m(\lambda, \mu)\, m(\lambda, \nu) = |W_\mu \backslash W / W_\nu|$ for all μ, ν and similarly for n in place of m. For m this follows from 3.12 (a) and 5.1. For n this has been observed in 3.8 (a).

(3) Now we can complete the proof, that $m(\lambda, \mu) = n(\lambda, \mu)$ by downward induction on λ in terms of the partial ordering $\geqq$. Fix λ_0 and assume the result known for higher λ's. By (2) we have $\sum m(\lambda, \mu)\, m(\lambda, \lambda_0) = \sum n(\lambda, \mu)\, n(\lambda, \lambda_0)$. By (1) only the terms for which $\lambda \geqq \lambda_0$ make a contribution, so that by induction $m(\lambda_0, \mu)\, m(\lambda_0, \lambda_0) = n(\lambda_0, \mu)\, n(\lambda_0, \lambda_0)$. With $\mu = \lambda_0$ this gives $m(\lambda_0, \lambda_0) = n(\lambda_0, \lambda_0)$, and then since this number is not 0 by (1), $m(\lambda_0, \mu) = n(\lambda_0, \mu)$ for all μ, as required.

Corollary 5.5. *If P and Q are associated parabolic subgroups, then $(G/P)_u$ and $(G/Q)_u$ have the same number of components of dimension d_u.*

Proof. In the preceding proof, each $P(\mu)$ can be replaced by an associate without changing things essentially.

Quite likely this result holds also when G is not SL_n, but it should be observed that G/P and G/Q here need not be isomorphic.

Corollary 5.6. *Let λ be a partition, $u \in C(\lambda)$.* (a) *$(G/P(\lambda))_u$ has just one component of dimension d_λ and on it G_u has a dense orbit.* (b) *If $n_\lambda = \mathrm{dg}\,\rho_\lambda$, then $(G/B)_u$ has exactly n_λ components of dimension d_λ.* (c) *$W = S_n$ contains exactly n_λ^2 elements w such that $C(\lambda)$ cuts $U \cap {}^wU$ densely, of which n_λ are involutions and n_λ satisfy $w\sum_{M(\lambda)}^+ > 0$. They represent, respectively, the attitudes of pairs of the components in* (b), *of the components in* (b) *with themselves, and of the components in* (b) *with the one in* (a).

Proof. We have $m(\lambda, \lambda) = n(\lambda, \lambda)$, the multiplicity of ρ_λ in σ_λ, which is 1 by 5.2. Further by 4.5 with $M = M(\lambda)$ and $P = P(\lambda)$ the component in question is seen to be the closure of $G_u^0 P/P$, whence (a). If $\mu = 1^n$ (the partition of n into n ones) in 5.4 then $P(\mu) = B$ and $W_\mu = \{1\}$ so that σ_μ is the regular representation of W. The multiplicity of ρ_λ there equals its degree, whence (b). Then the first two parts of (c) follow from 3.5 (b) and 5.1. For the third we use (a), (b) and 5.1 to get the n_λ attitudes. By 3.5 (b) with $P = B$ and $Q = P(\lambda)$ these are just the w's in W/W_λ for which $C(\lambda)$ cuts $U \cap {}^wP(\lambda)$ densely. If w is the unique element of its coset such that $w\sum_{M(\lambda)}^+ > 0$, the last item becomes $U \cap {}^wU$, as required.

We present now an alternate proof of 5.6 (b) which yields other results, among them the fact that the components of $(G/B)_u$ all have the same dimension. In this last result, incidentally, B may not be replaced by an arbitrary parabolic.

Let $u = u(\lambda)$ be unipotent in $SL(V)$ and identify G/B with the variety of flags on V. Let $W = W_0 \supseteq W_1 \supseteq W_2 \ldots$ be the filtration of $W = \ker(u-1)$ given by $W_i = W \cap \mathrm{im}\,(u-1)^i$. If μ is the dual of λ, then $\dim W_i = \mu_i$ for all i. Presently we prove: (*) G_u is transitive on the lines of $W_i - W_{i+1}$ for each i. Granted (*) we continue thus. Fix a line V_1 in $W_i - W_{i+1}$ (assuming one exists). For u acting on V/V_1 in place of V, μ gets replaced by μ^i, obtained by changing μ_i to $\mu_i - 1$ in μ. By induction on n, the flags on V/V_1 fixed by u form, say, $n'(\mu^i)$ components each of dimension, say, $d'(\mu^i)$. By 5.1 and (*) the flags of $(G/B)_u$ that project into the lines of $W_i - W_{i+1}$, a set dimension $\mu_i - 1$, split accordingly into $n'(\mu^i)$ components, each of dimension $d'(\mu^i) + (\mu_i - 1)$, which, by induction, works out to $\sum(\mu_j^2 - \mu_j)/2$, which is independent of i, and equal to $d(\lambda)$. The allowable i's here are those for which $W_i \neq W_{i+1}$, that is, $\mu_i > \mu_{i+1}$. The total number of components is thus (×) $n'(\mu) = \sum n'(\mu^i)$ summed over such i. Now recall that the diagram of μ consists of the n lattice points (i, j) with $1 \leqq j \leqq \mu_i$ and a Young tableau is what is obtained by placing $1, 2, 3, \ldots, n$ one at each of these points so that the numbers increase as i and j do. The number of such tableaux, which does not change when μ is replaced by its dual λ since this amounts to replacing the diagram by its transpose, also satisfies the recurrence (×) since the possible positions for n in the above tableaux are at the ends of i-th rows for which $\mu_i > \mu_{i+1}$. It follows that the number of components equals the number of such tableaux, for μ or for λ, which according to the theory of Young is just the degree of ρ_λ. It remains to prove (*). For any two lines in $W_i - W_{i+1}$ there exists x in $SL(W)$ mapping the first onto the second and stabilizing the filtration on W. It is enough to extend x to an element of G_u. From the Jordan normal form for u we see that there exists a direct sum decomposition $V = \sum V_i$ with $(u-1)\,V_i \subseteq V_{i-1}$ and $(u-1)^i\colon V_i \to W_i$ an isomorphism for all i. Let p_i be the inverse isomorphism. Then we define $x_i = p_i x (u-1)^i$ on V_i for all i to get the required extension of x.

Remarks 5.7. (a) One can also finish the proof by comparing ($\times$) with the "branching formula" $\rho(\mu)=\sum \rho(\mu^i)$, which expresses $\rho(\mu)$ restricted to S_{n-1} in terms of its irreducible components. We observe that ($\times$) is unchanged when expressed in terms of the dual λ of μ. (b) Directly from ($\times$) one may deduce the many properties of the n_λ's that have arisen in the representation theory of S_n, in particular the following formula for n_λ, which shows again that $n_\lambda=n_\mu$ if μ is the dual of λ, viz.: $n!/\prod(\lambda_i+\mu_j-i-j+1)$, the product over the points of the diagram of λ. The factor on the bottom is just the number of points in the "hook" of the diagram consisting of (i,j) and the points directly above it or directly to its right (see [12]). (c) In terms of the association that has been made above between the components of $(G/B)_{u(\lambda)}$ and the Young tableaux of shape λ, an alorithm can be given to yield the attitude between any two of these components, and vice versa, in particular to yield for each $w\in W$ the class $C(\lambda)$ that cuts $U\cap {}^wU$ densely. Further, analogous results can be obtained for parabolic subgroups other than B in terms of Young tableaux with possibly repeated entries. These matters, quite combinatorial in nature, will be left to a later paper. (d) J. Vargas in his thesis [18] has obtained, among other interesting results, an example in which a component of $(G/B)_u$ is singular.

We close this section with an example to show that G_u^0 does not always have a dense orbit on each component of $(G/B)_u$, even in SL_n. Let $n=8$, $\lambda=(44)$. Then $d_\lambda=4=\dim(\mathbb{P}^1)^4$, so that the flags $W_1\subset W_2\subset\cdots$ with $W_2=\ker(u-1)$, $W_4=\ker(u-1)^2$, $W_6=\ker(u-1)^3$ form an irreducible component. When G_u^0 acts, the cross-ratio of the four lines $W_1, uW_3, u^2W_5, u^3W_7$ is seen to be the only invariant. Thus there are infinitely many orbits, each of dimension 3. We also have here an example of B acting on $U\cap C(u)$, hence on U, with infinitely many orbits, for g_1B and g_2B are in the same G_u^0-orbit of $(G/B)_u$ if and only if $g_1^{-1}ug_1$ and $g_2^{-1}ug_2$ are in the same B-orbit of U.

6. Complements

In this section we prove 1.1 and indicate how the results obtained for unipotent elements can be extended to arbitrary elements.

We start with the proof of 1.1. First W is nonsingular. For let W' be the open subset defined by $gB\in U^-B$, with U^- the unipotent radical of the Borel subgroup containing T and opposite to B. Then W' equals $\{({}^gu, gB)$ with $g\in U^-, u\in U\}$, hence is isomorphic to $U^-\times U$, which is clearly nonsingular. We have used here the isomorphism between U^-B/B and U^-. Since W is covered by translates of W' (G acts on W via $x.\ (v, gB)=({}^xv, xgB)$), it also is nonsingular. Now the simple points of V are just the regular elements and they form a single class V' under G (see [16] or [13]). Thus to prove that π is a desingularization we must show that $\pi\colon \pi^{-1}V'\to V'$ is an isomorphism. Now it is bijective since each regular element fixes a unique point of G/B [16, Section 3], and it is a G-map of homogeneous G-spaces. Thus we are reduced to showing that it is also separable. If u is regular and in U, then restricting to the open set W' identified with $U\times U^-$ as above, this amounts to showing that the map $\alpha\colon U\times U^-\to V$, $\alpha(h,g)=ghg^{-1}$ is separable at

$(u, 1)$, i.e., that $d\alpha$ is injective since the two sides have the same dimension. At $(u, 1)$ we have $(d\alpha)(X, Y)=X+(u^{-1}-1)\,Y$ for $X\in L(U)$, $Y\in L(U^-)$. For simplicity we replace u by u^{-1} in what follows.

Lemma. *For u as above and $Y\in L(U^-)$ write $uY=Y'+H'+X'\in L(U^-)+L(T)+L(U)$. If $H'=0$, then $X'=0$.*

Granted this lemma, we see that if $X+(u-1)\,Y=0$ above then $(u-1)\,Y\in L(U^-)$, whence $X=0$ and $(u-1)\,Y=0$. But then $Y=0$ since the kernel of $u-1$ on $L(G)$ is in $L(B)$, because u is regular [16, Section 4]. Thus $d\alpha$ is injective as required.

Proof of Lemma. Choose a standard basis for $L(G)$ made up of root elements X_α, one for each root α, written $X_{\pm i}$ for $\pm$ the simple root α_i $(1\leqq i\leqq r)$, and the coroots $H_i=[X_i, X_{-i}]$ of the simple roots. By the separability assumption above the H_i's are linearly independent (and conversely), so that this really is a basis. Now take $u=u_r u_{r-1}\dots u_2 u_1$ in standard form [16, Section 4] with $u_i\in U_{\alpha_i}$ and $u_i\neq 1$. If $Z\in L(G)$ has weight β, then $u_i Z=Z_0+Z_1+Z_2+\cdots$ with $Z_0=Z$ and Z_j of weight $\beta+j\alpha_i$ for all j [17, p. 80]. It follows that $u_i X_\alpha\in L(U^-)$ if $\alpha<0$, $\alpha\neq-\alpha_i$, that $u_i H\in H+L(U)$ if $H\in L(T)$, and that $u_i L(U)=L(U)$. Further $u_i X_{-i}=X_{-i}+H_i-X_i$ by appropriate choice of u_i, by calculations within the group SL_2. Thus if we define $Z^0=Y$ (given in the lemma) and then $Z^i=u_i Z^{i-1}$ inductively, and write $Z^i=Y^i+H^i+X^i$ as in the lemma (so that $Z^r=uY$, $H^r=H'$ and $X^r=X'$), then (1) $H^i=H^{i-1}+c_i H_i$, and (2) $X^i=u_i X^{i-1}+(u_i-1)\,H_i-c_i X_i$, with c_i the coefficient of X_{-i} in Y^{i-1}. Assume now that $H^r=0$. By (1), $H^r=\sum c_i H_i$, whence every $c_i=0$ since the H_i's are linearly independent. By (1) (and induction) every $H^i=0$ and then by (2) every $X^i=0$. Thus $X^r=0$ and we are done.

Remark 6.1. If the H_i's above are linearly dependent with $\sum c_i H_i=0$ a nontrivial relation, then $(u-1)\,Y\in L(U)$ if $Y=\sum c_i X_{-i}$, by the above formulas, so that $d\alpha$ is not injective. Hence the separability assumption of 1.1 is also necessary.

Lemma 6.2. *Assume $t\in T$, P standard parabolic, and set $G_1=G_t^0$, $W_1=$ the Weyl group of G_1.*

(a) $(G/P)_t=G_1 WP/P$.

(b) *The irreducible components of $(G/P)_t$ are the sets $C_w=G_1 wP/P$ with $w\in W_1\backslash W/W_P$. Hence they are disjoint.*

(c) *C_w is isomorphic to $G_1/(G_1\cap {}^wP)$, a quotient by a parabolic subgroup.*

Proof. We observe first that G_1 is reductive (see, e.g., [15, Ch. II, Section 4]) and that t fixes each point of $G_1 wP/P$. Now assume that t fixes gP. Then $g^{-1}tg\in P$, so that $p^{-1}g^{-1}tgp$ is in T for some $p\in P$ and hence equals $w^{-1}tw$ for some $w\in W$. Thus $gpw^{-1}\in G_t$ and $g\in G_t wP$. Since $G_t=G_1.(G_t\cap W)$, this yields (a). Now $G_1\cap {}^wB$ is Borel in G_1. For ${}^wB$ is supported by a positive system of roots for G and this contains a positive system for G_1. Thus $G_1\cap {}^wP$ is parabolic in G_1 and $G_1/(G_1\cap {}^wP)$ is complete. Since C_w is isomorphic to the latter, by 2.1, it also is complete, hence closed in G/P, as well as irreducible. Since distinct double cosets are disjoint the C_w's must be the irreducible components of $(G/P)_t$. Finally, the correspondence between the C_w's and the elements of $W_1 \backslash W / W_P$ is a standard result. Alternatively, one can prove (b) and (c) by first showing that G_1 is transitive on each irreducible component of $(G/P)_t$, as is essentially done in [3, Section 11.18]. (The torus there can be replaced by any diagonalizable subgroup without invalidating the proof.)

Remark 6.3. If $x = tu$ is the Jordan decomposition of x in G with t in T, as may be assumed, and P is parabolic, then $(G/P)_x$ is isomorphic, by 6.2, to the disjoint union of the sets $(G_1/G_1 \cap {}^wP)_u$; if $P = B$ the situation is especially nice, for then these sets, $|W/W_1|$ in number, become isomorphic, each to $(G_1/B_1)_u$ with B_1 Borel in G_1. The extension of our earlier results to arbitrary elements of G is now clear. Here are two examples of this.

Corollary 6.4. (Grothendieck). *Let G be as in* 1.1, *V the closure of any regular class of elements, $t \in T \cap V$, and W the set of (v, gB) in $V \times G/B$ such that $g^{-1} v g \in tU$. Then $\pi: W \to V$ is a desingularization.*

The proof is a straight-forward modification of that of 1.1 in accordance with 6.2.

Theorem 6.5. *If G is very good as in* 4.6, *then* $\dim (G/B)_x = 1/2(\dim G_x - r)$ *for any $x \in G$.*

Proof. If $x = tu$ and the other notation is as above, then $\dim (G/B)_x = \dim (G_1/B_1)_u$ and $\dim G_x = \dim G_{1u}$. Thus 6.5 follows from 4.6.

References

1. Bala, P., Carter, R.W.: Classes of unipotent elements of simple algebraic groups. To appear.
2. Benard, M.: On the Schur indices of characters of the exceptional Weyl groups. Ann. of Math. **94**, 89–107 (1971)
3. Borel, A.: Linear algebraic groups. New York: Benjamin 1969
4. Borel, A., Tits, J.: Groupes réductifs. Publ. Math. I.H.E.S. **27**, 55–150 (1965)
5. Brieskorn, E.: Singular elements of semi-simple algebraic groups. Proc. Int. Congress Math., Nice (1970)
6. Dynkin, E.B.: Semisimple subalgebras of semisimple Lie algebras, Amer. Math. Soc. Translations **6**, (2) 111–244 (1957)
7. Elkington, G.B.: Centralizers of unipotent elements in semisimple algebraic groups. J. Algebra **23**, 137–163 (1972)
8. Frobenius, G., Schur, I.: Über die reellen Darstellungen der endlichen Gruppen. Sitzgsber. preuss. Akad. Wiss. 186–208 (1906)
9. Konstant, B.: The principal three-dimensional. Amer. J. Math. **81**, 973–1032 (1959)
10. Richardson, R.W.: Conjugacy classes in Lie algebras and algebraic groups. Ann. of Math. **86**, 1–15 (1967)
11. Richardson, R.W.: Conjugacy classes in parabolic subgroups of semisimple algebraic groups. Bull. London Math. Soc. **6**, 21–24 (1974)
12. Robinson, G. de B.: Representation theory of the symmetric group. University of Toronto Press 1961
13. Springer, T.A.: The unipotent variety of a semisimple group. Proc. Colloq. Alg. Geom. Tata Institute 373–391 (1969)
14. Springer, T.A.: Trigonometric sums. Green functions on finite groups and representations of Weyl groups. To appear

15. Springer, T. A., Steinberg, R.: Conjugacy classes. Lecture Notes in Math. 131 part E. Berlin-Heidelberg-New York: Springer 1970
16. Steinberg, R.: Regular elements of semisimple algebraic groups. Publ. Math. I. H. E. S. **25**, 49–80 (1965)
17. Steinberg, R.: Conjugacy classes in algebraic groups. Lecture Notes in Math. 366. Berlin-Heidelberg-New York: Springer 1974
18. Vargas, J.: Fixed points under the action of unipotent elements. Ph. D. thesis, U.C.L.A. (1976)

Received February 14, 1976

Robert Steinberg
University of California, Los Angeles
Department of Mathematics
Los Angeles, California 90024
USA

REPRINTED FROM
CONTRIBUTIONS TO ALGEBRA:
A collection of Papers Dedicated to Ellis Kolchin

ACADEMIC PRESS, INC.
NEW YORK SAN FRANCISCO LONDON

On theorems of Lie–Kolchin, Borel, and Lang

Robert Steinberg

*University of California
Los Angeles*

This somewhat expository note is concerned with two topics. The first one comes from the observation that Kolchin's proof of the Lie–Kolchin theorem (see Corollary C) as given in [5, §7, Th. 1] can be adapted to yield an extension (see Theorem A) which has the Borel subgroup conjugacy theorem and the Borel fixed point theorem as quick consequences. (See [1, §§15, 16] for the orginal proofs of these theorems.) The development is quite elementary. The notion of quotient space is not needed, and of completeness not until the very end. It is our understanding that M. Sweedler has also found an "easy proof" of the conjugacy theorem (see [4, p. 138, Notes]).

Our second concern is the theorem, apparently new, that if G is a connected algebraic group and σ an endomorphism such that G_σ is finite, then $1 - \sigma$: $G \to G$ is a finite (and dominant) morphism. This implies Lang's theorem (see [6], or [2, §16] for a different proof) that $1 - \sigma$ is surjective in case σ is the Frobenius map for some rational structure on G, and our development yields a new, especially simple, proof of this theorem in case G is affine.

As general references for the theory of algebraic groups we cite [2–4]. Here is the key to our first development.

A. *Theorem* *Let G be a connected solvable algebraic group acting linearly on a vector space V of finite dimension and stabilizing there a nonzero homogeneous closed cone C. Then G stabilizes some line in C.*

Remark This is a special case of Borel's fixed point theorem (Corollary E), but this will not emerge from our development since the completeness of projective space will not be used.

Proof We assume first that G is Abelian, and then using induction on dim V, that dim $V > 1$. Since G is Abelian, its elements have a common eigenvector, hence fix some line V_1 of V. If dim $V = 2$ then either C is all of V and hence contains V_1 or else is a finite union of lines each of which is fixed since G is connected. To reduce to this case in general we may assume $V_1 \nsubseteq C$. Consider the natural projection $p: V \to V/V_1$. We have $p(C)$ nonzero since $V_1 \nsubseteq C$. We claim that $p(C)$ is closed. To see this without using the completeness of projective space observe that p has the comorphism $k[X_2, X_3, \dots, X_n] \to k[X_1, X_2, \dots, X_n]$ if V_1 is taken as the first coordinate axis of V. Since $V_1 \nsubseteq C$, there exists a homogeneous polynomial f on V such that $f(V_1) \neq 0$ and $f(C) = 0$, whence X_1 is integral over $k[X_2, \dots, X_n]$ on C. Thus $p: C \to \overline{p(C)}$ is a finite morphism and $p(C)$ is closed. Now by induction G fixes a line V_2/V_1 of $p(C)$. We have V_2 fixed by G and $C \cap V_2 \neq 0$, hence the required reduction to dimension 2.

Assume now that G is not Abelian. By induction on dim G there exists a line $V_1 \subseteq C$ fixed by G'. The sum of all such lines yields a subspace W such that $C \cap W \neq 0$. Write $W = \sum W_i$ according to the various characters with which G' acts on the lines that it fixes. Since G normalizes G', it permutes the W_i's, hence fixes each of them since it is connected. On each W_i we thus have G' acting via scalars of determinant 1, hence trivially since G' is connected. Thus G is Abelian on W, and we are done.

B. *Corollary* *If G in Theorem A stabilizes a nonempty closed set of flags on V, then G fixes one of them.*

Proof We imbed the flag variety in the projective space $\mathbb{P}(V \otimes \wedge^2 V \otimes \wedge^3 V \otimes \cdots)$ in the usual way (see, e.g., [2, §10.3] for details) and apply Theorem A to the action of G on this space.

C. *Corollary* (Lie–Kolchin) *A connected solvable algebraic group acting linearly on a finite-dimensional vector space V fixes some flag there, i.e., can be put in upper triangular form.*

Proof We apply Corollary B to the set of all flags on V, or else we apply Theorem A and use induction on dim V.

D. *Corollary* (Borel) *Let G be a connected linear algebraic group. Then any two Borel subgroups of G are conjugate.*

Proof (Standard, adapted from [2, §11.1]) Let B be a Borel subgroup of G of maximum dimension. There exists a representation space V for G and a line L of V whose stabilizer is B (see [2, §5.1]), and by Corollary C a flag F extending L and fixed by B. Now G acts on the flag variety and the orbit through F has minimum dimension (each orbit has as dimension the codimension in G of the stabilizer of any of its elements, a solvable subgroup of G), hence is closed. If B' is another Borel subgroup, then B' has a fixed point on GF by Corollary B: $B'xF = xF$ for some $x \in G$. Then $x^{-1}B'x$ fixes F and is contained in B, as required.

E. *Corollary* (Borel fixed-point theorem) *A connected solvable linear algebraic group acting on a nonempty complete variety has a fixed point there.*

Proof Let G be the group, X the variety. By replacing X by an orbit which has minimum dimension and hence is closed and hence complete, we may assume that there is a single orbit. Fix $x \in X$ and choose a representation space V for G containing a line L whose stabilizer is G_x, the stabilizer of x in G. Let y be the point of $\mathbb{P}(V)$ corresponding to L, Y the orbit of y under G, and Z the orbit of $(x, y) \in X \times \mathbb{P}(V)$ under G. The natural projections yield bijective G-morphisms from Z to X and to Y. Now Z is closed in $X \times \mathbb{P}(V)$ since any G-orbit there projects onto X, hence has dimension $\geq \dim X = \dim Z$. Since X is complete, the projection of Z on the second factor, Y, is also closed. By Theorem A applied to Y, or rather to the cone in V over Y, we get a fixed point for G on Y (which thus reduces to a point), hence also for G on X by the above-mentioned bijectivity.

We continue with our second theme.

F. *Theorem* *Let G be a connected algebraic group and σ an endomorphism such that G_σ is finite. Then $1 - \sigma\colon x \to x\sigma(x^{-1})$ is a finite morphism.*

Throughout this part, finite morphisms are required to be dominant. Here $1 - \sigma$ is so since its nonempty fibers are the elements of G/G_σ and hence have dimension 0. The rest of the proof will be given in several steps, the first of which yields the simple proof of Lang's theorem mentioned in the introduction.

G. *Lemma* *If G is affine in Theorem F and σ is the Frobenius map for some rational structure on G, then $1 - \sigma$ is finite, and hence surjective.*

Proof Let k be the base field, $|k| = q$, and let σ_0 denote the comorphism of σ. Then $\sigma_0 f = f^q$ for every f in $A = k[G]$. For such an f and y, z in G

we have $f(yz) = \sum e_i(y)f_i(z)$ with $\{e_i\}$ (resp. $\{f_i\}$) a basis for the space of right (resp. left) translates of f, whence $f_j(yz) = \sum e_{ji}(y)f_i(z)$ with each e_{ji} in the space of left-right translates of f. With $y = (\sigma - 1)z$ this yields $f_j^q = \sigma_0 f_j = \sum (\sigma_0 - 1)e_{ji} \cdot f_i$. Thus $(\sigma_0 - 1)A[\{f_j\}]$ is finitely generated as a module over $(\sigma_0 - 1)A$, and it contains f. Hence A is integral over $(\sigma_0 - 1)A$ and $\sigma - 1$ is a finite morphism. Thus $1 - \sigma$ is also.

H. ***Lemma*** *If G is semisimple in Theorem F, then* $1 - \sigma$ *is finite.*

Proof If σ is surjective, then it is shown in [8, §10.5] that some σ^n $(n \geq 1)$ is the Frobenius map for a suitable rational structure on G. Then $1 - \sigma^n$ is finite by Lemma G and so are both of its factors in $1 - \sigma^n = (1 + \sigma + \sigma^2 + \cdots + \sigma^{n-1})(1 - \sigma)$. Assume next that G is simply connected, and also, as is permissible, that σ is not surjective. Let $K = \ker \sigma$ and let H be the product of the simple components of G that are not in K, so that $G = HK$ (direct). For $y \in H$, $z \in K$, we have $\sigma(yz) = \alpha(y)\beta(y)$ with α, β homomorphisms from H into H, K, and $(1 - \sigma)(yz) = (1 - \alpha)y \cdot z\beta(y^{-1})$. Since G_σ is finite, H_α is also, whence $1 - \alpha$ is a finite morphism, by induction on dim G. Then so is $1 - \sigma$, the composition of $(1 - \alpha) \times \mathrm{id}_K$ and the automorphism (of varieties) $yz \to yz\beta(y^{-1})$. Finally, in the general case let $p: \overline{G} \to G$ be the universal covering, $\overline{A}$ and A the affine algebras of $\overline{G}$ and G, and $\bar{\sigma}$ the lifting of σ to $\overline{G}$ (for this see, e.g., [8, §9.16]). We have $p\overline{G}_{\bar{\sigma}} \subseteq G_\sigma$, whence $\overline{G}_{\bar{\sigma}}$ is finite and $1 - \bar{\sigma}$ is a finite morphism since $\overline{G}$ is simply connected. Thus

$$\overline{A} \text{ is integral over } (1 - \bar{\sigma}_0)\overline{A}. \qquad (*)$$

But $\overline{A}$ is also integral over $p_0 A$. This follows since $\overline{A}$ is algebraic over $p_0 A$ and the latter consists of the elements of $\overline{A}$ that are fixed by $\ker p$ and annihilated by $\ker dp$ (differential of p). It also follows from Corollary M. Thus $(1 - \bar{\sigma}_0)\overline{A}$ is integral over $(1 - \bar{\sigma}_0)p_0 A = p_0(1 - \sigma_0)A$, and so is $\overline{A}$ by $(*)$, whence $p_0 A$ is also. Thus A is integral over $(1 - \sigma_0)A$, as required. We have used here the injectivity of $1 - \bar{\sigma}_0$ and of p_0.

I. ***Corollary*** *In Lemmas G and H if n is any positive integer, then* $1 + \sigma + \sigma^2 + \cdots + \sigma^{n-1}$ *is finite.*

For G_{σ^n} is finite in these cases (but not in general) and the above factorization of $1 - \sigma^n$ may be used.

J. ***Theorem*** *In Theorem F the map* $1 - \sigma$ *is surjective.*

Proof If G is Abelian, then $1 - \sigma$ is a homomorphism, whence $(1 - \sigma)G$ is closed and $1 - \sigma$ is surjective, while if G is semisimple, then $1 - \sigma$ is surjective by Lemma H. From this the general case readily follows since G has a normal series such that each term is fixed by σ and each factor group is either Abelian or semisimple (see, e.g., [8, §§10.3, 10.4] for the patching procedure).

K. ***Lemma*** *Let $f: X \to Y$ be a morphism of irreducible varieties which is dominant and has finite fibers. Then there exist affine open subsets U of X, V of Y such that $f^{-1}(V) = U$ and $f: U \to V$ is a finite morphism.*

Proof This well-known result is easily established by two localizations, the first yielding a reduction to the affine case (see, e.g., the reasoning of [7, pp. 99, 94, 95]).

L. ***Corollary*** *Let G be a connected algebraic group and $f: X \to Y$ a G-morphism of homogeneous G-spaces with finite fibers. Then f is finite.*

Proof Choose V and $U = f^{-1}V$ as in Lemma K. Then apply the conclusion of Lemma K to all pairs of sets gU, gV $(g \in G)$ to conclude that f is finite.

M. ***Corollary*** *A surjective homomorphism of connected algebraic groups with finite kernel is finite.*

N. *Completion of proof of Theorem F* We apply Corollary L with $X = G$, $Y = G$, $f = 1 - \sigma$, and G acting on X by $g \cdot x = gx$ and on Y by $g \cdot y = gy\sigma(g^{-1})$, which is permissible since Y is homogeneous by Theorem J.

O. ***Corollary*** *If $1 - \sigma$ in Theorem F is separable, then G/G_σ as a variety is isomorphic to G itself. This holds in the cases of Lemmas G and H.*

Proof The map $1 - \sigma: G \to G$ is surjective by Theorem J, has as its fibers the elements of G/G_σ, and is separable. Thus it defines a quotient by [2, §6.7]. In Lemma G $d\sigma$ is 0 and in Lemma H nilpotent (see the proof of Lemma H). Thus $d(1 - \sigma)$ is an isomorphism in both cases and $1 - \sigma$ is separable.

P. ***Corollary*** *The map $1 - \sigma$ of Theorem F is closed and open, as is the map $1 + \sigma + \sigma^2 + \cdots + \sigma^{n-1}$ of Corollary I.*

Proof The map $1 - \sigma$ is closed since it is finite (and dominant), hence, via complements, is open on sets consisting of complete fibers. Thus if U is any open subset of G, then $(1 - \sigma)U = (1 - \sigma)(UG_\sigma)$ is open. For $1 + \sigma + \sigma^2 + \cdots + \sigma^{n-1}$, the proof is similar.

References

1. A. Borel, Groupes algébriques linéaires, *Ann. of Math.* (2) **64** (1956), 20–82.
2. A. Borel, "Linear Algebraic Groups." Benjamin, New York, 1969.
3. C. Chevalley, *Sém. classification groupes Lie algébriques, Paris, 1956–1968.*

4. J. E. Humphreys, "Linear Algebraic Groups." Springer-Verlag, New York–Heidelberg–Berlin, 1975.
5. E. R. Kolchin, Algebraic matric groups and the Picard–Vessiot theory of homogeneous linear differential equations, *Ann. of Math.* (2) **49** (1948), 1–42.
6. S. Lang, Algebraic groups over finite fields, *Amer. J. Math.* **78** (1956), 555–563.
7. D. Mumford, Introduction to algebraic geometry, Harvard lecture notes.
8. R. Steinberg, Endomorphisms of linear algebraic groups, *Amer. Math. Soc. Mem.* **80** (1968).

AMS (MOS) 1970 subject classification: 20G15

Reprinted from JOURNAL OF ALGEBRA
All Rights Reserved by Academic Press, New York and London

Vol. 55, No. 2, December 1978
Printed in Belgium

Conjugacy in Semisimple Algebraic Groups

ROBERT STEINBERG

Department of Mathematics, University of California, Los Angeles, California 90024

Communicated by Walter Feit

Received February 17, 1978

1. INTRODUCTION

In [6, Problem (4)] we made the following conjecture.

Conjecture 1. If G is a semisimple algebraic group over an algebraically field k, then two elements a and a' are conjugate in G if and only if $f(a)$ and $f(a')$ are conjugate in $GL(V)$ for every irreducible rational representation (f, V) of G.

In this note we prove this result, at least when char k is 0 or sufficiently large (cf. Remark 1 at the end of this paper).

Recently Gauger [1, Theorem 1] has proved the analogous result for Lie algebras (in characteristic 0). His proof, unlike ours, is not adaptable to groups. In Section 2, below, we prove the result for Lie algebras and in Section 3 we give the modifications needed for algebraic groups. Until the remarks at the end of the paper k is assumed to be an algebraically closed field of characteristic 0.

2. LIE ALGEBRAS

In this section L is a semisimple Lie algebra over k and G is its adjoint group.

THEOREM 2. *The elements A and A' of L are conjugate under G if and only if $f(A)$ and $f(A')$ are conjugate under $GL(V)$ for every irreducible representation (f, V) of L.*

Proof. Since "only if" is obvious, we turn to the proof of "if." If A and A' are semisimple the result is known: If $f(A)$ and $f(A')$ are conjugate for every f, then $p(A) = p(A')$ for every invariant polynomial p on L, and hence A is conjugate to A' under G. The argument here is well known: By a theorem of Chevalley (see, e.g., [2, 126–129]) one may replace L by a Cartan subalgebra and G by the Weyl group in the result to be proved and then easily prove it since the latter group is finite.

348

0021-8693/78/0552–0348$02.00/0

LEMMA. *Let L_1 be a reductive Lie algebra over k as above, X a nilpotent element, and G_1 the adjoint group. Then X can be imbedded in a triple $\{X, H, Y\}$ in L_1 isomorphic to the standard basis of sl_2 (consisting of E_{12}, E_{11}–E_{22}, E_{21}, respectively, if $\{E_{ij}\}$ are the matrix units). Here H is necessarily semisimple. Further X determines H up to conjugacy under G_{1X} (the centralizer of X in G_1) and similarly with X and H interchanged. Finally, the relation between X and H is preserved by morphisms.*

For the proof of the lemma, see [3]. Now take A in Theorem 2 and decompose it into its semisimple and nilpotent parts S and X. Thus $X \in L_S$, the centralizer of S in L, a reductive algebra. Choose $H \in L_S$ as in the lemma with L_1 replaced by L_S. Similarly define S', X', H' in terms of A'. Since $f(A)$ and $f(A')$ are conjugate under $GL(V)$, so are their semisimple parts $f(S)$ and $f(S')$. Thus S and S' are conjugate under G by the semisimple case of Theorem 2. Hence we may assume from now on that $S = S'$. Thus $f(X)$ and $f(X')$ are conjugate under $GL(V)_{f(S)}$. Hence so are $f(H)$ and $f(H')$ by the lemma with L_1 replaced by $gl_{f(S)}$, and so also are $f(S + cH)$ and $f(S + cH')$ for every $c \in k$. Thus $S + cH$ and $S + cH'$ are conjugate under G by the semisimple case of Theorem 2. Replacing H' by a conjugate under G_S if necessary we may assume that H and H' belong to some Cartan subalgebra of L_S, thus that S, H, and H' belong to some Cartan sublagebra of L. Then $S + cH$ and $S + cH'$ are conjugate by an element $w = w(c)$ of the Weyl group, and a single w will work for all c since k is irreducible as an algebraic variety and the Weyl group is finite. Thus $wS = S$, so that w is represented in G_S, and $wH = H'$. Hence H and H' are conjugate under G_S, hence also X and X' by the lemma, and finally also A and A', as required.

3. ALGEBRAIC GROUPS

We now prove:

THEOREM 3. *Conjecture* 1 *holds if* char $k = 0$.

Proof. Here also the semisimple case is known, even if char $k \neq 0$: If $f(a)$ and $f(a')$ are conjugate for every f, then $\operatorname{tr} f(a) = \operatorname{tr} f(a')$ for every f, hence $g(a) = g(a')$ for every class function g, and a and a' are conjugate (see [5, Section 6]). In the general case with $a = sx$ and $a' = s'x'$ the Jordan decompositions we may assume that $s = s'$ as before. We now attach to x (which may be any unipotent element in any reductive group over k) a one-parameter subgroup $\alpha: k^* \to G_s$ as follows. Write $x = \exp X$ with X nilpotent, choose $\{X, H, Y\}$ in L_s as in the lemma, and then let $\alpha(t)$ be the image of $\operatorname{diag}(t, t^{-1})$ under the morphism of SL_2 into G whose differential maps the standard basis of sl_2 onto $\{X, H, Y\}$; another description (at least in characteristic 0): $\alpha: k^* \to G_s$ is the unique morphism whose differential maps $t(d/dt)$ onto H. Here x determines α up to $G_{s,x}$-conjugacy and similarly with x and α interchanged; and the relation between x and α is preserved by morphisms.

All of this follows easily from the lemma. It enables us to parallel the argument of Section 2. Define α' in terms of x' as α in terms of x. Since $f(a)$ and $f(a')$ are conjugate in $GL(V)_{f(s)}$, so are $f \circ \alpha$ and $f \circ \alpha'$, hence $f(\alpha(t))$ and $f(\alpha'(t))$ for every $t \in k^*$. Thus $s\alpha(t)$ and $s\alpha'(t)$, which are semisimple, are conjugate in G for every $t \in k^*$, by an element of the Weyl group once all of these elements are put in the same maximal torus, which is permissible as before, by an element independent of t, again as before. Thus α and α' are conjugate under G_s. Hence so are x and x', and finally a and a', as required.

Remarks. 1. The proof just given also works when char k is 0 or large enough, about 4 times the Coxeter number (see [4, Paragraph III, 4.3]).

2. Not all f's are needed in Theorem 3 (or in Theorem 2), just enough so that the coefficients of the characteristic polynomials of the $f(a)$'s for the f's used generate the algebra of all class functions, as our proof shows. For types A_r, C_r, the single standard f will do, for types B_r, D_r, the spin representations will do, and for any simply connected group the fundamental representations will do (see [5, Section 6]). Further details on this point, in the case of Lie algebras, may be found in [1].

References

1. M. Gauger, Conjugacy in a semisimple Lie algebra is determined by similarity under fundamental representations, *J. Algebra* **48** (1977), 382–389.
2. J. E. Humphreys, "Introduction to Lie Algebras and Representation Theory," Springer-Verlag, New York, 1970.
3. B. Kostant, The principal three-dimensional subgroup, *Amer. J. Math.* **81** (1959), 973–1032.
4. T. A. Springer and R. Steinberg, "Conjugacy Classes," pp. E1–E100, Springer Lecture Notes in Mathematics, No. 131, Springer, New York, 1970.
5. R. Steinberg, Regular elements of semisimple algebraic groups, *Publ. Math. I.H.E.S.* **25** (1965), 49–80.
6. R. Steinberg, Classes of elements of semisimple algebraic groups, *in* "Proceedings, I.C.M., Moscow, 1966."

Printed by the St Catherine Press Ltd., Tempelhof 37, Bruges, Belgium.

Proceedings of Symposia in Pure Mathematics
Volume 37, 1980

KLEINIAN SINGULARITIES AND UNIPOTENT ELEMENTS

ROBERT STEINBERG

1. Let F be a finite nontrivial subgroup of $SU_2(\mathbf{C})$. Then $\mathbf{C}^2/F$ is a surface with an isolated singularity at the origin. These singularities were classified and studied by Klein in 1872 (or so) in his work on the invariant theory of the regular solids in $\mathbf{R}^3$. However, they also arise in many other contexts which are not totally understood, including that of simple (Lie or algebraic) groups. The purpose of this talk was to discuss some of these connections, concentrating on the connection with simple groups, and then at the end to mention some areas that (we think) deserve further investigation. The main reference is [S] which contains a comprehensive treatment of the subject, including exact definitions, and an extensive bibliography.

2. The Kleinian singularities. Here is Klein's classification.

F	$\lvert F\rvert$	Degrees of invariants	Relator	Type
Cyclic	$r+1$	$2, r+1, r+1$	$X^{r+1}+YZ$	A_r
Bin. dihed.	$4(r-2)$	$4, 2(r-2), 2(r-1)$	$X^{r-1}+XY^2+Z^2$	D_r
Bin. tetra.	24	6, 8, 12	$X^4+Y^3+Z^2$	E_6
Bin. octa.	48	8, 12, 18	$X^3Y+Y^3+Z^2$	E_7
Bin. icosa.	120	12, 20, 30	$X^5+Y^3+Z^2$	E_8

Bin. = Binary, dihed. = dihedral, tetra. = tetrahedral,
octa. = octahedral, icosa. = icosahedral.

Each F here is just the stabilizer in SO_3 of a regular solid in $\mathbf{R}^3$, which may be degenerate, lifted from SO_3 to SU_2 via the spin map. In each case the polynomials on $\mathbf{C}^2$ invariant under F form an algebra with three generators X, Y, Z bound by a single relation as given in the table, and this yields $\mathbf{C}^2/F$ as a surface in $\mathbf{A}^3$ with a singularity at 0. For example, in the first case F is cyclic,

1980 *Mathematics Subject Classification*. Primary 20G20, 14J17; Secondary 20H15.

generated, in terms of the underlying coordinates, by the map $(u, v) \to (\alpha u, \alpha^{-1}v)$ with α a primitive $(r + 1)$th root of 1. Generating invariants are thus $X = uv$, $Y = u^{r+1}$, $Z = v^{r+1}$, and the relation $X^{r+1} - YZ = 0$ on them implies all others. In the second case we take $r + 1$ even in the first case, thus replace $r + 1$ by $2(r - 2)$, and adjoin $(u, v) \to (-v, u)$ to our group. The invariants that remain are generated by $X = u^2v^2$, $Y = u^{2r-4} + v^{2r-4}$, $Z = uv(u^{2r-4} - v^{2r-4})$ and these satisfy $-4X^{r-1} + XY^2 - Z^2 = 0$. The remaining cases are more complicated. The last column will be explained presently.

Our discussion of these singularities will be local. Thus only the germs of the surfaces, algebraic or analytic as the reader chooses, will be relevant, and similarly for various functions that occur along the way. As examples we note that the singularity represented by $X^2 + cX^3$ at 0 is isomorphic to that represented by X^2 under the local map $X \to X\sqrt{1 + cX}$, while that represented by $XY(X - Y)(X - cY)$ ($c \in \mathbf{C}$) depends essentially on c since any local isomorphism acts linearly mod terms of degree 5 and higher and hence fixes the cross-ratio of the 4 linear factors, which is c.

3. Other incarnations. (a) Let f: $\mathbf{C}^3 \to \mathbf{C}$, $f(0) = 0$, have a singularity at 0. Assume that in "the space of all such singularities" there is a neighborhood of f containing only finitely many singularities up to isomorphism (e.g. this condition fails for the example at the end of §2). Then f is a Kleinian singularity, i.e., is given by one of the polynomials in the table in §2.

This result of Arnold and of Siersma shows that the Kleinian singularities occupy an open part of "singularity space" and can be expected to occur frequently, as in fact they do.

(b) Let R be a nonregular two-dimensional analytic local ring in which unique factorization holds. Then R is of type E_8, i.e., isomorphic to the local ring of $\mathbf{C}\{X, Y, Z\}/(X^2 + Y^3 + Z^5)$ at 0 or its completion.

Brieskorn [$\mathbf{B_1}$] is responsible for this remarkable result.

(c) Theorem. *For an isolated normal surface singularity* (S, s_0) *the following are equivalent*: (1) *it is Kleinian*, (2) *it is rational with multiplicity* 2, (3) *it is rational with imbedding dimension* 3, (4) *the minimal resolution has as singular fibre a union of projective lines having intersection matrix minus that of the Cartan matrix of the root system of type* A_r, D_r, E_6, E_7 *or* E_8. *The correspondence between* (1) *and* (4) *is as in the table of* §2.

Du Val [**D**] and Artin [$\mathbf{A_r}$] get the credit here.

Explanation. A resolution of the singularity of S at s_0 is a map π: $S' \to S$ with S' a nonsingular surface and π an isomorphism from $\pi^{-1}(S - \{s_0\})$ to $S - \{s_0\}$. The singular fibre is $\pi^{-1}(s_0)$. "Rational" means, roughly, that S' necessarily has the same arithmetic genus as S.

Example. Following Brieskorn (see [$\mathbf{B_2}$] for the reference) we give a resolution of the singularity of type A_r. We have S: $X^{r+1} = YZ$, $s_0 = (0, 0, 0)$, all in $\mathbf{A}^3$. Let $U_1, U_2, \ldots, U_r$ be coordinates in $(\mathbf{P}^1)^r$. Define a surface S' in $\mathbf{A}^3 \times (\mathbf{P}^1)^r$ thus:

$$XU_1 = Y, \qquad XU_{i+1} = U_i \quad (1 \leqslant i \leqslant r - 1), \qquad X = U_rZ.$$

Then if π: $\mathbf{A}^3 \times (\mathbf{P}^1)^r \to \mathbf{A}^3$ is the natural projection, $\pi|_{S'}$: $S' \to S$ is a resolution of $(S, 0)$. The relevant points, that S' is nonsingular and π an isomorphism on $S' - \pi^{-1}(0)$, are easily checked (the first, e.g., since the Jacobian of the equations for S' has rank $r + 1$ everywhere). To get a typical point $(0, 0, 0; U_1, U_2, \ldots, U_r)$ of the singular fibre, we let j $(1 \leqslant j \leqslant r)$ be the smallest index for which $U_j \neq 0$. From the equations for S' it readily follows that $U_i = 0$ for $i < j$ and $U_i = \infty$ for $i > j$; and conversely. Thus the singular fibre is the union of r projective lines L_j $(1 \leqslant j \leqslant r)$ and these intersect in the A_r pattern as shown.

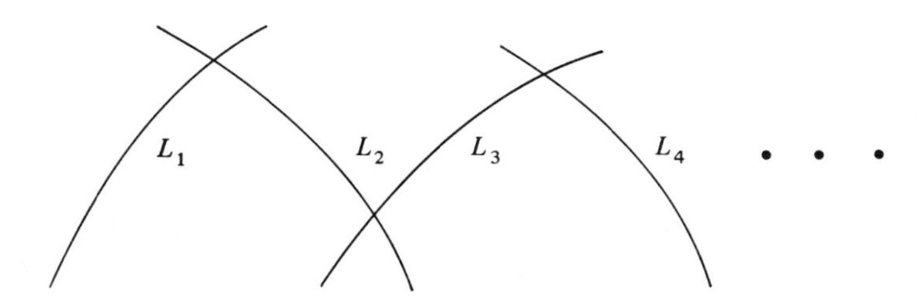

Brieskorn has also given ad hoc resolutions of the other Kleinian singularities, that of type E_8 being quite complicated indeed.

4. Universal deformation (or unfolding). For a surface singularity (S, s_0) this is (roughly) a map

$$\pi: (B, b_0) \to (T, t_0), \qquad \pi(b_0) = t_0$$

with B nonsingular and $(\pi^{-1}(t_0), b_0)$ isomorphic to (S, s_0). The fibres represent the stages that the original singularity (in the fibre over t_0) goes through as it gets deformed (unfolded). For universality it is required that every deformation factor uniquely through this one.

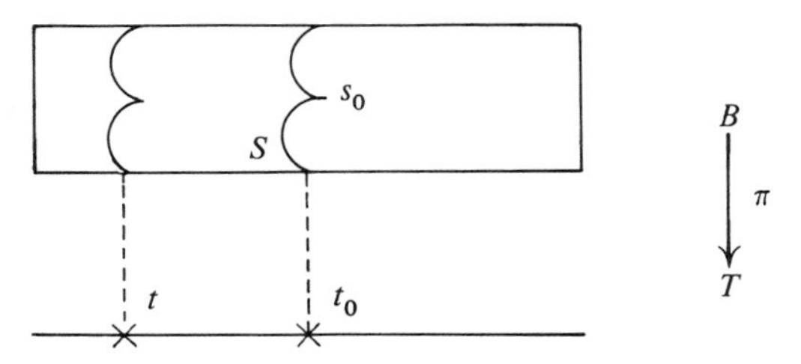

EXAMPLE. For type A_r the parameter space is $\mathbf{A}^r$ with coordinates T_0, $T_1, \ldots, T_{r-1}$, B is the zero set of $YZ + X^{r+1} + T_{r-1}X^{r-1} + T_{r-2}X^{r-2} + \cdots + T_0$, and π is projection onto the space of the first three coordinates X, Y, Z. The rule here, which applies quite generally, is to add on to the original polynomial $f(X, Y, Z)$ (as in the table of §2) the above linear combination formed from a set of polynomials $((X^{r-1}, X^{r-2}, \ldots, X, 1)$ for type $A_r)$ which projects into a basis for $\mathbf{C}[X, Y, Z]$ modulo the ideal generated by $\partial f/\partial X$, $\partial f/\partial Y$, $\partial f/\partial Z$.

5. Resolution (desingularization) of the unipotent variety. From now on G is a simple, simply connected Lie or algebraic group over $\mathbf{C}$ of type A_r, D_r, E_6, E_7 or E_8. V is the variety of unipotent elements of G, B is a Borel subgroup and G/B is the flag manifold.

THEOREM (a). *Let $\underline{W}$ be the subset of $V \times G/B$ formed by all (x, gB) such that $xgB = gB$. Then the natural projection π_1: $\underline{W} \to V$ is a resolution of V.*

According to this remarkable result of Springer [**Sp**] the singularity of a point x in V is thus measured by the variety of flags $(G/B)_x$ that if fixes, or, equivalently, by the variety of Borel subgroups in which it is contained. In accordance with (a) we have:

THEOREM (b) (SEE [**St$_1$**]). *The elements of V each fixing a single flag form a single conjugacy class V_{reg} which is dense and open in V. On $\pi_1^{-1}(V_{\text{reg}})$ the map π_1 is an isomorphism.*

DEFINITION. For each simple root α let P_α be the corresponding parabolic subgroup so that P_α/B is a projective line in G/B. A *Dynkin curve* is a union of translates L_α of such lines, one for each α, so that L_α and L_β intersect just when α and β are joined in the Dynkin diagram of G.

THEOREM (c). (1) *In $V - V_{\text{reg}}$ there exists a single dense open conjugacy class V_{subreg}. It has codimension* 2.

(2) *Dynkin curves exist and all are congruent under G.*

(3) *An element x of V is in V_{subreg} if and only if the fibre $(G/B)_x$ of Theorem* (a) *is a Dynkin curve.*

This result is due to Tits and the author. For a proof see [**St$_1$**].

6. The main theorems. In addition to the notation of §5 let T be a maximal torus of B, W the Weyl group, and T/W the corresponding affine space $\mathbf{A}^r$, coordinatized by the fundamental characters. Here r is the rank of G.

THEOREM. *Let x be a subregular unipotent element of G and S a transversal* (*of dimension $r + 2$*) *to V_{subreg} in G passing through x.*

(1) *x is an isolated Kleinian singularity of $V \cap S$ of the type of* (*the Dynkin diagram of*) *G.*

(2) *The restriction of π_1 of Theorem* 5(a) *to $\pi_1^{-1}(V \cap S)$ is a minimal resolution of $(V \cap S, x)$.*

(3) *The natural map $(S, V \cap S) \to^{\pi_2} (T/W, 1)$, given by y (in S) $\to y_s$ (semisimple part) $\to$ conjugate in T, is a universal deformation of $(V \cap S, 1)$.*

This result was originally conjectured by Grothendieck. The proof is due to Brieskorn (see [**B$_2$**]) in the analytic case, to Slodowy [**S**] in the algebraic case, with a significant contribution by H. Esnault.

Thus as t changes from 1 to a nearby value the singularity in the fibre above t changes from Kleinian of type G to Kleinian of type $Z_G(t)$ (one singular point for each simple component of $Z_G(t)$), a "simpler" singularity. If the map π_1 of (2) is extended so that G takes the place of V (see the start of §5), then the combination $\cdots \to^{\pi_1} S \to^{\pi_2} \cdots$ yields in effect a simultaneous resolution of the singularities of the whole deformation (Grothendieck, see [**B$_2$**]).

A FINAL EXAMPLE. We take $G = \mathrm{SL}_{r+1}$. The standard regular element (unipotent) is the identity with 1's filled in just above the diagonal. In the standard subregular x the (12)-entry is then replaced by 0. SL_n, and hence x, acts on the underlying space V_{r+1}, hence also on the space of flags: $\{V_1 \subset V_2 \subset \cdots \subset V_{r+1}, \dim V_i = i\}$. As is easily seen the flags fixed by x form r projective lines L_j $(1 \leqslant j \leqslant r)$ with L_j all flags of the form:

$$V_i = \begin{cases} \langle e_1, e_2, \ldots, e_i \rangle & \text{if } i < j, \\ \langle e_1, e_2, \ldots, e_{i-1}, ae_i + be_0 \rangle \quad (a, b \in \mathbf{C}) & \text{if } i = j, \\ \langle e_0, e_1, \ldots, e_i \rangle & \text{if } i > j, \end{cases}$$

in terms of a basis $e_0, e_1, \ldots, e_r$ of V_{r+1}. This is a Dynkin curve of type A_r and is in close analogy with the ad hoc desingularization given in §3. To exhibit a universal deformation (following Arnold [A]) we switch to the nilpotent element $N = x - 1$ in the Lie algebra sl_{r+1} and take as our cross-section S the set shown below, analogous to the general rational single Jordan block that it would be if N were regular. For the characteristic polynomial $f(t)$ we have in terms of that, $f_1(t)$, of the lower right hand block $f(t) = (t - X_1)f_1(t) - YZ$, and on setting $t = X_1$, $f(X_1) + YZ = 0$. This is just the deformation given in §4 for the singularity $X_1^{r+1} + YZ = 0$ of type A_r with $T_0, T_1, \ldots, T_{r-1}$ the coefficients of f, i.e., the fundamental characters, i.e., the coordinates on $\underline{t}/W$ as mentioned earlier.

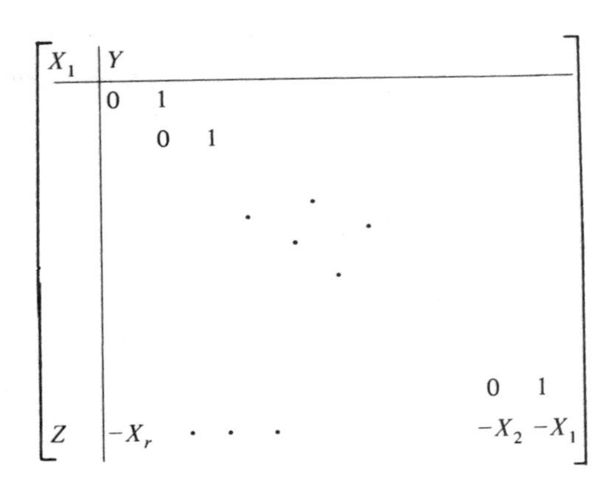

7. Remarks and problems. At several places in the known proofs of the theorems above, the classification is used (although large parts of the development are quite satisfactory). In particular, no one seems to have found a really direct connection between the finite group F used at the start and the algebraic group G used at the end. Nor is there yet an explanation for J. McKay's astonishing observation that the character table for F is an eigenmatrix for the extended Cartan matrix of G. Thus our first problem is:

(1) Repair this situation.

(2) Explain why the Kleinian singularities and the Arnold-Siersma singularities (see 3(a)) are the same.

(3) Study the fibres $(G/B)_x$ of the resolution 5(a) of V. Some nice properties are known: each is connected, with all irreducible components of the same dimension, which is known, as is the number of components. Further Springer has given a realization of the Weyl group representations in these fibres and I a connection between Weyl group elements and pairs of components. However, there is a lot about these fibres that is not known. This problem is included in the following one which I first posed about 13 years ago and which has, I like to think, led to some of the above developments.

(4) Study the unipotent variety thoroughly.

REFERENCES

[A] V. Arnold, *Normal forms for functions near degenerate critical points, the Weyl groups A_k, D_k and E_k and Lagrangian singularities*, Functional Anal. Appl. **6** (1972), 254–272.

[A_r] M. Artin, *On isolated rational singularities of surfaces*, Amer. J. Math. **88** (1966), 129–136.

[B_1] E. Brieskorn, *Rationale Singularitäten komplexer Flächen*, Invent. Math. **4** (1967/68), 336–358. 358.

[B_2] ______, *Singular elements of semi-simple algebraic groups*, Internat. Congr. Math. Nice (1970), vol. 2, pp. 279–284.

[D] P. Du Val, *On isolated singularities of surfaces which do not affect the conditions of adjunctions*, Proc. Cambridge Philos. Soc. **30** (1933/34), 453–465; 483–491.

[S] P. Slodowy, *Simple singularities and simple algebraic groups*, Lecture Notes in Math., vol. 815, Springer-Verlag, New York (1980),

[Sp] T. Springer, *The unipotent variety of a semisimple group*, Proc. Bombay Colloq. on Alg. Geom., Oxford Press, London, 1969, pp. 373–391.

[St_1] R. Steinberg, *Lectures on conjugacy classes*, Lecture Notes in Math., vol. 366, Springer-Verlag, Berlin and New York, 1974.

[St_2] ______, *Desingularization of the unipotent variety*, Invent. Math. **36** (1976), 209–224.

QUEEN MARY COLLEGE, ENGLAND

Reprinted from JOURNAL OF ALGEBRA
All Rights Reserved by Academic Press, New York and London

Vol. 71, No. 2, August 1981
Printed in Belgium

Generators, Relations and Coverings of Algebraic Groups, II

ROBERT STEINBERG*

Department of Mathematics, University of California, Los Angeles, California 90024

Communicated by Walter Feit

Received November 12, 1980

1. INTRODUCTION

Our purpose here is to fill the two remaining gaps in the determination of the Schur multipliers of the finite simple Chevalley groups and their twisted analogues by providing 1.1, 5.1 and 6.1 (see also 6.2) below.

THEOREM 1.1. *Let G be the group of rational points of a simply connected simple affine algebraic group defined and split over a finite field of q elements. Then the Schur multiplier M of G is trivial with the following exceptions*: (a) *If G is* $A_1(4)$, $A_2(2)$, $A_3(2)$, $C_3(2)$, $F_4(2)$ *or* $G_2(4)$ *then M is* $\mathbb{Z}_2$ (*cyclic of order* 2); *if G is* $A_1(9)$, $B_3(3)$ *or* $G_2(3)$ *it is* $\mathbb{Z}_3$. (b) *If G is* $A_2(4)$ (resp. $D_4(2)$) *then M is the direct product of* $\mathbb{Z}_4$ (*resp.* $\mathbb{Z}_2$) *with itself.*

This is proved in Schur [11, p. 119] for $G = SL_2$ and in [13, 4.1] for the oher types with $q \geqslant 5$. Most of the exceptional cases of 1.1 have been treated by other authors (see Sects. 2, 3 below). In Sects. 2, 3 we treat the case $q \leqslant 4$, limit M as indicated, and construct an appropriate covering group when this is easy, using mainly the spin covering which is developed in Section 7. In this regard we call attention to a forthcoming paper of Robert Griess, "Schur multipliers of the known finite simple groups III," in which such a construction is given for each of the simple groups (including the 26 sporadic groups) for which this has not been done here or elsewhere in the literature. The main idea in this part, as in [13], is to attempt to lift a certain presentation of G (see Section 2 below) to an arbitrary central extension.

In the second part of this paper, Sections 4–6, we obtain similar presentations for the finite quasisplit groups $SU_{2n+1}(k/k_0)$. From this, one can determine the Schur multipliers of these groups, and in fact Griess [9] has

* Part of this paper was written in 1979–1980 while the author was a guest at Queen Mary College. He acknowledges with pleasure the extreme hospitality shown him during this visit.

0021-8693/81/080527–17$02.00/0

already done this, among other things. His results together with those of Grover [10] and Alperin and Gorenstein [2] show that if G is twisted in 1.1 then M is trivial except for the following cases: $M({}^2A_3(2))=\mathbb{Z}_2$, $M({}^2A_3(3))=\mathbb{Z}_3\times\mathbb{Z}_3$, $M({}^2A_5(2))=\mathbb{Z}_2\times\mathbb{Z}_2$, $M({}^2B_2(8))=\mathbb{Z}_2\times\mathbb{Z}_2$, $M({}^2E_6(2))=\mathbb{Z}_2\times\mathbb{Z}_2$. Other papers that are relevant to this part are Abe [1] and Deodhar [7], where fragments of our calculations may be found.

Our results were obtained more than 10 years ago (that the number of exceptions in 1.1 is finite is mentioned in [13]) but remained unpublished because of a misreading of the literature on our part. We are indebted to Robert Griess for setting us straight on this point and for helping us otherwise in the preparation of this paper.

2. Start of the Proof of Theorem 1.1

For $G=SL_2$ the result is due to Schur [11, p. 119] and is also proved in [13]. The exceptional cases are covered by the sequences $SL_2(4)\sim PSL_2(5)\leftarrow SL_2(5)$ and $SL_2(9)\rightarrow PSL_2(9)\sim A_6$ (alternating group) which has a 6-fold covering group [12]. (Here and elsewhere "$\sim$" denotes an isomorphism.) Hence for the rest of the proof we may assume that the rank is at least 2. The group G is as in 1.1, thus is a universal Chevalley group in the language of [14] (this is a departure from [13], where G denotes the adjoint group), H a split Cartan subgroup, N, $W=N/H$ (Weyl group), $R=\{r,s,t,...\}$ the root system, $x_r(u)$ as in [13], as are $X_r=\{x_r(u)\mid u\in k\}$, $w_r(u)=x_r(u)\,x_{-r}(-u^{-1})\,x_r(u)$ $(u\neq 0)$ and $h_r=w_r(u)\,w_r(-1)$. The X_r's taken together generate G and they satisfy the following relations, which in case k is finite as is now being assumed form a complete set [13, 3.3].

$$\text{(A)}\qquad x_r(u)\,x_r(v)=x_r(u+v),$$

$$\text{(B)}\qquad (x_r(u),x_s(v))=\prod x_{ir+js}(C_{ijrs}u^iv^j).$$

Here u and v run over k and r and $s\neq -r$ over R. The term on the left of (B) is the commutator of the two factors and on the right is the product over all pairs of positive integers (i,j) taken in any order, the C_{ijrs} being certain integers which depend on the order chosen but not on u or v and are known, at least up to sign, in the various cases (see, e.g., [6], [13] or [14]). Now let $\pi\colon G_1\rightarrow G$ be a universal central extension of G (so that $\ker\pi$ is $M(G)$). As stated above our object is to lift relations (A) and (B) to G_1 if this is possible and to measure the obstruction if not. Since this is done for $q\geqslant 5$ in [13] we may assume from now on that $q\leqslant 4$. To compensate for the smallness of k, we shall have to be able to produce roots with suitable geometric properties. The basic result towards this end is as follows.

LEMMA 2.1. *Let R be a root system and S a subsystem closed under real (or rational) linear combinations. Then any basis, i.e., simple system, for S extends to one for R.*

Proof. Let A and B be the real spaces generated by R and by S, and C the orthogonal complement of B in A. Order C arbitrarily, B compatibly with a given basis for S, and then A lexicographically so that $b+c>0$ if $c>0$ or if $c=0$ and $b>0$. If now s is simple in S and $s=r_1+r_2$, a sum of positive roots in R, with $r_1=b_1+c_1$ and $r_2=b_2+c_2$ in $B+C$, then $s=b_1+b_2$ and $0=c_1+c_2$. Thus $c_1=c_2=0$, whence r_1 and r_2 are in S, and s remains simple in R, as required.

We turn now to the proof of 1.1 proper, considering in the rest of this section groups for which there is just one root length.

2.2. We assume first that $q>2$.

(1) In the present case (R simple, one root length) any two roots are contained in a subsystem of type A_1, A_2 or A_3. To see this, let B be a basis for the system generated over the reals by the given roots and A an extension to R as given by 2.1. If B is not of type A_1 or A_2 then it is of type A_1^2 and we may adjoin to B the root which is the sum of the intermediate roots to those of B in the Dynkin diagram of A (which is connected since R is simple) to get a basis of type A_3. Henceforth we shall omit proofs of this nature.

Now we define $f: X_r \to G_1$ as in [13, Sect. 9]: choose $h_r \in H$ so that $r(h_r) \neq 1$, i.e., $x \to (h_r, x)$ is a bijection on X_r, and then define $f(x)$ so that $f((h_r, x)) = (f(h_r), f(x))$ with $f(h_r)$ any lift of h_r. This is not circular since the right side is independent of f.

(2) $\pi^{-1}H$ transforms $f(X_r)$ into itself. More generally, if $n \in N$ corresponds to $w \in W$ then $\pi^{-1}n$ transforms $f(X_r)$ into $f(X_{wr})$. The proof is like that in [13, Sect. 9, step (1)].

(3) Each $f(X_r)$ is Abelian. Write $r=s+t$, the sum of two other roots and $x=(y,z) \in X_r \cap (X_s, X_t)$ accordingly. If $x' \in X_r$, then x' commutes with y and z since $r+s$ and $r+t$ are not roots. Thus $f(x')$ commutes with $f(y)$ and $f(z)$ up to central elements of G_1, thus with $(f(y), f(z))$ exactly, thus also with $f(x)$ which differs from this by a central element of G_1.

(4) The relations (A) hold (for the elements $f(X_r)$). The proof is as in [13, Sect. 9, step (3)].

(5) If r and s are independent roots there exists $h \in H$ such that $r(h)=1$ and $s(h) \neq 1$ (in k) unless $G=A_2(4)$ or $q=3$ and r and s are orthogonal. For this assume first that r and s are not orthogonal. If the rank is at least three we choose a root t with $\langle r,t \rangle = 0$ (Cartan integer) and $\langle s,t \rangle = 1$ (this is possible in A_3, hence in general by (1)) and set $h=h_t(c)$ with $c \neq 1$; if $G=A_2(3)$ set $h=h_r(c)$. If r and s are orthogonal and $q=4$ then $h=h_s(c)$ works.

(6) If $G \neq A_2(4)$ then relations (B) hold. We use induction on the number of roots involved. Let $c(t,u)$ denote the ratio of the two sides of (B) when the x's are replaced by the $f(x)$'s. By (4) and our inductive assumption it easily follows that c is "biadditive." In the favorable cases of (5) we conjugate our relation by $f(h)$, any lift of h. By (5) and the centrality of c we get $c(t, u) = c(t, us(h))$ whence $c(t, u(s(h) - 1)) = 1$ and c is identically 1. If $q = 3$ and r and s are orthogonal and also t and u are both 1, which we may assume since the additive group of k is cyclic, there exists $n \in N$ transforming $x = x_r(1)$ and $y = x_s(1)$ into each other (easily checked in SL_4, see (1).) Thus $(f(x), f(y)) = f(n)(f(x), f(y))\, f(n)^{-1} = (f(y), f(x))$ and $c^2 = 1$. But also $c^3 = 1$ since c is additive. Thus $c = 1$, as required.

(7) Now let $G = A_2(4)$. Everything is as before through (4) so that only relations (B) need be considered. In the 60° case ($r + s$ is not a root) the obstruction is biadditive, hence yields an elementary 2-group. In the 120° case if we write (B) (lifted) as $f(x_r(t))\, f(x_s(u))\, f(x_r(t))^{-1} = c_{rs}(t, u)\, f(x_{r+s}(tu))\, f(x_s(u))$, the same with u replaced by v, and then multiply, we get $(*)$ $c_{rs}(t, u + v) = c_{rs}(t, u)\, c_{rs}(t, v)\, c_{r,r+s}(u, tv)$. This shows (take $v = u$) that $c_{rs}^4 = 1$ and that the 60° obstructions are expressible in terms of the 120° ones. If we make H act and use (2) we get $c_{r,r+s}(t, u) = c_{r,r+s}(tv, uv^{-1})$ and $(**)$ $c_{rs}(t, u) = c_{rs}(tv, uv)$ for all $v \in k^*$, so that in particular $c_{r,r+s}$ is symmetric. Let a be a generator of k^*. Take the product of $(*)$ with $t = a$ and (u, v) running over all pairs of distinct elements of k^*. In view of the above remarks we get $c_{rs}(a, 1)\, c_{rs}(a, a)\, c_{rs}(a, a^2)$ equal to the same item with s replaced by $r + s$, which is 1 by additivity in the last position. Since also the Weyl group is transitive on 120° pairs of roots and $(**)$ holds, the obstruction to lifting all of relations (B), hence to lifting G itself, is reduced to a potential (4, 4) group generated by $c_{rs}(a, 1)$ and $c_{rs}(a, a^2)$. That this obstruction is real has been shown by Burgoyne and by Thompson, unpublished. One way to do this is to construct the central extension over B (the Borel subgroup corresponding to the positive roots) first and then to check certain compatibility conditions for the action of the Weyl group, sufficient for the extension of the construction to the whole group. (See [9, pp. 366–371], where this method is applied to the group $G_2(3)$, mentioned in Sect. 3.5 below.)

2.3. Now we assume that $q = 2$ (and still that there is just one root length). We set $x_r = x_r(1)$ and define $y_r = f(x_r)$ thus. Write $r = s + t$, a sum of two other roots, so that $x_r = (x_s, x_t)$ (one of the relations of (B)) and then $y_r = (y_s, y_t)$.

(1) If the rank is $\geqslant 3$ and $\langle r, s\rangle = 1$ (i.e., r, s make an angle of 60°), then y_r and y_s commute. To see this, write $s = t + u$ with r, t, u independent. Then $r + t$ and $r + u$ are not roots: if, e.g., $r + t$ were a root, we would have $\langle r, t\rangle < 0$, whence $\langle r, u\rangle > 1$, which is impossible since $r \neq u$. Thus y_r

commutes with y_t and y_u up to central elements and thus with y_s exactly (see step (3) of 2.2).

(2) If the rank is $\geqslant 3$, then (A) holds (for the y_r's), i.e., $y_r^2 = 1$. Write $y_r = (y_s, y_t)$ as at the start. Then y_r commutes with y_s and y_t by (1). Thus $y_r^2 = (y_s, y_t^2) = 1$ since y_t^2 is central.

(3) Assume that the rank is $\geqslant 4$ and that the type is A_n or E_n. If r and s are orthogonal roots, then y_r and y_s commute. By 2.2 (1) and our present assumptions we may imbed r and s into a root system of type A_3 and the latter into one of type A_4. There we write $s = t + u$ with all roots orthogonal to r. Then y_r commutes with y_s as in (1).

(4) Assume as in (3). Then the y_r's satisfy conditions (A) and (B). By (1), (2) and (3) it remains only to show that the normalization of y_r is independent of the choice of s and t. Let $r = u + v$ be another such choice. If $u = t$ and $v = s$, the result holds by (2). Otherwise u is linearly independent from s and t, and from $\langle s + t, u\rangle = \langle u + v, v\rangle = 1$, we get one of $\langle s, u\rangle$, $\langle t, u\rangle$, say, the first, equal to 1, the other equal to 0, and vice versa if u is replaced by v. Thus $s - u = v - t$ is a root, call it p, and it is orthogonal to $s + t$. If we conjugate (y_s, y_t) by $w_p \in W$, represented in N by an element of $X_p X_{-p} X_p$, we leave it unchanged by (2) and transform it into (y_u, y_v) since $w_p s = u$ and $w_p t = v$, whence (4).

(5) If the type is A_2 or A_3, then π is a double covering. For A_2 we get in (1) an obstruction $j = (y_r, y_s)$, $j^2 = 1$, independent of r and s (because of the action of W), and then each $x_r^2 = j$ in (2). We can realize the covering concretely via $SL_3(2) \sim PSL_2(7) \leftarrow SL_2(7)$. For type A_3 we have (1) and (2), but in (3) an obstruction $j = (y_r, y_s)$, $j^2 = 1$, independent of r and s, and in (4) an obstruction expressible in terms of j. Here we get our realization via $SL_4(2) \sim A_8$ (alternating group) $\subset SO_8 \leftarrow \mathrm{Spin}_8$ (See 7.9 below.)

(6) The groups $D_n(2)$ $(n \geqslant 4)$ remains to be considered. As above (1) and (2) hold; thus the crux of the matter is (3). In terms of coordinates the roots may be written $\pm v_i \pm v_j$ $(1 \leqslant i < j \leqslant n)$, more simply $\pm i \pm j$. Assume $n > 4$ first. Then for r and s of the form 12 and 34 we write $34 = 35 + (4 - 5)$ to get (3). For the only other possible form 12, $1 - 2$ we set $j = (y_r, y_s)$, which is independent of the indices involved. We conjugate y_{12} by $w_{1-2} w_{34} w_{3-4}$. By our result for pairs of the first form and the fact that w_{1-2} may be chosen in $X_{1-2} X_{2-1} X_{1-2}$ we get $j y_{12}$. If we conjugate in the same way (y_{13}, y_{-32}), equal to y_{12} up to a central element, we get (y_{2-3}, y_{13}), which is (y_{13}, y_{2-3}) by (2). Thus $j = 1$ and we have (3) and hence (4) in this case. If $n = 4$, there are three orbits of orthogonal pairs of roots, represented by $(12, 1 - 2)$, $(12, 3 - 4)$ and $(12, 34)$. These yield obstructions j, k and l in (3), each of square 1. We claim that $jkl = 1$. In fact the calculation just made shows that this is so. Further the obstructions in

(4) are expressible in terms of j, k and l. We thus have a potential (2, 2) covering which can be realized via $D_4(2) \leftarrow SW(E_8)$ (special Weyl group) $\subset SO_8 \leftarrow \mathrm{Spin}_8$. Here the first arrow is a double covering realized by reduction mod 2.

The proof of 1.1 in the single root case is now complete.

3. The Proof Concluded

3.1. We turn next to the groups $C_n(q)$. We assume first that $n \geqslant 2$ and $q = 4$. We normalize $y_r(u) \equiv f(x_r(u))$ as in 2.2 and see as there that

(1) $\pi^{-1}H$ normalizes $\{y_r(u)\}$ for each r.

(2) If $r \neq s$ and $x_r(u)$ and $x_s(v)$ commute, then so do $y_r(u)$ and $y_s(v)$. If r and s are orthogonal we remove any potential obstruction by conjugation by $h_r(c)$ $(c \neq 1)$ as in 2.2 (6). If r and s make an acute angle, there exists a long root t orthogonal to one of them and not to the other: if r is long and s short, then $t = r - 2s$ will do, while if both are short we imbed $\{r, -s\}$ in a basis of type C_3 (see 2.1) and use the remaining basis vector. In both cases $h_t(c)$ works as in the first case.

(3) Each $\{y_r(u)\}$ is Abelian. Consider a relation (B) of the form $(y_s, y_t) = cy_{s+t}y_{s+2t}$ with c central. Since y'_{s+t} commutes with y_s, y_t and y_{s+2t} by (2), it also commutes with y_{s+t}; and similarly for y'_{s+2t}. Since every y_r occurs, up to a central factor, in the role of y_{s+t} or y_{s+2t} above, we have (3).

(4) Relations (A) hold. By (1) and (3) as in 2.2.

(5) Relations (B) hold. By (2) we are left with the case in which r and s make an obtuse angle. By (2) again the obstruction c is biadditive, hence may be removed as in the second case of (2).

3.2. Next come the groups $C_n(3)$ $(n \geqslant 2)$. For $n = 2$ the central extension $Sp_4(3) \to SU_4(2)$ shows that $\ker \pi$ is both a 2-group and a 3-group (by analogues of 6.2 below which hold for these groups and are proved in the same way (see [13, 4.5]), hence is trivial. Thus we may assume that $n \geqslant 3$. For r short we normalize $y_r(u)$ as in 2.2, while for r long, in which case H is too small for this, we write $r = s + t$ with s and t short so that $x_r(u) = (x_s(u), x_t(\varepsilon))$ for some sign ε and then set $y_r(u) = (y_s(u), y_t(\varepsilon))$.

(1) Ir $r + s$ is not a root then relations (B) hold. If r and s are both short, the relation is within a subgroup of type $A_3(3)$, a case treated in 2.2. If both are long we can write s as a sum of two (short) roots orthogonal to r and proceed as in 2.3 (1), while if r is long and s short we can choose a long root t orthogonal to r and not to s and then use $h_t(-1)$ as in 2.2 (6).

(2) Relations (A) hold. If r is short we go to $A_3(3)$ as before. If r is long, then $(y_s(u), y_t(\varepsilon))$ (as above) commutes with both factors by (1). Thus it is "additive" in the first factor, thus also in u since central factors inside commutators are immaterial.

(3) Relations (B) hold. By (1) we may assume that $r+s$ is a root, say, t. If all roots are short we are inside $A_2(3)$ and so done. If r and s are short and t long, assume first that $t=r+s$ is the decomposition of t used in the normalization of the y_t's, so that (B) holds for the choice u, ε of the parameters. But then it holds for $\pm u$, $\pm\varepsilon$ since the commutator on the left commutes with both of its factors by (1), i.e., for all values of the parameters since $q=3$. If $t=r'+s'$ is another decomposition of t of the present type, there exists a root p orthogonal to t such that $w_p r=r'$ and $w_p s=s'$. Transforming $y_t=(y_r, y_s)$ by w_p represented in the form $y_p y_{-p} y'_p$ in G_1, we get y_t on the left by (1) and some $(y_{r'}, y_{s'})$ on the right, so that (B) holds in this case. Finally, if r is long and s short, we get in (B) an obstruction which is biadditive by what has already been proved and so may be removed as in the last case of (1).

3.3. For type C_n only the groups $C_n(2)$ ($n\geqslant 3$) remain to be treated. If r is short we define $y_r=y_r(1)=(y_s, y_t)$ with $r=s+t$ a sum of short roots. If r is long we write $r=s+2t$ and then define $y_r=y_{s+2t}$ by $(y_s, y_t)=y_{s+t}y_{s+2t}$. We assume first that $n\geqslant 5$.

(1) The relations (A) and (B) involving only short roots hold. This is because the short roots support a subgroup of G isomorphic to $D_n(2)$, a case treated in 2.3.

(2) y_r and y_s commute if r is long and $r+s$ is not a root. In terms of the usual coordinates for roots for C_n, the possibilities, up to the action of the Weyl group, are: $r=2v_1$ and $s=v_1-v_2$, v_2-v_3 or $2v_2$. In the first case we write $v_1-v_2=(v_1-v_3)+(v_3-v_2)$ and proceed as in 2.3 (1). In the second case there is a potential obstruction $j=(y_r, y_s)$, $j^2=1$, invariant under W. We apply $w_r w_s$ to $(y_a, y_b)=cy_{a+b}$ with $a=v_2-v_1$, $b=v_1+v_3$ and c central. The left side remains unchanged by (1) since the two terms there get interchanged up to central factors, while the right side gets multiplied by $j^3=j$ since w_s has no effect by (1) and w_r may be chosen as $y_r y_{-r} y'_r$. Thus $j=1$. In the third case we set $s=2v_3+2(v_2-v_3)=a+2b$, say. Then in $(y_a, y_b)=cy_{a+b}\,y_{a+2b}$ the left side commutes with y_r as does the second term on the right by the second case just treated; thus so does the final term, as required.

(3) If r is long relations (A) hold. We square the equation used to define y_r. By (1) and (2) we get $(y_s^2, y_t)=1$ on the left and $y_{s+2t}^2=y_r^2$ on the right.

(4) Relations (B) hold. Because of (1) and (2) this amounts to

showing that for r long our normalization of y_r is indepenent of s and t. If s' and t' form another possibility then up to the action of W there is just one case:

$$r = 2v_1 = 2v_2 + 2(v_1 - v_2) = 2v_3 + 2(v_1 - v_3).$$

If we apply the reflection corresponding to $v_1 - v_2$ to the first normalization we get the second one with y_r intact by (1) and (2), whence our result.

(5) We now assume that n is 3 or 4. We have arranged the steps above so that they all apply modulo the first, i.e., so that all obstructions are expressible in terms of those found for $D_n(2)$ in 2.3. If $n = 4$ the obstructions can be removed. For since the orbits through (12, 3–4) and (12, 34) near the end of 2.3 fuse, $k = l$, whence $j = 1$; and conjugating $(y_r, y_s) = cy_{r+s}y_{r+2s}$ with $r = 2v_4$ and $s = v_3 - v_4$ by y_t with $t = v_1 + v_2$ we get $l = 1$. This requires knowing the second case of (2) which can be proved directly for all $n \geqslant 4$ by writing $v_2 - v_3 = (v_2 - v_4) + (v_4 - v_3)$ and proceeding as in 2.3 (1). If $n = 3$, the apparent obstruction of order 2 is actual as is shown by the sequence $C_3(2) \sim SW(E_7) \subset SO_7 \leftarrow \mathrm{Spin}_7$.

(6) The case $n = 2$ is not covered by our theorem, but we observe that $C_2(2) \sim S_6$ (symmetric group), so that there is a double covering, as was shown by Schur himself in [12].

3.4. Because of the isomorphisms $B_n \sim C_n$ if q is even, $B_2 \sim C_2$ always, the groups $B_n(3)$ $(n \geqslant 3)$ are the only classical ones still to be considered. Here we can normalize all y_r's as in 2.2.

(1) The relations (A) and (B) that involve only long roots hold. For the long roots support a subgroup $D_n(3)$ for which the result is already known.

(2) If r is short the relations (A) hold. Since k is cyclic $\{y_r(u)\}$ is Abelian. Thus (2) follows as in 2.2 (4).

(3) Relations (B) hold if we exclude the single case: r long, s short and orthogonal to r, $n = 3$. By induction on the number of roots involved the obstruction is biadditive and hence may be removed as in 2.2 (6) by conjugation by $h_t(-1)$ with t a (long) root orthogonal to r but not to s.

(4) If $n = 3$ there is a potential obstruction of order 3 in the excluded case of (3), independent of (r, s) because of the action of W. The obstruction here is real (Fischer, unpublished): $B_3(3)$ can be imbedded in ${}^2E_6(2)$ (adjoint group) (this is the hard part) and the index is prime to 3. Thus we get a nontrivial triple covering of $B_3(3)$ by restricting that of ${}^2E_6(2)$ (adjoint group) given by the universal group of the same type.

3.5. We conclude with the two remaining types, first F_4. If $q \geqslant 3$, we normalize $y_r(u)$ as in 2.2; then each of the relations (A) and (B) occurs in a subgroup of type B_4 or C_3, hence holds by what has been done in 3.1, 3.2 and 3.4 (and an extra bit of argument because the normalization here is different from that in 3.2). If $q=2$, we can respect the outer automorphism interchanging long roots and short roots by setting $y_r = y_r(1) = (y_s, y_t)$ with all roots of the same length, whether r is long or short. Using our results for $D_4(2)$ as imbedded in $C_4(2)$ we get all relations in which only roots of one length are involved and also those in which the right side of (B) is 1. We are thus left with a potential obstruction j of order 2 coming from the relations of the form $(y_r, y_s) = jy_{r+s}y_{r+2s}$, a real obstruction as has been shown by Griess [9 pp. 374–379].

For the final type of group G_2 the situation is tighter and the relations of type (B) are more complicated. Since the details are available in [9, pp. 357–371], we omit them here. The results are: for $q=3$ a triple cover, for $q=4$ a double cover. Griess' existence proof in the latter case is very clever and is contained in Section 7. below.

The proof of 1.1 is hereby completed.

4. Split BN Pairs of Rank 1

In this section G is a group of rank 1 with split BN pair, $B = HX$, and $B^- = HY$ opposite to B satisfying:

(1) X and Y (together) generate G.

(2) $N' \equiv (N - B) \cap XYX$ generates N.

(3) $Y \cap B = 1$.

These properties, which are not independent, are enough to ensure uniqueness in the Bruhat normal form. They hold for SL_2, the quasisplit SU_3 (property (2) is verified just after 5.3 below) and the other rank 1 groups arising in the theory of algebraic groups, including the Suzuki and Ree groups. We consider the following relations on the elements of X and Y which hold in G.

(A) Those that hold in X; those in Y.

(B′) Those of the form $w = xyx'$ as in (2) above, thought of as a definition of w in terms of x, y, x'; those of the form ${}^{w}x'' = y''$.

(C) Those of the form $w = w'$ in N'.

(D) The further relations on the elements of N' needed for an abstract definition of N.

Remark 4.1. If $w \in N'$ then $w^{-1} \in N'$ by (B′), whence N' transforms Y into X and $w = {}^{w}w \in YXY$. Thus relations (B′) are entirely symmetric in X and Y.

PROPOSITION 4.2. (a) *Relations* (A) *to* (D) (*on the set* $X \cup Y$) *define* G *as an abstract group.* (b) *Relations* (A) *and* (B′) *define a group* G_1 *which is a central extension of* G.

Proof. Let H_1 be the subgroup of G_1 consisting of even products of elements of N'. Then H_1 transforms X and Y into themselves, in fact by the same equations as in G, and every $y \in Y - 1$ has the form xwx', both by (B′) and 4.1. It readily follows that if w_0 is a fixed element of N' then $XH_1 \cup XH_1 w_0 X$ is invariant under right multiplication by X and Y and hence equals G_1, and that the kernel of the natural map $\pi\colon G_1 \to G$ is contained in H_1. Now each $h \in \ker \pi$ transforms X and Y in G_1 as $\pi h = 1$ does in G, i.e., trivially, hence lies in the center of G_1, whence (b). The extra relations (C) and (D) make π injective on H_1, hence also on G_1 since $\ker \pi \subset H_1$, whence (a).

LEMMA 4.3. *In* G_1 (*or in* G) *let* (*) $w = xyx'$ *as in* (B′). (a) *Then also* $w = x'(w^{-1}xw)(w^{-1}yw)$ *and* $w = {}^w y {}^w x' x$ *in* XYX. (b) *Each of* x, y, x' *is different from* 1. (c) *Conversely for each* x, y *or* x' *different from* 1 *the other items in the equation* (*) *exist and are unique.* (d) *If we write* $w = w(x)$ *in* (c) *then* $w(x) = w(x')$. (e) *Further* $w(x^{-1}) = w(x)^{-1}$.

Proof. (a) It is easily checked that each equation here is formally equivalent to (*). (b) $y \neq 1$ since $w \notin X$ and similarly for x and x' by the equations in (a). (c) Given any $y \neq 1$ then $y \notin B$ and hence $y = xwx'$ (Bruhat form) for unique x, x' and w, whence $w = x^{-1}y(x')^{-1}$. Given $x \neq 1$ write $x = ywy'$, whence $w = x(y')^{-1}(w^{-1}y^{-1}w)$ with y, w, y' uniquely determined by x; similarly for x'. (d) By each equation in (a). (e) Take inverses in (*) and use (d).

LEMMA 4.4. *In* G_1 *assume that* x_1, x_2, x_3 *are in* X *and different from* 1 *and that* $x_2 = x_3 x_1$. *Write* $w(x_i) = x_i y_i x_i'$ *for* $i = 2, 3$ *as in* 4.3(*) *and* $w(x_1) = x_1'' y_1 x_1$ *as in the second equation of* 4.3 (a). *Then* $y_1 y_3 \neq 1$, *and if* x_4 *corresponds to* $y_4 \equiv y_1 y_3$ *as* x *does to* y *in* 4.3 (c) *then* $w(x_1)\, w(x_2)^{-1}\, w(x_3) = w(x_4)$.

Proof. Writing w_i for $w(x_i)$ and $x_5, x_6, \ldots$ for elements of X whose precise values are not needed we have:

$$
\begin{aligned}
w_1 w_2^{-1} w_3 &= w_1 y_2^{-1} x_2^{-1} (w_2 (x_2')^{-1} w_2^{-1})\, w_3 && \text{(def. of } w_2) \\
&= w_5 w_1 x_2^{-1} w_3 x_6 && \text{(with } x_5 = w_1 y_2^{-1} w_1^{-1}) \\
&= x_5 x_1'' y_1 x_1 x_2^{-1} x_3 y_3 x_3' x_6 && \text{(def. of } w_1, w_3) \\
&= x_7 y_1 y_3 x_8 && \text{(since } x_1 x_2^{-1} x_3 = 1) \\
&= x_7 y_4 x_8 && \text{(def. of } y_4).
\end{aligned}
$$

It follows that $x_7 y_4 x_8 \in N'$ in G, hence also in G_1. Then $y_4 \neq 1$ by 4.3 (b) and $x_7 = x_4$ by the definition of x_4, so that $x_7 y_4 x_8 = w(x_4) = w_4$, as required.

5. The Quasisplit SU_3

In this section G is as just described relative to a separable quadratic extension of fields k/k_0 and a split Hermitian form which may be taken as $u_1\bar{u}_3 - u_2\bar{u}_2 + u_3\bar{u}_1$ in terms of suitable coordinates. As usual X (resp. Y, N, H) is the subgroup of superdiagonal unipotent (resp. subdiagonal unipotent, monomial, diagonal) elements of G. Our goal is the following result.

THEOREM 5.1. *Assume that $G = SU_3$ in* 4.2 (a) *and that k is finite. Then relations* (C) *and* (D) *may be omitted. In other words,* (A) *and* (B′) *suffice to define G abstractly.*

The proof requires several steps. At the start k need not be finite.

For x to be in X it must be of the form $(1, a, b;\ 0, 1, \bar{a};\ 0, 0, 1)$ (with the rows of x written out in order) with

5.2. $a\bar{a} = b + \bar{b}$.

We write $x(a, b)$ for this element. Then $x(a, b)\,x(c, d) = x(a + c, b + a\bar{c} + d)$ and $x(a, b)^{-1} = x(-a, \bar{b})$. Further $x(a, b) \neq 1$ just when $(a, b) \neq (0, 0)$, i.e., when $b \neq 0$ by 5.2; and similarly for Y. For $x = x(a, b)$ equation 4.3 $(*)$ works out to

5.3. $w(a, b) = x(a, b)\, y(-\bar{a}\bar{b}^{-1}, \bar{b}^{-1})\, x(ab^{-1}\bar{b}, b)$.

This therefore is our definition of $w(a, b)$ in the abstract group G_1 (in which all of the calculations of this section are taking place). Now $w(a, b)$ in G is just the offdiagonal matrix $[b, -b^{-1}\bar{b}, \bar{b}^{-1}]$. Thus in view of 5.2 condition (2) at the start of Section 4 amounts to: every $b \in k^*$ is a product of such elements having traces that are norms. If k is finite this is clear since then every element of k_0 is a norm. If k is infinite two elements suffice: bj and j^{-1} with $j = b - \bar{b}$ or any nonzero skew element according as $b - \bar{b}$ is nonzero or not.

By 5.3 equation 4.3 (d) becomes

5.4. $w(a, b) = w(ab^{-1}\bar{b}, b)$.

And Lemma 4.4 now reads:

LEMMA 5.5. *In G_1 we have that $w(a_1, b_1)\, w(a_2, b_2)^{-1}\, w(a_3, b_3)$ is of the form $w(a_4, b_4)$ if b_1, b_2, b_3 are nonzero, $a_2 = a_1 + a_3$ and $b_2 = b_1 + b_3 + \bar{a}_1 a_3$. Then $a_4 = b_1 b_2^{-1} b_3(a_1\bar{b}_1^{-1} + a_3 b_3^{-1})$ and $b_4 = b_1 b_2^{-1} b_3$.*

Proof. The first set of equations follows from $x_2 = x_3 x_1$, the second from $y_4 = y_1 y_3$. (From 5.3 it follows that $y_3 = y(-\bar{a}_3 \bar{b}_3^{-1}, \bar{b}_3^{-1})$ and from 5.3 and a simple calculation that $y_1 = y(-\bar{a}_1 b_1^{-1}, \bar{b}_1^{-1})$, whence $y_4 = y_1 y_3 = y(-\bar{a}_1 b_1^{-1} - \bar{a}_3 \bar{b}_3^{-1}, \bar{b}_1^{-1} + \bar{b}_3^{-1} + a_1 \bar{b}_1^{-1} \bar{a}_3 \bar{b}_3^{-1})$. Then $x_4 = x(a_4, b_4)$ with $\bar{a}_4 \bar{b}_4^{-1} = \bar{a}_1 b_1^{-1} + \bar{a}_3 \bar{b}_3^{-1}$ and $\bar{b}_4^{-1} = \bar{b}_1^{-1} + \bar{b}_3^{-1} + a_1 \bar{b}_1^{-1} \bar{a}_3 \bar{b}_3^{-1}$ by 5.3, whence the expressions for a_4 and b_4 after some simplication using the equation for b_2 already established.)

We now specialize (a_2, b_2) to $(0, j)$ with j a fixed nonzero skew element of k and set $c = j^{-1}b$ and $h(a, c) = w(a, jc)\, w(0, j)^{-1}$. Observe that $h(0, 1) = 1$. Conditions 5.2 and 5.4 become:

5.6. $a\bar{a} = j(c - \bar{c})$.

5.7. $h(a, c) = h(-ac^{-1}\bar{c}, c)$.

If in 5.5 we solve for a_3, b_3, a_4, b_4 in terms of a_1, b_1 and drop the subscript 1 then that relation becomes

5.8. $h(a, c)\, h(-a, 1 - \bar{c}) = h(-ac\bar{c}^{-1}, c(1 - \bar{c}))$.

Now $h(a, c)$ in G works out to $\operatorname{diag}(c, c^{-1}\bar{c}, \bar{c}^{-1})$. Thus in view of 5.6 relations (C) become

(C′) $h(a, c)$ depends only on c, i.e., does not change when a is multiplied by an element of norm 1.

We now reinstate our assumption that k is finite in 5.1. Then each element of k_0 is a norm so that each $c \in k^*$ is allowable in 5.6. Thus the relations (D) require, in addition to (C′), that:

(D′) $h(\cdot, c)\, h(\cdot, d) = h(\cdot, cd)$ for all $c, d \in k^*$ and some choice of the dots.

It remains to show that (C′) and (D′) hold in G_1.

5.9. In a relation (D′) that holds in G_1 all of the dots may be multiplied by any $u \in k_0$ of norm 1.

Let σ be the automorphism of G given by conjugation by $\operatorname{diag}(u, 1, u^{-1})$. It acts on X, Y, N',..., hence also on relations (A) and (B′) and so yields an automorphism of G_1. Applying σ to (D′) we get the same relation with each dot multiplied by u.

5.10. We have $h(a, c)\, h(0, d\bar{d}) = h(ad^2\bar{d}^{-1}, cd\bar{d})$.

We conjugate $h(a, c) = w(a, jc)\, w(0, j)^{-1}$ by $h = h(\cdot, d)$. The left side remains unchanged since H_1 is a central extension of the cyclic group H (isomorphic to k^*) and hence is Abelian; on the right side we use ${}^{h}w(a, b) = w(ad^2\bar{d}^{-1}, bd\bar{d})$ which follows from 5.3 and the fact that h acts on X and Y

by the same formulas in G_1 as in G. The result is an equation which works out to 5.10.

5.11. (C′) holds if c is a generator of k^*.

We set $d = c^{-1}\bar{c}$; it generates the group of elements of norm 1 in k^*. Because of 5.10 we may multiply a by d^3 without changing $h(a, c)$, and because of 5.7 by $-d$, hence also by $d^3(-d)^{-2} = d$, hence by any element of norm 1.

5.12. Relations (D′) hold.

If c or d is a norm, i.e., is in k_0^*, then (D′) holds by 5.10. Assume next that none of c, d, cd is in k_0^*. Then the equation $pc + q\bar{d} = 1$ has a solution with $p, q \in k_0^*$. Set $p = r\bar{r}$ and $q = s\bar{s}$. Then $h(0, p)\, h(a, c)\, h(b, d)\, h(0, q) = h(ar^2\bar{r}^{-1}, cr\bar{r})\, h(bs^2\bar{s}^{-1}, ds\bar{s})$ by 5.10; write this as $h(A, C)\, h(B, D)$. We have $D = 1 - \bar{C}$ by the choice of p and q, and $A\bar{A} = j(C - \bar{C}) = j(D - \bar{D}) = B\bar{B}$ by 5.6, so that $-AB^{-1}$ has norm 1. Replacing b by $-AB^{-1}b$ at the start we achieve $B = -A$ at the end. Thus the original product equals $h(\cdot\,, CD)$ by 5.8, then $h(\cdot, pcqd)$ by the definition of C and D, and then $h(0, p)\, h(\cdot, cd)\, h(0, q)$ by 5.10, whence 5.12 in this case. Finally if $c, d \notin k_0$ and $cd \in k_0$ choose a generator e of k^* so that $de \notin k_0$: if e is any generator then either e or e^{-1} will work since otherwise $e^2 \in k_0^*$ and k^*/k_0^* has order at most 2, which is impossible. We also have e, $cde \notin k_0$. Thus by cases already done and 5.9 we have, for suitable choices of the dots, $h(\cdot, c)\, h(\cdot, d)\, h(\cdot, e) = h(\cdot, c)\, h(\cdot, de) = h(\cdot, cde) = h(\cdot\,, cd)\, h(\cdot, e)$. The two $h(\cdot, e)$'s here are equal by 5.11, whence 5.12 in this last case.

5.13. Relations (C′) hold.

Given $h(a, c)$, write $c = c_0^n$ with c_0 a fixed generator of k^*. Then $h(a, c)$ is a product of n elements of the form $h(\cdot, c_0)$ by 5.12 and 5.9, hence depends only on n by 5.11, hence only on $c = c_0^n$, as required.

Since relations (C′) and (D′), and hence also (C) and (D), have been shown to hold in G_1, the proof of 5.1 is now complete.

Remarks 5.14. Theorem 5.1 also holds for SL_2 (see [13, 3.3] whose proof provides a model for the present proof) and probably also holds for the Suzuki groups and the Ree groups. It does hold if $q \neq 2, 8$ in the first case and $q \neq 3$ in the second. For then it can be proved that G and G_1 are perfect (easy) and that G has trivial Schur multiplier (one prime at a time, as Schur did for $SL_2(q \neq 4, 9)$, see [2],), whence the central extension of 4.2 (b) is trivial. The same method works for the current groups SU_3 if $q \neq 4$. Unfortunately, most of the omitted cases above are needed in the treatment of groups of higher rank, as in the next section.

6. The Quasisplit SU_{2n+1} ($n > 1$)

In this section G is of this type. It is generated by unipotent subgroups X_r each isomorphic to k (additive group) or to the subgroup X of SU_3 of Section 5. Each X_r has an "opposite" X_{-r} which with it generate a group isomorphic to SL_2 or to SU_3 in these two cases. In G the following relations on the X_r's hold.

(A) For each r the relations in the group X_r.

(B) For $r \neq -s$ those of the form $(x, y) = \prod Z_{ij}$ analogous to (B) in Section 2.

(C), (D) As in Section 4 with (X, Y) replaced by (X_r, X_{-r}) for each r.

For the details of (B) the reader may consult [9, p. 388]. Our goal is the following result.

THEOREM 6.1. *Let G be as above, the quasisplit SU_{2n+1} with $n > 1$.* (a) *Relations* (A) *to* (D) *define G as an abstract group.* (b) *Relations* (A) *and* (*B*) *define a group G_1 which is a central extension of G.* (c) *If the base field is finite, i.e., if q is so, then* (A) *and* (B) *suffice in* (a), *i.e., they imply* (C) *and* (D).

Proof. Since the proof is close to that for split groups given in [13, 3.3] we shall be sketchy. Relations (A) and (B) restricted to positive (negative) roots define abstractly the maximal unipotent subgroups $X(Y)$ of G (see [13, 7.1]). They also imply the relations (B′) of Section 4 for each r and in fact that the w's in all of the N_r''s transform all of the X_s's in G_1 exactly as in G (see [13, 7.2] for the argument). If N_1 denotes the subgroup of G_1 generated by all of the N_r''s then $G_1 = XN_1X$ readily follows and from this (b), as in the proof of 4.2 (b) above. The extra relations (C) and (D) then permit the identification of N_1 with N, the corresponding subgroup of G, whence (a). Finally, if k is finite then for each rank 1 subgroup $\langle X_r, X_{-r}\rangle$ of G_1 we have the relations (A) and (B′) of Section 4, as already noted. By 5.1 and the corresponding result for SL_2 (see [13, 3.3]) we also have (C) and (D), whence (c).

COROLLARY 6.2. *If k is finite in* 5.1 *or in* 6.1 *and p is its characteristic then the p'-part of the Schur multiplier of G is trivial.*

Proof. Here also we essentially follow [13]. We use the easy fact that (*) a central extension of a finite p-group by a p'-group always spits (see, e.g., [15, Theorem 2.5] where a more general result of Schur is proved). Now let $\pi: G_1 \to G$ be a central extension with $\ker \pi$ a p'-group. In any one of the relations in (A) or in (B) the elements involved all lie in a p-subgroup of G.

Thus by $(*)$ that relation can be lifted from G to G_1. In the process each element of G involved is lifted to the unique p-element of G_1 above it. Thus all of the relations can be lifted together. By 5.1 or 6.1 this yields a splitting map for π, whence 6.2.

7. The Spin Covering

Throughout this section let V be a vector space of finite dimension $n \geqslant 2$ over $\mathbb{R}$ and $(\ , \ )$ a positive–definite inner product on it. Our object is to give a quick self-contained introduction to the spin group in this situation, enough to prove 7.7 below. Other, more comprehensive, treatments may be found in [3–5]. The Clifford algebra $C = C(V)$ is, by definition, the associative algebra (with 1) generated by the linear space V and the relations $(*)$ $v^2 = (v, v)$ for all $v \in V$. In terms of an orthonormal basis $\{v_i\}$ of V these relations become $(**)$ $v_i^2 = 1$ and $v_i v_j = -v_j v_i$ if $i \neq j$. If for each subset S of $\{v_i\}$ we take the product v_S of its elements with the subscripts in increasing order, we get a basis for C which thus has dimension 2^n. For, first by use of the relations every element of C can be reduced to a linear combination of v_S's. To prove linear independence we may do so over $\mathbb{Z}$ since the structural coefficients are all in $\mathbb{Z}$, hence over $\mathbb{F}_2$ on reduction mod 2. By $(**)$, C is now the tensor product of n subalgebras, the ith generated by v_i subject to the condition $v_i^2 = 1$, thus having 1 and v_i linearly independent. In the full algebra the 2^n products v_S are thus linearly independent, as required.

In C we have the subalgebra C^+ of even elements, generated by the products $v_i v_j$; it has dimension 2^{n-1}. For example, if $n = 2$, C^+ is the field of complex numbers, and if $n = 3$ it is the skew field of quaternions.

7.1. Center(C) $\cap\, C^+ = \mathbb{R}$.

If $x = \sum c_S v_S \in$ Center (C), we have $v_i x v_i^{-1} = \sum c_S v_i v_S v_i^{-1}$ with $v_i v_S v_i^{-1} = v_S$ or $-v_S$ according as $S - \{i\}$ is of even or odd size. For S even and nonempty the minus sign holds for $i \in S$, whence $c_S = 0$.

On C there is a unique antiautomorphism $x \to x^*$ fixing V, for there is certainly one on the tensor algebra $T(V)$ and it preserves the relations $(*)$ that define C. The product xx^* is generally not in $\mathbb{R}$ but it is so if x is decomposable ($x = u_1 u_2 \cdots u_k$ with each u_i in V) and then defines a norm which extends the given norm on V (imbedded in C in the obvious way) and is multiplicative, as is easily checked. We define Spin(V) (or Spin$_n$) to be the group of such x for which k is even and $xx^* = 1$, or, equivalently, each u_i is a unit vector.

7.2. If u is a unit vector in V, it preserves V by conjugation and acts there as minus the reflection corresponding to u.

For $uuu^{-1} = u$, while if v is orthogonal to u then $uvu^{-1} = -v$ by (**). We denote this action and its extension to $\mathrm{Spin}(V)$ by π.

7.3. Every element x of $O(V)$ is a product of at most n reflections.

If 1 is an eigenvalue of x then by induction x is a product of at most $n-1$ reflections. If 1 is not one then from $\det(x-1) = \det(-x)\det(x^t - 1) = \det(-x)\det(x-1)$, we get $\det(x) = (-1)^n$ so that replacing x by xr with r any reflection puts us back in the first case.

THEOREM 7.4. *If* $\pi: \mathrm{Spin}(V) \to GL(V)$ *is as above then* $\operatorname{im} \pi$ *is* $SO(V)$ *and* $\ker \pi$ *is the central subgroup* $\{\pm 1\}$.

The first point is by 7.2 and 7.3. If $x \in \ker \pi$ then x commutes with every $v \in V$, hence lies in $\mathrm{Center}(C) \cap C^+$, hence is a scalar by 7.1, ± 1 since its norm is 1; and both cases occur: $-1 = v(-v) \in \mathrm{Spin}(V)$ for any unit vector v.

Remarks 7.5. (a) By 7.3 and 7.4 every element of $\mathrm{Spin}(V)$ is a product of at most n elements of V. Since the unit vectors of V form a compact connected set it follows that $\mathrm{Spin}(V)$ is a compact connected Lie subgroup of C^*. That $\pi: \mathrm{Spin}(V) \to SO(V)$ is the universal covering requires further argument (see [4]). (b) It also follows that $\mathrm{Spin}(V)$ could have been defined as the commutator subgroup of the group of those elements of $(C^+)^*$ that conjugate V into itself.

COROLLARY 7.6. *An involution* x *of* $SO(V)$ *with* -1 *an eigenvalue of multiplicity* $2k$ *lifts to an element of order* 4 *in* $\mathrm{Spin}(V)$ *if* k *is odd, to one of order* 2 *if* k *is even.*

If $v_1, v_2, \ldots, v_{2k}$ are the first elements of an orthonormal basis of V then $\pi(v_1 v_2 \cdots v_{2k})$ may be taken as x. Since $(\pm v_1 v_2 \cdots v_{2k})^2 = (-1)^k$ by (**) we have 7.6.

COROLLARY 7.7. *If* G *is a subgroup of* $SO(V)$ *containing an involution as in* 7.6 *with* k *odd then* $\pi: \pi^{-1}G \to G$ *does not split. In particular this so for* $\pi: \mathrm{Spin}(V) \to SO(V)$.

This follows at once from 7.6.

COROLLARY 7.8. *If* n *is odd* $\mathrm{Spin}(V)$ *has center of order* 2. *If* n *is even the center has order* 4 *and it is cyclic just when* $n/2$ *is odd.*

This follows from 7.6 applied to the center of G which is 1 if n is odd, $\{\pm 1\}$ if n is even.

COROLLARY 7.9. *If* G *is a subgroup of* A_n *(alternating group) containing* *an involution which is the product of $2k$ disjoint transpositions with k odd then G has a nonsplit 2-fold central extension.*

We apply 7.7 to the obvious imbedding $A_n \subset SO_n$.

This result goes back to Schur [12]. He proved it by in effect explicitly constructing the fragment of the spin group lying above A_n (and of the pin group lying above S_n).

EXAMPLES 7.10. (a) Corollary 7.9 was used for $n = 8$ in 2.3 and could have been used for $n = 5$ in Section 2 since $SL_2(4) \sim A_5$. (b) Corollary 7.7 applied to the groups $SW(E_n)$ $(n = 7, 8)$ is used in 3.1 and 2.3. The case $n = 6$ figures indirectly in 3.2 since $SW(E_6) \sim SU_4(2) \sim PSp_4(3)$. (c) Here is the pretty way in which Griess [9, pp. 363–364] gets a nonsplit double covering of the group $G = G_2(4)$ of 3.5. Let b be the simple short root and P the corresponding parabolic subgroup. Then G acts on G/P and it turns out that $x_a(1)$ has 25 fixed points, thus acts as a product of $(1365 - 25)/2 = 670$ disjoint transpositions, so that 7.9 applies.

REFERENCES

1. E. ABE, Coverings of twisted Chevalley groups over commutative rings, *Tokyo Kyoiku Daigaku* (*A*) **13** (1977), 194–218.
2. J. L. ALPERIN AND D. GORENSTEIN, The multiplicators of certain simple groups, *Proc. Amer. Math. Soc.* **17**(1966), 515–519.
3. E. ARTIN, "Geometric Algebra," Interscience, New York, 1957.
4. C. CHEVALLEY, "Theory of Lie Groups," Princeton Univ. Press, Princeton, N. J., 1946.
5. C. CHEVALLEY, "The Algebraic Theory of Spinors," Columbia Univ. Press, New York, 1954.
6. C. CHEVALLEY,Sur certains groupes simples, *Tôkoku Math. J.* **7**(1955), 14–66.
7. V. V. DEODHAR, On central extensions of rational points of algebraic groups, *Amer. J. Math.* **100**(1978), 303–386.
8. R. L. GRIESS, Schur multipliers of the known finite simple groups, *Bull. Amer. Math. Soc.* **78** (1972), 68–71.
9. R. L. GRIESS, Schur multipliers of finite simple groups of Lie type, *Trans. Amer. Math. Soc.* 183 (1973), 355–421.
10. J. GROVER,Covering groups of groups of Lie type, *Pacific J. Math.* **30**(1969), 645–655.
11. I. SCHUR, Über die Darstellung der endlichen Gruppen durch gebrochene lineare Substitutionen, *J. Reine Angew. Math.* **127** (1904), 20–50; **132** (1907), 85–137.
12. I. SCHUR, Über die Darstellung der symmetrischen und der alternierenden Gruppe durch gebrochene lineare Substitutionen, *J. Reine Angew. Math.* **139** (1911), 155–250.
13. R. STEINBERG, Générateurs, relations, et revêtements de groupes algébriques, *in* "Colloq. Théorie des Groupes Algébriques, Bruxelles (1962)," pp. 113–127.
14. R. STEINBERG, "Lectures on Chevalley Groups," Yale Univ. Lecture Notes, New Haven Conn., 1967.
15. H. J. ZASSENHAUS, "The Theory of Groups," 2nd ed., Chelsea, New York, 1958.

Printed by the St. Catherine Press Ltd., Tempelhof 41, Bruges, Belgium

PACIFIC JOURNAL OF MATHEMATICS
Vol. 118, No. 2, 1985

FINITE SUBGROUPS OF SU_2, DYNKIN DIAGRAMS AND AFFINE COXETER ELEMENTS

ROBERT STEINBERG

Dedicated to the memory of my friend Ernst Straus

Using, among other things, some properties of affine Coxeter elements, for which we also present normal forms, we offer an explanation of the McKay correspondence, which associates to each finite subgroup of SU_2 an affine Dynkin diagram.

J. McKay [**M**] has observed that for each finite (Kleinian) subgroup G of SU_2 the columns of the character table of G, one column for each conjugacy class, form a complete set of eigenvectors for the corresponding affine Cartan matrix (of type A_n, D_n or E_n), the one that arises in connection with the resolution of the singularity of $\mathbf{C}^2/G$ at the origin (see 1(9) below). As he has observed, this follows at once from: if ρ is the two-dimensional representation by which G is defined, $\{\rho_i\}$ is the set of (complex) irreducible representations of G, and $\sum n_{ij}\rho_j$ denotes the decomposition of $\rho \otimes \rho_i$, then $C \equiv [c_{ij}] \equiv [2\delta_{ij} - n_{ij}]$ is the relevant Cartan matrix. Partial explanations have been given by several authors (see [**G**], [**H**], [**K**], [$\mathbf{S}_1$, Appendix III]). Here we shall give our own explanation of this and some related facts, including two normal forms for affine Coxeter elements which enter into our considerations. Section 1 deals mainly with McKay's correspondence, Section 2 mainly with affine Coxeter elements. As general references for Kleinian groups, Kleinian singularities and root systems, we cite [**C**, Chapters 7, 11], [$\mathbf{S}_1$, Section 6], [**B**], and the survey article [$\mathbf{S}_2$].

1. In this section G is a finite group, ρ is a faithful (complex) representation of G of finite dimension d, $\{\rho_i\}$ is the set of all irreducible representations of G with ρ_0 the trivial one, $\sum n_{ij}\rho_j$ denotes the decomposition of $\rho \otimes \rho_i$, and C is the matrix $[d\delta_{ij} - n_{ij}]$.

(1) The column $[\chi_j(x)$ (x in G fixed, $j = 1, 2, \ldots$)$]$ of the character table of G is an eigenvector of C with $d - \chi(x) = \chi(1) - \chi(x)$ as the corresponding eigenvalue. In particular $[d_1, d_2, \ldots]$ ($d_i = \dim \rho_i$) is an eigenvector corresponding to the eigenvalue 0.

We have $\chi(x)\chi_i(x) = \sum n_{ij}\chi_j(x)$, whence the first statement. Then $x = 1$ yields the second.

(2) The following equations hold.

(a) $n_{ij} = n_{\bar{j}\bar{i}}$ (bar denotes dual)

(b) $dd_i = \sum n_{ij} d_j$

(c) $dd_i = \sum n_{ji} d_j$

(d) $n_{ij} = n_{ji}$ for all i and j if and only if ρ is self-dual.

(a) This follows from $n_{ij} = (\chi\chi_i, \chi_j) = \text{Average}\, \chi\chi_i\bar{\chi}_j$.

(b) This is the second statement of (1). (c) $dd_i = dd_{\bar{i}} = \sum n_{\bar{i}j} d_j = \sum n_{\bar{j}i} d_{\bar{j}} = \sum n_{ji} d_j$. (d) If $n_{ij} = n_{ji}$ always then $n_{0j} = n_{j0} = n_{0\bar{j}}$ and ρ is self-dual. If ρ is self-dual then $n_{ij} = n_{\bar{i}\bar{j}} = n_{ji}$.

(3) Now form a real vector space V with a basis vector α_i for each ρ_i and a scalar product given by $(\alpha_i, \alpha_j) = c_{ij} \equiv d\delta_{ij} - n_{ij}$. Then the line through $\sum d_i\alpha_i$ is the radical of (,) from the left and from the right, and it is also the radical of the quadratic form (α, α) and this form is positive semidefinite.

By (2b) and (2c) the given line belongs to these radicals. It will be enough to prove the converse for the quadratic form since its radical contains the others. With $\alpha = \sum x_i\alpha_i$ arbitrary in V we have

$$\begin{aligned} 2(\alpha, \alpha) &= 2\sum c_{ij}x_ix_j \\ &= 2\sum (d - n_{ii})x_i^2 - 2\sum n_{ij}x_ix_j \qquad (i \neq j) \\ &= \sum (n_{ij} + n_{ji})d_i^{-1}d_jx_i^2 - 2\sum n_{ij}x_ix_j \qquad (i \neq j) \end{aligned}$$

by (2b) and (2c). For $i < j$ the pairs i, j and j, i together contribute

$$\begin{aligned} &(n_{ij} + n_{ji})\left(d_i^{-1}d_jx_i^2 + d_j^{-1}d_ix_j^2 - 2x_ix_j\right) \\ &\quad = (n_{ij} + n_{ji})d_i^{-1}d_j^{-1}\left(d_jx_i - d_ix_j\right)^2 \geq 0. \end{aligned}$$

Thus (α, α) is positive semidefinite. Now for each i there exists a sequence $i_1, i_2, \ldots, i_n$ with $i_1 = 0$ corresponding to the trivial representation, $i_n = i$, and $n(i_p, i_{p+1}) \neq 0$ for all p; this is because ρ_i is necessarily contained in some tensor power of the faithful representation ρ. It follows from this and the above inequalities that if $(\alpha, \alpha) = 0$ then $x_i = (x_0/d_0)d_i$ for all i, so that α is in the line of $\sum d_i\alpha_i$.

We now specialize to the case in which ρ imbeds G into SU_2. We assume that $G \neq \{1\}$.

(4) (a) $c_{ij} = c_{ji}$ (i.e. $n_{ij} = n_{ji}$) always.

(b) $c_{ii} = 2$ (i.e. $n_{ii} = 0$) always.

(c) If $G \neq \{\pm 1\}$ and $i \neq j$ then $c_{ij} = 0$ or -1 (i.e. $n_{ij} = 0$ or 1).

In other words (a) C is symmetric, (b) ρ_i is disjoint from $\rho \otimes \rho_i$, and (c) $\rho \otimes \rho_i$ is multiplicity-free.

(a) This is by (2d): if α, α^{-1} are the eigenvalues of $\rho(x)$ then $\alpha + \alpha^{-1} = \alpha + \bar{\alpha} \in \mathbf{R}$. (b) If ρ is reducible it has the form $\sigma \dotplus \bar{\sigma}$ with $\dim \sigma = 1$. Thus G is cyclic and all ρ_i have dimension 1. So, since G is nontrivial, $\sigma \otimes \rho_i$ and $\bar{\sigma} \otimes \rho_i$ are different from ρ_i and hence disjoint from it. If ρ is irreducible then $\{\pm 1\} \subseteq$ Center (G). For then 2 divides $|G|$ and -1 is the unique element of order 2 of SU_2. Now if -1 in G acts as (multiplication by) 1 (resp. -1) on ρ_i, it acts as -1 (resp. 1) on $\rho \otimes \rho_i$, hence also on its irreducible components, all of which must thus be different from ρ_i. (c) We have

$$\textstyle\sum_j n_{ij}^2 = (\rho \otimes \rho_i, \rho \otimes \rho_i) \equiv \mathrm{Av}|\chi(x)|^2 |\chi_i(x)|^2 \leq \mathrm{Av}\, 4|\chi_i(x)|^2 = 4,$$

since $|\chi(x)| \leq 2$ for all x in G with equality only if $x = \pm 1$. If the strict inequality holds then $\sum_j n_{ij}^2 < 4$ and each n_{ij} is 0 or 1. If equality holds then $\chi_i(x) = 0$ for all $x \neq \pm 1$. If also multiplicity occurs then $\rho \otimes \rho_i = 2\rho_j$ with ρ_j irreducible, and $\rho \otimes \rho_j = 2\rho_i$ because of the values of χ_i and χ_j. Now -1, if it is in G, acts trivially on ρ_i or on ρ_j, say on ρ_j. If H denotes G modulo its intersection with $\{\pm 1\}$, then ρ_j yields an irreducible representation of H with character value d_j at 1, 0 elsewhere, whence $|H| = d_j^2$. However $|H| = \sum' d_j^2$, summed over all irreducible representations of H. It follows that ρ_j is the unique irreducible representation of H, hence that H is trivial. Thus $G \subseteq \{\pm 1\}$, contradicting our assumptions.

We now introduce a diagram Γ with one vertex corresponding to each basis vector α_i of V (or to each irreducible representation ρ_i of G) and one edge for each pair i, j such that $n_{ij} = 1$ (i.e. $c_{ij} = -1$) in (4c), which is unambiguous by (4a). By (4b) no edge of Γ is a loop.

(5) Γ (resp. C) is the extended Dynkin diagram (resp. matrix) of a reduced, irreducible root system with all roots of one length (which we take to be $\sqrt{2}$) and $\Gamma' \equiv \{\alpha_i | i \neq 0\}$ is a simple system and $\sum_{i \neq 0} d_i \alpha_i$ is the corresponding highest root.

By the argument at the end of (3) Γ is connected. But then so is Γ': We have $\rho \otimes \rho_0 = \rho$. Thus if ρ is irreducible it yields the unique vertex of Γ joined to α_0 and Γ' is connected. If ρ is reducible then Γ is a loop, as may be checked; thus again Γ' is connected. Now by (3), if V' is the subspace of V generated by the α's other than α_0 and L is the line $\mathbf{R}\sum d_i \alpha_i$, then V' projects isometrically onto V/L and there (,) is positive definite. We identify the two spaces. Further $(\alpha_i, \alpha_i) = 2$ and for $i \neq j$ $(\alpha_i, \alpha_j) = -1$ or 0 according as α_i and α_j are or are not joined in Γ.

Thus $\{\alpha_i | i \neq 0\}$ is a simple system for an irreducible root system in which $(\alpha, \alpha) = 2$ for every root and Γ' is its Dynkin diagram. Further $-\alpha_0$ is a root since $(-\alpha_0, -\alpha_0) = 2$, and it is dominant and hence the highest root since also $(-\alpha_0, \alpha_i) = n_{0i} \geq 0$ for $i \neq 0$. Thus $\Gamma = \Gamma' \cup \{\alpha_0\}$ is the corresponding extended Dynkin diagram. On V' we have $-\alpha_0 = -d_0\alpha_0 = \sum_{i \neq 0} d_i \alpha_i$, whence the last point of (5).

(6) G/G' is isomorphic to F, the center of the simply connected complex Lie group L whose extended Cartan matrix is C.

First the orders of the two groups are equal: $|G/G'|$ is the number of 1-dimensional representations of G, i.e., the number of d_i's equal to 1, hence is 1 + the number of coefficients that are 1 in the highest root, which is known to be $|F|$. An isomorphism is given by $x \in G/G' \to \Pi \alpha_i^*(\det \rho_i(x))$. Here α_i^* denotes the coroot of α_i, viewed as a 1-parameter subgroup into L. All of this is relative to a choice of a maximal torus and an ordering of its character group. The proposed isomorphism is injective since if x is in the kernel then $\det \rho_i(x) = 1$ for all i, whence $\rho_i(x) = 1$ for all i with $d_i = 1$, and $x \in G'$. The image is in F since if α_j is any simple root then $\alpha_j(\text{image}) = \Pi(\det \rho_i(x))^{c(i,j)} = 1$, as we see by taking determinants in $\rho \otimes \rho_i = \sum n_{ij} \rho_j$ and using $\det \rho = 1$, $\det \rho_0 = 1$ and $c_{ij} = 2\delta_{ij} - n_{ij}$.

(7) The (unextended) Dynkin diagram for G' can be gotten from that for G by deleting all vertices α_i for which $d_i = 1$.

At present this is only an empirical observation.

Because of (6) and (7) the derived series for G can be written down easily in any given case. For example, $E_7 \supset E_6 \supset D_4 \supset A_1 \supset \{1\}$, with corresponding quotients $C_2, C_3, C_2 \times C_2, C_2$.

If G is reducible on $\mathbf{C}^2$ and hence cyclic, then Γ is a cycle, hence of type A_n, as is mentioned above. Conversely if Γ contains a cycle then by standard arguments Γ must be a cycle, α_0 (corresponding to the trivial representation) has two neighbors and $\rho = \rho \otimes \rho_0$ is reducible. Aside from these cases, the possibilities for G are classified by numbers $p_1 \geq q_1 \geq r_1 \geq 2$ with $p_1^{-1} + q_1^{-1} + r_1^{-1} > 1$, ($2p_1, 2q_1, 2r_1$ are the orders of the maximal cyclic subgroups of G, one subgroup for each conjugacy class), and so are the possibilities for Γ' (ordinary Dynkin diagram) (p, q, r are the branch lengths including the branch point).

(8) If $\Gamma'(p, q, r)$ is the diagram coming from $G(p_1, q_1, r_1)$ then $(p, q, r) = (p_1, q_1, r_1)$. Thus McKay's correspondence is bijective.

Let x, y, z in G be such that their eigenvalues on $\mathbf{C}^2$ are $\exp(\pm \pi i/p)$, $\exp(\pm \pi i/q)$, $\exp(\pm \pi i/r)$. The elements 1, -1, x^a $(1 \leq a < p_1)$, y^b $(1 \leq b < q_1)$, z^c $(1 \leq c < r_1)$ form a system of representatives of the conjugacy class of G, and on them χ, the character of ρ, has the values 2, -2, $2\cos \pi a/p_1$, etc., resp. By (1) these are the eigenvalues of $2 - C$ with C the corresponding extended Cartan matrix. Thus 2, 2, $2\cos 2\pi a/p_1$, etc., are the eigenvalues of $(2 - C)^2 - 2$. If we can show that this also holds with p_1, q_1, r_1 replaced by p, q, r we will be done. Consider an affine Coxeter element c, the product of the reflections corresponding to the affine simple roots; these are the ordinary simple roots with $1 - \mu$ adjoined (μ is the highest root). Since Γ has no circuits the conjugacy class of c is independent of the order of the factors and the affine simple roots may be so ordered that the first few are mutually orthogonal as are the rest of them. Then in partitioned form we have

$$2 - C = \begin{bmatrix} 0 & N \\ N' & 0 \end{bmatrix}, \quad \text{whence } (2 - C)^2 - 2 = \begin{bmatrix} NN' - 2 & 0 \\ 0 & N'N - 2 \end{bmatrix}.$$

On the other hand if $c = c_1 c_2$ in accordance with this partition of the roots, then

$$c_1 = \begin{bmatrix} -1 & N \\ 0 & 1 \end{bmatrix} \quad \text{and} \quad c_2 = \begin{bmatrix} 1 & 0 \\ N' & -1 \end{bmatrix}$$

in matrix form. It follows that $c + c^{-1} = (2 - C)^2 - 2$. Thus by the above formulas the eigenvalues of c are 1, 1, $\exp(2\pi i a/p_1)$ $(1 \le a < p_1)$, etc., and those of c', the linear part of c, are the same with the first 1 deleted. We now invoke a result which will be proved in the next section (after (10) there).

(*) c', the linear part of c, is conjugate in the Weyl group to c'', the product of the ordinary simple reflections with the one at the branch point excluded. From (*) it follows that c'' has the same eigenvalues as c' as given above. However c'' is the product of three Coxeter elements of types A_n $(n = p - 1, q - 1, r - 1)$ corresponding to the mutually orthogonal subsystems along the branches of Γ', and these, together with the branch root, contribute the eigenvalues 1, $\exp(2\pi i a/p)$ $(1 \le a < p)$, etc. Thus $(p, q, r) = (p_1, q_1, r_1)$, as required.

(9) Consider the minimal resolution of the singular surface $\mathbf{C}^2/G$. The singular fiber is a union of projective lines, and if we form a diagram by taking one node for each line and joining two nodes, by a simple bond, just when the corresponding lines intersect, we get an ordinary Dynkin diagram, of type A_n, D_n or E_n (see [$\mathbf{S}_1$]). It remains to show that this correspondence agrees with McKay's. Let p, q, r be the branch lengths of the diagram just obtained. Type A_n may be included by taking $q = r = 1$ in what follows. Let $C' = [c(i, j)]$ be the ordinary Cartan matrix. Then

the group G is isomorphic to the abstract group defined by n generators and the n relations $\Pi x_i^{c(i,j)} = 1$ $(j = 1, 2, \ldots)$. (Thus G/G', the Abelianized group, is just F, as given in (6) above.) This result is due to Mumford [**Mu**]. The relations yields, via an application of Van Kampen's Theorem, a presentation of the fundamental group of a "sphere" around the singular point of $\mathbf{C}^2/G$, and that group, quite clearly, is G itself. Now let $x_1, x_2, \ldots, x_p$ be the generators along a branch of length p towards the branch point x_p. The given relations yield $x_1^2 x_2^{-1} = 1$, $x_1^{-1} x_2^2 x_3^{-1} = 1, \ldots$, whence if $x_1 \equiv x$ then $x_a = x^a$ for $(1 \le a \le p)$. Similarly on the other branches $y_q = y^q$, $z_r = z^r$ and $x_p = y_q = z_r$. The relation at the branch point yields $(x^p)^2 = x^{p-1} y^{q-1} z^{r-1}$. Thus $xyz = x^p = y^q = z^r$. As is well known [**C**, 11.7, 7.4] this is a presentation of the Kleinian group of type (p, q, r). Thus G and the graph corresponding to it have the same type, as required.

(10) McKay's correspondence can be extended to yield Dynkin diagrams with multiple bonds in several different ways. One way, used in [**H**], is to start with representations over fields that are not algebraically closed. This yields most Dynkin diagrams, but not all of them. Another way, suggested in [**S**, App.III], which does yield all diagrams, is to start with certain pairs $G \triangleleft H$ of finite subgroups of SU_2, or, equivalently, with a single subgroup G and an automorphism σ of G which stabilizes the defining representation of G, and then to associate a node to each σ-orbit of irreducible representations G, or, dually, to each representation of $\langle G, \sigma \rangle$ induced by an irreducible representation of G. One can then carry out large parts of the above development in this new context (with weighted nodes, multiple bonds, etc.) or else notice that the coalescence of irreducible representations into σ-orbits corresponds exactly to the foldings of Dynkin diagrams according to their symmetries.

2. Affine Coexter elements. Our purpose is to develop the principal properties of these elements, including two normal forms and a proof of the property $(*)$ used in the proof of (8) above. Π will be a simple system for an irreducible root system. We write λ_α for the fundamental weight corresponding to α and α^* for the coroot $2\alpha/(\alpha, \alpha)$. We can decompose Π into disjoint parts Π_1 and Π_2 so that each is an orthogonal set of roots. We exclude type A_n mostly. Then μ, the highest root, is orthogonal to all roots of Π but one, so that the notation can be chosen so that μ is orthogonal to Π_2. We write w_1 (resp. w_2) for the product of the simple reflections in Π_1 (resp. Π_2), then $w_i = w_1$ (resp. w_2) when i is odd (resp. even) (and similarly for Π_i), and finally $w^1 = w_1$, $w^2 = w_2 w_1$,

$w^3 = w_3w_2w_1, \ldots$. We observe that w_1, w_2 and each odd w^i is an involution. Since type A_n has been excluded, the Coxeter number, the order of w_1w_2, is even: $h = 2g$. For as can be easily proved or read off from the classification, h is odd only for type A_{2n}.

(1) We have $1 < w^1 < w^2 \cdots < w^{2g}$ and $w^{2g} = w_0$, the element of the Weyl group that makes all positive roots negative.

Here $w < w'$ means that w is the terminal segment of a minimal expression for w' as a product of simple reflections, i.e., that the length of w' is the sum of those of w and $w'w^{-1}$. The more general Bruhat order could also be used in all that follows. The fact that $w^{2g} = w_0$ is proved in [**St**]. In the expression $w_0 = w_{2g} \cdots w_2w_1$ with w_i written as the product of the reflections for the roots in Π_i, the number of roots listed is $g|\Pi| = (h/2)|\Pi|$, which, as is known [**St**], is equal to the number of positive roots. It follows that the expression is minimal, as in each terminal segment, whence (1).

(2) Let λ be a dominant (integral) weight, and $w < w'$ in the Weyl group. Then $w\lambda \geq w'\lambda$. Hence $\lambda \geq w^1\lambda \geq w^2\lambda \geq \cdots \geq w^{2g}\lambda$.

It is enough, by induction, to prove this when $w' = w_\beta w$, with $\beta > 0$ and $w^{-1}\beta > 0$. Here and elsewhere w_β is the reflection relative to β. Now $w'\lambda = w_\beta w\lambda = w\lambda - (w\lambda, \beta^*)\beta$, and $(w\lambda, \beta^*) = (\lambda, w^{-1}\beta^*) \geq 0$ since λ is dominant and $w^{-1}\beta$ is positive, whence (2).

(3) Assume that λ is dominant, $w_0\lambda = -\lambda$, and $\operatorname{Supp}\lambda \subseteq \Pi_1$. Then $w^i\lambda = -w^{2g-1-i}\lambda$ for $0 \leq i < g$.

Here the third condition on λ is that in its expression in terms of the fundamental weights λ_α only those with $\alpha \in \Pi_1$ are needed. We have

$$\begin{aligned} w^i\lambda &= w_i \cdots w_2w_1\lambda \\ &= w_{i+1} \cdots w_{2g} \cdot w_{2g} \cdots w_{i+1}w_i \cdots w_1\lambda = -w_{i+1} \cdots w_{2g}\lambda \quad \text{(by (1))} \\ &= -w^{2g-i-1}\lambda \quad \text{since } w_{2g} = w_2 \text{ fixes } \lambda. \end{aligned}$$

(4) In (3) $w^{g-1}\lambda$ is a nonnegative combination of roots in Π_g.

We have $w^{g-1}\lambda = (w^{g-1}\lambda - w^g\lambda)/2$ by (3). This is ≥ 0 by (2), and, since it equals $(1 - w_g)w^{g-1}\lambda/2$, it involves only the simple roots in Π_g.

(5) $w^{g-1}\mu$ is a simple root, an element of Π_g. It is the unique long simple root b at which there is a branch point or a multiple bond. (Recall that μ is the highest root).

First, by (4), which is applicable since μ as a dominant weight has its support in Π_1, $w^{g-1}\mu = b$ is a nonnegative combination of roots in Π_g. Since b is a root and the elements of Π_g are mutually orthogonal, it easily follows that b is an element of Π_g. And since μ is a long root, so is b.

Since type A_n is being excluded, μ is connected to a unique simple root α. Assume first that α is shorter than μ. To prove (5) in this case we show that there is only one long simple root. In the extended Dynkin diagram $1 - \mu$ is joined only to α, by a multiple bond, and in the ordinary diagram α is connected to a nearest long root by a chain C, ending with a multiple bond. It is enough to show that C is the full Dynkin diagram. If it were not, then $C \cup \{1 - \mu\}$ would be a proper connected subdiagram of the extended Dynkin diagram, hence a Dynkin diagram in its own right, but one with two multiple bonds, namely those at its two ends, which is impossible. Now assume that α has the same lengths as μ. We have

$$(w^g\mu, w^{g+1}\mu) = (w^{2g+1}\mu, \mu) = -(w_1\mu, \mu) \quad (\text{by } (3)) = -|\mu|^2/2$$

since $w_1\mu = w_\alpha\mu$ and α and μ have equal length and form an angle of $60°$ in the present case. Thus $|w^g\mu - w^{g+1}\mu|^2 = 3|\mu|^2$. However $w^g\mu - w^{g+1}\mu = (1 - w_{g+1})w^g\mu = (1 - w_{g+1})(-b) = \Sigma(b, \gamma^*)\gamma$, summed over the elements of Π_{g+1} that are not orthogonal to b, i.e., over the neighbors of b. Since μ and b have the same length, the last two equations imply that $\Sigma(\gamma, b^*)(b, \gamma^*) = 3$. Thus 3 bonds come together at b, which is therefore a branch point or a point with a multiple bond.

In the development in this section so far we have borrowed ideas from Kostant **[Ko]**, who in turn has borrowed ideas from an earlier version of this paper. In that version, the transition from μ to b was effected differently, namely by alternate applications of (1) the reflection corresponding to b, (2) the product of the other simple reflections ordered so as to move away from b. That method brings up other points of interest, but we shall not pursue them here.

(6) Assume as in (3) and (4) except that $\operatorname{Supp} \lambda \subseteq \Pi_2$. Then $w^i\lambda = -w^{2g+1-i}\lambda$ and $w^g\lambda$ is a nonnegative combination of the roots in Π_{g+1}.

This easily follows from (3) and (4) with the roles of Π_1 and Π_2 interchanged.

(7) Let λ be a weight such that $w_0\lambda = -\lambda$, and λ_1 (resp. λ_2) the parts supported by the λ_α with $\alpha \in \Pi_1$ (resp. Π_2). Then $w^g\lambda_1$ (resp. $w^g\lambda_2$) is the part of $w^g\lambda$ supported by the roots of Π_g (resp. Π_{g+1}).

This follows from (4) and (6).

(8) Write $\mu = \Sigma_1 n_\alpha\alpha + \Sigma_2 n_\beta\beta$, the sums over Π_1 and Π_2. Then, with b as in (5),

(a) $w^g\Sigma_1 n_\alpha\alpha = -2\lambda_b$.

(b) $w^g\Sigma_2 n_\beta\beta = 2\lambda_b - b$.

By the orthogonality relations between the simple coroots and the fundamental weights we have $b = 2\lambda_b + \Sigma(b, \gamma^*)\lambda_\gamma$. Here b is in Π_g and the sum, equal to $b - 2\lambda_b$, is over the neighbors of b, all in Π_{g+1}. Using (4) we get $-\mu = (w^g)^{-1}(2\lambda_b) + (w^g)^{-1}\Sigma(b, \gamma^*)\lambda_\gamma$. By (7) with w_g, w_{g+1}, $(w^g)^{-1}$ in the roles of w_1, w_2, w_g, the first term on the right has support in Π_1, the second in Π_2, whence (a) and (b).

(9) With the notation as in (8)

$$\sum\nolimits_1 n_\alpha \alpha = 2\sum\nolimits_1 n_\alpha \lambda_\alpha - 2\sum\nolimits_2 n_\beta \lambda_\beta.$$

If we dot with α^* we get $2n_\alpha$ on both sides. If we dot with β^* instead we get $-2n_\beta$ since the left side equals $\mu - \Sigma_2 n_\beta \beta$ and μ is orthogonal to all elements of Π_2.

(10) With the notation as in (8)

(a) $w^g \Sigma_1 n_\alpha \lambda_\alpha = -\lambda_{b,g}$

(b) $w^g \Sigma_2 n_\beta \lambda_\beta = \lambda_{b,g+1}$

(c) $w^g(\Sigma_1 n_\alpha \lambda_\alpha - \Sigma_2 n_\beta \lambda_\beta) = -\lambda_b$

Here $\lambda_{b,g}$, for example, denotes the part of λ_b supported by Π_g, that one of Π_1 and Π_2 that contains b. By (8) and (9) we have (c). Then (a) and (b) follow from (7).

We turn now to our discussion of affine Coxeter elements. One of these is $c = ww_2w_1$ with w_1 and w_2 as above and $w = w_{1-\mu}$ the reflection corresponding to $1 - \mu$. First we give the proof of (∗) of 1(8), thus completing the proof of that result. We have to show that c', the linear part of c, is conjugate to the product of the reflections other than that at the branch point b. By (5), which is all that is needed, we have $w^g w_\mu (w^g)^{-1} = w_b$. This can be written as $w^g(w_\mu w_2 w_1)(w^g)^{-1} = w_b w_g w_{g+1}$. The left side is conjugate to c' and the right side equals the stated product since the w_b in front cancels the w_b that occurs as a factor of w_g.

Next we present a normal form for the affine Coxeter element $c = ww_2w_1$. Let F denote the standard fundamental domain for the affine Weyl group, defined as the region (a simplex) where $\alpha \geq 0$ for all $\alpha \in \Pi$, and $1 - \mu \geq 0$. Then w_1 is the reflection across the facet F_2 of F where all $\alpha \in \Pi_1$ are 0 (since the elements of Π_1 are mutually orthogonal), and ww_2 is the reflection across the opposite facet F_1 where $1 - \mu$ and all $\beta \in \Pi_2$ are 0. We seek points γ_1 and γ_2 in F_1 and F_2 such that the line L joining them is orthogonal to F_1 and to F_2. Then c will be a screw displacement along L (in 3 dimensions the motion of a screw whose axis is L): translation by the vector $2(\gamma_1 - \gamma_2)$ in the direction of L composed with a rotation around L (i.e. an isometry fixing the points of L), the two factors necessarily commuting and being determined by the stated conditions. Since c', the linear part of c, has 1 as an eigenvalue of multiplicity 1 (with corresponding eigenvector in the direction of L), L, in the present case, may also be described as the set of points moved the least distance by c.

(11) Write $\mu = \Sigma_1 n_\alpha \alpha + \Sigma_2 n_\beta \beta$ as in (8) set $\delta = \Sigma_1 n_\alpha \alpha$, so that $(\delta, \delta) = \Sigma_1 n_\alpha^2(\alpha, \alpha)$. Then the solution to our problem is $\gamma_1 = 2\Sigma_1 n_\alpha \lambda_\alpha/(\delta, \delta)$ and $\gamma_2 = 2\Sigma_2 n_\beta \lambda_\beta/(\delta, \delta)$. Thus c is a rotation around the line joining these points composed with the translation by the vector $2(\gamma_1 - \gamma_2) = 2\delta/(\delta, \delta)$, of length $2/|\delta|$, along the line. Further γ_2 is the point on L closest to the origin.

First δ is orthogonal to F_2 clearly and also to F_1 since $\delta = \mu - \Sigma_2 n_\beta \beta$. Write the equation of (9) as $\delta = \delta_1 - \delta_2$. Then by the definitions $\delta_2 \in F_2$ and $\delta_1 \in F_1$ except for the condition $(\mu, \delta_1) = 1$. Now $(\mu, \delta_1) = 2\Sigma_1 n_\alpha^2(\alpha, \lambda_\alpha) = \Sigma_1 n_\alpha^2(\alpha, \alpha) = (\delta, \delta)$. It follows that $\gamma_1 = \delta_1/(\delta, \delta)$ is in F_1 and γ_2 is in F_2, and that $2(\gamma_1 - \gamma_2) = 2(\delta_1 - \delta_2)/(\delta, \delta) = 2\delta/(\delta, \delta)$ is orthogonal to F_1 and to F_2. Finally γ_2 is the point of L closest to the origin since it is orthogonal to the vector δ along L.

In the normal form for c just given, the axis L and the translational part can be calculated quite explicitly in any given case, but the same can not be said of the rotational part. Here is another normal form which remedies this deficiency.

(12) Let b be the root in (5) above, and write $\lambda_b = \lambda_{b,g} + \lambda_{b,g+1}$ with $\lambda_{b,g}$ the part supported by Π_g (which includes b) and $\lambda_{b,g+1}$ the part supported by Π_{g+1}. Let L be the line through $\varepsilon = \lambda_{b,g+1}/2(\lambda_b, \lambda_b)$ in the direction of λ_b. Then ε is the point of L closest to the origin. Form the rotation around L whose linear part is the product of the simple reflections other than that for b and compose it with the translation by $-\lambda_b/(\lambda_b, \lambda_b)$ in the direction of L. Then the result is an affine Coxeter element c expressed in standard form as a screw displacement.

With $\gamma_1, \gamma_2, \delta$ as in (11) we have from (8) and (10) that $w^g\delta = -2\lambda_b$, $w^g\gamma_1 = -2\lambda_{b,g}/(\delta, \delta)$ and $w^g\gamma_2 = 2\lambda_{b,g+1}/(\delta, \delta)$. Thus (12) follows from (11).

We observe that since $\Pi - \{b\}$ is a union of systems of type A_n, the eigenvalues and eigenvectors of the linear part of c above, as well as its order, can be easily determined. So can λ_b, hence the other items of (12) also, especially when all roots have the same length, so that b is a branch point:

(13) If $\lambda_b = \Sigma m_\alpha \alpha$ $(\alpha \in \Pi)$, then $m_b = (p^{-1} + q^{-1} + r^{-1} - 1)^{-1}$ in terms of the branch lengths, and along the branch of length p, for example, starting at the end point the m_α's are $p^{-1}m_b, 2p^{-1}m_b, \ldots$.

For, as is easily seen, the scalar product of the proposed vector with b^* is 1, with all other simple coroots is 0.

We conclude our paper with some further remarks about the McKay correspondence that arise from the ideas of this section. G will be a

Kleinian group, as in §1, Π the corresponding simple root system, and the other notations as above.

(14) $2\Sigma_1 n_\alpha^2 = g$, the order of $|G|$.

Regarding the n_α's as the degrees of the irreducible representations of G we have $\Sigma_1 n_\alpha^2 + \Sigma_2 n_\beta^2 + 1$ (trivial representation) $= g$. The same for the group $G/\{\pm 1\}$ yields $\Sigma_2 + 1 = g/2$ (see the proof of 1(4b)), whence (14).

(15) $2\Sigma_1 n_\alpha^2 = 4(p^{-1} + q^{-1} + r^{-1} - 1)^{-1}$ in terms of the branch lengths.

For, $(\delta, \delta) = \Sigma_1 n_\alpha^2 (\alpha, \alpha)$ in (11) and $(2\lambda_b, 2\lambda_b) = 4(\lambda_b, \Sigma m_\alpha \alpha) = 2m_b(b, b)$. These are equal by (8a), and then (15) follows from (13) and the equality of all root lengths.

These equations show that m_b in (13) is just $g/4$. They also lead to another, nontrigonometric, proof of §1(8), with which we close our paper. We have $g = 4(p_1^{-1} + q_1^{-1} + r_1^{-1} - 1)$ since the decomposition of G into conjugacy classes yields $g = 1 + 1 + (p_1 - 1)g/2p_1 + (q_1 - 1)g/2q_1 + (r_1 - 1)g/2r_1$. Since $r = r_1 = 2$, this, (14) and (15) yield $p^{-1} + q^{-1} = p_1^{-1} + q_1^{-1}$. But also $p + q = p_1 + q_1$ since $p + q + r - 1$ is the number of irreducible representations of G and $p_1 + q_1 + r_1 - 1$ is the number of conjugacy classes. Dividing one equation by the other we get $pq = p_1 q_1$. Thus $(p, q) = (p_1, q_1)$, as required.

References

[B] N. Bourbaki, *Groupes et Algèbres de Lie*, vol. IV, V, VI, Hermann and Co., Paris (1968).

[C] H. S. M. Coxeter, *Regular Complex Polytopes*, Cambridge University Press, Cambridge (1974).

[G] G. Gonzalez-Sprinberg and J. L. Verdier, *Construction géométrique de la correspondance de McKay*, Ann. Sci. E.N.S., **16** (1983), 409–449.

[H] D. Happel, U. Preiser and C. M. Ringel, *Binary polyhedral groups and Euclidean diagrams*, Manuscripta Math., **31** (1980), 317–329.

[K] H. Knörrer, *Group representations and resolution of rational double points*, Bonn University preprint.

[Ko] B. Kostant, *On finite subgroups of SU(2), simple Lie algebras and the McKay correspondence*, M.S.R.I. preprint.

[M] J. McKay, *Graphs, singularities, and finite groups*, Amer. Math. Soc. Proc. Sym. Pure Math., **37** (1980), 183–186.

[Mu] D. Mumford, *The topology of normal singularities of an algebraic surface and a condition for simplicity*, Publ. Math. I.H.E.S., **9** (1961), 5–22.

[S$_1$] P. Slodowy, *Simple Singularities and Simple Algebraic Groups*, Lecture Notes in Math. 815, Springer-Verlag, New York (1980).

[S_2] _____, *Platonic Solids, Kleinian Singularities and Lie Groups*, Lecture Notes in Math. 1008, Springer-Verlag, New York (1980), 102–138.
[St] R. Steinberg, *Finite reflection groups*, Trans. Amer. Math. Soc., **91** (1959), 493–504.

Received October 29, 1984.

UNIVERSITY OF CALIFORNIA
LOS ANGELES, CA 90024

Some Consequences of the Elementary Relations in SL_n

Robert Steinberg

1. The philosophy of this paper (and talk) is that significant results can be obtained for various classes of groups (all of the split and quasisplit Chevalley groups of rank > 1, which include SL_n, Sp_n, some Spin_n, SO_n and SU_n) by using certain obvious relations among the simplest elements of these groups in rather simple-minded ways. The results obtained are not always the most general of their kind that are now known, especially towards the end when discrete subgroups are considered, but the proofs are extremely simple. For expository purposes and in order to reach a larger audience I shall confine my discussion to the canonical example of a split group, SL_n. The proofs for the other groups are not essentially different. The following notation will be used: $x_{ij} = 1 + E_{ij}$, $x_{ij}^t = 1 + tE_{ij}$ ($1 \leq i, j \leq n$; $i \neq j$; $\{E_{ij}\}$ the matrix units, $t \in \mathbf{C}$). Observe that for $t \in \mathbf{Z}$ this really is the tth power. The elements x_{ij}^t of SL_n are all unipotent.

2. We start with a classical theorem of Nielsen and Magnus (stated differently, but equivalently, by them; see [**M**]).

THEOREM 1. *The group* $\Gamma \equiv SL_n(\mathbf{Z})$ $(n \geq 3)$ *is generated by the elements* $x_{ij} (i \neq j)$ *subject to the relations*

$$\text{(B)} \qquad (x_{ij}, x_{k\ell}) = \begin{cases} x_{i\ell} & \text{if } i \neq \ell,\ j = k, \\ 1 & \text{if } i \neq \ell,\ j \neq k. \end{cases}$$

$$\text{(C)} \qquad (x_{12} x_{21}^{-1} x_{12})^4 = 1.$$

Here (x, y) denotes the commutator $xyx^{-1}y^{-1}$. The proof is elementary and may be found, somewhat simplified by Silvester, in [**Mi**]. The result has been extended to other split groups by Wardlaw and Humphreys, but not yet to G_2 as far as I know.

1991 *Mathematics Subject Classifications.* Primary 20-xx; Secondary 22-xx.

Reprinted from Some consequences of the elementary relations in SL_n, *Cont. Math.* **45** (1985), 335–350.

3. In about 1965 Christofides [**C**], and somewhat later I, independently proved:

THEOREM 2. *Theorem 1 holds with* $\mathbf{Z}$ *replaced by* $\mathbf{Z}/q\mathbf{Z}$ $(q \geq 1)$ *provided that one adds the relation:*

$$\text{(D)} \qquad x_{12}^q = 1.$$

It is further true that the relation (C) is needed only when 4 divides q. The proof runs as follows. First the Chinese Remainder Theorem reduces us to the case in which $q = p^a$, a prime power. Here there is a normal form: if L, U, M denote the subgroup of $SL_n(\mathbf{Z}/p^a\mathbf{Z})$ consisting of those matrices that are strictly lower triangular, strictly upper triangular, monomial, respectively, then $SL_n(\mathbf{Z}/p^a\mathbf{Z}) = LUM$, with uniqueness of expression for the element 1. This normal form is Christofides'; mine is somewhat different. The proof of the theorem is completed by showing that an arbitrary element of the group, given as a product of x_{ij}'s, can be reduced to the above normal form as a consequence of the relations.

In proving this theorem the two of us had the same objective in mind, namely, an elementary proof of the congruence subgroup theorem over $\mathbf{Z}$, proved slightly earlier by Mennicke [**Me**] and by Bass-Lazard-Serre [**B-L-S**]. One version of this result is:

THEOREM 3. *Let* Γ *denote* $SL_n(\mathbf{Z})$ $(n \geq 3)$ *and* Γ_q *(respectively* Γ_q^**) the subgroup of elements that are* 1 *(respectively scalar)* $\bmod q$. *If* N *is any normal subgroup of* Γ *then there exists a unique* $q \geq 0$ *such that* $\Gamma_q \subseteq N \subseteq \Gamma_q^*$. *Hence either* N *consists of scalars and is finite* $(q = 0)$ *or* N *is of finite index* $(q > 0)$.

The proof runs as follows. If $q \equiv$ g.c.d. $\{a_{ij},\ a_{ii} - a_{jj}\ (i \neq j\},\ a = (a_{ij}) \in N\}$ then by elementary matrix calculations due to Brenner [**Br**] $x_{12}^q \in N$. If E_q denotes the normal subgroup of Γ generated by x_{12}^q it follows that $E_q \subseteq N \subseteq \Gamma_q^*$. It is enough now to show that $E_q = \Gamma_q$. For $q = 0$ this is obvious. Assume $q > 0$. From $E_q \subseteq \Gamma_q \subseteq \Gamma$ we have a homomorphism $f : \Gamma/E_q \to \Gamma/\Gamma_q$. Since $\Gamma/\Gamma_q \simeq SL_n(\mathbf{Z}/q\mathbf{Z})$ and $x_{12}^q = 1$ in Γ/E_q, Theorem 2 produces a map in the other direction. Thus f is an isomorphism and $E_q = \Gamma_q$.

4. Earlier, in 1962 [**St1**], I had obtained a result like Theorem 1 but with $\mathbf{Z}$ replaced by a field.

THEOREM 4. $G \equiv SL_n(k)$ $(n \geq 3,\ k$ *any field) is generated by the* x_{ij}^t*'s subject to*

$$(A) \qquad x_{ij}^t x_{ij}^u = x_{ij}^{t+u}$$

$$(B) \qquad (x_{ij}^t, x_{k\ell}^u) = \begin{cases} x_{i\ell}^{tu} & \text{if } i \neq \ell,\ j = k, \\ 1 & \text{if } i \neq \ell,\ j \neq k \end{cases}$$

$$(C) \qquad h_{12}(t)h_{12}(u) = h_{12}(tu) \text{ with}$$

$$h_{12}(t) \equiv x_{12}^t x_{21}^{-t^{-1}} x_{12}^t \cdot x_{12}^{-1} x_{21} x_{12}^{-1} \text{ for each } t \in k^*.$$

If k *is finite, the relations* (C) *may be omitted.*

Observe that when k is replaced by $\mathbf{Z}$ and the generating set $\{x_{ij}^t\}$ by $\{x_{ij}\}$ the relations (A) become redundant, (B) reduces to the earlier (B), and (C) with $t, u = -1$, the only nontrivial units of $\mathbf{Z}$, reduces to the earlier (C).

This result has had some bearing on the classification problem for finite simple groups just completed, mainly in the recognition part of it. But when I first proved it I was concerned with something else, namely the possibility that every projective representation of G can be lifed to an ordinary one, or, quivalently, that every central extension is trivial, or, again, that the Schur multiplier is trivial. This (as in the original paper) turns out to be the case for SL_n and the other split and quasisplit finite groups, with a handful of exceptions, by a program initiated by me in [**St1**] and completed by Grover, Griess and me in [**Gro**], [**Gr**] and [**St2**]. The idea is, given a central extension $\bar{G} \to G$, to lift the relations (A) and (B) from G to $\bar{G}$ and thus to split the extension. If k has enough elements (≥ 5) this is relatively easy; if not, it takes a bit of a scramble. I should add that for some of these groups results of this kind were obtained earlier by H. Zassenhaus, but as far as I know he never published them.

5. If k is infinite the relations (A) and (B) can also be lifted (more easily, in fact) and they then define a universal central extension $\bar{G}$ of G. The kernel, $K_2(G)$, coming from the relations (C) (and therefore trivial if k is finite by Theorem 4), has been analyzed by myself, Moore [**Mo**] and, finally, Matsumoto [**Ma**] who has proved:

THEOREM 5. *For $G = SL_n(k)$ ($n \geq 3$, k any field) $K_2(G)$ is generated by symbols $(t, u) \in k^{*2}$, in correspondence with the relations (C), subject to:*
(1) (t, u) is bimultiplicative.
(2) $(t, 1-t) = 1$ if $t \neq 0, 1$.

If k is a number field this has connections with reciprocity laws via the norm residue symbol which also satisfies these relations. In particular, if $k = \mathbf{Q}$ the result of the analysis is a very pretty proof of the classical quadratic reciprocity law for $\mathbf{Z}$. The proof, due to Tate and closely related to Gauss' first proof, may be found in [**Mi**]. If k is a local field (e.g., $\mathbf{R}$ or some $\mathbf{Q}_p$) and we restrict to continuous symbols (t, u), the group $K_2(G)$ works out to be $\mu(k)$, the (finite, cyclic) group of roots of 1 in k which may be interpreted: the fundamental group of $SL_n(k)$, which is now a topological group, is $\mu(k)$. For $k = \mathbf{R}$, $\mu(k) = \mathbf{Z}/2\mathbf{Z}$, this result is classical, but for $k = \mathbf{Q}_p$ this was done for the first time by Moore and it provides a reasonable definition for the fundamental group of G, which in this case is a totally disconnected group. These results, it should be mentioned, are closely related to the congruence subgroup problem (see Theorem 3) for $\mathbf{Z}$ and number fields [**B-M-S**]. It should also be mentioned that if k is replaced by R, any associative ring with 1, and $n \to \infty$, the relations (A) and (B) still define a central extension of the group generated by all x_{ij}^t's, which yields a kernel, $K_2(R)$, on which a great deal of work has been done, with applications to topology and algebraic K-theory.

6. What I would like to discuss from now on are some results on the representations of arithmetic subgroups of (some) Lie groups that can be proved rather simply with the aid of the relations that we have been discussing. More general results have been obtained by Borel [**Bo**], Margulis (see [**Ti**]), Weil [**W**] and others, but by methods that are considerably harder.

THEOREM 6. *Let $\Gamma = G(\mathbf{Z}) = SL_n(\mathbf{Z})$ ($n \geq 3$) and $f : \Gamma \to GL_m(\mathbf{C})$ be any representation. Then there exist representations g and $h : \Gamma \to GL_m(\mathbf{C})$ such that:*
(a) $f = gh$ and $g(\Gamma)$ and $h(\Gamma)$ commute elementwise. Further $g = u \circ f$ and $h = s \circ f$ ($u =$ unipotent part, $s =$ semisimple part) on each x_{ij}^t.

(b) *g is the restriction of a polynomial representation of $SL_n(\mathbf{C})$ and agrees with f on a subgroup E of finite index in Γ, on some Γ_q as in Theorem 3, so that h factors through the finite quotient Γ/E.*

(c) *The conditions in* (a) *or those in* (b) *determine g and h uniquely.*

The proof of this basic result will be given in several steps.

(1) *Let x, y, z in $GL_m(C)$ be such that $(x, y) = z$ and z commutes with x and y.*

(a) *$(x_u, y) = z_u$, $(x, y_s) = z_s$ and $(x_u, y_s) = 1$.*

(b) *$z_s^q = 1$, that is, z^q is unipotent, for some q dividing $m!$.*

To prove the second equation in (a), for example, we write $xyx^{-1} = zy$ and take semisimple parts. Then, taking the commuting semisimple elements z_s and y_s in diagonal form as we may, we see that x acts on y_s via a permutation of the diagonal entries. If q is the order of this permutation then $y_s = x^q y_s x^{-q} = z_s^q y_s$, whence $z_s^q = 1$.

(2) (a) *$f(x_{ij})_s$ and $f(x_{k\ell})_u$ commute for all i, j, k, ℓ.*

(b) *$f(x_{ij}^q)$ is unipotent for all i, j and q as in* (1).

If $i \neq \ell$ or $j \neq k$ then (a) follows from the relations (B) (see Theorem 1) and (1a) above. In the remaining case we choose an index $p \neq i, j$ (recall that $n \geq 3$). Then $f(x_{ij})_u$ commutes with $f(x_{jp})_s$ and $f(x_{pi})_s$, hence also with their commutator, which is $f(x_{ji})_s$ by (B) and (1a) above. For (b) we use (1b) with $x = f(x_{ip})$, $y = f(x_{pj})$, $z = f(x_{ij})$.

(3) *Set $Y_{ij} = \log f(x_{ij})_u$ (using the usual series for* $\log(1 + T)$, *a polynomial when $1 + T$ is unipotent and hence T is nilpotent). Then the following relations hold:*

$$(\mathrm{B}') \qquad [Y_{ij}, Y_{k\ell}] = \begin{cases} Y_{i\ell} & \text{if } i \neq \ell,\ j = k \\ 0 & \text{if } i \neq \ell,\ j \neq k \end{cases}$$

Consider the first case. From (B) it readily follows that $(x_{ij}^r, x_{j\ell}^s) = x_{i\ell}^{rs}$ for all $r, s \in \mathbf{Z}$. This relation continues to hold if we apply f and take unipotent parts, by (1a). If we then write $f(x_{ij}^r)_u = f(x_{ij})_u^r = (\exp Y_{ij})^r = \exp rY_{ij} = 1 + rY_{ij} + \cdots$ and similarly for the other terms and then equate the coefficients of rs on the two sides we get $[Y_{ij}, Y_{j\ell}] = Y_{i\ell}$. The proof of the second case is similar but easier.

(4) *The relations (B′) of (3) form an abstract definition of the Lie algebra $s\ell_n$ $(n \geq 3)$ with Y_{ij} corresponding to the matrix unit E_{ij}.* Set $Z_i = [Y_{i,i+1}, Y_{i+1,i}]$ $(i = 1, 2, \ldots, n-1)$. In $s\ell_n$ the corresponding elements $E_{ii} - E_{i+1,i+1}$ together with the E_{ij} $(i \neq j)$ form a basis. Thus we need only show that the product of any $Y_{k\ell}$ with any Y_{ij} or Z_i is a linear combination of elements of these types as a consequence of the relations (B′), and we may assume that $\ell = k \pm 1$ since these Y's generate the others. The only products not obviously covered by (B′) in conjunction with Jacobi's identity are those of the form $[Y_{k,k+1}, Z_k]$ and $[Y_{k+1,k}, Z_k]$. In the first case, for example, we choose a third index p, replace $Y_{k,k+1}$ by $[Y_{k,p}, Y_{p,k+1}]$ (a relation in (B′)) and use Jacobi's identity to get to cases already considered.

(5) Proof of (a) and (b). As an algebraic group $SL_n(\mathbf{C})$ is simply connected. Hence by (3) and (4) there exists a polynomial representation $g : SL_n(\mathbf{C}) \to GL_m(\mathbf{C})$ such that $dg : E_{ij} \to Y_{ij}$. We have $g(x_{ij}) = g(\exp E_{ij}) = \exp Y_{ij} = f(x_{ij})_u$. (The exp series here, like the log series earlier, is finite). We can also get to this point by using the simpleconnectedness of $SL_n(\mathbf{C})$ as a Lie group. Set $h(x_{ij}) =$

$f(x_{ij})_s$ $(= f(x_{ij})g(x_{ij})^{-1})$ for all i and j. Then h extends a representation of Γ into $GL_m(\mathbf{C})$: if $R \equiv R(x_{ij}) = 1$ is any relation on the x_{ij}'s in Γ, then $R(f(x_{ij})_s) = R(f(x_{ij})_s f(x_{ij})_u) R(f(x_{ij})_u)^{-1}$ (by (2a)) $= f(R) \cdot g(R)^{-1} = 1$ since f and g are homomorphisms. Clearly $f = gh$ and $g(\Gamma)$ and $h(\Gamma)$ commute elementwise. Finally $h(x_{ij}^q) = f(x_{ij})_s^q = 1$ by (2b) so that $\ker h \supseteq \Gamma_q$, of finite index in Γ as in Theorem 3, and $f = g$ on Γ_q.

(6) Proof of (c). The conditions in (a) determine g and h since $\{x_{ij}\}$ generates Γ and those in (b) do because each Γ_q is dense in $SL_n(\mathbf{C})$ in the Zariski topology.

Remarks. (a) The results are the same over any field of char 0. (b) The map g above is defined over the field $\mathbf{Q}(f(\Gamma))$. (c) The proof can be adapted to show that any representation of Γ in char $p \neq 0$ factors through a finite quotient Γ/E. (d) A result quite close to Theorem 6 occurs in [**B-M-S**], but with a proof that is not easy. (e) These results suggest that the group Γ is rather rigidly embedded in $SL(\mathbf{C})$. For a related aspect of this situation see [**W**].

We consider next a result like Theorem 6, but over $\mathbf{Q}$, which we deduce, as is done in [**B-M-S**], from that theorem.

THEOREM. *Every representation* $f : SL_n(\mathbf{Q}) \to GL_m(\mathbf{C})$ *is polynomial.*

Unlike most of the other results of this section this result holds also for $n - 2$. We assume first that $n \geq 3$. Let g be as in Theorem 6 relative to the restriction of f to $SL_n(\mathbf{Z})$. If g_d (d diagonal in $SL_n(\mathbf{Q})$) is defined likewise for $dSL_n(\mathbf{Z})d^{-1}$, then g and g_d agree on all x_{ij}^a for a suitable positive integer a, whence $g = g_d$ since these elements generate a dense subgroup of $SL_n(\mathbf{C})$ in the Zariski topology. Now if r is any positive rational number and the indices i and j are fixed, d may be chosen so that $x_{ij}^r = dx_{ij}^s d^{-1}$ with s some multiple of q as in Theorem 6. It follows that $f = g$ on all x_{ij}^r, and since these elements generate $SL_n(\mathbf{Q})$ that $f = g$. Assuming now that $n = 2$, we set $x_{12}^t = x(t)$, $x_{21}^t = y(t)$, $\operatorname{diag}(t, t^{-1}) = h(t)$. Using the relations $h(r)x(t)h(r)^{-1} = x(r^2 t)$ we get, as in the proof of steps (1b) and (2b) of Theorem 6, that $f(x(qt))$ is unipotent for all $t \in \mathbf{Q}$ and some q dividing $m!^2$, that is, $f(x(t))$ is always unipotent, and likewise for $f(y(t))$. Set $X = \log f(x(1))$ and $Y = \log f(y(1))$. Then (*) $(ad\ X)^2 Y = -2X$ and $(ad\ Y)^2 X = -2Y$. To see this, apply f to the relation $y(-t^{-1})x(t)y(t^{-1}) = x(t)y(-t^{-1})x(t)$, take logs, and then work out the coefficients of t and t^{-1} on the two sides. Since these relations with X, Y corresponding to E_{12}, E_{21} form a presentation of $s\ell_2$, a polynomial map g may be defined as in step (5) of the proof of Theorem 6 and the proof completed as above.

We return to our main development.

THEOREM 7. *The main conclusions of Theorem 6 hold if Γ is any arithmetic subgroup of $SL_n(\mathbf{C})$, that is, any subgroup of $SL_n(\mathbf{Q})$ commensurable with $SL_n(\mathbf{Z})$, namely* (a), (b) *and* (c) *with the second sentence of* (a) *and the reference to Γ_q in* (b) *omitted.*

First we show that f agrees with g, polynomial on $SL_n(\mathbf{C})$, on a subgroup E of Γ of finite index. If this holds for some Γ and if Γ' is such that $\Gamma' \supseteq \Gamma$ and $[\Gamma' : \Gamma] < \infty$, then clearly it holds for Γ'. If it holds for Γ' instead and f is a representation for Γ, then it holds for the induced representation $f(\Gamma \to \Gamma')$ of Γ', whose restriction to Γ is a multiple of f. By Theorem 6 we thus have our first assertion. Now define $h(x) = f(x)g(x)^{-1}$ on Γ. Then for $x \in \Gamma$, $y \in E$, which we suppose to be normal in Γ as we may, we have

$$h(x)g(y)h(x)^{-1} = f(x)g(x)^{-1}g(y)g(x)f(x)^{-1} = f(x)g(x^{-1}yx)f(x)^{-1} =$$
$$f(x)f(x^{-1}yx)f(x)^{-1} = f(y) = g(y)$$

since y and $x^{-1}yx$ are in E, so that $h(x)$ and $g(y)$ commute. This continues to hold for $y \in \Gamma$ since E is dense in Γ, in fact in all of $SL_n(\mathbf{C})$, in the Zariski topology, and g is polynomial. It follows that h is also a representation and that $\ker h \supseteq E$, whence (a) and (b). This calculation is taken from Serre [**S**] who used it for a similar purpose. Uniqueness in (c) is as before.

THEOREM 8. *Let Γ be any arithmetic subgroup of $SL_n(\mathbf{Q})$ $(n \geq 3)$ and $f : \Gamma \to GL_m(\mathbf{C})$ any representation.*

(a) *f is semisimple (completely reducible).*

(b) *If f is irreducible then f is uniquely the tensor product of a representation g_1 that is polynomial on a subgroup E of finite index in Γ and another one h_1 that has E in its kernel, both necessarily irreducible.*

(c) *Further the tensor product of two representations g_1 and h_1 as in* (b) *is always irreducible.*

In (a) let $V = C^m$, V_1 a Γ-submodule. We restrict to E as in Theorem 7 to get a polynomial representation and hence a complementary E-submodule V_2 by Weyl's theorem that polynomial representations of semisimple groups in characteristic 0 are semisimple. Let p_1 be the projection of V onto V_1 along V_2. Then $f(x)p_1f(x)^{-1}$ averaged over the finite group Γ/E works out to another projection of V onto V_1 commuting with the action of Γ, and its kernel is a Γ-submodule complementary to V_1. For (b) let g and h be as in Theorem 7. They yield an irreducible representation of $\Gamma \times \Gamma/E$ via $(x, y) \to g(x)h(y)$, which is necessarily of the form $g_1 \otimes h_1$ with g_1 irreducible on Γ and h_1 on Γ/E. On $\Gamma \times 1$ we have $g = g_1 \otimes$ (trivial action on the space of h) and similarly for h on $1 \times \Gamma/E$. It follows that g_1 and h_1 have the required properties and that via the diagonal action of Γ, $f = g_1 \otimes h_1$. Let g_1' and h_1' also work, relative to E' which may be taken equal to E. Since h_1 and h_1' are trivial on E, $rg_1 = r'g_1'$ $(r = \dim h_1,\ r' = \dim h_1')$ on E. Since E is dense in $SL_n(\mathbf{C})$ it follows that $g_1 = g_1'$ and thus $h_1 = h_1'$. Now let $f = g_1 \otimes h_1$ as in (c), so that f is irreducible on $\Gamma \times \Gamma/E$. Any Γ-submodule (Γ acting diagonally) is an $E \times 1$ submodule, hence also a $\Gamma \times 1$ submodule since E is dense in Γ and g_1 is polynomial, hence also is a $\Gamma \times \Gamma/E$ submodule since in $\Gamma \times \Gamma$ we have $(x, y) = (xy^{-1}, 1) \cdot (y, y)$. Thus f is irreducible on Γ as required.

Remarks. (a) A consequence of (a) is that $H^1(\Gamma, V) = 0$ for any Γ-module V finite-dimensional over any field of characteristic 0, a result related to the rigidity mentioned at the end of Theorem 6. (b) In principle the (irreducible) polynomial representations of Γ are known (highest weight classification). Thus there remains the problem of determining those with kernel of finite index which is mostly unsolved even in simple cases. For $\Gamma = SL_n(\mathbf{Z})$, $E = \Gamma_q$ for some $q > 0$ by Theorem 3, the problem is to determine the representations of $\Gamma/E \simeq SL_n(\mathbf{Z}/q\mathbf{Z})$, and $q = p^a$ (p prime) may be assumed. The case $n = 2$ has been solved by Shalika [**Sh**] and Tanaka [**T**] independently, in difficult work, but for $n \geq 3$ it is unsolved although for $a = 1$ a good deal is known [**G**], [**Sr**]. These representations are needed for the harmonic analysis of the topological group $SL_n(\mathbf{Q}_p)$.

THEOREM 9. *Let* Γ, Γ' *be arithmetic subgroups of* $SL_n(\mathbf{Q})$, $SL_m(\mathbf{Q})$ $(n, m \geq 3)$ *and* $f : \Gamma - \Gamma'$ *a homomorphism.*

(a) f *is polynomial on a subgroup of finite index.*

(b) *If* f *is surjective then* f *is the product of a polynomial homomorphism and one into the (finite) center of* Γ.

To get (a) we embed Γ' in $SL_m(\mathbf{C})$ and use Theorem 6. For (b) let g and h be as in Theorem 6. Then $g(\Gamma) \cap \Gamma'$ has finite index in Γ', hence is dense in $SL_m(\mathbf{C})$, hence acts irreducibly on $\mathbf{C}^m$. Since $h(\Gamma)$ commutes with this action, it consists of scalars, whence (b).

COROLLARY. (a) *In* (b) *we necessarily have* $n = m$. *Hence an arithmetic subgroup of* SL_n *can never be isomorphic to one of* SL_m *if* $n \neq m$.

(b) *Any automorphism of* Γ *above is of the form* gh *with* g *induced by an automorphism of the algebraic group* $SL_n(\mathbf{C})$ *and* h *a homomorphism into the center of* $SL_n(\mathbf{C})$.

The map g above induces an isomorphism of the corresponding algebraic groups since both are simple (over their centers), whence both have the same dimension and $n = m$. In fact $PSL_n(k)$ and $PSL_m(K)$ are isomorphic as abstract groups only if $n = m$ and k is isomorphic to K, with a few exceptions and k and K are small, by a classical theorem of Schreier and van der Waerden [**S-W**]. Part (b) follows from Theorem 9(b).

Remarks and Examples. If $\Gamma = SL_n(\mathbf{Z})$ $(n \geq 3)$ then every automorphism of Γ is induced by conjugation by an element x of $GL_n(\mathbf{Z})$ or else is such composed with the inverse transpose. By (B) Γ equals its own derived group. Thus $h = 1$ above and $f = g$ on Γ. As an automorphism of the algebraic group (or complex Lie group) $SL_n(\mathbf{C})$, g has the required form with $x \in GL_n(\mathbf{C})$. Since g preserved Γ it readily follows that x may be taken in $GL_n(\mathbf{Z})$. The result of (a) is substantially true for $n = 2$ also [**H-R**], with an elementary proof based on the fact that $PSL_2(\mathbf{Z})$ is the free product of a group of order 2 and one of order 3. As remarked earlier the above methods can be adapted to arbitrary split and quasisplit groups of rank greater than 1 so that, for example: the automorphisms of $Sp_{2n}(\mathbf{Z})$ can be determined without trouble [**R**], and arithmetic subgroups of Sp_{2n} and SL_m $(m \geq 3)$ can never be isomorphic. As also remarked earlier these results are not new but the above proofs are considerably simpler than any that we have seen elsewhere.

These methods can also be extended quite easily to apply to number fields other than $\mathbf{Q}$ to prove, for example, the following result (also given in [**S**]).

THEOREM 10. *Let* R *be the ring of integers of an algebraic number field* k *and* $\Gamma = SL_n(R)$ $(n \geq 3)$. *Let* $f : \Gamma \to GL_m(\mathbf{C})$ *be any representation.*

(a) f *is semisimple.*

(b) *Let* f *be irreducible. Let* $\sigma_1, \sigma_2, \ldots, \sigma_r$ $(r = [k : \mathbf{Q}])$ *denote the embeddings of* k *into* $\mathbf{C}$. *Then* f *is expressible uniquely as* $h\Pi(g_i \circ \sigma_i)$ *(tensor product;* i *goes from* 1 *to* r*) with each* g_i *polynomial and* h *factoring through some finite quotient* Γ/E, *and all factors are irreducible.*

(c) *Conversely, any tensor product of factors as in* (b) *is irreducible.*

It is an easy exercise to deduce this from Theorem 8.

References

[B-L-S] A. Bass, M. Lazard, J.-P. Serre, *Sous-groupes d'indice fini dans* $SL(n, \mathbf{Z})$, Bull. Amer. Math. Soc. **70** (1964), 385–392.

[B-M-S] A. Bass, J. Milnor, J.-P. Serre, *Solution of the congruence subgroup problem*, Publ. Math. I.H.E.S. **33** (1967), 21–137.

[Bo] A. Borel, *On the automorphisms of certain subgroups of semisimple groups*, Procedings Bombay Colloq. Alg. Geom. (1968), 43–73.

[Bo-T] A. Borel, J. Tits, *Homomorphismes "abstraits" de groupes algébriques simples*, Ann. of Math. **97** (1973), 499–571.

[Br] J.L. Brenner, *The linear homogeneous group III*, Ann. of Math. **71** (1960), 210–223.

[C] A. Christofides, Thesis, Queen Mary College, U. London (1966).

[G] J.A. Green, *The characters of the finite general linear groups*, Trans. Amer. Math. Soc. **80** (1955), 402–447.

[Gr] R. Griess, *Schur multipliers of finite simple groups of Lie type*, Trans. Amer. Math. Soc. **183** (1973), 355–421.

[Gro] J. Grover, *Covering groups of groups of Lie type*, Pacific J. Math. **30** (1969), 645–655.

[H-R] L. Hua, I. Reiner, *Automorphisms of the unimodular group*, Trans. Amer. Math. Soc. **71** (1951), 331–348.

[M] W. Magnus, *Über n-dimensionale Gittertransformationen*, Acta Math. **64** (1935), 353–367.

[Ma] H. Matsumoto, *Sur les sous-groupes arithmétiques des groupes semisimples déployés*, Ann. Ecole Norm. Sup. **2** (1969), 1–62.

[Me] J. Mennicke, *Finite factor groups of the unimodular group*, Ann. of Math. **81** (1965), 31–37.

[Mi] J. Milnor, *Introduction to Algebraic K-theory*, Ann. of Math. Studies #72, Princeton U. Press, 1971.

[Mo] C. Moore, *Group extensions of p-adic and adelic linear groups*, Publ. Math. I.H.E.S. **35** (1969), 5–74.

[R] I. Reiner, *Automorphisms of the symplectic modular group*, Trans. Amer. Math. Soc. **80** (1955), 35–50.

[S-W] O. Schreier, B. L. Van der Waerden, *Die Automorphismen der projectiven Gruppen*, Hamburg Univ. Math. Seminar **6** (1928), 303–322.

[S] J.-P. Serre, *Le problème des groupes de congruence pour* SL_2, Ann. of Math. **92** (1970), 489–527.

[Sh] J. Shalika, *Representations of the two by two unimodular group over local fields, I, II*, Seminar at Institute for Advanced Study (1966).

[Sr] B. Srinivasan, *Representations of finite Chevalley groups*, vol. 764, Springer Lecture Notes, 1979.

[St1] R. Steinberg, *Générateurs, relations et revêtements de groupes algébriques*, Colloque Théorie des Groupes Algébriques, Bruxelles (1962), 113–127.

[St2] R. Steinberg, *Generators, relations and coverings of algebraic groups II*, J. Algebra **71** (1981), 527–543.

[T] S. Tanaka, *Irreducible representations of the binary modular congruence groups* mod p^λ, J. Math. Kyoto Univ. **7** (1967), 123–132.

[Ti] J. Tits, *Travaux de Margulis*, No. 482, Sém. Bourbaki, 1975-6.

[W] A. Weil, *Remarks on the cohomology of groups*, Ann. of Math. **80** (1964), 149–157.

University of California Los Angeles

Proceedings of Symposia in Pure Mathematics
Volume **47** (1987)

Tensor Product Theorems

ROBERT STEINBERG

Introduction. The following fundamental, although elementary, theorem dates from the time of Burnside (see, e.g., [**C-R**, Theorem 10.33]).

THEOREM 0. *If $G = H \times K$, a direct product of groups, then every (absolutely) irreducible (finite-dimensional) representation of G is expressible uniquely as a tensor product of irreducible representations of H and of K.*

We have seen this theorem used many times at this conference, applied to abstract representations, polynomial representations, and so on. What we want to discuss in this paper are some cases in which groups which are not direct products nevertheless turn out to have irreducible representations which are tensor products. Other results of this nature have been obtained by Clifford [**Cl**], updated in [**C-R**, §11], Gajendragadkar [**G**], and Ferguson and Turull [**F-T**], among others, but, unlike those results, the theorems that we shall discuss in this paper have the striking feature that they apply to all irreducible representations (continuous, abstract, ..., as the case may be) of the groups involved.

Continuous representations. The oldest theorem of this kind that we are aware of is the following one.

THEOREM 1. *Let G be a simply connected complex Lie group. Then every continuous representation of G is expressible uniquely as the tensor product of one that is holomorphic and one that is antiholomorphic.*

We shall establish only the existence of the decomposition since that is our main theme. The following proof is taken from [**Se2**, Exp. 22]. Let $f\colon G \to \mathrm{GL}_m(\mathbf{C})$ be the given irreducible representation of G. With G and $\mathrm{GL}_m(\mathbf{C})$ viewed as Lie groups over $\mathbf{R}$ the continuity of f implies its differentiability, and then the differential yields a homomorphism of $\mathbf{R}$-algebras: $(*)$ $df\colon g \to gl_m(\mathbf{C})$. To keep track of the fact that g is also an algebra over $\mathbf{C}$ we write J for the $\mathbf{R}$-linear map of multiplication by i. We extend the scalars from $\mathbf{R}$ to $\mathbf{C}$ to

1980 *Mathematics Subject Classification* (1985 *Revision*). Primary 20G05, 22E46; Secondary 22E40.

get a homomorphism of **C**-algebras: (**) $(df)^{\mathbf{C}}\colon g^{\mathbf{C}} \to gl_m(\mathbf{C})$. Let g_1 and g_2 denote the i and $-i$ eigenspaces of J acting on $g^{\mathbf{C}}$. These are **C**-algebras, $g^{\mathbf{C}}$ is their direct sum, and they are, respectively, isomorphic and complex-conjugate isomorphic to g via the natural imbedding of g into $g^{\mathbf{C}}$. If G_1 and G_2 denote the simply connected complex Lie groups whose Lie algebras are g_1 and g_2, the factorization $g \to g^{\mathbf{C}} = g_1 + g_2 \to \mathrm{gl}_m(\mathbf{C})$ of df leads to a factorization $G \to G_1 \times G_2 \to \mathrm{GL}_m(\mathbf{C})$ of f. By the above remarks the components of the first map are respectively holomorphic and antiholomorphic while the second map is holomorphic and hence is a tensor product of holomorphic representations of G_1 and G_2 by Theorem 0. The required result follows.

The process of going from G to $G_1 \times G_2$ above is called restriction of the base field (from **C** to **R**). If $L \supset K$ is a finite extension of fields, the process produces from a variety V defined over L a variety of $[L\colon K]$ times the dimension defined over K. The new variety $R_{L/K}V$ has the property that there is a natural bijection between $V(L)$ and $R_{L/K}V(K)$. If V is defined over K, then $R_{L/K}V$ is isomorphic over L to V^n ($n = [L\colon K]$) and the bijection becomes $x \to (\sigma_1(x), \sigma_2(x), \dots, \sigma_n(x))$ in terms of the set $\{\sigma_i\}$ of distinct K-imbeddings of L into $\overline{K}$. In the case of Theorem 1 we have $K = \mathbf{R}$, $\overline{K} = L = \mathbf{C}$, $\{\sigma_i\} = \{\text{identity, complex conjugation}\}$, and the principal result may be restated as follows.

THEOREM 1′. *Every continuous representation of $G(\mathbf{C})$ factors through a holomorphic representation of $(R_{\mathbf{C}/\mathbf{R}}G)(\mathbf{C})$.*

Polynomial representations in positive characteristics. The basic result here is the following one.

THEOREM 2. *Let G be a simply connected semisimple algebraic group defined over the field of p elements, and let X_p denote the set of dominant weights λ that satisfy $0 \le (\lambda, \check{\alpha}) \le p-1$ for all simple coroots $\check{\alpha}$ and M_p the corresponding set of irreducible polynomial representations of G. Then every irreducible polynomial representation of G is expressible uniquely as a tensor product $\bigotimes_{i=0}^{\infty}(g_i \circ F^i)$ (finite product, $g_i \in M_p$, F =Frobenius). The corresponding result for the finite group $G(\mathbf{F}(q))$ $(q = p^n)$ is also true with the product truncated to $\bigotimes_{i=0}^{n-1}$.*

This result was first proved for the type $G = \mathrm{SL}_2$, by Brauer and Nesbitt [**B-N**] in the finite case and by Chevalley [**Se1**, Exp. 20] in the algebraic case, and then for arbitrary types of groups by us in [**St1**]. Simultaneously, Wong treated the types SL_n and Sp_n and later in [**Wo**] extended his methods to apply to all types, with the emphasis throughout on the finite groups. Since then other proofs and expositions have been given by Ballard [**B**], Borel [**B-A**], Cartier [**C**], Cline, Parshall, and Scott [**C-P-S**], Kempf [**K**], and Donkin and Jantzen, the last two unpublished.

The [**C-P-S**] proof is probably the simplest one for the algebraic groups, but it does not apply to the finite groups. We sketch here our original proof since it applies to both kinds of groups and also to converses to Theorems 1, 3, and 5 of the present paper, since it depends mainly on the fact that the endomorphisms F^i of the base field are distinct.

By the classification of the irreducible polynomial representations of G according to their highest weights each of which is expressible uniquely in the form $\sum \lambda_i p^i$ ($\lambda_i \in X_p$), it is enough to show that the tensor product given in the theorem is irreducible. Now the action of G on this product leads, via a twisted version of the differential, to an action of g in which tX acts on the ith factor as $F^i(t)X$ whenever X is defined over $\mathbf{F}(p)$ and t is any scalar, and it is enough to show that this action is irreducible. Since the endomorphisms F^i are distinct, they are linearly independent. It easily follows that we need only show that g is irreducible on each factor and this is a theorem of Curtis [**Cu**]: each element of M_p is infinitesimally irreducible; i.e., the corresponding representation of g given by the differential is also irreducible.

In view of the last-mentioned result we have the following reformulation of Theorem 2 in terms of restriction of the base field.

THEOREM 2′. *Every irreducible polynomial representation of G factors through an infinitesimally irreducible representation of $R_{\mathbf{F}(p^n)/\mathbf{F}(p)}G$ for some (sufficiently large) n depending on the given representation.*

Finally, we note that under very special circumstances the elements of M_p can themselves be expressed as tensor products. This occurs, in the case of simple groups, when p is expressible as the square of the ratio of the lengths of two roots ($p = 2$ for type B_n, C_n, or F_4; $p = 3$ for type G_2). Then each f in M_p is uniquely expressible as $f' \otimes f''$ with the highest weight of f' (respectively f'') supported by the long (respectively short) simple roots [**St1**, Theorem 11.1]. According to a result of Seitz [**S**], for irreducible polynomial representations of semisimple algebraic groups, no further decompositions are possible. This includes the case of characteristic 0, where no decompositions at all are possible.

Abstract representations. The principal result here, not stated in its most general form, is due to Borel and Tits [**B-T**].

THEOREM 3. *Let G be a simple algebraic group defined over an infinite field k and of positive k-rank. Let $f\colon G(k) \to \mathrm{PGL}_m(k')$ be an irreducible projective representation with $m \geq 2$ and k' an algebraically closed field. Then there exist distinct imbeddings σ_i of k into k', finite in number, and corresponding irreducible polynomial representations g_i of $\sigma_i^\circ(G)$ such that $f = \bigotimes_i (g_i \circ \sigma_i^\circ)$ on $G(k)$.*

Here σ_i° denotes the isomorphism of groups induced by the change of base field σ_i.

If G is simply connected, then in many cases Theorem 3 is true with projective representations replaced by linear representations throughout [**Ti2**]. In any case it is true up to multiplication by a representation of $G(k)$ by scalars, as easily follows from Theorem 3. Thus we have the remarkable fact (in the present case) that if G is quasisplit, i.e., contains a Borel subgroup B defined over k, then the principal result about irreducible polynomial representations of $G(k)$ holds also for abstract ones: $B(k)$ fixes a unique line and the representation is determined by the action of $B(k)$ on that line.

The proof of Theorem 3 depends on the following result (see [**B-T**, Theorem 8.1]).

THEOREM A. *In Theorem* 3 *replace* PGL_m *by any simple adjoint k'-group G' and f by any homomorphism of $G(k)$ into $G'(k')$ whose image is dense. Then there exists an imbedding σ of k into k' and a k'-isogeny of algebraic groups g of $\sigma^\circ(G)$ into G' with $dg \neq 0$, both unique, such that $f = g \circ \sigma^\circ$ on $G(k)$.*

We sketch proofs for the above theorems in the quasisplit case. Let $B = TU$ be a Borel subgroup of G with T a maximal torus and U a maximal unipotent subgroup, all defined over k. These are characterized abstractly, up to conjugacy, as follows. B is maximal solvable and has only itself as a subgroup of finite index. U is the derived group of B. T is maximal nilpotent and every subgroup of finite index of T is of finite index in its normalizer. It follows that $\overline{f(B(k))}$ (Zariski closure) is a Borel subgroup of G' in Theorem A and that it equals $\overline{f(T(k))} \cdot \overline{f(U(k))}$, as B equals TU in G. The group $U(k)$ captures the additive group of k and $T(k)$ the multiplicative group, and the action of T on U binds the two together as in the distributive law. From this we get the required imbedding of k into k' and then the other parts of Theorem A (for more details see [**B-T**] or [**St2**]).

Now $\overline{f(G(k))}$ in Theorem 3 is connected by what has been said above, and it acts irreducibly on $P(k'^m)$. Hence it is reductive, hence also semisimple and adjoint (by Schur's lemma), hence it is a product of simple algebraic groups $\prod G_i$. The identity map yields an irreducible polynomial representation of this group which is therefore, by Theorem 0, a tensor product of similar representations h_i of the components G_i. Theorem A applied to the components of the factorization $G \to \prod G_i(k') \xrightarrow{\text{id}} \mathrm{PGL}_m(k')$ of f then yields the required conclusions of Theorem 3.

We note that Theorem 1 follows from Theorem 3. Further, if $k = \mathbf{R}$ and the values of f are all in $\mathrm{PGL}_m(\mathbf{R})$, or if $k = \mathbf{Q}$, then the only possibility for σ_i in Theorem 3 is the obvious one, so that we get the strong conclusion that f itself is polynomial.

An interesting problem is to extend the results of this section to groups of k-rank 0, where the known results, for general k, are fragmentary (cf. [**B-T**, §9], where such results over locally compact fields are discussed).

Representations of arithmetic subgroups. The principal result here, which we do not present in its most general form, is due to Margulis [**M**, Theorem 11]. (For an exposition of Margulis's results see [**Ti1**] or [**Z**].)

THEOREM 4. *Let G be a simply connected simple algebraic group defined over $\mathbf{Q}$ of $\mathbf{R}$-rank ≥ 2 and Γ an arithmetic subgroup of $G(\mathbf{Q})$. Let $f\colon \Gamma \to \mathrm{GL}_m(C)$ be an irreducible representation. Then f is expressible uniquely as a tensor product $g \otimes h$ such that g is polynomial and h has kernel of finite index in Γ.*

Here "arithmetic" means that, for some faithful realization of G by matrices (as a group defined over $\mathbf{Q}$), Γ equals $G(\mathbf{Z})$ or some subgroup of $G(\mathbf{Q})$ commensurable to this group. Actually, Margulis states his results for discrete subgroups and it is one of his major results that, under very general conditions, such subgroups are always arithmetic.

As is well known to people working in this area, Theorem 4 is essentially equivalent to:

THEOREM B. *In Theorem 4 the representation f is polynomial on a subgroup of finite index of Γ, even if f is not irreducible.*

Clearly Theorem 4 implies Theorem B in case f is irreducible. For a proof of the converse we follow Serre [**Ser**, p. 502]. Let g be the rational representation of Γ given by Theorem 4 and Γ' the subgroup of finite index, which may be taken to be normal, on which g agrees with f. Set $h(x) = g(x)^{-1}f(x)$ for all $x \in \Gamma$. We claim that $h(x)$ and $g(y)$ commute for all $x, y \in \Gamma$. Granted the claim, we see that h is also a representation Γ, and its kernel contains Γ'. It follows that the map $(x, y) \to g(x)h(y)$ defines a representation of $\Gamma \times \Gamma$ which, by Theorem 0, can be expressed as a tensor product of representations of the two factors. Since f factors through this representation via the diagonal imbedding of Γ in $\Gamma \times \Gamma$, the tensor product decomposition of f in Theorem 4 and its required properties easily follow. Returning to the claim, we have for $x \in \Gamma$ and $y \in \Gamma'$ that

$$\begin{aligned} h(x)^{-1}g(y)h(x) &= f(x)^{-1}g(x)g(y)g(x)^{-1}f(x) = f(x)^{-1}g(xyx^{-1})f(x) \\ &= f(x)^{-1}f(xyx^{-1})f(x) = f(y) = g(y), \end{aligned}$$

so that $h(x)$ and $g(y)$ commute. This continues to hold for $y \in \Gamma$ since g is a polynomial map and Γ' is dense in G, and hence also in Γ, in the Zariski topology, whence the claim and the equivalence of the two theorems.

Theorem B is a very hard theorem, but a relatively simple proof can be given when G is a split or quasisplit group. We now sketch such a proof, in several steps, taken from [**St3**, Theorem 6], in case $G = \mathrm{SL}_n (n \geq 3)$ and $\Gamma = \mathrm{SL}_n(\mathbf{Z})$.

(1) Let x, y, z in $\mathrm{GL}_m(\mathbf{C})$ be such that $z = (x, y)$ (commutator) and z commutes with x and y.

(a) $\quad (x_u, y) = z_u, (x, y_s) = z_s \quad$ and $\quad (x_u, y_s) = 1.$

(b) $\quad z_s^q = 1, \quad$ i.e., z^q is unipotent, for some q dividing $m!$.

To prove the second equation in (a), for example, write $xyx^{-1} = zy$ and take semisimple parts. Then, taking the commuting semisimple elements z_s and y_s in diagonal form as we may, we see that x acts on y_s via a permutation of the diagonal entries. If q is the order of this permutation then $y_s = x^q y_s x^{-q} = z_s^q y_s$, whence $z_s^q = 1$.

Now for $i \neq j$ let E_{ij} denote the corresponding $n \times n$ matrix unit and x_{ij} the unipotent element $1 + E_{ij}$ of $\Gamma = \mathrm{SL}_n(\mathbf{Z})$.

(2) Set $Y_{ij} = \log f(x_{ij})_u$ (using the usual series for $\log(1+T)$, a polynomial when $1+T$ is unipotent and hence T is nilpotent). Then the following relations

hold:

$$[Y_{ij}, Y_{kl}] = \begin{cases} Y_{il} & \text{if } i \neq l,\ j = k, \\ 0 & \text{if } i \neq l,\ j \neq k. \end{cases}$$

For the first case we use the relation $(x_{ij}^r, x_{jl}^s) = x_{il}^{rs}$ $(r, s \in \mathbf{Z})$. By (1a) this relation continues to hold if we apply f and take unipotent parts. If we then replace $f(x_{ij}^r)_u$ by $\exp rY_{ij} = 1 + rY_{ij} + \cdots$, and similarly for the other terms, and equate the coefficient of rs on the two sides of the resulting relation, we get the first case of (2). The proof of the second case is similar but easier.

It is not difficult to prove that the relations of (2) form an abstract definition of sl_n under the correspondence $E_{ij} \Leftrightarrow Y_{ij}$. Since the group SL_n is simply connected as an algebraic group there exists a polynomial representation $g\colon \mathrm{SL}_n(\mathbf{C}) \to \mathrm{GL}_m(\mathbf{C})$ such that $dg\colon E_{ij} \to Y_{ij}$ for all i and j. We have $g(x_{ij}) = g(\exp E_{ij}) = \exp Y_{ij} = f(x_{ij})_u$. The exp series here, like the log series earlier, is finite.) We also have $f(x_{ij}) = (f(x_{ik}), f(x_{kj}))$ $(k \neq i, j; k$ exists since $n \geq 3)$, so that $f(x_{ij}^q)$ is unipotent by (1b). It follows that f agrees with g on the elements x_{ij}^q and similarly on their conjugates and hence on the subgroup Γ' of Γ generated by all of these elements. By the congruence subgroup theorem [**B-L-S**] Γ' is the kernel of reduction $\mathrm{mod}\, q$, hence is of finite index in Γ, as required.

It follows from Theorem B that every representation of Γ is completely reducible, and hence also that the irreducible representations of Γ are known once those with kernels of finite index are known, since the polynomial ones are known in terms of the highest weight classification, Weyl's character formula, and so on. However, the former are unknown even in simple cases like $\Gamma = \mathrm{SL}_n(\mathbf{Z})$, where the irreducible representations of $\mathrm{SL}_n(\mathbf{Z}/p^a\mathbf{Z})$ (p prime) have to be determined. These representations are needed for the harmonic analysis of the topological group $\mathrm{SL}_n(\mathbf{Q}_p)$ For $a = 1$ we have one of the classical essentially simple finite groups and a good deal is known about its representations (see [**L**], where earlier references may be found). For $a \geq 2$, however, only the case $n = 2$ has been solved, by Shalika [**Sh**] and Tanaka [**T**] independently. Therefore we pose here the problem of determining the representations in all cases, and similarly for the other simple algebraic groups.

We close with an extension of Theorem 4 to number fields. Again the general case is due to Margulis [**M**] but a special case occurred in the work of Serre [**Ser**, Theorem 5, Corollary 3]. For split or quasisplit groups the theorem can be proved along the lines of the proof of Theorem 4 for $G = \mathrm{SL}_n$ sketched above.

THEOREM 5. *In Theorem 4 let* $\mathbf{Q}$ *be replaced by* k, *a number field, and* Γ *by an arithmetic subgroup of* $G(k)$. *Let* $\sigma_1, \sigma_2, \ldots, \sigma_r$ *be the distinct imbeddings of* k *into* $\mathbf{C}$. *Then there exist unique representations* $g_1, g_2, \ldots, g_r$ *and* h *of* Γ *such that each* g_i *is polynomial,* h *has kernel of finite index in* Γ, *and* $f = (g_1 \circ \sigma_1) \otimes \cdots \otimes (g_r \circ \sigma_r) \otimes h$. *In other words,* f *factors through* $g \otimes h$ *with* h *as above and* g *a polynomial representation of* $R_{k/\mathbf{Q}}G$.

References

[B] J. Ballard, *Clifford's theorem for algebraic groups and Lie algebras*, Pacific J. Math. **106** (1983), 1–15.

[B-A] A. Borel et al, *Seminar on algebraic groups and related finite groups*, Lecture Notes in Math., vol. 131, part A, Springer-Verlag, 1970.

[B-L-S] H. Bass, M. Lazard, and J.-P. Serre, *Sous-groupes d'indice fini dans* $SL(n, \mathbf{Z})$, Bull. Amer. Math. Soc. **70** (1964), 385–392.

[B-N] R. Brauer and C. Nesbitt, *On the modular characters of groups*, Ann. of Math. (2) **42** (1941), 556–590.

[B-T] A. Borel and J. Tits, *Homomorphismes "abstraits" de groupes algébriques simples*, Ann. of Math. (2) **97** (1973), 499–571.

[C] P. Cartier, *Représentations linéaires des groupes algébriques semi-simples en caractéristique non nulle*, Sém. Bourbaki **255**, (1962–63).

[Cl] A. Clifford, *Representations induced in an invariant subgroup*, Ann. of Math. (2) **38** (1937), 533–550.

[C-P-S] E. Cline, B. Parshall, and L. Scott, *On the tensor product theorem for algebraic groups*, J. Algebra **63** (1980), 264–267.

[C-R] C. Curtis and I. Reiner, *Methods of representation theory with applications to finite groups and orders.* Vol. I, Wiley-Interscience, New York, 1981.

[Cu] C. Curtis, *Representations of Lie algebras of classical type with applications to linear groups*, J. Math. and Mech. **9** (1960), 307–326.

[F-T] P. Ferguson and A. Turull, *Prime characters and factorization of quasi-primitive characters*, Math. Z. **190** (1985), 583–604.

[G] D. Gajendragadkar, *A characteristic class of characters of finite* π*-separable groups*, J. Algebra **59** (1979), 237–259.

[K] G. Kempf, *Representations of algebraic groups in prime characteristics*, Ann. Sci. École Norm. Sup. (4) **14** (1981), 61–76.

[L] G. Lusztig, *Characters of reductive groups over a finite field*, Ann. of Math. Studies, vol. 107, Princeton Univ. Press, Princeton, N. J., 1984.

[M] G. Margulis, *Discrete subgroups of motions of manifolds of non-positive curvature*, Proc. Internat. Congr. Math. (Vancouver, 1974), Vol. 2, Canad. Math. Congress, Montreal, 1975, pp. 21–34.

[S] G. Seitz, *The maximal subgroups of classical algebraic groups*, Mem. Amer. Math. Soc. no. 365 (1987), 1–286.

[Se1] Séminare C. Chevalley, *Classification des groupes de Lie algébriques*, 2 vol., Inst. Henri Poincaré, Paris, 1956–58.

[Se2] Séminare "Sophus Lie," Inst. Henri Poincaré, Paris, 1954–55.

[Ser] J.-P. Serre, *Le problème des groupes de congruence pour* SL_2, Ann. of Math. (2) **92** (1970), 489–527.

[Sh] J. Shalika, *Representations of the two by two unimodular group over local fields.* I, II, Seminar at the Institute for Advanced Study, Princeton, 1966.

[St1] R. Steinberg, *Representations of algebraic groups*, Nagoya Math. J. **22** (1963), 33–56.

[St2] ——, *Abstract homomorphisms of simple algebraic groups* (after A. Borel and J. Tits), Sém. Bourbaki **435** (1972–73).

[St3] ——, *Some consequences of the elementary relations in* SL_n, Contemp. Math., no. 45, Amer. Math. Soc., Providence, R. I., 1985, pp. 335–350.

[T] S. Tanaka, *Irreducible representations of the binary modular congruence groups* $\bmod p^\lambda$, J. Math. Kyoto Univ. **7** (1967), 123–132.

[Tl1] J. Tits, *Travaux de Margulis*, Sém. Bourbaki **482** (1975–76).

[**Ti2**] ____, *Groupes de Whitehead de groupes algébriques simples sur un corps*, Sém. Bourbaki **505** (1976–77).

[**W**] A. Weil, *Adèles and algebraic groups*, Prog. Math, vol. 23, Birkhäuser, 1982.

[**Wo**] W. Wong, *Irreducible modular representations of finite Chevalley groups*, J. Algebra **20** (1972), 355–367.

[**Z**] R. Zimmer, *Ergodic theory and semisimple groups*, Birkhäuser, Boston, 1984.

UNIVERSITY OF CALIFORNIA AT LOS ANGELES

J. Fac. Sci. Univ. Tokyo
Sect. IA, Math.
34 (1987), 699-707.

On Dickson's theorem on invariants

To my friend Nagayoshi Iwahori

By Robert STEINBERG

1. Introduction.

The theorem in question, first proved in [3], is as follows.

THEOREM A. *Let* $G=GL_n(k)$ $(k=\boldsymbol{F}_q)$ *act on* $k[X_1, X_2, \cdots, X_n]$, *the algebra of formal polynomials, in the usual way. Then* $k[X_1, X_2, \cdots, X_n]^G$, *the algebra of invariants, is a polynomial algebra on the generators*

$$I_r=[01 \cdots \hat{r} \cdots n]/[01 \cdots n-1] \qquad (0 \leqq r \leqq n-1).$$

Here, for any nonnegative integers, $[e_1e_2 \cdots e_n]$ denotes the determinant of the matrix whose ij entry is the q^{e_j}th power of X_i. This is nonzero if the e's are distinct, since the main diagonal term is not cancelled by any other term. Also since it is reproduced by each T in $GL_n(k)$ with the scalar factor $\det(T)$, by an easy calculation using Fermat's theorem that $c^q=c$ for each c in k, it follows that every $[e_1 \cdots e_n]$ is $SL_n(k)$-invariant and that the ratio of any two such is $GL_n(k)$-invariant. The I_r's are called the Dickson invariants.

Simultaneously with Theorem A we shall consider the following result which we believe to be new.

THEOREM B. $k[X_1, X_2, \cdots, X_n]$ *is free as a module over* $k[X_1, X_2, \cdots, X_n]^G$ *with a basis consisting of the monomials* $X_1^{i_1}X_2^{i_2} \cdots X_n^{i_n}$ $(0 \leqq i_r < q^n - q^{r-1}$ *for all* r).

Dickson's original proof of Theorem A was very complicated. Since then a number of other proofs have appeared (see Bourbaki [2, p. 137–8], Ore [7, p. 566–8] and Wilkerson [11]), some of them quite simple. In Section 2 we present a proof which, we believe, is simpler than any of these and has the additional advantage that it yields Theorems A and B together. It should be mentioned that the polynomial nature of $k[X]^G$ and the freeness of $k[X]$ over $k[X]^G$ are equivalent under quite general conditions, but the proof involves some serious commutative algebra (see [2]). The essential step of our proof is a relative version of the above results (Theorem E below) which at once yields Theorems A and B and analogous results for many of its subgroups.

In [3] Dickson also proved the following result.

THEOREM C. *If $G=SL_n(k)$ in Theorem* A *then the conclusion there holds with I_0 replaced by $I_0^{1/(q-1)}=[01\cdots n-1]$.*

The equality here follows from $I_0=[12\cdots n]/[01\cdots n-1]=[01\cdots n-1]^q/[01\cdots n-1]$. As is known Theorem C follows easily from Theorem A (and vice versa), and we have included a proof of this fact.

The invariants that depend on several formal vectors have not yet been determined, except in case $n=2$, q is a prime, and there are just two vectors, by Krathwohl [5]. In Section 3 we present some contributions to this problem which lead to yet another proof of Dickson's Theorem.

Finally we consider the situation over $\boldsymbol{Z}/p^r\boldsymbol{Z}$. Here the invariants have been obtained by Feldstein [4], following Turner [10] who did the case $r=2$. In Section 4 we obtain more general results in a simpler way which makes transparent the transition from $\bmod p$ invariants to $\bmod p^r$ invariants for $GL_n(\boldsymbol{Z}/p^r\boldsymbol{Z})$ and many of its subgroups.

2. Dickson's theorem.

In this section we shall prove Theorems A, B and C. Our approach is that of Artin [1, p. 39-41] who considered the invariants of the symmetric group S_n.

THEOREM D. *For $0\leqq r\leqq n$ let $G(r)$ be the subgroup of $GL_n(k)$ consisting of the matrices that agree with the identity in the first r rows.*

(a) *$k[X]^{G(r)}$ is generated by $X_1,\cdots,X_r,I_r,\cdots,I_{n-1}$, hence is a polynomial algebra over k.*

(b) *For $r\geqq 1$ $k[X]^{G(r)}$ is free as a module over $k[X]^{G(r-1)}$ with $\{X_r^i\mid i<q^n-q^{r-1}\}$ as a basis.*

Observe that $G(n)=1$ and $G(0)=GL_n(k)$ and that the proposed generators in (a) all belong to $k[X]^{G(r)}$. Part (a) with $r=0$ yields Theorem A, while part (b) for the values of r from 1 to n combined yields Theorem B as well as an analogous theorem for each of the groups $G(r)$. We start our proof with two simple lemmas.

LEMMA 1. *The polynomial $T^{q^n}-I_{n-1}T^{q^{n-1}}+I_{n-2}T^{q^{n-2}}-\cdots+(-1)^nI_0T$ has as its roots the q^n distinct linear forms in $X_1,X_2,\cdots,X_n$.*

Let $[01 \cdots n]$ be defined as $[01 \cdots n-1]$ above but with an extra variable X_{n+1} and an extra exponent q^n. As a polynomial in X_{n+1} its degree is q^n since the highest coefficient $[01 \cdots n-1]$ is, as noted above, nonzero. If X_{n+1} is replaced in $[01 \cdots n]$ by any linear combination of the first n X's then the $(n+1)$th row is replaced by the same linear combination of the first n rows. It follows that the roots are just the q^n linear forms in the first n X's. If we expand $[01 \cdots n]$ along the $(n+1)$th row, divide through by $[01 \cdots n-1]$ and then replace X_{n+1} by T, we get the polynomial of Lemma 1, as required.

An immediate consequence of this lemma is that the I's are all polynomials.

LEMMA 2. *For each r, X_r is a root of a monic polynomial of degree $q^n - q^{r-1}$ over $k[X_1, \cdots, X_{r-1}, I_{r-1}, \cdots, I_{n-1}]$.*

If F_n is the polynomial of Lemma 1 and F_{r-1} is defined similarly so that its roots are the q^{r-1} linear forms in the first $r-1$ X's, then F_n/F_{r-1} is the required polynomial. First it is monic of degree $q^n - q^{r-1}$ and has X_r as a root. Further its coefficients are in $k[X_1, \cdots, X_{r-1}, I_1, \cdots, I_n]$ since those of F_{r-1} are in $k[X_1, \cdots, X_{r-1}]$ and those of F_n are in $k[I_1, \cdots, I_n]$, by Lemma 1. Finally, the I_s $(s<r-1)$ can be dropped since their degrees are all larger than that of F_n/F_{r-1}.

We come now to the heart of the proof, the deduction of Theorem D from Lemma 2. Since the argument is valid in a number of other cases of interest to us, we carry it out in a more general context which focuses attention on the essential features of the situation.

THEOREM E. *Let G and H be subgroups of $GL_n(k)$ with H contained in G. Let S be a subset of $k[X]^H$ and T a subset of $k[X]^G$ such that* (1) *S generates $k[X]^H$,* (2) *S contains just one element, Y, which is not in T, and* (3) *Y is a root of a polynomial over $k[T]$ which is monic and of degree at most $|G/H|$. Then T generates $k[X]^G$, and $k[X]^H$ is free as a module over $k[X]^G$ with $\{Y^i \mid 0 \leq i < |G/H|\}$ as a basis.*

Here k can be any field as long as G is finite. Let A be the polynomial of (3) and F any element of $k[X]^H$. Then F is in $k[S]$ by (1), and by using the equation $A(Y)=0$ we can write F as a polynomial in Y of degree less than $|G/H|$, with coefficients in $k[T]$ because of the assumption (2). Thus the given Y^i's generate $k[X]^H$ as a module over $k[T]$, hence also $k[X]^H$ over $k[X]^G$, and $k(X)^H$ over $k(X)^G$, the last because for any $P(X)/Q(X)$ in $k(X)^H$ the denominator can always be converted to one in $k[X]^G$, namely,

$\Pi g\cdot Q(X)$ (g in G). However the number of Y^i's is at most $|G/H|$, which by the first theorem of Galois theory equals the dimension of $k(X)^H$ as a vector space over $k(X)^G$ (see [1, Th. 14]). It follows that the Y^i's form a basis for this space and hence a free generating set for $k[X]^H$ over $k[X]^G$. It remains to show that each F in $k[X]^G$ belongs to $k[T]$. Write $F=\sum C_i Y^i$ $(i<|G/H|)$ as a linear combination over $k[T]$, and $F=FY^0$, over $k[X]^G$. Since the Y^i's are free over $k[X]^G$, we get $F=C_0$, an element of $k[T]$, as required.

Labelling the two parts of Theorem D (a_r) and (b_r), we shall now show that for $r\geqq 1$ (a_r) implies (b_r) and (a_{r-1}). Then since (a_n) is obviously true, the theorem will follow. The hypotheses of Theorem E are satisfied with $G=G(r-1)$, $H=G(r)$, $S=\{X_1,\cdots,X_r,I_r,\cdots,I_{n-1}\}$, $T=\{X_1,\cdots,X_{r-1},I_{r-1},\cdots,I_{n-1}\}$ and $Y=X_r$, in view of Lemma 2 and the assumption (a_r). We conclude that (b_r) and (a_{r-1}) hold, except for the last point of (a_{r-1}). Since $k[X]$ is algebraic over $k[X]^{G(r-1)}$, the transcendence degree of the latter over k equals that of the former, which is n, so that that point also holds and the proof of Theorem D is complete.

REMARKS. (a) If X_1 is replaced by 1 the group G_1 above becomes the affine group on the remaining coordinates. It follows that for this group also the invariants form a polynomial algebra, with generators obtained from the I_r's $(r\neq 0)$ by the same replacement. (b) The above method, as embodied in Theorem E, also works for all parabolic subgroups of $GL_n(k)$ and their unipotent radicals. To indicate the results obtained we state them in a simple, but typical, case. Let $n=4$ and let P be the parabolic subgroup in which the 2 by 2 block in the upper right hand corner is required to be 0. Then $k[X]^P$ is generated by I_0', I_1', I_2, I_3, with the first two I's calculated on the space of the first two coordinates, and a basis for $k[X]/k[X]^P$ consists of the monomials with $0\leqq i_r<q^2-q^{r-1}$ for $r=1$ and 2 and $0\leqq i_r<q^4-q^{r-1}$ for $r=3$ and 4.

We come now to the proof of Theorem C. For this we need another (well-known) lemma.

LEMMA 3. *$[01\cdots n-1]$ equals the product of all of the nonzero linear forms in the X's for which the last nonzero coefficient is* 1, *and it divides every $SL_n(k)$-invariant which X_1 divides.*

Any such invariant is divisible by all of the transforms of X_1 under $SL_n(k)$, hence by all of the linear forms as above and hence also by their product. But this product and $[01\cdots n-1]$ have the same lowest terms (lexicographically). Thus they are equal and the lemma follows.

It also follows that $[01\cdots n-1]$ divides every $[e_1e_2\cdots e_n]$, which provides another proof that the I_r's are polynomials.

Now let F be any invariant for $SL_n(k)$. We must show that it is expressible in terms of the n invariants given by Theorem C. The elements $T_c = \mathrm{diag}(c, 1, 1, \cdots, 1)$ (c in k^*) form a system of representatives for the cosets of $GL_n(k)$ over $SL_n(k)$. If we write $F = -\sum T_c F + \sum (T_c F - F) = F_1 + F_2$, say, it follows that F_1 is an invariant for $GL_n(k)$ and hence by Theorem A is expressible as required by Theorem C. From the form of F_2 it is divisible by X_1 and hence also by $[01 \cdots n-1]$ by Lemma 3. We can now finish by induction since the degree of $F_2/[01 \cdots n-1]$ is less than that of F.

3. Several vectors.

In this section we consider $GL_n(k)$ acting on several vectors. The results are fragmentary, but the ideas introduced may be of further use.

THEOREM F. *If the group $G = GL_n(k)$ acts on a set of $m \geqq n$ independent formal vectors whose coordinates are viewed as functions on $\bar{k}^{mn}$ and written in matrix form as $v_1 v_2 \cdots v_m = XY$ with X consisting of the first n v's and Y of the others, then on the set where* $\det X = [X] \neq 0$, *or more generally on any locally closed G-invariant subset of this set, the map $f: XY \to (X^{-1}X^{(q)})(X^{-1}Y)$ defines a quotient for the action.*

Here $X^{(q)}$ denotes the matrix of qth powers of the entries of X. The Lang-Speiser Theorem states that the map $X \to X^{-1}X^{(q)}$ on $GL_n(\bar{k})$ is surjective. In [9] a refinement is proved, that this map is a finite morphism. From this the theorem readily follows.

THEOREM G. *In addition to the above notation let $[X_{ij}]$ denote the quantity obtained from $[X]$ by replacing v_i by v_j^q if $j \leqq n$, by v_j if $j > n$. Then on the set in $\bar{k}^{mn}$ where $[X] \neq 0$ the algebra of polynomials invariant under G is generated by the mn elements $[X_{ij}]/[X]$ together with $[X]^{1-q}$.*

By Theorem F we need only work out the coordinates of $X^{-1}X^{(q)}$, $X^{-1}Y$ and $[X^{-1}X^{(q)}]^{-1}$ in terms of those of X and Y. By Cramer's rule for X^{-1} the results are as stated.

This result also holds for $SL_n(k)$ provided that $[X]^{1-q}$ is replaced by $[X]^{-1}$.

COROLLARY. *If $m \geqq n$, then $k(v_1 v_2 \cdots v_m)^G$ is purely transcendental over k with the mn $[X_{ij}]/[X]$'s as a generating set.*

This follows at once from Theorem G.

For $m<n$ analogous results may be obtained as follows. We first expand $v_1 \cdots v_m$ to a square matrix X by adjoining the q^jth powers of v_m for $j=1, \cdots, n-m$. (We could also adjoin powers of several v's.) With this X and the resulting n^2+1 functions of Theorem G which can be brought down to $mn+1$ since those with $m \leqq j<n$ are constant, the result there holds. Finally, dropping the function $[X]^{1-q}$, we see that the corollary also holds, all for $m<n$.

To solve our main problem, the determination of $k[v_1 \cdots v_m]^G$, we would, according to the present approach, have to remove the condition $[X] \neq 0$, that is, determine the polynomials in the $[X_{ij}]$'s that are divisible by $[X]$, and we can not do this. We can, however, do this in case $m=1$ and thus obtain another proof of Dickson's Theorem, with which we close this section.

Let v be any formal vector. We expand it to a square matrix X by adjoining $n-1$ of its powers as in the preceding proof. The $n+1$ nonconstant invariants that result from Theorem G in this case are just the Dickson invariants $I_0, \cdots, I_{n-1}$ together with I_0^{-1}, in terms of the coordinates $X_1, \cdots, X_n$ of v. These therefore generate the invariant polynomials on the open set in $\bar{k}^n$ where $I_0 \neq 0$. Let F be any element of $k[X]^G$. Then F is a polynomial in the I's and I_0^{-1}, so that for some $m \geqq 0$, $I_0^m F$ is a polynomial in the I's and is thus expressible as $\sum H_r(I_1, \cdots, I_{n-1}) I_0^r$ with each H_r a polynomial. Assuming m to be minimal we must show that $m=0$. Suppose not. Then $H_0 \neq 0$. Since X_n divides I_0, H_0 vanishes at $X_n=0$. However, as easily follows from the definitions, the substitution $X_n=0$ converts the I_r's $(1 \leqq r<n-1)$ into the qth powers of the $(n-1)$-dimensional Dickson invariants, and these are algebraically independent over k, by what has been said above (or else by a direct proof by induction). Thus $H_0=0$, a contradiction. This shows that $r=0$ and thus that F is a polynomial in the I's, as required.

4. **Invariants** mod p^r.

One of our main theorems in this area, extended by the remarks that follow its proof, is the following.

THEOREM H. *Let H be the subgroup of $GL_n(\boldsymbol{Z}/p^r\boldsymbol{Z})$ $(n \geqq 2, r \geqq 1)$ consisting of the matrices that are unimodular and congruent to the unit matrix* mod p. *Then F in $\boldsymbol{Z}/p^r\boldsymbol{Z}[X_1, X_2, \cdots, X_n]$ is invariant under H if and only if it is expressible in one of the following forms.*

(a) *$F=\sum p^i F_i$ with F_i a polynomial in the p^{r-i-1}th powers of*

$X_1, X_2, \cdots, X_n$ for each $i=0, 1, 2, \cdots, r-1$.

(b) *$F=\sum c_I X^I$ with p^{r-1} dividing $c_I I$ for every multi-index $I=(i_1, \cdots, i_n)$.*

Since the sufficiency in (a) and (b) are easily verified, we turn to the proof of necessity, the one in (b), since that yields the one in (a) at once. Let F be an invariant for H written as the sum of its homogeneous parts relative to the total degree in X_1 and X_2. If we apply $T_{12}: (X_1 \to X_1 + pX_2, X_j \to X_j \text{ if } j \neq 1)$ to F then each of these parts is kept fixed. If $F_m = \sum A_i X_1^i X_2^{m-i}$ (with each A_i a polynomial in $X_3, X_4, \cdots$) is the part of degree m then $T_{12}F_m = F_m$ yields, when like terms are grouped together, $\sum_{j=0}^{m-1} X_1^j X_2^{m-j} \sum_{0<k\leqq m-j} p^k \binom{j+k}{j} A_{j+k} = 0$. Here each of the inner sums must be 0. The first term in the jth sum is $p(j+1)A_{j+1}$ and the later terms may be written as $(p^{k-1}/k)\binom{j+k-1}{j} p(j+k)A_{j+k}$. Here each p^{k-1}/k is an integer mod p since $p^{k-1} \geqq k$ so that p can divide k at most $k-1$ times. It follows by downward induction that p^r divides piA_i, that is, p^{r-1} divides iA_i, for all i. In terms of F written as in (b) this means that p^{r-1} divides every $c_I i_1$. Similarly this holds with i_1 replaced by $i_2, i_3, \cdots$, which proves the necessity in (b) and hence the theorem.

REMARK. The group H in its entirety is not needed for the above proof of necessity. If T_{ij} $(i \neq j)$ is defined as above then one could replace H by the subgroup generated by any set of T_{ij}'s such that the index i takes on all values from 1 to n, for example, T_{12} and the set T_{i1} $(2 \leqq i \leqq n)$.

REMARK. A similar theorem holds for the group of unimodular matrices that are congruent to the unit matrix mod p^s $(0<s<r)$. It is only necessary to replace p^{r-1} by p^{r-s} in (b) and make an analogous change in (a). The proof is essentially the same, with the previous remark still applying.

THEOREM I. *Let G be any subgroup of $GL_n(\mathbf{Z}/p^r\mathbf{Z})$ $(n \geqq 2, r \geqq 1)$ which contains the subgroup H of Theorem H, or, more generally, any subgroup of H for which the conclusion there holds (see the first remark above). Let $\{J_1, \cdots, J_m\}$ be a generating set for the algebra of polynomial invariants in the induced action of G on $(\mathbf{Z}/p\mathbf{Z})^n$. Then F is an invariant for G in the original action if and only if it can be written $F=\sum p^i F_i$ with F_i a polynomial in the p^{r-i-1}th powers of the J's for each $i=0, 1, \cdots, r-1$.*

Again the sufficiency is easily verified; thus we turn to the proof of necessity. Let F be a G-invariant polynomial. Then by Theorem H it can be written, mod p, as a polynomial in the p^{r-1}th powers of the X's. Now G acts, mod p, on these powers of the X's and on the X's themselves by exactly the same formulas. It follows that, mod p, F is a polynomial in the J's evaluated at the p^{r-1}th powers of the X's, which, mod p, is the same polynomial in the p^{r-1}th powers of the J's. In other words, $F=F_0+pF'$ with F_0 as in the conclusion of the theorem and F' a polynomial. Further by the sufficiency, which has already been established, F_0 is an invariant and hence pF' is also. Thus F' is an invariant mod p^{r-1}, and the proof may be completed by induction.

THEOREM J. *A polynomial is invariant under* $GL_n(\boldsymbol{Z}/p^r\boldsymbol{Z})$ *if and only if it can be written* $\sum p^i F_i$ *with each* F_i *a polynomial in the* I's, *as in Theorem* A *but with* q *replaced by* p, *and similarly for* SL_n *with* I_0 *replaced by* $[01 \cdots n-1]$.

This result follows at once from Theorems A, C and I.

As they are stated above, Theorems H and I also apply to most parabolics and, in case n is even, to the symplectic group. With minor modifications they apply to all parabolics and their unipotent radicals and the groups G_r of Section 2.

Using the same methods, we have obtained a version of the above results for a class of local rings which includes $\boldsymbol{F}_q$ and $\boldsymbol{Z}/p^r\boldsymbol{Z}$ as special cases and thus a common generalization of Theorems A and J. Many other cases that we have worked out lead us to the conjecture that such results should hold for arbitrary finite local rings.

References

[1] Artin, E., Galois theory, Notre Dame Math. Lectures 2, Second Edition, NAPCO Inc., 1959.

[2] Bourbaki, N., Groupes et Algèbres de Lie, Ch. 4, 5 et 6, Hermann, Paris, 1968.

[3] Dickson, L. E., A fundamental system of invariants of the general modular linear group with a solution of the form problem, Trans. Amer. Math. Soc. **12** (1911), 75-98.

[4] Feldstein, M. M., Invariants of the linear group modulo p^k, Trans. Amer. Math. Soc. **25** (1923), 223-238.

[5] Krathwohl, W. C., Modular invariants of two pairs of cogredient variables, Amer. J. Math. **36** (1914), 449-460.

[6] Mui, H., Modular invariant theory and the cohomology algebras of symmetric groups, J. Fac. Sci. Univ. Tokyo Sect. IA Math. **22** (1975), 319-369.

[7] Ore, O., On a special class of polynomials, Trans. Amer. Math. Soc. **35** (1933),

559-584.

[8] Serre, J.-P., Groupes finis d'automorphismes d'anneaux locaux réguliers, Colloque d'Algèbre, ENSJF (1967), Exp. 8.

[9] Steinberg, R., On theorems of Lie-Kolchin, Borel and Lang, Contributions to Algebra (dedicated to Ellis Kolchin), Academic Press, New York, 1977.

[10] Turner, J. S., A fundamental system of invariants of a modular group of transformations, Trans. Amer. Math. Soc. **24** (1922), 129-134.

[11] Wilkerson, C., A primer on Dickson invariants, Amer. Math. Soc. Contemp. Math. **19** (1983), 421-434.

(Received May 1, 1987)

MSRI
1000 Centennial Drive
Berkeley, CA 94720
U. S. A.

and

Department of Mathematics
University of California
Los Angeles, CA 90024
U. S. A.

Reprinted from JOURNAL OF ALGEBRA
Vol. 113, No. 2, March 1988
Printed in Belgium

An Occurrence of the Robinson–Schensted Correspondence

ROBERT STEINBERG

Department of Mathematics, University of California, Los Angeles, California 90024

Communicated by Walter Feit

Received October 3, 1986

1. INTRODUCTION

Let V be an n-dimensional vector space over an infinite field, $\mathscr{F}$ the flag manifold of V, u a unipotent transformation of V, and $\lambda = (\lambda_1 \geqslant \lambda_2 \geqslant \cdots)$ the type of u, a partition of n whose parts are the sizes of the Jordan blocks for u. As is shown in [4, 7, 8], and also below, the components of $\mathscr{F}_u$, the variety of flags fixed by u, correspond naturally to the standard tableaux of shape λ. The purpose of this note is to show that the "relative position" of any two components of $\mathscr{F}_u$ (in general an element of the Weyl group, in the present case an element of S_n (see [5] or [7])) is given, in terms of the corresponding tableaux, by the Robinson–Schensted correspondence.

1	2
3	5
4	

We proceed to a fuller explanation. Let $F = (V_0 \subset V_1 \subset \cdots)$ be a flag fixed by u. As is easily seen (Lemma 2.3, below), for each $k \geqslant 1$ the type of $u \mid V_{k-1}$ is obtained from that of $u \mid V_k$ by decreasing some part by 1. Thus we may associate to F a tableau T of shape λ filled in with the numbers from 1 to n, each used just once, so that for each k the subtableau T_k supported by the numbers from 1 to k has the shape of the type of $u \mid V_k$. It is clear from the construction that T is standard: the numbers increase across the rows and down the columns. In the tableau shown, for example, $u \mid V_4$

0021-8693/88 $3.00

is of type 211 since T_4 has rows of lengths 2, 1, and 1. As was mentioned earlier, the map $F \to T$ yields a bijection between the irreducible components of $\mathscr{F}_u$ and the standard tableaux of shape λ, each fibre of the map being a dense open part of the corresponding component.

For each pair of flags (F, F') the position of F' relative to F, or, equivalently, the $GL(V)$ orbit of the pair (F, F'), may be defined as the unique permutation $w = w(F, F')$ with the following property. If $F = (V_0 \subset V_1 \subset \cdots)$ and $F' = (V'_0 \subset V'_1 \subset \cdots)$ then there exists a basis $\{v_1, v_2, ..., v_n\}$ of V such that $\{v_1, v_2, ..., v_i\}$ is a basis of V_i and $\{v_{w1}, v_{w2}, ..., v_{wj}\}$ is a basis of V'_j for all i and j. The existence and uniqueness of such a w, which is the essence of the Bruhat lemma for $GL(V)$, are easily proved by induction on n [6, Lemma 2.1].

On the other hand, the permutation $w = w(T, T')$ associated to a pair of standard tableaux of the same shape by the Robinson–Schensted correspondence is defined by the following algorithm (see, e.g., [1, 3], where other references can be found). From T' remove the number n (and the box that contains it). Then take the number which is in the same position in T as n was in T' and move it up one row to displace the largest number in that row that is smaller than it; use the displaced number to displace a number in the next higher row according to the same rule, and so on, until a number, say r, is displaced from the first row; set $w(n) = r$. Repeat the process on the two tableaux of size $n-1$ produced by the first step to get $w(n-1)$, and so on, to get the required w, usually written as a word $w(1)\,w(2) \cdots w(n)$. For example, if T is the tableau in the diagram above and T' is obtained from T by interchanging 4 and 5, then, as the reader may verify, the word works out to 45132.

We can now state our theorem precisely.

THEOREM 1.1. *Let $\mathscr{F}$ be the flag manifold on V, and $u = u(\lambda)$ a unipotent transformation as above. Let T and T' be standard tableaux of shape λ and C and C' the corresponding irreducible components of $\mathscr{F}_u$. Then for generic elements F and F' of these components, i.e., for elements in suitable dense open subsets, we have $w(F, F') = w(T, T')$.*

The main lemma in our proof of Theorem 1.1 runs as follows.

LEMMA 1.2. *Let $F = (V_0 \subset V_1 \subset \cdots)$ and $F' = (V'_0 \subset V'_1 \subset \cdots)$ be as in 1.1. Let F'_1 be the subflag $(V'_0 \subset V'_1 \subset \cdots \subset V'_{n-1})$ of F', and F_1 the flag on V'_{n-1} defined thus: if r is the smallest number such that $V_r \not\subset V'_{n-1}$, then the components of F_1 are $W_i \equiv V_i \cap V'_{n-1}$ $(i \neq r)$ (and hence are labeled by the numbers $\leqslant n$ with r excluded, rather than by the numbers $\leqslant n-1$). Then the tableaux corresponding to F_1 and F'_1 are obtained from those corresponding to F and F' by the first step of the Robinson–Schensted algorithm.*

The theorem was obtained by us in 1976, as is mentioned in [7, p. 221]. We had hoped to consider the situation in the other classical groups, but time has passed us by with the appearance, among other things, of the comprehensive treatise [5]. A proof of the theorem can also be found there (as Proposition II9.8), but in that proof the Robinson–Schensted process occurs only implicitly through some of its derived properties rather than in the direct and explicit manner provided by our main lemma. Therefore we feel that the publication of our proof is still worthwhile.

2. Preliminaries

The material in this section can also be found in [4, 7, 8]. It is given here to make the present paper as self-contained as possible. We let $u = u(\lambda)$ be as in Section 1 and set $N = u - 1$, a nilpotent transformation of V.

Lemma 2.1. *A hyperplane W of V is stable under N (or u) if and only if $W \supset NV$.*

For, if W is stable then N acts nilpotently, hence as 0, on V/W.

Lemma 2.2. *For each $k \geqslant 1$ we have:*

(a) *The codimension of* $\ker N^{k-1}$ *in* $\ker N^k$ *equals the number of parts of λ of size $\geqslant k$.*

(b) *The codimension of* $NV + \ker N^{k-1}$ *in* $NV + \ker N^k$ *equals the number of parts of λ of size k.*

Both parts are easily seen to hold on each Jordan block for N and hence on all of V.

Lemma 2.3. *Let W be an N-stable hyperplane and j the unique index such that $W \supset \ker N^{j-1}$ and $W \not\supset \ker N^j$. Then some part, λ_i,* of *λ equals j. If i is the largest index with this property, i.e., if i is the number of parts of λ of size $\geqslant j$, then the type λ' of $u \mid W$ is obtained from λ by decreasing λ_i by* 1.

By Lemma 2.1 the left side of part (b) of Lemma 2.2 is positive when $k = j$ so that λ contains a part of size j. On intersection with W the codimension in Lemma 2.2(a) changes only if $k = j$ and then it goes down by 1. Thus λ' is obtained from λ by decreasing some part of size j by 1, the last such part since the parts of λ' are to be written in decreasing order.

Now if T is any standard tableau of shape λ we can see how the subset $C(T)$ of $\mathscr{F}_u$ consisting of the flags whose corresponding tableau is T is to be constructed. Let n occur in the (i, j) position of T, so that λ_i is the last part of λ of size j. Then by Lemma 2.3 the component V_{n-1} of our flag can be any hyperplane which contains $NV+\ker N^{j-1}$ but does not contain $NV+\ker N^j$; then V_{n-2} is to be chosen to satisfy the same conditions with V, N,... replaced by V_{n-1}, $N|V_{n-1}$, ..., and so on. The possibilities for V_{n-1}, for example, correspond to the points of a dense open part of the projective space $P(V/(NV+\ker N^{j-1}))$ whose dimension is $i-1$ by Lemma 2.2(b). It follows that $C(T)$ is an irreducible algebraic variety with a continued fibration by dense open parts of projective spaces of total dimension $\sum(i-1)\lambda_i$ summed over i, since this equals $\sum(i-1)$ summed over the squares of T. Since this depends only on λ and not on T, it follows that $C(T)$ is a dense open part of an irreducible component of $\mathscr{F}_u$ and that we have produced a bijection between the standard tableaux of shape λ and the irreducible components of $\mathscr{F}_u$, as was mentioned in the Introduction.

3. Proof of Theorem 1.1 and Lemma 1.2

Two further lemmas precede the proof proper.

Lemma 3.1. *For fixed k with $1\leqslant k\leqslant\lambda_1$ let W be a hyperplane generic relative to: W is N-stable and $W\supset\ker N^{k-1}$. Then $N|W$ is of the type specified by Lemma 2.3 with $\lambda_i=j$ the last part of λ of size $\geqslant k$.*

Let j be the number just specified. It exists because $k\leqslant\lambda_1$. It follows from Lemma 2.3 and the choice of j that $NV+\ker N^{k-1}=NV+\ker N^{j-1}\subsetneqq NV+\ker N^j$. Thus W contains the second term, but generically does not contain the third, so that Lemma 2.3 applies.

Lemma 3.2. *Let W be an N-stable hyperplane with $N|W$ of the type λ' specified by $\lambda_i\to\lambda_i-1$ as in Lemma 2.3, and X an N-stable hyperplane which is generic relative to this property.*

(a) *If $i=1$ then $X=W$; i.e., there is a unique hyperplane of this type.*

(b) *If $i>1$ then $X\neq W$, and if we set $Y=X\cap W$, then the type λ'' of $N|Y$ is obtained from λ' by decreasing λ_{i-1} by 1.*

The hyperplanes containing $NV+\ker N^{j-1}$ form a projective space of dimension $i-1$, as was mentioned towards the end of Section 2. Thus (a) and the first point of (b) hold. Now we have $(*)$ Y is a hyperplane in W and $Y\supset NW+\ker N^{j-1}|W$. Since the last part of λ' of size $\geqslant j$ is λ_{i-1} it is enough by 3.1 to show that Y is generic relative to $(*)$, hence enough to show that each Y which satisfies $(*)$ is the intersection with W of some X which satisfies the original conditions. Now $Y\not\supset W$, and $Y\not\supset\ker N^j$ since $W\not\supset\ker N^j$. Thus there exists a hyperplane X in V such that $X\supset Y$,

$X \neq W$, and $X \not\supset \ker N^j$. The required properties of X are all immediate, except possibly for the N-stability which follows by Lemma 2.1 from

$$NV = N(W + \ker N^j) \subseteq NW + \ker N^{j-1} \subseteq Y \subseteq X \qquad \text{and} \qquad \ker N^{j-1} \subset W.$$

We come now to the proof of our main lemma, Lemma 1.2. Let T_1 be the tableau corresponding to F_1, thus labeled, as F_1 is, with the numbers from 1 to n with r missing. The key point here is the following. (*) In the transition from T to T_1 let some number m, originally in the (i, j) position, move. (a) If $i = 1$ then $m = r$ and m disappears from the tableau. (b) If $i > 1$ then m moves up one row and displaces a smaller number there. Recall that r is the smallest index such that $V_r \not\subset V'_{n-1}$. It follows that $W_k = V_k$ for all $k < r$, so that each such k occupies the same position in T_1 as in T. Thus $m \geqslant r$. Assume $m > r$. Since $W_m \subset V_m$ it follows that m remains in the subtableau T_m and hence moves into T_{m-1} where it must displace a smaller number. Similarly each number less than $m-1$, other than r, stays in T_{m-1}. Hence W_m is a hyperplane in V_m of the type specified by Lemma 2.3 with (i, j) as above and $V, W, \ldots$ in Lemma 2.3 replaced by $V_m, W_m, \ldots$. But clearly V_{m-1} is also of this type, and $V_{m-1} \neq W_m$ since $V_{m-1} \not\subset V'_{n-1}$, because $m - 1 \geqslant r$. By Lemma 3.2(a) we have $i > 1$, under our current assumption that $m > r$. In other words, if $i = 1$ then $m = r$. Now assume that $i > 1$ and that V_{m-1} is chosen generically in V_m, assuming, by induction, that V_m has already been chosen, with V'_{n-1} kept fixed throughout the discussion. Then by Lemma 3.2(b), since $W_{m-1} = V_{m-1} \cap W_m$, the type of $N | W_{m-1}$ is obtained from that of $N | W_m$ by decreasing the $(i-1)$th part by 1. Thus m lies in $(i-1)$th row of T_1 and (*) is proved.

Now let n occupy the (i, j) position in T' and let m be the number in this position in T. It follows from (*) that, in the transition from T to T_1, the number m moves forward one row to displace a smaller number, which in turn displaces a smaller number in the row just above it, and so on, until finally the number r is displaced from the first row and removed from the tableau. No other number of T can move, for if one did it would create another chain of displaced numbers leading to the first row and hence including r by (*) (a), which is impossible since different numbers in T cannot move to the same position in T_1. Finally, since the tableau T_1 is standard, at each stage of the chain the number displaced is the largest number in that row smaller than the displacing number, so that the first step of the Robinson–Schensted algorithm has been achieved, and Lemma 1.2 is proved.

It seems that Theorem 1.1 now follows at once, but this is not quite the case since, if $n \geqslant 3$, then F_1 in Lemma 1.2 need not be generic relative to its tableau even though each of F and F' is so. However, we can finish the proof as follows. For each permutation $w = w(1)\, w(2) \cdots w(n)$ and for all i

and j from 1 to n let $d_{ij}(w)$ denote the number of p's from 1 to n that satisfy $p \leqslant j$ and $wp \leqslant i$. Here w can be recovered from the matrix $(d_{ij}(w))$ since, for each j, wj is the smallest i such that $d_{ij} \neq d_{ij-1}$. We introduce a partial order in $S_n : w \geqslant w'$ if and only if $d_{ij}(w) \leqslant d_{ij}(w')$ for all i and j. Now let F and F' be any two flags in V and w their relative position, as defined in Section 1. Using a corresponding basis of V, we see that $(*)$ $d_{ij}(w) = \dim(V_i \cap V'_j)$. Now the pairs of subspaces of V which satisfy $(*)$ with i, j, and d_{ij} arbitrary fixed numbers form an irreducible variety whose closure consists of those which satisfy $d_{ij}(w) \leqslant \dim(V_i \cap V'_j)$. It follows that for fixed w the pairs of flags (F, F') which satisfy $(**)$ $w(F, F') \leqslant w$ form a closed set. This is actually the closure of the set for which equality holds [2, Proposition 7.1] but we shall not need this fact. Returning to the proof of Theorem 1.1, we see by Lemma 1.2, the inductive assumption in Theorem 1.1, and $(**)$ above that $w(F_1, F'_1) \leqslant w(T_1, T'_1)$, and then putting the number r at the ends of these words, that $w(F, F') \leqslant w(T, T')$. But in the latter inequality each element of S_n occurs exactly once on the left as λ runs over the partitions on n and (C, C') over the pairs of irreducible components of $\mathscr{F}_{u(\lambda)}$ (see [7, Theorems 3.5, 3.6]) and exactly once on the right as (T, T') runs over the pairs of standard tableaux of the same shape (see [3]). Thus $w(F, F') = w(T, T')$. The theorem is proved.

References

1. D. E. Knuth, Permutation matrices and generalized Young tableaux, *Pacific J. Math.* **34** (1970), 709–727.
2. R. Proctor, Classical Bruhat orders and lexicographic shellability, *J. Algebra* **77** (1982), 104–126.
3. M.-P. Schützenberger, "La correspondance de Robinson," *in* Lecture Notes in Mathematics, Vol. 579, Springer-Verlag, Berlin/New York, 1977.
4. N. Spaltenstein, On the fixed point set of a unipotent transformation on the flag manifold, *Proc. K. Ned. Akad. Wet.* **79**(5) (1976), 452–458.
5. N. Spaltenstein, "Classes unipotentes de sous-groupes de Borel," *in* Lecture Notes in Mathematics, Vol. 946, Springer-Verlag, Berlin/New York, 1982.
6. R. Steinberg, A geometric approach to the representations..., *Trans. Amer. Math. Soc.* **71** (1951), 274–282.
7. R. Steinberg, On the desingularization of the unipotent variety, *Invent. Math.* **36** (1976), 209–224.
8. J. Vargas, Fixed points under the action of unipotent elements of SL_n in the flag variety, *Bol. Soc. Mat. Mexicana* (2) **24** (1979), 1–14.

Printed by Catherine Press, Ltd., Tempelhof 41, B-8000 Brugge, Belgium

Invent Math. 110, 649–671 (1992)

Inventiones mathematicae

Parabolic subgroups with Abelian unipotent radical

Roger Richardson[1], **Gerhard Röhrle**[2], **and Robert Steinberg**[3]

[1] Centre for Mathematics and its Applications, Australian National University, Canberra, ACT 2601, Australia
[2] Department of Mathematics, University of Southern California, Los Angeles, CA 90089, USA
[3] Department of Mathematics, University of California, Los Angeles, CA 90024, USA

Oblatum 12-II-1992

1 Introduction

Let G be a (connected) reductive algebraic group over an algebraically closed field k and P a parabolic subgroup, with V its unipotent radical and L a Levi subgroup. We wish to study the orbits of L on V and on G/P in the case that V is Abelian. One of our principal results is the following.

Theorem 1.1 *Let G be a reductive algebraic group, $P=LV$ a parabolic subgroup whose unipotent radical V is Abelian, and $P^-=LV^-$ the opposite parabolic subgroup relative to some maximal torus T of L. Assume that char $k \neq 2$ in case some simple component of G has roots of different lengths.*

(a) *Each (P, P) double coset of G meets V^- in a single L-orbit. Hence $L \cdot x = PxP \cap V^-$ for every x in V^-, and the number of (P, P) double cosets equals the number of L-orbits of V^- (or of V).*

(b) *Any P-orbit and any P^--orbit of G/P are disjoint or else intersect in a single L-orbit. Hence $LxP = PxP \cap P^- xP$ for every x in G.*

Part (a) here turns out to be an easy consequence of part (b). We present two proofs of these results. The first one, given in Sects. 2 and 3, produces further information about the orbits including systems of representative elements as follows.

Theorem 1.2 *Assume as in Theorem 1.1 and also that G is simple (over its center). Let $(\beta_1, \beta_2, \dots, \beta_r)$ be a maximal sequence of long roots which are mutually orthogonal and contained in the root system for V. For each i let $u_{-\beta_i}$ be a nontrivial element of the one-parameter unipotent subgroup $U_{-\beta_i}$ corresponding to the root $-\beta_i$, and let w_{β_i} be the reflection in the Weyl group W corresponding to β_i, realized as an element of $N_G(T)$. For $0 \leqq t \leqq s \leqq r$, let $x_s = \prod_{i=1}^{s} u_{-\beta_i}$, $w_s = \prod_{i=1}^{s} w_{\beta_i}$, and $x_{st} = \prod_{i=1}^{t} w_{\beta_i} \prod_{j=t+1}^{s} u_{-\beta_j}$.*

(a) $\{x_s | 0 \leqq s \leqq r\}$ *is a system of representatives for the L-orbits of* V^-.
(b) $\{w_s | 0 \leqq s \leqq r\}$ *is a system of representatives for the* (P, P) *double cosets of* G.
(c) $\{x_{st} P | 0 \leqq t \leqq s \leqq r\}$ *is a system of representatives for the L-orbits of* G/P.

Thus there are $r+1$ *L-orbits in* V^-, $r+1$ (P, P) *double cosets in* G, *and* $(r+1)(r+2)/2$ *L-orbits in* G/P.

We note in passing that it easily follows from the Bruhat lemma that $\{x_s\}$ is also a system of representatives in (b) (from which 1.1(a) follows at once).

In the main step of our first proof (Theorem 2.1 below, which embodies 1.1(a) and 1.2(a, b)) we establish natural bijections among the following sets: the L-orbits of V^- (or of V), the $N_L(T)$-orbits of elements of V^- of the form $\prod_{i=1}^{s} u_{-\beta_i}$ with $(\beta_1, \beta_2, \ldots, \beta_s)$ as in 1.2 but not necessarily maximal, the W_L-orbits of such sequences of roots, the W_L-orbits of involutions of W expressible in the form $\prod_{i=1}^{s} w_{\beta_i}$, the (W_L, W_L) double cosets of W, and the (P, P) double cosets of G; and we show that each of the orbits of the second, third or fourth kinds is determined by the number s.

Our second proof of Theorem 1.1, given in Sect. 4 (under a mild extra assumption), depends on a tangent space computation. This proof does not yield the detailed information of the first proof, but it is quite short and it brings up other ideas that we feel are of interest.

Along the way we present other information including the closures and dimensions of the orbits, and in an appendix we consider cases in which the assumptions that V is Abelian and that k is algebraically closed can be relaxed.

If $k=\mathbb{C}$, the homogeneous space G/P is a simply connected Hermitian symmetric space of compact type and, conversely, every simply connected Hermitian symmetric space of compact type is of this form. There exists a real form G_0 of G such that the intersection of G_0 with L is a maximal compact subgroup of both G_0 and L. The orbits of G_0 on G/P have been extensively studied within the framework of Hermitian symmetric spaces by a number of authors, including Takeuchi [15], Korànyi and Wolf [6], Wolf [17, 18] and Lasalle [7]. They obtain results on these orbits which are, in a sense, dual to the ones we obtain for the L-orbits. If we combine the results of Wolf [18] with the results of Theorem 1.2, we obtain a particularly nice set $\{y_{st} | 0 \leqq t \leqq s \leqq r\}$ of representatives for the L-orbits on G/P which are also representatives for the G_0-orbits. Thus we obtain a bijection between the L-orbits and the G_0-orbits. This bijection reverses the partial orders on the sets of orbits given by inclusion of the orbit closures.

In [10], Muller et al. study the L-orbits on $\mathfrak{v}$, the Lie algebra of V. They obtain results which are roughly equivalent to those of Theorem 1.2(a), assuming that char k is 0.

2 The orbits of L on V^- and on V

We start with proofs of Theorems 1.1(a) and 1.2(a, b) and, simultaneously, of the following closely related result. In its statement and in all that follows the notation $G, P, L, V, T, \ldots$ is as in Theorems 1.1 and 1.2 and in addition $W, W_L, \ldots$

denote the Weyl groups of $G, L, \ldots$ and $\Psi, \Psi_P, \Psi_V, \ldots$ the root systems of $G, P, V, \ldots$. A sequence $(\beta_1, \ldots, \beta_r)$ of roots is *orthogonal* if β_i is orthogonal to β_j whenever $i \neq j$. For the standard results about root systems and Weyl groups we refer the reader to [2].

Theorem 2.1 *Let G be a simple algebraic group and $P = LV$ a parabolic subgroup with V Abelian. Assume that the characteristic is not 2 in case roots of different lengths occur. Then there are natural bijections among:*

(a) *The L-orbits of V^- (or of V).*
(b) *The orbits of $N_L(T)$, the normalizer of T in L, on the set of elements of V^- of the form $\prod u_{-\beta_i}$ with $(\beta_1, \beta_2, \ldots, \beta_s)$ an orthogonal sequence of long roots in Ψ_V and $u_{-\beta_i} \in U_{-\beta_i} \setminus \{1\}$ for every i.*
(c) *The W_L-orbits of sequences $(\beta_1, \ldots, \beta_s)$ as in* (b).
(d) *The W_L-orbits of involutions $w \in W$ expressible in the form $\prod w_{\beta_i}$ with $(\beta_1, \ldots, \beta_s)$ as in* (b).
(e) *The (W_L, W_L) double cosets of W.*
(f) *The (P, P) double cosets of G.*

Further the orbits in (b), (c) *and* (d) *are determined by the cardinality s of the sequence in* (b), *so that if r is the maximal length of any such sequence then the cardinalities in* (a) *to* (f) *are all equal to $r+1$. The bijections are given as follows. The maps among* (b), (c), (d) *and* (e) *are the obvious ones and the bijection between* (e) *and* (f) *is an easy consequence of the Bruhat lemma. Finally, the maps from* (b) *to* (a) *and from* (a) *to* (f) *are just the inclusions.*

The bijectivity of this last inclusion amounts to the fact that every (P, P) double coset of G contains a unique L-orbit of V^-, which is just 1.1(a). This is in case G is simple, but the general case follows almost at once since every reductive group is, modulo its center, a direct product of its simple components. Theorems 1.2(a) and 1.2(b) are also formal consequences of 2.1. However, we shall prove 1.2(b) directly as a step in the proof of 2.1.

The proof of Theorem 2.1 will be given in a series of lemmas. First we need some more notation. We assume henceforth that G is a simple group and that the characteristic is not 2 if there are two root lengths. Let T be a maximal torus of G and let B be a Borel subgroup containing T. Let $\Psi^+ = \Psi_B$ be the set of positive roots determined by B and let Δ be the corresponding set of simple roots. If $\beta \in \Psi$, we may write $\beta = \sum_{\alpha \in \Delta} n_\alpha(\beta)\, \alpha$, where the coefficients $n_\alpha(\beta)$ are integers. If $\alpha \in \Delta$ and $\beta \in \Psi$, then $n_\alpha(\beta)$ is the *α-height* of β. Let ρ denote the highest root.

Lemma 2.2 *Let G be a simple group and P a proper parabolic subgroup containing B. Then the unipotent radical V of P is Abelian if and only if P is maximal and the corresponding simple root α (such that $-\alpha \notin \Psi_P$) occurs in the highest root ρ with coefficient* 1, *i.e., the α-height of ρ is* 1.

Proof. If the last condition holds, then for β and γ in Ψ_V the sum $\beta + \gamma$ is never a root (and vice versa). Thus the commutator relations between U_β and U_γ show that these groups centralize each other. If the condition fails then $\beta + \gamma$ can be a root and then U_β and U_γ fail to commute, unless the commutator relations collapse because of the value of char k. If G is not a group of type G_2, this collapse can only occur if β and $\beta + \gamma$ have different lengths and char k

$=2$, which is ruled out by our assumptions. The case $G=G_2$ requires further discussion which we omit. As general references for specific information about the commutator relations among the root subgroups of G used here and later on, we cite [3] and [13]. □

If $\alpha \in \Delta$, we let P_α denote the corresponding maximal parabolic subgroup.

Remark 2.3 Let $\Delta=\{\alpha_1, \ldots, \alpha_n\}$ and let the simple roots α_i be indexed as in [2, Planches I–IX]. We list below all simple roots α such that $n_\alpha(\rho)=1$ or, equivalently, such that the maximal parabolic subgroup P_α has abelian unipotent radical:

(1) Type A_n. $\alpha=\alpha_1, \alpha_2, \ldots, \alpha_n$.
(2) Type B_n. $\alpha=\alpha_1$.
(3) Type C_n. $\alpha=\alpha_n$.
(4) Type D_n. $\alpha=\alpha_1, \alpha_{n-1}, \alpha_n$.
(5) Type E_6. $\alpha=\alpha_1, \alpha_6$.
(6) Type E_7. $\alpha=\alpha_7$.

For types E_8, F_4 and G_2, there are no maximal parabolic subgroups with abelian unipotent radical.

We assume henceforth that $P=P_\alpha$ is a maximal parabolic subgroup containing B with abelian unipotent radical, so that we have $n_\alpha(\rho)=1$.

Lemma 2.4 *The α-height of a root β is* 1, 0 *or* -1, *and, accordingly, β belongs to Ψ_V, Ψ_L or Ψ_V^-.*

This is clear.

Lemma 2.5 *The roots α and ρ (the highest root) belong to Ψ_V and they are both long.*

Proof. The first point is clear and it is well known that ρ is always long. Since the α-height of ρ is 1 it follows that in the dual root system the $\alpha^\vee$-height of $\rho^\vee$ is $(\alpha, \alpha)/(\rho, \rho)$, which can only be an integer if α is also a long root. (Here $\alpha^\vee$, e.g., denotes the coroot $2\alpha/(\alpha, \alpha)$.) □

Lemma 2.6 *W_L is transitive on all roots of Ψ_V of the same length. More precisely the orbits are represented by ρ, the highest root, and ρ_s the highest short root (in case there are two root lengths).*

Proof. This is a special case of a more general result in which the coefficients of several simple roots are prescribed [1]. In the present case, if $\gamma \in \Psi_V$ is arbitrary and $(\gamma, \beta)<0$ for some simple $\beta \in \Psi_L$ then $w_\beta \cdot \gamma$ has greater height than γ. Thus if γ has the greatest height in its orbit then $(\gamma, \beta) \geqq 0$ for all $\beta \in \Psi_L$. Since also $(\gamma, \alpha) \geqq 0$, because $\gamma+\alpha$ is not a root, γ is dominant in Ψ and hence uniquely determined by its length, as required. □

Corollary 2.7 *If two root lengths occur and $\beta \in \Psi_V$ is short, then there exists a long $\gamma \in \Psi_V$ such that $(\beta, \gamma)>0$.*

Proof. By 2.6 β may be taken to be ρ_s and then γ to be ρ. □

More generally than 2.6 we have:

Proposition 2.8 *For each $s=1, 2, \ldots$ the group W_L is transitive on all orthogonal sequences $(\beta_1, \beta_2, \ldots, \beta_s)$ of s roots of Ψ_V of the same length.*

Proof. For convenience of exposition we assume that the above roots are all long. Given a sequence as above we normalize using 2.6 so that $\beta_1=\rho$. Let Ψ_1 be the system of all roots orthogonal to ρ. Since ρ is dominant $\Psi_1\cap\Delta$ is a simple system for Ψ_1 and it contains α if $s\geqq 2$, as we may assume. For, from $(\rho,\beta_2)=(\beta_1,\beta_2)=0$, the dominance of ρ, and the fact that the α-height of β_2 is positive, it follows that α is orthogonal to ρ. Further $\beta_2,\beta_3,\dots,\beta_s$ all lie in $\Psi_1\cap\Psi_V$ which is again Abelian and is the unipotent radical of the parabolic subsystem of Ψ_1 corresponding to the simple root α. Here Ψ_1 need not be irreducible, but its irreducible component containing α also contains $\beta_2,\beta_3,\dots,\beta_s$ since all of these roots have positive α-height. By induction it follows that the Weyl group corresponding to $\Psi_1\cap\Psi_L$ is transitive on all possibilities for $(\beta_2,\beta_3,\dots,\beta_s)$ and hence that W_L is transitive on all possibilities for $(\beta_1,\beta_2,\dots,\beta_s)$. □

Remark. Similarly one can show that, for any s and t, W_L is transitive on all orthogonal sequences from Ψ_V containing s long roots and t short ones in prescribed positions.

Corollary 2.9 *Any two maximal orthogonal sequences of long roots in Ψ_V are W_L-conjugate and hence have the same length.*

Lemma 2.10 *Let $(\beta_1,\beta_2,\dots,\beta_s)$ be an orthogonal set of long roots in Ψ_V and let $w=w_{\beta_1}\cdots w_{\beta_s}$.*

(a) *For each i, $w\cdot\beta_i=-\beta_i$.*

(b) *For $\beta\in\Psi_V$ not equal to any β_i, the set of β_i's not orthogonal to β has cardinality* 0, 1 *or* 2 *and, accordingly, $w\cdot\beta\in\Psi_V$, Ψ_L or Ψ_{V^-}.*

(c) *If s is maximal, the first case of* (b) *does not occur, i.e., β can not be orthogonal to all of the β_i's.*

Proof. (a) This is obvious.

(b) If β is not orthogonal to β_i then $(\beta,\beta_i)>0$ since $\beta+\beta_i$ is not a root, and hence $(\beta,\beta_i^\vee)=1$, since β_i is long. Thus $w\cdot\beta=\beta-\sum\beta_i$, the sum over all such β_i. From this the α-height of $w\cdot\beta$ is 1 minus the number of such β_i. Using 2.4 we get (b).

(c) We may choose, as before, β_1 to be ρ, the highest root. If $(\rho,\alpha)>0$, then $(\rho,\beta)>0$ for every $\beta\in\Psi_V$, since ρ is dominant and α-height$(\beta)>0$. Thus (c) holds in this case with $\max s=1$. If $(\rho,\alpha)=0$, we use induction applied to the system Ψ_1 of the proof of 2.8 to get our conclusion. □

Proposition 2.11 *Let $(\beta_1,\beta_2,\dots,\beta_r)$ be a maximal orthogonal sequence of long roots in Ψ_V, and for each $s\leqq r$ let $w_s=w_{\beta_1}w_{\beta_2}\cdots w_{\beta_s}$. Then $\{w_s|0\leqq s\leqq r\}$ is a system of representatives for $W_L\backslash W/W_L$.*

Proof. First we show that each double coset contains some w_s. We start with an arbitrary element w of the given double coset and write it as a product of reflections relative to roots, some in Ψ_L and some in Ψ_V. Since the former normalize the latter, we may move all reflections of the first kind to the left, conjugating those of the second kind into others of like kind and length as we do so, and then drop those of the first kind. Since the latter all belong to W_L, we remain in the same double coset and thus get a representative expressed as a product of reflections relative to the roots in Ψ_V. Now if a reflection w_β (β short) occurs in this product, then by 2.7 there exists a long $\gamma\in\Psi_V$

with $(\beta, \gamma)>0$ and hence $w_\gamma \cdot \beta=\beta-\gamma=\delta$, say, a root in Ψ_L (see 2.4). We now replace w_β by $w_\gamma^2 w_\beta = w_\gamma w_\gamma w_\beta w_\gamma^{-1} w_\gamma = w_\gamma w_\delta w_\gamma$; the only short root here is δ which is in Ψ_L. Thus, as before, w_δ may be moved to the left and then dropped. Thus all w_β with β short may be removed, resulting in a product involving only long reflections. If two of these, w_β and w_γ, are not orthogonal, we take them as close to each other as possible in the product; then all intermediate reflections are orthogonal to both, hence commute with both, and the two may be brought together in the product. If $\beta=\gamma$, the pair may be cancelled since $w_\beta^2=1$, while if $\beta\neq\gamma$, then $(\beta, \gamma)>0$ (since $(\beta, \gamma)<0$ can not occur because $\beta+\gamma$ is not a root), and then $w_\beta w_\gamma = w_\beta w_\gamma w_\beta^{-1} w_\beta = w_\delta \cdot w_\beta$ with $\delta=\gamma-\beta\in\Psi_L$ and w_δ may be removed as before. In both cases, the product has been shortened so that this process ends after a finite number of steps. When it does, we have as our coset representative a product of reflections in a sequence of long orthogonal roots in Ψ_V, which by 2.8 is conjugate to some w_s under W_L. The given double coset then contains this w_s, as required.

To complete the proof we show that w_s and w_t belong to the same double coset only if $s=t$. Let $u w_s v = w_t$ with $u, v\in W_L$. Here v may be moved to the left and absorbed in u at the expense of conjugating w_s into a like element w_s' but for a possibly changed sequence of roots $(\alpha_1, \alpha_2, \ldots, \alpha_s)$ and we must show that in this situation $s=t$. If some α_i equals some β_j we cancel the corresponding reflections in the relation $u w_s' = w_t$ and finish the proof by induction. Hence we may assume that no α_i equals any β_j. For each $k\leqq s$ we have $w_t(\alpha_k) = u w_s'(\alpha_k) = u(-\alpha_k)$ which is in Ψ_V^-. It follows from 2.10(b) that $(\alpha_k, \beta_l)\neq 0$ for exactly two values of $l\leqq t$. Since this is so for every k, the number of pairs (k, l) with $(\alpha_k, \beta_l)\neq 0$ equals $2s$. Similarly it equals $2t$. Thus $s=t$, as required. □

Remark 2.12 The representative w_s above may be characterized up to W_L-conjugacy as follows. It is an involution in W whose (-1)-eigenspace is spanned by a set of s orthogonal long roots in Ψ_V.

We observe that 2.11 is just 1.2(b), in view of the Bruhat lemma.

We now turn to the elements of V and V^-.

Proposition 2.13 *Every L-orbit of V contains an element of the form* $\prod_{j=1}^{s} u_{\beta_j}$ *with* $(\beta_1, \beta_2, \ldots, \beta_s)$ *as in* 2.11 *and* $u_{\beta_j}\in U_{\beta_j}\setminus\{1\}$. *Two such elements lie in the same L-orbit if and only if their supports have the same cardinality or, equivalently, are* W_L*-conjugate, or yet again, if and only if the elements themselves are* $N_L(T)$*-conjugate. Analogous results hold for* V^- *in place of V.*

Proof. First we show that every orbit has an element of the stated form. We start with an arbitrary element v of the given orbit. If $v=1$, then v has the required form with $s=0$. Assume that $v\neq 1$. Thus $v=\prod u_\beta (\beta\in\Psi_V)$, and some $u_\beta\neq 1$. By applying some $w\in W_L$ (realized in G) to v, we may, by 2.6, assume that this β is dominant. We want β to also be long, i.e., to be the highest root ρ. If this is not the case, then $\beta=\rho_s$ and $(\rho, \beta)>0$, and then $\rho, \rho-\beta, \rho-2\beta$ are roots, while $\rho-3\beta$ is not, since the group G_2 has been ruled out. Then $\rho-\beta\in\Psi_L$ so that $U_{\rho-\beta}\subseteq L$, and in terms of suitable parametrizations of the root subgroups, the commutator relations yield $u_{\rho-\beta}(c)\, u_\beta(d)\, u_{\rho-\beta}(c)^{-1} = u_\beta(d)\, u_\rho(2cd)$ (since $(\rho-\beta)-\beta$ is a root and $(\rho-\beta)-2\beta$ is not). Since we are assuming in this case that the characteristic is not 2, it follows that if we conjugate v by $u_{\rho-\beta}(1)$, the result has ρ in its support. Now that we have

$u_\rho \neq 1$ in v, we apply elements $u_{-\beta}(\beta \in \Psi_L^+)$ to remove from the support of v all γ's (other than ρ) which are not orthogonal to ρ. This can be done inductively: Assume that all such γ's of height $\geqq j$ have been removed, and that β of height $j-1$, not orthogonal to ρ, remains in the support of v. Since ρ is dominant, $(\rho, \beta) > 0$. Thus $\rho - \beta$ is a root, and it lies in Ψ_L^+. The commutator relation between u_ρ and $u_{\beta-\rho}$ takes the form

$$u_{\beta-\rho}(c)\, u_\rho(d)\, u_{\beta-\rho}(c)^{-1} = u_\rho(d)\, u_\beta(cd)\, u_{2\beta-\rho}(c^2 d).$$

The last term occurs only if $2\beta - \rho = 2(\beta - \rho) + \rho$ is a root, and then, although it lies in Ψ_V, its height is less than that of β. In this relation the structural constants can all be made 1, as shown, because $\rho - (\beta - \rho) = \rho + (\rho - \beta)$ is not a root. For a given value of $d \neq 0$, any value of cd can be achieved by an appropriate choice of c. It follows that $u_{\beta-\rho} \in U_{\beta-\rho}$ can be chosen so that conjugation by $u_{\beta-\rho}$ removes the u_β term from v, and without affecting the higher terms already removed, since any new terms, aside from those already discussed, correspond to roots of the form $\gamma + (\beta - \rho)$ and $\gamma + 2(\beta - \rho)\ (\gamma \neq \rho)$ and these have lower height than β. After removing all $u_\beta((\beta, \rho) \neq 0)$ from v we arrive at an orbit representative $v = u_\rho \cdot v_1$, with v_1 supported by the roots of Ψ_V orthogonal to ρ. As in 2.8 let Ψ_1 be the system of all roots of Ψ orthogonal to ρ and G_1 the corresponding reductive subgroup of G containing T. Although the group G_1 may not be simple $\Psi_1 \cap \Psi_V$ (which contains the support of v_1), if nonempty, is contained in the root system of one of its simple components. By induction on the rank of G, we can conjugate v_1 by an element of $L \cap G_1$ to make its support a set of long orthogonal roots in $\Psi_1 \cap \Psi_V$. Then $v = u_\rho v_1$ has as its support a set of long orthogonal roots of Ψ_V, as required. The analogous result for the L-orbits of V^- then follows.

We now consider questions of uniqueness. If two products as in the statement of 2.13 have the same length, then by 2.8 their supports in Ψ_V are W_L-conjugate and may thus be taken to be the same. The products then agree up to a choice of the elements $u_{\beta_j} \in U_{\beta_j} \setminus \{1\}$. In terms of a parametrization, T acts on U_{β_j} thus: $t \cdot u_{\beta_j}(c) = u_{\beta_j}(\beta_j(t)\, c)$. Since $\{\beta_j\}$ is an orthogonal and hence linearly independent set of characters on T, an arbitrary sequence of nonzero values $(\beta_1(t), \beta_2(t), \ldots)$ can be achieved by appropriate choice of t. It follows that the last two products are T-conjugate, and hence that the original ones are $N_L(T)$-conjugate. It remains to show that if two products are L-conjugate they have the same length. We may take the products $v_1 = \prod u_{-\beta_i}$ and $v_2 = \prod u_{-\gamma_j}$ to be in V^-. Let $y_1 = \prod w_{\beta_i}$ and $y_2 = \prod w_{\gamma_j}$. Since v_1 and v_2 are L-conjugate, they lie in the same (P, P) double coset of G. If we apply the Bruhat lemma $u_{-\beta_i} = u'_{\beta_i} w_{\beta_i} u''_{\beta_i}$ to each $u_{-\beta_i}$ in the rank 1 simple group corresponding to β_i and use the fact that these groups centralize each other we see that the first double coset is $P y_1 P$; similarly the second is $P y_2 P$. Since these are equal it follows, again from the Bruhat lemma, that $W_L y_1 W_L$ and $W_L y_2 W_L$ are equal, and hence from 2.11 that the products have the same length, as required. □

Remark 2.14 In the situation we are studying we may regard V as a vector space and L as a group acting linearly on it. The procedure carried out in the first part of the proof of 2.13 is similar to that of completing squares to write an arbitrary quadratic form as a sum of squares, to which it in fact reduces in case G is a group of type C_n. As in this special case, the general algorithm works just when the characteristic is not 2. Of course, in the single root length case this restriction is not needed, nor has it been used.

In view of the above results, notably 2.11 and 2.13, the proofs of Theorems 1.1(a), 1.2(a), 1.2(b), and 2.1 are now complete.

We continue with some consequences.

Proposition 2.15 *Let* $L\cdot x_s (0 \leqq s \leqq r)$ *be the collection of L-orbits of* V^- *as in Theorem* 1.2.

(a) $\overline{L\cdot x_s} = \bigcup_{s' \leqq s} L\cdot x_{s'}$, *for each* $s \leqq r$.

(b) $L\cdot x_r$ *is the unique open L-orbit and* $L\cdot x_0 = \{1\}$ *the unique closed orbit.*

Proof. $L\cdot x_s$ contains $\prod (U_{-\beta_i} \backslash \{1\})(i \leqq s)$, by 2.1(b) or directly. Thus its closure contains every $x_{s'}(s' \leqq s)$, and the left side of (a) contains the right side. From this, if $s' > s$, then $L\cdot x_s$ is contained in the closure of $L\cdot x_{s'}$ and hence can not contain the latter in its closure. This proves (a), and then (b) follows. □

We next determine the dimensions of these orbits. For each $0 \leqq s \leqq r$ we write d_s for the number of roots of Ψ_V that are orthogonal to $\{\beta_1, \beta_2, \ldots, \beta_s\}$. Thus $d_0 = \dim V$, and if G_s denotes the reductive subgroup of G containing T and supported by the roots orthogonal to $\{\beta_1, \ldots, \beta_s\}$ and V_s is the group playing the same role in G_s as V does in G, then $\dim V_s = d_s$.

Proposition 2.16 *Let* $L\cdot x_s$ *and* d_s *be as above.*

(a) $\dim L\cdot x_s = \dim P w_s P/P = d_0 - d_s$.

(b) $\dim G\cdot x_s = 2(d_0 - d_s)$.

Proof. First, for $x \in V^-$ the L-orbits $L\cdot x$ in V^- (with L acting by conjugation) and LxP/P in G/P (with L acting from the left) are isomorphic since $(h\cdot x)P = hxP$ for each $h \in L$, as is easily checked, and V^- and P are transversal. Thus $\dim L\cdot x_s = \dim Lx_s P/P$. Now $L\cdot x_s = Pw_s P \cap V^-$ by 1.1(a), and hence $Lx_s P = Pw_s P \cap V^- P$. Since $V^- P$ is dense in G it follows that $Lx_s P$ is dense in $Pw_s P$ and hence has the same dimension as it. Thus $\dim Lx_s P/P = \dim P w_s P/P = \dim P\,{}^wP/{}^wP = \dim P/(P \cap {}^wP)$, with $w = w_s$. Here and elsewhere ${}^wP$ stands for wPw^{-1}. Since each of P, ${}^wP$, $P \cap {}^wP$ is a subgroup of G containing the maximal torus T, its dimension equals $\dim T$ plus the number of roots in its root system. Thus the required dimension equals the number of roots $\beta \in \Psi_P$ such that $w^{-1}\cdot\beta \notin \Psi_P$, i.e., $w^{-1}\cdot\beta \in \Psi_V^-$, or, setting $w^{-1}\cdot\beta = -\gamma$, the number of roots $\gamma \in \Psi_V$ such that $w\cdot\gamma \in \Psi_P^-$. By 2.10(b) these are just the roots in Ψ_V not orthogonal to $\{\beta_1, \beta_2, \ldots, \beta_s\}$, and from the definitions their number is $d_0 - d_s$, as required for part (a). □

For part (b) we use the following lemma.

Lemma 2.17 (a) *For* $x \in V^-$ *and* $y \in V$, *if* $xyx^{-1} \in V$ *then* x *and* y *commute.*

(b) $\dim Z_V(x_s) = d_s$.

Proof. (a) We may assume that $x = x_s$ for some s and then replace x_s by w_s as in the proof of 2.13, since the extra factors u'_{β_i} and u''_{β_i} both lie in V. Then w_s maps the support of y into Ψ_V. By 2.10(b) the support of y is orthogonal to $\{\beta_1, \beta_2, \ldots, \beta_s\}$, hence to the support of x. Since the latter consists of long roots it follows that x and y commute.

(b) From the discussion in (a) it follows that $Z_V(x_s)$ is the subgroup of V supported by the roots of Ψ_V orthogonal to $\{\beta_1, \beta_2, \ldots, \beta_s\}$. Thus its dimension is d_s. □

From 2.17(a) it follows that if $h \in L$, $y \in V$, and if hy commutes with x_s then both h and y do. Since $V^- L V$ is dense in G and V^- is Abelian it then follows that $\dim Z_G(x_s) = d_0 + \dim Z_L(x_s) + d_s$, the last term by 2.17(b), and this equals $\dim L + 2d_s$, since $\dim Z_L(x_s) = \dim L - \dim L \cdot x_s = \dim L - (d_0 - d_s)$ by part (a). Since also $\dim G = \dim V^- + \dim L + \dim V = \dim L + 2d_0$ we get $\dim G \cdot x_s = \dim G - \dim Z_G(x_s) = 2(d_0 - d_s)$, as required.

Corollary 2.18 *Each G-orbit that meets V^- does so in a single L-orbit. In other words, $G \cdot x \cap V^- = L \cdot x$ for each $x \in V^-$.*

Proof. If $s \neq t$ then by 2.16(b) the orbits $G \cdot x_s$ and $G \cdot x_t$ have different dimensions and hence can not be equal. Or, we can argue as follows. First $G = PV^- P = PV^- \cdot V^- P$, since in the final expression both factors are dense and open in G. Thus if $x, y \in V$ are in the same G-orbit then $(pzq) \cdot x = y$ with $p, q \in P$ and $z \in V^-$, whence $z \cdot (q \cdot x) = p^{-1} \cdot y$ with $z \in V^-$ and $q \cdot x$, $p^{-1} \cdot y \in V$. It follows from part (a) of the lemma that $q \cdot x = p^{-1} \cdot y$, whence x and y are in the same P-orbit, hence in the same L-orbit, as required. □

Another proof of 2.18 may be found in [12] where the following more general result due to G. Seitz is also proved. If P is an arbitrary parabolic subgroup of G then two elements of the center of V (or of V^-) conjugate under G are also conjugate under L.

We illustrate 2.16(a) with an example.

Example 2.19

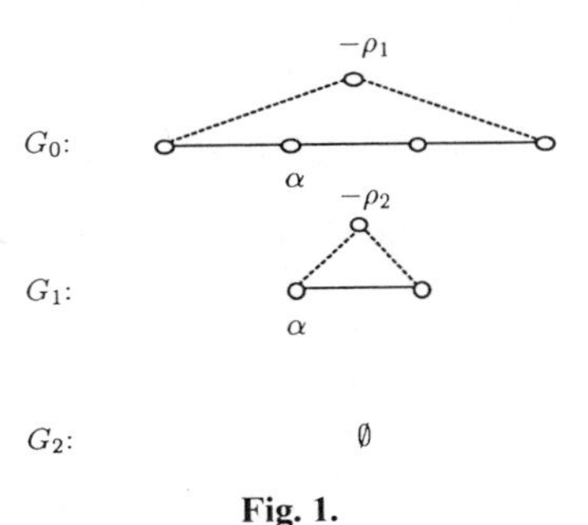

Fig. 1.

Here $G = G_0 = A_4$ is the original group, α is the root defining P, and ρ_1 is the highest root. Then G_1 is the subgroup supported by the roots orthogonal to ρ_1, and ρ_2 is its highest root. And so on. From the formula $d_i = N(G_i) - N(L_i)$ (N = number of positive roots), we get $d_0 = 10 - 4 = 6$, $d_1 = 3 - 1 = 2$, $d_2 = 0$. Clearly, $r = 2$. Thus there are three orbits and their dimensions are 0, 4, and 6.

In the same way all of the cases can be worked out (another example is given in 3.11 below) and, in particular, all of the values of r found. The results, in the notation of 2.3, are as follows.

Table 1

Type of G	The root α	r
A_n	$\alpha_i (1 \leqq i \leqq n/2)$	i
B_n	α_1	2
C_n	α_n	n
D_n	α_1	2
D_n	α_{n-1} or α_n	$\left[\frac{n}{2}\right]$
E_6	α_1 or α_6	2
E_7	α_7	3

3 The orbits of L on G/P

In this section we obtain the principal results about these orbits, using mainly the results of the last section. Our previous assumptions and notations continue. In particular, G is simple, $P = P_\alpha = LV$ with V abelian, $(\beta_1, \beta_2, \ldots, \beta_r)$ is a maximal orthogonal sequence of long roots in Ψ_V, and $w_s, x_{st}, \ldots$ are as in Theorem 1.2.

Lemma 3.1 *$P = LU_{\beta_1} U_{\beta_2} \cdots U_{\beta_r} L$, even if G is not simple.*

Proof. If G is simple, this follows from 2.1 (see (a) and (b)). If G is not simple, we apply this first case to each of its simple components and collect the results. □

Lemma 3.2 *For each $s \leqq r$ we have $P w_s P = LU_{\beta_1} U_{\beta_2} \cdots U_{\beta_s} w_s P$.*

Proof. We write $w = w_s$ and consider first the case in which ${}^wP = P^-$. This condition is equivalent to the pair of conditions ${}^wL = L$ and ${}^wV = V^-$ (and also to the same conditions with w replaced by the longest element of W, which will not be used). It holds only when $s = r$, since otherwise w_s would stabilize U_{β_r} and hence not map it into V^-. We have

$$\begin{aligned} P w_r P &= LU_{\beta_1} \cdots U_{\beta_r} L w_r P \quad \text{(by 3.1)} \\ &= LU_{\beta_1} \cdots U_{\beta_r} w_r P, \end{aligned}$$

where the second equality holds since $Lw_r P = w_r {}^wLP = w_r LP = w_r P$ (because $w = w_r$ is an involution). This proves the first case, which is valid also when G is not simple since 3.1 is. In the general case let V' be the subgroup of V supported by the roots $\beta \in \Psi_V$ such that $w_s \cdot \beta \in \Psi_{V^-}$. Observe that $\beta_1, \beta_2, \ldots, \beta_s$ are among these roots and that ${}^wV'$ is the subgroup of V^- supported by the negatives of these roots since w_s is an involution. Similarly let L' be the reductive subgroup of L containing T whose root system consists of the roots $\gamma \in \Psi_L$ such that $w_s \cdot \gamma \in \Psi_L$. Specifically, $V' = V \cap {}^w V^-$ and $L' = (L \cap {}^w L)^0$. It can be checked that $\Psi_{V'}$, $-\Psi_{V'}$ and $\Psi_{L'}$ together form a closed semisimple subsystem of Ψ. Let G' be the corresponding reductive subgroup of G containing T. Another straightforward check shows that $P' = L'V'$ is a parabolic subgroup of G' and that V' is Abelian. We apply the first case to G', which, as noted above, need not be simple. Let V'' be the subgroup of V with support $\Psi_V \setminus \Psi_{V'}$. Then $V = V'V''$ and $w_s \Psi_{V''} \subseteq \Psi_P$. Thus $P w_s P = LV'V'' w_s P = LV' w_s P$ (since ${}^w V'' \subseteq P$) $= L(L'V' w_s P') P = L(P' w_s P') P$. We finish by applying the first case to the inner expression and then simplifying the result. □

For each root β, let G_β be the three dimensional simple subgroup of G generated by $\{U_\beta, U_{-\beta}\}$, let $T_\beta = G_\beta \cap T$ and let $B_\beta = T_\beta U_\beta$. Then B_β is a Borel subgroup of G_β, T_β is a maximal torus of G_β and $Y_\beta = G_\beta / B_\beta$ is a projective line.

Lemma 3.3 (a) $P w_s P = \bigcup_{0 \leqq t \leqq s} L x_{st} P$.

(b) $G = \bigcup_{0 \leqq t \leqq s \leqq r} L x_{st} P$.

Proof. For each root β, $T_\beta U_\beta w_\beta B_\beta$ is the union of $T_\beta w_\beta B_\beta$ and $T_\beta U_{-\beta} B_\beta$, all taken in the simple group G_β corresponding to β. For, in the group SL_2 or PSL_2 isomorphic to G_β, these three sets are, respectively, the set of matrices $\begin{bmatrix} a & b \\ c & d \end{bmatrix}$ for which $c \neq 0$, the subset for which $a = 0$ and the subset for which $a \neq 0$. Here the standard choices for $T, B, \ldots$ (diagonal, superdiagonal, ...) are being used. It follows from 3.2 that each (L, P) coset contained in $P w_s P$ has a representative of the form $\prod_{i=1}^{s} y_i$, where, for each i, $y_i = w_{\beta_i}$ or $y_i = u_{-\beta_i}$. Since any permutation of the β_i's can be achieved by an element of W_L, by 2.8, the i's for which $y_i = w_{\beta_i}$ may be taken to be $1, 2, \ldots, t$ for some $t \leqq s$. These representatives thus become $\{x_{st} | t \leqq s\}$, as required for (a). Since $G = \cup P w_s P$, part (b) follows. □

Lemma 3.4 *$\{w_s | 0 \leqq s \leqq r\}$ is not only a system of representatives for $P \setminus G / P$ but also for $P^- \setminus G / P$.*

This follows from a more general result whose proof we postpone to Sect. 5 so as not to interrupt the present flow.

Lemma 3.5 (a) *The cosets $L x_{st} P$ $(0 \leqq t \leqq s \leqq r)$ are distinct.*

(b) $P w_s P = \bigcup_{0 \leqq t \leqq s} L x_{st} P$.

(c) $P^- w_t P = \bigcup_{t \leqq s \leqq r} L x_{st} P$.

(d) *$P w_s P$ and $P^- w_t P$ intersect in $L x_{st} P$ if $t \leqq s$ and they are disjoint if $t > s$.*

Proof. If $L x_{s't'} P = L x_{st} P$, then $P w_{s'} P = P w_s P$ and hence $s' = s$ by 1.1(b). Similarly $t' = t$ by 3.4. Thus (a) holds. Since each of $P w_s P$ and $P^- w_t P$ is a union of (L, P) double cosets, (b), (c), and (d) follow from (a), 3.3(a), and 3.4. □

We return now to Theorems 1.1 and 1.2. Note that 1.1(a), 2.1(a) and 2.1(b) have already been proved and that 1.1(b) has been reduced to the case in which G is simple. By 3.3(b) and 3.5(d) every L-orbit of G/P is the intersection of a P-orbit and a P^--orbit, which is 1.1(b), and by 3.3(a) and 3.5(a) $\{x_{st}P\}$ is a system of representatives for the L-orbits of G/P. The proof of Theorems 1.1 and 1.2. is complete.

We can now prove a nice geometric corollary of the above results. Since β_i and β_j are orthogonal long roots if $i \neq j$, the groups G_{β_i} and G_{β_j} commute elementwise. Thus the product $\prod_{i=1}^{r} G_{\beta_i}$ is a closed algebraic subgroup of G.

Corollary 3.6 (a) $G = L \prod_{i=1}^{r} G_{\beta_i} P$.

(b) *The L-orbits of G/P correspond bijectively to the* $\left(S_r \times \prod_{i=1}^{r} T_{\beta_i}\right)$*-orbits of* $\prod_{i=1}^{r} G_{\beta_i} P/P$: *each of the former contains a unique one of the latter.*

Proof. Here S_r is the symmetric group, acting by permuting the G_{β_i}'s. Observe that $\prod G_{\beta_i}P/P$ is naturally isomorphic to $\prod G_{\beta_i}/B_{\beta_i}$, a product of projective lines. Part (a) easily follows from 3.3(b). As for (b) it is straightforward to verify that $\{x_{st}\}$ is a system of representatives for $S_r \times \prod T_{\beta_i} \backslash G_{\beta_i} P/P$. □

This reduction to a product of SL_2's acting on a product of projective lines is a theme that recurs in the earlier papers on this subject (see [7, 18]; see also 5.6).

Corollary 3.7 (a) $\overline{Pw_s P} = \bigcup_{s' \leqq s} P w_{s'} P$.

(b) $\overline{P^- w_t P} = \bigcup_{t' \geqq t} P^- w_{t'} P$.

(c) $\overline{Lx_{st} P} = \bigcup_{t \leqq t' \leqq s' \leqq s} Lx_{s't'} P$.

(d) *$Lx_{st}P$ is closed (in G or in G/P) if and only if $t = s$. There are thus $r+1$ closed L-orbits on G/P.*

(e) *$Pw_s P$ and $P^- w_t P$ contain $Lx_{so} P$ and $Lx_{rt} P$ as their unique open L-orbits and $Lx_{ss} P$ and $Lx_{tt} P$ as their unique closed or relatively closed L-orbits.*

Proof. For each $i \leqq s$, $Pw_s P$ contains $B_{\beta_i} w_{\beta_i} B_{\beta_i}$, which is dense in G_{β_i} and hence contains 1 in its closure. Thus $\overline{Pw_s P}$ contains every $Pw_{s'} P$ $(s' \leqq s)$ and then (a) follows as in the proof of 2.15(a). Part (b) is proved similarly. The set $T_{\beta_s} u_{-\beta_s} B_{\beta_s}$ contains both 1 and w_{β_s} in its closure. Thus, if $t < s$ then $\overline{Lx_{st} P}$ contains $Lx_{s-1,t} P$ and $Lx_{s,t+1} P$ and hence, by induction, it contains $\cup Lx_{s't'} P$ $(t \leqq t' \leqq s' \leqq s)$. By (a), (b) and 3.5(d) the reverse containment holds, whence (c). Then (d) and (e) easily follow. □

Remarks. (a) The closure in 3.7(d) holds much more generally. In fact LwP is closed in G if G is any reductive group, P any parabolic subgroup, w any element of W, and L any reductive subgroup of G containing a maximal torus of P. The proof goes as follows: For each β in Ψ_L at least one of $w^{-1}(\pm\beta)$ is in Ψ_P. Thus the stabilizer, Q, of wP in L is a parabolic subgroup of L. Thus in the natural morphism from L/Q onto LwP/P the first term is a complete variety and hence the second one closed. In each of the cases of 3.7(d) the parabolic subgroup Q can be worked out. An amusing fact is that *its* unipotent radical is always Abelian since it is conjugate under w_s to a subgroup of V.

(b) The closure of $Lx_{st}P$ may be written as $L\cdot\prod_{i=1}^{t} w_{\beta_i} \prod_{j=t+1}^{s} G_{\beta_j} P$.

Next we determine the dimensions of the orbits.

Proposition 3.8 *Let G be an algebraic group and A, B, C closed subgroups. Assume that G, $A\cap B$ and C are connected and that AB is dense in G. Then $AC\cap BC$ is irreducible and* $\dim(AC\cap BC)=\dim(AC)+\dim(BC)-\dim G$.

Proof. Let $X=AC\cap BC$, Y the (algebraic) set of all $(a, b, c_1, c_2)\in A\times B\times C\times C$ such that $ac_1=bc_2$ and Z the set of all $(a, b, c)\in A\times B\times C$ such that $ab=c$. The projection of Z on C has image $C'=AB\cap C$, which is dense in C, since AB is dense in G, and the nonempty fibres are isomorphic to $A\cap B$. Since C' is irreducible and all fibres are irreducible and of the same dimension it follows that Z is irreducible and that its dimension is $\dim C+\dim(A\cap B)$. Applying a similar argument to the morphism of Y onto Z sending (a, b, c_1, c_2) to $(a^{-1}, b, c_1c_2^{-1})$ we see that Y is irreducible and that its dimension is $2\dim C+\dim(A\cap B)$. Finally, the morphism $(a, b, c_1, c_2)\mapsto ac_1$ from Y to X yields that X is irreducible and that its dimension is $2\dim C+\dim(A\cap B)-\dim(A\cap C)-\dim(B\cap C)$, which agrees with the required formula since $\dim(E\cap F)=\dim E+\dim F-\dim(EF)$ for any two subgroups of G (and $\dim(AB)=\dim G$). □

Remark. It follows similarly that if A and B intersect transversally then AC and BC do also and hence that $AC\cap BC$ is smooth.

Corollary 3.9 *Using our earlier notations we have*

$$\dim Lx_{st}P/P=\dim Pw_sP/P+\dim P^-w_tP/P-\dim G/P.$$

Proof. We apply 3.8 with $A=P$, $B=P^-$ and $C={}^xP$ $(x=x_{st})$. Since $Lx_{st}P=Pw_sP\cap P^-w_tP$, the corollary follows. □

Remark. The above results, as they apply to sets of the form $PR\cap QR$ with P, Q and R parabolic subgroups of G, as well as our earlier results on closures in Corollary 3.7, are treated in great generality in [11].

As in Sect. 2 we write d_s $(s=0, 1, \ldots, r)$ for the number of roots in Ψ_V orthogonal to $\{\beta_1, \beta_2, \ldots, \beta_s\}$.

Theorem 3.10 *For all* $0\leqq t\leqq s\leqq r$ *we have*

$$\dim Lx_{st}P/P=(d_t-d_s)+t(d_{t-1}-d_t-1).$$

Proof. First we have $\dim Pw_sP/P=d_0-d_s$. This restatement of 2.16(a) is also the case $t=0$ of the present theorem. Secondly we have $\dim P^-w_tP/P=d_t+t(d_{t-1}-d_t-1)$, the case $s=r$ of the theorem. For, paralleling the proof of 2.16(a) we see that $\dim P^-w_tP/P$ equals the number of roots $\gamma\in\Psi_V$ with $w_t\cdot\gamma\in\Psi_P$. By 2.10(b) this equals the number of roots in Ψ_V each orthogonal to at most one of $\{\beta_1, \ldots, \beta_t\}$, excluding these roots themselves. There are d_t such roots orthogonal to all of them and, for each $i=1, 2, \ldots, t$, exactly $d_{t-1}-d_t-1$ such roots other than β_i orthogonal to $\{\beta_1, \ldots, \hat{\beta}_i, \ldots, \beta_t\}$ but not to β_i. This holds for $i=t$ by the definitions and thus for every i by symmetry. From this our second case follows. Since also $\dim G/P = \dim V^- P/P = \dim V^- = d_0$, the theorem follows from 3.9. □

As special cases we have the two occuring in the proof of the theorem which apply to the open L-orbits of the various P-orbits and the various P^--orbits of G/P, and also the equation $\dim Lx_{ss} P/P = \dim Lw_s P/P = s(d_{s-1} - d_s - 1)$ which applies to the various closed L-orbits.

Example 3.11 As in the example at the end of Sect. 2 we present the situation in terms of extended Dynkin diagrams.

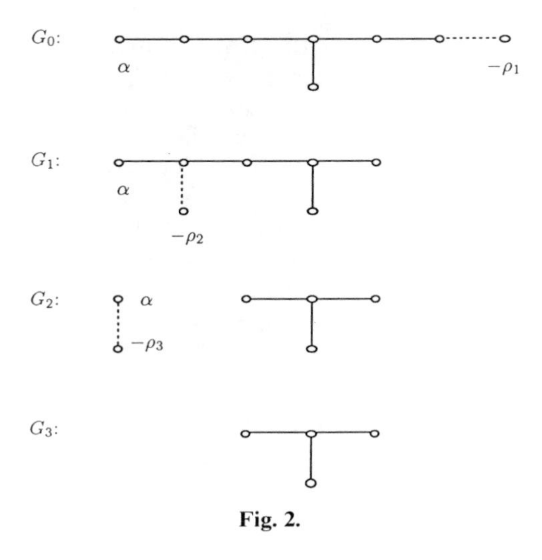

Fig. 2.

Here $r=3$ and the d_i's work out as follows

$$d_0 = 63 - 36 = 27, \quad d_1 = 30 - 20 = 10, \quad d_2 = 1 - 0 = 1, \quad d_3 = 0.$$

From 3.10 we get the following table of values of the dimensions of the L-orbits.

Table 2.
Table of values of $\dim Lx_{st} P/P$ for E_7

$t\backslash s$	0	1	2	3
0	0	17	26	27
1		16	25	26
2			16	17
3				0

Here the blanks represent cases that do not occur. The first row represents the L-orbits that are open in the various P-orbits of G/P, the last column those that are open in the various P^--orbits, and the main diagonal those that are closed. Observe the symmetry in the second diagonal and the two 0's which represent fixed points. If the reader works out the table for the example of 2.19, he will notice that this symmetry fails and that there is only one fixed point.

We proceed to an explanation of these discrepancies.

Proposition 3.12 *The following conditions are equivalent.*

(a) *P and P^- are conjugate in G.*
(b) *If $w=w_r$, then ${}^wP=P^-$.*
(c) *Each root in Ψ_V other than $\beta_1, \beta_2, \ldots, \beta_r$ is not orthogonal to exactly two of these roots.*
(d) *$d_{r-1}=1$.*

Proof. If (a) holds then ${}^{w'}P=P^-$ for some $w'\in W$, by the Bruhat lemma. Then ${}^{w'}V=V^-$ and $w'\cdot\Psi_V=\Psi_{V^-}$, and hence also $w'\cdot\Psi_L=\Psi_L$. By modifying w' by an element of W_L we may assume that $w'\Psi_L^+=\Psi_L^-$, i.e., that w' is the longest element of W. Now Pw_rP is dense in G by 3.7(a) and $Pw'P$ is also since it contains the big cell $Bw'B$ of the Bruhat lemma. Thus $Pw'P=Pw_rP$ and w' may be taken to be w_r. Thus if (a) holds then (b) does also; the converse is obvious. If (b) fails then w_r maps some root in Ψ_V into Ψ_P and, by 2.10(b), (c) fails; and vice versa. If (c) fails there is a root $\beta\in\Psi_V$ not orthogonal to exactly one of the β_i's, by 2.10(c), which may be taken to be β_r. Then β and β_r are orthogonal to $\{\beta_1, \ldots, \beta_{r-1}\}$, $d_{r-1}\geqq 2$, and (d) fails. This argument also may be reversed. □

Lemma 3.13 (a) *The map sending each connected subgroup of G to its Lie algebra is injective when it is restricted to the set of subgroups containing a fixed maximal torus.*
(b) *If Q and R are two subgroups containing a common maximal torus of G, then $T_1(QR)=\underline{q}+\underline{r}$.*

Proof. Here T_1 denotes the tangent space at 1 and $\underline{q}$ and $\underline{r}$ the Lie algebras of Q and R; and G can be taken to be any semisimple algebraic group. We may assume the maximal torus to be the standard one, T. If $S=S_Q=\{\alpha\in\Psi \mid U_\alpha\subset Q\}$, then $T\cdot\prod_{\alpha\in S} U_\alpha$ is a dense open part of Q (the big cell) and, as an algebraic variety, it is the direct product of its factors. Thus $\underline{q}=\underline{t}+\sum_{\alpha\in S}\underline{u}_\alpha$ (direct) and (a) follows. Similar considerations applied to R and QR, with S_Q replaced by $S_Q\cup S_R$ in the latter case, then yield (b). □

Proposition 3.14 *If the conditions of 3.12 hold then L has two fixed points on G/P, viz., P and w_rP. If they fail there is only one fixed point, P.*

Proof. Let xP be a fixed point: $LxP=xP$, i.e., $L\subseteq{}^xP$. Now it is known that L acts on $\underline{v}$ and on $\underline{v}^-$ (the Lie algebras of V and V^-) irreducibly (as follows easily from 2.6), via inequivalent representations since the weights of T on the first are all positive and these on the second are all negative. It follows that $\underline{l}$, $\underline{p}$, $\underline{p}^-$, and $\underline{g}$ are the only subalgebras of $\underline{g}$ containing $\underline{l}$, thus by 3.13(a) P and P^- are the only subgroups of their dimension containing L, and thus, since $L\subseteq{}^xP$, that ${}^xP=P$ or P^-. If the conditions of 3.12 hold we conclude that $x\in P$ or $x\in w_rP$ (using also the fact that P equals its own normalizer), while if they fail then the second case can not occur and $x \in P$ is the only possibility.

Observe that the two fixed points are the L-orbits for which $(s, t)=(0, 0)$ and $(s, t)=(r, r)$, respectively. □

Proposition 3.15 *The conditions of* 3.12 *hold if and only if*

$$\dim Lx_{st} P/P = \dim Lx_{r-t, r-s} P/P \quad \textit{for all } 0 \leqq t \leqq s \leqq r,$$

i.e., the table of dimensions is symmetric in the second diagonal.

Proof. Assume that the conditons hold. Then

$$w_r P w_s P = w_r P w_r^{-1} \cdot w_r w_s \cdot P = P^- w_{\beta_{s+1}} \cdots w_{\beta_r} P = P^- w_{r-s} P,$$

the last step since $\{\beta_{s+1}, \ldots, \beta_r\}$ is conjugate to $\{\beta_1, \ldots, \beta_{r-s}\}$ under W_L. Similarly $w_r P^- w_t P = P w_{r-t} P$. Since $Lx_{st} P = P w_s P \cap P^- w_t P$, we get $w_r Lx_{st} P = Lx_{r-t, r-s} P$, and the required symmetry follows. If the conditions of 3.12 fail then $\dim Lx_{00} P/P = 0$ while $\dim Lx_{rr} P/P \neq 0$ by 3.14 and the stated symmetry fails. □

Proposition 3.16 *Assume that the conditions of* 3.12 *hold. Then the dimensions of* $P w_s P/P$, $P^- w_t P/P$, *and* $Lx_{st} P/P$ *are, respectively,* $d_0 - d_s$, $d_0 - d_{r-t}$, and $d_0 - d_s - d_{r-t}$.

Proof. The first formula is just 2.16(a). Since $w_r P^- w_t P = P w_{r-t} P$, the second formula follows. The third formula then follows from 3.9. □

Remark. The formulas given here and earlier in terms of the d_i's are not unique since, e.g., $d_s = d_0 - s(d_0 - d_1) + \binom{s}{2}(d_0 - 2d_1 + d_2)$ for all $s \leqq r$. Since $d_0 - d_1$ counts the roots in Ψ_V not orthogonal to a specified root in $\{\beta_1, \beta_2, \ldots, \beta_s\}$ and $d_0 - 2d_1 + d_2$ counts those not orthogonal to (each of) two specified roots in $\{\beta_1, \beta_2, \ldots, \beta_s\}$, the equation follows from the principal of inclusion and exclusion. The special case $s=3$, $d_0 - 3d_1 + 3d_2 - d_3 = 0$, expresses the fact that there is no root not orthogonal to three of the β_i's.

4 Second proof of Theorem 1.1

For this proof we must assume that char $k \neq 2$ even though, as we have seen in the first proof, this assumption is not always needed. As we have also seen we may assume that G is simple. Our second proof depends mainly on the following result.

Proposition 4.1 *Assume as in Theorem* 1.1 *and also that G is simple. Then*

$$T_x(PxP) \cap T_x(P^- xP) = T_x(LxP) \quad \textit{for every } x \in G.$$

Granted 4.1 we proceed as follows. We have

$$T_x(LxP) \subseteq T_x(PxP \cap P^- xP) \subseteq T_x(PxP) \cap T_x(P^- xP) = T_x(LxP).$$

Thus $T_x(LxP)=T_x(PxP\cap P^-xP)$. Since LxP is smooth (as an orbit of $L\times P$ acting on G) and $PxP\cap P^-xP$ is irreducible (by 3.8 with $A=P$, $B=P^-$, $A\cap B=L$, and $C={}^xP$) we get $\dim LxP\geqq\dim(PxP\cap P^-xP)$, and clearly equality must hold. Using the irreducibility again we see that LxP is a dense open subset of $(PxP\cap P^-xP)$. If LyP is another double coset contained in $PxP\cap P^-xP$ it too must be a dense open subset of $PxP\cap P^-xP$ and hence must meet LxP, a contradiction. Thus $PxP\cap P^-xP$ consists of the single double coset LxP, as required for 1.1(b). Since $PV^-P=G$ (see the proof of 2.18), every (P, P) double coset of G meets V^-, and it remains, for the proof of 1.1(a), to show that $L\cdot x=PxP\cap V^-$ for every $x\in V^-$. Assume $y\in PxP\cap V^-$. Then $y\in PxP\cap P^-xP=LxP$ by 1.1(b), and $y=hxp$ with $h\in L$ and $p\in P$. Thus $(hxh^{-1})^{-1}y=hp\in V^-\cap P=\{1\}$, and $y=hxh^{-1}\in L\cdot x$, as required.

Returning to 4.1, if we right translate by x^{-1} the result to be proved becomes

$$T_1(P^xP)\cap T_1(P^{-\,x}P)=T_1(L^xP).$$

We have $T_1(P^xP)=\underline{p}+x\cdot\underline{p}$ and $T_1(P^{-\,x}P)=\underline{p}^-+x\cdot\underline{p}$ by 3.13(b) since any two parabolic subgroups contain a common maximal torus of G by the Bruhat lemma, and $T_1(L^xP)\supseteq\underline{l}+x\cdot\underline{p}$. Thus it is enough to prove

$$(1)\qquad (\underline{p}+x\cdot\underline{p})\cap(\underline{p}^-+x\cdot\underline{p})\subseteq\underline{l}+x\cdot\underline{p})\qquad \text{for every } x\in G.$$

We may write $x=p_1wp_2$ with $p_1, p_2\in P$ and $w\in W$, and then $p_1=hv$ with $h\in L$ and $v\in V$. We may then drop h and p_2 since h stabilizes $\underline{p}$ and $\underline{p}^-$ and v stabilizes $\underline{p}$ and thus take $x=v\,w$. Then writing $v=v_1v_2$ with $v_1\in V\cap{}^wV^-$ and $v_2\in V\cap{}^wP$ we may drop v_2 since $v_2w=w(w^{-1}v_2w)$ and $w^{-1}v_2w\in P$ stabilizes $\underline{p}$. Applying v_1^{-1} to (1) and writing v in place of v_1^{-1}, we get as the result to be proved:

$$(2)\quad (\underline{p}+w\cdot\underline{p})\cap(v\underline{p}^-+w\cdot\underline{p})\subseteq v\underline{l}+w\cdot\underline{p}\qquad \text{for all } w\in W \text{ and } v\in V\cap{}^wV^-.$$

Finally we may drop the second $w\cdot\underline{p}$ and replace $\underline{p}^-$ by $\underline{v}^-$ since $v\cdot\underline{p}^-=v\cdot\underline{v}^-+v\cdot\underline{l}$ and $v\cdot\underline{l}\subseteq\underline{p}$ to reduce (2) to:

$$(3)\qquad (\underline{p}+w\cdot\underline{p})\cap(v\cdot\underline{v}^-)\subseteq v\cdot\underline{l}+w\cdot\underline{p}\qquad \text{for all } w\in W \text{ and } v\in V\cap{}^wV^-.$$

Lemma 4.2 *For each $v\in V\cap{}^wV^-$ there exists $X\in\underline{v}\cap w\cdot\underline{v}^-$ such that for all $Y\in\underline{v}^-$*

(a) $v\cdot Y=Y+[X, Y]+\frac{1}{2}[X,[X, Y]]$.

(b) $v\cdot[X, Y]=[X, Y]+[X,[X, Y]]$.

Proof. Assume first that $k=\mathbb{C}$. Choose a basis $\mathscr{B}$ of $\underline{g}$ made up of root elements X_β $(\beta\in\Psi)$ and elements T_i $(1\leqq i\leqq n)$ of $\underline{t}$. Then the parametrization of the root subgroups U_β may be chosen so that $u_\beta(x)=\exp(xX_\beta)$ for all $x\in\mathbb{C}$. Thus if v as in the lemma is written as $\prod u_\beta(x_\beta)$ $(\beta\in\Psi_V\cap w\cdot\Psi_{V^-})$, then $v=\exp X$ with $X=\sum x_\beta X_\beta$ and $\mathrm{Ad}(v)=\exp(\mathrm{ad}\,X)=I+\mathrm{ad}\,X+\frac{1}{2}(\mathrm{ad}\,X)^2+\ldots$. Since $\mathrm{ad}\,X$ increases α-heights by 1 it maps $\underline{v}^-$ to $\underline{l}$, $\underline{l}$ to $\underline{v}$, and $\underline{v}$ to $\{0\}$, and our result follows. In other words,

$$(*)\qquad \textit{if } v=\prod u_\beta(x_\beta),\qquad \textit{then (a) and (b) hold with } X=\sum x_\beta X_\beta.$$

Since this is true for arbitrary $x_\beta\in\mathbb{C}$, the relations (a) and (b) may be regarded as formal polynomial identities in the coordinates $\{x_\beta\}$ of v and the coordinates, say $\{y_\gamma\}$, of Y. Now if $\mathscr{B}$ is taken to be a Chevalley basis of $\underline{g}$ the coefficients of the polynomials are all integers (see [3, p. 27, formula (5)] or [13, p. 22, Lemma 15]), and thus the identities persist over any field. Thus (*) with (a) multiplied by 2 is true for any field k, and (*) itself is true as long as char $k \neq 2$. □

Now with v as in (3) and $Y \in \underline{v}^-$ we have $v \cdot Y = Y + A$ with $A \in \underline{p}$ by 4.2(a). Also $\underline{p} + w \cdot \underline{p} = \underline{p} + (w \cdot \underline{p} \cap \underline{v}^-)$ since w stabilizes $\underline{t}$ and permutes the root spaces. Thus if $v \cdot Y \in \underline{p} + w \cdot \underline{p}$ it follows that $Y \in w \cdot \underline{p} \cap \underline{v}^-$. Thus to prove (3) we need only prove the following lemma.

Lemma 4.3 *If* $v \in V \cap {}^w V^-$ *and* $Y \in w \cdot \underline{p} \cap \underline{v}^-$ *then* $v \cdot Y \in v \cdot \underline{l} + w \cdot \underline{p}$.

Proof. Choose X as in 4.2. Since $Y \in w \cdot \underline{p}$ it is enough to show that $v \cdot \underline{l} + w \cdot \underline{p}$ contains $[X, Y] + \frac{1}{2}[X, [X, Y]]$, i.e., that it contains $Z + \frac{1}{2}[X, Z]$ if we write Z for $[X, Y]$. We have $Z \in \underline{l}$ and also $Z \in [w \cdot \underline{v}^-, w \cdot \underline{p}] \subseteq w\underline{p}^- = w \cdot \underline{v}^- + w \cdot \underline{l}$. Thus we may write $Z = Z_1 + Z_2$, with $Z_1 \in \underline{l} \cap w \cdot \underline{v}^-$ and $Z_2 \in \underline{l} \cap w \cdot \underline{l}$, and it is enough to prove the required containment for $Z = Z_1$ and $Z = Z_2$. We have $[X, Z_1] \in [w \cdot \underline{v}^-, w \cdot \underline{v}^-] = 0$. Thus $Z_1 + \frac{1}{2}[X, Z_1] = Z_1 + [X, Z_1] = v \cdot Z_1 \in v \cdot \underline{l}$, the last equality by 4.2(b). We also have $v \cdot Z_2 = Z_2 + [X, Z_2]$. Thus $Z_2 + \frac{1}{2}[X, Z_2] = \frac{1}{2}(Z_2 + v \cdot Z_2)$ with $Z_2 \in w \cdot \underline{p}$ and $v \cdot Z_2 \in v \cdot \underline{l}$. □

Lemma 4.3 is thus established and with it Proposition 4.1 and Theorem 1.1.

5 Appendix

We present some complements to our main development.

Proposition 5.1 *If G is a reductive group, T a maximal torus, and Q and R parabolics containing T, then a system of representatives for $W_Q \backslash W / W_R$ is also one for $Q \backslash G / R$.*

This yields 3.4 which was used in the proof of 3.5.

Proof. First 5.1 is true when Q and R are standard, i.e., contain a common Borel subgroup containing T, by the Bruhat lemma. In the general case there exist $a, b \in W$ such that ${}^aQ$ and ${}^bR$ are standard. If S is a system of representatives for $W_Q \backslash W / W_R$, then aSb^{-1} is one for ${}^aW_Q \backslash W / {}^bW_R$, hence it is also one for ${}^aQ \backslash G / {}^bR$ by the first case, and hence $S = a^{-1}(aSb^{-1})b$ is one for $Q \backslash G / R$. □

Next we use a construction from [1] to extend our results partially to the case in which V is not necessarily Abelian. Assume this to be the case. By the shape of a root $\beta = \sum n_\alpha(\beta)\alpha$ $(\alpha \in \Delta)$ we mean the subsum over the elements of $\Delta_V = \Psi_V \cap \Delta$. The condition for V to be Abelian given in 2.2 may be restated: there exists $\alpha \in \Delta$ such that every root in Ψ_V has shape α. We now fix a nonzero shape S occurring for some positive root β. Then the roots of shape a positive (integral) multiple of S form a closed subsystem Ψ_{V_1} of Ψ_V. The corresponding subgroup V_1 of V is invariant under L and we want to consider its decomposition into L-orbits. Let Ψ_1 be the set of all roots whose shapes are multiples of S. It is closed and semisimple, and it contains Ψ_L (the set of roots of shape 0) as a subsystem of corank 1 since its elements are all multiples of S modulo (the integral span of) Ψ_L. Thus the positive simple system for Ψ_1 consists of Δ_L and one other root α_1 of shape exactly S. Thus if G_1 is the reductive subgroup of G corresponding to Ψ_1 (and containing T) then $P_1 = LV_1$ is the maximal parabolic subgroup of G_1 corresponding to the simple root α_1. The group G_1 need not be simple (over its center), but the simple component containing α_1 as a root contains all of Ψ_{V_1} and hence also the group V_1, while the other simple components occur passively. We now assume that V_1 is Abelian, i.e., that $2S$ is not a shape, which is so, e.g. if $2S$ exceeds the highest root ρ in some coefficient. Then our previous results apply to $P_1 = LV_1$ in G_1 to yield the L-orbits of V_1, their closures, representative elements, etc., and the following connection with double cosets of G.

Proposition 5.2 *The natural map from V_1^-/L to $P\backslash G/P$ given by inclusion is injective. Thus for each $x \in V_1^-$, $PxP \cap V_1^- = L \cdot x$.*

Proof. Let $x, y \in V_1^-$ lie in different L-orbits. Then by 1.1(a) they lie in different elements of $P_1\backslash G_1/P_1$. These correspond to different elements of $W_L\backslash W_1/W_L$ by the Bruhat lemma in G_1, and hence to different elements of $P\backslash G/P$ by the Bruhat lemma in G. □

Then incursions may be made into G/P since the L-orbit structure of $G_1\ P/P$ is naturally isomorphic to that of G_1/P_1.

It follows from 5.2 that the number of L-orbits of V_1 is bounded above by the number of (P, P) double cosets of G. Parts of the above argument can be combined with results from [12] to show that this result is true if V_1 is replaced by $V_1/[V_1, V_1]$ in case V_1 is not Abelian, as long as the simple component of G_1 containing it is a classical group.

Here is a simple example to illustrate the above construction. The group is E_6, and S, α_1 and G_1 are as shown. Since $G_1 = A_4 \times A_1$ with A_1 appearing passively the L-orbit structure here is the same as that of the example worked out partially in 2.19.

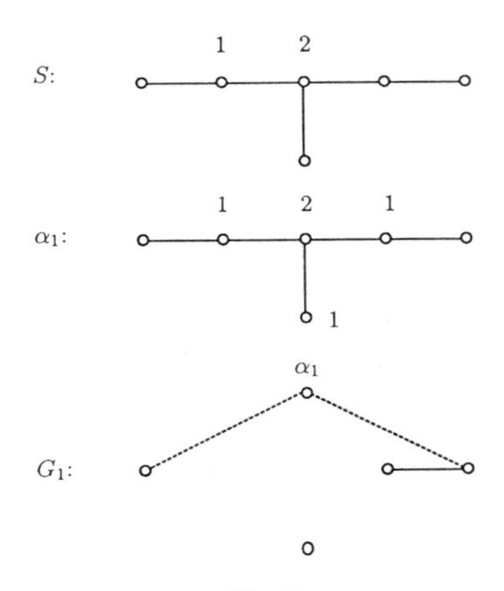

Fig. 3.

We next consider a partial extension of our results to the case in which k is not necessarily algebraically closed.

Theorem 5.3 *Assume that G, L, V, ..., as in 1.1 and 1.2, are all defined over a field k and that G and T split over k. Assume further that G is simple and of* *adjoint type. Then the results of* 1.1 *and* 1.2 *are true with* $G, L, \ldots$ *replaced by* $G(k), L(k), \ldots$ *in each of the following cases.*

(a) *Only one root length occurs.*

(b) *There are two root lengths,* char $k \neq 2$, *and every element of* k *is a square.*

Proof. Consider (a). If we check the proofs of 2.1 and the first part of Sect. 3, we see that every $L(k)$-orbit of $G(k)/P(k)$ has a representative of the required form. The extra point to be checked is that, with $\{\beta_1, \ldots, \beta_s\}$ fixed, any two choices of $\{u_{-\beta_i}\}$ and $\{w_{\beta_i}\}$ in $G(k)$ are $T(k)$-conjugate. This follows from:

Lemma 5.4 *In case* (a) *above* $\{\beta_1, \beta_2, \ldots, \beta_s\}$ *can be extended to a simple system* Δ' *for* Ψ.

Assume this for a moment. Then Δ' is a basis for the character group of T since G is adjoint. Since also G and T are split over k, the map $T(k) \to (k^*)^s$ given by $t \mapsto (\beta_1(t), \beta_2(t), \ldots)$ is surjective and the required $T(k)$-conjugacy easily follows.

Returning to 5.4 we note first $(*)$ if β is any root rationally dependent on $\{\beta_1, \beta_2, \ldots, \beta_s\}$ then $\beta = \pm\beta_i$ for some i. For, let $\beta = \sum n_i \beta_i$ be a root. By applying suitable w_{β_i}'s we may assume that every $n_i \geqq 0$. We have $2n_i = (\beta, \beta_i^\vee) = 0$, 1 or 2. Since the α-height of β is $\sum n_i > 0$, it must be 1. Thus either one n_i is 1 or two n_i's are $\frac{1}{2}$, and the rest are 0. In the first case $\beta = \beta_i$. In the second case $(\beta, \beta) = \frac{1}{2}(\beta_i, \beta_i)$ and there are two root lengths, a contradiction. In the $\mathbb{Q}$-space generated by the roots we form a basis starting with a basis of the orthogonal complement of $\{\beta_i\}$ and finishing with $\beta_1, \beta_2, \ldots, \beta_s$ and use the lexicographic ordering in which a nonzero vector is positive if its first nonzero coefficient is positive. Because of $(*)$, the positive simple system of roots corresponding to this ordering includes all of the β_i's, which proves 5.4.

Case (b) of 5.3 is proved similarly, with the assumption that every element of k is a square enough to overcome the failure of 5.4, as is indicated in our discussion of $(*)$. □

With a little extra work 5.3 can be extended to the case in which G is quasi-split, i.e., has a Borel subgroup defined over k.

5.5 Hermitian symmetric spaces. We let $k = \mathbb{C}$ and we continue with the earlier notation. We assume that G is simple and simply connected. In this case it is known that the homogeneous space $X = G/P$ is an irreducible simply connected Hermitian symmetric space of compact type. The argument goes as follows. Let $c \in T$ be chosen so that $\alpha(c) = -1$ and $\beta(c) = 1$ for every $\beta \in \Delta \setminus \{\alpha\}$ (c is uniquely determined modulo the center of G). The inner automorphism $\mathrm{Int}(c) = \theta$ of G is of order 2 and the fixed point subgroup G^θ is equal to L(we note that G^θ is connected since G is simply connected [14, 8.1]). In order to conform more closely with standard notation in the theory of symmetric spaces, we henceforth denote $G^\theta = L$ by K. Let C denote the (unique) maximal compact subgroup of T. Clearly $c \in C$. Let U be a maximal compact subgroup of G containing C. Then there exists an $\mathbb{R}$-structure on G such that U is the set of $\mathbb{R}$-rational points of G (this follows easily from the results of [5, Chap. XVII, Sect. 5]). Let τ denote the corresponding complex conjugation of G (that is, τ denotes the action of the non-trivial element of the Galois group $\mathrm{Gal}(\mathbb{C}/\mathbb{R})$ on G). Then τ is an automorphism of the real Lie group G and U is equal to G^τ, the fixed point subgroup of τ. Since $c \in U$, $\theta = \mathrm{Int}(c)$ and τ commute. Let $\sigma = \tau\theta = \theta\tau$ and let $G_0 = G^\sigma$. (The group G_0 is the set of $\mathbb{R}$-points of G with respect to the

$\mathbb{R}$-structure on G defined by σ, but we will not use that here.) Since G_0 and K are τ-stable, the restrictions $\tau|_{G_0}$ and $\tau|_K$ are Cartan involutions of G_0 and K. Thus $K_0 = G_0 \cap K$ is a maximal compact subgroup of both G_0 and K. Since G is simply connected, the groups G_0, K and K_0 are all connected. The homogeneous space G_0/K_0 is an irreducible Riemannian symmetric space of non-compact type and U/K_0 is the compact dual of G_0/K_0 (see [4, Chaps. 6–7]). Since K (and hence K_0) has a center of positive dimension, it follows that G_0/K_0 and U/K_0 are Hermitian symmetric spaces [4, Chap. 8]. It is not difficult to show that $UP = G$ and that $U \cap P = K_0$, so that $U/K_0 \simeq G/P$. One can choose the complex structure on U/K_0 so that this is an isomorphism of complex manifolds.

Conversely, let $X = U/K_0$ be a simply connected irreducible Hermitian symmetric space of compact type. We may assume that U is simply connected [4, Chap. 8]. Let G denote the universal complexification of U (see [5, Chap. XVII, Sect. 5]). Then G is a simple, simply connected complex algebraic group. Furthermore there exists a maximal parabolic subgroup P of G such that $U/K_0 \simeq G/P$. The quickest way to see this is to compare the list in remark 2.3 with the list for the classification of Hermitian symmetric spaces in Helgason [4, p. 518]. One can also give a direct proof, but we won't do that here.

5.6 G_0-orbits on G/P. We continue with the notation of 5.5. The G_0-orbits on G/P have been classified by Takeuchi [15]. He shows that these orbits can be parametrized by the set $\{(s, t) | 0 \leqq t \leqq s \leqq r\}$, where r is the rank of the symmetric space G/P. One can show that the rank of G/P is the length of a maximal orthogonal sequence of long roots in Ψ_V. Hence his parameter set for the G_0-orbits is the same as ours for the K-orbits.

In [18], Wolf gives a new proof of Takeuchi's result using ideas similar to those of Corollary 3.6. If we combine the results of Wolf's Orbit Structure Theorem [18, p. 303] with Corollary 3.6, we can find a common set of representatives for the K-orbits and the G_0-orbits.

First we consider a particular case. Let $G = \mathrm{SL}(2, \mathbb{C})$ act on $\mathbb{P}^1(\mathbb{C}) = \mathbb{C} \cup \{\infty\}$ by fractional linear transformations. Let B_0 denote the Borel subgroup of $\mathrm{SL}(2, \mathbb{C})$ consisting of all lower triangular matrices in $\mathrm{SL}(2, \mathbb{C})$ and let $T_0 = \{\mathrm{diag}(z, z^{-1}) | z \in \mathbb{C}^*\}$. Then B_0 is the isotropy subgroup of $0 \in \mathbb{C}$ in $\mathrm{SL}(2, \mathbb{C})$ and T_0 is a maximal torus of B_0. Define $\tau: \mathrm{SL}(2, \mathbb{C}) \to \mathrm{SL}(2, \mathbb{C})$ by $\tau(g) = ({}^t\bar{g})^{-1}$. Then $U = \mathrm{SL}(2, \mathbb{C})^\tau = \mathrm{SU}(2)$ is a maximal compact subgroup of $\mathrm{SL}(2, \mathbb{C})$. Let $c = \mathrm{diag}(\sqrt{-1}, -\sqrt{-1})$ and let $\theta = \mathrm{Int}(c) \in \mathrm{Aut}(\mathrm{SL}(2, \mathbb{C}))$. Then $\mathrm{SL}(2, \mathbb{C})^\theta = T_0$ and θ and τ commute. Let $\sigma = \theta\tau = \tau\theta$. Then $\mathrm{SL}(2, \mathbb{C})^\sigma = \mathrm{SU}(1, 1)$. Furthermore $\mathrm{SL}(2, \mathbb{C})/B_0 = \mathbb{P}^1(\mathbb{C})$ is a Hermitian symmetric space. The subgroup $\mathrm{SU}(1, 1) = \mathrm{SL}(2, \mathbb{C})^\sigma$ has three orbits on $\mathbb{P}^1(\mathbb{C})$, namely the unit disk $D_0 = \{z \in \mathbb{C} | \, |z| < 1\}$, the unit circle $S^1 = \{z \in \mathbb{C} | \, |z| = 1\}$, and $D_\infty = \{z \in \mathbb{C} | \, |z| > 1\} \cup \{\infty\}$. The group $T_0 = \mathrm{SL}(2, \mathbb{C})^\theta$ also has three orbits, $\{0\}$, $\mathbb{C}^*$ and $\{\infty\}$. It is clear that $\{0, 1, \infty\}$ is a set of representatives for both the $\mathrm{SU}(1, 1)$-orbits and the T_0-orbits on $\mathbb{P}^1(\mathbb{C}) = \mathrm{SL}(2, \mathbb{C})/B_0$.

We return to the notation of 5.5. We can now use Corollary 3.6 and the analogous result of Wolf for the G_0-orbits to carry over the results above to the general case. Since G is simply connected, for each root $\beta \in \Psi$, the subgroup G_β is isomorphic to $\mathrm{SL}(2, \mathbb{C})$. Moreover, each G_β is both τ-stable and θ-stable. If $\beta \in \Psi_K (= \Psi_L)$, then $G_\beta^\theta = G_\beta$ and G_β^τ is isomorphic to $\mathrm{SU}(2)$. If $\beta \in (\Psi_V \cup \Psi_{V^-})$, then $G_\beta^\theta = T_\beta$ and G_β^τ is isomorphic to $\mathrm{SU}(1, 1)$.

Now we choose a maximal sequence $(\beta_1, \ldots, \beta_r)$ of strongly orthogonal roots in Ψ_V, constructed as in [18, p. 279, (3.2)]. (The construction is originally due to Harish Chandra.) One can check that all of the roots in this sequence are long roots, so that we have a maximal orthogonal sequence of long roots in Ψ_V. For each $i=1, \ldots, r$, we define an isomorphism v_i: $\mathrm{SL}(2, \mathbb{C}) \to G_{\beta_i}$ such that $v_i(T_0)=T_{\beta_i}$, $v_i(B_0)=B_{\beta_i}$ and $v_i(\mathrm{SU}(1,1))=G^\sigma_{\beta_i}=G_0 \cap G_{\beta_i}$. The isomorphism v_i induces an isomorphism (of varieties) of $\mathbb{P}^1(\mathbb{C})=\mathrm{SL}(2, \mathbb{C})/B_0$ onto $G_{\beta_i}/B_{\beta_i}=Y_{\beta_i}$. Let $H=\prod_{i=1}^{r} G_{\beta_i}$ and define an isogeny v: $\mathrm{SL}(2, \mathbb{C})^r \to H$ by $v(g_1, \ldots, g_r) = v_1(g_1)\, v_2(g_2) \cdots v_r(g_r)$. Let $v(\mathrm{SU}(1,1)^r)=H^*$. The isogeny v induces an isomorphism η of $\mathbb{P}^1(\mathbb{C})^r$ onto $Y=\prod_{i=1}^{r} Y_{\beta_i}=HP/P$. In order to simplify our notation, we will identify $\mathbb{P}^1(\mathbb{C})^r$ with Y by means of η. As in Corollary 3.6, we let the symmetric group S_r act on $Y=\mathbb{P}^1(\mathbb{C})^r$ by permuting the factors.

The following result is (part of) Wolf's Orbit Structure Theorem for the G_0-orbits [18, p. 303].

Theorem 5.6.1 *Every G_0-orbit on G/P meets Y in a unique $(S_r \times H^*)$-orbit.*

Thus we obtain a bijection between the G_0-orbits on G/P and the $(S_r \times H^*)$-orbits on Y.

Using Corollary 3.6 and Theorem 5.6.1, it is now easy to get a set of common representatives for the K-orbits and the G_0-orbits.

Theorem 5.6.2 (a) *For each (s, t) with $0 \leqq t \leqq s \leqq r$, define $y_{st} \in Y=\mathbb{P}^1(\mathbb{C})^r$ by*

$$y_{st}=(\infty, \ldots, \infty, 1, \ldots, 1, 0, \ldots, 0),$$

with t terms equal to ∞, $s-t$ terms equal to 1 *and $r-s$ terms equal to* 0. *Then*

$$G/P = \coprod_{0 \leqq t \leqq s \leqq r} G_0 \cdot y_{st} = \coprod_{0 \leqq t \leqq s \leqq r} K \cdot y_{st},$$

where the symbol $\coprod$ *denotes the disjoint union.*

(b) $\overline{G_0 \cdot y_{st}} = \bigcup_{t' \leqq t \leqq s \leqq s'} G_0 \cdot y_{s't'}$

(c) $\overline{K \cdot y_{st}} = \bigcup_{t \leqq t' \leqq s' \leqq s} K \cdot y_{s't'}$.

Proof. The proof of (a) follows from Corollary 3.6 and Theorem 5.6.1. Part (b) is part of Wolf's Orbit Structure Theorem and (c) is equivalent to 3.7(c). □

Remarks. (a) The y_{st}'s of 5.6.2 are just the $x_{st}P$'s of 1.2 with the w_{β_i}'s and the $u_{-\beta_j}$'s appropriately chosen relative to the $\mathbb{R}$-structure of G.

(b) We see from Theorem 5.6.2 that the map $K \cdot y_{st} \to G_0 \cdot y_{st}$ is an order-reversing bijection from the K-orbits on G/P to the G_0-orbits. For the case of orbits on the flag manifold G/B, a theorem of Wolf [17] and Matsuki [8] (see also [16, p. 108]) gives an order-reversing bijection between the K-orbits and the G_0-orbits on G/B. Matsuki [9] has generalized this to give an order reversing bijection between G_0-orbits and K-orbits on G/Q, where Q is an arbitrary parabolic subgroup of G. Using this result, one can obtain the classification of the

G_0-orbits of G/P and the order relations between these orbits from the results of this paper.
(c) The proof of Wolf's Orbit Structure Theorem makes use of analysis on Hermitian symmetric spaces. Wolf's proof has been simplified by Lasalle [7], but his proof still requires the theory of the Shilov boundary.

References

1. Azad, H., Barry, M., Seitz, G.: On the structure of parabolic subgroups. Commun. Algebra **18**, 551–562 (1990)
2. Bourbaki, N.: Groupes et algèbres de Lie, Chaps. 4, 5 et 6. Paris: Hermann 1975
3. Chevalley, C.: Sur certains groupes simples. Tôhoku Math. J. **7**, 14–66 (1955)
4. Helgason, S.: Differential Geometry, Lie Groups, and Symmetric Spaces. New York London: Academic Press 1978
5. Hochschild, G.: The Structure of Lie Groups. San Francisco: Holden-Day 1965
6. Korànyi, A., Wolf, J.: Generalized Cayley transforms of bounded symmetric domains. Am. J. Math. **87**, 899–939 (1965)
7. Lassalle, M.: Les orbites d'un espace hermitien symétrique compact. Invent. Math. **52**, 199–239 (1979)
8. Matsuki, T.: The orbits of affine symmetric spaces under the action of minimal parabolic subgroups. J. Math. Soc. Japan **31** (no. 2), 331–357 (1979)
9. Matsuki, T.: Orbits on affine symmetric spaces under the action of parabolic subgroups. Hiroshima Math. J. **12**, 307–320 (1983)
10. Muller, I., Rubenthaler, H., Schiffmann, G.: Structure des espaces prèhomogènes associès á certaines algèbres de Lie graduèes. Math. Ann. **274**, 95–123 (1986)
11. Richardson, R.W.: Intersections of Double Cosets in Algebraic Groups. Indag. Math. (to appear)
12. Röhrle, G.: On the Structure of Parabolic Subgroups in Algebraic Groups. J. Algebra (to appear)
13. Steinberg, R.: Lectures on Chevalley Groups. Yale University (1967)
14. Steinberg, R.: Endomorphisms of linear algebraic groups. Mem. Am. Math. Soc. **80** (1968)
15. Takeuchi, M.: On orbits in a compact hermitian symmetric space. Am. J. Math. **90**, 657–680 (1968)
16. Vogan, D.: Representations of Real Reductive Lie Groups. Boston Basel Stuttgart: Birkhäuser 1981
17. Wolf, J.A.: The Action of a Real Semisimple Group on a complex flag manifold. I: Orbit Structure and Holomorphic Arc Components. Bull. Am. Math. Soc. **75**, 1121–1237 (1969)
18. Wolf, J.A.: Fine structure of Hermitian symmetric spaces. In: Boothby, W., Weiss, G.L. (eds.) Symmetric Spaces, Short Courses Presented at Washington University, pp. 271–357. New York: Marcel Dekker 1972

Nagata's Example

Robert Steinberg

To the memory of my friend Roger Richardson

1

At the International Congress of Mathematicians held in Edinburgh in 1958 M. Nagata [**N**$_1$] presented an example of a group acting linearly on a finite-dimensional vector space such that the algebra of polynomials on the space invariant under the group is not finitely generated. He thereby answered in the negative Problem 14 of Hilbert's famous list presented to the Congress of 1900. Our object in this article is to present other examples which are simpler and easier to establish than Nagata's and also yield a better result. We hasten to add that our development is close to his, with one twist which produces the improved examples.

Here is our first example.

1.1 THEOREM. *Let k be a field of characteristic 0 and $V_2 = k^2$, viewed as a vector space over k. Let a_i $(1 \le i \le 9)$ be distinct elements of k such that $\Sigma a_i \neq 0$. Let G_a denote the additive group of k. Let G_a^9 act on $V = V_2^9$, given the coordinates (x_i, t_i) $(1 \le i \le 9)$, thus: $t_i \to t_i$ and $x_i \to x_i + c_i t_i$ $(1 \le i \le 9)$ for each $(c_1, c_2, \ldots, c_9)$ in $k^9 \simeq G_a^9$. Finally, let $G \simeq G_a^6$ be the subgroup of G_a^9 consisting of all $(c_1, c_2, \ldots, c_9)$ such that (1) $\Sigma c_i = 0$, (2) $\Sigma a_i c_i = 0$, (3) $\Sigma a_i^3 c_i = 0$. Then $k[V]^G = k[x_1, t_1, x_2, t_2, \ldots]^G$, the algebra of polynomials on V invariant under G, is not finitely generated.*

In Nagata's example (his first one, given in [**N**$_1$]) 16 copies of V_2 figure rather than 9, and accordingly the group acting is G_a^{13} rather than G_a^6.

Following Nagata, we prove instead of Theorem 1.1 the following result.

1.2 THEOREM. *Enlarge G in Theorem 1.1 to $H = GT$ with T the torus $\{(d_1, d_2, \ldots, d_9) \in k^{*9} \mid \Pi d_i = 1\}$ acting on V via $x_i \to d_i x_i$, $t_i \to d_i t_i$ $(1 \le i \le 9)$. Then $k[V]^H$ is not finitely generated.*

As will be seen below the proof of 1.2 to be given becomes one of 1.1 once some obvious changes are made. Alternatively, one can note that 1.1 follows from 1.2

1991 *Mathematics Subject Classifications.* Primary 20-xx; Secondary 14-xx.

Reprinted with the permission of Cambridge University Press, from Nagata's example, in Algebraic groups and Lie groups, *Austral. Math. Soc. Lecture Series 9*, Cambridge University Press, 1997, 375–384.

and the fact that if a torus acts on an algebra which is finitely generated, then the algebra of invariants is also finitely generated (see, e.g., Ch. 2 of [**S**] for this).

To get an example in char $k \neq 0$ we choose the a_i's in 1.1 so that Πa_i is not 0 or any nth root of 1 and replace a_i^3 in (3) by $a_i^2 - a_i^{-1}$. As is easily seen, such a_i's exist if k is any infinite field which is not locally finite, i.e., not an algebraic extension of a finite field, hence, in particular, if char $k = 0$. In this sense our second example is more general than our first. However, both examples are contained in a more general example which we now explain.

As is known (for this, see Ch. 5 of [**Fu**]), the nonsingular points of an irreducible cubic curve C in the projective plane form an Abelian group A in which every line not containing any singular point meets C in 3 points (counting multiplicities) whose sum in A is 0. The last condition requires only that one of the flexes of C (whose coordinates need not lie in k) is chosen as the neutral element of A. We choose one of them and call A "the" group of C. For example, if C is $y - x^3$, in affine coordinates, then the only singular point is the point at infinity in the direction of the y-axis, and every line not through this point has the form $y - (ax + b)$. Since such a line always meets C in 3 points for which the sum of the x-coordinates is 0 (since they satisfy $x^3 - (ax + b)$), it follows that A is, in this case, the additive group of k. Similarly, for $y - (x^2 - x^{-1})$, or, equivalently, $xy - x^3 + 1$, the product of the x-coordinates is 1, so that A is the multiplicative group of k. Thus our two examples are contained in the following result.

1.3 THEOREM. *In the construction of G in Theorem* 1.1 *let the* 9 *points* $(1 : a_i : a_i^3)$ *used in the conditions* (1), (2) *and* (3) *be replaced by* 9 *nonsingular points on an arbitrary irreducible cubic curve. If the sum of these points has infinite order in the group of the curve, then the conclusions of Theorem* 1.1 *and* 1.2 *remain true.*

Other examples have been given (see [**A**], [**D**], [$\mathbf{N}_{1,2,3}$], [**NO**], [**R**], [**Ro**]), some of them for other versions of Hilbert's problem. The one by A' Campo [**A**], based on Roberts [**Ro**], is quite simple. The group there is G_a^{12} and the field must be of characteristic 0. Also, many positive results have been obtained; for example, for finite or reductive groups the algebra of invariants is always finitely generated. For discussions of some of these matters we refer the reader to the survey articles [**Hu**] and [**M**].

In our main development we establish our first two examples because this can be done using only elementary algebra, and then we indicate the changes needed for Theorem 1.3. This is done in the next three sections. In Section 5 we present yet other examples which involve reducible cubic curves and figure in a converse to our main lemma. In a final section we consider some problems suggested by our development.

2

In this section the two main lemmas are presented and Theorem 1.2 is deduced from them. The lemmas themselves are proved in the following two sections.

2.1 LEMMA (Nagata). *With the notations as in Theorems* 1.1 *and* 1.2, *set* $t = \Pi t_i$, $z_1 = \Sigma x_i t/t_i$, $z_2 = \Sigma a_i x_i t/t_i$, $z_3 = \Sigma a_i^3 x_i t/t_i$. *Then* t, z_1, z_2 *and* z_3 *are algebraically independent over* k *and each of them is invariant under* H.

Further $k[V]^H$ *consists of the elements of* $k[z_1, z_2, z_3, t, t^{-1}]$ *that are expressible as sums of elements of the form* $f(z_1, z_2, z_3)/t^m$ *in which* f *is a nonzero homogeneous polynomial such that the corresponding plane projective curve has multiplicity at least* m *at each of the points* $P_i = (1 : a_i : a_i^3)$ $(1 \leq i \leq 9)$.

This elementary lemma, which relates the structure of $k[V]^H$ to an interpolation problem in the projective plane, is the heart of Nagata's method. Actually, he proved a more general result which will be formulated and proved in the next section.

2.2 LEMMA. *Let* k *be a field of characteristic* 0 *and* $a_1, a_2, \ldots, a_9$ *distinct elements of* k *such that* $\Sigma a_i \neq 0$. *For each* i *let* P_i *be the point* (a_i, a_i^3) *on the plane affine cubic curve* $f_0 : y - x^3$.

(a) *For each* $m \geq 0$ *there exists, up to scalar multiplication, a unique curve (or polynomial) of degree* $\leq 3m$ *with multiplicity* $\geq m$ *at every* P_i, *namely,* f_0^m, *the cubic counted* m *times.*

(b) *For every* $d \geq 3m$ *the multiplicity conditions of* (a) *are linearly independent on the space of all polynomials of degree* $\leq d$.

(c) *There exists a polynomial of degree* $3m + 1$ *which has multiplicity* $\geq m$ *at every* P_i *and is not divisible by* f_0.

It is here that we have deviated from Nagata in his development in [**N**$_1$] and [**N**$_2$]. He uses instead of 2.2 at this point the result that if k has a large enough transcendence degree over some subfield and if $r \geq 4$, then, for each $m \geq 1$, there does not exist a curve of degree rm with multiplicity $\geq m$ at each of r^2 general points of the plane. Nagata's ingenious proof of this is a tour de force but the results from algebraic geometry that he uses are by no means elementary. Since his result is false for $r = 3$, his minimal example has $r = 4$ and $G = G_a^{16-3} = G_a^{13}$.

In order to be able to combine these two lemmas we imbed $\mathbb{A}^2$ in $\mathbb{P}^2$ and match up the coordinates so that $x = z_2/z_1$ and $y = z_3/z_1$, and accordingly match up the nonzero polynomials in x and y of degree $\leq d$ with the homogeneous ones in z_1, z_2 and z_3 of degree exactly d via $f(x, y) \leftrightarrow z_1^d f(z_2/z_1, z_3/z_1)$. The point P_i of 2.2 gets identified with that of 2.1, and the cubic $y - x^3$, viewed as a polynomial of degree ≤ 3, takes on the homogeneous form $z_1^2 z_3 - z_2^3$.

We now give the proof of Theorem 1.2. Assuming that $k[V]^H$ is finitely generated, we shall come to a contradiction. By Lemmas 2.1 and 2.2 there exists a finite generating set of the form $S = \{f_0/t, f_1/t^{m_1}, \ldots, f_r/t^{m_r}\}$, where $f_0 = z_1^2 z_3 - z_2^3$ is our basic cubic and each later generator has the form specified by Lemma 2.1, and where it can be further arranged that no f_j $(j \geq 1)$ is divisible by f_0. Set $d_j = \deg f_j$. Then $d_j > 3m_j$, so that $d_j - 3m_j$ is positive, for every $j \geq 1$. This is so for $d_j > 0$ by Lemma 2.2(a) and also for $d_j = 0$ since then $m_j \leq 0$ and we can rule out the case of equality by dropping every generator that is a constant. Let m be a positive integer larger than every m_j $(j \geq 1)$. By Lemma 2.2(c) there exists a polynomial f of degree $3m + 1$ which has multiplicity $\geq m$ at every P_i and is not divisible by f_0. We show that f/t^m is not expressible as a polynomial in the elements of S and thus reach the desired contradiction. Assume that such a polynomial expression does exist. Then there exists one such that all of the terms appearing in it have the same degree in $\{z_1, z_2, z_3\}$, and also in t, as does f/t^m.

This matching of degrees makes sense because t, z_1, z_2 and z_3 are algebraically independent over k. Since f_0 does not divide f there exists at least one of the terms, $c\Pi(f_j/t^{m_j})^{e_j}$, which does not involve f_0/t. Equating the degrees of this term and of f/t^m, we get

(a) $\Sigma d_j e_j = 3m + 1$ (degree in $\{z_1, z_2, z_3\}$)

(b) $\Sigma m_j e_j = m$ (degree in t)

The combination (a) - 3(b) yields

(c) $\Sigma(d_j - 3m_j)e_j = 1$.

Since every $d_j - 3m_j$ is positive, as noted above, it follows from (c) that exactly one e_j, say e_{j_0}, is 1 while all the others are 0, and then from (b) that $m = m_{j_0}$. This contradiction to the choice of m shows that $k[V]^H$ is not finitely generated and thus completes the proof that Theorem 1.2 follows from Lemmas 2.1 and 2.2.

3

In this section we present a proof of Lemma 2.1 which is essentially Nagata's own but is changed in one or two places so that it can be given entirely in the language of elementary algebra. Because of the utility of this result we prove it in full generality. Thus k is replaced by an arbitrary infinite field and the points P_i $(1 : a_i : a_i^3)$ by an arbitrary number n of points P_i $(a_{i1} : a_{i2} : a_{i3})$ not all on one line, a condition which holds for the original points since at most 3 points of the cubic can lie on one line. The conditions (1), (2) and (3) defining G in Theorem 1.1 and the definition of the z_j's given in Lemma 2.1 are then changed accordingly.

First, it is routine to verify that t, z_1, z_2 and z_3 are invariant under H. For example, if the element $(c_1, c_2, \ldots, c_n)$ of G acts on $z_j = \Sigma a_{ij} x_i t/t_i$ the result is $\Sigma a_{ij}(x_i + c_i t_i)t/t_i$, which is again z_j in view of condition (j) of Theorem 1.1.

The rest of the proof is given in three steps.

(1) It is shown first that t, z_1, z_2 and z_3 are algebraically independent over k and that $k[V]^H = k[V] \cap k[z_1, z_2, z_3, t, t^{-1}]$.

Since the P_i's are not all on one line we can, by interchanging P_3 and one of the later P_i's if necessary, arrange that P_1, P_2 and P_3 are not on a line and thus that the matrix of their coordinates, (a_{ij}) $(1 \leq i, j \leq 3)$, is invertible. It follows that if we set $w_i = x_i t/t_i$ $(1 \leq i \leq n)$ so that $z_j = \Sigma a_{ij} w_i$ $(1 \leq j \leq 3)$, then the latter equations can be solved for w_1, w_2 and w_3 as linear functions of the other w_i's and z_1, z_2 and z_3. If we then use the definition of the w_i's we get that $k[V][t^{-1}]$ is the polynomial ring over $k[t^{-1}]$ in the variables z_1, z_2, z_3, $x_4, \ldots, x_n$, t_1, $t_2, \ldots, t_n$. Now for an arbitrary c in k the equations (1), (2) and (3) of 1.1 have a solution with $c_4 = c$ and $c_i = 0$ for every $i > 4$, because, once these values are assigned, (1), (2) and (3) can be solved for c_1, c_2 and c_3, again by the invertibility of the above matrix. If the corresponding element of G acts on an element of $k[V][t^{-1}]$, then only the variable x_4 in the above list changes and it goes to $x_4 + ct_4$. Since c takes on infinitely many values here it follows that if the element is invariant then it does not involve x_4; and similarly for $x_5, \ldots, x_n$. Thus we see that $k[V][t^{-1}]^G = k[z_1, z_2, z_3, t_1, t_2, \ldots, t_n, t^{-1}]$, and, if we then make T act, that $k[V][t^{-1}]^H = k[z_1, z_2, z_3, t, t^{-1}]$, and finally that $k[V]^H = k[V] \cap k[z_1, z_2, z_3, t, t^{-1}]$. Also, since k is infinite, the x_i's are algebraically independent over $k[t_1, t_2, \ldots, t_n]$, hence so are z_1, z_2, z_3, $x_4, \ldots, x_n$ by the above, and thus t, z_1, z_2 and z_3 are algebraically independent over k.

(2) It is next shown that $k[V]^H$ is spanned by its elements of the form $f(z_1, z_2, z_3)/t^m$, in which f is a homogeneous polynomial in z_1, z_2 and z_3 which is divisible by t^m in $k[V]$.

If F is any element of $k[V]^H$, then by (1) it may be written as a Laurent polynomial in t with coefficients which are polynomials in z_1, z_2 and z_3: (*) $F = \Sigma f_m(z_1, z_2, z_3)/t^m$. Assume that F is homogeneous of degree d in the union of the x_i's and the t_i's. Since, from the definitions, each z_j is homogeneous in the x_i's and also in the t_i's, of degrees 1 and $n-1$, respectively, and t is homogeneous in the t_i's, of degree n, it follows that each term $pz_1^a z_2^b z_3^c$ of f_m in (*) satisfies $d = (a+b+c)n - mn$. This yields $a+b+c = m + d/n$, which depends only on m. Thus f_m is a homogeneous polynomial in the z_j's, of degree $m + d/n$. By the above f_m is also homogeneous of this degree when viewed as a polynomial in the x_i's, with coefficients polynomials in the t_i's. Since also this degree changes as m does it follows that (*) gives the decomposition of F into its x-homogeneous components. Finally, since F is in $k[V]$ so is each of these components, which proves (2).

(3) By (2) it remains to show that a nonzero homogeneous polynomial $f(z_1, z_2, z_3)$ is divisible by t^m in $k[V]$ if and only if (the curve in $\mathbb{P}^2$ represented by) f has multiplicity at least m at all n of the points P_i.

Consider first the point P_1 which by a projective change of coordinates (e.g., $z_1' = z_1/a_{11}$, $z_2' = a_{11}z_2 - a_{12}z_1$, $z_3' = a_{11}z_3 - a_{13}z_1$ in case $a_{11} \neq 0$) may be taken to be $(1:0:0)$. Then (from the definition of the z_j's) (*) $z_1 = x_1 t/t_1 + t_1 u_1$, $z_2 = t_1 u_2$, $z_3 = t_1 u_3$ with $u_j = a_{2j}x_2 t/t_1 t_2 + a_{3j}x_3 t/t_1 t_3 + \cdots$ for $j = 1$, 2 and 3. Observe that t_1 is not involved in any u_j. Since P_1, P_2 and P_3 are not all on one line, the sums for u_2 and u_3 above are linearly independent in the terms involving x_2 and x_3, and hence they are algebraically independent over $k(t_2, \dots, t_n)$. Now if $f(z_1, z_2, z_3)$ is expanded as $f_m(z_2, z_3)z_1^{d-m} + f_{m+1}(z_2, z_3)z_1^{d-m-1} + \cdots$ with $f_m \neq 0$ and each f_k homogeneous of degree k, then $f = t_1^m f_m(u_2, u_3) + t_1^{m+1}(\cdot)$ by (*) above, and $f_m(u_2, u_3) \neq 0$ by the algebraic independence just noted. Thus if m is the multiplicity of f at P_1 then the highest power of t_1 dividing f in $k[V]$ is t_1^m; and similarly for every P_j and t_j since in the above development P_1 can be any of the n given points. Since $t = \Pi t_i$, (3) is proved, and with it Lemma 2.1.

4

In this section we prove Lemma 2.2, thus establishing our first example, and then consider our second example and Theorem 1.3.

We are again in the affine plane and x and y are the coordinates, and now the multiplicity of a curve, or of a polynomial $f(x, y)$ representing it, at a point P is the degree of the first nonzero term in the (finite) power series expansion of f around P. The following facts will be used repeatedly.

Over an arbitrary infinite field the (linear) space of all polynomials (which we shall also call curves) of degree $\leq d$ has dimension $\binom{d+2}{2}$ (binomial coefficient). The (linear) conditions that specify that a curve of degree $\geq m-1$ should have multiplicity $\geq m$ at a given point P are independent and in number equal to $\binom{m+1}{2}$.

The first formula comes from counting the monomials in $f = \Sigma c_{ij} x^i y^j$ $(i+j \leq d)$ (one of degree 0, two of degree 1, and so on), while the second counts the number of coefficients in the power series expansion around P which must be set equal to 0 to achieve the required multiplicity (e.g., all c_{ij} $(i+j<m)$ in case $P=(0,0)$).

We turn now to the proof of Lemma 2.2.

(a) Let $f(x,y)=c_0(x)y^{3m}+c_1(x)y^{3m-1}+\cdots+c_{3m}(x)$ have degree $\leq 3m$ and multiplicity $\geq m$ at every $P_i=(a_i,a_i^3)$. Here c_j is, for each j, a polynomial of degree $\leq j$. The substitution $y\to x^3$ converts f into a polynomial in x of degree $\leq 9m$ which is just the remainder when f is divided by $y-x^3$. This polynomial has multiplicity $\geq m$ at every a_i because f has multiplicity $\geq m$ at P_i and, in terms of the coordinates there, $y-a_i^3$ has been replaced by $x^3-a_i^3$ which is a multiple of $x-a_i$. Thus it is divisible by $\Pi(x-a_i)^m$, whose degree is exactly $9m$. We thus have the equation

(*) $c_0x^{9m}+c_1(x)x^{9m-3}+\cdots+c_{3m}(x)=c_0\Pi(x-a_i)^m$.

The coefficient of x^{9m-1} on the left is 0 since $dg\ c_j\leq j$ for every j, while it is $-c_0m\Sigma a_i$ on the right, if we assume that $m>0$. Since char $k=0$, and $\Sigma a_i\neq 0$ by assumption, we get $c_0=0$. Thus the left side of (*), i.e., the remainder when f is divided by $y-x^3$, is 0, so that $f_0=y-x^3$ divides f. Since f_0 has multiplicity exactly 1 at every P_i, f/f_0 has degree $\leq 3(m-1)$ and multiplicity $\geq m-1$ at every P_i, and (a) follows by induction, the case $m=0$ being obvious.

(b) The polynomials of degree $\leq 3m$ form a vector space of dimension $\binom{3m+2}{2}=(9m^2+9m+2)/2$, while the multiplicity conditions at the 9 points number $9\binom{m+1}{2}=(9m^2+9m)/2$, which is exactly 1 less. Thus the conditions are independent on the space if and only if they define a subspace of dimension 1 exactly, and that is the case by part (a). Since for every $d\geq 3m$ the polynomials of degree $\leq 3m$ are included in those of degree $\leq d$, the conditions remain independent on the space of polynomials of degree $\leq d$. (Linear functions on a vector space are linearly independent if their restrictions to some subspace are so.)

(c) The space of polynomials of degree $\leq 3m+1$ with multiplicity $\geq m$ at every P_i has, by (b), the dimension $\binom{3m+3}{2}-9\binom{m+1}{2}$, which is equal to $3m+3$. The subspace of those that are also divisible by f_0 has the dimension $3m$, obtained by replacing m by $m-1$ in the preceding formula. Therefore there exists a polynomial in the space and not in the subspace (in fact, a punctured 3-dimensional subspace of such polynomials), whence (c).

This completes the proof of Lemma 2.2 and hence also of Theorem 1.2.

Our second example, the one mentioned soon after Theorem 1.2 in Section 1 can be established in essentially the same way. The only proof that has to be changed is that of (a) given above. If we replace y by x^2-x^{-1} instead of by x^3 and then multiply by x^{3m} to clear away the negative powers of x, we get an equation like (*) which yields $c_0((\Pi a_i)^m-1)=0$ and thus that $c_0=0$ because of the assumptions made on the a_i's, and that is all that is needed.

We turn now to the proof of Theorem 1.3. Again the only proof that has to be changed is that of Lemma 2.2(a). Let C be the curve and A its group of nonsingular points. We use the following known result: (*) If G is a curve of degree n $(n\geq 1)$ which does not contain C or any of its singular points then the intersection G.C (in which each point is counted with its intersection multiplicity) consists of $3n$ points whose sum in A is 0. Assume the result (*) for a moment. Let F be a curve of degree $3m$ with multiplicity $\geq m$ $(m\geq 1)$ at each of the 9 given points. To prove our required result by induction as before it is enough to show that F contains C as a component. Suppose that it does not. By Bezout's Theorem F.C consists of $3m.3 = 9m$ points, while by the multiplicity assumptions F.C contains each of the 9 points at least m times. Thus F. C consists of the 9 points, each counted exactly m times. By (*) this yields that the sum of these points in A has finite order ($\leq m$), in contradiction to one of our assumptions. It remains to prove (*). Since F is a line if $n = 1$, we assume $n \geq 2$ and start with the case n even. Set F.C = S. Partition the $3n$ points of S into $3n/2$ pairs. Let G be the union of the lines through these pairs, a curve of degree $3n/2$, and T the set of third points in their intersections with C. We have $G.C = S \cup T$. By Max Noether's Fundamental Theorem it follows from $G.C \supset F.C$ that there exists a curve E of degree deg $G -$ deg $F = n/2$ such that $E.C = T$, the residual intersection. By induction the sum of the points of T is 0, and because G is a union of lines the sum of those of $S \cup T$ is also; thus the same holds for those of S, as required. If n is odd, the same argument with $(3n-1)/2$ lines through pairs of points of S and one extra line through the remaining point produces a curve E of degree $(n+1)/2$. Thus we are done. For an account of the results about plane curves used here we refer the reader to Chapter 5 of Fulton [**Fu**].

5

It does not seem to be widely known that reducible cubic curves also have groups attached to them. This fact leads to other negative examples for Hilbert's problem and to a converse of our main interpolation result, Lemma 2.2. We start with the examples.

5.1 Examples. If each of the sets of 9 points below is used in Lemma 2.2 and Theorem 1.1 then the conclusions there remain true.

(a) (char $k = 0$) 3 points on each of the lines $y-1$, y and $y+1$ in the affine plane, $(a_i, 1)$, $(b_j, 0)$ and $(c_k, -1)$, such that $\Sigma a_i + \Sigma c_k - 2\Sigma b_j \neq 0$.

(b) 3 points on each of the coordinate lines z_1, z_2 and z_3 of the projective plane, $(0 : 1 : -a_i)$, $(-b_j : 0 : 1)$ and $(1 : -c_k : 0)$ such that $\Pi a_i \cdot \Pi b_j \cdot \Pi c_k \neq 0$ or any root of 1.

(c) (char $k = 0$) 6 points (a_i, a_i^2) on the parabola $y - x^2$ and 3 points at infinity (on the lines with slopes) m_j such that $\Sigma a_i - \Sigma m_j \neq 0$.

(d) 6 points (a_i, a_i^{-1}) on the hyperbola $xy - 1$ and 3 points m_j at infinity such that $\Pi a_i \cdot \Pi m_j \neq 0$ or any root of 1.

(e) ($k = \mathbb{R}$) 6 points on the unit circle $x^2 + y^2 - 1$ in the Euclidean plane and 3 points at infinity specified by their angular coordinates α_i (mod 2π) and β_j (mod π), all measured from the positive x-axis, such that $\Sigma\alpha_i - 2\Sigma\beta_j$ is not a rational multiple of π.

The interested reader is invited to establish (a) by restricting the polynomial $f(x, y)$ used in the proof of 2.2(a) in Section 4 to the lines $y-1$, y and $y+1$ and comparing the results; and similarly for (b) and the lines z_1, z_2 and z_3.

5.2 Proposition. *Let C be a cubic curve without any multiple component. Then, with the singular points of C removed, the various irreducible components of C can be made into algebraic groups which can be identified by suitable isomorphisms so that (*) every transversal meets C in 3 points whose sum is 0.*

Here and later, "transversal" means a line not through any singular point of C. The group involved in 5.1(a), for example, is the additive group since every transversal meets the cubic $y^3 - y$ in points $(a, 1)$, $(b, 0)$ and $(c, -1)$ such that the weighted sum $a + c - 2b$ is 0. Similarly in (b) we have the multiplicative group since the points $(0 : 1 : -a)$, $(-b : 0 : 1)$ and $(1 : -c : 0)$ on the cubic $z_1 z_2 z_3$ are in a line if and only if $abc = 1$. (A version of this in Euclidean geometry is known as Menelaus' Theorem.) This actually proves 5.2 in case C is the union of 3 lines since then C can, by appropriate choice of coordinates, be put in the form $y^3 - y$ or $z_1 z_2 z_3$ according as the lines are concurrent or not. To slightly shorten the discussion we have assumed in the first case that char $k \neq 2$. Similarly (c) and (d) take care of the cases in which C is the union of a conic Q and a line L which meets Q in 1 or 2 distinct points. Left, among the reducible cubics, are those for which L fails to meet Q. An example occurs in (e) where the group is the circle group. In general, when Q contains rational points, we get the group of units of a definite quadratic form $x^2 + axy + by^2$ over k.

We now given an outline of a proof of 5.2 in which all cubic curves are treated together. Let C' be the set of nonsingular points of C. Let Z be the additive group of divisors $\Sigma n_i P_i$ ($n_i \in \mathbb{Z}$, $P_i \in C'$) on C' such that the sum of the n_i's for every irreducible component of C' is 0. Let B be the subgroup generated by all divisors $C \cdot L_1 - C \cdot L_2$ with L_1 and L_2 transversals to C. Let $D = Z/B$. The key point of our development is that if P_0 is any point of C' then every element of D has a unique representation $P - P_0$ with P a point of the irreducible component of C' that contains P_0. The uniqueness is proved as follows. Let $P' - P_0$ be another representation. Then, by the definitions, there exist curves F and F', each the union of n transversals, for some n, such that the divisors $F \cdot C + P$ and $F' \cdot C + P'$ are equal. Let L be any transversal through P, meeting C again in Q and R. Then $(F \cup L) \cdot C = F \cdot C + P + Q + R = F' \cdot C + P' + Q + R \supset F' \cdot C$. By Max Noether's Theorem there exists a line L' containing the residual intersection $P' + Q + R$. Since P' is the third point of intersection of QR with C, and P is also, it follows that $P' = P$. We next choose a transversal L_0 which meets each component of C' in a unique point. (If C is irreducible, for example, then L_0 must be the tangent at a flex, while if C is the union of 3 lines then L_0 can be any transversal.) We then make every component A of C' into a group thus: if Q and R are points of A and P_0 is the unique point of L_0 in A, then $Q + R$ is defined to be the point S of A such that $(Q - P_0) + (R - P_0)$ is equivalent to $S - P_0 \mod B$. This group is seen to be isomorphic to D via the map $Q \to Q - P_0$ $(Q \in A)$. We thus get isomorphisms of the groups of the various components of C' with D and hence with each other for which the condition (*) of 5.2 holds because, for any transversal L, the divisor $C \cdot L - C \cdot L_0$ represents O in D.

This brings us to our main interpolation result.

5.3 THEOREM. *For the distinct points P_i $(1 \leq i \leq 9)$ in the projective plane, the conditions* (a) *and* (b) *below are equivalent.*

(a) *For every $m \geq 1$ there exists a unique curve of degree $\leq 3m$ with multiplicity $\geq m$ at every P_i, i.e., the multiplicity conditions being imposed here are linearly independent.*

(b) *There exists a cubic curve C with the following properties.*

(1) *C has no multiple component.*

(2) *C contains every P_i as a nonsingular point.*

(3) *The number of P_i's on any irreducible component A of C is 3 times the degree of A.*

(4) *ΣP_i is an element of infinite order of the group of C.*

Because of 5.2 the argument in the proof of Theorem 1.3 given at the end of Section 4, slightly extended, can be used to prove that (b) implies (a). Since (a) implies the conclusions of Theorem 1.1, the five examples of 5.1 are thus established. The proof that (a) implies (b), which proceeds mainly by contradiction, is omitted.

6

We conclude with some problems. The first one is to decide whether the group G_a^6 of our examples can be made even smaller. In char 0 it is known (Weitzenbock's Theorem, proved in [**W**] and in [**F**]) that the group $G_a = G_a^1$ always produces a finitely generated algebra of invariants, but in char $p \neq 0$ not even this is known. There is also the problem of finding a negative example over every infinite locally finite field or showing that one does not exist. One can not expect one of the type presented above since the group of every cubic curve over a locally finite field is itself locally finite.

Finally, the geometric results discussed in our development focus attention on the following fundamental problem, hardly a new one. Find the dimension of the space of all polynomials (or curves) of a given degree with prescribed multiplicities at the points of a given finite set in general position in the plane, thus also determine if there is a curve, i.e., a nonzero polynomial, in the space and if the multiplicity conditions are independent. If the set has 9 points or fewer, the problem has been solved, by Nagata himself (see [**N**$_4$]). His algorithm, which has been rediscovered by others (including ourselves), is short and elementary and it uses a weak version of our main lemma.

For 10 points or more the problem becomes considerably more difficult. Hirschowitz [**H**] has solved it in case every $m_i \leq 3$ and he and Alexander [**AH**] have extended his results for $m_i \leq 2$ to higher dimensions, but results of this generality are rare. Chudnovsky [**C**] has tried to establish Nagata's geometric lemma for $r = \sqrt{10}$ (and thus to bring his group down to G_a^7), but his arguments are inconclusive. Thus our final problem is to prove it, i.e., to show that there is no curve of degree $d \leq m\sqrt{10}$ with multiplicity $\geq m$ at each of 10 general points, or to do related special cases such as $(d,m) = (19,6)$, $(38,12)$, $(177,55), \ldots$ which should just fail to exist because $\binom{d+2}{2} - 10\binom{m+1}{2} = 0$ in each case. Here the first two cases have been done (cf. [**H**]), with a number of ingenious but ad hoc constructions, but the third case remains undone.

References

[A] A. A'Campo-Neuen, *Note on a counterexample to Hilbert's fourteenth problem given by P. Roberts*, Indag. Math., N.S. **5** (1994), 253-257.

[AH] J. Alexander and A. Hirschowitz, *La méthode d'Horace éclatée: application à l'interpolation en degré quatre*, Invent. Math. **107** (1992), 585-602.

[C] G. V. Chudnovsky, *Sur la construction de Rees et Nagata pour le 14e problème de Hilbert*, C. R. Acad. Sci. Paris **286** (1978), A1133-1135.

[D] J. Dixmier, *Solution négative du problème des invariants (d'après Nagata)*, Sém. Bourbaki 1958/59, Exp. 175, also published by Benjamin, 1966.

[F] J. Fogarty, *Invariant Theory*, Benjamin, 1969.

[Fu] W. Fulton, *Algebraic Curves*, Benjamin, 1969.

[H] A. Hirschowitz, *La méthode d'Horace pour l'interpolation à plusieurs variables*, Manuscr. Math. **50** (1985), 337-388.

[Hu] J. E. Humphreys, *Hilbert's fourteenth problem*, Amer. Math. Monthly **85** (1978), 341-353.

[M] D. Mumford, *Hilbert's fourteenth problem - the finite generation of subrings such as the ring of invariants*, AMS Proc. Symp. Pure Math., vol. 28, 1976, pp. 431-444.

[N$_1$] M. Nagata, *On the fourteenth problem of Hilbert*, Proc. I.C.M. 1958, 459-462, Cambridge University Press, 1960.

[N$_2$] ———, *On the fourteenth problem of Hilbert*, Amer. J. Math. **81** (1959), 766-772.

[N$_3$] ———, *Lectures on the Fourteenth Problem of Hilbert*, Tata Institute, 1965.

[N$_4$] ———, *On rational surfaces II,*, Mem. Coll. Sci. Kyoto **33** (1960), 271-293.

[NO] M. Nagata and K. Otsuka, *Some remarks on the 14th problem of Hilbert*, J. Math. Kyoto Univ. **5** (1965), 61-66.

[R] D. Rees, *On a problem of Zariski*, Illinois J. Math. **2** (1958), 145-149.

[Ro] P. Roberts, *An infinitely generated symbolic blow-up in a power series ring and a new counterexample to Hilbert's fourteenth problem*, J. Algebra **132** (1990), 461-473.

[S] T. A. Springer, *Invariant Theory*, Springer Lect. Notes in Math. 585, Springer, 1977.

[W] R. Weitzenbock, *Über die Invarianten von linearen Gruppen*, Acta Math. **58** (1932), 231-293.

University of California, Los Angeles, CA 90095

Comments on the Papers

Robert Steinberg

In this section (a) refers to a paper (numbered a) which is in this volume and [b] to a paper which is not.

1. A geometric approach to the representations of the full linear group over a Galois field

As the reader may have noticed, the work of Section 2 constitutes the decomposition of the induced character 1_B^G ($G = GL(n,q)$, $B =$ upper triangular subgroup) into its irreducible components, in terms of the 2^{n-1} subgroups P containing B. This is so because the $e(\nu)$'s being permuted in the representation $C(\nu)$ are just the flags of type ν and the stabilizer of one of these is the subgroup $P(\nu)$ consisting of the elements of G that have the block upper triangular form relative to the decomposition $n = \nu_1 + \nu_2 + \cdots$. Previously I had done the cases $n = 2, 3$ and 4 in an ad hoc manner. Then I learned by chance about Frobenius' remarkable paper [**3**] on S_n. This led at once to the decomposition of 1_B^G in terms of that of the regular representation of S_n (which is just the Weyl group of $GL(n)$, a fact that I did not know at the time). By now there is a vast literature on the subject, for arbitrary reductive groups, with the Hecke algebra, i.e., the commuting algebra of 1_B^G, playing a large role. The reader may also have noticed that Lemma 2.1 is in essence the Bruhat lemma for $GL(n)$, which states that the elements of S_n form a system of representatives for the double cosets of G relative to B.

The larger problem broached in this article, that of determining *all* of the complex irreducible characters of $GL(n,q)$, has been solved brilliantly by J. A. Green in his influential article, *The characters of the finite general linear groups*, Trans. Amer. Math. Soc. **80** (1955), 402–447. Subsequently, the same problem for all of the finite simple (or almost simple) groups arising from the Lie theory (which include the general linear, orthogonal, symplectic and unitary groups and also the exceptional groups) has been treated with notable success by G. Lusztig, P. Deligne, T. A. Springer, D. Kazhdan and others. We cite here the book by B. Srinivasan, *Representations of finite Chevalley groups*, Lecture Notes in Math., Vol. 764, Springer-Verlag, 1979, which contains a very readable account of the situation as of 1979, including the necessary background material on algebraic groups and the étale cohomology.

3. Prime power representations of finite linear groups

I was now enjoying, really enjoying, my first sabbatical leave. I made use of the extra time by constructing a representation to go along with the interesting character of $GL(n,q)$ that I had found in (1), and likewise for the other classical groups. Once again the Weyl groups are present and the Bruhat decomposition hovers in the background. Since the representation for $GL(n,q)$ has gained some currency in algebraic topology, I want to call attention here to a presentation for it, one which is implicit in the paper. The generators of the underlying vector space are the nondegenerate oriented simplices $[P_1, P_2, P_3, \dots, P_n]$ of $\mathbb{P}^{n-1}$ and the relations are $[P_1, P_2, P_3, \dots, P_n] = [P_1, P, P_3, \dots, P_n] + [P, P_2, P_3, \dots, P_n]$ for every point P on the line P_1P_2.

My next paper, Part II of this one, contains my final construction, for all finite reductive groups together.

4. Prime power representations of finite linear groups II

With this paper I was becoming a semisimple person, strongly influenced by the material in [**1**] and in [**3**] which I had been reading at the time.

I felt that I had finally found a proper setting for the representation that I had studied in (1) and in (3), first as a character for $GL(n,q)$ and then as a representation for all of the classical groups. Nowadays the representation R is called the Steinberg representation and it is denoted St, as are a number of related representations. In view of the ubiquitous nature of these representations I have decided, with some encouragement from several of my colleagues, to include here a brief essay on St discussing other constructions as well as situations in which it figures prominently. This follows some comments on the present paper itself.

First, the reader may have noticed that in this paper, which I shall call Part II, I put together the two main ingredients of my previous paper, Part I, in which I treated the classical groups. These are $B = UH$ and $H\sum \varepsilon(w)w$, which for the group $GL(n)$ are, respectively, the stabilizer of a flag (which is called a composition sequence (c.s.) in Part I) and the oriented stabilizer of a simplex. (Recall that H, e.g., denotes not only a certain subgroup of G but also the sum of the elements of that subgroup in the group algebra $F(G)$.) To my surprise and delight the product of these two ingredients, which is essentially the idempotent $|U|/|G| \cdot e$, works perfectly, as do the main arguments of Part I.

Next, the axioms, and hence also the conclusions, of Part II hold not only for those of the twisted Chevalley groups that were known at the time, as we expected they would, but also for those that were discovered soon after by me, M. Suzuki and R. Ree, in chronological order (see item (8) in this volume and references [**29**] and [**16**] of the bibliography of (23)). Also, as suggested in the paper, the list of axioms can be made shorter. The ultimate in this direction is undoubtedly the theory of BN-pairs, a brilliant invention of J. Tits (see, e.g., his notes *Buildings of Spherical Type and Finite BN-Pairs*, Lecture Notes in Math., Vol. 386, Springer-Verlag, 1974).

This may also be an appropriate place to point out that, in case the characteristic of the base field F is 0 or p, if V denotes the group $w_0Uw_0^{-1}$ and $v_1, v_2, \dots$ is a list of its elements, then $\{E_{ij} = |U|/|G| \cdot v_i e v_j^{-1} | i, j = 1, 2, \dots\}$ is a set of matrix units for the subalgebra of $F(G)$ corresponding to St, i.e., we have the relations $(*)\ E_{ij}E_{kl} = \delta_{jk}E_{il}$ for all i, j, k and l. As a consequence of this, the central idempotent of $F(G)$ corresponding to St may be written as $|U|/|G| \cdot \sum_{v \in V} vev^{-1}$. In view of Lemma 2 of the paper $(*)$ follows from: $eve = 0$ for every $v \in V$, $v \neq 1$, which we now prove. If ev is written as $B \sum \varepsilon(w)wv$, then the coset B does not occur in the sum since from $Bwv = B$ and $v = w_0 u w_0^{-1}$ $(u \in U)$ it would follow that $Bww_0u = Bw_0$, then $w = 1$ by (7) of the paper, then $u = 1$ by (6) and (7), and finally $v = 1$ in contradiction to $v \neq 1$. It follows from the corollary to Theorem 1 that if ev is written as $\sum c(u)eu$, then $\sum c(u) = 0$, whence $evU = 0$ and $eve = 0$, as required.

Besides e, St has another very important generator, namely, $f = eU$, which I want to discuss briefly. It is bi-invariant under B and hence represents an element of the Hecke algebra of G relative to B, i.e., of the commuting algebra of 1_B^G. As such, it is in fact, up to a multiplicative scalar which is easily computed, just the G-projection of 1_B^G onto its component St. In terms of f, a basis of St consists of the elements fv $(v \in V)$. From the point of view of algebraic groups, f is a highest weight vector whereas e, more democratically, is the sum of a complete set of extreme weight vectors, a highest one and the rest of its images under W.

What has been said above about the Hecke algebra H and its "idempotent" element f may be formulated as a construction of St as an "induced representation" of G from a 1-dimensional representation of H as follows, using the group algebra whose elements are the F-valued functions on G with convolution $(f \cdot g)(x) = \sum f(xy^{-1})g(y)$, sometimes normalized, as multiplication. Accordingly H consists of the functions which are bi-invariant under B, i.e., are linear combinations of the characteristic functions of the various double cosets BwB $(w \in W)$, and $|B|^{-1}\chi_B$ is its unit element. Finally, f corresponds to the 1-dimensional H-module F on which $|B|^{-1}\chi_{BwB}$ acts as multiplication by 1 or -1 according as $w \in W$ has even or odd length, and St is realized on $F[G/B] \otimes_H F$ with G acting naturally on the first factor and trivially on the second.

There are other constructions of St which have gained currency and I want to discuss two of these next. G is still a finite Chevalley group, $B = UH, W, \dots$ as before, and the language of semisimple groups will be used. The first construction involves the *Tits complex* (or *building*) of G, another of his far-reaching discoveries (see loc. cit.). This is, in the present context, the finite simplicial complex K in which there is one simplex S_P, of dimension $r - r(P) - 1$, corresponding to each parabolic subgroup P of G, with S_P a face of S_Q just when $P \supset Q$. Here r is the rank of G, P runs through the conjugates of the 2^r subgroups of G containing B, and $r(P)$ is the semisimple rank of P. The (simplices corresponding to the) Borel subgroups of G, i.e., the conjugates of B, have the maximum dimension, $r - 1$, which is thus the dimension of K, while for $P = G$, S_P is the empty simplex, considered to have the dimension -1. (The inclusion of the empty simplex in the definition of the Tits complex, or of any simplicial complex for that matter, is not customary, but in the present situation it works perfectly. Observe that one of its effects is the replacement of $H_0(K)$ by its reduced version.) In case $G = GL(n)$, for example, $r = n - 1$, and if $P(\nu)$ $(n = \nu_1 + \nu_2 + \cdots)$ is the parabolic group discussed in our commentary on (1), $r(P(\nu)) = \sum(\nu_i - 1)$. In any case G acts simplicially on K and hence also on its homology groups (or spaces). The result of Solomon-Tits (see L. Solomon, *The Steinberg character of a finite group with a BN-pair*, Theory of Finite Groups (Harvard Symposium), Benjamin, New York, 1969, 213–221) is that the representation of G on the highest homology space $H_{r-1}(K)$ is isomorphic to St, provided of course that the same base field (or even ring, since everything is done over $\mathbb{Z}$) is used in both cases. In fact the two constructions themselves are very closely related, as I shall now explain.

First, the $(r-1)$-simplices of K correspond to the conjugates of B, hence also to the cosets Bx ($x \in G$), because B equals its own normalizer, as does every parabolic subgroup of G. With this identification, the simplices Bw ($w \in W$) fit together in K to form an $(r-1)$-sphere S. This is so because, in the action of W as a reflection group on $\mathbb{R}^r$ and hence also on $S^{r-1} \subset \mathbb{R}^r$, the reflecting hyperplanes divide S^{r-1} into $|W|$ fundamental spherical simplices which are the images Fw ($w \in W$) of one of them (with the group continuing to act on the right) and these fit together to form S^{r-1} exactly as the corresponding simplices Bw ($w \in W$) fit together in K to form S. The same remarks apply with Bw replaced by Bwx for each fixed $x \in G$ so that K is a union of $(r-1)$-spheres. Since S is the union of the $(r-1)$-simplices Bw, we have, in the chain group $C_{r-1}(K)$, $S = \sum \varepsilon(w)Bw$, which is just the element e of our paper. Thus the spheres correspond to the elements ex ($x \in G$) of our original representation space. In this interpretation, the main relation (17) of our paper states that the sphere $e\overline{u}_\alpha$ is homologically the sum of the spheres e and $eu_\alpha w_\alpha^{-1}$ (exactly as the outer cycle of the figure θ is the sum of the two inner cycles, which is just the case $r = 2$), and Lemma 4 of Part I has a similar interpretation. Then, as in the paper, the spheres Sx ($x \in G$) are, in $H_{r-1}(K)$, linear combinations of the spheres Su ($u \in U$) and the latter ones are linearly independent. Therefore to make our comparison complete it remains only to show that every element of $H_{r-1}(K)$ is a signed sum of these spheres. This is contained in the following proposition which in fact proves everything ab initio, except for the irreducibility of St.

PROPOSITION. *For each rank-1 parabolic subgroup P_i containing B ($i = 1, 2, \ldots, r$) let $\partial_i : F[B \setminus G] \to F[P_i \setminus G]$ be the linear map defined by $Bx \to P_i x$ for all $x \in G$. Then the following subspaces of $F[B \setminus G]$ are equal.*

(a) *The right ideal of $F[G]$ generated by e.*

(b) *The linear span of $\{eu | u \in U\}$.*

(c) $H_{r-1}(K)$.

(d) $\bigcap \ker \partial_i$.

Here $F[B \setminus G]$, for example, denotes the span of $\{Bx | x \in G\}$ and the P_i are just the subgroups mentioned in Axiom (11) of the paper. Because of our identification of $F[B \setminus G]$ with the chain group $C_{r-1}(K)$, the boundary operator on this group is just $\partial = \sum (-1)^i \partial_i$. Since $\partial(Bx) = \sum (-1)^i P_i x$ and the P_i's for different i's are never conjugate it follows that $\ker \partial = \bigcap \ker \partial_i$, so that (the spaces in) (c) and (d) are equal. Clearly (a) contains (b), while (d) contains (a) because $\partial_i(ex) = \sum \varepsilon(w) P_i x$ and this is 0 since the terms for w and $w_i w$ in the sum cancel each other. Finally, to show that an element $f = \sum c_x Bx$ of (d) is necessarily in (a), in fact in (b), we may achieve the normalization $c_x = 0$ for every $x \in w_0 U$ by subtracting from f a suitable linear combination of the eu's, all of which are by now known to be in (d). If $f \neq 0$, let $c_x Bx$ ($x \in wU_w$) be one of its terms such that $l(w)$, the length of w, is maximal. Since $w \neq w_0$, there exists an i such that $l(ww_i) > l(w)$. From the properties of the Bruhat decomposition (Axiom (7) of the paper), it follows that $P_i x \neq P_i x'$ for any other term $c_{x'} B x'$ of f and thus that the term $c_x P_i x$ of $\partial_i f$ is not canceled by any other term, so that $\partial_i f \neq 0$ in contradiction to $f \in$ (d). Thus $f = 0$, which is in (a), as required.

It follows from this discussion, especially from the equation $P_i e = 0$ proved above, that St has a presentation in which the generators of the underlying vector space are the Borel subgroups of G and the relations are $\sum B = 0$ $(B \in P)$, one relation for each rank-1 parabolic subgroup P of G.

The third construction of St that we want to consider, one which appeared before that of Solomon and Tits, is due to C. W. Curtis, and, independently, to W. Feit. (See C. W. Curtis, *The Steinberg character of a finite group with a BN-pair*, J. Algebra **4** (1966), 433–441.) It states that, as a character, St is given by the formula $(*)$ $\sum(-1)^{r(P)} 1_P^G$, in which P runs over the 2^r subgroups of G that contain B. Curtis also shows that St is the unique irreducible component of 1_B^G which is not a component of any other 1_P^G. Thus $(*)$ is a sort of inclusion-exclusion formula for St (although Curtis does not obtain it in this way), but the simplest way to relate it to our previous considerations is to use the Euler-Poincaré principle: If a group G acts simplicially on a finite simplicial complex and hence also on its (co)homology, and if C_i is the ith chain group and H_i is the ith homology group $(i \geq -1)$, then the characters of G on $\sum(-1)^i C_i$ and on $\sum(-1)^i H_i$ are equal. We apply this to the Tits complex K, interpreted as above so that the simplices are the cosets Px. From the definitions it follows that $\sum(-1)^i C_i$ is isomorphic to $\sum(-1)^{r-r(P)-1} 1_P^G$ as a G-module and, since K is a "bouquet", i.e., a wedge, of $(r-1)$-spheres, that $H_i = 0$ unless $i = r-1$ in which case it is isomorphic to St, as in the above proposition. Here the H_0 of the bouquet of spheres is the reduced one. Thus $\sum(-1)^{r-r(P)-1} 1_P^G$ is isomorphic to $(-1)^{r-1} St$, as required.

We discuss next some of the advantages of each of the three approaches to St that we have given above. The one in terms of the Tits complex leads to a very pleasant conceptual determination of the character values of St. (See C. W. Curtis, G. I. Lehrer and J. Tits, *Spherical buildings and the character of the Steinberg representation*, Invent. Math. **58** (1980), 201–220, and the discussion of (18), as well as the papers (18) and (23) themselves where other determinations, with the sign not worked out, are given.) The result is that $St(x)$ is 0 unless x is a semisimple element of G (i.e., one whose order is prime to p) in which case it is $(-1)^s|P|$, where P is a maximal unipotent subgroup (i.e., a Sylow p-subgroup) of $Z_G(x)$ and s is the difference between the ranks of G and of $Z_G(x)$. Another plus for the Solomon-Tits construction is that by now suitable (co)homology spaces related to G have yielded constructions for *all* of the complex irreducible representations of G (see the discussion of (1)).

Curtis' approach to St leads to a very interesting duality among the complex irreducible representations of G given as follows. Let P be a (parabolic) subgroup of G containing B. There exists a decomposition $P = M_P U_P$ (semidirect) with U_P a normal subgroup of P called the unipotent radical and M_P a complementary (reductive) subgroup called a Levi subgroup. Given a representation of G on a space V, we get one of M_P by restricting the action of M_P to the subspace consisting of the points of V that are kept fixed by U_P, while given a representation of M_P, we get one for G by composing it with the map $P \to M_P$ to get a representation of P and then inducing the result to G. Let D_P denote the composition of these operations, from G back to G. Then $D = \sum(-1)^{r(P)} D_P$ $(P \supset B)$ is the operator that we are after. It is involutary and acts as an isometry on the character ring of G; hence it pairs each irreducible representation of G with another one. It follows from Curtis' formula that St is paired with 1, the trivial representation, and that fact plays a significant role in the further development of the theory. For a full treatment of the subject and many relevant references we cite the book by C. W. Curtis and I. Reiner, *Methods of Representation Theory with Applications to Finite Groups and Orders* II, John Wiley and Sons, New York, 1987.

As to our own approach to St, it has the following advantages. It yields a quick proof of irreducibility, even in characteristic p, its construction in terms of the generator f yields an important connection with the Hecke algebra of G relative to B, and finally the generator e which is at the center of our development has recently resurfaced in algebraic topology, in particular in connection with the solutions of conjectures of G. Segal (the Burnside conjecture) and G. W. Whitehead. For this, see N. J. Kuhn, S. A. Mitchell and S. B. Priddy, *The Whitehead conjecture and splitting* $B(\mathbb{Z}/2)^k$, Bull. Amer. Math. Soc. (N.S.) **7** (1982), 255–258, J. F. Adams, J. H. Gunawardena and H. Miller, *The Segal conjecture for elementary Abelian p-groups*, Topology **24** (1985), 435–460, and the other papers mentioned there.

The most significant applications of St, however, occur in representation theory itself, in connection with the problems of determining *all* of the irreducible characters of G, those in characteristic 0 and also those in characteristic p.

In the first case (which is briefly discussed in our commentary on (1)) St plays a useful but limited role. The character values (which we have just given above) come into play of course, but the simpler fact that these values are 0 away from the set of semisimple elements of G often carries one a long way.

In the second case, that of representations in characteristic p, where considerable progress has also been made (see, e.g., J. C. Jantzen, *Representations of Algebraic Groups*, Pure Appl. Math., Vol. 131, Academic Press, 1987, and H. H. Andersen, J. C. Jantzen and W. Soergel, *Representations of Quantum Groups at a pth Root of Unity and of Semisimple Groups in Characteristic p: Independence of p*, Astérisque 220, 1994), St plays what seems to be an indispensable role. Among its significant properties are the following. It is both irreducible and projective and it is the only such representation of G. As an irreducible representation it is the largest (in dimension), and as a projective one the smallest since it is also a tensor factor of *every* projective representation of G. It follows that it is also self-dual and that every projective representation of G is injective and vice versa.

It turns out, as is suggested by the references that have just been given, that the problem in this second case, that of determining the irreducible characteristic p representations of the finite group G, is equivalent to that of determining the irreducible rational representations of the algebraic group $\overline{G}$ obtained by replacing by its algebraic closure the field F used in the definition of G. This is so because each irreducible representation of G extends to a rational representation of $\overline{G}$. (This result, due in part to C. W. Curtis, is proved in our article (18) in this volume.) The irreducible rational representations of $\overline{G}$ are in a sense well known: they are classified by their highest weights (see (18)). The problem is to determine their characters, or even just their dimensions, in all cases.

In particular, the representation St of G extends to the representation of $\overline{G}$ whose highest weight is $(q-1)\sum \omega_\alpha$ (in terms of the fundamental weights as in (18)). The latter representation is also denoted St, or sometimes St_q since there is one representation of $\overline{G}$ corresponding to each $q = p, p^2, p^3, \dots$ because the finite fields of these orders all have the same algebraic closure, and it plays the same role in the representation theory of $\overline{G}$ that the earlier St did in that of G.

This representation of $\overline{G}$, used for a sufficiently high power of p, enters significantly into the solution of Mumford's conjecture (see W. J. Haboush, *Reductive groups are geometrically reductive*, Ann. of Math. **102** (1975), 67–83). The result, Haboush's Theorem, states that if V is any rational $\overline{G}$-module and if v is any nonzero element of V invariant under $\overline{G}$, then there exists a polynomial function f on V which is homogeneous, of positive degree, invariant under $\overline{G}$, and such that $f(v) = 1$. This property of reductive algebraic groups plays the same role in characteristic p that complete reducibility does in characteristic 0 (in which case there exists an f which is linear), for example, in the proof that for rational $\overline{G}$-modules the algebras of polynomial invariants are always finitely generated.

There also exist infinite-dimensional representations which are sometimes called Steinberg representations, not for the finite group G or the algebraic group $\overline{G}$, but for the topological group $\widehat{G}$ obtained by taking the base field to be a p-adic field. All of the constructions of St in the finite case (via the Hecke algebra, the top cohomology of the Tits complex or building, Curtis' formula) and many of its properties (e.g., that its restriction to U is the regular representation of U) carry over to $\widehat{G}$ if they are properly formulated, but a second double coset decomposition of $\widehat{G}$ also has to be used, along with the Hecke algebra and the building (called the Bruhat-Tits building) that are associated with it. In this decomposition the role of the Borel subgroup B is taken by a certain compact open subgroup of $\widehat{G}$ called an Iwahori subgroup and that of W by the corresponding affine Weyl group. If the choices that have to be made (e.g., of $L^2(\widehat{G}, \mathbb{C})$ as the group algebra of $\widehat{G}$) are made suitably, the three constructions used in the finite case can be carried out here also and, as in the finite case, they all produce the same representation of $\widehat{G}$. Parts of the proof of this are quite complicated and perhaps have not yet attained their final form. In any case, the resulting representation, which is called *the special representation* if the Hecke algebra construction is used, but *the Steinberg representation* if Curtis' formula is used, can be realized on an infinite-dimensional complex Hilbert space so that it is continuous, unitary, square-integrable and again irreducible, and it is characterized by these properties and the further one that its continuous cohomology is nonzero. The main contributors here are A. Borel, W. Casselman, Harish-Chandra, H. Matsumoto, J-P. Serre and J. Shalika. We cite here the following papers whose bibliographies should also be consulted: A. Borel and J-P. Serre, *Cohomologie d'immenbles et de groupes S-arithmétiques*, Topology **15** (1976), 211–232 (where both types of buildings (and others as well), Curtis' formula, and even the idempotent e of our paper, in the guise of a sphere, come into play, as do applications to the cohomology of arithmetic subgroups of algebraic groups defined over number fields), A. Borel, *Admissible representations of a semi-simple group over a local field with vectors fixed under an Iwahori subgroup*, Invent. Math. **35** (1976), 233–259 (in which the Hecke algebra construction is done in great generality, much of it purely algebraically), A. Borel and N. Wallach, *Continuous Cohomology, Discrete Subgroups, and Representations of Reductive Groups*, Ann. of Math. Studies, no. 94, Princeton University Press, 1980 (especially Section XI.3.8), and M. Reeder, *The Steinberg module and the cohomology of arithmetic groups*, J. Algebra **141** (1991), 287–315.

This essay has become considerably longer than I expected it to be when I started it. Therefore I shall stop it here by calling attention to another essay on the same subject: J. E. Humphreys, *The Steinberg representation*, Bull. Amer. Math. Soc. (N.S.) **16** (1987), 237–263. This article covers roughly the same ground as mine, but it has further information on some of the topics that we have gone into and also on some of those that we have not; and it has an extensive bibliography which contains many items beyond those that we have chosen to mention here.

5. On the number of sides of a Petrie polygon

The main objective of this paper was a quick classification of the regular polytopes in 4-dimensional Euclidean space. Since the left side of (1) is positive it easily follows that $(*)$ $p+2q+r<15$. Since also each of p, q and r is at least 3, the possibilities can be checked out quickly. Of the resulting 7 solutions of $(*)$ only $\{4,3,4\}$ has to be ruled out since it makes the right side of (1) 0 (and hence corresponds to a tesselation of 3-dimensional space). A comprehensive account of these matters and of regular polytopes in general can be found in the second edition of [**1**].

6. Eigenvalues of the unitary part of a matrix (with A. Horn)

It seems to me likely that the theorem of this paper, suitably formulated, can be carried over to arbitrary complex reductive Lie groups. The analogous result of my coauthor Alfred Horn in [**2**], which concerns the positive definite part of a matrix, *has* been carried over in this way. See B. Kostant, *On convexity, the Weyl group and the Iwasawa decomposition*, Ann. Sci. Ecole Norm. Sup. **6** (1973), 413–455, where another of Horn's theorems, which determines the possible diagonals of a rotation matrix, is also considered.

7. Finite reflection groups

I had decided that in order to become a semisimple mathematician I would have to learn all about finite reflection groups. The result was the present paper.

I should have included in the assumptions made on the simplex F at the start of Section 3 that it is required to have only nonobtuse dihedral angles (which are the interior angles between neighboring pairs of faces). This is used, via Lemma 2.4, in the last paragraph of the proof of Lemma 3.2, and it is a property which holds in the main result Theorem 4.2 since there the dihedral angles are proper submultiples of π.

The calculation in the proof of 8.4 is a bit cryptic and might be amplified thus: at the third step the term for $\rho=\mu$ increases by exactly 1 since $2\cdot 1/2$ is being replaced by $2^2 \cdot 1/2$, while the other terms remain unchanged since in each of them either 0 or 1 is being replaced by its square.

Finally, the "unexplained" fact in Section 9 has been explained, quite satisfactorily, by B. Kostant in *The principal three-dimensional subgroup and the Betti numbers of a complex simple Lie group*, Amer. J. Math. **81** (1959), 973–1032. His attention was drawn to it by A. Shapiro who had noticed it independently of me.

8. Variations on a theme of Chevalley

A strong impetus for this paper was my desire to extend the results of my earlier paper (4) to the (finite) unitary groups and the so-called second orthogonal groups, those groups of the present paper that were known at the time that it was written. The discovery of new families of simple groups was, of course, a very pleasant surprise. I overlooked the fact that the special isogenies that exist for groups of type C_2, F_4 and G_2 in characteristics $2, 2$ and 3, respectively, could be used to construct yet other families of simple groups in very much the same way. This oversight was soon rectified by M. Suzuki (who found the first family in a different way) and R. Ree (see items [**29**] and [**16**] of the bibliography of (23)), thus completing the list in the eventual classification of the finite simple groups except for the final 21 sporadic groups which were yet to come.

On the 11th line from the end of Section 13, *and sign changes* should be added after *permutations*. The example that follows is faulty because b is not in L. However the method of calculating $n(w)$ is correct, as is its use in the calculation of $P(t)$.

A simpler and more conceptual proof of simplicity than the ones offered in this paper and in our next one has been given by J. Tits, following an idea of K. Iwasawa, in the framework of his theory of B-N pairs (see J. Tits, *Algebraic and abstract simple groups*, Ann. of Math. **80** (1964), 313–329). It applies to all of the Chevalley groups, twisted and untwisted, and goes as follows. Let $B = UH$ (this is U^1H^1 in the present paper) be the standard Borel subgroup and K any nontrivial normal subgroup of G. Then (1) B is solvable, (2) G equals its own derived group, and (3) $G = BK$. Here (1) is easy because U is nilpotent and $B/U \cong H$ is Abelian, while (2) and (3) require more serious proofs. G/K equals its own derived group by (2), $B/(B \cap K)$ is solvable by (1), and these groups are isomorphic by (3). It follows that both groups are trivial and hence that $K = G$, as required.

10. Automorphisms of finite linear groups

Regarding the remarks at the beginning of Section 8, the results of this paper *do* hold when the base field K is infinite, as was first shown to me by A. Borel in 1961. He, in collaboration with J. Tits, has obtained even more general results, which involve a larger class of groups, homomorphisms instead of automorphisms, etc. Our Séminaire Bourbaki account of their work appears as item (26) of this volume.

12. Automorphisms of classical Lie algebras

The exceptional cases of Section 7, in which G_2, F_4 and C_2 in characteristics $3, 2$ and 2, respectively, figure, are closely related to the existence of special isogenies (in which long roots are mapped onto short ones and vice versa) for the corresponding algebraic groups, first found by C. Chevalley in his famous seminar on the classification of the semisimple algebraic groups. (See reference [**8**] of (20), especially Exp. 23 and Exp. 24.)

13. A general Clebsch-Gordon theorem

In the original version of this paper I had also considered the known result that $m_\lambda(\mu) \neq 0$, i.e., μ is a weight of the irreducible representation whose highest weight is λ, if and only if $\lambda - \mu$ is in the lattice generated by the roots and μ is in the convex hull of the extreme weights $s\lambda$ ($s \in W$); thus 0 is a weight if and only if λ is in the lattice generated by the roots. The interested reader might want to prove for himself that this result, like Kostant's formula and the theorem of this paper, is also a formal consequence of Weyl's character formula.

14. Generators for simple groups

It is now known that, not only can one of the two generators of G be chosen to be an involution (cf. the first paragraph of the paper), it can be chosen to be any nontrivial element of G. Further this holds for all finite simple groups, including those not discussed in our paper (see R. Guralnick and W. M. Kantor, to appear).

15. A closure property of sets of vectors

I had noticed that, in my own work on semisimple groups and Lie algebras, the property of root systems that I was using most frequently was property P and so decided to pursue it. I was especially pleased to find that among its consequences was 1.4 about the decomposition of space into simplexes, generally of many shapes and sizes.

A consequence of the results, especially of 1.5, is that there are, up to linear isomorphisms, only finitely many systems with property P of any given dimension. It might be worthwhile to find out what they all are.

16. Complete sets of representations of algebras

R. Brauer, in *A note on theorems of Burnside and Blichfeldt*, Proc. Amer. Math. Soc. **15** (1964), 31–34, has refined Burnside's theorem to say that only the first r tensor powers of R are needed in (1) in case the character of R takes on only r distinct values on G. The same conclusion holds in our main result (2) if we add to the assumptions there first that B is a group (so that A is its group algebra) and second that, if V is the vector space underlying R, there exists a linear function λ on $\mathrm{End}(V)$ such that $\lambda(1) = 1$, $\lambda(x) \neq \lambda(1)$ for any $x \neq 1$ ($x \in B$), and λ takes on only r distinct values on B.

17. Générateurs, relations et revêtements de groupes algébriques

In current terminology, our main result 4.1 states that (π, Δ) is *the universal central extension* of G.

For the validity of this result and our others, not all of the relations in (A), (B) and (C) are needed: some of them are consequences of the others. In particular, (C) is needed only for a single root—any long one will do—and (B) for those pairs of roots which, taken together, are supported by pairs of simple roots. The last point follows from the result of C. Chevalley (see reference [**8**], Exp. 23 and Exp. 24, of our later paper (20)) that G is isomorphic to the free product of its standard semisimple subgroups of rank 2 amalgamated at their intersections in G.

The kernel of the map $\pi : \Delta \to G$ has been determined by H. Matsumoto following work by C. Moore (see references [**Ma**] and [**Mo**] of (35)). Recall that Δ is the group defined by the relations (A) and (B), while G is the (semisimple) group obtained when the relations (C) are added (in case G is simply connected, which we shall henceforth assume). The result is that $\operatorname{Ker}\pi$ is generated by the elements $(t, u) \equiv h(t)h(u)h(tu)^{-1}$ $(t, u \in K^*)$ of Δ (which are used near the end of Section 7), subject only to the relations:

(a) $(t, uv) = (t, u)(t, v)$ and $(tu, v) = (t, v)(u, v)$,

(b) $(t, 1-t) = 1$.

It must be assumed here that G is not a symplectic group, in which case the relations have to be modified. Moore and Matsumoto were motivated by the fact that the norm residue symbol also satisfies the relations (a) and (b) and they established a deep connection between the structure of semisimple algebraic groups over number fields and their completions on the one hand and the reciprocity laws of algebraic number theory on the other. An interesting biproduct of their work is a double covering of the symplectic group called the metaplectic group which had been constructed earlier by A. Weil, in a different way, and used by him in his retake of C. L. Siegel's work on quadratic forms.

Next we want to discuss Milnor's K_2, a functor in algebraic K-theory devised by him to go along with K_0 and K_1 (see J. Milnor, *Introduction to Algebraic K-Theory*, Ann. of Math. Studies, no. 72, Princeton Univ. Press, 1971). For any ring R let $G_n(R)$ be the group of $n \times n$ matrices over R generated by the elementary matrices, and $G(R)$ the limit as $n \to \infty$ (relative to the imbedding(s) of $G_n(R)$ in the upper-left-hand corner of $G_{n+1}(R)$). The elementary matrices $x_{ij}(t) = 1 + tE_{ij}$ $(i \neq j$, E_{ij} matrix unit, $t \in R)$ satisfy (A) and (B) (which are written down for this case near the end of Section I3). If $\Delta(R)$ is the abstract group defined by these generators and relations, it turns out that the natural map $\pi : \Delta \to G(R)$ is the universal central extension of $G(R)$ (it dominates all others). Its kernel is, by definition, $K_2(R)$. In case R is a field it follows from our earlier discussion that $K_2(R)$ is defined by the relations (a) and (b), but in general this is not so because R might not have enough units. Since the ensuing work in this area is vast and quite diverse, I shall not give any further references. Anyone can, of course, consult the section on Algebraic K-theory in Mathematical Reviews and trace back any of the results or subjects that strike her or his fancy.

Since the relation (b) $(t, 1-t) = 1$ has turned out to be such an important feature of this paper, I have decided to include here a more transparent proof of it than the one given in the paper, at the end of Section 7. (An analogous proof in a different setting occurs in (33), which is Part II of this paper.) Since $h(t) = w(t)w(1)^{-1}$ we have to show that $w(t)w(1)^{-1}w(u) = w(tu)$ if $t + u = 1$ (and

$t, u \in K^*$). We have

$$(*) \qquad \begin{aligned} w(t)w(1)^{-1}w(u) &= w(t)y(1)x(-1)y(1)w(u) \\ &= x(\cdot)w(t)x(-1)w(u)x(\cdot) \\ &= x(\cdot)x(\cdot)y(\cdot)x(t)x(-1)x(u)y(\cdot)x(\cdot)x(\cdot) \\ &= x(\cdot)y(\cdot)x(\cdot). \end{aligned}$$

Here dots indicate elements of K whose exact values are not needed. In the first and third equalities we use the definition of $w(\cdot)$, in the second that $w(\cdot)$ conjugates $y(\cdot)$ into $x(\cdot)$, and in the fourth that $t + u = 1$. Since after projection to G (via π) the first item of (*) equals $w(tu)$, which can be written uniquely in the form $x(\cdot)y(\cdot)x(\cdot)$ in G, we see that the last item is $x(tu)y(-t^{-1}u^{-1})x(tu)$, which is the definition of $w(tu)$ in Δ.

Finally, we call attention to our paper (35) where the relations that we have been discussing are put to other good uses.

18. Representations of algebraic groups

Theorem 6.1, the tensor product theorem, is the principal result of this paper, especially the version of it that deals with linear representations rather than projective ones. It focuses attention on the finite set consisting of the p^l rational irreducible representations of Γ and of Γ_q (which can be any simply connected semisimple algebraic group defined over the finite field $\mathbb{F}_q$ and its subgroup of points that are rational over $\mathbb{F}_q$) for which the coefficients of the highest weights are all less than p. Despite this reduction, the main problem here, that of the determination of the characters, remains open. There has, however, been considerable progress towards its solution, as we mentioned in our discussion of (4), where references, especially to the book by Jantzen, can be found. There are by now many different proofs of Theorem 6.1. References to some of these are given in our later article (36). Also there is an analogous theorem for quantum groups. And there are such theorems for GL_n in which the fields defining the groups and the representation spaces have different prime characteristics. For a discussion of these and many references see R. Dipper, *Polynomial representations of finite general linear groups in non-describing characteristic*, Progr. Math., Vol. 95, Birkhäuser-Verlag, 1991, 343–370.

The representation M_q of Section 8 is just the representation St_q discussed in our comments on (4). Observe that the Brauer character of Theorem 8.4 equals the ordinary character of St_q viewed as a representation in characteristic 0. The method that we have used to determine it has produced as a bonus all of the eigenvalues of the representing linear transformations, not just their traces. A more conceptual method is the one by Curtis-Lehrer-Tits which we mentioned in our commentary on (4). It runs as follows. Let x be any element of G, $x = su$ its semisimple-unipotent decomposition, $G' = Z_G(s)$, which is essentially also a reductive group, and K and K' the Tits complexes of G and G'. From the properties of K and the action of G on it, it follows that the trace of x on $\sum(-1)^i H_i(K)$ equals the trace of u on $\sum(-1)^i H_i(K')$, and thus (since all but the top homology groups are 0) that $St_G(x) = (-1)^{r-r'} St_{G'}(u)$, if r and r' are the ranks of G and G'. Finally, $St_{G'}(u)$ is 0 unless $u = 1$, i.e., unless x is semisimple, in which case it equals $\dim St_{G'}$ which is the order of a Sylow p-subgroup of G'. The Tits complexes that are used here are actually slight modifications of the ones that were defined earlier.

We conclude with some further remarks. Theorem 1.1 is, of course, also true for linear representations in case G is simply connected. I regret my usage of the term *high weight* (instead of the standard *highest weight*) here and later on, which, fortunately, has not been taken up by very many other authors. Conjecture 1.2, in a more general form, has been proved by A. Borel and J. Tits (see (26) which is our Séminaire Bourbaki account of their work). A complete proof of Step (2) of Lemma 3.9, originally worked out in conversation with T. A. Springer over tea at the Institute for Advanced Study in Princeton, is given in our joint survey article (25), in Section II.4. A more general result, which states that G_σ is connected for any semisimple automorphism σ of a simply connected semisimple algebraic group G, is also proved later in this volume, in Section 8 of (23). In the statement of 7.1 of the present paper the relations $x_r(t)x_r(u) = x_r(t+u)$ $(r \in \Sigma; t, u \in K)$ should have been added, and in the first part of 7.5 it should have been stated that the extension can be made rational. Both 9.4 and 9.5 hold for groups of type A_{2n} also, as follows from results in our later paper (33). In 10.2 the isomorphism is meant to be one of abstract groups. As algebraic groups, the groups there are only isogenous. Theorem 11.1, for the group Γ, can be proved more simply by applying Theorem 6.1 to the representations ρ and $\rho \circ \theta$ (with θ as in 10.1 and ρ any irreducible rational representation of Γ) and comparing the results. It is not hard to show that for Γ there are no further decompositions of its rational irreducible representations into tensor products, and that for a semisimple group of characteristic 0 there are no such decompositions at all. Finally, as suggested just after the statement of 12.4, that result is indeed true for the groups of type G_2.

19. Differential equations invariant under finite reflection groups

In the statement of Theorem 1.3 I should have included the following condition, which is easily seen to be equivalent to 1.3(e) and 1.3(f): S is free as a module over I. The fact that this condition implies 1.3(g), which states that I is a polynomial algebra, holds under quite general conditions, in which groups need not be involved. The proof requires serious commutative algebra. See, e.g., reference [**4**] of (28). Also, writing $d_L = d_L(G)$ to indicate the dependence on G, I should have stated in 1.2 that $d_L(G) = |G/G_L| \cdot d_0(G_L)$. Thus $d_L(G) = |G|$ if and only if $d_0(G_L) = |G_L|$, i.e., by 1.3, if and only if G_L is a reflection group.

20. Regular elements of semisimple algebraic groups

As mentioned in the paper, 1.4 yields normal forms for the regular elements of G, which for $G = SL(r+1)$ are just the elements which consist of a single block in the rational Jordan canonical form. As far as I know, the problem of extending this to arbitrary elements remains open, at least for general semisimple algebraic groups.

The proof of the last assertion of 2.10, which implies the one just before it, has appeared. It is given in Sections 8 and 9 of (23). That $\dim G_x - r$ in 3.10 is always even is now known to be true, and it is in fact equal to twice the dimension of the variety whose points are the Borel subgroups of G that contain x. For discussions of these matters see III.3.24 of (25) and also (29) in this volume, and for updates and a comprehensive bibliography see the book by J. E. Humphreys, cited in our commentary on (25). In these matters there are general proofs for the classical groups and also when char k is *good*, i.e., it is not equal to any coefficient of the highest root of any simple component of G. When G is an exceptional group, the conjugacy classes can be listed and checked out on a case-by-case basis.

Also when char k is good, J. F. Kurtzke has shown that the abstract characterization of regular elements suggested in 3.13 is indeed true. See J. F. Kurtzke, *Centralizers in reductive algebraic groups*, Comm. Algebra **19** (1991), 3393–3410, and his earlier article in Pacific J. Math. **104** (1983), 133–154. The other abstract characterizations mentioned in the paragraph after 3.14 are proved in our article (26).

G. Lusztig, *On the finiteness of the number of unipotent classes*, Invent. Math. **34** (1976), 201–213, has shown that this finiteness does indeed hold in all semisimple groups. As suggested in our paper, this yields at once the existence of regular unipotent elements. Lusztig's proof, however, is by no means simple, and our proof in Section 4 remains, we believe, the simplest known proof of the existence of these elements.

Theorem 9.8 has been extended by R. E. Kottwitz, *Rational conjugacy classes in reductive groups*, Duke Math. J. **49** (1982), 785–806, as follows. If G is reductive with a simply connected derived group and, as before, it contains a Borel subgroup defined over the perfect field k, then every semisimple class defined over k contains an element defined over k. The same holds for *every* conjugacy class in case also char k is 0 or a large enough prime so that the Jacobson-Morosov theory can be applied.

The condition which appears in 9.3(b) and elsewhere should be corrected to read: *There exists a finite Galois extension K of k such that $x_\gamma = 1$ for each $\gamma \in \Gamma$ which acts trivially on K.*

Finally, the reference to Serre in [**13**] should be updated. It is now a book *Cohomologie Galoisienne*, Lecture Notes in Math., Vol. 5, Springer-Verlag, 1964, of which the latest edition should be sought.

22. Classes of elements of semisimple algebraic groups

For further information on the results and problems mentioned in this paper the reader should consult our survey article (25), written jointly with T. A. Springer, and the recent book of J. E. Humphreys mentioned in our commentary on (25).

23. Endomorphisms of linear algebraic groups

I start with a correction. The definition of Σ' in the paragraph just before 3.7 should read $\{k + \alpha | k \textit{ integral}, \alpha \in \Sigma\}$. Incidentally, this marks, as far as I know, the introduction of affine root systems into the mathematical literature. In earlier articles in this area the roles of the roots $k + \alpha$ were played by the corresponding half-spaces of V.

As suggested in the introduction, the twisted version of Bott's formula given in Theorem 3.10 *has* turned out to have applications to linear algebraic groups defined over local fields. Here are two instances of this: R. Kottwitz, *Tamagawa numbers*, Ann. of Math. **127** (1988), 629–646 (the result here is that these numbers are always 1 for simply connected semisimple algebraic groups defined over number fields); A. Borel and G. Prasad, *Sous-groupes discrets de groupes p-adiques à covolumes borné*, C. R. Acad. Sci. Paris Sér. I Math. **305** (1987), 357–362.

Since this paper was written alternate proofs for the formula given in Theorem 15.1 (for the number of unipotent elements, i.e., p-elements, of G) have appeared, and we next want to discuss some of them. Let S denote the set of unipotent elements of G_σ, and B_σ and U_σ its standard Borel and maximal unipotent subgroups. In his paper *Unipotent elements and characters of finite Chevalley groups*, Osaka J. Math. **12** (1975), 523–554, N. Kawanaka proves, as a special case of a more general result, that if D is any (B_σ, B_σ) double coset of G_σ, then $|S \cap D| = |U_\sigma^- \cap D| \cdot |U_\sigma|$ (actually his proof works for any (U_σ, U_σ) double coset). From this $|S| = |U_\sigma^-| \cdot |U_\sigma|$ follows at once. Like me, he makes use of the character St in his proof, specifically through its pairing with the trivial character 1 in the Curtis duality (discussed in our commentary on (4)). Unlike me, however, he only needs the value of St at each unipotent element, which is easy: it is $|U_\sigma|$ at the unit element of G_σ and 0 at every other unipotent element. Kawanaka also shows that, if χ is any character contained in $1_{B_\sigma}^{G_\sigma}$ and $\hat{\chi}$ is its dual, then $\sum_{x \in S} \chi(x) = \hat{\chi}(1)|U_\sigma|$. This, for $\chi = 1, \hat{\chi} = St$, again yields $|S| = |U_\sigma|^2$. Yet another proof of this formula has been given by G. I. Lehrer, *Rational tori, semisimple orbits and the topology of hyperplane complements*, Comment. Math. Helv. **67** (1992), 226–251. It does not use St or any representation theory at all and runs as follows. First Lehrer shows that the number of pairs (T, x) consisting of a maximal torus T of G and an element x of T, both stable under σ, is just $|G_\sigma|$; this fact, previously known, is quite easy to establish. It follows that if $M(x)$ denotes the number of maximal tori of G_x that are stable under σ, then $|G_\sigma| = \sum M(x)$, summed over all semisimple elements x of G_σ. Combined with the formula $|G_\sigma| = \sum P(x)$ in the paragraph following 15.5 (in which $P(x)$ is the number of unipotent elements of G_x stable under σ), this yields, again by induction, that $P(x) = M(x)$ for every semisimple x in G_σ. Thus $|S| = P(1) = M(1)$, which by 14.14 is $|U_\sigma|^2$. Lehrer's method works equally well for nilpotent elements in the Lie algebra of G, the first step being simpler since σ acts semilinearly on the Lie algebra of T. Independently of each other, Kawanaka (loc. cit.) and Lehrer (in collaboration with J. A. Green) have proved the following result, under the necessary assumption that p is a good prime. If S' is the set of *regular* unipotent elements of G_σ, then $|S' \cap Bx|$ is independent of $x \in G_\sigma$. In this respect, I offer here the following conjecture. Let $g, b, u, \ldots$ denote the Lie algebras of $G, B, U, \ldots$ and s the set of nilpotent elements of g. Then $|s \cap (b + X)|$ is independent of $X \in g$ (and hence is always equal to $|u_\sigma| = |U_\sigma|$, the case $X = 0$).

24. Algebraic groups and finite groups

As mentioned in the discussion of (1) great progress has been made towards a solution of the problem given at the end of (D), that of determining the complex irreducible characters (or representations) of G_σ.

25. Conjugacy classes (with T. A. Springer)

We start with some minor corrections and clarifications. In I.3.2, σ is required to be surjective. In I.4.8(a) (resp. (b)) the Lie algebra involved is that of the simply connected (resp. adjoint) group of the family. In line 4 of II.1.14 *compact* should be replaced by *minisotropic*. In the definition of the universal covering of a semisimple algebraic group given at the start of II.2 the condition on unipotent subgroups is equivalent to: the kernel of $d\pi$ (as well as that of π) is central.

Some of the progress that has been made on the problems posed in our article is discussed in the supplement which follows it in this volume. The paper by S. Keny mentioned there has in the meantime appeared, as *Existence of regular nilpotent elements in the Lie algebra of a simple algebraic group in bad characteristics*, J. Algebra **108** (1987), 195–201. One of her results, that $Z_G(X)$ is always connected when G is a simple adjoint group and $X = \sum X_a$, summed over the simple roots, is a regular nilpotent element in its Lie algebra, can also be proved as follows. Let $S \subset T$ be the 1-dimensional torus $(\bigcap_{a,b \text{ simple}} \operatorname{Ker}(a-b))^0$. For each $c \in k^*$ there exists $s \in S$ such that $\operatorname{Ad}(s)(X_a) = cX_a$ for every simple a. Since also $Z_G(X) = Z_G(cX)$, it follows that s normalizes $Z_U(X)$ and acts on it thus: if $x = \prod_{r>0} x_r(t_r)$, then $s \cdot x = \prod x_r(c^{h(r)} t_r)$ ($h(r)$ = height of r, a positive integer). Thus x and 1 are both contained in the connected subset $\overline{S \cdot x}$ of $Z_U(X)$. Since x is arbitrary that group is connected, and since G is adjoint so is $Z_G(X)$.

Finally, we call attention to the book by J. E. Humphreys, *Conjugacy Classes in Semisimple Algebraic Groups*, Math. Surveys Monographs, Vol. 43, Amer. Math. Soc., 1995, a survey which is more comprehensive than ours, and of course more up-to-date.

The conjecture 1.4(b), which states that $G(k) = G^+$ always (when G is simply connected), is false, as has been shown by V. P. Platonov. For an account of his work and of the area in general, we cite J. Tits, *Groupes de Whitehead de groupes algébriques simples sur un corps* (d'après V. Platonov et al.), Séminaire Bourbaki, Lecture Notes in Math., Vol. 677, Springer-Verlag, 1978, Exp. 505.

28. On a theorem of Pittie

In the equation six lines from the end of Section 2, the right side should be $2\sum \lambda_a n_a$. The anomalous situations concerning SO_{2r+1} in Theorem 1.2(e) and Spin_{2r+1} in Lemma 3.1(d) occur because the action of W on T for the group Sp_{2r}, which is simply connected, is isomorphic to that for SO_{2r+1}, which is not. Just as Theorem 1.1 has a significant application to topological K-theory, Theorem 2.2 has one to algebraic K-theory, in I. A. Panin, *On the algebraic K-theory of twisted flag varieties*, to appear, where it plays a central role. In case $G = SU_n$ and G' is a maximal torus of G, Theorem 2.2 may be reformulated as follows. For each σ in the symmetric group $W = S_n$, acting on $\mathbb{Z}[x_1, x_2, \ldots, x_n]$ by permutations of the coordinates, let $\beta_\sigma = \prod x_{\sigma(1)} x_{\sigma(2)} \cdots x_{\sigma(i)}$, the product over all i such that $\sigma(i) > \sigma(i+1)$. Then the monomials β_σ ($\sigma \in W$) form a free basis for $\mathbb{Z}[x_1, x_2, \ldots, x_n]$ as a module over $\mathbb{Z}[x_1, x_2, \ldots, x_n]^W$. This special case of 2.2 has been rediscovered and put to good use by A. Garsia, in *Combinatorial methods in the theory of Cohen-Macaulay rings*, Adv. in Math. **38** (1980), 229–266.

29. On the desingularization of the unipotent variety

Let V be the variety of unipotent elements of G, and let A be the subvariety of $V \times G/B \times G/B$ made up of all (v, g_1B, g_2B) such that v fixes both g_1B and g_2B. Observe that A is the union of the principal varieties of our paper, one for each G-orbit of V. The main idea of the paper may be viewed as a comparison of this decomposition of A with the one obtained by considering the G-orbits of $G/B \times G/B$ instead. The variety A has since been used by D. Kazhdan and G. Lusztig, *A topological approach to Springer's representations*, Adv. in Math. **38** (1980), 222–228, in a treatment of Springer's construction of the irreducible representations of Weyl groups which does not require the étale cohomology. The reason that I used two copies of B in my main construction in the first place was to get into play the term $2\dim(G/B)_u$ in my principal result 4.6 written as $\dim G_u = r + 2\dim(G/B)_u$. To my delight, this idea worked. The result 4.6 is now known to be true in all characteristics since the unipotent classes for the exceptional groups have been classified (in terms of lists) and checked out case-by-case. For this see, again, J. E. Humphreys' book mentioned in regard to (25). Finally, the algorithm mentioned in 5.7(c) is worked out in our later paper (38).

30. On theorems of Lie-Kolchin, Borel and Lang

This paper always provokes fond remembrances because I worked out most of it while lying in my sleeping bag looking up at the stars on a backpack trip in the high Sierras.

31. Conjugacy in semisimple algebraic groups

The main question left open in this paper, as to whether Conjecture 1 also holds when $\operatorname{char} k \neq 0$, is, as far as I know, still open.

32. Kleinian singularities and unipotent elements

This is the only one of my articles for which I got any money, 23 dollars from Uspekhi Mat. Nauk. **38** (1983), 165–170, where it was republished in a Russian translation. My Yale notes *Lectures on Chevalley groups* were also published in a Russian translation, but in the days before glasnost.

33. Generators, relations and coverings of algebraic groups II

The letter N, as it appears near the beginning of Section 2, denotes the normalizer of H in G.

36. Tensor product theorem

In addition to the examples offered here we call attention to those mentioned in our commentary on (18).

37. On Dickson's theorem on invariants

In the statement of Theorem J near the end of the paper, *in the I's* should be replaced by *in the* p^{r-i-1}*th powers of the I's*.

40. Nagata's example

It should have been mentioned in the paper that in Theorem 1.3, and again in Theorem 5.3 where the situation is more general, the order of the group element which is the sum of the 9 points involved is independent of the group structure put on the curve, i.e., of the choice of the 0-point for the group. We omit the proofs, which are short and easy.

Acknowledgments

The American Mathematical Society gratefully acknowledges the kindness of these institutions in granting the following permissions:

Academic Press, Inc.

Torsion in reductive groups, Advances in Math. **15** (1975), 63–92.

On the theorems of Lie-Kolchin, Borel and Lang, Contributions to Algebra, Academic Press (1977), 349–354.

Conjugacy in semisimple algebraic groups, J. Algebra **55** (1978), 348–350.

Generators, relations and coverings of algebraic groups II, J. Algebra **71** (1981), 527–543.

An occurrence of the Robinson-Schensted correspondence, J. Algebra **113** (1988), 523–528.

Cambridge University Press

Nagata's example, Algebraic groups and Lie groups, Australian Math. Soc. Lecture Series 9, Cambridge University Press, 1997, 375–384.

Canadian Mathematical Society

The representations of $GL(3,q), GL(4,q), PGL(3,q)$, *and* $PGL(4,q)$, Canad. J. Math. **3** (1951), 225–235.

Prime power representations of finite linear groups, Canad. J. Math. **8** (1956), 580–591.

Prime power representations of finite linear groups II, Canad. J. Math. **9** (1957), 347–351.

On the number of sides of a Petrie polygon, Canad. J. Math. **10** (1958), 220–221.

Automorphisms of finite linear groups, Canad. J. Math. **12** (1960), 606–615.

Invariants of finite reflection groups, Canad. J. Math. **12** (1960), 616–618.

Generators for simple groups, Canad. J. Math. **14** (1962), 277–283.

Elsevier Science, LTD.

On a theorem of Pittie, Topology **14** (1975), 173–177.

Gordon & Breach

On the Galois cohomology of linear algebraic groups, Proc. International Conf. on the Theory of Groups, Canberra (1965), Gordon and Breach (1967), 315–319.

Illinois Journal of Mathematics

Algebraic groups and finite groups, Illinois J. Math. **13** (1969), 81–86.

Nagoya Mathematical Journal

Representations of algebraic groups, Nagoya Math. J. **22** (1963), 33–56.

Pacific Journal of Mathematics

(with A. Horn) *Eigenvalues of the unitary part of a matrix*, Pacific J. Math. **9** (1959), 541–550.

Variations on a theme of Chevalley, Pacific J. Math. **9** (1959), 875–891.

The simplicity of certain groups, Pacific J. Math. **10** (1960), 1039–1041.

Automorphisms of classical Lie algebras, Pacific J. Math. **11** (1961), 1119–1129.

Subgroups of SU_2, Dynkin diagrams and affine Coxeter elements, Pacific J. Math. **118** (1985), 587–598.

Presses Universitaires de France

Regular elements of semisimple algebraic groups, Publ. Sci. I.H.E.S. **25** (1965), 49–80.

Springer-Verlag Gmbh & Co. KG

On the desingularization of the unipotent variety, Invent. Math. **36** (1976), 209–224. Copyright © 1976 Springer-Verlag

(with G. Rohrle and R. Richardson) *Parabolic subgroups with Abelian unipotent radical*, Invent. Math. **110** (1992), 649–671. Copyright © 1992 Springer-Verlag

Springer-Verlag New York, Inc.

(with T. A. Springer) *Conjugacy classes, Seminar in algebraic groups and related finite groups*, Lecture Notes in Math., Springer-Verlag **131** (1970), 167–266; Addendum to second corrected printing (1986), 322–323. Copyright © 1970 Springer-Verlag New York, Inc.

Abstract homomorphisms of simple algebraic groups, Séminaire Bourbaki, Lecture Notes in Math., Springer-Verlag **383** (1972), 307–326. Copyright © 1972 Springer-Verlag New York, Inc.

University of Tokyo

On Dickson's theorem on invariants, J. Faculty of Sciences, Univ. of Tokyo **34** (1987), 699–707.

ISBN 0-8218-0576-2

GETTYSBURG COLLEGE
3326800 0310252 2